REAL VARIABLES:
An Introduction To The Theory of Functions

By John M. H. Olmsted

Late Professor Emeritus at

The University of Illinois

With A New Preface And Bibliography for the 2019 Edition by Karo Maestro

Blue Collar Scholar LLC/Kindle Publishing

Copyright © 1959 by
APPLETON-CENTURY-CROFTS, INC.

All rights reserved. This book, or parts thereof, must not be reproduced in any form without permission of the publisher.

5119-1

Library of Congress Card Number: 60-5076

All New Material@2019 Karo Maestro

3

3

4

4

PREFACE

For a student of mathematical analysis the importance of a thorough training in the techniques of critical reasoning is becoming increasingly recognized and accepted. This training should include extensive experience and intimate acquaintance with the construction of proofs and counter-examples and the formulations of negations, complete with "deltas and epsilons." An elementary course in logic, although important, is inadequate, and a standard course in calculus is usually limited to the introduction of subject matter and the development of manipulative techniques. Perhaps the most proper time for a student of mathematics to begin his serious study of the tools of analysis is immediately after the completion of a first course in calculus. It is at this time that the student should be encouraged to prove (in full detail) statements which he has previously been persuaded to accept because of their immediate obviousness.

The purpose of this book is to present the basic ideas and techniques of analysis, for real-valued functions of an arbitrary number of real variables, in such a way that students with at least standard calculus background can proceed at whatever pace and degree of intensity seem suitable. The organization of the book is designed for maximum flexibility. Topics of a relatively difficult or theoretical nature are starred (*) for possible omission or postponement. For example, the discussion of uniform continuity is starred while that of ordinary continuity is not. Frequently, if a theorem is easy to understand whereas its proof is more difficult, only the proof is starred. Occasionally such a proof is presented in a later section or chapter. An instance is the intermediate value property of a function continuous on an interval. Some topics which can be omitted from the starred portions without affecting the continuity of what remains receive double stars (**). This system of starring and double starring applies to the exercises as well as the text. The resultant arrangement of the contents makes the book unusually adaptable for courses of varying lengths and levels. The starring system

PREFACE

is used in a consistent pattern throughout the book. Prerequisite to any unstarred sections of the book are the preceding unstarred sections, and prerequisite to any singly starred sections are the preceding sections that are not doubly starred. As a corollary, omission of any starred material may necessitate omission of later starred material.

A dominant philosophy, dictated by the needs of the average as well as the superior student, is to proceed from the simple to the complex and from the concrete to the abstract. In this way a maximum body of subject matter is made available to the student of modest ability, and a strong groundwork is established for the student who is able to undertake the generalizations that follow. A few examples will illustrate this principle. In integration, the standard Riemann integral of a function of a single real variable is considered first; the Riemann-Stieltjes integral is treated as a generalization, in a starred section, with bounded variation doubly starred; multiple integrals are taken up later as extensions to higher dimensional spaces. The boundedness of a real-valued function continuous on a closed interval is extended to permit the domain to be an arbitrary compact set of real numbers, and then a compact set in n dimensions; after a generalization to mappings in higher dimensional spaces comes the *final* abstraction permitting both the domain and the range to be in general metric spaces. Improper integrals are introduced in Chapter 5, and more intensively studied with the aid of uniform convergence in Chapter 14. The implicit function theorem is proved *after* the student is given an opportunity to become familiar with its significance. Often, when a theorem has an elementary and a sophisticated proof, *both* are given; such is the case with the standard theorems concerning the convergence of an alternating series, the convergence of an absolutely convergent series, and the convergence of a product series to the appropriate sum.

The unstarred portions of the book include most of the substance of the classical intermediate and advanced calculus courses and are suitable for a one-year sequence at that level, meeting three times a week. Other courses, with considerable variation in content and depth, can be fashioned by different choices of sections. One possibility is to skip from the middle of Chapter 3 to Chapter 10, which gives an introduction to higher dimensional spaces and abstractions including both metric and topological spaces. Another possibility is to omit most of the material on functions of several variables for the sake of including Chapter 15 on Fourier series and linear function spaces. The doubly starred sections of the book, particularly in the exercises, permit the more advanced student to explore well into the graduate curriculum. Instances are the criterion for Riemann integrability in terms of continuity almost everywhere, the Lebesgue dominated convergence theorem for Riemann and

PREFACE

improper integrals, a complete proof of the transformation theorem for multiple integrals, and a thorough discussion of closedness and completeness in function spaces.

Special attention should be called to the abundant sets of exercises (there are over 2200 exercises!). These include routine drills for practice, intermediate exercises that extend the material of the text while retaining its character, and advanced exercises that go beyond the standard textual subject matter. Whenever guidance seems desirable, generous hints are included. In this manner the student is led to such items of interest as Weierstrass's theorem on uniform approximation of continuous functions by polynomials, the construction of a continuous nondifferentiable function, and the second category nature of the set of irrational numbers. Analytic treatment of the logarithmic, exponential, and trigonometric functions is presented in the exercises, where sufficient hints are given to make these topics available to all. The same is true for the standard theorems concerning line and surface integrals. Answers to all problems are given in the back of the book. Illustrative examples abound throughout.

A few words regarding notation should be given. The equal sign $=$ is used for equations, both conditional and identical, and the triple bar $\equiv$ is reserved for definitions. For simplicity, if the meaning is clear from the context, neither symbol is restricted to the indicative mood as in "$(a + b)^2 = a^2 + 2ab + b^2$," or "where $f(x) \equiv x^2 + 5$." Examples of subjunctive uses are "let $x = n$," and "let $\epsilon \equiv 1$," which would be read "let x be equal to n," and "let ϵ be defined to be 1," respectively. A similar freedom is granted the inequality symbols. For instance, the symbol $>$ in the following constructions "if $\epsilon > 0$, then $\cdots$," "let $\epsilon > 0$," and "let $\epsilon > 0$ be given," could be translated "is greater than," "be greater than," and "greater than," respectively. A relaxed attitude is also adopted regarding functional notation, and the tradition $(y = f(x))$ established by Dirichlet has been followed. When there can be no reasonable misinterpretation the notation $f(x)$ is used both to indicate the value of the function f corresponding to a particular value x of the independent variable and also to represent the function f itself (and similarly for $f(x, y)$, $f(x, y, z)$, and the like). This permissiveness has two merits. In the first place it indicates in a simple way the number of independent variables and the letters representing them. In the second place it avoids such awkward circumlocutions as "the function f defined by the equation $f(x) = \sin 2x$ is periodic," by permitting simply, "$\sin 2x$ is periodic." There is an appealing simplicity (and no confusion) to such a statement as "the functions $1, x, \cdots, x^n$ are linearly independent on every interval."

In a few places parentheses are used to indicate alternatives. The principal instances of such uses are heralded by announcements or foot-

PREFACE

notes in the text. Here again it is hoped that the context will prevent any ambiguity. Such a sentence as "The function $f(x)$ is integrable from a to b $(a > b)$" would mean that "$f(x)$ is integrable from a to b, where it is assumed that $a > b$," whereas a sentence like "A function having a positive (negative) derivative over an interval is strictly increasing (decreasing) there" is a compression of two statements into one, the parentheses indicating an alternative formulation.

One note of caution is in order. Because of the rich abundance of material available, complete coverage is difficult. Anybody using the book as a text should therefore be advised to use it selectively, in order that tempting stopovers may not preclude attractive explorations in other directions.

The present book is an expansion of the author's earlier *Intermediate Analysis* which, with minor alterations, comprises the first nine chapters of *Real Variables*. The author wishes to express his deep appreciation of the aid and suggestions given by Professor R. W. Brink of the University of Minnesota in the preparation of the manuscripts of both texts. He is also indebted to many others for their many helpful comments concerning both the first nine chapters, from *Intermediate Analysis*, and the succeeding chapters when they were used in mimeographed form.

J. M. H. O.

Minneapolis, Minn.

CONTENTS

	PAGE
PREFACE	v

Chapter 1
THE REAL NUMBER SYSTEM

SECTION
101.	Introduction	1
102.	Axioms of the basic operations	2
103.	Exercises	4
104.	Axioms of order	5
105.	Exercises	6
106.	Positive integers and mathematical induction	7
107.	Exercises	10
108.	Integers and rational numbers	13
109.	Exercises	13
110.	Geometrical representation and absolute value	16
111.	Exercises	18
112.	Some further properties	19
113.	Exercises	20
114.	Axiom of completeness	22
★115.	Further remarks on mathematical induction	24
116.	Exercises	25

Chapter 2
FUNCTIONS, SEQUENCES, LIMITS, CONTINUITY

201.	Functions and sequences	28
202.	Limit of a sequence	31
203.	Exercises	33
204.	Limit theorems for sequences	34
205.	Exercises	38
206.	Limits of functions	40
207.	Limit theorems for functions	44
208.	Exercises	45
209.	Continuity	49
210.	Types of discontinuity	51
211.	Continuity theorems	52

CONTENTS

212. Exercises	53
213. More theorems on continuous functions	54
214. Existence of $\sqrt{2}$ and other roots	55
215. Monotonic functions and their inverses	55
216. Exercises	57
★217. Further remarks on functions	60

★Chapter 3
SOME THEORETICAL CONSIDERATIONS

★301. A fundamental theorem on bounded sequences	61
★302. The Cauchy Criterion for convergence of a sequence	62
★303. Sequential criteria for continuity and existence of limits	64
★304. The Cauchy Criterion for functions	65
★305. Exercises	66
★306. Proofs of some theorems on continuous functions	69
★307. Uniform continuity	71
★308. Exercises	73
★★309. Point sets: open, closed, compact, connected sets	74
★★310. Point sets and sequences	78
★★311. Some general theorems	79
★★312. Exercises	81

Chapter 4
DIFFERENTIATION

401. Introduction	85
402. The derivative	85
403. One-sided derivatives	88
404. Exercises	91
405. Rolle's Theorem and the Law of the Mean	93
406. Consequences of the Law of the Mean	97
407. The Extended Law of the Mean	98
408. Exercises	100
409. Maxima and minima	105
410. Differentials	107
411. Approximations by differentials	109
412. Exercises	111
413. L'Hospital's Rule. Introduction	115
414. The indeterminate form $0/0$	116
415. The indeterminate form ∞/∞	117
416. Other indeterminate forms	120
417. Exercises	121
418. Curve tracing	123
419. Exercises	127
★420. Without loss of generality	129
★421. Exercises	129

Chapter 5
INTEGRATION

501. The definite integral	131

CONTENTS

503.	Exercises	143
504.	The Fundamental Theorem of Integral Calculus	154
505.	Integration by substitution	156
506.	Exercises	156
★507.	Sectional continuity and smoothness	159
★508.	Exercises	160
★509.	Reduction formulas	161
★510.	Exercises	162
511.	Improper integrals, introduction	164
512.	Improper integrals, finite interval	164
513.	Improper integrals, infinite interval	167
514.	Comparison tests. Dominance	168
515.	Exercises	171
★★516.	Bounded variation	175
★517.	The Riemann-Stieltjes integral	179
★518.	Exercises	184

Chapter 6

SOME ELEMENTARY FUNCTIONS

★601.	The exponential and logarithmic functions	188
★602.	Exercises	188
★603.	The trigonometric functions	191
★604.	Exercises	192
605.	Some integration formulas	194
606.	Exercises	196
607.	Hyperbolic functions	197
608.	Inverse hyperbolic functions	199
609.	Exercises	200
★610.	Classification of numbers and functions	201
★611.	The elementary functions	203
★612.	Exercises	204

Chapter 7

INFINITE SERIES OF CONSTANTS

701.	Basic definitions	206
702.	Three elementary theorems	207
703.	A necessary condition for convergence	207
704.	The geometric series	208
705.	Positive series	208
706.	The integral test	209
707.	Exercises	211
708.	Comparison tests. Dominance	212
709.	The ratio test	215
710.	The root test	218
711.	Exercises	219
★712.	More refined tests	221
★713.	Exercises	223

CONTENTS

714. Series of arbitrary terms	225
715. Alternating series	225
716. Absolute and conditional convergence	227
717. Exercises	229
718. Groupings and rearrangements	231
719. Addition, subtraction, and multiplication of series	233
★720. Some aids to computation	235
721. Exercises	237

Chapter 8
POWER SERIES

801. Interval of convergence	240
802. Exercises	243
803. Taylor series	244
804. Taylor's Formula with a Remainder	246
805. Expansions of functions	248
806. Exercises	249
807. Some Maclaurin series	250
808. Elementary operations with power series	253
809. Substitution of power series	255
810. Integration and differentiation of power series	259
811. Exercises	261
812. Indeterminate expressions	264
813. Computations	264
814. Exercises	266
★815. Analytic functions	267
★816. Exercises	268

★Chapter 9
UNIFORM CONVERGENCE

★901. Uniform convergence of sequences	270
★902. Uniform convergence of series	273
★903. Dominance and the Weierstrass M-test	274
★904. Exercises	275
★905. Uniform convergence and continuity	278
★906. Uniform convergence and integration	279
★907. Uniform convergence and differentiation	281
★908. Exercises	283
★909. Power series. Abel's Theorem	288
★★910. Proof of Abel's Theorem	289
★911. Exercises	290
★912. Functions defined by Power Series. Exercises	292

Chapter 10
FUNCTIONS OF SEVERAL VARIABLES

1001. Introduction	293
1002. Neighborhoods in the Euclidean plane	293

CONTENTS

1003. Point sets in the Euclidean plane	294
1004. Sets in higher dimensional Euclidean spaces	298
1005. Exercises	299
1006. Functions and limits	301
1007. Iterated limits	304
1008. Continuity	306
1009. Limit and continuity theorems	307
1010. More theorems on continuous functions	308
1011. Exercises	309
★1012. Uniform limits	311
★1013. Three theorems on uniform limits	312
★★1014. The Moore-Osgood theorem	313
★1015. Exercises	314
1016. More general functions. Mappings	315
★1017. Uniform continuity	318
★1018. Some theoretical considerations	319
★★1019. Two fundamental theorems on mappings	320
★★1020. Proofs of some theorems on continuous functions	321
1021. Exercises	322
★1022. Point set algebra	324
★1023. Exercises	327
★1024. Set algebra and operations involving neighborhoods	328
★1025. Exercises	330
★★1026. Metric spaces	333
★★1027. Topological spaces	335
★★1028. Exercises	337

Chapter 11
ARCS AND CURVES

1101. Duhamel's Principle for integrals	339
★1102. A proof with continuity hypotheses	340
1103. Arcs and curves	340
★★1104. Space-filling arcs	342
1105. Arc length	343
★1106. Existence of arc length	345
★1107. Independence of parametrization	346
1108. Integral form for arc length	347
★1109. Remark concerning the trigonometric functions	351
1110. Exercises	352
1111. Cylindrical and spherical coordinates	353
1112. Arc length in rectangular, cylindrical, and spherical coordinates	356
1113. Exercises	357
1114. Curvature and radius of curvature in two dimensions	357
1115. Circle of curvature	359
★1116. Evolutes and involutes	360
1117. Exercises	363

CONTENTS

Chapter 12
PARTIAL DIFFERENTIATION

1201.	Partial derivatives	366
1202.	Partial derivatives of higher order	367
★1203.	Equality of mixed partial derivatives	368
1204.	Exercises	369
1205.	The fundamental increment formula	371
1206.	Differentials	373
1207.	Change of variables. The chain rule	374
★1208.	Homogeneous functions. Euler's theorem	377
1209.	Exercises	378
★1210.	Directional derivatives. Tangents and normals	380
★1211.	Exercises	384
1212.	The Law of the Mean	385
1213.	Approximations by differentials	387
1214.	Maxima and minima	388
1215.	Taylor series	391
1216.	Exercises	392
1217.	Differentiation of an implicit function	394
1218.	Some notational pitfalls	396
1219.	Exercises	397
1220.	Envelope of a family of plane curves	398
1221.	Exercises	400
1222.	Several functions defined implicitly. Jacobians	401
1223.	Coordinate transformations. Inverse transformations	405
1224.	Functional dependence	410
1225.	Exercises	411
★1226.	Differentiation under the integral sign. Leibnitz's Rule	416
★1227.	Exercises	418
★1228.	The Implicit Function Theorem	419
★1229.	Existence theorem for inverse transformations	424
★1230.	Sufficiency conditions for functional dependence	426
★1231.	Exercises	428

Chapter 13
MULTIPLE INTEGRALS

1301.	Introduction	429
1302.	Double integrals	429
1303.	Area	431
1304.	Second formulation of the double integral	432
★1305.	Inner and outer area. Criterion for area	433
★1306.	Theorems on double integrals	436
★★1307.	Proof of the second formulation	441
1308.	Iterated integrals, two variables	444
★1309.	Proof of the Fundamental Theorem	446
1310.	Exercises	448
1311.	Triple integrals. Volume	451
1312.	Exercises	453

CONTENTS

1313.	Double integrals in polar coordinates	455
1314.	Volumes with double integrals in polar coordinates	457
1315.	Exercises	459
1316.	Mass of a plane region of variable density	459
1317.	Moments and centroid of a plane region	460
1318.	Exercises	461
1319.	Triple integrals, cylindrical coordinates	462
1320.	Triple integrals, spherical coordinates	463
1321.	Mass, moments, and centroid of a space region	465
1322.	Exercises	466
1323.	Mass, moments, and centroid of an arc	468
1324.	Attraction	468
1325.	Exercises	471
1326.	Jacobians and transformations of multiple integrals	472
1327.	General discussion	474
★★1328.	Proof of the transformation theorem	478
★1329.	Improper multiple integrals	486
1330.	Exercises	488
★★1331.	Line and surface integrals. Exercises	490

★Chapter 14
IMPROPER INTEGRALS

★1401.	Introduction. Review	495
★1402.	Alternating integrals. Abel's test	496
★1403.	Exercises	499
★1404.	Uniform convergence	500
★1405.	Dominance and the Weierstrass M-test	502
★1406.	The Cauchy criterion and Abel's test for uniform convergence	503
★1407.	Three theorems on uniform convergence	505
★1408.	Evaluation of improper integrals	508
★1409.	Exercises	510
★1410.	Iterated improper integrals	513
★1411.	Improper integrals of infinite series	515
★1412.	Exercises	517
★1413.	The gamma function	518
★1414.	The beta function	521
★1415.	Exercises	523
★1416.	Infinite products	524
★1417.	Wallis's infinite product for π	526
★1418.	Euler's constant	527
★1419.	Stirling's formula	528
★1420.	Weierstrass's infinite product for $1/\Gamma(\alpha)$	530
★1421.	Exercises	531
★★1422.	Improper Riemann-Stieltjes integrals	532
★★1423.	Exercises	535

CONTENTS

★Chapter 15

FOURIER SERIES AND ORTHOGONAL FUNCTIONS

★1501. Introduction	537
★1502. Linear function spaces	538
★1503. Distance in a function space	539
★1504. Inner product. Orthogonality. The space R^2	541
★1505. Least squares. Fourier coefficients	545
★1506. Generalized Fourier series. Bessel's inequality	547
★1507. Periodic functions	549
★1508. Linear spaces of periodic functions	550
★1509. Trigonometric Fourier series	551
★1510. A special convergence theorem	554
★1511. Exercises	556
★1512. Parseval's equation	558
★1513. Cosine series. Sine series	559
★1514. Other intervals	561
★1515. Exercises	562
★★1516. Partial sums of Fourier series	564
★★1517. Functions with one-sided limits	565
★★1518. The Riemann-Lebesgue Theorem	565
★★1519. Functions of bounded variation. The space $BV_{2\pi}$	566
★★1520. Jordan convergence theorem	567
★★1521. Fejér's summability theorem	568
★★1522. Uniform summability	570
★★1523. Weierstrass's Theorem	570
★★1524. Density of polynomials	571
★★1525. Uniform convergence	572
★★1526. Integration of Fourier series	573
★★1527. The Gibbs phenomenon	575
★★1528. Exercises	576
★1529. Applications of Fourier series. The vibrating string	577
★1530. A heat conduction problem	580
★1531. Exercises	581
★★1532. Bases. Closedness. Completeness	583
★★1533. Linear dependence and independence	586
★★1534. The Gram-Schmidt Process	587
★★1535. Legendre polynomials	589
★★1536. Orthogonality with respect to a weight function	590
★★1537. Other orthogonal systems	591
★★1538. Exercises	591
ANSWERS TO ALL PROBLEMS	595
INDEX	613

17

Preface To The 2019 Edition

During the last half-century, number and measurable quantity have been separated... the idea of number alone has been recognized as the foundation upon which Mathematical Analysis rests, and the theory of extensive magnitude is now regarded as a separate department in which the methods of Analysis are applicable, but as no longer forming part of the foundation upon which Analysis itself rests.- E. W. Hobson

The analyst, who pursues a purely esthetic aim, helps create, just by that, a language more fit to satisfy the physicist.-Henri Poincaré

Calculus is the easiest subject in the world to get an A in-as long as you're smart enough not to ask why anything's true in it. We're smart enough not to ask. You're not. That's why we'll be rich and you'll die at 30 after I foreclose on your house. -the President of the Pre-Med Society to me when I was an undergraduate (whose teenage son overdosed last year)

When I decided to republish John Olmsted's undergraduate analysis text, it was very apparent to me that unlike the situation when the book was first published, there is no shortage of analysis textbooks in English nowadays. [1] Indeed, unlike most subjects in mathematics and physics, today there's no shortage of *cheap and affordable quality* textbooks on mathematical analysis at many levels. Mattuck, Hoffman, Ash, Rosenlicht, Cummings and Lebl-even The Great Granddaddy Analysis Text Of Them All, Hardy's *A*

---
[1] Yes, this is a bit of an understatement on my part, but bear with me.

Course In Pure Mathematics- all range in quality from solid to outstanding and they're all available rather inexpensively.

And those are just the inexpensive ones. If one is able to dig deeper into one's pocket, the great standard classics and successor texts of equal stature and range of difficulty are also available: Apostol, "baby" Rudin, Ross, Burns, Strichartz, Abbott, Pugh, Mardsen, Bartle and Shubert, Korner, Binmore, Tao, Burkill, Krantz, etc, etc, etc.

So it's clear that if you want to publish-or republish, in this case-yet another real analysis text, then dammit, you better have a convincing rationale for the prospective buyer to even look at it, let alone buy it.

Well, first of all, I happen to like the book immensely. I think it should be made available for new generations for whom it's been forgotten. As I've said in the introductions to other books published by Blue Collar Scholar, one of our main missions will be to uncover out of print texts that we believe are too good and relevant in subject matter to languish in publishing limbo. These texts can still serve mathematics and science students very well if made available again inexpensively. I believe Olmsted's book certainly fits those criteria.

Why?

Well, my publishing company and unless I'm breaking some copyright law, I'll publish any book I feel like. So nyah nyah.

More specifically, I believe several aspects of the book's unusual organization and content make it very deserving of low cost republication.

Firstly, while it covers just about all the usual topics in any undergraduate analysis text-number systems, functions, limits of functions and sequences of one and several variables in $\mathbb{R}^n$, continuity, differentiation and integration of functions in $\mathbb{R}$, bounded sequences, metric spaces, basic point set topology, infinite series, power series, convergence tests, improper integrals, partial and total derivatives and multiple integrals- it has a number of unique aspects to the presentation that distinguish it from other textbooks. For example, it has a substantial final chapter on Fourier series and their applications to both approximation theory and differential equations requiring only Riemann integration as prerequisites. It also covers a number of important concepts of analysis in the starred sections and exercises that are not usually covered in general introductions, such as point set topology, Riemann-Steijles integration, the Moore-Osgood theorem for infinite series of several variables, vector analysis and differential forms. Yet another example- it contains an entire opening chapter giving a far more detailed axiomatic description of the number systems without explicitly constructing them. While most analysis texts have such an opening section, Olmstead's is longer and more detailed then the ones found in most books with many substantial exercises.

Another positive quality of the book is its' unusual midway level of difficulty. Calculus courses today are far weaker than they were when the standard textbooks such as Walter Rudin's *Principles of Mathematical Analysis* were published. As a result, a number of students beginning analysis today need a bit more foundational training in rigorous calculus before tackling functions in Euclidean spaces and abstract metric spaces. So usually students have to begin with a "baby real variables" text-such as Mattuck or Ross-before moving on to analysis on metric spaces. Olmsted does a fine job in his early chapters of presenting the properties of the real numbers and a precise presentation of calculus on the real line before moving on to functions of several variables. Metric spaces are covered in an optional section late in the book that builds on these earlier concepts. This allows the first half of the text to act as a "baby real variables" book i.e. a bridge between today's calculus courses and hard-edged classical analysis courses on metric spaces. Students can then either move on to a more abstract text or learn the basic elements of metric spaces as a generalization of functions in $\mathbb{R}^n$. Therefore, students will need only one low-priced textbook instead of two.

It also makes the book very well suited for self- study-*particularly since the solutions to many of the exercises can be found in the back.* While detailed solutions are not given, the final answers will be quite helpful as "hints", so that students may work backwards from them on exercises they've become stuck on. There are very few analysis textbooks you can say this about.

Another unique quality is the "pragmatic" sections that discuss the computational aspects of analysis that would be of great interest to applied mathematics, physics and engineering students. For example, Chapter 7 contains several useful sections expounding on grouping and rearrangements of terms in infinite series computations, something that's usually left to basic calculus classes and not discussed in the higher theoretical context. There is also an entire chapter on the so-called elementary functions-exponential, logarithmic and hyperbolic functions with their inverses-that are of great use in the physical sciences. While these functions were very standard in classical advanced calculus/real analysis texts from the turn of the 20th century onward, most modern texts omit them or shunt them to the exercises.

But the final and most important aspect of the book is the many unusual and richly detailed examples and counterexamples, which are considerably more diverse and sophisticated than one usually finds in most undergraduate analysis textbooks. These surprising examples make themselves known and felt throughout the book to dramatically demonstrate where many results would either fail altogether or not be as general as the beginner would expect them to be without their precise conditions of validity. There are a host of beautiful examples of functions that on subsets of $\mathbb{R}^n$ that are more subtle then usually encountered in the first course. For example, step functions are used throughout chapter 5 to give surprising examples of integrable functions-such as an example of a function on [0,1] *which has infinitely many discontinuities and yet is Riemann-integrable on [0,1]!* It is the large collection of challenging examples that allows Olmsted's book to actually be more sophisticated then it initially looks, yet not lose any clarity. They are probably the deepest and most useful parts of the book and what makes it so worthy of republication. I believe it's this aspect of the book in particular that will make it incredibly useful to both teachers and students of analysis as either a main text or a supplement.

It's clear to me that Olmsted's book is an extraordinarily versatile textbook for undergraduate analysis courses at all levels. It will be a wonderful addition to the undergraduate analysis textbook literature. This versatility of use will be particularly true because of its' low price.

In the first stages of preparing the book for republication, I considered adding a considerable amount of new material to make the new edition of the book even more valuable and versatile. I fully intended on adding 2 original chapters to Olmstead's original text: a fairly complete construction of the number systems and, more importantly, an introduction to the generalized Riemann integral. Indeed, there are currently substantial drafts of both of these chapters on my computer in both Microsoft Word and LaTeX formats. But in the end, I opted not to add them. This is because it feels to me like tampering with the original works in a self-defeating manner. If you're going to write a substantial amount of new material, then what's the point of digging this book up in the first place? The entire point of republishing these works is to make them available to a new generation. If they need this much improvement, then why bother to republish them?

Therefore, once again, I have decided to be conservative in this republication and limit the new material to the standard opinionated preface and recommended references which have been the unique characteristics of this book's predecessors in BCS. [2]

   A few words about the author, whose name is probably at least vaguely familiar to most mathematics acolytes, whether professional or amateur. It is familiar because John Olmsted-then a well-known and respected analyst in the 1960's-co-authored with Bernard Gelbuam one of the great classics of mathematics, *Counterexamples In Analysis*, which of course is available from Dover Books. If there is one reference which has proven itself to be a required possession for all serious students, teachers and researchers of both classical and modern analysis, it's this one. Counterexamples to theorems are just as important in learning mathematics as examples-understanding a theorem only gets you so far if you don't see what kinds of aberrations can be constructed when the conditions of the theorem fail. Before OG's publication, finding such counterexamples required either days or weeks of a painstaking search of the mathematical literature or an even more lengthy and uncertain journey with pen in hand through reams of scrap paper attempting to construct them from scratch. OG collected in one volume just about all the relevant major counterexamples in analysis in level from basic calculus all the way to second year graduate functional analysis. As such, it has become an absolutely indispensable reference.

---

2 Whether or not this was a wise decision, time will tell.

What most people don't know is that many of these examples were first class-tested in Olmsted's advanced calculus and real variables courses in the preceding 2 decades-many of which appear in the example and problem sections which I praised earlier. Olmsted was a gifted teacher whose classes were filled with the most hungry mathematics students-and his emphasis on counterexamples was one of the stronger indicators of his passion for analysis.

Olmsted (1911-1997) was part of the pre-WW II generation of American mathematics students who were the first native teachers of modern mathematics trained by the European-trained missionaries at top programs in such then-frontier subjects as topology and abstract algebra. He was a graduate student at Princeton University in the 1930's, mentioned by Paul Halmos in his reminisces of his time there as one of his 2 important guides to the esteemed university when he arrived there (the other was Arthur Brown)[3]. Olmsted got his PhD in 1940 under Salomon Bochner. His research was mostly focused on integration theory and function spaces. But even as a graduate student, what impressed most of the faculty and his classmates about Olmsted was his ability as a gifted teacher. He lectured clearly and creatively with unusual observations most professors wouldn't bring up. He also had a visible passion for proselytization which was contagious. It was really this skill as a teacher for which he was known and which his many textbooks are monuments to.

Olmsted taught most of his career at the University of Southern Illinois At Carbondale with periods at the University of Minnesota and a fellowship at the Princeton Institute for Advanced Study. He

---

[3] This bibliographic material is drawn from 2 sources: Halmos, Paul, *I Want To Be A Mathematician*, Springer Verlag, 1985 and the mathematics department webpages of the University of Southern Illinois At Carbondale.

was Department Chair at USIC for nearly 15 years and finished as Professor Emeritus. He was known there for personally designing and teaching virtually all analysis courses there ranging from freshman calculus to graduate analysis. USI is a small liberal arts college without a doctoral program in mathematics. As such, it focused on undergraduate training. Olmsted's goal was to prepare the strongest possible undergraduates so they could then move on to top tier graduate programs for their PhD. In his memory, USIC has named a department award for the best masters' student in the department-The John M. H. Olmsted Memorial Award for Graduate Students in Mathematics.

In addition to his teaching and research, Olmstead authored a number of textbooks sharing the same range as the courses they're based on i.e. from basic freshman to first year graduate level. Many of these textbooks became standard texts for undergraduates and graduate students in the 1970's onwards: *Solid Analytic Geometry, Calculus With Analytic Geometry, Prelude To Calculus and Linear Algebra, A Second Course In Calculus, The Real Number System, Advanced Calculus, Intermediate Analysis,* the aforementioned *Counterexamples In Analysis* as well as the more general follow up volume with the same co-author, *Theorems And Counterexamples In Mathematics*-and of course, the volume you hold in your hands. Sadly, all of them, with the exception of the last 2, were out of print until recently. Dover and several small publishing houses reissued *The Real Number System* last year, which is good news for students and teachers alike.[4]

---

[4] A word about *The Real Number System*: The title suggests this book contains a full construction of the real numbers from the natural numbers a la Landau's *Foundations of Analysis*. As I was, you would be quite mistaken. This is a book length axiomatic development of the number systems through axioms and examples, basically a very expanded version of first chapter of this book. While it doesn't give an explicit construction of the number systems, it does give a very detailed and informative study of the formal structure of these number systems and is very readable. It would make a very nice companion to the text in your hands.

My hope is that the republication of both these strong textbooks will revive interest in Olmsted's other works as ageless works on foundational subjects that will find renewed life in new generations of students. After all, that is one of my primary goals with BCS and I'd be remiss if I allowed these wonderful books to remain in limbo. I also like to think Professor Olmsted himself would have approved-especially republishing the book more or less as it appeared in 1959.

I've also written an extended recommended reading section which gives an overview and reviews of most of the undergraduate real analysis texts available today, particularly those that can be gotten dirt cheap. This one is longer then my usual ones because I feel analysis is the course where most mathematics majors lose their way and become overwhelmed by new concepts and techniques. As a result, they need all the help they can get in choosing sources to help them. I also hope this section will be helpful to prospective teachers of advanced calculus/ honors calculus/ real analysis in not only picking a textbook, but sources in writing their course materials.

I commend this classic to the care and use of new generations of teachers and students in analysis. I hope all the new users will be as impressed with it as I have-and learn as much from it as previous generations have.

I've dedicated each of my so-far published works to one of my former teachers. This is one is no exception. I first learned serious real analysis, after a smattering learned in various classes-under Gerald Itzowitz, a prominent analyst who specialized in abstract harmonic analysis. I audited a number of his courses because I was too broke at the time to actually register for them. He was kind enough to allow me to, including giving me his notes for his more advanced courses, which I couldn't attend full time with my advanced biochemistry courses and helping my cancer-stricken father. I still have his wonderfully clear and concise old school photocopy lecture notes for his full year undergraduate real analysis course as well as semester-long graduate courses in measure theory, functional analysis and topological groups.[5] They were filled with both examples and strong problem sets, which he required his students to work hard on as handed in solutions for them were the basis for most of the grade in the course. Like me, he hated exams because he felt they were easy to "beat" with tricks and didn't really prove anything about how much the student learned, whereas substantial problem sets demonstrated much more about a student's ability and understanding. That's why only the best and/or most committed mathematics students took his courses.[6] We didn't agree on everything, but I learned an enormous amount from him, as did all his students who were serious about mathematics. He was a wonderful teacher and mentor.

I'm very sure that Dr. Itzkowitz would approve highly of assigning Olmsted's text for his year-long undergraduate real analysis sequence, especially at these prices for his impoverished students at CUNY.[7]

---

5 I actually would love to retype all these notes in Tex at some point and republish them-but one step at a time. Here's hoping.

6 Unfortunately, towards the end of his tenure, there was a cheating scandal in his courses where students began to hire other students to either do the exercises or look up the answers for them-which they then handed in for perfect grades. Dr. Itzkowitz was far too experienced to fall for the con, but it forced him to begin giving in-class exams-much to the sadness of those of us who'd taken him earlier and loved the system. Which demonstrates what I've always said-cheating is not a victimless crime, never is.

7 To my knowledge, at this writing (early 2019), Professor Itzkowitz is still alive, although he has now been retired for well over a decade. If he by some random cosmic coincidence sees this book and the preface therein, I hope he'll confirm my assessment of his approval.

I dedicate this book in memory of his wonderful courses and our many discussions in his office where I arrived utterly confused and emerged far less so every time.

Karo Maestro
New York City
February 2019

1

The Real Number System

101. INTRODUCTION

The reader is already familiar with many of the properties of real numbers. He knows, for example, that $2 + 2 = 4$, that the product of two negative numbers is a positive number, and that if x, y, and z are any real numbers, then $x(y + z) = xy + xz$. The average reader at the level of Calculus usually knows these things because he has been told that they are true—probably by people who know such things simply because they have been told. He knows pretty well why some of these familiar facts (like $2 \cdot 3 = 3 \cdot 2$) are true, but is entitled to be unsure of the reasons for others (like $\sqrt{2} \cdot \sqrt{3} = \sqrt{3} \cdot \sqrt{2}$).

It is important to know that the properties of real numbers which we use almost daily are true, not by *fiat* or *decree*, but by rigorous mathematical *proof*. The situation is not unlike that of Euclidean Plane Geometry. In either case (numbers or geometry), all properties within one particular mathematical system follow by logical inference from a few basic *assumptions*, or *hypotheses*, or *axioms*, which are properties that are "true" only in the relative sense that they are assumed as a working basis for that particular mathematical system. What is true in one system may be false or meaningless in another, but within one logical system "true" means "implied by the axioms."

For the system of real numbers the axioms, as usually given,† are five in number, and are concerned only with the **natural numbers**, 1, 2, 3, ⋯ . These five axioms were given in 1889 by the Italian mathematician G. Peano (1858-1932), and are called the *Peano axioms for the natural numbers*. In the book by Landau, referred to in the footnote, the reader is led by a sequence of steps (consisting of 73 definitions and 301 theorems, most of which are very easy) through the entire construction of the real and complex number systems. The natural numbers are used to build the

---

† For a detailed discussion of the real number system see E. Landau, *Foundations of Analysis* (New York, Chelsea Publishing Company, 1951) or E. W. Hobson, *Theory of Functions* (Washington, Harren Press, 1950).

larger system of positive rational numbers (fractions); these in turn form the basis for constructing the positive real numbers; the positive real numbers then lead to the system of all real numbers; finally, the complex numbers are constructed from the real numbers. In each successive class the basic operations of addition and multiplication and (through the class of real numbers) the relation of order are defined, and shown to be consistent with those of the preceding class considered as a subclass. The result is the number system as we know it.

In this book we shall begin our discussion of the number system at a point far along the route just outlined. We shall, in fact, assume that we already have the entire real number system, and *describe* this system by means of a few fundamental properties which we shall accept without further question. These fundamental properties are so chosen that they give a *complete* description of the real number system, so that all properties of the real numbers are deducible from them.† For this reason, and to distinguish them from other properties, we call these fundamental properties *axioms*, and refer to them as such in the text. These axioms are arranged according to three categories: *basic operations*, *order*, and *completeness*.

The present chapter is devoted exclusively to the real number system. The first part of the chapter is elementary in nature, and includes the axioms for the basic operations and order, a review of mathematical induction, a brief treatment of the integers and rational numbers, and a discussion of geometrical representation and absolute value. The last part, of a more advanced character, is starred for possible omission or postponement, and includes the axiom of completeness and a further treatment of topics introduced earlier.

Except for passing mention, complex numbers are not used in this book. Unless otherwise specified, the word *number* should be interpreted to mean *real number*.

The reader should be advised that many of the properties of real numbers that are given in the Exercises of this chapter are used throughout the book without specific reference. In many cases (like the laws of cancellation and the unique factorization theorem) such properties, when desired, can be located by means of the index.

102. AXIOMS OF THE BASIC OPERATIONS

The basic operations of the number system are *addition* and *multiplication*. As shown below, *subtraction* and *division* can be defined in terms of these two. The axioms of the basic operations are further subdivided into three subcategories: *addition*, *multiplication*, and *addition and multiplication*.

† For a proof of this fact and a discussion of the axioms of the real number system, see G. Birkhoff and S. MacLane, *A Survey of Modern Algebra* (New York, The Macmillan Company, 1944). Also cf. Ex. 13, § 116.

I. Addition

(i) Any two numbers, x and y, have a unique sum, $x + y$.
(ii) **The associative law** holds. That is, if x, y, and z are any numbers,
$$x + (y + z) = (x + y) + z.$$
(iii) There exists a number 0 such that for any number x, $x + 0 = x$.
(iv) Corresponding to any number x there exists a number $-x$ such that
$$x + (-x) = 0.$$
(v) *The* **commutative law** *holds*. That is, if x and y are any numbers,
$$x + y = y + x.$$

The number 0 of axiom (iii) is called **zero**.
The number $-x$ of axiom (iv) is called the **negative** of the number x. The difference between x and y is defined:
$$x - y = x + (-y).$$
The resulting operation is called **subtraction**.

Some of the properties that can be derived from Axioms I alone are given in Exercises 1-7, § 103.

II. Multiplication

(i) Any two numbers, x and y, have a unique product, xy.
(ii) **The associative law** holds. That is, if x, y, and z are any numbers,
$$x(yz) = (xy)z.$$
(iii) There exists a number $1 \neq 0$ such that for any number x, $x \cdot 1 = x$.
(iv) Corresponding to any number $x \neq 0$ there exists a number x^{-1} such that $x \cdot x^{-1} = 1$.
(v) *The* **commutative law** *holds*. That is, if x and y are any numbers,
$$xy = yx.$$

The number 1 of axiom (iii) is called **one** or **unity**.
The number x^{-1} of axiom (iv) is called the **reciprocal** of the number x. The **quotient** of x and y ($y \neq 0$) is defined:
$$\frac{x}{y} = x \cdot y^{-1}.$$
The resulting operation is called **division**.

Some of the properties that can be derived from Axioms II alone are given in Exercises 8-11, § 103.

III. Addition and Multiplication

(i) *The* **distributive law** *holds*. That is, if x, y, and z are any numbers,
$$x(y + z) = xy + xz.$$

The distributive law, together with Axioms I and II, yields further familiar relations, some of which are given in Exercises 12-33, § 103.

103. EXERCISES

In Exercises 1-33, prove the given statement or establish the given equation.

1. There is only one number having the property of the number 0 of Axiom I (iii). *Hint:* Assume that the numbers 0 and $0'$ both have the property. Then simultaneously $0' + 0 = 0'$ and $0 + 0' = 0$.

2. The **law of cancellation for addition** holds: $x + y = x + z$ implies $y = z$. *Hint:* Let $(-x)$ be a number satisfying Axiom I (iv). Then
$$(-x) + (x + y) = (-x) + (x + z).$$
Use the associative law.

3. The negative of a number is unique. *Hint:* If y has the property of $-x$ in Axiom I (iv), § 102, $x + y = x + (-x) = 0$. Use the law of cancellation, given in Exercise 2.

4. $-0 = 0$.

5. $-(-x) = x$.

6. $0 - x = -x$.

7. $-(x + y) = -x - y; -(x - y) = y - x$. *Hint:* By the uniqueness of the negative it is sufficient for the first part to prove that
$$(x + y) + [(-x) + (-y)] = 0.$$
Use the commutative and associative laws.

8. There is only one number having the property of the number 1 of Axiom II (iii).

9. The **law of cancellation for multiplication** holds: $xy = xz$ implies $y = z$ if $x \neq 0$.

10. The reciprocal of a number ($\neq 0$) is unique.

11. $1^{-1} = 1$.

12. $x \cdot 0 = 0$. *Hint:* $x \cdot 0 + 0 = x \cdot 0 = x(0 + 0) = x \cdot 0 + x \cdot 0$. Use the law of cancellation on the extreme members.

13. Zero has no reciprocal.

14. If $x \neq 0$ and $y \neq 0$, then $xy \neq 0$. Equivalently, if $xy = 0$, then either $x = 0$ or $y = 0$. *Hint:* Assume $x \neq 0$, $y \neq 0$, and $xy = 0$. Then
$$x^{-1}(xy) = x^{-1} \cdot 0 = 0.$$
Use the associative law to infer $y = 0$.

15. If $x \neq 0$, then $x^{-1} \neq 0$ and $(x^{-1})^{-1} = x$.

16. $\dfrac{x}{y} = 0$ ($y \neq 0$) if and only if $x = 0$.

17. $\dfrac{1}{x} = x^{-1}$ ($x \neq 0$).

18. If $x \neq 0$ and $y \neq 0$, then $(xy)^{-1} = x^{-1}y^{-1}$, or $\dfrac{1}{xy} = \dfrac{1}{x} \cdot \dfrac{1}{y}$.

19. If $b \neq 0$ and $d \neq 0$, then $\dfrac{a}{b} = \dfrac{ad}{bd}$.

Hint: $(ad)(bd)^{-1} = (ad)(d^{-1}b^{-1}) = a[(dd^{-1})b^{-1}] = ab^{-1}$.

20. If $b \neq 0$ and $d \neq 0$, then $\dfrac{a}{b} \cdot \dfrac{c}{d} = \dfrac{ac}{bd}$.

21. If $b \neq 0$ and $d \neq 0$, then $\dfrac{a}{b} + \dfrac{c}{d} = \dfrac{ad + bc}{bd}$.

Hint: $(bd)^{-1}(ad + bc) = (b^{-1}d^{-1})(ad) + (b^{-1}d^{-1})(bc)$.

22. $(-1)(-1) = 1$. *Hint:* $(-1)(1 + (-1)) = 0$. The distributive law gives $(-1) + (-1)(-1) = 0$. Add 1 to each member.

23. $(-1)x = -x$.

Hint: Multiply each member of the equation $1 + (-1) = 0$ by x.

24. $(-x)(-y) = xy$.

Hint: Write $-x = (-1)x$ and $-y = (-1)y$.

25. $-(xy) = (-x)y = x(-y)$.

26. $-\dfrac{x}{y} = \dfrac{-x}{y} = \dfrac{x}{-y}$ $(y \neq 0)$.

27. $x(y - z) = xy - xz$.

★**28.** $(x - y) + (y - z) = x - z$.

★**29.** $(a - b) - (c - d) = (a + d) - (b + c)$.

★**30.** $(a + b)(c + d) = (ac + bd) + (ad + bc)$.

★**31.** $(a - b)(c - d) = (ac + bd) - (ad + bc)$.

★**32.** $a - b = c - d$ if and only if $a + d = b + c$.

33. The general linear equation $ax + b = 0$, $a \neq 0$, has a unique solution $x = -b/a$.

104. AXIOMS OF ORDER

In addition to the basic *operations*, the real numbers have an *order relation* subject to certain axioms. One form of these axioms expresses order in terms of the primitive concept of *positiveness*:

(i) *Some numbers have the property of being positive.*
(ii) *For any number x exactly one of the following three statements is true:* $x = 0$; x *is positive;* $-x$ *is positive.*
(iii) *The sum of two positive numbers is positive.*
(iv) *The product of two positive numbers is positive.*

Definition I. *The symbols $<$ and $>$* (read "less than" and "greater than," respectively) *are defined by the statements*

$$x < y \text{ if and only if } y - x \text{ is positive};$$
$$x > y \text{ if and only if } x - y \text{ is positive}.$$

Definition II. *The number x is **negative** if and only if $-x$ is positive.*

Definition III. *The symbols $\leq$ and $\geq$* (read "less than or equal to" and "greater than or equal to," respectively) *are defined by the statements*

$$x \leq y \text{ if and only if either } x < y \text{ or } x = y;$$
$$x \geq y \text{ if and only if either } x > y \text{ or } x = y.$$

NOTE 1. The two statements $x < y$ and $y > x$ are equivalent. The two statements $x \leq y$ and $y \geq x$ are equivalent.

NOTE 2. The *sense* of an inequality of the form $x < y$ or $x \leq y$ is said to be the **reverse** of that of an inequality of the form $x > y$ or $x \geq y$.

NOTE 3. The simultaneous inequalities $x < y$, $y < z$ are usually written $x < y < z$, and the simultaneous inequalities $x > y$, $y > z$ are usually written $x > y > z$. Similar interpretations are given the compound inequalities $x \leq y \leq z$ and $x \geq y \geq z$.

105. EXERCISES

In Exercises 1-24, prove the given statement or establish the given equation.

1. $x > 0$ if and only if x is positive.

2. The **transitive law** holds for $<$ and for $>$: $x < y$, $y < z$ imply $x < z$; $x > y$, $y > z$ imply $x > z$. (Cf. Ex. 19.)

3. The **law of trichotomy** holds: *for any x and y, exactly one of the following holds*: $x < y$, $x = y$, $x > y$.

4. Addition of any number to both members of an inequality preserves the order relation: $x < y$ implies $x + z < y + z$. A similar fact holds for subtraction. *Hint*: $(y + z) - (x + z) = y - x$.

5. $x < 0$ if and only if x is negative.

6. The sum of two negative numbers is negative.

7. The product of two negative numbers is positive. *Hint*: $xy = (-x)(-y)$. (Cf. Ex. 24, § 103.)

8. The square of any nonzero real number is positive.

9. $1 > 0$.

10. The equation $x^2 + 1 = 0$ has no real root.

11. The product of a positive number and a negative number is negative.

12. The reciprocal of a positive number is positive. The reciprocal of a negative number is negative.

13. $0 < x < y$ imply $0 < \frac{1}{y} < \frac{1}{x}$.

14. Multiplication or division of both members of an inequality by a positive number preserves the order relation: $x < y$, $z > 0$ imply $xz < yz$ and $x/z < y/z$.

15. Multiplication or division of both members of an inequality by a negative number reverses the order relation: $x < y$, $z < 0$ imply $xz > yz$ and $x/z > y/z$.

16. $a < b$, $c < d$ imply $a + c < b + d$.

17. $0 < a < b$, $0 < c < d$ imply $ac < bd$ and $a/d < b/c$.

18. If x and y are nonnegative numbers, then $x < y$ if and only if $x^2 < y^2$ (cf. Ex. 10, § 107).

19. The transitive law holds for $\leq$ (also for $\geq$): $x \leq y$, $y \leq z$ imply $x \leq z$. (Cf. Ex. 2.)

20. $x \leq y$, $y \leq x$ imply $x = y$.

21. $x + x = 0$ implies $x = 0$, $x + x + x = 0$ implies $x = 0$.

22. $x^2 + y^2 \geq 0$; $x^2 + y^2 > 0$ unless $x = y = 0$.

§ 106] POSITIVE INTEGERS 7

23. If x is a fixed number satisfying the inequality $x < \epsilon$ for every positive number ϵ, then $x \leq 0$. *Hint:* If x were positive one could choose $\epsilon = x$.

24. There is no largest number, and therefore the real number system is infinite. (Cf. Ex. 11, § 109.) *Hint:* $x + 1 > x$.

25. If $x < a < y$ or if $x > a > y$, then a is said to be **between** x and y. Prove that if x and y are distinct numbers, their arithmetic mean $\frac{1}{2}(x + y)$ is between them. (The number 2 is defined: $2 = 1 + 1$; cf. § 106.)

26. If $x^2 = y$, then x is called a **square root** of y. By Exercise 8, if such a number x exists, y must be nonnegative, and if $y = 0$, $x = 0$ is the only square root of y. Show that if a positive number y has square roots, it has exactly two square roots, one positive and one negative. The unique positive square root is called **the square root** and is written $\sqrt{y}$. It is shown in § 214 that such square roots do exist. (Cf. Ex. 14, § 116.) *Hint:* Let $x^2 = z^2 = y$. Then
$$x^2 - z^2 = (x - z)(x + z) = 0.$$
Therefore $x = z$ or $x = -z$.

106. POSITIVE INTEGERS AND MATHEMATICAL INDUCTION

In the development of the real number system, as discussed briefly in the Introduction, the natural numbers become absorbed in successively larger number systems, losing their identity but not their properties, and finally emerge in the real number system as **positive integers**, $1, 2, 3, \cdots$.

The *positive integers* have certain elementary properties that follow directly from the role of the natural numbers in the construction of the number system (cf. § 115 and Ex. 16, § 116, for further remarks):

(i) The "positive integers" are positive. (Cf. Ex. 9, § 105.)
(ii) If n is a positive integer, $n \geq 1$.
(iii) $2 = 1 + 1, 3 = 2 + 1, 4 = 3 + 1, 5 = 4 + 1, \cdots$.
(iv) $0 < 1 < 2 < 3 < 4 < \cdots$.
(v) The sum and product of two positive integers are positive integers.
(vi) If m and n are positive integers, and $m < n$, then $n - m$ is a positive integer.
(vii) If n is a positive integer, there is no positive integer m such that $n < m < n + 1$.

An important property of the positive integers, one which validates mathematical induction, is a rewording of the fifth Peano axiom for the natural numbers:

Axiom of Induction. *If S is a set of positive integers with the two properties (i) S contains the number 1, and (ii) whenever S contains the positive integer n it also contains the positive integer $n + 1$, then S contains all positive integers.*

An immediate consequence is the theorem:

Fundamental Theorem of Mathematical Induction. *For every positive integer n let $P(n)$ be a proposition which is either true or false. If (i) $P(1)$ is true and (ii) whenever this proposition is true for the positive integer n it is also true for the positive integer $n + 1$, then $P(n)$ is true for all positive integers n.*

Proof: Let S be the set of positive integers for which $P(n)$ is true, and use the Axiom of Induction.

Another consequence (less immediate) of the Axiom of Induction is the principle:

Well-ordering Principle. *Every nonempty set of positive integers (that is, every set of positive integers which contains at least one member) contains a smallest member.*

Proof. Let T be an arbitrary set of positive integers which contains at least one member, and assume that T has no smallest member. We shall obtain a contradiction by letting S be the set of all positive integers n having the property that every member of T is greater than n. Clearly 1 is a member of S, since every member of T is at *least* equal to 1, and if a member of T were *equal* to 1 it would be the smallest member of T (property (ii) above). Now suppose n is a member of S. Then every member of T is greater than n and therefore (property (ii) above) is at least $n + 1$. But by the same argument as that used above, any member of T equal to $n + 1$ would be the smallest member of T. Therefore every member of T is greater than $n + 1$ and $n + 1$ belongs to S. Consequently every positive integer is a member of S, and T must be empty.

We state two general laws which are familiar to all, whose detailed proofs by mathematical induction are given in § 115.

General Associative Laws. *Any two sums (products) of the n numbers $x_1, x_2, \cdots, x_n$ in the same order are equal regardless of the manner in which the terms (factors) are grouped by parentheses.*

Illustration. Let $a = x_1(x_2(x_3(x_4 x_5)))$ and $b = ((x_1 x_2)(x_3 x_4))x_5$. We shall show that $b = a$ by using the associative law of § 102 to transform b step by step into a. A similar sequence of steps would transform any product of the five numbers into the "standard" product a, and hence justify the theorem for $n = 5$. We start by thinking of the products $(x_1 x_2)$ and $(x_3 x_4)$ as single numbers and use the associative law to write $b = (x_1 x_2)((x_3 x_4) x_5)$. Repeating this method we have: $b = (x_1 x_2)(x_3(x_4 x_5)) = x_1(x_2(x_3(x_4 x_5))) = a$.

NOTE. As a consequence of the Axiom of Induction any sum or product of n numbers exists, and by the general associative laws any such sum or product can be written without parentheses, thus: $x_1 + x_2 + \cdots + x_n$ and $x_1 x_2 \cdots x_n$.

§ 106]   POSITIVE INTEGERS

General Commutative Laws. *Any two sums (products) of the n numbers $x_1, x_2, \cdots, x_n$ are equal regardless of the order of the terms (factors).*

Illustration. Let $a = x_1x_2x_3x_4x_5$ and $b = x_4x_1x_5x_2x_3$. We shall show that $b = a$ by using the commutative law of § 102 to transform b step by step into a. We first bring x_1 to the left-hand end: $b = (x_4x_1)(x_5x_2x_3) = (x_1x_4)(x_5x_2x_3) = x_1x_4x_5x_2x_3$. Next we take care of x_2: $b = x_1x_4(x_5x_2)x_3 = x_1(x_4x_2)x_5x_3 = x_1x_2x_4x_5x_3$. Finally, after x_3 is moved two steps to the left, the form a is reached.

The following examples illustrate the principles and uses of mathematical induction:

Example 1. Establish the formula

(1) $$1^2 + 3^2 + 5^2 + \cdots + (2n - 1)^2 = \frac{n(4n^2 - 1)}{3}$$

for every positive integer n.

Solution. Let $P(n)$ be the proposition (1). Direct substitution shows that $P(1)$ is true. We wish to show that whenever $P(n)$ is true for a *particular* positive integer n it is also true for the positive integer $n + 1$. Accordingly, we assume (1) and wish to establish

(2) $$1^2 + 3^2 + 5^2 + \cdots + (2n - 1)^2 + (2n + 1)^2 = \frac{(n + 1)[4(n + 1)^2 - 1]}{3}.$$

On the assumption that (1) is correct (for a particular value of n), we can rewrite the left-hand member of (2) by grouping:

$$[1^2 + 3^2 + \cdots + (2n - 1)^2] + (2n + 1)^2 = \frac{n(4n^2 - 1)}{3} + (2n + 1)^2.$$

Thus verification of (2) reduces to verification of

(3) $$\frac{n(4n^2 - 1)}{3} + (2n + 1)^2 = \frac{(n + 1)(4n^2 + 8n + 3)}{3},$$

which, in turn, is true (divide by 3) by virtue of

(4) $\quad 4n^3 - n + 3(4n^2 + 4n + 1) = (4n^3 + 8n^2 + 3n) + (4n^2 + 8n + 3).$

By the Fundamental Theorem of Mathematical Induction, (1) is true for all positive integers n.

Example 2. Prove the *general distributive law*:

(5) $\quad x(y_1 + y_2 + \cdots + y_n) = xy_1 + xy_2 + \cdots + xy_n.$

Solution. Let $P(n)$ be the proposition (5). $P(1)$ is a triviality, and $P(2)$ is true by the distributive law III(i), § 102. We wish to show now that the truth of (5) for a particular positive integer n implies the truth of $P(n + 1)$:

(6) $\quad x(y_1 + y_2 + \cdots + y_n + y_{n+1}) = xy_1 + xy_2 + \cdots + xy_n + xy_{n+1}.$

By using the distributive law of § 102 and the assumption (5), we can rewrite the left-hand member of (6) as follows:

$$x[(y_1 + \cdots + y_n) + y_{n+1}] = x(y_1 + \cdots + y_n) + xy_{n+1}$$
$$= (xy_1 + \cdots + xy_n) + xy_{n+1}.$$

Since this last expression is equal to the right-hand member of (6), the truth of $P(n)$, or (5), is established for all positive integers n by the Fundamental Theorem of Mathematical Induction.

107. EXERCISES

1. Prove that $2 + 2 = 4$. *Hint:* Use the associative law with $4 = 3 + 1 = (2 + 1) + 1$.

2. Prove that $2 \cdot 3 = 6$.

3. Prove that $6 + 8 = 14$ and that $3 \cdot 4 = 12$.

4. Prove that the sum and product of n positive integers are positive integers.

5. Prove that any product of nonzero numbers is nonzero.

6. Prove that if $x_1 \neq 0, x_2 \neq 0, \cdots, x_n \neq 0$, then $(x_1 x_2 \cdots x_n)^{-1} = x_1^{-1} x_2^{-1} \cdots x_n^{-1}$.

7. Prove that if $x_1 < x_2, x_2 < x_3, \cdots, x_{n-1} < x_n$ (usually written $x_1 < x_2 < \cdots < x_n$), then $x_1 < x_n$.

8. Prove that any sum or product of positive numbers is positive.

9. Use mathematical induction to prove that if n is any positive integer, then $x^n - y^n = (x - y)(x^{n-1} + x^{n-2}y + \cdots + xy^{n-2} + y^{n-1})$. *Hint:* $x^{n+1} - y^{n+1} = x^n(x - y) + y(x^n - y^n)$.

10. Prove that if x and y are nonnegative numbers and n is a positive integer, then $x > y$ if and only if $x^n > y^n$. *Hint:* Use Ex. 9. (Cf. Ex. 18, § 105.)

11. Establish the inequality $2^n > n$, where n is a positive integer.

12. Establish the formula $1 + 2 + \cdots + n = \frac{1}{2}n(n + 1)$. (Cf. Ex. 38.)

13. Establish the formula (cf. Ex. 39)
$$1^2 + 2^2 + \cdots + n^2 = \tfrac{1}{6}n(n + 1)(2n + 1).$$

14. Establish the formula (cf. Ex. 40)
$$1^3 + 2^3 + \cdots + n^3 = \tfrac{1}{4}n^2(n + 1)^2 = (1 + 2 + \cdots + n)^2.$$

15. Establish the formula (cf. Ex. 41)
$$1^4 + 2^4 + \cdots + n^4 = \tfrac{1}{30}n(n + 1)(2n + 1)(3n^2 + 3n - 1).$$

★16. Let A be a nonempty set of real numbers with the property that whenever x is a member of A and y is a member of A then $x - y$ is a member of A. Prove that whenever $x_1, x_2, \cdots, x_n$ are members of A then $x_1 + x_2 + \cdots + x_n$ is a member of A.

17. Establish the law of exponents: $a^m a^n = a^{m+n}$, where a is any number and m and n are positive integers. *Hint:* Hold m fixed and use induction on n.

18. Establish the law of exponents:
$$\frac{a^m}{a^n} = \begin{cases} a^{m-n} & \text{if } m < n, \\ \dfrac{1}{a^{n-m}} & \text{if } n < m, \end{cases}$$
where a is any nonzero number and m and n are positive integers.

19. Establish the law of exponents: $(a^m)^n = a^{mn}$, where a is any number and m and n are positive integers.

20. Establish the law of exponents: $(ab)^n = a^n b^n$, where a and b are any numbers and n is a positive integer. Generalize to m factors.

21. Establish the law of exponents $\left(\dfrac{a}{b}\right)^n = \dfrac{a^n}{b^n}$, where a and b are any numbers ($b \neq 0$) and n is a positive integer.

§ 107] EXERCISES 11

22. A positive integer m is a **factor** of a positive integer p if and only if there exists a positive integer n such that $p = mn$. A positive integer p is called **composite** if and only if there exist integers $m > 1$ and $n > 1$ such that $p = mn$. A positive integer p is **prime** if and only if $p > 1$ and p is not composite. Prove that if m_i, $i = 1, 2, \cdots, n$, are integers > 1, then $m_1 m_2 \cdots m_n > n$. Hence prove that any integer > 1 is either a prime or a product of primes. (Cf. Ex. 29.)

23. Two positive integers are **relatively prime** if and only if they have no common integral factor greater than 1. A fraction p/q, where p and q are positive integers, is **in lowest terms** if and only if p and q are relatively prime. Prove that any quotient of positive integers is equal to such a fraction in lowest terms. (Cf. Ex. 33.)

★24. Prove that the positive integers are **Archimedean** (cf. § 114): If a and b are positive integers, then there exists a positive integer n such that $na > b$. *Hint:* $ba \geq b$.

★25. Prove the **Fundamental Theorem of Euclid**: *If a and b are positive integers, there exist unique numbers n and r, each of which is either 0 or a positive integer and where $r < a$, such that*

$$b = na + r.$$

Hint: For existence, let $n + 1$ be the smallest positive integer such that $(n + 1)a > b$ (cf. Ex. 24).

★26. Prove that a prime number p cannot be a factor of the product of two positive integers, a and b, each of which is less than p. *Hint:* Assuming that p is a factor of ab, let c be the smallest positive integer such that p is a factor of ac, and let n be a positive integer such that $nc < p < (n + 1)c$ (cf. Ex. 24). Then $p - nc$ is a positive integer less than c having the property assumed for c!

★27. Prove that if a prime number p is a factor of the product of two positive integers, a and b, then p is a factor of either a or b (or both). *Hint:* Use Exs. 25 and 26.

★28. Prove that if a prime number p is a factor of the product of n positive integers $a_1, a_2, \cdots, a_n$, then p is a factor of at least one of these numbers. (Cf. Ex. 27.)

★29. Prove the **Unique Factorization Theorem**: *Every positive integer greater than 1 can be represented in one and only one way as a product of primes.* *Hint:* Assume $p_1 p_2 \cdots p_m$ and $q_1 q_2 \cdots q_n$ are two nonidentical factorizations of a positive integer. Cancel all identical prime factors from both members of the equation $p_1 p_2 \cdots p_m = q_1 q_2 \cdots q_n$. A prime number remains which is a factor of both members but which violates Ex. 28. (Cf. Ex. 22.)

★30. Prove that if a, b, and c are positive integers such that a and c are relatively prime and b and c are relatively prime, then ab and c are relatively prime. (Cf. Ex. 23.)

★31. Prove that if a, b, and c are positive integers such that a and c are relatively prime and c is a factor of ab, then c is a factor of b. (Cf. Ex. 23.)

★32. Prove that two positive integers each of which is a factor of the other are equal.

★33. Prove that if a fraction p/q, where p and q are positive integers, is in lowest terms, then p and q are uniquely determined. (Cf. Ex. 23.)

34. If n is a positive integer, **n factorial**, written $n!$, is defined: $n! = 1 \cdot 2 \cdot 3 \cdots n$. **Zero factorial** is defined: $0! = 1$. If r is a positive integer or zero and if $0 \leq r \leq n$, the binomial coefficient $\binom{n}{r}$ (also written $_nC_r$) is defined:

$$\binom{n}{r} = \frac{n!}{(n-r)!\,r!}.$$

Prove that $\binom{n}{r}$ is a positive integer. *Hint:* Establish the law of *Pascal's Triangle* (cf. any College Algebra text): $\binom{n+1}{r} = \binom{n}{r-1} + \binom{n}{r}$.

35. Prove the **Binomial Theorem** for positive integral exponents (cf. V, § 807): *If x and y are any numbers and n is a positive integer,*

$$(x+y)^n = \binom{n}{0}x^n + \binom{n}{1}x^{n-1}y + \cdots + \binom{n}{r}x^{n-r}y^r + \cdots + \binom{n}{n}y^n.$$

(Cf. Ex. 34.)

36. The **sigma summation notation** is defined:

$$\sum_{k=m}^{n} f(k) \equiv f(m) + f(m+1) + \cdots + f(n), \text{ where } n \geq m. \text{ Prove:}$$

(i) k is a **dummy variable**: $\sum_{k=m}^{n} f(k) = \sum_{i=m}^{n} f(i)$.

(ii) $\sum$ is **additive**: $\sum_{k=m}^{n} [f(k) + g(k)] = \sum_{k=m}^{n} f(k) + \sum_{k=m}^{n} g(k)$.

(iii) $\sum$ is **homogeneous**: $\sum_{k=m}^{n} cf(k) = c \sum_{k=m}^{n} f(k)$.

(iv) $\sum_{k=m}^{n} 1 = n - m + 1$.

★37. A useful summation formula is

(1) $$\sum_{k=1}^{n} [f(k) - f(k-1)] = f(n) - f(0).$$

Establish this by mathematical induction.

★38. By means of Exercises 36 and 37 *derive* the formula of Exercise 12. *Hint:* Let $f(n) \equiv n^2$. Then (1) becomes

$$\sum_{k=1}^{n} [k^2 - (k-1)^2] = \sum_{k=1}^{n} (2k-1) = n^2, \text{ or}$$

$$2 \sum_{k=1}^{n} k = n^2 + n.$$

★39. By means of Exercises 36–38, *derive* the formula of Exercise 13. *Hint:* Let $f(n) \equiv n^3$ in (1).

★40. By means of Exercises 36–39, *derive* the formula of Exercise 14.

★41. By means of Exercises 36–40, *derive* the formula of Exercise 15.

★42. Use mathematical induction to prove that $\sum_{k=1}^{n} k^m$ is a polynomial in n of degree $m+1$ whose leading coefficient (coefficient of the term of highest degree) is $1/(m+1)$. (Cf. Exs. 36–41.)

§ 108] INTEGERS AND RATIONAL NUMBERS

★43. If $a_1, a_2, \cdots, a_n$ and $b_1, b_2, \cdots, b_n$ are real numbers, show that
$$\left(\sum_{i=1}^{n} a_i^2\right)\left(\sum_{i=1}^{n} b_i^2\right) - \left(\sum_{i=1}^{n} a_i b_i\right)^2 = \sum_{1 \leq i < j \leq n} (a_i b_j - a_j b_i)^2.$$
Hence establish the **Schwarz** (or **Cauchy**) **inequality**:
$$\left(\sum_{i=1}^{n} a_i b_i\right)^2 \leq \sum_{i=1}^{n} a_i^2 \sum_{i=1}^{n} b_i^2.$$
(Cf. Ex. 29, § 503; Ex. 26, § 711; Ex. 14, § 717; Exs. 38, 40, § 1225.)

★44. By use of Exercise 43, establish the **Minkowski inequality**:
$$\sqrt{\sum_{i=1}^{n} (a_i + b_i)^2} \leq \sqrt{\sum_{i=1}^{n} a_i^2} + \sqrt{\sum_{i=1}^{n} b_i^2}.$$
Hint: Square both members of the Minkowski inequality, expand each term on the left, cancel identical terms that result, divide by 2, and reverse steps. (Cf. Ex. 30, § 503; Ex. 14, § 717; Exs. 39, 40, § 1225.)

108. INTEGERS AND RATIONAL NUMBERS

In order to rely as little as possible on the constructive process behind real numbers, we shall formulate definitions of integers and rational numbers in terms of the more primitive concept of *positive integer*. Having done this in order to achieve an economy in basic concepts, we must suffer the consequence of being forced to prove "obvious" statements. For instance, we all know that the sum of two integers is an integer, yet the student is asked to prove this elementary fact in Exercise 1 of the following section! Inspection of the following two definitions and of the hint accompanying that exercise should dispel any confusion.

Definition I. *A number x is a **negative integer** if and only if $-x$ is a positive integer. A number x is an **integer** if and only if it is 0 or a positive integer or a negative integer.*

Definition II. *A number x is a **rational number** if and only if there exist integers p and q, where $q \neq 0$, such that $x = p/q$. The real numbers that are not rational are called **irrational**.*

109. EXERCISES

1. Prove that the sum of two integers is an integer. *Hint:* Consider all possible cases of signs. For the case $m > -n > 0$, use property (vi), § 106.

2. Prove that the product of two integers is an integer.

3. Prove that the difference between two integers is an integer.

4. Prove that the quotient of two integers need not be an integer.

5. Prove that the integers are rational numbers.

6. Prove that the sum, difference, and product of two rational numbers are rational numbers, and that the quotient of two rational numbers the second of which is nonzero is a rational number.

7. Define integral powers a^n, where $a \neq 0$ and n is any integer, and establish the laws of exponents of Exercises 17-21, § 107, for integral exponents.

8. The square root of a positive number was defined in Exercise 26, § 105. Existence is proved in § 214. Prove that $\sqrt{2}$ is irrational. That is, prove that there is no positive rational number whose square is 2. *Hint:* Assume $\sqrt{2} = p/q$, where p and q are relatively prime positive integers (cf. Ex. 23, § 107). Then $p^2 = 2q^2$, and p is even and of the form $2k$. Repeat the argument to show that q is also even! (Cf. Ex. 14, § 116.)

★9. Assume that $f(x) = a_0 x^n + a_1 x^{n-1} + \cdots + a_{n-1} x + a_n$ is a polynomial with integral coefficients $a_0, \cdots, a_n$, of which the leading coefficient a_0 and the constant term a_n are nonzero. Prove that if p/q is a rational root of the equation $f(x) = 0$, where p and q are relatively prime integers (that is, they are nonzero and they have no common integral factor greater than 1; cf. Ex. 23, § 107), then p is a factor of a_n and q is a factor of a_0. *Hint:* By assumption $a_0 p^n + a_1 p^{n-1} q + \cdots + a_n q^n = 0$. Use Ex. 31, § 107, and the fact that each of p and q is a factor of n of these terms, and therefore also of the remaining term.

★10. Prove that a positive integer m cannot have a rational nth root (n is a positive integer), unless m is a perfect nth power (of a positive integer), and hence generalize Exercise 8. *Hint:* Use Ex. 9.

★11. Prove that a positive rational number p/q, where p and q are relatively prime positive integers, cannot have a rational nth root (n is a positive integer) unless both p and q are perfect nth powers. (Cf. Ex. 23, § 107; Ex. 10 above.)

★12. Let a and b be any two nonzero integers, and consider the set
$$G(a, b) = \{ma + nb\}$$
consisting of all numbers of the form $ma + nb$, where m and n are arbitrary integers. Prove that $G(a, b)$ consists precisely of all integral multiples of a positive integer k. *Hint:* Show first that whenever c and d belong to $G(a, b)$, so do $c + d$ and $c - d$, and that whenever c belongs to $G(a, b)$ and h is an arbitrary integer, then ch belongs to $G(a, b)$. Then let k be the smallest positive integer in $G(a, b)$. If there were a member of $G(a, b)$ not a multiple of k, use the Fundamental Theorem of Euclid (Ex. 25, § 107) to obtain a contradiction.

★13. Prove that the number k of Exercise 12 is unique. Prove that k is a factor of both a and b, and that any common factor of a and b is a factor of k. (In other words, k is the *highest common factor* of a and b.)

★14. Prove that if a and b are relatively prime integers, there exist integers m and n such that
$$ma + nb = 1.$$
(Cf. Exs. 12-13.)

NOTE. In the following problems, **nonzero polynomial** means any polynomial different from the **zero polynomial** all of whose coefficients are zero.

★15. Prove the *Fundamental Theorem of Euclid for polynomials:* If $f(x)$ and $g(x)$ are nonzero polynomials, there exist unique polynomials $Q(x)$ and $R(x)$, where $R(x)$ either is 0 or has degree less than that of $g(x)$, such that
$$f(x) = Q(x)g(x) + R(x).$$

The polynomials $Q(x)$ and $R(x)$ are called the **quotient** and **remainder**, respectively, when $f(x)$ is divided by $g(x)$. (Cf. Ex. 25, § 107.)

★16. Let $f(x)$ and $g(x)$ be any two nonzero polynomials, and consider the set
$$G(f, g) = \{\phi(x)f(x) + \psi(x)g(x)\}$$
consisting of all polynomials of the form $\phi f + \psi g$, where ϕ and ψ are arbitrary polynomials. Prove that $G(f, g)$ consists precisely of all polynomial multiples of a nonzero polynomial $P(x)$. *Hint:* See Ex. 12, observing that $P(x)$ is not uniquely determined, but is any nonzero polynomial in $G(f, g)$ of lowest degree.

★17. State and prove the analogue of Exercise 13 for Exercise 16. *Hint:* $P(x)$ is unique except for a nonzero constant factor.

★18. Prove that if $f(x)$ and $g(x)$ are relatively prime polynomials (that is, they are nonzero and have only constants as common polynomial factors), there exist polynomials $\phi(x)$ and $\psi(x)$ such that
$$\phi(x)f(x) + \psi(x)g(x) = 1.$$
(Cf. Exs. 16–17.)

★19. Prove that any fraction of the form $N(x)/f(x)g(x)$, where $N(x)$ is a polynomial and $f(x)$ and $g(x)$ are relatively prime polynomials, can be written in the form
$$\frac{N(x)}{f(x)g(x)} = \frac{A(x)}{f(x)} + \frac{B(x)}{g(x)},$$
where $A(x)$ and $B(x)$ are polynomials. *Hint:* Take the equation of Ex. 18, divide by $f(x)g(x)$, and multiply by $N(x)$.

★20. Prove that if $f(x)$ and $g(x)$ are nonconstant polynomials and if the degree of $N(x)$ is less than that of $f(x)g(x)$, then the polynomials $A(x)$ and $B(x)$ of Exercise 19 can be chosen so that their degrees are less than those of $f(x)$ and $g(x)$, respectively. *Hint:* Write
$$\frac{N(x)}{f(x)g(x)} = \frac{\Phi(x)}{f(x)} + \frac{\Psi(x)}{g(x)},$$
where $\Phi(x)$ and $\Psi(x)$ are polynomials, and write $\Phi(x) = Q(x)f(x) + A(x)$, where the degree of $A(x)$ is less than that of $f(x)$, and define $B(x) = \Psi(x) + Q(x)g(x)$. Under the assumption that the degree of $B(x)$ is at least that of $g(x)$, obtain a contradiction from the equation $N(x) = A(x)g(x) + B(x)f(x)$.

★21. Prove that any fraction of the form $A(x)/[p(x)]^n$, where $A(x)$ and $p(x)$ are nonconstant polynomials and n is a positive integer, and where the degree of the numerator $A(x)$ is less than that of the denominator $[p(x)]^n$, can be written in the form
$$\frac{A(x)}{[p(x)]^n} = \frac{B(x)}{[p(x)]^{n-1}} + \frac{C(x)}{[p(x)]^n},$$
where $B(x)$ and $C(x)$ are polynomials such that $B(x)$ either is 0 or has degree less than that of $[p(x)]^{n-1}$, and $C(x)$ either is 0 or has degree less than that of $p(x)$. *Hint:* Use Ex. 15 to write $A(x) = B(x)p(x) + C(x)$.

★22. Recall the fact from College Algebra that any real polynomial (any polynomial with real coefficients) of positive degree can be factored into real linear and quadratic factors; more precisely, as a constant times a product of factors of the form $(x + a)$ and $(x^2 + bx + c)$, where the discriminant $b^2 - 4c$ is negative. Use this fact and those contained in Exercises 19-21 to prove the

Fundamental Theorem on Partial Fractions (used in Integral Calculus): *Any quotient of real polynomials, where the degree of the numerator is less than that of the denominator, can be expressed as a sum of fractions of the form*

$$\frac{A}{(x+a)^m} \quad \text{and} \quad \frac{Bx+C}{(x^2+bx+c)^n},$$

where A, B, and C are constants and m and n are positive integers.

110. GEOMETRICAL REPRESENTATION AND ABSOLUTE VALUE

The reader has doubtless made use of the standard representation of real numbers by means of points on a straight line. It is conventional, when considering this line to lie horizontally as in Figure 101, to adopt a

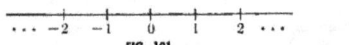

FIG. 101

uniform scale, with numbers increasing to the right and decreasing to the left. With an appropriate axiomatic system for Euclidean geometry† there is a one-to-one correspondence between real numbers and points on a line. That is, to any real number there corresponds precisely one point of the line, and to any point of the line, there corresponds precisely one real number. For this reason it is often immaterial whether one speaks of *numbers* or *points*. In this book we shall frequently use these two words interchangeably, and feel free, for example, to speak of the *point* 3. In this sense, in Figure 101, positive numbers lie to the right of the point 0, and $x < y$ if and only if the point x is to the left of the point y. Again, if $x < z$, then the number y satisfies the simultaneous inequalities $x < y < z$ if and only if the point y is between the points x and z (cf. Ex. 25, § 105). Properties of the real numbers, axiomatized and obtained in this chapter, lend strength to our intuitive conviction that a straight line with a number scale furnishes a reliable picture of the real number system.

Definition I. *If a and b are any two real numbers such that $a < b$, the* **open interval** *from a to b, written (a,b), is the set of all numbers x between a and b, $a < x < b$. The* **closed interval** *from a to b, written [a,b], includes the points a and b and is the set of all x such that $a \leq x \leq b$. The* **half-open intervals** *(a,b] and [a,b) are defined by the inequalities $a < x \leq b$ and $a \leq x < b$, respectively. In any of these cases the interval is called a* **finite interval** *and the points a and b are called* **end-points**. **Infinite intervals** *are denoted and defined as follows, the point a, where it appears, being the*

† Cf. D. Hilbert, *The Foundations of Geometry* (La Salle, Ill., The Open Court Publishing Company, 1938).

end-point of the interval: $(a, +\infty), x > a; [a, +\infty), x \geq a; (-\infty, a), x < a;$ $(-\infty, a], x \leq a; (-\infty, +\infty),$ all x.† *Any point of an interval that is not an end-point is called an **interior point** of the interval.*

Definition II. *The **absolute value** of a number x, written $|x|$, is defined:*

$$|x| \equiv \begin{cases} x \text{ if } x \geq 0, \\ -x \text{ if } x < 0. \end{cases}$$

The absolute value of a number can be thought of as its *distance* from the **origin** 0 in Figure 101. Similarly, the absolute value of the difference between two numbers, $|x - y|$, is the distance between the two points x and y. Some of the more useful properties of the absolute value are given below. For hints for some of the proofs, see § 111.

Properties of Absolute Value

  I. $|x| \geq 0$; $|x| = 0$ *if and only if* $x = 0$.
 II. $|xy| = |x| \cdot |y|$.
III. $\left|\dfrac{x}{y}\right| = \dfrac{|x|}{|y|}$ $(y \neq 0)$.
 IV. *If $\epsilon > 0$, then*
    (i) *the inequality $|x| < \epsilon$ is equivalent to the simultaneous inequalities $-\epsilon < x < \epsilon$;*
    (ii) *the inequality $|x| \leq \epsilon$ is equivalent to the simultaneous inequalities $-\epsilon \leq x \leq \epsilon$.*
  V. *The **triangle inequality**‡ holds*: $|x + y| \leq |x| + |y|$.
 VI. $|-x| = |x|$; $|x - y| = |y - x|$.
VII. $|x|^2 = x^2$; $|x| = \sqrt{x^2}$. (Cf. Ex. 26, § 105.)
VIII. $|x - y| \leq |x| + |y|$.
 IX. $||x| - |y|| \leq |x - y|$.

Definition III. *A **neighborhood** or **epsilon-neighborhood** of a point a is an open interval of the form $(a - \epsilon, a + \epsilon)$, where ϵ is a positive number.*

By Property IV, the neighborhood $(a - \epsilon, a + \epsilon)$ consists of all x satisfying the inequalities $a - \epsilon < x < a + \epsilon$, or $-\epsilon < x - a < \epsilon$, or $|x - a| < \epsilon$. It consists, therefore, of all points whose distance from a is less than ϵ. The point a is the midpoint of each of its neighborhoods.

---

† These infinite intervals are also sometimes designated directly by means of inequalities: $a < x < +\infty$, $a \leq x < +\infty$, $-\infty < x < a$, $-\infty < x \leq a$, and $-\infty < x < +\infty$, respectively.

‡ Property V is called the *triangle inequality* because the corresponding inequality for complex numbers states that any side of a triangle is less than or equal to the sum of the other two. (Cf. Ex. 22, § 1005; §§ 1026, 1503.)

18   THE REAL NUMBER SYSTEM   [§ 111

111. EXERCISES

1. Prove Property I, § 110.
2. Prove Property II, § 110.
3. Prove Property III, § 110. *Hint:* Use Property II with $z = x/y$, so that $x = yz$.
4. Prove Property IV, § 110.
5. Prove Property V, § 110. *Hint:* For the case $x > 0$ and $y < 0$,
$$x + y < x + 0 < x - y = |x| + |y|,$$
$$-(x + y) = -x - y < x - y = |x| + |y|.$$
Use Property IV. Consider all possible cases of sign.
6. Prove Property VI, § 110.
7. Prove Property VII, § 110.
8. Prove Property VIII, § 110.
9. Prove Property IX, § 110. *Hint:* The inequality $|x| - |y| \leq |x - y|$ follows by Property V from $|(x - y) + y| \leq |x - y| + |y|$.
10. Prove that $|x_1 \cdot x_2 \cdots x_n| = |x_1| \cdot |x_2| \cdots |x_n|$.
11. Prove the *general triangle inequality*:
$$|x_1 + x_2 + \cdots + x_n| \leq |x_1| + |x_2| + \cdots + |x_n|.$$
12. Replace by an equivalent single inequality:
$$x > a + b, \quad x > a - b.$$

In Exercises 13-24, find the values of x that satisfy the given inequality or inequalities. Express your answer without absolute values.

13. $|x - 2| < 3$.   14. $|x + 3| \geq 2$.
15. $|x - 5| < |x + 1|$. *Hint:* Square both members. (Cf. Ex. 18, § 105.)
★16. $|x - 4| > x - 2$.   ★17. $|x - 4| \leq 2 - x$.
★18. $|x - 2| > x - 4$.   ★19. $|x^2 - 2| \leq 1$.
★20. $x^2 - 2x - 15 < 0$. *Hint:* Factor and graph the left-hand member.
★21. $x^2 + 10 < 6x$.   ★22. $|x + 5| < 2|x|$.
★23. $x < x^2 - 12 < 4x$.   ★24. $|x - 7| < 5 < |5x - 25|$.

In Exercises 25-28, solve for x, and express your answer in a simple form by using absolute value signs.

★25. $\dfrac{x - a}{x + a} > 0$.   ★26. $\dfrac{a - x}{a + x} \geq 0$.

★27. $\dfrac{x - 1}{x - 3} > \dfrac{x + 3}{x + 1}$.   ★28. $\dfrac{x - a}{x - b} > \dfrac{x + a}{x + b}$.

In Exercises 29-38, sketch the graph. (These problems presuppose Analytic Geometry.)

★29. $y = |x|$.   ★30. $y = \dfrac{x}{|x|}$.

★31. $y = x \cdot |x|$.   ★32. $y = \sqrt{|x|}$.

★33. $y = |\,|x| - 1|$.   ★34. $|y| = |x|$.

★37. $|x| + |y| < 1$. ★38. $|x| - |y| \leq 1$.

112. SOME FURTHER PROPERTIES

In this section we give a few relations which exist between *certain* real numbers (integers and rational numbers, to be precise) and real numbers *in general*. In the constructive development of the real number system, discussed in the introduction, these properties follow easily and naturally from the definitions. From the point of view of a set of axioms descriptive of the real numbers, it is of interest that the properties of this section are implied by the properties listed as axioms in other sections of this chapter, in particular the axiom of completeness (cf. § 114 for statements and proofs not given in this section).

(i) *If x is any real number there exists a positive integer n such that $n > x$.*

(ii) *If x is any real number there exist integers m and n such that*
$$m < x < n.$$

(iii) *If x is any real number there exists a unique integer n such that*
$$n \leq x < n + 1.$$

(iv) *If ϵ is any positive number there exists a positive integer n such that*
$$\frac{1}{n} < \epsilon.$$

(v) *The rational numbers are **dense** in the system of real numbers. That is, between any two distinct real numbers there is a rational number (in fact, there are infinitely many).*

We defer to § 114 the proof of (i) and present here the proofs of the remaining properties on the assumption that (i) is true.

Proof of (ii). According to (i), let n be a positive integer such that $n > x$, let p be a positive integer such that $p > -x$, and let $m = -p$.

Proof of (iii). Existence: By (ii) there exist integers r and s such that $r < x < s$. If x is an integer, let $n = x$. If x is not an integer, x must lie between two consecutive integers of the finite set $r, r+1, r+2, \cdots, s$. *Uniqueness:* If m and n are distinct integers such that $m \leq x < m+1$ and $n \leq x < n+1$, assume $n < m$. Then $n < m \leq x < n+1$, in contradiction to property (vii), § 106.

Proof of (iv). Let $n > 1/\epsilon$, by (i).

Proof of (v). Let a and b be two real numbers, where $a < b$, or $b - a > 0$. Let q be a positive integer such that $\frac{1}{q} < b - a$, by (iv). We now seek an integer p so that $\frac{p}{q}$ shall satisfy the relation

This will hold if p is chosen so that $aq < p \leq aq + 1$, or $p - 1 \leq aq < p$, according to (iii). With this choice of p and q we have found a rational number $r_1 = p/q$ between a and b. A second rational number r_2 must exist between r_1 and b, a third between r_2 and b, etc.

113. EXERCISES

1. Prove that if r is a nonzero rational number and x is irrational, then $x \pm r$, $r - x$, xr, x/r, r/x are all irrational. *Hint for $x + r$:* If $x + r = s$, a rational number, then $x = s - r$.

2. Prove that the irrational numbers are dense in the system of real numbers. *Hint:* Let x and y be any two distinct real numbers, assume $x < y$, and find a rational number p/q between $\sqrt{2}\,x$ and $\sqrt{2}\,y$. Then divide by $\sqrt{2}$. (Cf. Ex. 8, § 109, and Ex. 1, above.)

3. Prove that the sum of two irrational numbers may be rational. What about their product?

★4. Prove that the **binary numbers**, $p/2^n$, where p is an integer and n is a positive integer, are dense in the system of real numbers. (Cf. Ex. 11, § 107, and Ex. 5, below.)

★5. Prove that the **terminating decimals** $\pm d_{-m}d_{-m+1}\cdots d_{-1}d_0.d_1d_2\cdots d_n$, where d_i is an integral digit ($0 \leq d_i \leq 9$), are dense in the system of real numbers. *Hint:* Each such number is of the form $p/10^q$, where p is an integer and q is a positive integer. (Cf. Ex. 4.)

NOTE. Decimal expansions are treated in Exercises 28-30, § 711.

6. Prove that if x and y are two fixed numbers and if $y \leq x + \epsilon$ for every positive number ϵ, then $y \leq x$. Prove a corresponding result as a consequence of an inequality of the form $y < x + \epsilon$; of the form $y \geq x - \epsilon$; of the form $y > x - \epsilon$. (Cf. Ex. 23, § 105.)

★7. Axioms I (i), (ii), (iii), and (iv), § 102, define a **group**. That is, any set of objects with one operation (in this case addition) satisfying these axioms is *by definition* a group. Show that the set of all nonzero numbers with the operation of multiplication (instead of addition) is a group.

★8. Axioms I, II, and III, § 102, define a **field**. That is, any set of objects with two operations satisfying these axioms is *by definition* a field. Show that the set of five elements 0, 1, 2, 3, 4, with the following addition and multiplication tables, is a field:

	0	1	2	3	4
0	0	1	2	3	4
1	1	2	3	4	0
2	2	3	4	0	1
3	3	4	0	1	2
4	4	0	1	2	3

Addition

	0	1	2	3	4
0	0	0	0	0	0
1	0	1	2	3	4
2	0	2	4	1	3
3	0	3	1	4	2
4	0	4	3	2	1

Multiplication

§ 113] EXERCISES 21

Hint: The sum (product) of any two elements is the remainder left after division by 5 of the real-number sum (product) of the corresponding two numbers.

★9. Prove that the integers form a group but not a field. (Cf. Exs. 7 and 8.)

★10. Prove that the rational numbers form a field. (Cf. Ex. 8.)

★11. Axioms I, II, and III, § 102, and the axioms of § 104 define an **ordered field**. That is, any set of objects with two operations and an order relation satisfying these axioms is *by definition* an ordered field. Prove that the rational numbers form an ordered field. Prove that any ordered field is infinite. (Cf. Ex. 8, above, and Ex. 24, § 105.)

★12. Any set whose members can be put into a one-to-one correspondence with the natural numbers is called **denumerable** or **enumerable**. Thus the set of positive integers is denumerable. Prove that the set of all integers is denumerable. *Hint:* Arrange the integers as follows: $0, 1, -1, 2, -2, 3, -3, \cdots$, and count them off: first, second, third, etc.

★13. Prove that the set of positive rational numbers is denumerable. (Cf. Ex. 12.) *Hint:* Represent each positive rational as a quotient of relatively prime positive integers (cf. Ex. 23, § 107), and arrange them as follows, and count them off as indicated:

$$
\begin{array}{cccc}
1 \rightarrow & 2 & 3 \rightarrow & 4 \quad \cdots \\
1/2 \leftarrow & 3/2 \nearrow & 5/2 \swarrow & 7/2 \quad \cdots \\
1/3 & 2/3 \leftarrow & 4/3 & 5/3 \quad \cdots \\
1/4 & 3/4 & 5/4 & 7/4 \quad \cdots \\
\cdots & & & \cdots
\end{array}
$$

★14. Prove that the set of all rational numbers is denumerable. (Cf. Ex. 12.) *Hint:* Let the positive rationals be (cf. Ex. 13) $r_1, r_2, r_3, \cdots$, and arrange the rationals: $0, r_1, -r_1, r_2, -r_2, \cdots$.

NOTE. It is shown in Exercise 30, § 711, that not all infinite sets are denumerable, and that, in fact, the set of real numbers is not.

★★15. An expression of the form

$$a_1 + \cfrac{1}{a_2 + \cfrac{1}{a_3 + \cfrac{1}{a_4}}}, \quad \text{written} \quad a_1 + \frac{1}{a_2+} \frac{1}{a_3+} \frac{1}{a_4},$$

and, more generally, an expression of the form

$$(1) \qquad a_1 + \frac{1}{a_2+} \frac{1}{a_3+} \cdots \frac{1}{a_n},$$

where a_1 is a nonnegative integer and $a_2, \cdots, a_n$ are positive integers with $a_n > 1$, is called a **simple continued fraction**.† Prove that any positive rational number can be expressed in one and only one way as a simple continued fraction.

† For a treatment of continued fractions, see George Chrystal, *Algebra* (New York, Chelsea Publishing Company, 1952) or B. M. Stewart, *Number Theory* (New York, The Macmillan Company, 1952).

★★16. Show that the numbers

$$1,\ 1+\tfrac{1}{2}=\tfrac{3}{2},\ 1+\cfrac{1}{2+\cfrac{1}{2}}=\tfrac{7}{5},\ 1+\cfrac{1}{2+\cfrac{1}{2+\cfrac{1}{2}}}=\tfrac{17}{12},\ 1+\cfrac{1}{2+\cfrac{1}{2+\cfrac{1}{2+\cfrac{1}{2}}}}=\tfrac{41}{29},\ \ldots$$

satisfy the *recursion relation:* if p/q is any one, the next is $(p+2q)/(p+q)$. Also show that they are alternately less than and greater than $\sqrt{2}$. Finally, prove that each is closer to $\sqrt{2}$ than its predecessor and that if ϵ is an arbitrary positive number there exists a number r in this set such that $|r-\sqrt{2}|<\epsilon$. *Hint:* If r is any member of the set, the next member is $(r+2)/(r+1)$, and $\left(\dfrac{r+2}{r+1}-\sqrt{2}\right)/(\sqrt{2}-r)=\dfrac{\sqrt{2}-1}{r+1}<\tfrac{1}{2}$. Make use of Ex. 11, § 107, and Property (*iv*), § 112.

★★17. If x is an irrational number, under what conditions on the rational numbers a, b, c, and d, is $(ax+b)/(cx+d)$ rational?

114. AXIOM OF COMPLETENESS

The remaining axiom of the real number system can be given in any of several forms, all of which involve basically the ordering of the numbers. We present here the one that seems to combine most naturally a fundamental simplicity and an intuitive reasonableness. It is based on the idea of an **upper bound** of a set of numbers; that is, a number that is at least as large as anything in the set. For example, the number 10 is an upper bound of the set consisting of the numbers -15, 0, 3, and 7. It is also an upper bound of the set consisting of the numbers -37, 2, 5, 8, and 10. The number 16 is an upper bound of the open intervals $(-3, 6)$ and $(8, 16)$, and also of the closed intervals $[-31, -2]$ and $[15, 16]$. However, 16 is *not* an upper bound of the set of numbers -6, 13, and 23, nor of the open interval $(8, 35)$, nor of the set of all integers. There is *no* number x that is an upper bound of the set of all positive real numbers, for x is not greater than or equal to the real number $x+1$. (Cf. Ex. 24, § 105.)

If, for a given set A of numbers, there is some number x that is an upper bound of A, the set A is said to be **bounded above.** This means that the corresponding set B of points on a number scale does not extend indefinitely to the right but, rather, that there is some point x at least as far to the right as any point of the given set (Fig. 102).

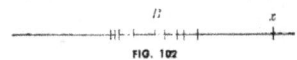
FIG. 102

Whenever a set is bounded above, it has *many* upper bounds. For example, if x is an upper bound, so is $x+1$ and, indeed, so is any number greater than x. An appropriate question: "Is there any upper bound

§ 114] AXIOM OF COMPLETENESS 23

less than x?" If the answer to this question is "No," then x is the smallest of all possible upper bounds† and is called the **least upper bound**† of the set. Since two numbers cannot be such that each is less than the other, there cannot be *more* than one such *least* upper bound. When a number x is a least upper bound of a set, it is therefore called *the* least upper bound of the set. Furthermore, in this case, we say that the set *has x* as its least upper bound, whether x is a member of the set or not. Thus the open interval (a, b) and the closed interval $[a, b]$, where $a < b$, both have b as their least upper bound, whereas b is a member of only the *closed* interval.

The essential question now is whether a given set has a least upper bound. Of course, if a set is not bounded above it has no upper bound and *a fortiori* no *least* upper bound. Suppose a set *is* bounded above. Then does it have a least upper bound? The final axiom gives the answer.

Axiom of Completeness. *Any nonempty set of real numbers that is bounded above has a least upper bound.*

This axiom can be thought of geometrically as stating that there are no "gaps" in the number scale. For instance, if a single point x were removed from the number scale (Fig. 103), the remaining numbers would

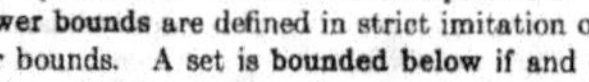

FIG. 103

no longer be complete, since the set consisting of all numbers less than x would be bounded above (e.g., by $x + 1$), but would have no least upper bound (if y is an upper bound, then $x < y$ and $x < \frac{1}{2}(x + y) < y$, so that $\frac{1}{2}(x + y)$ is a smaller upper bound). (Cf. Ex. 25, § 105.)

It might seem that the axiom of completeness is biased or one-sided. Why should we speak of *upper* bounds and *least* upper bounds, when we might as naturally consider *lower* bounds and *greatest* lower bounds? The answer is that it is immaterial, as far as the axiomatic system is concerned, whether we formulate completeness in terms of upper bounds or lower bounds. This fact is illustrated by Theorem I, below, whose statement is actually *equivalent* to the axiom of completeness. **Lower bounds** and **greatest lower bounds** are defined in strict imitation of upper bounds and least upper bounds. A set is **bounded below** if and only if it has a lower bound. A set is **bounded** if and only if it is bounded both below and above; in other words, if and only if it is contained in some finite interval.

Theorem I. *Any nonempty set of real numbers that is bounded below has a greatest lower bound.*‡

---

† The least upper bound of a set A is also called the **supremum** of A, abbreviated $\sup(A)$ (pronounced "supe of A").

‡ The greatest lower bound of a set A is called the **infimum** of A, abbreviated $\inf(A)$ (pronounced "inf of A").

Proof. The set obtained by changing every member of the given set to its negative is bounded above by the negative of any lower bound of the original set. (Look at the number scale with a mirror.) By the axiom of completeness the new set then has a least upper bound, whose negative must be the greatest lower bound of the original set. (Look at the number scale directly again, without the mirror.)

We shall now show that the axiom of completeness implies the first property of § 112 (Corollary, below). In essence, this property is an algebraic formulation of a basic principle of Euclidean geometry known as the **Archimedean property**.† This principle states that any length (however large) can be exceeded by repeatedly marking off a given length (however small), each successive application starting where the preceding one stopped. (A midget ruler, if used a sufficient number of times, can measure off an arbitrarily large distance.) For real numbers this principle, again called the **Archimedean property**, has the following formal statement and proof.

Theorem II. *If a and b are positive numbers, there is a positive integer n such that $na > b$.*

Proof. If the theorem were false, the inequality $na \leq b$ would hold for all positive integers n. That is, the set $a, 2a, 3a, \cdots$ would be bounded above. Let c be the least upper bound of this set. Then $na \leq c$ for all n, and hence $(n + 1)a \leq c$ for all n. Therefore $na + a \leq c$, or $na \leq c - a$, for all n. Thus $c - a$ is an upper bound that is *less* than the *least* upper bound c. This is the desired contradiction.

Corollary (*(i)*, § 112). *If x is any real number there exists a positive integer n such that $n > x$.*

Proof. If $x \leq 0$, let $n = 1$. If $x > 0$, use Theorem II with $a = 1$ and $b = x$.

★115. FURTHER REMARKS ON MATHEMATICAL INDUCTION

If the real number system is considered as defined by the descriptive axioms of this chapter rather than by the constructive development outlined in the Introduction, the simple properties of positive integers given in § 106 are no longer obvious. However, they are still true if we superimpose on the axiomatized real number system the structure of the natural numbers given by the Peano axioms. The positive integers greater than 1 are then defined inductively, $2 \equiv 1 + 1$, $3 \equiv 2 + 1$, $\cdots$, and the remaining properties of § 106 are proved by mathematical induction. An

† Cf. D. Hilbert, *op. cit.*

alternative uses set theory to construct the *set of positive integers.* (For this method of establishing the properties of § 106, cf. Ex. 16, § 116.)

We present now the formal detailed proofs by mathematical induction of the general associative and commutative laws stated in § 106. For simplicity of notation we restrict ourselves to the multiplicative form.

Proof of the general associative law. Let $P(n)$ be the proposition: "Any product of the m numbers $x_1, x_2, \cdots, x_m$, in that order, is equal to the special product $x_1(x_2(x_3(\cdots(x_{m-1}x_m)\cdots)))$, whenever $m \leq n$." For $n = 1$ and $n = 2$ the proposition is trivial, and for $n = 3$ it follows from the associative law, $(x_1x_2)x_3 = x_1(x_2x_3)$. Assume now the truth of $P(n)$, for a fixed n, and consider any possible form for the product of the $n + 1$ numbers $x_1, x_2, \cdots, x_{n+1}$, in that order. Such a product must have the form ab, where a and b are products of at most n of the x's. By the induction assumption that $P(n)$ is true, each of these two factors can be rewritten, if necessary, in the form $a = x_1y$ and $b = x_{k+1}z$, where y is either 1 or a product of the factors $x_2, \cdots, x_k$, and z is either 1 or a product of the factors $x_{k+2}, \cdots, x_{n+1}$. By the associative law of § 102, $ab = (x_1y)(x_{k+1}z) = x_1(y(x_{k+1}z))$. Again using the induction hypothesis, we can write the product $y(x_{k+1}z)$ in the special form $x_2(x_3(\cdots(x_nx_{n+1})\cdots))$. This fact, with the aid of the Fundamental Theorem of Mathematical Induction, § 106, establishes the truth of $P(n)$ for every positive integer n. Finally, since any two products of n numbers in a given order are equal to the same special product, they must be equal to each other.

Proof of the general commutative law. Let $P(n)$ be the proposition: "Any two products of n numbers are equal regardless of the order of the factors." For $n = 1$ the proposition is trivial and for $n = 2$ it follows from the commutative law of § 102: $x_2x_1 = x_1x_2$. Assume now the truth of $P(n)$ for a particular n and consider any possible product of the $n + 1$ numbers $x_1, x_2, \cdots, x_{n+1}$. This product must have the form xx_1y, where x is either 1 or the product of some of the x's and y is either 1 or the product of some of the x's. By the commutative and associative laws, $xx_1y = (xx_1)y = (x_1x)y = x_1(xy)$. The product xy contains the n factors $x_2, x_3, \cdots, x_{n+1}$ which, by the induction assumption that $P(n)$ is true, can be rearranged according to the order of the subscripts. Therefore $P(n + 1)$ follows from $P(n)$, and application of the Fundamental Theorem completes the proof.

116. EXERCISES

1. Prove that the system of integers satisfies the axiom of completeness. (Cf. Ex. 2, below.)

2. Prove that the system of rational numbers does not satisfy the axiom of completeness. *Hint:* Consider the set S of all rational numbers less than $\sqrt{2}$. Then S has an upper bound in the system of rational numbers (the rational number 2 is one such). Assume that S has a least upper bound r. Use the density of the rational numbers to show that if $r < \sqrt{2}$ then r is not even an upper bound of S, and that if $r > \sqrt{2}$ then r is not the *least* upper bound of S.

3. Let x be a real number and let S be the set of all rational numbers less than x. Show that x is the least upper bound of S.

★**4.** Prove that if S is a bounded nonempty set there is a smallest closed interval I containing S. That is, I has the property that if J is any closed interval containing S, then J contains I.

★**5.** Prove by counterexample that the statement of Exercise 4 is false if the word *closed* is replaced by the word *open*.

★**6.** Prove that if S is a set of numbers dense in the system of real numbers, and if any finite number of points are deleted from S, the remaining set is still dense.

7. Let S be a nonempty set of numbers bounded above, and let x be the least upper bound of S. Prove that x has the two properties corresponding to an arbitrary positive number ϵ: (i) every element s of S satisfies the inequality $s < x + \epsilon$; (ii) at least one element s of S satisfies the inequality $s > x - \epsilon$.

★**8.** Prove that the two properties of Exercise 7 characterize the least upper bound. That is, prove that a number x subject to these two properties is the least upper bound of S.

★**9.** Prove the analogue of Exercise 7 for greatest lower bounds.

★**10.** Prove the analogue of Exercise 8 for greatest lower bounds.

★★**11.** Prove **Dedekind's Theorem:** *Let the real numbers be divided into two nonempty sets A and B such that (i) if x is an arbitrary member of A and if y is an arbitrary member of B then $x < y$ and (ii) if z is an arbitrary real number then either z is a member of A or z is a member of B. Then there exists a number c (which may belong to either A or B) such that any number less than c belongs to A and any number greater than c belongs to B.*

★★**12.** Prove that the real number system is both minimal and maximal in the sense that it is impossible to have two systems R and S both satisfying the axioms of this chapter, where every member of R belongs to S but not every member of S belongs to R, and where the members of R are combined algebraically and related by order in the same way whether they are thought of as members of R or as members of S. *Hint:* The multiplicative unit 1 of S must be the multiplicative unit of R, and therefore R must contain all rational members of S. Since R is dense in S and complete it must be the same as S.

★★**13.** Discuss the essential uniqueness of the real number system in the following sense: Prove that if R and S are any two "real number systems" subject to the axioms of this chapter, then it is possible to establish a one-to-one correspondence between their members which preserves algebraic operations and order. *Hint:* Let correspond, first, the additive units 0 and 0′, then the multiplicative units 1 and 1′, then the positive integers n and n', then the integers n and n', and then the rational numbers p/q and p'/q', and show that these correspondences preserve operations and order. Finally, since any real num-

★★14. Prove the existence of $\sqrt{2}$ as follows: Let A be the set of all positive rational numbers r such that $r^2 < 2$, and define $a = \sup(A)$. Prove that $a^2 = 2$ by disproving both $a^2 < 2$ and $a^2 > 2$. *Hints:* Assuming $a^2 < 2$, let s be a positive rational number less than a suitably small fraction of $2 - a^2$, let t be a member of A greater than $a - s$, and square $s + t$. Assuming $a^2 > 2$, let s be small compared with $a^2 - 2$, and let $a - s < r < a$.

★★15. Prove the statement of Exercise 2 by means of the technique of Exercise 14, *without going beyond the system of rational numbers*.

★★16. Prove that every ordered field F (cf. Ex. 11, § 113) contains a subset PI of "positive integers" having the properties $(i) - (vii)$, § 106, and satisfying the Axiom of Induction, as follows: We borrow from set theory the notation $x \in A$ to mean that x is a member of the set A, and the concept of *intersection* of a family of sets A_α to mean the set B such that $b \in B$ if and only if $b \in$ every A_α. Define an **inductive subset** of F to be a subset A such that (i) $1 \in A$, and (ii) whenever $x \in A$, then $x + 1 \in A$. Prove:

a. The set of all $x \geq 1$ is inductive.
b. The intersection of any family of inductive sets is inductive.
c. There exists a *smallest* inductive set PI equal to the intersection of *all* inductive sets.
d. Properties (i) and (ii), § 106, hold (by parts a and c).
e. If $n > 1$ and $n \in PI$, then $n - 1 \in PI$ (since otherwise n could be deleted to give a *smaller* inductive set).
f. The Axiom of Induction holds (by definition of PI).
g. Properties (iii) and (iv), § 106, hold. (By part e, every member of PI greater than 1 is obtained by adding 1 to a member of PI.)
h. Property (v), § 106, holds. (Addition: For given m, let $n > 1$ be the smallest member of PI such that $m + n$ is *not* in PI. Then $m + (n - 1) \in PI$!)
i. Property (vii), § 106, holds. (Let n be the *smallest* member of PI such that $n < m < n + 1$ is possible. Then $n > 1$. Subtract 1.)
j. Property (vi), § 106, holds. (For given m let n be the smallest member of $PI > m$ such that $n - m$ is not in PI. Then $(n - 1) - m \in PI$!)

Furthermore, show that any inductive subset of an ordered field F for which the Axiom of Induction holds is uniquely determined, and thus prove that if F is the real number system, then the set PI is identical with the set of positive integers described in § 106.

§ 116] EXERCISES 27

ber is the least upper bound of all rational numbers less than it (Ex. 3), the correspondence can be extended to all elements of R and S.

NOTE. By virtue of Exercise 13 it is possible to describe completely the real number system by the definition: *The real numbers are a complete ordered field.* (Cf. Ex. 11, § 113.)

★★14. Prove the existence of $\sqrt{2}$ as follows: Let A be the set of all positive rational numbers r such that $r^2 < 2$, and define $a = \sup(A)$. Prove that $a^2 = 2$ by disproving both $a^2 < 2$ and $a^2 > 2$. *Hints:* Assuming $a^2 < 2$, let s be a positive rational number less than a suitably small fraction of $2 - a^2$, let t be a member of A greater than $a - s$, and square $s + t$. Assuming $a^2 > 2$, let s be small compared with $a^2 - 2$, and let $a - s < r < a$.

★★15. Prove the statement of Exercise 2 by means of the technique of Exercise 14, *without going beyond the system of rational numbers*.

★★16. Prove that every ordered field F (cf. Ex. 11, § 113) contains a subset PI of "positive integers" having the properties $(i) - (vii)$, § 106, and satisfying the Axiom of Induction, as follows: We borrow from set theory the notation $x \in A$ to mean that x is a member of the set A, and the concept of *intersection* of a family of sets A_α to mean the set B such that $b \in B$ if and only if $b \in$ every A_α. Define an **inductive subset** of F to be a subset A such that (i) $1 \in A$, and (ii) whenever $x \in A$, then $x + 1 \in A$. Prove:

  a. The set of all $x \geq 1$ is inductive.
  b. The intersection of any family of inductive sets is inductive.
  c. There exists a *smallest* inductive set PI equal to the intersection of *all* inductive sets.
  d. Properties (i) and (ii), § 106, hold (by parts a and c).
  e. If $n > 1$ and $n \in PI$, then $n - 1 \in PI$ (since otherwise n could be deleted to give a *smaller* inductive set).
  f. The Axiom of Induction holds (by definition of PI).
  g. Properties (iii) and (iv), § 106, hold. (By part e, every member of PI greater than 1 is obtained by adding 1 to a member of PI.)
  h. **Property (v), § 106, holds.** (Addition: For given m, let $n > 1$ be the smallest member of PI such that $m + n$ is *not* in PI. Then $m + (n - 1) \in PI$!)
  i. Property (vii), § 106, holds. (Let n be the *smallest* member of PI such that $n < m < n + 1$ is possible. Then $n > 1$. Subtract 1.)
  j. Property (vi), § 106, holds. (For given m let n be the smallest member of $PI > m$ such that $n - m$ is not in PI. Then $(n - 1) - m \in PI$!)

Furthermore, show that any inductive subset of an ordered field F for which the Axiom of Induction holds is uniquely determined, and thus prove that if F is the real number system, then the set PI is identical with the set of positive integers described in § 106.

2

Functions, Sequences, Limits, Continuity

201. FUNCTIONS AND SEQUENCES

Whenever one says that y is a function of x, one has in mind some mechanism that assigns values to y corresponding to given values of x. The most familiar examples are real-valued functions of a real variable given by formulas, like $y = 3x^2 - 12x$ or $y = \pm\sqrt{x^2 - 4}$. These and other examples (where the variables x and y need not be related by formula, or even be real numbers) are given below.

Definition I. *Let D and R be two sets of objects. Then $y = f(x)$ is called a **function** with **domain** (of definition) D and **range** (of values) R if and only if to each member x of D there corresponds at least one member y of R, and for each member y of R there is at least one member x of D to which y corresponds. The general member of D and the general member of R are called the **independent** and the **dependent variable**, respectively. In case no two members of R correspond to the same member of D, $f(x)$ is called **single-valued**. In case R consists of just one object, $f(x)$ is called a **constant function**. In case D consists of real numbers, $f(x)$ is called a function of a **real variable**. In case R consists of real numbers, $f(x)$ is called **real-valued**. The symbol $f(x_0)$ denotes the members of R that correspond to the member x_0 of D.* (Cf. § 217 for further discussion.)

Example 1. The function $y = 3x^2 - 12x$ is defined for all real numbers. If we take D to be the set of all real numbers, R consists of all real numbers $\geqq -12$, since the function has an absolute minimum (cf. § 409) when $x = 2$. The function is a single-valued real-valued function of a real variable.

Example 2. The function of Example 1 restricted to the domain $D = (1, 5)$ (the open interval from 1 to 5) has range R equal to the half-open interval $[-12, 15)$. This function is not the same as that of Example 1, since it has a different domain. It is also, however, a single-valued real-valued function of a real variable.

Example 3. The function $y = \sqrt{x^2 - 4}$ with domain D consisting of all real numbers x such that $|x| \geqq 2$ is a single-valued real-valued function of a

§ 201]   FUNCTIONS AND SEQUENCES   29

real variable, with range R consisting of all nonnegative real numbers. (Cf. § 214.)

Example 4. The function $y = \pm\sqrt{x^2 - 4}$ with the same domain as the function of Example 3 is real-valued, but is single-valued only for $x = \pm 2$. Otherwise it is double-valued. Its range is the set of all real numbers.

Example 5. The **bracket function** or **greatest integer function**, $f(x) = [x]$, is defined to be the largest integer less than or equal to x, with domain all real numbers (Fig. 201). It is a single-valued real-valued function of a real variable. Its range is the set of all integers.

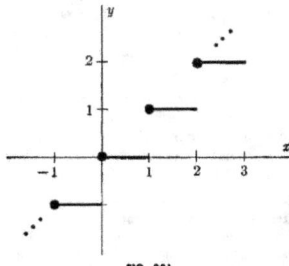

FIG. 201

Example 6. Let D be the closed interval $[0, 1]$, and define $f(x)$, for x in D, to be 1 if x is rational and 0 if x is irrational. (See Fig. 202.) Then R consists of the two numbers 0 and 1.

Example 7. Let D be the contestants in a radio quiz show and define $f(x)$ as follows: If x is a contestant who has correctly answered all questions, then

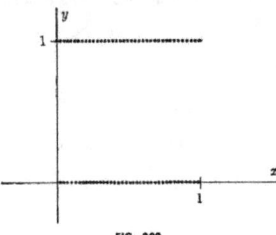

FIG. 202

30 FUNCTIONS AND SEQUENCES [§ 201

$f(x) = n$ where n is the number of dollars of prize money; if x is a contestant who has incorrectly answered a question, then $f(x)$ is a box of breakfast food. In this example $f(x)$ is a single-valued function. In some cases its value is a real number, and in some cases its value is a box of breakfast food.

A type of function of particular importance in mathematics is specified in the following definition.

Definition II. *An* **infinite sequence** *is a single-valued function whose domain of definition is the positive integers.*

This means that corresponding to any positive integer there is a unique value or **term** determined. In particular, there is a first term a_1 corresponding to the number 1, a second term a_2 corresponding to the number 2, etc. An infinite sequence can thus be represented:

$$a_1, a_2, \cdots, a_n, \cdots, \text{ or } \{a_n\}.$$

The nth term, a_n, is sometimes called the **general term** of the infinite sequence. Since it is a function of n, $(a_n = f(n))$, it must be prescribed by some rule. If the terms are numbers, this rule may sometimes be expressed as a simple algebraic formula. Such a formula may be impractical, but a definite rule must exist.

NOTE 1. Frequently an infinite sequence is indicated by an explicit listing of only the first few terms, in case the general rule for procedure is clear *beyond reasonable doubt*. For instance, in part (c) of the following Example 8 the rule that is clearly implied by alternating 1's and 0's for the first six terms is alternating 1's and 0's for all terms, although the ingenious artificer could construct any number of infinite sequences that start with alternating 1's and 0's (the terms could continue by being identically 0, or with alternating 6's and 7's, for example). Such interpretations, we hold, are not only unnatural, but deliberately mischievous.

NOTE 2. For convenience, if the meaning is clear, the single word *sequence* will be used henceforth to mean *infinite sequence*.

Example 8. Give a rule for obtaining the general term for each of the following sequences:

(a) $\frac{1}{2}, -\frac{2}{5}, \frac{3}{8}, -\frac{4}{11}, \cdots$;   (b) $1, \frac{1}{2}, \frac{1}{3}, \frac{1}{4}, \cdots$;
(c) $1, 0, 1, 0, 1, 0, \cdots$;   (d) $1, 2, 3, 1, 2, 3, 1, 2, 3, \cdots$;
(e) $\frac{1}{2}, \frac{1}{3}, \frac{1}{4}, \frac{1}{9}, \frac{1}{8}, \frac{1}{27}, \cdots$.

Solution. (a) The factor $(-1)^n$ or $(-1)^{n+1}$ is a standard device to take care of alternating signs. The general term is $(-1)^{n+1} \dfrac{n}{3n-1}$. (b) If $n = 1$, $a_n = 1$; if $n > 1$, $a_n = \dfrac{1}{n-1}$. (c) First formulation: if n is odd, $a_n = 1$; if n is even, $a_n = 0$. Second formulation: $a_{2n-1} = 1$; $a_{2n} = 0$. Third formulation: $a_n = \frac{1}{2}[(-1)^{n+1} + 1]$. (d) $a_{3n-2} = 1$; $a_{3n-1} = 2$; $a_{3n} = 3$. (e) $a_{2n-1} = \dfrac{1}{2^n}$; $a_{2n} = \dfrac{1}{3^n}$.

202. LIMIT OF A SEQUENCE

A sequence is said to **tend toward**, or **converge to**, a number if and only if the absolute value of the difference between the general term of the sequence and this number is less than any preassigned positive number (however small) whenever the subscript n of the general term is sufficiently large.

Symbolically, this is written

$$\lim_{n \to +\infty} a_n = a \quad \text{or} \quad \lim_{n \to \infty} a_n = a \quad \text{or} \quad a_n \to a,$$

where a_n is the nth term of the sequence and a is the number to which it converges. If $\{a_n\}$ converges to a, a is called the **limit** of the sequence.

A more concise form of the definition given above is the following:

Definition I. *The sequence $\{a_n\}$ has the **limit** a, written $\lim_{n \to +\infty} a_n = a$, if and only if corresponding to an arbitrary positive number ϵ there exists a number $N = N(\epsilon)$ such that $|a_n - a| < \epsilon$ whenever $n > N$.*

NOTE 1. In conformity with the discussion following Definition III, § 110, the statement that $\{a_n\}$ converges to a is equivalent to the statement that every neighborhood of a contains all of the terms of $\{a_n\}$ *from some point on*, and is also equivalent to the statement that every neighborhood of a contains all but a finite number of the terms of $\{a_n\}$ (that is, all of the terms except for a finite number of the subscripts).

If a sequence converges to some number, the sequence is said to be **convergent**; otherwise it is **divergent**.

The concept of an *infinite limit* is important, and will be formulated in precise symbolic form. As an exercise, the student should reformulate the following definition in his own words, without the use of mathematical symbols.

Definition II. *The sequence $\{a_n\}$ has the limit $+\infty$, written*

$$\lim_{n \to +\infty} a_n = +\infty, \quad \text{or} \quad a_n \to +\infty,$$

if and only if corresponding to an arbitrary number B (however large) there exists a number $N = N(B)$ such that $a_n > B$ whenever $n > N$; the sequence $\{a_n\}$ has the limit $-\infty$, written

$$\lim_{n \to +\infty} a_n = -\infty, \quad \text{or} \quad a_n \to -\infty,$$

if and only if corresponding to an arbitrary number B (however large its negative) there exists a number $N = N(B)$ such that $a_n < B$ whenever $n > N$; the sequence has the limit ∞ (unsigned infinity), written

32   FUNCTIONS AND SEQUENCES   [§ 202

$$\lim_{n \to +\infty} a_n = \infty, \text{ or } a_n \to \infty,$$

if and only if $\lim_{n \to +\infty} |a_n| = +\infty$.

NOTE 2. Although the word *limit* is applied to both the finite and infinite cases, the word *converge* is used only for finite limits. Thus, a sequence tending toward $+\infty$ diverges.

NOTE 3. In any extensive treatment of limits there are numerous statements which can be interpreted to apply to both finite and infinite cases, and which are of such a nature that the proofs for the finite and infinite particularizations are in essence identical. In such instances these proofs can be combined into a single proof by appropriate extensions of the word *neighborhood*. We define "neighborhoods of infinity" as follows: (i) a **neighborhood of** $+\infty$ is an open interval of the form $(a, +\infty)$; (ii) a **neighborhood of** $-\infty$ is an open interval of the form $(-\infty, b)$; (iii) a **neighborhood of** ∞ is the set of all x satisfying an inequality of the form $|x| > a$. With these conventions, for example, all cases of Definitions I and II can be included in the following single formulation for $\lim_{n \to +\infty} a_n = a$ (where a may be a number, or $+\infty$, $-\infty$, or ∞): *Corresponding to every neighborhood N_a of a there exists a neighborhood $N_{+\infty}$ of $+\infty$ such that whenever n belongs to $N_{+\infty}$, a_n belongs to N_a.* In the sequel we formulate theorems and proofs separately for the finite and infinite forms, but suggest that the student interested in exploring the simplifying techniques available with general neighborhoods try his hand at combining the separate formulations into unified ones. A word of warning is in order, however: Do not confuse infinite symbols with numbers, and write such nonsense as $|a_n - \infty| < \epsilon$ when dealing with an infinite limit! It is to avoid such possible confusion of ideas that we have adopted the policy of maintaining (in the main) the separation of the finite and infinite.

Definition III. *A **subsequence** of a sequence is a sequence whose terms are terms of the original sequence arranged in the same order. That is, a subsequence of a sequence $\{a_n\}$ has the form $a_{n_1}, a_{n_2}, a_{n_3}, \cdots$, where*

$$n_1 < n_2 < n_3 < \cdots.$$

It is denoted by $\{a_{n_k}\}$.

Example 1. The sequence $\frac{1}{2}, \frac{1}{4}, \frac{1}{6}, \cdots$ is a subsequence of the sequence $\frac{1}{2}, \frac{1}{3}, \frac{1}{4}, \frac{1}{5}, \frac{1}{6}, \cdots$ of Example 8, § 201. The sequence $0, 1, 0, 1, 0, 1, \cdots$ is a subsequence of the sequence $1, 0, 1, 0, 1, 0, 0, 0, 1, \cdots$.

Example 2. Show that the following sequences converge to 0:

(a) $1, \frac{1}{2}, \frac{1}{3}, \cdots, \frac{1}{n}, \cdots$; (b) $\frac{1}{2}, \frac{1}{4}, \frac{1}{8}, \cdots, \frac{1}{2^n}, \cdots$; (c) $\frac{1}{2}, -\frac{1}{4}, \frac{1}{8}, \cdots, \frac{(-1)^{n+1}}{2^n}, \cdots$.

Solution. (a) Since $|a_n - a| = \left|\frac{1}{n} - 0\right| = \frac{1}{n}$, and since $\frac{1}{n} < \epsilon$ whenever $n > 1/\epsilon$, we can choose as the function $N(\epsilon)$ of Definition I the expression $1/\epsilon$. (Cf. § 112.) (b) By Ex. 11, § 107, $2^n > n$ for all positive integers, so that we can choose $N(\epsilon) = 1/\epsilon$. (Cf. Ex. 14, § 205.) (c) This reduces immediately to (b).

§ 203]   EXERCISES   33

Example 3. Find the limit of each sequence: (a) $\frac{1}{2}, \frac{3}{4}, \frac{7}{8}, \cdots, 1 - \frac{1}{2^n}, \cdots$.
(b) 3, 3, 3, $\cdots$, 3, $\cdots$; (c) 1, $\frac{1}{2}$, 1, $\frac{3}{4}$, 1, $\frac{7}{8}$, $\cdots$.

Solution. (a) The expression $|a_n - 1|$ is equal to $\frac{1}{2^n}$, which is less than any preassigned positive number whenever n is sufficiently large, as shown in Example 2, (b). Therefore the limit is 1. (b) The absolute value of the difference between the general term and 3 is identically zero, which is less than any preassigned positive number for *any* n, and certainly for n sufficiently large. Therefore the limit is 3. (c) By combining the reasoning in parts (a) and (b) we see that the general term differs numerically from 1 by less than any preassigned positive number if n is sufficiently large. The odd-numbered terms form a subsequence identically 1, while the even-numbered terms form a subsequence which is the sequence of part (a). The limit is 1.

Example 4. Show that each of the following sequences diverges:
   (a) 1, 2, 1, 2, 1, 2, $\cdots$;   (b) 1, 2, 4, 8, 16, $\cdots$;
   (c) 1, 2, 1, 3, 1, 4, $\cdots$;   (d) 1, -2, 4, -8, 16, $\cdots$.

Solution. (a) If $\{a_n\}$ converges to a, *every* neighborhood of a must contain all terms from some point on, and therefore must contain both numbers 1 and 2. On the other hand, no matter what value a may have, a neighborhood of a of length less than 1 cannot contain both of these points! (b) No finite interval about any point can contain all terms of this sequence, from some point on. The limit is $+\infty$. (c) The comment of part (b) applies to this sequence, since there is a subsequence tending toward $+\infty$. This sequence has no limit, finite or infinite. (d) The subsequence of the odd-numbered terms tends toward $+\infty$, and that of the even-numbered terms tends toward $-\infty$. The sequence of absolute values tends toward $+\infty$, so that the sequence itself has the limit ∞.

203. EXERCISES

In Exercises 1-10, draw the graph of the given function, assuming the domain of definition to be as large as possible. Give in each case the domain and the range of values. The bracket function $[x]$ is defined in Example 5, § 201, and square roots are discussed in § 214. (Also cf. Exs. 5-10, § 216.)

1. $y = \sqrt{x^2 - 9}$.   2. $y = \pm\sqrt{25 - x^2}$.
3. $y = \sqrt{-x}$.   4. $y = \pm\sqrt{[x]}$.
5. $y = \sqrt{4x - x^2}$.   6. $y = \sqrt{[x^2 - 16]}$.
7. $y = x - [x]$.   8. $y = (x - [x])^2$.
9. $y = \sqrt{x - [x]}$.   10. $y = [x] + \sqrt{x - [x]}$.

In Exercises 11-18, give a rule for finding the general term of the sequence.

11. 2, $\frac{4}{3}$, $\frac{8}{5}$, $\frac{16}{7}$, $\cdots$.   12. $\frac{1}{3}$, $-\frac{1}{6}$, $\frac{1}{11}$, $-\frac{1}{18}$, $\frac{1}{27}$, $\cdots$.
13. 1, -1, $\frac{1}{3}$, $-\frac{1}{16}$, $\frac{1}{120}$, $\cdots$.   14. 1, 2, 24, 720, 40320, $\cdots$.
15. $1 \cdot 3$, $1 \cdot 3 \cdot 5$, $1 \cdot 3 \cdot 5 \cdot 7$, $1 \cdot 3 \cdot 5 \cdot 7 \cdot 9$, $\cdots$.
16. 1, 2, 3, 2, 1, 2, 3, 2, 1, $\cdots$.

34   FUNCTIONS AND SEQUENCES   [§ 204

17. $-1, 1, 1, -2, 2, 2, -3, 3, 3, -4, 4, 4, \cdots$.
18. $1, 2\cdot 4, 1\cdot 3\cdot 5, 2\cdot 4\cdot 6\cdot 8, 1\cdot 3\cdot 5\cdot 7\cdot 9, \cdots$.

In Exercises 19-24, find the limit of the sequence and justify your contention (cf. Exs. 25-30).

19. $2, 2, 2, 2, 2, \cdots$.   **20.** $\frac{3}{2}, \frac{5}{4}, \frac{7}{6}, \cdots, \frac{2n+1}{2n}, \cdots$.
21. $\frac{3}{5}, \frac{4}{7}, \frac{5}{9}, \frac{6}{11}, \frac{7}{13}, \frac{8}{15}, \cdots$.   **22.** $1, 4, 9, 16, \cdots, n^2, \cdots$.
23. $\frac{3}{2}, -\frac{5}{4}, \frac{7}{8}, -\frac{9}{16}, \cdots$.   **24.** $9, 16, 21, 24, \cdots, 10n - n^2, \cdots$.

In Exercises 25-30, give a simple explicit function $N(\epsilon)$ or $N(B)$, in accord with Definition I or II, for the sequence of the indicated Exercise.

★**25.** For Ex. 19.   ★**26.** For Ex. 20.   ★**27.** For Ex. 21.
★**28.** For Ex. 22.   ★**29.** For Ex. 23.   ★**30.** For Ex. 24.

In Exercises 31-34, prove that the given sequence has no limit, finite or infinite.

31. $1, 5, 1, 5, 1, 5, \cdots$.   **32.** $1, 2, 3, 1, 2, 3, 1, 2, 3, \cdots$.
33. $1, 2, 1, 4, 1, 8, 1, 16, \cdots$.   **34.** $2^1, 2^{-2}, 2^3, 2^{-4}, 2^5, 2^{-6}, \cdots$.

204. LIMIT THEOREMS FOR SEQUENCES

Theorem I. *The alteration of a finite number of terms of a sequence has no effect on convergence or divergence or limit. In other words, if $\{a_n\}$ and $\{b_n\}$ are two sequences and if M and N are two positive integers such that $a_{M+n} = b_{N+n}$ for all positive integers n, then the two sequences $\{a_n\}$ and $\{b_n\}$ must either both converge to the same limit or both diverge; in case of divergence either both have the same infinite limit or neither has an infinite limit.*

Proof. If $\{a_n\}$ converges to a, then every neighborhood of a contains all but a finite number of the terms of $\{a_n\}$, and therefore all but a finite number of the terms of $\{b_n\}$. Proof for the case of an infinite limit is similar.

Theorem II. *If a sequence converges, its limit is unique.*

Proof. Assume $a_n \to a$ and $a_n \to a'$, where $a \neq a'$. Take neighborhoods of a and a' so small that they have no points in common. Then each must contain all but a finite number of the terms of $\{a_n\}$. This is clearly impossible.

Theorem III. *If all terms of a sequence, from some point on, are equal to a constant, the sequence converges to this constant.*

Proof. Any neighborhood of the constant contains the constant and therefore all but a finite number of the terms of the sequence.

Theorem IV. *Any subsequence of a convergent sequence converges, and its limit is the limit of the original sequence.* (Cf. Ex. 12, § 205.)

§ 204] LIMIT THEOREMS FOR SEQUENCES 35

Proof. Assume $a_n \to a$. Since every neighborhood of a contains all but a finite number of terms of $\{a_n\}$ it must contain all but a finite number of terms of any subsequence.

Definition I. *A sequence is **bounded** if and only if all of its terms are contained in some interval. Equivalently, the sequence $\{a_n\}$ is bounded if and only if there exists a positive number P such that $|a_n| \leq P$ for all n.*

Theorem V. *Any convergent sequence is bounded.* (Cf. Ex. 2, § 205.)

Proof. Assume $a_n \to a$, and choose a definite neighborhood of a, say the open interval $(a - 1, a + 1)$. Since this neighborhood contains all but a finite number of terms of $\{a_n\}$, a suitable enlargement will contain these missing terms as well.

Definition II. *If $\{a_n\}$ and $\{b_n\}$ are two sequences, the sequences $\{a_n + b_n\}$, $\{a_n - b_n\}$, and $\{a_n b_n\}$ are called their **sum, difference,** and **product,** respectively. If $\{a_n\}$ and $\{b_n\}$ are two sequences, where b_n is never zero, the sequence $\{a_n/b_n\}$ is called their **quotient.** The definitions of sum and product extend to any finite number of sequences.*

Theorem VI. *The sum of two convergent sequences is a convergent sequence, and the limit of the sum is the sum of the limits:*

$$\lim_{n \to +\infty} (a_n + b_n) = \lim_{n \to +\infty} a_n + \lim_{n \to +\infty} b_n.$$

(Cf. Ex. 4, § 205.) *This rule extends to the sum of any finite number of sequences.*

Proof. Assume $a_n \to a$ and $b_n \to b$, and let $\epsilon > 0$ be given. Choose N so large that the following two inequalities hold *simultaneously* for $n > N$:

$$|a_n - a| < \tfrac{1}{2}\epsilon, \quad |b_n - b| < \tfrac{1}{2}\epsilon.$$

Then, by the triangle inequality, for $n > N$

$$(a_n + b_n) - (a + b)| = |(a_n - a) + (b_n - b)|$$
$$\leq |a_n - a| + |b_n - b| < \tfrac{1}{2}\epsilon + \tfrac{1}{2}\epsilon = \epsilon.$$

The extension to the sum of an arbitrary number of sequences is provided by mathematical induction. (Cf. Ex. 3, § 205.)

Theorem VII. *The difference of two convergent sequences is a convergent sequence, and the limit of the difference is the difference of the limits:*

$$\lim_{n \to +\infty} (a_n - b_n) = \lim_{n \to +\infty} a_n - \lim_{n \to +\infty} b_n.$$

Proof. The details are almost identical with those of the preceding proof. (Cf. Ex. 6, § 205.)

Theorem VIII. *The product of two convergent sequences is a convergent sequence and the limit of the product is the product of the limits:*

$$\lim_{n \to +\infty} (a_n b_n) = \lim_{n \to +\infty} a_n \cdot \lim_{n \to +\infty} b_n.$$

(Cf. Ex. 5, § 205.) *This rule extends to the product of any finite number of sequences.*

Proof. Assume $a_n \to a$ and $b_n \to b$. We wish to show that $a_n b_n \to ab$ or, equivalently, that $a_n b_n - ab \to 0$. By addition and subtraction of the quantity ab_n and by appeal to Theorem VI, we can use the relation

$$a_n b_n - ab = (a_n - a)b_n + a(b_n - b)$$

to reduce the problem to that of showing that both sequences $\{(a_n - a)b_n\}$ and $\{a(b_n - b)\}$ converge to zero. The fact that they do is a consequence of the following lemma:

Lemma. *If $\{c_n\}$ converges to 0 and $\{d_n\}$ converges, then $\{c_n d_n\}$ converges to 0.*

Proof of lemma. By Theorem V the sequence $\{d_n\}$ is bounded, and there exists a positive number P such that $|d_n| \leq P$ for all n. If $\epsilon > 0$ is given, choose N so large that $|c_n| < \epsilon/P$ for $n > N$. Then for $n > N$,

$$|c_n d_n| = |c_n| \cdot |d_n| < (\epsilon/P) \cdot P = \epsilon.$$

This inequality completes the proof of the lemma, and hence of the theorem.

The extension to the product of an arbitrary number of sequences is provided by mathematical induction. (Cf. Ex. 3, § 205.)

Theorem IX. *The quotient of two convergent sequences, where the denominators and their limit are nonzero, is a convergent sequence and the limit of the quotient is the quotient of the limits:*

$$\lim_{n \to +\infty} \frac{a_n}{b_n} = \frac{\lim_{n \to +\infty} a_n}{\lim_{n \to +\infty} b_n}.$$

Proof. Assume $a_n \to a$, $b_n \to b$, and that b and b_n are nonzero for all n. Inasmuch as $a_n/b_n = (a_n) \cdot (1/b_n)$, Theorem VIII permits the reduction of this proof to that of showing that $1/b_n \to 1/b$ or, equivalently, that

$$\frac{1}{b_n} - \frac{1}{b} = \frac{b - b_n}{b} \cdot \frac{1}{b_n} \to 0.$$

Let $c_n \equiv (b - b_n)/b$ and $d_n \equiv 1/b_n$ and observe that the conclusion of the Lemma of Theorem VIII is valid (with no change in the proof) when the sequence $\{d_n\}$ is assumed to be merely bounded (instead of convergent). Since the sequence $\{c_n\} = \{(b - b_n) \cdot (1/b)\}$ converges to zero (by this same lemma), we have only to show that the sequence $\{d_n\} = \{1/b_n\}$ is bounded. We proceed now to prove this fact. Since $b \neq 0$, we can choose neighborhoods of 0 and b which have no points in common. Since $b_n \to b$, the

§ 204] LIMIT THEOREMS FOR SEQUENCES

neighborhood of b contains all but a finite number of the terms of $\{b_n\}$, so that only a finite number of these terms can lie in the neighborhood of 0. Since b_n is nonzero for all n, there is a (smaller) neighborhood of 0 that excludes *all* terms of the sequence $\{b_n\}$. If this neighborhood is the open interval $(-\epsilon, \epsilon)$, where $\epsilon > 0$, then for all n, $|b_n| \geq \epsilon$, or $|d_n| = |1/b_n| \leq 1/\epsilon$. The sequence $\{d_n\}$ is therefore bounded, and the proof is complete.

Theorem X. *Multiplication of the terms of a sequence by a nonzero constant k does not affect convergence or divergence. If the original sequence converges, the new sequence converges to k times the limit of the original, for any constant k:*

$$\lim_{n \to +\infty} (k\, a_n) = k \cdot \lim_{n \to +\infty} a_n.$$

Proof. This is a consequence of Theorems III and VIII.

Theorem XI. *If $\{a_n\}$ is a sequence of nonzero numbers, then $a_n \to \infty$ if and only if $1/a_n \to 0$; equivalently, $a_n \to 0$ if and only if $1/a_n \to \infty$.*

Proof. If $|a_n| \to +\infty$ and if $\epsilon > 0$ is given, there exists a number N such that for $n > N$, $|a_n| > 1/\epsilon$, and therefore $|1/a_n| < \epsilon$. Conversely, if $1/a_n \to 0$ and B is any given *positive* number, there exists a number N such that for $n > N$, $|1/a_n| < 1/B$, and therefore $|a_n| > B$.

Theorem XII. *If $a > 1$, $\lim_{n \to +\infty} a^n = +\infty$.*

Proof. Let $p \equiv a - 1 > 0$. Then $a = 1 + p$, and by the Binomial Theorem (cf. Ex. 35, § 107), if n is a positive integer,

$$a^n = (1 + p)^n = 1 + np + \frac{n(n-1)}{2} p^2 + \cdots \geq 1 + np.$$

Therefore, if B is a given positive number and if $n > B/p$, then

$$a^n \geq 1 + np > 1 + B > B.$$

Theorem XIII. *If $|r| < 1$, $\lim_{n \to +\infty} r^n = 0$.*

Proof. This is a consequence of the two preceding theorems.

Definition III. *A sequence $\{a_n\}$ is **monotonically increasing** (**decreasing**),† written $a_n \uparrow$ ($a_n \downarrow$), if and only if $a_n \leq a_{n+1}$ ($a_n \geq a_{n+1}$) for every n. A sequence is **monotonic** if and only if it is monotonically increasing or, monotonically decreasing.*

Theorem XIV. *Any bounded monotonic sequence converges. If $a_n \uparrow$ ($a_n \downarrow$) and if $a_n \leq P$ ($a_n \geq P$) for all n, then $\{a_n\}$ converges; moreover, if $a_n \to a$, then $a_n \leq a \leq P$ ($a_n \geq a \geq P$) for all n.*

† Parentheses are used here to indicate an alternative statement. For a discussion of the use of parentheses for alternatives, see the Preface.

Proof. We give the details only for the case $a_n \uparrow$ (cf. Ex. 7, § 205). Since the set A of points consisting of the terms of the sequence $\{a_n\}$ is bounded above, it has a least upper bound a (§ 114), and since P is an upper bound of A, the following inequalities must hold for all n: $a_n \leq a \leq P$. To prove that $a_n \to a$ we let ϵ be a given positive number and observe that there must exist a positive integer N such that $a_N > a - \epsilon$ (cf. Ex. 7, § 116). Therefore, for $n > N$, the following inequalities hold:

$$a - \epsilon < a_N \leq a_n \leq a < a + \epsilon.$$

Consequently $a - \epsilon < a_n < a + \epsilon$, or $|a_n - a| < \epsilon$, and the proof is complete.

NOTE. As a consequence of Theorem XIV we can say in general that *any monotonic sequence has a limit* (finite, $+\infty$, or $-\infty$), and that the limit is finite if and only if the sequence is bounded. (The student should give the details in Ex. 7, § 205.)

Theorem XV. *If $a_n \leq b_n$ for all n, and if $\lim_{n \to +\infty} a_n$ and $\lim_{n \to +\infty} b_n$ exist (finite, $+\infty$, or $-\infty$), then $\lim_{n \to +\infty} a_n \leq \lim_{n \to +\infty} b_n$.*

Proof. If the two limits are finite we can form the difference

$$c_n \equiv b_n - a_n$$

and, by appealing to Theorem VII, reduce the problem to the special case: *if $c_n \geq 0$ for all n and if $C = \lim_{n \to +\infty} c_n$ exists and is finite then $C \geq 0$.* By the definition of a limit, for any positive ϵ we can find values of n (arbitrarily large) such that $|c_n - C| < \epsilon$. Now if $C < 0$, let us choose $\epsilon = -C > 0$. We can then find arbitrarily large values of n such that $|c_n - C| = |c_n + \epsilon| < \epsilon$, and hence $c_n + \epsilon < \epsilon$. This contradicts the nonnegativeness of c_n. On the other hand, if it is assumed that $a_n \to +\infty$ and $b_n \to B$ (finite), we may take $\epsilon = 1$ and find first an N_1 such that $n > N_1$ implies $a_n > B + 1$, and then an N_2 such that $n > N_2$ implies $b_n < B + 1$. Again the inequality $a_n \leq b_n$ is contradicted (for n greater than both N_1 and N_2). The student should complete the proof for the cases $a_n \to A$ (finite), $b_n \to -\infty$ and $a_n \to +\infty$, $b_n \to -\infty$.

205. EXERCISES

1. Prove that if two subsequences of a given sequence converge to distinct limits, the sequence diverges.

2. Show by a counterexample that the converse of Theorem V, § 204, is false. That is, a bounded sequence need not converge.

3. Prove the extensions of Theorems VI and VIII, § 204, to an arbitrary finite number of sequences.

4. Prove that if $a_n \to +\infty$ and either $\{b_n\}$ converges or $b_n \to +\infty$, then $a_n + b_n \to +\infty$.

5. Prove that if $a_n \to +\infty$ and either $b_n \to b > 0$ or $b_n \to +\infty$, then $a_n b_n \to +\infty$. Prove that if $a_n \to \infty$ and $b_n \to b \neq 0$, then $a_n b_n \to \infty$.

6. Prove Theorem VII, § 204.

7. Prove Theorem XIV, § 204, for the case $a_n \downarrow$, and the statement of the Note that follows. *Hint:* Let $b_n = -a_n$ and use Theorem XIV, § 204, for the case $b_n \uparrow$. Cf. Ex. 23.

8. Show by counterexamples that the sum (difference, product, quotient) of two divergent sequences need not diverge.

9. Prove that if the sum and the difference of two sequences converge, then both of the sequences converge.

10. Prove that $a_n \to 0$ if and only if $|a_n| \to 0$.

11. Prove that $a_n \to a$ implies $|a_n| \to |a|$. Is the converse true? Prove, or give a counterexample. *Hint:* Use Property IX, § 110.

12. Prove that if a sequence has the limit $+\infty$ ($-\infty, \infty$) then any subsequence has the limit $+\infty$ ($-\infty, \infty$).

13. Prove that if $a_n \leq b$ ($a_n \geq b$) and $a_n \to a$, then $a \leq b$ ($a \geq b$). Show by an example that from the strict inequality $a_n < b$ ($a_n > b$) we cannot infer the strict inequality $a < b$ ($a > b$).

14. Prove that if $0 \leq a_n \leq b_n$ and $b_n \to 0$, then $a_n \to 0$. More generally, prove that if $a_n \leq b_n \leq c_n$ and $\{a_n\}$ and $\{c_n\}$ converge to the same limit, then $\{b_n\}$ also converges to this same limit.

15. Prove that if x is an arbitrary real number, there is a sequence $\{r_n\}$ of rational numbers converging to x. *Hint:* By the density of the rationals ((v), § 112), there is a rational number r_n in the open interval $\left(x - \frac{1}{n}, x + \frac{1}{n}\right)$.

★★16. If $\{s_n\}$ is a given sequence, define $\sigma_n \equiv \frac{1}{n}(s_1 + s_2 + \cdots + s_n)$. Prove that if $\{s_n\}$ converges to 0 then $\{\sigma_n\}$ also converges to 0. (Cf. Ex. 17.) *Hint:* Let m be a positive integer $< n$, and write

$$\sigma_n = \frac{1}{n}(s_1 + \cdots + s_m) + \frac{1}{n}(s_{m+1} + \cdots + s_n).$$

If $\epsilon > 0$, first choose m so large that whenever $k > m$, $|s_k| < \frac{1}{2}\epsilon$. Holding m fixed, choose $N > m$ and so large that $|s_1 + \cdots + s_m|/N < \frac{1}{2}\epsilon$. Then choose $n > N$, and consider separately the two groups of terms of σ_n, given above.

★★17. With the notation of Exercise 16, prove that if $\{s_n\}$ converges, then $\{\sigma_n\}$ also converges and has the same limit. Show by the example $0, 1, 0, 1, \cdots$ that the convergence of $\{\sigma_n\}$ does not imply that of $\{s_n\}$. Can you find a divergent sequence $\{s_n\}$ such that $\sigma_n \to 0$? *Hint:* Assume $s_n \to l$, let $t_n \equiv s_n - l$, and use the result of Ex. 16.

★★18. With the notation of Exercise 16, prove that if $\lim_{n \to +\infty} s_n = +\infty$, then $\lim_{n \to +\infty} \sigma_n = +\infty$. Show by the example $0, 1, 0, 2, 0, 3, \cdots$ that the reverse implication is not valid. *Hint:* If B is a given positive number, first choose m so large that whenever $k > m$, $s_k > 3B$. Then choose $N > 3m$ and so large that $|s_1 + \cdots + s_m|/N < B$. Then follow the hint of Ex. 16.

★★19. Show by the example $1, -1, 2, -2, 3, -3, \cdots$ that with the notation of Exercise 16, $\lim_{n\to+\infty} s_n = \infty$ does not imply $\lim_{n\to+\infty} \sigma_n = \infty$. Can you find an example where $\lim_{n\to+\infty} s_n = \infty$ and $\lim_{n\to+\infty} \sigma_n = 0$?

★★**20.** A number x is called a **limit point** of a sequence $\{a_n\}$ if and only if there exists some subsequence of $\{a_n\}$ converging to x. Prove that x is a limit point of $\{a_n\}$ if and only if corresponding to $\epsilon > 0$ the inequality $|a_n - x| < \epsilon$ holds for infinitely many values of the subscript n. Show by an example that this does not mean that $|a_n - x| < \epsilon$ must hold for infinitely many *distinct* values of a_n.

★★**21.** Prove that a bounded sequence converges if and only if it has exactly one limit point. (Cf. Ex. 20. If necessary, refer to § 301.)

★★**22.** Explain what you would mean by saying that $+\infty$ $(-\infty)$ is a limit point of a sequence $\{a_n\}$. Prove that a sequence is unbounded above (below) if and only if $+\infty$ $(-\infty)$ is a limit point of the sequence. Show how the word "bounded" can be omitted from Exercise 21. (Cf. Ex. 20.)

★**23.** Prove Theorem XIV, § 204, for the case $a_n \downarrow$, directly (without reference to the case $a_n \uparrow$), using the principle of *greatest lower bound*.

★★**24.** Formulate the results of Exercises 12-14, 16, § 113, in terms of the language of sequences.

206. LIMITS OF FUNCTIONS

In this section we recall and extend some of the basic limit concepts of elementary calculus. Before formalizing the appropriate definitions for such limits as $\lim_{x\to a} f(x)$ and $\lim_{x\to+\infty} f(x)$, let us agree on two things: Henceforth, unless explicit statement to the contrary is made, it will be implicitly assumed that *all functions considered are single-valued* and, whenever a limit of a function is concerned, that *the quantities symbolized exist for at least some values of the independent variable neighboring the limiting value of that variable*. For example, when we write $\lim_{x\to a} f(x)$ we shall assume that every neighborhood of the point a contains at least one point x different from a for which the function $f(x)$ is defined;† and when we write $\lim_{x\to+\infty} f(x)$ we shall assume that for any number N, $f(x)$ exists for some $x > N$.

A function $f(x)$ is said to **tend toward** or **approach** or **have** a limit L as x approaches a number a if and only if the absolute value of the difference between $f(x)$ and L is less than any preassigned positive number (however small) whenever the point x belonging to the domain of definition of $f(x)$ is sufficiently near a but not equal to a. This is expressed symbolically:

$$\lim_{x\to a} f(x) = L.$$

---

† In the terminology of the next chapter (§ 309), a is a *limit point* of the domain of definition D of $f(x)$. It can be shown that every neighborhood of a contains *infinitely many* points of D.

If in this definition the independent variable x is restricted to values greater than a, we say that x approaches a from the **right** or from **above** and write
$$\lim_{x \to a+} f(x) = L.$$
Again, if x is restricted to values less than a, we say that x approaches a from the **left** or from **below** and write
$$\lim_{x \to a-} f(x) = L.$$
The terms *undirected limit* or *two-sided limit* may be used to distinguish the first of these three limits from the other two in case of ambiguity arising from use of the single word *limit*.

A more concise formulation for these limits is given in the following definition:

Definition I. *The function $f(x)$ has the limit L as x approaches a, written*
$$\lim_{x \to a} f(x) = L, \quad \text{or} \quad f(x) \to L \text{ as } x \to a,$$
if and only if corresponding to an arbitrary positive number ϵ there exists a positive number $\delta = \delta(\epsilon)$ such that $0 < |x - a| < \delta$ implies $|f(x) - L| < \epsilon$, for values of x for which $f(x)$ is defined;† *$f(x)$ has the limit L as x approaches a from the right (left),*‡ *written*
$$\lim_{x \to a+} f(x) = L, \quad \text{or} \quad f(x) \to L \text{ as } x \to a+$$
$$(\lim_{x \to a-} f(x) = L, \quad \text{or} \quad f(x) \to L \text{ as } x \to a-),$$
if and only if corresponding to an arbitrary positive number ϵ there exists a positive number $\delta = \delta(\epsilon)$ such that $a < x < a + \delta$ $(a - \delta < x < a)$ implies $|f(x) - L| < \epsilon$, for values of x for which $f(x)$ is defined. These one-sided limits (if they exist) are also denoted:
$$f(a+) \equiv \lim_{x \to a+} f(x), \quad f(a-) \equiv \lim_{x \to a-} f(x).$$

Since the definition of limit employs only values of x different from a, it is completely immaterial what the value of the function is at $x = a$ or, indeed, whether it is defined there at all. Thus a function can fail to have a limit as x approaches a only by its misbehavior for values of x near a but not *equal* to a. Since $\lim_{x \to a} f(x)$ exists if and only if $\lim_{x \to a+} f(x)$ and

---

† An open interval with the midpoint removed is called a **deleted neighborhood** of the missing point. The inequalities $0 < |x - a| < \delta$, then, define a deleted neighborhood of the point a.

‡ Parentheses are used here and in the following two definitions to indicate an alternative statement. For a discussion of the use of parentheses for alternatives, see the Preface.

42 FUNCTIONS AND SEQUENCES [§ 206

$\lim_{x \to a} f(x)$ both exist and are equal (cf. Exs. 13-14, § 208), $\lim_{x \to a} f(x)$ may fail to exist either by $\lim_{x \to a+} f(x)$ and $\lim_{x \to a-} f(x)$ being unequal or by either or both of the latter failing to exist in one way or another. These possibilities are illustrated in Example 1 below.

Limits as the independent variable becomes infinite have a similar formulation:

Definition II. *The function $f(x)$ has the limit L as x becomes positively (negatively) infinite, written*

$$f(+\infty) \equiv \lim_{x \to +\infty} f(x) = L, \quad \text{or} \quad f(x) \to L \text{ as } x \to +\infty$$

$$(f(-\infty) \equiv \lim_{x \to -\infty} f(x) = L, \quad \text{or} \quad f(x) \to L \text{ as } x \to -\infty),$$

if and only if corresponding to an arbitrary positive number ϵ there exists a number $N = N(\epsilon)$ such that $x > N$ ($x < N$) implies $|f(x) - L| < \epsilon$, for values of x for which $f(x)$ is defined.

In an analogous fashion, infinite limits can be defined. Only a sample definition is given here, others being requested in the Exercises of § 208.

Definition III. *The function $f(x)$ has the limit $+\infty$ ($-\infty$) as x approaches a, written*

$$\lim_{x \to a} f(x) = +\infty \ (-\infty), \quad \text{or} \quad f(x) \to +\infty \ (-\infty) \text{ as } x \to a,$$

if and only if corresponding to an arbitrary number B there exists a positive number $\delta = \delta(B)$ such that $0 < |x - a| < \delta$ implies $f(x) > B$ ($f(x) < B$), for values of x for which $f(x)$ is defined.

As with limits of sequences it is often convenient to use an *unsigned infinity*, ∞. When we say that a variable, dependent or independent, tends toward ∞, we shall mean that its absolute value approaches $+\infty$. Thus $\lim_{x \to \infty} f(x) = L$ is defined as in Definition II, with the inequality $x > N$ replaced by $|x| > N$, and $\lim_{x \to a} f(x) = \infty$ is equivalent to $\lim_{x \to a} |f(x)| = +\infty$.

Example 1. Discuss the limits of each of the following functions as x approaches 0, $0+$, and $0-$, and in each case sketch the graph: (a) $f(x) = \frac{x^2 + x}{x}$ if $x \neq 0$, undefined for $x = 0$; (b) $f(x) = |x|$ if $x \neq 0$, $f(0) = 3$; (c) the **signum function**, $f(x) = \operatorname{sgn} x = 1$ if $x > 0$, $f(x) = \operatorname{sgn} x = -1$ if $x < 0$, $f(0) = \operatorname{sgn} 0 = 0$; (d)† $f(x) = \sin \frac{1}{x}$ if $x \neq 0$, $f(0) = 0$; (e) $f(x) = \frac{1}{x}$ if $x \neq 0$, undefined if $x = 0$; (f) $f(x) = \frac{1}{x^2}$ if $x \neq 0$, undefined if $x = 0$.

† For illustrative examples and exercises the familiar properties of the trigonometric functions will be assumed. An analytic treatment is given in §§ 603-604.

§ 206] LIMITS OF FUNCTIONS 43

Solution. The graphs are given in Figure 203. In part (a) if $x \neq 0$, $f(x)$ is identically equal to the function $x + 1$, and its graph is therefore the straight line $y = x + 1$ with the single point $(0, 1)$ deleted; $\lim_{x \to 0} f(x) = f(0+) = f(0-) = 1$. In part (b) $\lim_{x \to 0} f(x) = f(0+) = f(0-) = 0$. The fact that $f(0) = 3$ has no bearing on the statement of the preceding sentence. For the signum function (c), $f(0+) = 1$, $f(0-) = -1$, and $\lim_{x \to 0} f(x)$ does not exist. In part (d) all three limits fail to exist. In part (e) $f(0+) = +\infty$, $f(0-) = -\infty$, and $\lim_{x \to 0} f(x) = \infty$ (unsigned infinity) (cf. Exs. 31-32, § 208). In part (f), $f(0+) = f(0-) = \lim_{x \to 0} f(x) = +\infty$ (cf. Ex. 32, § 208).

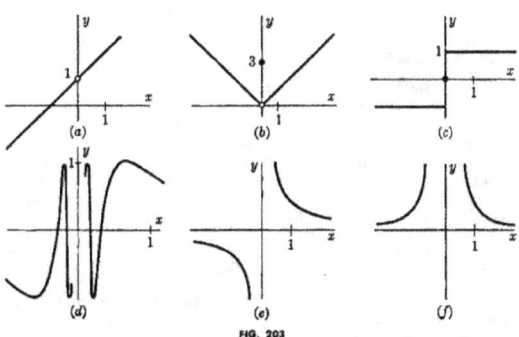

FIG. 203

Example 2. Show that $\lim_{x \to 2} \dfrac{x^2 - x + 18}{3x - 1} = 4$. Find an explicit function $\delta(\epsilon)$ as demanded by Definition I.

Solution. Form the absolute value of the difference:

(1) $\qquad \left| \dfrac{x^2 - x + 18}{3x - 1} - 4 \right| = \left| \dfrac{x^2 - 13x + 22}{3x - 1} \right| = |x - 2| \cdot \left| \dfrac{x - 11}{3x - 1} \right|.$

We wish to show that this expression is small if x is near 2. The first factor, $|x - 2|$, is certainly small if x is near 2; and the second factor, $\left| \dfrac{x - 11}{3x - 1} \right|$, is not dangerously large if x is near 2 and at the same time not too near $\tfrac{1}{3}$. Let us make this precise by first requiring that $\delta \leq 1$. If x is within a distance less than δ of 2, then $1 < x < 3$, and hence also $-10 < x - 11 < -8$ and $2 < 3x - 1 < 8$, so that $|x - 11| < 10$ and $|3x - 1| > 2$. Thus the second factor is less than $\tfrac{10}{2} = 5$. Now let a positive number ϵ be given. Since the

44　　FUNCTIONS AND SEQUENCES　　[§ 207

expression (1) will be less than ϵ if simultaneously $|x - 2| < \frac{\epsilon}{5}$ and $\left|\frac{x-1}{3x-1}\right| < 5$, we have only to take $\delta = \delta(\epsilon)$ to be the smaller of the two numbers 1 and $\frac{\epsilon}{5}$, $\delta(\epsilon) = \min\left(1, \frac{\epsilon}{5}\right)$. The graph of this function $\delta(\epsilon)$ is shown in Figure 204.

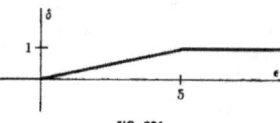

FIG. 204

207. LIMIT THEOREMS FOR FUNCTIONS

Many of the theorems of § 204 are special cases of more general limit theorems which apply to real valued functions, whether the independent variable tends toward a (finite) number, toward a number from one side only, or toward $+\infty$, $-\infty$ or ∞. Some of these are stated below. Since in each case the statement is essentially the same regardless of the manner in which the independent variable approaches its limit,† this latter behavior is unspecified. Furthermore, since the proofs are mere reformulations of those given in § 204, only one sample is given here. Others are requested in the Exercises of § 208.

Theorem I. *If $\lim f(x)$ exists it is unique.*

Theorem II. *If the function $f(x)$ is equal to a constant k, then $\lim f(x)$ exists and is equal to k.*

Theorem III. *If $\lim f(x)$ and $\lim g(x)$ exist and are finite, then $\lim [f(x) + g(x)]$ exists and is finite, and*

$$\lim [f(x) + g(x)] = \lim f(x) + \lim g(x).$$

In short, the limit of the sum is the sum of the limits. This rule extends to the sum of any finite number of functions.

Theorem IV. *Under the hypotheses of Theorem III, the limit of the difference is the difference of the limits:*

$$\lim [f(x) - g(x)] = \lim f(x) - \lim g(x).$$

---

† The word *limit* has been given specific meaning in this chapter only for functions, but it is convenient to extend its use to apply to *both* dependent and independent variables. The reader should recognize, however, that such an isolated statement as "the limit of the variable v is l" is meaningless, and takes on meaning only if v is associated with another variable, thus: $\lim_{x \to \infty} v(x) = l$, or $\lim_{x \to a} f(v) = k$.

Theorem V. *Under the hypotheses of* Theorem III, *the limit of the product is the product of the limits:*
$$\lim [f(x)\, g(x)] = \lim f(x) \cdot \lim g(x).$$
This rule extends to the product of any finite number of functions.

Proof for the case $x \to a$. Assume $f(x) \to L$ and $g(x) \to M$ as $x \to a$. We wish to show that $f(x)\, g(x) \to LM$ or, equivalently, that
$$f(x)\, g(x) - LM \to 0, \text{ as } x \to a.$$
By addition and subtraction of the quantity $L \cdot g(x)$ and by appeal to Theorem III, we can use the relation
$$f(x)\, g(x) - LM = [f(x) - L]g(x) + L[g(x) - M]$$
to reduce the problem to that of showing that $[f(x) - L]g(x) \to 0$ and $L[g(x) - M] \to 0$ as $x \to a$. The fact that they do is a consequence of the following lemma:

Lemma. *If* $\phi(x) \to 0$ *and* $\psi(x) \to \mu$ *as* $x \to a$, *then* $\phi(x)\psi(x) \to 0$ *as* $x \to a$.

Proof of lemma. First, by letting $\epsilon = 1$ in the definition of $\lim_{x \to a} \psi(x) = \mu$, we observe that there is a positive number δ_1 such that for x in the deleted neighborhood $0 < |x - a| < \delta_1$ the values of the function $\psi(x)$ lie in the neighborhood $(\mu - 1, \mu + 1)$, and are therefore bounded. Let P be a positive number such that $0 < |x - a| < \delta_1$ implies $|\psi(x)| \leq P$. Now let ϵ be an arbitrary positive number, and let δ be a positive number $\leq \delta_1$ such that $0 < |x - a| < \delta$ implies $|\phi(x)| < \epsilon/P$. Then $0 < |x - a| < \delta$ implies
$$|\phi(x)\psi(x)| < (\epsilon/P) \cdot P = \epsilon,$$
and the proof of the theorem is complete.

Theorem VI. *Under the hypotheses of* Theorem III *and the additional hypothesis that* $\lim g(x) \neq 0$, *the limit of the quotient is the quotient of the limits:*
$$\lim \frac{f(x)}{g(x)} = \frac{\lim f(x)}{\lim g(x)}.$$

Theorem VII. *If* $f(x) \leq g(x)$ *and if* $\lim f(x)$ *and* $\lim g(x)$ *exist (finite, or* $+\infty$, *or* $-\infty$), *then* $\lim f(x) \leq \lim g(x)$.

208. EXERCISES

In Exercises 1-7, prove the indicated limit theorem for the specified behavior of the independent variable.

1. Theorem I, § 207; $x \to a$, finite or infinite limit.
2. Theorem II, § 207; $x \to \infty$.

3. Theorem III, § 207; $x \to a+$.
4. Theorem III, § 207; $x \to +\infty$.
5. Theorem IV, § 207; $x \to a-$.
6. Theorem V, § 207; $x \to -\infty$.
7. Theorem VII, § 207; $x \to a$.

8. Prove that if $\lim_{x \to a} \phi(x) = 0$, and if $\psi(x)$ is bounded in some deleted neighborhood of a (that is, if there exist positive numbers P and η such that $|\psi(x)| \leq P$ for $0 < |x - a| < \eta$), then $\lim_{x \to a} \phi(x)\psi(x) = 0$. Find $\lim_{x \to 0} x \sin \frac{1}{x}$. (cf. Example 2, § 403).

9. Prove that if $\lim_{x \to a} f(x)$ exists and is positive (negative), then $f(x)$ is positive (negative) in some deleted neighborhood of a. Prove, in fact, that if $\lim_{x \to a} f(x) = m \neq 0$, then for all x within some deleted neighborhood of a, $f(x) > \frac{1}{2}m$ if $m > 0$ and $f(x) < \frac{1}{2}m$ if $m < 0$. Consequently show that the reciprocal of a function is bounded in some deleted neighborhood of any point at which the function has a nonzero limit. Show that this last statement is true even if the nonzero limit is infinite. *Hint:* In case $\lim_{x \to a} f(x) = m > 0$, let δ be a positive number such that $0 < |x - a| < \delta$ implies $|f(x) - m| < \frac{1}{2}m$, so that $m - f(x) < \frac{1}{2}m$.

10. Prove Theorem VI, § 207, for the case $x \to a$. (Cf. Exs. 8 and 9.)

11. Prove Theorem VI, § 207, for the case $x \to +\infty$. (Cf. Exs. 8-10.)

12. Prove that if $f(x) \leq g(x) \leq h(x)$ and if $\lim f(x)$ and $\lim h(x)$ are finite and equal (for the same behavior of the independent variable subject to the restrictions of the first paragraph of § 207), then $\lim g(x)$ exists and is equal to their common value. Extend this result to include infinite limits.

In Exercises 13-20, prove the given statement.

13. $\lim_{x \to a} f(x)$ exists and is finite if and only if $\lim_{x \to a+} f(x)$ and $\lim_{x \to a-} f(x)$ exist and are finite and equal.

14. $\lim_{x \to a} f(x)$ exists (in the finite or infinite sense) if and only if $\lim_{x \to a+} f(x)$ and $\lim_{x \to a-} f(x)$ exist and are equal.

15. Theorems III and V, § 207, hold for any finite number of functions.

16. If k is a constant and $\lim f(x)$ exists and is finite, then $\lim k f(x)$ exists and is equal to $k \lim f(x)$, whatever the behavior of the independent variable x, subject to the restrictions of the first paragraph of § 207.

17. $\lim_{x \to a} x = a$.

18. If n is a positive integer, $\lim_{x \to a} x^n = a^n$. *Hint:* Use Theorem V, § 207, and Ex. 17.

19. If $f(x)$ is a polynomial,
$$f(x) = a_0 x^n + a_1 x^{n-1} + \cdots + a_{n-1} x + a_n,$$
then $\lim_{x \to a} f(x) = f(a)$.

§ 208]   EXERCISES   47

20. If $f(x)$ is a rational function,
$$f(x) = g(x)/h(x),$$
where $g(x)$ and $h(x)$ are polynomials, and if $h(a) \neq 0$, then $\lim_{x \to a} f(x) = f(a)$.

In Exercises 21-26, find the indicated limit.

21. $\lim_{x \to 3} (2x^2 - 5x + 1)$.

22. $\lim_{x \to -2} \dfrac{3x^2 - 5}{2x + 17}$.

23. $\lim_{x \to -2} \dfrac{3x^2 - x - 10}{x^2 + 5x - 14}$. *Hint:* Reduce to lowest terms.

24. $\lim_{x \to -3} \dfrac{x^3 + 27}{x^4 - 81}$. (Cf. Ex. 23.)

25. $\lim_{x \to a} \dfrac{x^2 - a^2}{x - a}$. (Cf. Ex. 23.)

26. $\lim_{x \to a} \dfrac{x^m - a^m}{x - a}$, where m is an integer. (Cf. Ex. 23, above, and Ex. 9, § 107.)

In Exercises 27-30, give a precise definition for the given limit statement.

27. $\lim_{x \to a+} f(x) = -\infty$.

28. $\lim_{x \to -\infty} f(x) = +\infty$.

29. $\lim_{x \to \infty} f(x) = \infty$.

30. $\lim_{x \to a} f(x) = \infty$.

31. Prove that $\lim_{x \to 0} \dfrac{1}{x} = \infty$ and $\lim_{x \to \infty} \dfrac{1}{x} = 0$. More generally, assuming $f(x)$ to be nonzero except possibly for the limiting value of the independent variable, prove that $\lim f(x) = 0$ if and only if $\lim \dfrac{1}{f(x)} = \infty$. Discuss Theorem VI, § 207, if $\lim f(x) \neq 0$ and $\lim g(x) = 0$. (Cf. Theorem XI, § 204.)

32. Prove that $\lim_{x \to 0+} \dfrac{1}{x} = +\infty$. More generally, assuming $f(x)$ to be positive except possibly for the limiting value of the independent variable, prove that $\lim f(x) = +\infty$ if and only if $\lim \dfrac{1}{f(x)} = 0$.

In Exercises 33 and 34, assuming the standard facts regarding trigonometric functions (cf. §§ 603-604 for analytic definitions of the trigonometric functions), find the specified limits, or establish their nonexistence.

33. (a) $\lim_{x \to \frac{1}{2}\pi-} \tan x$; (b) $\lim_{x \to \frac{1}{2}\pi+} \tan x$;

(c) $\lim_{x \to \frac{1}{2}\pi} \tan x$; (d) $\lim_{x \to 0+} \cot x$;

(e) $\lim_{x \to 0-} \cot x$; (f) $\lim_{x \to 0} \cot x$.

34. (a) $\lim_{x \to +\infty} \sin x$; (b) $\lim_{x \to -\infty} \cos x^2$;

48 FUNCTIONS AND SEQUENCES [§ 208]

(c) $\lim\limits_{x\to\infty} \dfrac{\sin x}{x}$; (d) $\lim\limits_{x\to+\infty} \dfrac{\sec x}{x}$;

(e) $\lim\limits_{x\to-\infty} \dfrac{x - \cos x}{x}$; (f) $\lim\limits_{x\to\infty} \dfrac{x \sin x}{x^2 - 4}$.

35. Let $f(x)$ be a rational function
$$f(x) = \frac{a_0 x^m + a_1 x^{m-1} + \cdots + a_{m-1} x + a_m}{b_0 x^n + b_1 x^{n-1} + \cdots + b_{n-1} x + b_n},$$
where $a_0 \neq 0$ and $b_0 \neq 0$. Show that $\lim\limits_{x\to\infty} f(x)$ is equal to 0 if $m < n$, to a_0/b_0 if $m = n$, and to ∞ if $m > n$. In particular, show that if $f(x)$ is any nonconstant polynomial, $\lim\limits_{x\to\infty} f(x) = \infty$. *Hint:* Divide every term in both numerator and denominator by the highest power of x present.

36. Discuss the result of Exercise 35 for the case $m > n$ if (i) $x \to +\infty$; (ii) $x \to -\infty$; (iii) $x \to \infty$ and m and n are either both even or both odd. Consider in particular the special case where $f(x)$ is a polynomial.

In Exercises 37–42, find the indicated limit. (Cf. Exs. 35–36.)

37. $\lim\limits_{x\to\infty} \dfrac{5x^2 - 3x + 1}{6x^2 + 5}$. **38.** $\lim\limits_{x\to+\infty}(2x^4 - 350x^2 - 10{,}000)$.

39. $\lim\limits_{x\to\infty}(-6x^4 - 9x^3 + x)$. **40.** $\lim\limits_{x\to\infty} \dfrac{2x^3 - 5}{3x + 7}$.

41. $\lim\limits_{x\to+\infty} \dfrac{150x + 2000}{x^2 + 3}$. **42.** $\lim\limits_{x\to-\infty} \dfrac{8x^3 + 13x + 6}{5x^3 + 11}$.

In Exercises 43–50, interpret and prove each relation. For these exercises p designates a positive number, n a negative number, m a nonzero number, and q any number. *Hint for* Ex. 43: This means: If $\lim f(x) = p$ and $\lim g(x) = +\infty$, then $\lim f(x) g(x) = +\infty$. For simplicity let $x \to a$.

43. $p \cdot (+\infty) = +\infty$. **44.** $n \cdot (+\infty) = -\infty$.
45. $q - (-\infty) = +\infty$. **46.** $q + (\infty) = \infty$.
47. $(-\infty) - q = -\infty$. **48.** $(+\infty) + (+\infty) = +\infty$.
49. $\dfrac{0}{\infty} = \dfrac{m}{\infty} = 0$. **50.** $\dfrac{\infty}{0} = \dfrac{m}{0} = \infty$.

In Exercises 51–56, show by examples that the given expression is indeterminate. (See Hint for Ex. 43.)

51. $\infty + \infty$. **52.** $(+\infty) - (+\infty)$.

Hint for Ex. 52: Consider the examples (i) $x - x$ and (ii) $x^2 - x$ as $x \to +\infty$.

53. $(+\infty) + (\infty)$. **54.** $0 \cdot \infty$.
55. $\dfrac{0}{0}$. **56.** $\dfrac{\infty}{\infty}$.

★★**57.** Give an example of a function $f(x)$ satisfying the following three conditions: $\lim\limits_{x\to 0} |f(x)| = 1$, $\lim\limits_{x\to 0-} f(x) = -1$, $\lim\limits_{x\to 0+} f(x)$ does not exist. (Cf. Ex. 39, § 216.)

§ 209] CONTINUITY

★★58. Give an example of a function $f(x)$ satisfying the following three conditions: $\lim_{x\to 0+} f(x) = \infty$, $\lim_{x\to 0+} f(x) \neq +\infty$, $\lim_{x\to 0+} f(x) \neq -\infty$. (Cf. Ex. 57.)

In Exercises 59-64, find the required limit, prove that it is the limit by direct use of Definition I, § 206, and obtain explicitly a function $\delta(\epsilon)$ as demanded by that definition. (Cf. Exs. 32-37, § 216.)

★59. $\lim_{x\to 2} 3x$.

★60. $\lim_{x\to -3} x^2$.

★61. $\lim_{x\to 4} (3x^2 - 5x)$.

★62. $\lim_{x\to -5} \frac{1}{x}$.

★63. $\lim_{x\to 1} \frac{4x^2 - 1}{5x + 2}$.

★64. $\lim_{x\to 2} \frac{3x}{4x - 7}$.

In Exercises 65-70, find the required limit, prove that it is the limit by direct use of Definition II or Definition III, § 206, and obtain explicitly a function $N(\epsilon)$ or $\delta(B)$ as demanded by the definition.

★65. $\lim_{x\to +\infty} \frac{5}{x}$.

★66. $\lim_{x\to -\infty} \frac{1}{x^2}$.

★67. $\lim_{x\to +\infty} \frac{3x - 2}{x + 5}$.

★68. $\lim_{x\to +\infty} \frac{5x^2 + 1}{3x^2}$.

★69. $\lim_{x\to 0} \frac{1}{x^2}$.

★70. $\lim_{x\to 1} \frac{2x - 5}{x^3 - 2x^2 + x}$.

209. CONTINUITY

Continuity of a function at a point a can be defined either in terms of limits (Definition I, below), or directly by use of the type of δ-ϵ formulation in which the original limit concepts are framed (Definition II, below). When continuity is couched in terms of limits, we shall make the same implicit assumption that was stated in the first paragraph of § 206, namely, that every neighborhood of the point a contains at least one point x different from a for which the function $f(x)$ is defined. (Cf. Note 2, below.)

Definition I. *A function $f(x)$ is **continuous** at $x = a$ if and only if the following three conditions are satisfied:*

(i) $f(a)$ exists; that is, $f(x)$ is defined at $x = a$;

(ii) $\lim_{x\to a} f(x)$ exists and is finite;

(iii) $\lim_{x\to a} f(x) = f(a)$.

By inspection of the definition of $\lim_{x\to a} f(x)$, it is possible (cf. Ex. 11, § 212) to establish the equivalence of this definition and the following (in case the implicit assumption of the first paragraph, above, is satisfied).

Definition II. *A function $f(x)$ is* **continuous** *at $x = a$ if and only if it is defined at $x = a$ and corresponding to an arbitrary positive number ϵ, there exists a positive number $\delta = \delta(\epsilon)$ such that $|x - a| < \delta$ implies*

$$|f(x) - f(a)| < \epsilon,$$

for values of x for which $f(x)$ is defined.

NOTE 1. Definition I is sometimes called the *limit definition of continuity* and Definition II the *δ-ϵ definition of continuity*.

NOTE 2. The δ-ϵ definition is applicable even when the function is not defined at points neighboring $x = a$ (except at a itself). In this case a is an *isolated point* of the domain of definition, and $f(x)$ is continuous there, although $\lim_{x \to a} f(x)$ has no meaning.

NOTE 3. Each of the following limit statements is a formulation of continuity of $f(x)$ at $x = a$ (Ex. 3, § 212):

(1) $\qquad\qquad f(a + h) - f(a) \to 0 \quad \text{as} \quad h \to 0;$

(2) $\qquad\qquad \text{if} \quad \Delta y = f(a + \Delta x) - f(a), \quad \text{then}$
$$\Delta y \to 0 \quad \text{as} \quad \Delta x \to 0.$$

A function is said to be **continuous on a set** if and only if it is continuous at every point of that set. In case a function is continuous at every point of its domain of definition it is simply called **continuous**, without further modifying words.

Continuity from the right, or **right-hand continuity**, is defined by replacing, in Definition I, $\lim_{x \to a} f(x)$ by $\lim_{x \to a+} f(x)$. Similarly, **continuity from the left**, or **left-hand continuity**, is obtained by replacing $\lim_{x \to a} f(x)$ by $\lim_{x \to a-} f(x)$. Thus $f(x)$ is continuous from the right at $x = a$ if and only if $f(a+) = f(a)$, and $f(x)$ is continuous from the left if and only if $f(a-) = f(a)$, it being assumed that the expressions written down exist.

A useful relation between continuity and limits is stated in the following theorem:

Theorem. *If $f(x)$ is continuous at $x = a$ and if $\phi(t)$ has the limit a, as t approaches some limit (finite or infinite, one-sided or not), then*

$$f(\phi(t)) \to f(a).$$

In short, the limit of the function is the function of the limit:

$$\lim f(\phi(t)) = f(\lim \phi(t)).$$

Proof. We shall prove the theorem for the single case $t \to +\infty$. (Cf. Exs. 12-13, § 212.) Accordingly, let ϵ be an arbitrary positive number, and let δ be a positive number such that $|x - a| < \delta$ implies

$$|f(x) - f(a)| < \epsilon.$$

Then choose N so large that $t > N$ implies $|\phi(t) - a| < \delta$. Combining

§ 210] TYPES OF DISCONTINUITY 51

these two implications by setting $x = \phi(t)$ we have the result: $t > N$ implies $|f(\phi(t)) - f(a)| < \epsilon$. This final implication is the one sought.

Example 1. A function whose domain of definition is a closed interval $[a, b]$ is continuous there if and only if it is continuous at each interior point, continuous from the right at $x = a$, and continuous from the left at $x = b$. (See Fig. 205.)

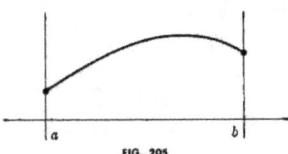

FIG. 205

Example 2. The function $[x]$ (Example 5, § 201) is continuous except when x is an integer. It is everywhere continuous from the right. (See Fig. 206.)

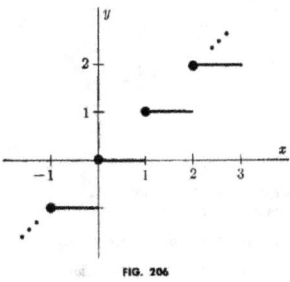

FIG. 206

210. TYPES OF DISCONTINUITY

The principal types of discontinuity are the following four:

(i) *The limit of the function exists, but the function either is not defined at the point or has a value different from the limit there.* (Figure 207, (a) and (b); cf. Example 1, § 206.) Such a discontinuity is called a **removable discontinuity** because if the function is redefined to have the value $f(a) = \lim_{x \to a} f(x)$ at $x = a$, it becomes continuous there.

52   FUNCTIONS AND SEQUENCES   [§ 211

(ii) *The two one-sided limits exist and are finite, but are not equal.* An example is the signum function (**Figure 207**, (c); cf. Example 1, § 206). Such a discontinuity is called **a jump discontinuity**.

(iii) *At least one one-sided limit fails to exist.* An example is $\sin \frac{1}{x}$ (Figure 207, (d); cf. Example 1, § 206).

(iv) *At least one one-sided limit is infinite* (Figure 207, (e) and (f); cf. Example 1, § 206).

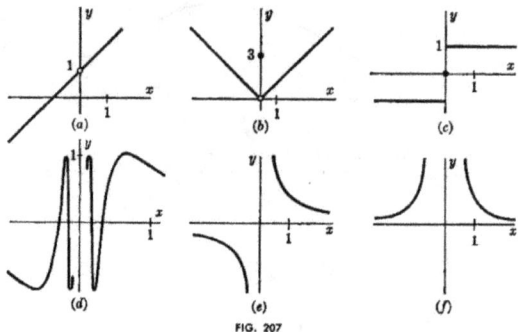

FIG. 207

211. CONTINUITY THEOREMS

The limit theorems II-VI of § 207 have (as immediate corollaries) counterparts in terms of continuity:

Theorem I. *Any constant function is continuous.*

Theorem II. *If $f(x)$ and $g(x)$ are continuous at $x = a$, then their sum $f(x) + g(x)$ is also continuous at $x = a$. In short, the sum of two continuous functions is a continuous function. This rule applies to any finite number of functions.*

Theorem III. *Under the hypotheses of Theorem II, $f(x) - g(x)$ is continuous at $x = a$: the difference of two continuous functions is a continuous function.*

Theorem IV. *Under the hypotheses of Theorem II, $f(x) g(x)$ is continuous at $x = a$: the product of two continuous functions is a continuous function. This rule applies to any finite number of functions.*

Theorem V. *Under the hypotheses of* Theorem II *and the additional hypothesis that* $g(a) \neq 0$, $f(x)/g(x)$ *is continuous at* $x = a$: *the quotient of two continuous functions is continuous where the denominator does not vanish.*

A direct consequence of the Theorem of § 209 is the following:

Theorem VI. *A continuous function of a continuous function is a continuous function. More precisely, if* $f(x)$ *is continuous at* $x = a$, *if* $g(y)$ *is continuous at* $y = b = f(a)$, *and if* $h(x) \equiv g(f(x))$, *then* $h(x)$ *is a continuous function of* x *at* $x = a$.

212. EXERCISES

1. Prove that any polynomial is continuous for all values of the independent variable. (Cf. Ex. 19, § 208.)

2. Prove that any rational function is continuous except where the denominator vanishes. (Cf. Ex. 20, § 208.)

3. Prove the statements in Note 3, § 209.

4. Redefine the signum function (Example 1, § 206) at $x = 0$ so that it becomes everywhere continuous from the right; so that it becomes everywhere continuous from the left.

5. Establish continuity at $x = 0$ for each of the two functions:

(a) $f(x) = x \sin \frac{1}{x}$ if $x \neq 0$, $f(0) = 0$; (b) $g(x) = x^2 \sin \frac{1}{x}$ if $x \neq 0$, $= 0$ if $x = 0$. (Cf. Ex. 8, § 208. Also see Examples 2 and 3, § 403, for graphs and further discussion.)

6. Define the function $x \cos \frac{1}{x}$ at $x = 0$ so that it becomes everywhere continuous.

In Exercises 7-10, state the type of discontinuity at $x = 0$.

7. $f(x) = x^2 - 8x$ if $x \neq 0$, $f(0) = 6$.

8. $f(x) = x^2 \cos \frac{1}{x^2}$ if $x \neq 0$, undefined if $x = 0$.

9. $f(x) = \frac{1}{x}$ if $x > 0$, $f(x) = 0$ if $x \leq 0$.

10. $f(x) = x + 1$ if $x > 0$, $f(x) = -x - 1$ if $x < 0$, $f(0) = 0$.

11. Prove the equivalence of the two definitions of continuity, § 209, under the assumptions of the first paragraph of that section.

12. Prove the Theorem of § 209 (a) for the case $t \to c$; (b) for the case $t \to c-$.

13. Give an example to show that the Theorem of § 209 is false if the continuity assumption is omitted.

14. Prove that the **negation of continuity** of a function at a point of its domain can be expressed: $f(x)$ *is discontinuous at* $x = a$ *if and only if there is a positive number* ϵ *having the property that corresponding to an arbitrary positive number* δ *(however small), there exists a number* x *such that* $|x - a| < \delta$ *and* $|f(x) - f(a)| \geq \epsilon$.

54   FUNCTIONS AND SEQUENCES   [§ 213

15. Show that the function defined for all real numbers, $f(x) = 1$ if x is rational, $f(x) = 0$ if x is irrational, is everywhere discontinuous. (Cf. Example 6, § 201.)

16. Prove that x is everywhere continuous.

★17. Prove that $f(x)$ is continuous wherever $f(x)$ is. Give an example of a function defined for all real numbers which is never continuous but whose absolute value is always continuous. *Hint:* Consider a function like that of Ex. 15, with values ± 1.

18. Prove that if $f(x)$ is continuous and positive at $x = a$, then there is a neighborhood of a in which $f(x)$ is positive. Prove that, in fact, there exist a positive number ϵ and a neighborhood of a such that in this neighborhood $f(x) > \epsilon$. State and prove corresponding facts if $f(a)$ is negative. *Hint:* Cf. Ex. 9, § 208.

★19. Prove that if $f(x)$ is continuous at $x = a$, and if $\epsilon > 0$ is given, then there exists a neighborhood of a such that for any two points in this neighborhood the values of $f(x)$ differ by less than ϵ.

★★20. Show that the function with domain the closed interval $[0, 1]$, $f(x) = x$ if x is rational, $f(x) = 1 - x$ if x is irrational, is continuous only for $x = \frac{1}{2}$.

★★21. Show that the following example is a counterexample to the following False Theorem, which strives to generalize the Theorem of § 209, and Theorem VI, § 211: If $\lim \phi(t) = a$, where a is finite and t approaches some limit, and if $\lim_{x \to a} f(x) = b$, where b is finite, then $\lim f(\phi(t)) = b$, as t approaches its limit.

Counterexample: $\phi(t) \equiv 0$ for all t, $f(x) \equiv 0$ if $x \neq 0$, $f(0) \equiv 1$; $t \to 0$, $x \to 0$. Prove that this False Theorem becomes a True Theorem with the additional hypothesis that $\phi(t)$ is never equal to a, except possibly for the limiting value of t.

213. MORE THEOREMS ON CONTINUOUS FUNCTIONS

We list below a few important theorems on continuous functions, whose proofs depend on certain rather sophisticated ideas discussed in Chapter 3, which is starred for possible omission or postponement. The proofs of these particular theorems are given in § 306.

Theorem I. *A function continuous on a closed interval is bounded there.* That is, if $f(x)$ is continuous on $[a, b]$, there exists a number B such that $a \leq x \leq b$ implies $|f(x)| \leq B$.

Theorem II. *A function continuous on a closed interval has a maximum and a minimum there.* That is, if $f(x)$ is continuous on $[a, b]$, there exist points x_1 and x_2 in $[a, b]$ such that $a \leq x \leq b$ implies $f(x_1) \leq f(x) \leq f(x_2)$.

Theorem III. *If $f(x)$ is continuous on the closed interval $[a, b]$ and if $f(a)$ and $f(b)$ have opposite signs, there is a point x_0 between a and b for which $f(x_0) = 0$.*

214. EXISTENCE OF $\sqrt{2}$ AND OTHER ROOTS

Theorem IV. *A function continuous on an interval assumes (as a value) every number between any two of its values.* (Cf. Ex. 47, § 408.)

NOTE. For other properties of a function continuous on a closed interval, see § 307 and § 501.

214. EXISTENCE OF $\sqrt{2}$ AND OTHER ROOTS

In Chapter 1 (Ex. 8, § 109) $\sqrt{2}$ was mentioned as an example of an irrational number, but proof of its existence was deferred. We are able now, with the aid of the last theorem of the preceding section, to give a simple proof that there exists a positive number whose square is 2. The idea is to consider the function $f(x) = x^2$, which is continuous everywhere (Ex. 1, § 212) and, in particular, on the closed interval from $x = 0$ to $x = 2$. Since the values of the function at the end-points of this interval are $0^2 = 0$ and $2^2 = 4$, and since the number 2 is between these two extreme values, there must be a (positive) number between these end-points for which the value of the function is 2. That is, there is a positive number whose square is 2. The following theorem generalizes this result, and establishes the uniqueness of positive nth roots:

Theorem. *If p is a positive number and n is a positive integer, there exists a unique positive number x such that $x^n = p$. This number is called the nth root of p and is written $x = \sqrt[n]{p}$.*

Proof. We establish uniqueness first. If $x^n = y^n = p$, where x and y are positive, then (Ex. 9, § 107) $x^n - y^n = (x - y)(x^{n-1} + x^{n-2}y + \cdots + y^{n-1}) = 0$. Since the second factor is positive (Ex. 8, § 107) the first factor must vanish (Ex. 14, § 103): $x - y = 0$, or $x = y$.

For the proof of existence, we note first that since $x^n \geq x$ for $x \geq 1$, $\lim_{x \to +\infty} x^n = +\infty$. Therefore there exists a number b such that $b^n > p$. The number p is thus between the extreme values assumed by x^n on the closed interval $[0, b]$, and therefore, since x^n is continuous, this function must assume the value p at some point x between 0 and b: $x^n = p$.

215. MONOTONIC FUNCTIONS AND THEIR INVERSES

Definition. *A function $f(x)$ is monotonically increasing (decreasing), written $f(x) \uparrow (\downarrow)$, on a set A if and only if whenever a and b are elements of A and $a < b$, then $f(a) \leq f(b)$ ($f(a) \geq f(b)$). In either case it is called monotonic. Whenever $a < b$ implies $f(a) < f(b)$ ($f(a) > f(b)$), $f(x)$ is called strictly increasing (decreasing), and in either case strictly monotonic.* (Cf. Fig. 208.)

56   FUNCTIONS AND SEQUENCES   [§ 215

Theorem XIV, § 204, states facts about monotonic sequences. If we permit the inclusion of infinite limits we can drop the assumption of boundedness and state that any monotonic sequence has a limit (cf. the Note, § 204). In Exercise 27, § 216, the student is asked to generalize this fact and prove that a monotonic function $f(x)$ always has one-sided limits. Consequently the only type of discontinuity that a monotonic function can have (at a point where it is defined) is a finite jump (Ex. 28, § 216).

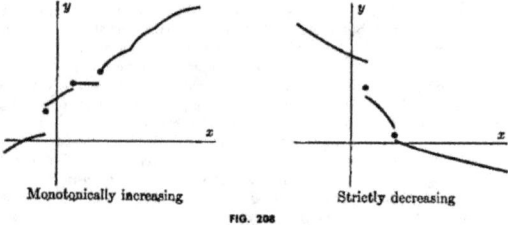

Monotonically increasing   Strictly decreasing

FIG. 208

Consider now the case of a function that is continuous and strictly monotonic on a closed interval $[a, b]$. For definiteness assume that $f(x)$ is strictly *increasing* there, and let $c = f(a)$ and $d = f(b)$. Then $c < d$, and (Theorem IV, § 213) $f(x)$ assumes in the interval $[a, b]$ every value between c and d. Furthermore, it cannot assume the same value twice, for if $\alpha < \beta$, then $f(\alpha) < f(\beta)$. (Cf. Fig. 209.) The function $f(x)$, there-

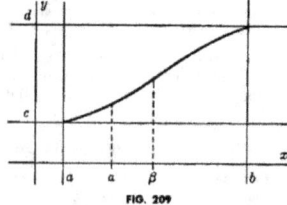

FIG. 209

fore, establishes a one-to-one correspondence between the points of the two closed intervals $[a, b]$ and $[c, d]$. Thus to each point y of the closed interval $[c, d]$ corresponds a unique point x of $[a, b]$ such that $y = f(x)$. Since y determines x uniquely, x can be considered as a single-valued func-

§ 216] EXERCISES 57

tion of y, $x = \phi(y)$, and is called the **inverse function** of $y = f(x)$, the latter being referred to, then, as the **direct function**. The nth roots obtained in the preceding section are inverse functions: if $y = f(x) = x^n$, then $x = \phi(y) = \sqrt[n]{y}$. We now state an important fact regarding the continuity of inverse functions. The proof is given in § 306.

Theorem. *If $y = f(x)$ is continuous and strictly monotonic on a closed interval $[a, b]$, its inverse function $x = \phi(y)$ is continuous and strictly monotonic on the corresponding closed interval $[c, d]$.*

Corollary. *The function $y = \sqrt[n]{x}$, where n is a positive integer, is a continuous and strictly increasing function of x for $x \geq 0$.* (Cf. Ex. 18, § 216.)

216. EXERCISES

1. Show that the equation $x^4 + 2x - 11 = 0$ has a real root. Prove that if $f(x)$ is a real polynomial (real coefficients) of odd degree, the equation $f(x) = 0$ has a real root.

2. Prove that the maximum (minimum) value of a function on a closed interval is unique, but show by examples that the point ξ at which $f(x)$ is a maximum (minimum) may or may not be unique.

3. Prove that $\sqrt{f(x)}$ is continuous wherever $f(x)$ is continuous and positive, and that $\sqrt[3]{f(x)}$ is continuous wherever $f(x)$ is continuous.

4. Assuming continuity and other standard properties of $\sin x$ and $\cos x$ (cf. §§ 603-604), discuss continuity of each function.

(a) $\sqrt{1 + \sin x}$;   (b) $\dfrac{1}{\sqrt{1 - \sin x}}$;

(c) $\sqrt{\cos x^2}$;   (d) $\dfrac{1}{\sqrt{1 + \cos^2 x - \sin^2 x}}$.

In Exercises 5-10, discuss the discontinuities of each function, and draw a graph in each case. The bracket function $[x]$ is defined in Example 5, § 201. (Also cf. Exs. 7-10, § 203.)

5. $[-x]$.   **6.** $[x] + [-x]$.
7. $[\sqrt{x}]$.   **8.** $[x^2]$.
9. $[x] + \sqrt{x - [x]}$.   **10.** $[x] + (x - [x])^2$.

11. Prove the two laws of radicals:

(i) $\sqrt[n]{ab} = \sqrt[n]{a} \sqrt[n]{b}$,

(ii) $\sqrt[n]{\dfrac{a}{b}} = \dfrac{\sqrt[n]{a}}{\sqrt[n]{b}}$, $b \neq 0$,

where n is a positive integer, and in case n is even a and b are nonnegative.

12. Prove that $\lim\limits_{x \to +\infty} \sqrt[n]{x} = +\infty$, where n is a positive integer.

★13. Prove that $\lim_{x\to+\infty} \sqrt{x}(\sqrt{x+a} - \sqrt{x}) = \tfrac{1}{2}a$. *Hint:* Multiply by $(\sqrt{x+a} + \sqrt{x})/(\sqrt{x+a} + \sqrt{x})$.

★14. Let $f(x)$ be a polynomial of degree m and leading coefficient $a_0 > 0$, and let $g(x)$ be a polynomial of degree n and leading coefficient $b_0 > 0$, and let k be a positive integer. Prove that the limit
$$\lim_{x\to+\infty} \sqrt[k]{f(x)} / \sqrt[k]{g(x)}$$
is equal to 0 if $m < n$, to $\sqrt[k]{a_0/b_0}$ if $m = n$, and to $+\infty$ if $m > n$. (Cf. Ex. 35, §208.)

★15. Give an example of a function which is continuous and bounded on the open interval $(0, 1)$, but which has neither maximum nor minimum there.

★16. Give an example of a function which is defined and single-valued on the closed interval $[0, 1]$, but which is not bounded there.

★17. Give an example of a function which is defined, single-valued, and bounded on the closed interval $[0, 1]$, but which has neither maximum nor minimum there.

★18. Prove that $\sqrt[n]{x}$ is an increasing function of n for any fixed x between 0 and 1, and a decreasing function of n for any fixed $x > 1$. *Hint:* Show that the desired order relation between $\sqrt[n]{x}$ and $\sqrt[n+1]{x}$ follows from Ex. 10, §107, by taking $n(n+1)$th powers.

★19. Prove that if $x > 0$ then $\lim_{n\to+\infty} \sqrt[n]{x} = 1$. *Hint:* Assume for definiteness that $0 < x < 1$. Then (Ex. 18) $\sqrt[n]{x} \uparrow$ as $n \uparrow$. Since $\sqrt[n]{x} < 1$ for all n, $\lim_{n\to+\infty} \sqrt[n]{x} = L$ exists (Theorem XIV, §204) and $L \leq 1$. If $L < 1$, then $\sqrt[n]{x} \leq L$ for all n, and $x \leq L^n$ for all n. Use Theorem XIII, §204.

★20. Define $\sqrt[2n-1]{x}$ for all x, and show that $\lim_{n\to+\infty} \sqrt[2n-1]{x} = \operatorname{sgn} x$ (cf. Example 1 (c), §206).

★21. If r is a positive rational number, represented as the quotient of two positive integers, $r = p/q$, prove that the two definitions of x^r, where $x \geq 0$, are equivalent: $x^r = \sqrt[q]{x^p}$; $x^r = (\sqrt[q]{x})^p$. Also show that either definition is independent of the representation of r as a quotient of positive integers. Discuss the function $f(x) = x^r$ as a strictly increasing continuous function of x for $x \geq 0$.

★★22. Define x^r, where $x > 0$ and r is any rational number, and establish the laws of exponents of Exercises 17–21, §107, for rational exponents. (Cf. Ex. 7, §109, and Exs. 11 and 21, above.)

★★23. Discuss the function $f(x) = x^r$, where r is any rational number and $x > 0$, as a continuous monotonic function of x. Find $\lim_{x\to 0+} f(x)$ and $\lim_{x\to+\infty} f(x)$. (Cf. Ex. 21.)

★24. If $f(x)$ is continuous on $[a, b]$, if $a < c < d < b$, and $K = f(c) + f(d)$, prove that there exists a number ξ between a and b such that $K = 2f(\xi)$. More generally, if m and n are positive numbers, show that $mf(c) + nf(d) = (m + n)f(\xi)$, for some ξ between a and b. Finally, extend this result to the formula

§ 216] EXERCISES 59

$$m_1 f(c_1) + \cdots + m_k f(c_k) = (m_1 + \cdots + m_k) f(\xi).$$

Hint: Show that $\frac{m}{m+n} f(c) + \frac{n}{m+n} f(d)$ is between $f(c)$ and $f(d)$, and use Theorem IV, § 213.

★★25. Discuss the function $h(r) = x^r$, where r is any rational number and $x > 0$, as a monotonic function of r. Show that for $0 < x < 1$, $h(r) \downarrow$ as $r \uparrow$ and that for $x > 1$, $h(r) \uparrow$ as $r \uparrow$. Find $\lim_{r \to \infty} h(r)$ and $\lim_{r \to -\infty} h(r)$. (Cf. Ex. 18.)

★★26. Prove that the function $h(r)$ of Exercise 25 is continuous, if only rational values of r are considered. *Hint:* Write $x^r - x^{r_0} = x^{r_0}(x^{r-r_0} - 1)$, and use Exs. 19 and 25.

★27. Prove that a monotonic function always has one-sided limits (finite or infinite). *Hint:* Use suprema or infima (cf. Th. XIV, § 204; Ex. 23, § 205).

★28. Prove that the only discontinuities that a monotonic function can have at points where it is defined are finite jumps. (Cf. Ex. 27.)

★29. Prove that the set of points of discontinuity of a monotonic function is either finite or denumerable. (Cf. Ex. 12, § 113. Also cf. Ex. 25, § 717.) *Hint:* There are at most two values of x at which $y = f(x)$ can become infinite. Any other point of discontinuity, on the x-axis, corresponds to a finite interval of jump on the y-axis, and to distinct points of jump discontinuity correspond open intervals having no point in common. In each of these intervals of jump lies a rational number ((e), § 112), and the rational numbers are denumerable (Ex. 14, § 113).

★30. Prove that any monotonic function can be redefined at the points of discontinuity at which it is defined so that it becomes everywhere continuous from the right (or from the left).

★31. Prove the following converse of the statement of § 215 that a continuous strictly monotonic function has a single-valued inverse: If $f(x)$ is continuous on a closed interval $[a, b]$ and if $f(x)$ does not assume there any value twice (that is, at distinct points of $[a, b]$ $f(x)$ has distinct values), then $f(x)$ is strictly monotonic there.

In Exercises 32-37, find a specific function $\delta = \delta(\epsilon)$ as specified by Definition II, § 209, for the given function at the prescribed point. (Cf. Exs. 59-64, § 208.)

★32. $\sqrt{x}$, $a = 3$. *Hint:* If $x > 0$,
$$|\sqrt{x} - \sqrt{3}| = \left|\frac{\sqrt{x} - \sqrt{3}}{1} \cdot \frac{\sqrt{x} + \sqrt{3}}{\sqrt{x} + \sqrt{3}}\right| = \frac{|x - 3|}{\sqrt{x} + \sqrt{3}} < |x - 3|.$$

★33. $\sqrt{x^2 - 4}$, $a = 2$. ★34. $\sqrt{3x^2 - x}$, $a = -1$.

★35. $\sqrt[3]{x}$, $a = 5$. *Hint:* If $x > 0$,
$$|\sqrt[3]{x} - \sqrt[3]{5}| = \left|\frac{\sqrt[3]{x} - \sqrt[3]{5}}{1} \cdot \frac{\sqrt[3]{x^2} + \sqrt[3]{5x} + \sqrt[3]{25}}{\sqrt[3]{x^2} + \sqrt[3]{5x} + \sqrt[3]{25}}\right| < |x - 5|.$$

★36. $\frac{1}{\sqrt{x}}$, $a = 6$. ★37. $\frac{1}{\sqrt[3]{x}}$, $a = 1$.

★★38. If a is a fixed positive rational number prove that the set of all numbers of the form $r + s\sqrt{a}$, where r and s are arbitrary rational numbers, form a field (cf. Ex. 8, § 113).

★★**39.** Prove that each example demanded by Exercises 57 and 58, § 208, must be discontinuous and that, in fact, it must have infinitely many points of discontinuity.

★★**40.** Prove that $\lim_{n \to +\infty} \left(1 + \frac{1}{n}\right)^n$ exists and is between 2 and 3. (This is one definition of e. For another treatment see §§ 601-602.) *Hint:* First express the binomial expansion of $a_n = \left(1 + \frac{1}{n}\right)^n$ in the form
$$1 + 1 + \frac{1}{2!}\left(1 - \frac{1}{n}\right) + \frac{1}{3!}\left(1 - \frac{1}{n}\right)\left(1 - \frac{2}{n}\right) + \cdots,$$
and thereby show that $a_n \uparrow$ and, furthermore, that
$$a_n < 1 + 1 + \frac{1}{2} + \frac{1}{2^2} + \frac{1}{2^3} + \cdots + \frac{1}{2^{n-1}}.$$

★217. FURTHER REMARKS ON FUNCTIONS

In a strict sense, Definition I, § 201, is more an explanation of what is meant by the statement *y is a function of x* than a definition of a *function*. A more precise statement rests on the concept of *ordered pair* (x, y) of two objects, x and y. We use the notation $a \in A$ to mean that a is a member of the set A. (Cf. § 1022.)

Definition. *Let D and R be two sets of objects. A **function** with domain of definition D and range of values R is a set F of ordered pairs (x, y), where $x \in D$ and $y \in R$, having the two properties:*

(i) *If $x \in D$ there exists a $y \in R$ such that $(x, y) \in F$.*
(ii) *If $y \in R$ there exists an $x \in D$ such that $(x, y) \in F$.*

*The **inverse function** of a given function F is the set of all pairs (y, x), such that $(x, y) \in F$.*

*A function F is **single-valued** if and only if the two statements $(x, y) \in F$ and $(x, z) \in F$ imply the equality $y = z$.*

*A **one-to-one correspondence** is a single-valued function with a single-valued inverse.*

NOTE. If the **product space** of D and R, denoted $D \times R$, is defined to be the set of *all* ordered pairs (x, y), where $x \in D$ and $y \in R$, then functions merely become certain subsets of product spaces. In essence, the preceding Definition identifies a function with its "graph."

3

★Some Theoretical Considerations

★301. A FUNDAMENTAL THEOREM ON BOUNDED SEQUENCES †

One of the properties of the real number system that was assumed as an axiom in Chapter 1 is that of *completeness*. A useful consequence (which is actually one of several alternative formulations of the concept of completeness) is stated in the following theorem:

Theorem. Fundamental Theorem on Bounded Sequences. *Every bounded sequence (of real numbers) contains a convergent subsequence.*

In order to prove this theorem we observe first that we already know (Theorem XIV, § 204) that every bounded *monotonic* sequence converges. Therefore the proof will be complete as soon as we establish the following lemma:

Lemma I. *Every sequence (of real numbers) contains a monotonic subsequence.*

Before proceeding with the details of the proof of this lemma we discuss briefly the idea of *largest term* of a sequence, and give a few illustrative examples.

By a **largest term** of a sequence $a_1, a_2, a_3, \cdots$ we mean any term a_k with the property $a_k \geq a_n$ for every positive integer n. The examples below show that a sequence may not have a largest term, and that if it does have a largest term it may have several. In case a sequence has a largest term, there must be a *first* largest term (first in the order of the terms according to the subscripts). This particular largest term will be called for convenience **the largest term**.

Example 1. The sequence $0, 1, 0, 1, \cdots$ contains the two monotonic constant subsequences $0, 0, 0, \cdots$ and $1, 1, 1, \cdots$. It has a largest term $a_k = 1$,

---

† Unless otherwise qualified the word *sequence* should be interpreted in this chapter to mean *infinite sequence of real numbers*. (Cf. the footnote, p. 74.)

where k may be any even positive integer. *The largest term of the sequence is the second.*

Example 2. The sequence $1, -1, 2, -2, 3, -3, \cdots$ contains the monotonic subsequences $1, 2, 3, \cdots$ and $-1, -2, -3, \cdots$. It has no largest term, but the subsequence $-1, -2, -3, \cdots$ has the largest term -1.

Example 3. The sequence $1, -1, \frac{1}{2}, -\frac{1}{2}, \frac{1}{3}, -\frac{1}{3}, \cdots$ contains the monotonic subsequences $1, \frac{1}{2}, \frac{1}{3}, \cdots$ and $-1, -\frac{1}{2}, -\frac{1}{3}, \cdots$, and has the largest term 1. The subsequence $-1, -\frac{1}{2}, -\frac{1}{3}, \cdots$ has no largest term.

Example 4. The sequence $1, \frac{1}{2}, \frac{2}{3}, \frac{3}{4}, \frac{4}{5}, \cdots$ contains the monotonic subsequence $\frac{1}{2}, \frac{2}{3}, \frac{3}{4}, \cdots$. The sequence has the largest term 1, but no monotonic subsequence has a largest term.

As an aid in the proof of Lemma I we establish a simple auxiliary lemma:

Lemma II. *If a sequence S has a largest term equal to x, and if a subsequence T of S has a largest term equal to y, then $x \geq y$.*

Proof of Lemma II. Let $S = \{a_n\}$. Since T is a subsequence of S, y is equal to some term a_n of S. If a_k is the largest term of S, then $a_k \geq a_n$, or $x \geq y$.

We are now ready to prove Lemma I:

Proof of Lemma I. Let $S = a_1, a_2, a_3, \cdots$ be a given sequence of real numbers, and let certain subsequences of S be denoted as follows: $S_0 \equiv S$, $S_1 \equiv a_2, a_3, a_4, \cdots$, $S_2 \equiv a_3, a_4, a_5, \cdots$, etc. There are two cases to consider:

Case I. Every sequence $S_0, S_1, S_2, \cdots$ contains a largest term. Denote by a_{n_1} the largest term of S_0, by a_{n_2} the largest term of S_{n_1}, by a_{n_3} the largest term of S_{n_2}, etc. By construction, $n_1 < n_2 < n_3 \cdots$, so that $a_{n_1}, a_{n_2}, a_{n_3}, \cdots$ is a subsequence of S, and by Lemma II it is a monotonically decreasing subsequence.

Case II. There exists a sequence S_N containing no largest term. Then every term of S_N is followed ultimately by some larger term (for otherwise there would be some term of S_N which could be exceeded only by its predecessors, of which there are at most a finite number, so that S_N would have a largest term). We can therefore obtain inductively an increasing subsequence of S by letting $a_{n_1} \equiv a_{N+1}$, $a_{n_2} \equiv$ the first term following a_{n_1} that is greater than a_{n_1}, $a_{n_3} \equiv$ the first term following a_{n_2} that is greater than a_{n_2}, etc. This completes the proof of Lemma I, and therefore that of the Fundamental Theorem.

★302. THE CAUCHY CRITERION FOR CONVERGENCE OF A SEQUENCE

Convergence of a sequence means that there is some number (the limit of the sequence) that has a particular property, formulated in terms of ϵ and N. In order to test the convergence of a given sequence, then, it might seem that one is forced to obtain its limit first. That this is not *always* the case was seen in Theorem XIV, § 204, where the convergence of a *monotonic* sequence is guaranteed by the simple condition of *boundedness*. A natural question to ask now is whether there is some way of testing an *arbitrary* sequence for convergence without knowing in advance what its limit is. It is the purpose of this section to answer this question by means of the celebrated Cauchy Criterion which, in a crude way, says that the terms of a sequence get arbitrarily close to something fixed, for sufficiently large subscripts, if and only if they get arbitrarily close to each other, for sufficiently large subscripts.

Definition. *If $\{a_n\}$ is a sequence (of real numbers), the notation $\lim_{m,n \to +\infty} (a_m - a_n) = 0$ means that corresponding to an arbitrary positive number ϵ there exists a number N such that $m > N$ and $n > N$ together imply $|a_m - a_n| < \epsilon$. A sequence $\{a_n\}$ satisfying the condition $\lim_{m,n \to +\infty} (a_m - a_n) = 0$ is called a* **Cauchy sequence.**

Theorem. Cauchy Criterion. *A sequence (of real numbers) converges if and only if it is a Cauchy sequence.*

Proof. The "only if" part of the proof is simple. Assume that the sequence $\{a_n\}$ converges, and let its limit be a: $a_n \to a$. We wish to show that $\lim_{m,n \to +\infty} (a_m - a_n) = 0$. Corresponding to a preassigned $\epsilon > 0$, let N be a number such that $n > N$ implies $|a_n - a| < \frac{1}{2}\epsilon$. Then if m and n are both greater than N, we have simultaneously

$$|a_m - a| < \tfrac{1}{2}\epsilon \quad \text{and} \quad |a_n - a| < \tfrac{1}{2}\epsilon,$$

so that by the triangle inequality (§ 110)

$$|a_m - a_n| = |(a_m - a) - (a_n - a)| \leq |a_m - a| + |a_n - a| < \epsilon.$$

For the "if" half of the proof we assume that a sequence $\{a_n\}$ satisfies the Cauchy condition, and prove that it converges. The first step is to show that it is bounded. To establish boundedness we choose N so that $n > N$ implies $|a_n - a_{N+1}| < 1$. (This is the Cauchy Criterion with $\epsilon = 1$ and $m = N + 1$.) Then, by the triangle inequality, for $n > N$,

$$|a_n| = |(a_n - a_{N+1}) + a_{N+1}| \leq |a_n - a_{N+1}| + |a_{N+1}| < |a_{N+1}| + 1.$$

Since the terms of the sequence are bounded for all $n > N$, and since there are only a finite number of terms with subscripts less than N, the entire sequence is bounded. According to the Fundamental Theorem (§ 301), then, there must be a convergent subsequence $\{a_{n_k}\}$. Let the limit of this subsequence be a. We shall show that $a_n \to a$. Let ϵ be an arbitrary positive number. Then, by the Cauchy condition, there exists a number N such that for $m > N$ and $n > N$, $|a_m - a_n| < \frac{1}{2}\epsilon$. Let n be an *arbitrary* positive integer greater than N. We shall show that $|a_n - a| < \epsilon$. By the triangle inequality,

$$|a_n - a| \leq |a_n - a_{n_k}| + |a_{n_k} - a|.$$

This inequality holds for all positive integers k and, in particular, for values of k so large that (i) $n_k > N$ and (ii) $|a_{n_k} - a| < \frac{1}{2}\epsilon$. If we choose any such value for k, the inequality above implies $|a_n - a| < \frac{1}{2}\epsilon + \frac{1}{2}\epsilon = \epsilon$, and the proof is complete.

NOTE. Without the Axiom of Completeness, the Cauchy condition is not a criterion for convergence. For example, in the system of rational numbers, where the Axiom of Completeness fails (Ex. 2, § 116), a sequence of rational numbers converging to the irrational number $\sqrt{2}$ (Ex. 15, § 205) satisfies the Cauchy condition, but does not converge *in the system of rational numbers*.

At this point, the student may wish to do a few of the Exercises of § 305. Some that are suitable for performance now are Exercises 1-6 and 15-22.

★303. SEQUENTIAL CRITERIA FOR CONTINUITY AND EXISTENCE OF LIMITS

For future purposes it will be convenient to have a necessary and sufficient condition for continuity of a function at a point, in a form that involves only sequences of numbers, and a similar condition for existence of limits. (Cf. the Theorem, § 209.)

Theorem I. *A necessary and sufficient condition for a function $f(x)$ to be continuous at $x = a$ is that whenever $\{x_n\}$ is a sequence of numbers which converges to a (and for which $f(x)$ is defined), then $\{f(x_n)\}$ is a sequence of numbers converging to $f(a)$; in short, that $x_n \to a$ implies $f(x_n) \to f(a)$.*

Proof. We first establish necessity. Assume that $f(x)$ is continuous at $x = a$, and let $x_n \to a$. We wish to show that $f(x_n) \to f(a)$. Let $\epsilon > 0$ be given. Then there exists a positive number δ such that $|x - a| < \delta$ implies $|f(x) - f(a)| < \epsilon$ (for values of x for which $f(x)$ is defined). Since $x_n \to a$, there exists a number N such that, for $n > N$, $|x_n - a| < \delta$. Therefore, for $n > N$, $|f(x_n) - f(a)| < \epsilon$. This establishes the desired convergence.

Next we prove sufficiency, by assuming that $f(x)$ is discontinuous at $x = a$, and showing that we can then obtain a sequence $\{x_n\}$ converging to a such that the sequence $\{f(x_n)\}$ does not converge to $f(a)$. By Exercise 14, § 212, the discontinuity of $f(x)$ at $x = a$ means that there exists a positive number ϵ such that however small the positive number δ may be, there exists a number x such that $|x - a| < \delta$ and $|f(x) - f(a)| \geq \epsilon$. We construct the sequence $\{x_n\}$ by requiring that x_n satisfy the two inequalities $|x_n - a| < 1/n$ and $|f(x_n) - f(a)| \geq \epsilon$. The former guarantees the con-

§ 304]   CAUCHY CRITERION FOR FUNCTIONS   65

vergence of $\{x_n\}$ to a, while the latter forbids the convergence of $\{f(x_n)\}$ to $f(a)$. This completes the proof of Theorem I.

The formulation and the proof of a sequential criterion for the existence of a limit of a function are similar to those for continuity. The statement in the following theorem does not specify the manner in which the independent variable approaches its limit, since the result is independent of the behavior of the independent variable, subject to the restrictions of the first paragraph of § 207. The details of the proof, with hints, are left to the student in Exercises 8 and 9, § 305.

Theorem II. *The limit, $\lim f(x)$, of the function $f(x)$ exists (finite or infinite) if and only if, for every sequence $\{x_n\}$ of numbers having the same limit as x but never equal to this limit (and for which $f(x)$ is defined), the sequence $\{f(x_n)\}$ has a limit (finite or infinite); in short, if and only if $x_n \to \lim x$, $x_n \neq \lim x$ imply $f(x_n) \to \lim it$.*

As might be expected, the Cauchy Criterion for convergence of a sequence has its application to the question of the existence of a limit of a function. With the same understanding regarding the behavior of the independent variable as was assumed for Theorem II, we have as an immediate corollary of that theorem the following:

Theorem III. *The limit, $\lim f(x)$, of the function $f(x)$ exists and is finite if and only if, for every sequence $\{x_n\}$ of numbers approaching the same limit as x but never equal to this limit (and for which $f(x)$ is defined), the sequence $\{f(x_n)\}$ is a Cauchy sequence; in short, if and only if $x_n \to \lim x$, $x_n \neq \lim x$ imply $\{f(x_n)\}$ is a Cauchy sequence.*

★304. THE CAUCHY CRITERION FOR FUNCTIONS

The Cauchy Criterion for sequences gives a test for the convergence of a sequence that does not involve explicit evaluation of the limit of the sequence. Similar tests can be formulated for the existence of finite limits for more general functions, where explicit evaluation of the limits is not a part of the test. Such criteria are of great theoretical importance and practical utility whenever direct evaluation of a limit is difficult. Owing to the latitude in the behavior granted the independent variable, we have selected for specific formulation in this section only two particular cases, and the proof of one. The remaining proof and other special cases are treated in the exercises of § 305.

Theorem I. *Assume that every deleted neighborhood of the point a contains at least one point of the domain of definition of the function $f(x)$. Then the limit $\lim f(x)$ exists and is finite if and only if corresponding to an arbitrary*

66 SOME THEORETICAL CONSIDERATIONS [§ 305

positive number ϵ there exists a positive number δ such that $0 < |x' - a| < \delta$ and $0 < |x'' - a| < \delta$ imply $|f(x') - f(x'')| < \epsilon$, for values of x' and x'' for which $f(x)$ is defined.

Proof. "*Only if*": Assume $\lim_{x \to a} f(x) = L$ and let $\epsilon > 0$. Then there exists $\delta > 0$ such that $0 < |x - a| < \delta$ implies $|f(x) - L| < \frac{1}{2}\epsilon$. If x' and x'' are any two numbers such that $0 < |x' - a| < \delta$ and $0 < |x'' - a| < \delta$, the triangle inequality gives

$$|f(x') - f(x'')| = |(f(x') - L) - (f(x'') - L)|$$
$$\leq |f(x') - L| + |f(x'') - L| < \tfrac{1}{2}\epsilon + \tfrac{1}{2}\epsilon = \epsilon.$$

"*If*": Let $\{x_n\}$ be an arbitrary sequence of numbers (for which $f(x)$ is defined) such that $x_n \to a$ and $x_n \neq a$. By Theorem III, § 303, we need only show that the sequence $\{f(x_n)\}$ is a Cauchy sequence:

$$\lim_{m, n \to +\infty} [f(x_m) - f(x_n)] = 0.$$

If ϵ is a preassigned positive number, let δ be a positive number having the assumed property that $0 < |x' - a| < \delta$ and $0 < |x'' - a| < \delta$ imply $|f(x') - f(x'')| < \epsilon$. Since $x_n \to a$ and $x_n \neq a$, there exists a number N such that $n > N$ implies $0 < |x_n - a| < \delta$. Accordingly, if $m > N$ and $n > N$, we have simultaneously $0 < |x_m - a| < \delta$ and $0 < |x_n - a| < \delta$, so that $|f(x_m) - f(x_n)| < \epsilon$. Thus the sequence $\{f(x_n)\}$ is a Cauchy sequence, and the proof is complete.

Theorem II. *The limit $\lim_{x \to +\infty} f(x)$ exists and is finite if and only if corresponding to an arbitrary positive number ϵ there exists a number N such that $x' > N$ and $x'' > N$ imply $|f(x') - f(x'')| < \epsilon$ (for values of x' and x'' for which $f(x)$ is defined).*

★305. EXERCISES

★1. Prove that a sequence $\{a_n\}$ (of real numbers) converges if and only if corresponding to an arbitrary positive number ϵ there exists a positive integer N such that for all positive integers p and q, $|a_{N+p} - a_{N+q}| < \epsilon$.

★2. Prove that a sequence $\{a_n\}$ (of real numbers) converges if and only if corresponding to an arbitrary positive number ϵ there exists a number N such that $n > N$ implies $|a_n - a_N| < \epsilon$.

★3. Prove that a sequence $\{a_n\}$ (of real numbers) converges if and only if corresponding to an arbitrary positive number ϵ there exists a positive integer N such that for all positive integers p, $|a_{N+p} - a_N| < \epsilon$.

★4. Prove that the condition $\lim_{n \to +\infty} (a_{n+p} - a_n) = 0$ for every positive integer p is necessary but not sufficient for the convergence of the sequence $\{a_n\}$. *Hint:* Consider a sequence suggested by the terms $1, 2, 2\frac{1}{2}, 3, 3\frac{1}{3}, 3\frac{2}{3}, 4, 4\frac{1}{4}, 4\frac{1}{2}, 4\frac{3}{4}, 5, 5\frac{1}{5}, \cdots$.

★5. Find the error in the "theorem" and "proof": *If $a_n \to a$, then $a_n = a$ for sufficiently large n.* Proof. Any convergent sequence is a Cauchy sequence. Therefore, if $\epsilon > 0$ there exists a positive integer N such that for all positive integers m and n, with $m > N$, $|a_m - a_{N+n}| < \epsilon$. But this means that $\lim_{n \to +\infty} a_{N+n} = a_m$, which in turn implies that $\lim_{n \to +\infty} a_n = a = a_m$ for all $m > N$.

★6. If $\{b_n\}$ is a convergent sequence and if $\{a_n\}$ is a sequence such that $|a_m - a_n| \leq |b_m - b_n|$ for all positive integers m and n, prove that $\{a_n\}$ converges.

★7. If $f(x)$ and $g(x)$ are functions defined for $x > 0$, if $\lim_{x \to 0+} g(x)$ exists and is finite, and if $|f(b) - f(a)| \leq |g(b) - g(a)|$ for all positive numbers a and b, prove that $\lim_{x \to 0+} f(x)$ exists and is finite.

★8. Prove Theorem II, § 303, for the case $x \to a$. *Hint:* First show that if $x_n \to a$, $x_n \neq a$ imply that the sequence $\{f(x_n)\}$ has a limit then this limit is unique, by considering, for any two sequences $\{x_n\}$ and $\{x_n'\}$ which converge to a, the compound sequence $x_1, x_1', x_2, x_2', x_3, x_3', \cdots$. For the case $f(x_n) \to +\infty$, assume that $\lim_{x \to a} f(x) \neq +\infty$, and show that there must exist a constant B and a sequence $\{x_n\}$ such that $0 < |x_n - a| < 1/n$ and $f(x_n) \leq B$. For other cases of $\lim f(x_n)$, proceed similarly.

★9. Prove Theorem II, § 303, for the case $x \to a+$; $x \to a-$; $x \to +\infty$; $x \to -\infty$; $x \to \infty$. (Cf. Ex. 8.)

★10. Assuming only Theorem I, § 303, and the limit theorems for sequences (§ 204), prove the continuity theorems of § 211. *Hint for Theorem IV:* Let $\{x_n\}$ be an arbitrary sequence converging to a. Then $f(x_n) \to f(a)$ and $g(x_n) \to g(a)$, and therefore $f(x_n)g(x_n) \to f(a)g(a)$.

★11. Assuming only Theorem II, § 303, and the limit theorems for sequences (§ 204), prove the limit theorems of § 207. (Cf. Ex. 10.)

★12. Reformulate and prove Theorem I, § 304, for the case $x \to a+$; for the case $x \to a-$.

★13. Prove Theorem II, § 304.

★14. Reformulate and prove Theorem II, § 304, for the case $x \to -\infty$; for the case $x \to \infty$.

★15. Let $\{a_n\}$ be a sequence of real numbers, and let A_n be the least upper bound of the set $\{a_n, a_{n+1}, a_{n+2}, \cdots\}$ ($A_n = +\infty$ if this set is not bounded above). Prove that either $A_n = +\infty$ for every n, or A_n is a monotonically decreasing sequence of real numbers, and that therefore $\lim_{n \to +\infty} A_n$ exists ($+\infty$, finite, or $-\infty$). Prove a similar result for the sequence $\{B_n\}$, where B_n is the greatest lower bound of the set $\{a_n, a_{n+1}, a_{n+2}, \cdots\}$.

★16. The **limit superior** and the **limit inferior** of a sequence $\{a_n\}$, denoted $\overline{\lim}_{n \to +\infty} a_n$, or $\limsup a_n$, and $\underline{\lim}_{n \to +\infty} a_n$, or $\liminf a_n$, respectively, are defined as the limits of the sequences $\{A_n\}$ and $\{B_n\}$, respectively, of Exercise 15. Justify the following formulations:

68 SOME THEORETICAL CONSIDERATIONS [§ 305

$$\overline{\lim_{n \to +\infty}} a_n = \lim_{n \to +\infty} A_n = \inf_{n=1}^{+\infty} \left[\sup_{m=n}^{+\infty} (a_m) \right],$$

$$\underline{\lim_{n \to +\infty}} a_n = \lim_{n \to +\infty} B_n = \sup_{n=1}^{+\infty} \left[\inf_{m=n}^{+\infty} (a_m) \right].$$

(Cf. § 114.)

★17. Prove that a number L is the limit superior of a sequence $\{a_n\}$ if and only if it has the following two properties, where ϵ is an arbitrary preassigned positive number:

(i) The inequality $a_n < L + \epsilon$ holds for all but a finite number of terms.
(ii) The inequality $a_n > L - \epsilon$ holds for infinitely many terms.

State and prove a similar result for the limit inferior.

Prove that the limit superior and limit inferior of a bounded sequence are limit points of that sequence (cf. Exs. 20-22, § 205), and that any other limit point of the sequence is between these two. Extend this result to unbounded sequences. (Cf. Exs. 15-16, and Ex. 29, § 312.)

★18. Prove that for any sequence $\{a_n\}$, bounded or not, $\underline{\lim_{n \to +\infty}} a_n \leq \overline{\lim_{n \to +\infty}} a_n$.

Prove that a sequence converges if and only if its limit superior and limit inferior are finite and equal, and that in the case of convergence, $\lim_{n \to +\infty} a_n = \underline{\lim_{n \to +\infty}} a_n = \overline{\lim_{n \to +\infty}} a_n$. Extend these results to include infinite cases. (Cf. Exs. 15-17.)

In Exercises 19-22, find the limit superior and the limit inferior. (Cf. Exs. 15-18.)

★19. $0, 1, 0, 1, 0, 1, \cdots$.
★20. $1, -2, 3, -4, \cdots, (-1)^{n+1} n, \cdots$.
★21. $\frac{2}{3}, \frac{1}{3}, \frac{3}{4}, \frac{1}{4}, \frac{4}{5}, \frac{1}{5}, \cdots$.
★22. $\frac{2}{3}, -\frac{1}{3}, \frac{3}{4}, -\frac{1}{4}, \frac{4}{5}, -\frac{1}{5}, \cdots$.

★23. Let $f(x)$ be a real-valued function, defined for at least some values of x neighboring the point $x = a$ (except possibly at a itself). If δ is an arbitrary positive number, let $\phi(\delta)$ and $\psi(\delta)$ be the least upper bound and the greatest lower bound, respectively, of the values of $f(x)$ for all x such that $0 < |x - a| < \delta$ (and for which $f(x)$ is defined). Prove that, in a sense that includes infinite values (cf. Ex. 15), $\phi(\delta)$ and $\psi(\delta)$ are monotonic functions and therefore have limits as $\delta \to 0+$. (Cf. § 215.)

★24. The limit superior and the limit inferior of a function $f(x)$ at a point $x = a$, denoted $\overline{\lim_{x \to a}} f(x)$, or lim sup $f(x)$, and $\underline{\lim_{x \to a}} f(x)$, or lim inf $f(x)$, respectively, are defined as the limits of the functions $\phi(\delta)$ and $\psi(\delta)$, respectively, of Exercise 23. Justify the following formulations (cf. Exs. 17-19, § 1021):

$$\overline{\lim_{x \to a}} f(x) = \lim_{\delta \to 0+} \phi(\delta) = \inf_{\delta > 0} \left[\sup_{0 < |x - a| < \delta} f(x) \right],$$

$$\underline{\lim_{x \to a}} f(x) = \lim_{\delta \to 0+} \psi(\delta) = \sup_{\delta > 0} \left[\inf_{0 < |x - a| < \delta} f(x) \right].$$

★25. Prove that a number L is the limit superior of a function $f(x)$ at $x = a$ if and only if it has the following two properties, where ϵ is an arbitrary preassigned positive number:

§ 306] CONTINUOUS FUNCTIONS 69

(i) The inequality $f(x) < L + \epsilon$ holds for all x in some deleted neighborhood of $a : 0 < |x - a| < \delta$.

(ii) The inequality $f(x) > L - \epsilon$ holds for some x in every deleted neighborhood of $a : 0 < |x - a| < \delta$.

State and prove a similar result for the limit inferior. (Cf. Exs. 23-24.)

★26. Prove that for any function $f(x)$, $\underline{\lim}_{x \to a} f(x) \leq \overline{\lim}_{x \to a} f(x)$ (cf. Ex. 18). Prove that $\lim_{x \to a} f(x)$ exists if and only if $\underline{\lim}_{x \to a} f(x) = \overline{\lim}_{x \to a} f(x)$, and in case of equality, $\lim_{x \to a} f(x) = \underline{\lim}_{x \to a} f(x) = \overline{\lim}_{x \to a} f(x)$. (Cf. Exs. 23-25.)

★27. Formulate definitions and state and prove results corresponding to those of Exercises 23-26 for the case $x \to a+$; $x \to a-$.

★28. Formulate definitions and state and prove results corresponding to those of Exercises 23-26 for the case $x \to +\infty$; $x \to -\infty$; $x \to \infty$.

In Exercises 29-34, find the limit superior and the limit inferior. Draw a graph. (Cf. Exs. 23-28.)

★29. $x \sin x$, as $x \to +\infty$. ★30. $\cos \frac{1}{x}$, as $x \to 0$.

★31. $\frac{x+1}{x} \sin x$, as $x \to +\infty$. ★32. $\frac{x+1}{x} \sin x$, as $x \to -\infty$.

★33. $\frac{x-1}{x} \cos x$, as $x \to +\infty$. ★34. $(x^2 + 1) \sin \frac{1}{x}$, as $x \to 0+$.

In Exercises 35-37, establish the inequalities, including all cases except $(-\infty) + (+\infty)$ and $0 \cdot (+\infty)$ (cf. Exs. 43-56, § 208). State and prove corresponding relations for functions. (The notation $n \to +\infty$ is omitted for conciseness.)

★35. $\underline{\lim} (-a_n) = -\overline{\lim} a_n$; if $a_n > 0$, $\underline{\lim} (1/a_n) = 1/\overline{\lim} a_n$.

★36. $\underline{\lim} a_n + \underline{\lim} b_n \leq \underline{\lim} (a_n + b_n) \leq \overline{\lim} (a_n + b_n) \leq \underline{\lim} a_n + \overline{\lim} b_n$.

★37. If $a_n, b_n \geq 0$, $\underline{\lim} a_n \underline{\lim} b_n \leq \underline{\lim} a_n b_n \leq \overline{\lim} a_n b_n \leq \underline{\lim} a_n \overline{\lim} b_n$.

★★38. A function $f(x)$ is **upper semicontinuous** at a point $x = a$ if and only if $\overline{\lim}_{x \to a} f(x) = f(a)$, and is **lower semicontinuous** there if and only if $\underline{\lim}_{x \to a} f(x) = f(a)$. Prove that a function is continuous at a point if and only if it is both upper and lower semicontinuous at the point.

★306. PROOFS OF SOME THEOREMS ON CONTINUOUS FUNCTIONS

The first four sections of this chapter provide us with the means of proving the theorems on continuous functions given in §§ 213 and 215. Some important generalizations of these theorems are obtained in § 311.

Proof of Theorem I, § 213. In order to prove that a function continuous on a closed interval is bounded there, we assume the contrary and seek a contradiction. Let $f(x)$ be continuous on $[a, b]$, and assume it is unbounded there. Under these conditions, for any positive integer n there is a point x_n of the interval $[a, b]$ such that $|f(x_n)| > n$. By the Fundamental Theorem on bounded sequences (§ 301), since the sequence $\{x_n\}$ is bounded, it contains a convergent subsequence $\{x_{n_k}\}$, converging to some point x_0, $x_{n_k} \to x_0$. By Exercise 13, § 205, $a \leq x_{n_k} \leq b$ implies $a \leq x_0 \leq b$, so that x_0 also belongs to the closed interval $[a, b]$. Since $f(x)$ is continuous at x_0, $f(x_{n_k}) \to f(x_0)$ (Theorem I, § 303). But this means that the sequence $\{f(x_{n_k})\}$ is bounded, a statement inconsistent with the inequality

$$|f(x_{n_k})| > n_k.$$

This contradiction establishes Theorem I, § 213.

Proof of Theorem II, § 213. Let $f(x)$ be continuous on the closed interval $[a, b]$. We shall show that it has a maximum value there (the proof for a minimum value is similar). By the preceding theorem, $f(x)$ is bounded on $[a, b]$. Let M be the least upper bound of its values there. We wish to show that there is some number x_0 such that $f(x_0) = M$. By the definition of M, there are values of $f(x)$ arbitrarily close to M. For each positive integer n, let x_n be a point in the interval $[a, b]$ such that $|f(x_n) - M| < \frac{1}{n}$. Then $\lim_{n \to +\infty} f(x_n) = M$. Furthermore, since the sequence $\{x_n\}$ is bounded, it contains a convergent subsequence $\{x_{n_k}\}$ (by the Fundamental Theorem of § 301) converging to some point x_0 of the closed interval (cf. proof of Theorem I, § 213), $x_{n_k} \to x_0$. Since $f(x)$ is continuous at x_0, $f(x_{n_k}) \to f(x_0)$ (Theorem I, § 303). Since $\{f(x_{n_k})\}$ is a subsequence of the sequence $\{f(x_n)\}$ and $f(x_n) \to M$, $f(x_{n_k}) \to M$ (Theorem IV, § 204). Finally, by the uniqueness of the limit of the convergent sequence $\{f(x_{n_k})\}$ (Theorem II, § 204), $f(x_0) = M$, and the proof is complete.

Proof of Theorem III, § 213. Assume for definiteness that $f(a) < 0$ and $f(b) > 0$. We wish to find a number x_0 between a and b such that $f(x_0) = 0$. Let A be the set of points x of the interval $[a, b]$ where $f(x) < 0$ (this set contains at least the point a and is bounded above by b) and *define* x_0 to be the least upper bound of the set A: $x_0 = \sup (A)$. We shall prove that $f(x_0) = 0$ by showing that *each* of the inequalities $f(x_0) < 0$ and $f(x_0) > 0$ is impossible. First assume that $f(x_0) < 0$. By Exercise 18, § 212, any function negative and continuous at a point is negative (where defined) in some neighborhood of that point. If $f(x_0) < 0$, therefore, there must exist some neighborhood of x_0 throughout which $f(x) < 0$, or since $x_0 < b$, this means that $f(x) < 0$ for some values of x *greater* than x_0, in contradiction to the fact that x_0 is an upper bound of A. Finally, by the same argument, if $f(x_0) > 0$ then $f(x) > 0$ for values of x throughout some neighborhood of x_0, so that some number *less* than x_0 is an upper bound of A. This fact is inconsistent with the definition of x_0 as the *least* upper bound of A. The only alternative left is the one sought: $f(x_0) = 0$.

Proof of Theorem IV, § 213. Assume that c lies between the two numbers $f(a)$ and $f(b)$, and consider the function $g(x) \equiv f(x) - c$. Since $g(x)$ is continuous throughout the closed interval $[a, b]$ and has opposite signs at a and b, there must exist, by the theorem just established, a number x_0 between a and b at which $g(x)$ vanishes, so that $g(x_0) = f(x_0) - c = 0$, or $f(x_0) = c$.

Proof of the Theorem of § 215. The only part of the proof that presents any difficulty is the continuity of the inverse function. We shall seek a contradiction to the assumption that $x = \phi(y)$ is discontinuous at some point y_0 of the interval $[c, d]$. Accordingly, by Exercise 14, § 212, there must be a positive number ϵ such that however small the positive number δ may be, there exists a number y of the interval $[c, d]$ such that $|y - y_0| < \delta$ and $|\phi(y) - \phi(y_0)| \geqq \epsilon$. We construct the sequence $\{y_n\}$ such that if $x_n \equiv \phi(y_n)$ and $x_0 \equiv \phi(y_0)$, then $|y_n - y_0| < 1/n$ and $|x_n - x_0| \geqq \epsilon$. The former inequality guarantees the convergence of $\{y_n\}$ to y_0. The boundedness of the sequence $\{x_n\}$ (each x_n belongs to the interval $[a, b]$), by the Fundamental Theorem of § 301, ensures the convergence of some subsequence $\{x_{n_k}\}$ to some number x_0' of the interval $[a, b]$ (cf. proof of Theorem I, § 213). The inequality $|x_n - x_0| \geqq \epsilon$ implies $|x_{n_k} - x_0| \geqq \epsilon$, for all k, so that the subsequence $\{x_{n_k}\}$ cannot converge to x_0. In other words, $x_0' \neq x_0$. On the other hand, the direct function $y = f(x)$ is assumed to be continuous at x_0', so that $x_{n_k} \to x_0'$ implies $y_{n_k} = f(x_{n_k}) \to f(x_0')$. But $y_n \to y_0 = f(x_0)$ implies $y_{n_k} \to f(x_0)$ (Theorem IV, § 204). Finally, the uniqueness of limits (Theorem II, § 204) means that $f(x_0) = f(x_0')$, in contradiction to the assumption that $f(x)$ is strictly monotonic.

★307. UNIFORM CONTINUITY

The function $\frac{1}{x}$ is continuous in the open interval $(0,1)$ (Ex. 2, § 212). Let us consider this statement from the point of view of the δ-ϵ definition of continuity. Let x_0 be a point in $(0, 1)$, and let ϵ be a preassigned positive number. We wish to find δ as a function of ϵ so that $|x - x_0| < \delta$ implies $\left|\frac{1}{x} - \frac{1}{x_0}\right| < \epsilon$. In other words, by making the numerator of the fraction $\frac{|x - x_0|}{|x| \cdot x_0}$ small, we wish to ensure the smallness of the fraction itself. It is clear, however, that mere smallness of the numerator alone is not going to be enough. If the denominator is also small, the fraction is not restricted. The trick (cf. Exs. 59-64, § 208) is to pin down the denominator first by not permitting x to come too close to 0, and then to tackle the numerator. Without going through any of the details, however, it can be seen by inspection of Figure 301 that δ is going to depend quite essentially not only

72   SOME THEORETICAL CONSIDERATIONS   [§ 307

on ϵ but on the point x_0 as well. If x_0 is near 1, and an ϵ is given, we can be fairly generous in the size of δ. However, if x_0 is near 0, and the same ϵ is given, the δ required must be considerably smaller.

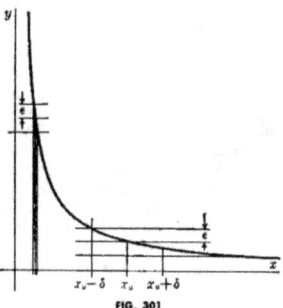

FIG. 301

In some cases it is possible, for a given interval or other set, and for a given ϵ, to choose δ so that the inequalities concerned will hold for all points of the set *without varying the number δ*. When this state exists we have a sort of *uniformity* to the amount of squeezing on $|x - x_0|$ that has to be imposed. This leads to the definition:

Definition. *A function $f(x)$, defined on a set A, is **uniformly continuous** on the set A if and only if corresponding to an arbitrary positive number ϵ there exists a positive number $\delta = \delta(\epsilon)$ such that for any x' and x'' belonging to A, $|x' - x''| < \delta$ implies $|f(x') - f(x'')| < \epsilon$.*

Let us contrast the definitions of continuity and uniform continuity. The most obvious distinction is that continuity is defined *at a point*, whereas uniform continuity is defined *on a set*. These concepts are also distinguished by the order in which things happen. In the case of continuity we have (1) the point x_0, (2) the positive number ϵ, and (3) the positive number δ, which depends on both x_0 and ϵ. In the case of uniform continuity we have (1) the positive number ϵ, (2) the positive number δ, which depends only on ϵ, and (3) the points x' and x''.

In spite of these contrasts, there are certain relations between the two ideas. The simplest one is stated in the following theorem, whose proof is requested in Exercise 1, § 308.

Theorem I. *A function uniformly continuous on a set is continuous at each point of that set. In brief, uniform continuity implies continuity.*

The converse, as we have seen by an example, is false. The function $\frac{1}{x}$ is continuous on the open interval $(0, 1)$, but it is not uniformly continuous there. (Cf. Ex. 9, § 308.) However, it is an important fact that for *closed intervals* there is a valid converse, as given in Theorem II below. (For a more general theorem see Theorem IV, § 311.) As an aid in the proof of Theorem II, we first formulate the *negation* of uniform continuity. The student is asked in Exercise 2, § 308, to validate this formulation. (Cf. Ex. 14, § 212.)

Negation of Uniform Continuity. *A function $f(x)$, defined on a set A, fails to be uniformly continuous on A if and only if there exists a positive number ϵ having the property that for any positive number δ there exist points x' and x'' of A such that $|x' - x''| < \delta$ and $|f(x') - f(x'')| \geq \epsilon$.*

Theorem II. *A function continuous on a closed interval is uniformly continuous there.*

Proof. Assume that $f(x)$ is continuous on the closed interval I but fails to be uniformly continuous there. Then (by the Negation above) for any positive integer n there exist $\epsilon > 0$ and points x_n' and x_n'' of I such that $|x_n' - x_n''| < \frac{1}{n}$ and $|f(x_n') - f(x_n'')| \geq \epsilon$. The bounded sequence $\{x_n'\}$ contains a convergent subsequence $\{x_{n_k}'\}$, converging to some point x_0 of I (cf. proof of Theorem I, § 213, given in § 306). We show that the corresponding sequence $\{x_{n_k}''\}$ also converges to x_0 by use of the triangle inequality:

$$|x_{n_k}'' - x_0| = |(x_{n_k}'' - x_{n_k}') + (x_{n_k}' - x_0)|$$
$$\leq |x_{n_k}'' - x_{n_k}'| + |x_{n_k}' - x_0| < 1/n_k + |x_{n_k}' - x_0|.$$

As k becomes infinite each of the last two terms approaches zero, and therefore so does the quantity $|x_{n_k}'' - x_0|$. By continuity of $f(x)$ at x_0, $\lim_{k \to +\infty} f(x_{n_k}') = \lim_{k \to +\infty} f(x_{n_k}'') = f(x_0)$, and therefore $\lim_{k \to +\infty} [f(x_{n_k}') - f(x_{n_k}'')] = 0$. This last statement is inconsistent with the inequality $|f(x_{n_k}') - f(x_{n_k}'')| \geq \epsilon$, which must hold for all values of k, and the proof is complete.

★308. EXERCISES

★1. Prove Theorem I, § 307.
★2. Establish the Negation of Uniform Continuity, § 307.

In Exercises 3–8, find an explicit function $\delta = \delta(\epsilon)$ in conformity with the definition of uniform continuity.

★3. $y = x^2$, for $0 \leq x \leq 1$. *Hint:*
$$|x''^2 - x'^2| = |x'' - x'| \cdot |x'' + x'| \leq 2|x'' - x'|.$$
★4. $y = x^2$, for $0 \leq x \leq 2$.
★5. $y = \sqrt{x}$, for $1 \leq x \leq 2$. *Hint:* $\sqrt{x''} - \sqrt{x'} = \dfrac{x'' - x'}{\sqrt{x''} + \sqrt{x'}}$.
★6. $y = \sqrt{x}$, for $0 \leq x \leq 1$. *Hint:*
$$\sqrt{x''} + \sqrt{x'} \geq \sqrt{x'' - x'} \quad \text{(cf. Ex. 5)}.$$
★7. $y = \dfrac{1}{x}$, for $x \geq 1$.
★8. $y = \sqrt{1 - x^2}$, for $|x| \leq 1$. *Hint:* For $0 \leq x' < x'' \leq 1$,
$$\sqrt{1 - x'^2} - \sqrt{1 - x''^2} = \frac{(x'' + x')(x'' - x')}{\sqrt{1 + x'}\sqrt{1 - x'} + \sqrt{1 + x''}\sqrt{1 - x''}}$$
$$\leq \frac{2(x'' - x')}{\sqrt{1 - x'} + \sqrt{1 - x''}} \leq \frac{2(x'' - x')}{\sqrt{(1 - x') - (1 - x'')}} \quad \text{(cf. Ex. 6)}.$$

In Exercises 9-12, use the Negation of § 307 to show that the given function is not uniformly continuous on the given interval. (Cf. Exs. 13-14.)

★9. $\dfrac{1}{x}$, $0 < x < 1$.   **★10.** x^2, $x \geq 1$.

★11. $\sin \dfrac{1}{x}$, $0 < x < 1$.   **★12.** $x \sin x$, $x > 0$.

In Exercises 13-14, find an explicit function $\delta = \delta(\epsilon, x_0)$, in conformity with the δ-ϵ definition of continuity, for the given function at a given point x_0 of the specified interval. Observe that δ depends essentially on x_0. (Cf. Exs. 9-10; also Exs. 59-64, § 208.)

★13. $\dfrac{1}{x}$, $0 < x < 1$.   **★14.** x^2, $x \geq 1$.

★15. If $f(x)$ is uniformly continuous on an open interval (a, b), prove that the two limits $f(a+)$ and $f(b-)$ exist and are finite. *Hint:* Use the Cauchy Criterion for functions.

★16. Prove the following converse to Exercise 15: If $f(x)$ is continuous on an open interval (a, b) and if the two limits $f(a+)$ and $f(b-)$ exist and are finite, then $f(x)$ is uniformly continuous on (a, b). *Hint:* Extend $f(x)$ to a function continuous on the closed interval $[a, b]$.

★★309.† POINT SETS: OPEN, CLOSED, COMPACT, CONNECTED SETS

Many of the results obtained in the preceding portions of this chapter are special cases of more general theorems which can be formulated for Euclidean spaces of any number of dimensions. In the remaining sections of this chapter we shall establish some of these general theorems for the particular one-dimensional space of real numbers. The statements of the

---

† Students wishing to study spaces of an arbitrary number of dimensions or, more generally, "metric" or "topological" spaces may find it profitable at this point to read the introductory section of Chapter 10.

§ 309]   POINT SETS   75

theorems and the concepts and techniques involved are of such a nature that their analogues in higher dimensional spaces are immediately available, with only a few minor adjustments in basic definitions. In this chapter the single word *set* means *set of real numbers*, and the word *point* means *real number*.

Definition I. *If A is a set, the **complement** of A, written A', is the set of all real numbers that are not members of A.*

Examples. If $A = (-\infty, 23]$, then $A' = (23, +\infty)$. If $A = [1, 3)$, then A' consists of all real numbers less than 1, or greater than or equal to 3.

Definition II. *A set A is **open** if and only if every member of A has some neighborhood contained entirely in A.*

Examples. The set $A = (2, 7)$ is open, because if $2 < x < 7$ and ϵ is a positive number less than both $7 - x$ and $x - 2$, then the open interval $(x - \epsilon, x + \epsilon)$ is a neighborhood of x lying within A. More generally, any open interval, finite or infinite, is open (Ex. 1, § 312). The set $B = (-1, 8]$ is not open, since every neighborhood of the point 8 extends to the right, outside the set B.

Definition III. *A point p is a **limit point** of a set A if and only if every deleted neighborhood of p contains at least one point of A.*

NOTE 1. If p is a limit point of a set A we say that the set A *has p as a limit point*, whether p is a member of A or not. For example, the two sets $B = [2, 5]$ and $C = (2, 5)$ both have 5 as a limit point, but only B contains 5 as a member.

NOTE 2. If p is a limit point of a set A, every neighborhood of p contains *infinitely many* points of A (Ex. 2, § 312).

Definition IV. *A set A is **closed** if and only if it contains all of its limit points.*

Examples. The set $A = [1, 6]$ is closed. More generally, any finite closed interval is closed, and the infinite intervals $[a, +\infty)$ and $(-\infty, b]$ are closed (Ex. 5, § 312). The set $B = (3, 5]$ is not closed, since the point 3 is a limit point of B but not a member of B. It is neither open nor closed. The set C of all integers is a closed set since it has *no* limit points and therefore contains all of its limit points. For the same reason any *finite* set is closed. The set D of all rational numbers is not closed since every real number, rational or irrational, is a limit point of D. It is neither open nor closed.

NOTE 3. The **empty set**, denoted $\emptyset$, is both open and closed. (If it were *not open* there would be *at least one member* every neighborhood of which contains a point of the complement, and if it were *not closed* there would be a limit point and, by Definition III, *at least one member*!) The entire space S of real numbers is also both open and closed. The student is asked to show in Exercise 11, § 312, that ∅ and S are the *only* sets of real numbers that are both open and closed. Any set with at least one member is called **nonempty**.

Definition V. *A set A is **compact** if and only if it is closed and bounded.*†

Examples. Any finite closed interval is compact. Any finite set is compact. The set consisting of the reciprocals of the positive integers together with the number 0 is compact. The set of all integers is not compact because it is unbounded. The open interval $(0, 1)$ is not compact because it is not closed.

The preceding examples show that a set may be neither open nor closed (a half-open interval and the set of rational numbers are two instances). On the other hand, as explained in the Note above, a set may be both open and closed. One might wonder whether there is any relation between these two concepts of openness and closedness. The following theorem gives the answer.

Theorem I. *A set is open if and only if its complement is closed. Equivalently, a set is closed if and only if its complement is open.*

Proof. "*Only if*": Let A be an open set. We wish to show that the complement A' contains all of its limit points. Assume that x is a limit point of A' that does *not* belong to A'. Then x belongs to A. Since A is open, x has a neighborhood lying entirely in A and therefore containing *no* points of A'. Therefore x is not a limit point of A'. Contradiction. "*If*": Let A' be a closed set. We wish to prove that A is open. Let x be any point of A. Then x is *not* a limit point of A'. Therefore x has a neighborhood that contains no points of A' and thus lies entirely in A. Therefore A is open, and the proof is complete.

A useful relation between least upper bounds (or greatest lower bounds) and limit points is the following, whose proof is requested in Exercise 8, § 312:

Theorem II. *If p is the least upper bound (or the greatest lower bound) of a set A and if p is not a member of A, then p is a limit point of A.*

An immediate consequence (Ex. 9, § 312) is the following:

Theorem III. *Every nonempty compact set of real numbers has a greatest member and a least member.*

Definition VI. *Two sets are **disjoint** if and only if they have no point in common.*

† This formulation of compactness is suitable for finite dimensional Euclidean spaces, but not for general metric or topological spaces. For a brief introduction to metric and topological spaces, and a further reference, the reader is directed to §§ 1026-1028, which are based only on Chapters 1-3 and parts of Chapter 10.

Examples. The intervals (0, 1) and (1, 2) are disjoint. The intervals (0, 1] and (1, 2] are disjoint. The intervals [0, 1] and [1, 2] are not disjoint, since they have the point 1 in common.

Definition VII. *Two sets are **separated** if and only if they are disjoint and neither contains a limit point of the other.*

Examples. The intervals (0, 1) and (1, 2) are separated. The intervals (0, 1] and (1, 2] are not separated, since the point 1 belongs to (0, 1] and is a limit point of (1, 2]. The intervals [0, 1] and [1, 2] are not separated, since they are not disjoint. The set of rational numbers and the set of irrational numbers are disjoint, but are about as far from being separated as two disjoint sets of real numbers can be!

Definition VIII. *A set A is said to be **split into two parts** B and C if and only if B and C are disjoint and every point of A belongs either to B or to C.*

Definition IX. *A set A is **connected** if and only if it cannot be split into two separated nonempty parts.*

The following theorem shows that for real numbers, connectedness is not a sufficiently rich concept to excite anyone. Our reason for presenting the subject here is that connectedness, along with openness, closedness, and compactness, is a very significant idea in the theory of spaces of more than one dimension. In the study of functions of several variables it is important to know the substance of the theorem that we now state and prove.

Theorem IV. *A nonempty set of real numbers is connected if and only if it is an interval or consists of one point.*

Proof. The "only if" half of this proof is fairly easy. Let A be any connected set consisting of more than one point. We divide the proof that A is an interval into two parts, the second of which is left to the student, with hints (Ex. 10, § 312): (i) if a and b are any two points belonging to A, then any point between a and b must also belong to A; (ii) any set having the property just specified in (i) is an interval. To prove (i), we seek a contradiction to the assumption that there exist two points a and b ($a < b$) which are members of A and a third point c between a and b ($a < c < b$) which is not a member of A. The point c provides a splitting of A into two parts (one consisting of all points of A less than c and the other consisting of all points of A greater than c) neither of which contains a limit point of the other. Since the set A is assumed to be connected, the desired contradiction has been obtained.

The "if" half is more difficult. We wish to show that every interval is connected. Let us assume the contrary, and let A be an interval which is not connected, and let A consist of the two nonempty parts B and C neither of which contains a limit point of the other. Let b be any point

78 SOME THEORETICAL CONSIDERATIONS [§ 310

in B and c any point in C and assume for definiteness that $b < c$. Since A is an interval, every point between b and c must belong either to B or to C. Denote by D the set of points of the closed interval $[b, c]$ which belong to B, and let $d = \sup (D)$. There are two cases: (i) d belongs to B. In this case $d < c$ since c belongs to C, and every point of the half-open interval $(d, c]$ belongs to C (by the definition of d). But d, being a limit point of $(d, c]$ must thereby be a limit point of C. But a member of B cannot be a limit point of C! (ii) d belongs to C. In this case d must be a limit point of D (Theorem II), and therefore of B. But a member of C cannot be a limit point of B! In either case we obtain a contradiction, and the theorem is proved.

★★310. POINT SETS AND SEQUENCES

Much of our earlier work has been based on sequential arguments resting ultimately on the Fundamental Theorem for sequences. In order to exploit the techniques already used, we obtain in this section some useful facts about certain types of sets, phrased in terms of sequences. It turns out that the key concept that permits the useful extension of the Fundamental Theorem for sequences to more general sets than intervals is *compactness*. This is shown in Theorem IV of this section.

Definition. *A sequence $\{a_n\}$ of points is called a **sequence of distinct points** if and only if no two terms are equal; that is, if and only if $m \neq n$ implies $a_m \neq a_n$.*

Theorem I. *If a sequence $\{a_n\}$ (of real numbers) has a finite limit a, then either all but a finite number of the terms are equal to a or there exists a subsequence of distinct terms converging to a.*

Proof. Assume that $a_n \to a$ and that for every N there exists an $n > N$ such that $a_n \neq a$. The subsequence sought can be obtained inductively. Let a_{n_1} be the first term different from a. Take $\epsilon = |a_{n_1} - a| > 0$ and let a_{n_2} be the first term a_n satisfying the inequalities $0 < |a_n - a| < \epsilon = |a_{n_1} - a|$, let a_{n_3} be the first term a_n satisfying the inequalities

$$0 < |a_n - a| < |a_{n_2} - a|, \text{ etc.}$$

By construction, $n_1 < n_2 < n_3 < \cdots$, so that $\{a_{n_k}\}$ is a subsequence, and no two terms are equal.

Theorem II. *A point p is a limit point of a set A if and only if there exists a sequence $\{a_n\}$ of distinct points of A converging to p.*

Proof. If $\{a_n\}$ is a sequence of distinct points of A converging to p, then every neighborhood of p contains all points of the sequence from some index on, and therefore infinitely many points of A, so that p is a limit point

§ 311] SOME GENERAL THEOREMS 79

of A. On the other hand, if p is a limit point of A, we can find, for each positive integer n, a point p_n of A such that $0 < |p_n - p| < 1/n$. The sequence $\{p_n\}$ therefore converges to p, and since none of the terms are equal to p it contains (by Theorem I) a subsequence $\{p_{n_i}\}$ no two terms of which are equal, such that $p_{n_k} \to p$. Let $a_k = p_{n_k}$.

Theorem III. *A set A of real numbers is closed if and only if the limit of every convergent sequence $\{a_n\}$ of points of A is a point of A.*

Proof. "If": Assume that the limit of every convergent sequence of points of A is a point of A and let p be a limit point of A. We wish to show that p is a point of A. Since p is a limit point of A, Theorem II guarantees the existence of a sequence of points of A converging to p, so that p must belong to A.

"Only if": Let A be a closed set, and let $\{a_n\}$ be a sequence of points of A converging to the point a. We wish to show that a belongs to A. According to Theorem I there are two possibilities. Either $a_n = a$ for all but a finite number of n (in which case a must belong to A) or the sequence $\{a_n\}$ contains a subsequence of distinct terms (in which case, by Theorem II, a must be a limit point of A). In either case, since A contains all of its limit points, a must belong to A.

Theorem IV. *A set A of real numbers is compact if and only if every sequence $\{a_n\}$ of points of A contains a subsequence converging to a point of A.*

Proof. "If": Assume that every sequence $\{a_n\}$ of points of A contains a convergent subsequence whose limit belongs to A. We wish to show that A is bounded and closed. If A were unbounded there would exist a sequence $\{a_n\}$ of points of A such that $|a_n| > n$, so that no subsequence could converge to *any* point. If A were not closed, there would exist (by Theorem III) a sequence $\{a_n\}$ of points of A converging to a point p not a member of A. Since every subsequence would also converge to p, no subsequence could converge to a point of A. These contradictions show that A must be both bounded and closed, and therefore compact.

"Only if": Assume that A is compact and let $\{a_n\}$ be an arbitrary sequence of points of A. Since A is bounded, $\{a_n\}$ contains a convergent subsequence $\{a_{n_k}\}$, and since A is closed, by Theorem III the limit of this subsequence must belong to A.

★★311. SOME GENERAL THEOREMS

With the aid of the theorems on sets and sequences given in the preceding section, we can now establish four of the most important general theorems on continuous functions. We remind the reader that in this chapter only real-valued functions of a real variable are considered, but the theorems are stated in general terms, so that they may be taken without change for more general use. Joining *compactness* as a key concept, now, is *connectedness*. It is assumed that every domain of definition is nonempty.

Theorem I. *A function continuous on a compact domain has a compact range.*

Proof. Let the function be $y = f(x)$, with compact domain A and range B. If B is *not* compact, there is a sequence $\{b_n\}$ of points of B such that $\{b_n\}$ contains no subsequence converging to a point of B (Theorem IV, § 310). For each n, let a_n be a point of A such that $b_n = f(a_n)$. Since A is assumed to be compact, the sequence $\{a_n\}$ contains a subsequence $\{a_{n_i}\}$ converging to some point a of A. But from the continuity of $f(x)$ at $x = a$ we can infer that $a_{n_i} \to a$ implies $f(a_{n_i}) \to f(a)$ (Theorem I, § 303). In other words, the subsequence $\{b_{n_i}\}$ of the sequence $\{b_n\}$ converges to the point $b = f(a)$ of B. This contradiction completes the proof.

Since any compact set is bounded and since any compact set of real numbers has a greatest member and a least member (Theorem III, § 309), we have immediately the following two corollaries, of which the first two theorems of § 213 are special cases where the domain is a closed interval:

Corollary I. *A function continuous on a compact domain is bounded there.*

Corollary II. *A real-valued function continuous on a compact domain has a maximum value and a minimum value there.*

Theorem II. *A function continuous on a connected domain has a connected range.*

Proof. Let the function be $y = f(x)$, with connected domain A and range B. If B is *not* connected, then B can be split into two disjoint nonempty parts, B_1 and B_2, neither of which contains a limit point of the other. Denote by A_1 the points x of A such that $f(x)$ is a point of B_1, and by A_2 the points x of A such that $f(x)$ is a point of B_2. Then A is split into the two disjoint nonempty subsets, A_1 and A_2. Since A is connected, one of these sets must contain a limit point of the other. For definiteness, assume that p belongs to A_1 and is a limit point of A_2, and let $\{a_n\}$ be a sequence of points of A_2 such that $a_n \to p$. Since $f(x)$ is assumed to be continuous at $x = p$, $a_n \to p$ implies $f(a_n) \to f(p)$. But this means that a sequence of points of B_2 converges to a point of B_1, in contradiction to the assumption that *no* point of B_1 is a limit point of B_2, and the proof is complete.

Since any connected set of real numbers is an interval or a one-point set, we have the following corollary, of which Theorems III and IV, § 213, are special cases where the domain is an interval:

§ 312] EXERCISES 81

Corollary. *A real-valued function continuous on a connected domain assumes (as a value) every number between any two of its values.*

The Theorem of § 215 on the continuity of the inverse function of a strictly monotonic function finds its generalization again based on compactness, with the monotonic property replaced by the mere existence of a single-valued inverse:

Theorem III. *If a function $y = f(x)$ is continuous on a compact domain A and never assumes the same value at distinct points of A, then the inverse function $x = \phi(y)$ is continuous on the range B of $f(x)$.*

Proof. We observe first that since $f(x)$ always has distinct values at distinct points of A, $f(x)$ establishes a one-to-one correspondence between the points of A and the points of B, so that the inverse function $x = \phi(y)$ exists on B. To establish continuity of $\phi(y)$ we wish to show that $b_n \to b$ (where b_n and b are points of B) implies $a_n \to a$, where $a_n \equiv \phi(b_n)$ and $a \equiv \phi(b)$ (a_n and a are points of A). Let us assume that $a_n \not\to a$, so that there exists a neighborhood of a outside of which lie infinitely many terms of the sequence $\{a_n\}$. Since these infinitely many terms form a subsequence of $\{a_n\}$, and since A is assumed to be compact, there must be a subsequence of this subsequence which converges to some point of A (Theorem IV, § 310) different from a. Denote by $\{a_{n_k}\}$ this new convergent subsequence, and denote by a' its limit: $a' \equiv \lim_{k \to +\infty} a_{n_k}$, where $a' \neq a$. Since $f(x)$ is assumed to be continuous at a', $a_{n_k} \to a'$ implies $f(a_{n_k}) = b_{n_k} \to f(a')$. On the other hand, $b_n \to b$ implies $b_{n_k} \to b = f(a)$. By the uniqueness of the limit of a sequence (Theorem II, § 204), as applied to the subsequence $\{b_{n_k}\}$, we infer that $f(a) = f(a')$, in contradiction to our assumption that $f(x)$ never assumes the same value at distinct points.

Finally, the proof given in § 307 that a function continuous on a closed interval is uniformly continuous there generalizes with only trivial notational changes (Ex. 23, § 312) to a function continuous on any compact set:

Theorem IV. *A function continuous on a compact domain is uniformly continuous there.*

★★312. EXERCISES

★★1. Prove that any open interval, finite or infinite, is open. Give some examples of open sets (of real numbers) that are not intervals.

★★2. Prove the statement of Note 2, § 309.

★★3. Prove that every nonempty open set (of real numbers) contains both rational and irrational numbers. Show, in fact, that it contains infinitely many of each type.

★★4. Find an example of a pair of nonempty open sets (of real numbers) such that every member of each set is exceeded by some member of the other

★★5. Prove that any finite closed interval and the infinite intervals $[a, +\infty)$ and $(-\infty, b]$ are closed.

★★6. Give some more examples of closed sets.

★★7. Give some more examples of sets (of real numbers) that are neither open nor closed.

★★8. Prove Theorem II, § 309.

★★9. Prove Theorem III, § 309.

★★10. Prove that if A is a set of real numbers containing more than one point, with the property that whenever two points belong to A every point between these two also belongs to A, then A is an interval (finite or infinite). *Hint:* If A is bounded, show that any point c between inf (A) and sup (A) is flanked by two members, a and b, of A: $a < c < b$. If A is unbounded proceed similarly.

★★11. Prove that the only sets of real numbers both open and closed are the empty set and the entire space S. *Hint:* Assume that A is a nonempty, open, and closed set of real numbers not containing all real numbers. Then its complement $B = A'$ is also nonempty, open, and closed. But this means that the connected set S is split into two nonempty separated parts, A and B.

★★12. Prove that if A and B are two nonempty disjoint open sets (of real numbers) there exist numbers a, b, and c, where c is between a and b, such that a belongs to A, b belongs to B, and c belongs to neither A nor B. *Hint:* Show that the contrary assumption implies that the set C made up of all of the points of A together with all of the points of B is connected (cf. Exs. 10-11).

★★13. Prove the **Bolzano-Weierstrass Theorem**: *Any infinite bounded set of real numbers has a limit point.* *Hint:* If A is an infinite bounded set, let $\{a_n\}$ be a sequence of distinct points of A, and let $\{a_{n_k}\}$ be a convergent subsequence of $\{a_n\}$ converging to a point p. Show that p is a limit point of A.

★★14. Give some examples of bounded sets (of real numbers) that have no limit points.

★★15. Give some examples of infinite sets that have no limit points.

★★16. Give some examples of sets having the property that each point of the set is a limit point of the set.

★★17. If A is an arbitrary set of real numbers, prove that the set B of all limit points of A is closed.

★★18. The **closure** of a set A, denoted $\bar{A}$, is the set made up by adjoining to A all limit points of A. Prove that $\bar{A}$ is always closed.

★★19. The **distance between a point p and a nonempty set A**, written $\delta(p, A)$, is defined to be the greatest lower bound of the set of distances $|p - a|$ between p and arbitrary points a of A. Prove that the distance between p and A is 0 if and only if p is either a point of A or a limit point of A. Prove that if p is not a point of a nonempty closed set A, then $\delta(p, A) > 0$.

★★20. Prove that if p is a point that is not a member of a nonempty closed set A, then there is a point a of A such that $\delta(p, A) = |p - a|$. (Cf. Ex. 19.) *Hint:* Let $\{a_n\}$ be a sequence of points of A such that $|p - a_n| < \delta(p, A) + \frac{1}{n}$ for every positive integer n, and let $a_{n_k} \to a$.

★★21. The **distance between two nonempty sets A and B**, written $\delta(A, B)$, is

arbitrary points a of A and b of B. Prove that $\delta(A, B) \geq 0$, and is always zero unless A and B are disjoint. Show that $\delta(A, B)$ may be zero if A and B are disjoint, and even if A and B are disjoint closed sets. *Hint:* Consider the example: A is the set of positive integers and B is the set of numbers of the form $n + \frac{1}{n}$, where n is a positive integer.

★★**22.** Prove that if A and B are nonempty disjoint closed sets at least one of which is bounded (compact), then there exist points a of A and b of B such that $\delta(A, B) = |a - b|$. Hence prove that $\delta(A, B) > 0$. (Cf. Exs. 20-21.) *Hint:* Assume that A is compact, and choose points a_n of A and b_n of B such that $|a_n - b_n| < \delta(A, B) + \frac{1}{n}$. Let $a_{n_k} \to a$, and choose a convergent subsequence of $\{b_{n_k}\}$.

★★**23.** Prove Theorem IV, § 311.

★★**24.** Let $f(x)$ be continuous for all real numbers x, and let c be a constant. Prove that the following sets are open: (i) all x such that $f(x) > c$; (ii) all x such that $f(x) < c$. Prove that the following sets are closed: (iii) all x such that $f(x) \geq c$; (iv) all x such that $f(x) \leq c$; (v) all x such that $f(x) = c$. If $f(x)$ is bounded, must any of these sets be bounded? (Prove or give a counterexample.)

★★**25.** Give an example of a function, defined for all real numbers x, such that the set of all points of continuity is (i) open but not closed; (ii) closed but not open; (iii) neither open nor closed.

★★**26.** A sequence $\{A_n\}$ of sets is called **monotonically decreasing** if and only if every set A_n of the sequence contains its successor A_{n+1}. This property is symbolized $A_n \downarrow$. (A **constant sequence**, where $A_n = A$, for all n, is an extreme example.) Prove the theorem: *If $\{A_n\}$ is a monotonically decreasing sequence of nonempty compact sets there exists a point x common to every set of the sequence.* The **nested intervals theorem** is the special case of this theorem where every compact set A_n is a (finite) closed interval. *Hint:* For every positive integer n let a_n be a point of A_n, and let $a_{n_k} \to x$. For any N, a_{n_k} belongs to A_N for sufficiently large k, so that x also belongs to A_N.

★★**27.** Show that the theorem of Exercise 26 is false if the assumption of compactness is replaced by either boundedness or closedness alone. *Hint:* Consider the sequences $\left\{\left(0, \frac{1}{n}\right)\right\}$ and $\{[n, +\infty)\}$.

★★**28.** A collection of open sets is said to **cover** a set A if and only if *every* point of A is a member of *some* open set of the collection. Such a collection of open sets is called an **open covering** of A. An open covering of a set A is said to be **reducible to a finite covering** if and only if there exists some finite subcollection of the open sets of the covering which also covers A. Prove the **Heine-Borel Theorem**: *Any open covering of a compact set is reducible to a finite covering.* *Hint:* First prove the theorem for the special case where the compact set is a (finite) closed interval I, as follows: Assume that F is a collection of open sets (each *member* of F is an open set) which covers I and which is not reducible to a finite covering of I. Consider the two closed intervals into which I is divided by its midpoint. At least one of these two subintervals cannot be

84 SOME THEORETICAL CONSIDERATIONS [§ 312]

and relabel $I = I_1$. Let I_2 be a closed half of I_1 that is not covered by any finite collection of sets of F, and repeat the process to obtain a decreasing sequence $\{I_n\}$ of closed intervals whose lengths tend toward zero. If x is a point common to every interval I_n (Ex. 26), let B be an open set of the family F which contains x. Then B contains a neighborhood $(x - \epsilon, x + \epsilon)$ of x which, in turn, must contain one of the intervals I_n. But this means that I_n is covered by a finite collection of sets of F, namely, the single set B. This contradiction establishes the special case. Now let A be an arbitrary compact set and let F be an arbitrary open covering of A. Let I be a closed interval containing A, and adjoin to the family F the open set A' (A' is the complement of A). This larger collection is an open covering of I, and accordingly is reducible to a finite covering of I, by the first part of the proof. Those sets of F which belong to this finite covering of I cover A.

★★29. A *limit point* of a sequence $\{a_n\}$ was defined in Exercise 20, § 205, to be a number x to which some subsequence converges. Prove that the set of all limit points of a given sequence is closed, and that the set of all limit points of a bounded sequence is compact. Hence show that any bounded sequence has a largest limit point and a smallest limit point. Prove that these are the limit superior and limit inferior, respectively, of the sequence. (Cf. Exs. 16-17, § 305.) Extend these results to unbounded sequences.

★★30. Let A be a set contained in the domain of definition of a bounded function $f(x)$. The **oscillation** of $f(x)$ on the set A, $\omega(A)$, is defined to be the difference between the least upper bound of its values there and the greatest lower bound of its values there,
$$\omega(A) \equiv \sup_{x \text{ in } A} (f(x)) - \inf_{x \text{ in } A} (f(x)).$$
Prove that if A is contained in B, then $\omega(A) \leq \omega(B)$. Hence show that if $\delta > 0$, and if $A_\delta \equiv (x_0 - \delta, x_0 + \delta)$, then $\omega(A_\delta)$ is a monotonically increasing function of δ, so that $\lim_{\delta \to 0+} \omega(A_\delta)$ exists (cf. § 215). Make appropriate modifications to take care of the possibility that A or A_δ may be only partly contained in the domain of definition of $f(x)$.

★★31. Let $f(x)$ be defined on a closed interval I. The **oscillation** $\omega(x_0)$ of $f(x)$ **at a point** x_0 of I is defined to be the limit of the function $\omega(A_\delta)$ of Exercise 30:
$$\omega(x_0) \equiv \lim_{\delta \to 0+} \omega(A_\delta).$$
Prove that $\omega(x_0) \geq 0$ and that $f(x)$ is continuous at $x = x_0$ if and only if $\omega(x_0) = 0$.

★★32. Let $f(x)$ be defined on a closed interval I. If $\epsilon > 0$, let D_ϵ be the set of all points x such that $\omega(x) \geq \epsilon$. Prove that D_ϵ is closed. (Cf. Ex. 31.) *Hint:* If $\omega(x_0) < \epsilon$, let $\eta \equiv \epsilon - \omega(x_0)$ and show that there exists a neighborhood N of x_0 for which $\omega(N) < \omega(x_0) + \eta$, so that for x in N, $\omega(x) < \omega(x_0) + \eta = \epsilon$.

★★33. Let $f(x)$ be defined on a closed interval I, and consider the sequence of closed sets $D_1, D_{\frac{1}{2}}, D_{\frac{1}{3}}, \cdots, D_{\frac{1}{n}}, \cdots$ (cf. Ex. 32). Prove that each of these sets is contained in its successor, and that the set of points x such that x is a member of some $D_{\frac{1}{n}}$ is the set D of points of discontinuity of $f(x)$. (Cf. Ex. 41, § 1025.)

4

Differentiation

401. INTRODUCTION

This and the following chapter contain a review and an amplification of certain topics from a first course in Calculus. Some of the theorems that are usually stated without proof in a first introduction are established here. Other results are extended beyond the scope of a first course. On the other hand, many definitions and theorems with which the student can be assumed to be familiar are repeated here for the sake of completeness, without the full discussion and motivation which they deserve when first encountered. For illustrative material we have felt free to use calculus formulas which either are assumed to be well known or are established in later sections. For example, the trigonometric, exponential, and logarithmic functions provide useful examples and exercises for these chapters, but their analytic treatment is deferred to Chapter 6. The only inverse trigonometric functions used in this book are the inverse sine, tangent, and cosine, denoted Arcsin x, Arctan x, and Arccos x, with values restricted to the principal value ranges $-\frac{\pi}{2} \leq \text{Arcsin } x \leq \frac{\pi}{2}$, $-\frac{\pi}{2} < \text{Arctan } x < \frac{\pi}{2}$, and $0 \leq \text{Arccos } x \leq \pi$ (the upper case A indicates principal values).

402. THE DERIVATIVE

We shall consider only single-valued real-valued functions of a real variable defined in a neighborhood of the particular value of the independent variable concerned (or possibly just for values of the independent variable neighboring the particular value on one side).

Definition. *A function $y = f(x)$ is said to have a derivative or be **differentiable** at a point x if and only if the following limit exists and is finite; the function $f'(x)$ defined by the limit is called its **derivative**:*

$$(1) \qquad \frac{dy}{dx} = f'(x) = \lim_{h \to 0} \frac{f(x+h) - f(x)}{h}.$$

NOTE. On occasion it is convenient to speak of an infinite derivative in the sense that the limit in the definition above is either $+\infty$ or $-\infty$. For simplicity we shall adopt the convention that the word *derivative* refers to a finite quantity unless it is preceded by the word *infinite*.

Let us observe first that any differentiable function is continuous. More precisely, if $f(x)$ has a (finite) derivative at $x = x_0$, it is continuous there. We show this by taking limits of both members of the equation

$$f(x_0 + h) - f(x_0) = h \cdot \frac{f(x_0 + h) - f(x_0)}{h},$$

as $h \to 0$, the limit of the right-hand member being $0 \cdot f'(x_0) = 0$ (cf. Theorem V, § 207, and Note 3, § 209). The finiteness of the derivative is essential. For example, the signum function (Example 1, § 206) has an infinite derivative at $x = 0$, but is not continuous there.

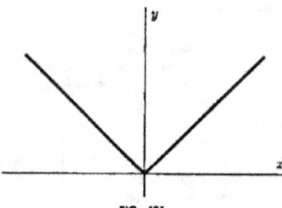

FIG. 401

The converse of the statement of the preceding paragraph is false. A continuous function need not have a derivative at every point. For example, the function $|x|$ (Fig. 401) is everywhere continuous, but has no derivative at $x = 0$. Example 2, § 403, contains a more pathological function. Even more startling is the renowned example of Weierstrass of a function which is everywhere continuous and nowhere differentiable. Although we shall not present this particular example in this book, every student at the level of Advanced Calculus should know of its existence. A discussion is given in E. W. Hobson, *The Theory of Functions of a Real Variable* (Washington, Harren Press, 1950). Another example of a continuous nondifferentiable function is presented in Exercise 41, § 908.

With the notation $\Delta y \equiv f(x + \Delta x) - f(x)$, the definition of a derivative takes the form

(2) $$\frac{dy}{dx} = \lim_{\Delta x \to 0} \frac{\Delta y}{\Delta x}.$$

This fact can be written $\lim_{\Delta x \to 0} \left(\frac{\Delta y}{\Delta x} - \frac{dy}{dx} \right) = 0$. In other words, the expression

$$\epsilon = \frac{\Delta y}{\Delta x} - \frac{dy}{dx},\quad(3)$$

as a function of Δx, is an infinitesimal (that is, its limit as $\Delta x \to 0$ is 0). With the aid of this infinitesimal, equation (3) can be rewritten in either of the following two ways:

$$\frac{\Delta y}{\Delta x} = \frac{dy}{dx} + \epsilon,\quad(3)$$

$$\Delta y = \frac{dy}{dx}\Delta x + \epsilon\Delta x.\quad(4)$$

Under the assumption that the difference Δx between the values of the independent variable is numerically small, equation (3) states that the difference quotient $\Delta y/\Delta x$ differs but slightly from the derivative, and equation (4) states that the difference Δy can be approximated closely by an expression involving the derivative. Use of this fact is made in § 411.

The derivations of the formulas for the derivative of a constant function and of the sum, product, and quotient of functions will be omitted since they are available in any Calculus text. Two standard formulas, however, are often not completely proved in a first course, and we supply their proofs now. The first is the *chain rule* for differentiation of a *composite function* (a function of a function).

Theorem I. Chain Rule. *If y is a differentiable function of u and if u is a differentiable function of x, then y, as a function of x, is differentiable and*

$$\frac{dy}{dx} = \frac{dy}{du} \cdot \frac{du}{dx}.\quad(5)$$

Proof. Let $u = f(x)$ be differentiable at $x = x_0$, $y = g(u)$ be differentiable at $u = u_0 = f(x_0)$, and let $h(x) \equiv g(f(x))$. With the customary notation,

$$\Delta u = f(x_0 + \Delta x) - f(x_0) = f(x_0 + \Delta x) - u_0$$
$$\Delta y = g(u_0 + \Delta u) - g(u_0)$$
$$= g(f(x_0 + \Delta x)) - g(f(x_0)) = h(x_0 + \Delta x) - h(x_0).$$

The usual simple device to make formula (5) seem plausible is the following: Write

$$\frac{\Delta y}{\Delta x} = \frac{\Delta y}{\Delta u}\frac{\Delta u}{\Delta x}\quad(6)$$

and take limits as $\Delta x \to 0$. Since u is a continuous function of x at $x = x_0$, $\Delta x \to 0$ implies $\Delta u \to 0$ (Note 3, § 209), and the desired formula is obtained:

$$\lim_{\Delta x \to 0}\frac{\Delta y}{\Delta x} = \lim_{\Delta u \to 0}\frac{\Delta y}{\Delta u} \cdot \lim_{\Delta x \to 0}\frac{\Delta u}{\Delta x}$$

—unless in this process Δu vanishes and makes equation (6) meaningless!

To avoid this difficulty we define a new function of the independent variable Δu:

(7) $$\epsilon(\Delta u) \equiv \begin{cases} \dfrac{\Delta y}{\Delta u} - \dfrac{dy}{du}, & \text{if } \Delta u \neq 0, \\ 0, & \text{if } \Delta u = 0. \end{cases}$$

Then $\lim_{\Delta x \to 0} \epsilon(\Delta u) = \lim_{\Delta u \to 0} \epsilon(\Delta u) = \dfrac{dy}{du} - \dfrac{dy}{du} = 0.$

From the formulation (7), if $\Delta u \neq 0$,

(8) $$\Delta y = \frac{dy}{du} \cdot \Delta u + \epsilon(\Delta u) \cdot \Delta u.$$

In fact, equation (8) holds whether Δu is zero or not! All that remains is to divide by Δx (which is *nonzero*) and take limits:

$$\lim_{\Delta x \to 0} \frac{\Delta y}{\Delta x} = \frac{dy}{du} \cdot \lim_{\Delta x \to 0} \frac{\Delta u}{\Delta x} + \lim_{\Delta x \to 0} \epsilon(\Delta u) \cdot \lim_{\Delta x \to 0} \frac{\Delta u}{\Delta x},$$

or $$\frac{dy}{dx} = \frac{dy}{du} \cdot \frac{du}{dx} + 0 \cdot \frac{du}{dx}.$$

Theorem II. *If $y = f(x)$ is strictly monotonic (§ 215) and differentiable in an interval, and if $f'(x) \neq 0$ in this interval, then the inverse function $x = \phi(y)$ is strictly monotonic and differentiable in the corresponding interval, and*

$$\frac{dx}{dy} = \frac{1}{\dfrac{dy}{dx}}.$$

Proof. By the Theorem of § 215, $\phi(y)$ is a continuous function, and therefore $\Delta x \to 0$ if and only if $\Delta y \to 0$ (Note 3, § 209). By Theorem VI, § 207, therefore,

$$\lim_{\Delta y \to 0} \frac{\Delta x}{\Delta y} = \lim_{\Delta x \to 0} \frac{1}{\dfrac{\Delta y}{\Delta x}} = \frac{1}{\lim_{\Delta x \to 0} \dfrac{\Delta y}{\Delta x}},$$

and the proof is complete.

It will be assumed that the reader is familiar with the definitions and notations for derivatives of higher order than the first.

403. ONE-SIDED DERIVATIVES

It is frequently important to consider one-sided limits in relation to derivatives. There are three principal ways in which this can be done. We give the three definitions here and call for examples and properties in the exercises of § 404. It happens that in many applications the most useful of these three definitions is the second, and accordingly we reserve

§ 403] ONE-SIDED DERIVATIVES 89

the term *right-hand* or *left-hand derivative* for that concept rather than for the first, for which it might at first seem more natural. For Definitions I and III we create names for distinguishing purposes.

Definition I. *The **derivative from the right** (left) of a function $f(x)$ at the point $x = a$ is the one-sided limit*

$$\lim_{h \to 0+} \frac{f(a+h) - f(a)}{h} \quad \left(\lim_{h \to 0-} \frac{f(a+h) - f(a)}{h} \right).$$

Definition II. *The **right-hand** (left-hand) derivative of a function $f(x)$ at the point $x = a$ is the one-sided limit*

$$\lim_{h \to 0+} \frac{f(a+h) - f(a+)}{h} \quad \left(\lim_{h \to 0-} \frac{f(a+h) - f(a-)}{h} \right),$$

where $f(a+) \equiv \lim_{x \to a+} f(x) \ \left(f(a-) \equiv \lim_{x \to a-} f(x) \right)$.

Definition III. *The **right-hand** (left-hand) limit of the derivative of a function $f(x)$ at the point $x = a$ is the one-sided limit*

$$f'(a+) \equiv \lim_{x \to a+} f'(x) \ \left(f'(a-) \equiv \lim_{x \to a-} f'(x) \right).$$

A function has a derivative at a point if and only if it has equal derivatives from the right and from the left at the point (cf. Ex. 14, § 208). The function $|x|$ (Fig. 401) is an example of a function which has unequal derivatives from the right and from the left. Definitions I, II, and III are related as follows: In case of one-sided continuity, Definitions I and II coincide, and hence if the limit in Definition I exists and is finite, the limit in Definition II exists and is equal to it. If the limit in Definition III exists and is finite, it can be proved (Ex. 54, § 408) that the limit in Definition II exists and is equal to it. Therefore, in case of differentiability, Definitions I and II are consistent with each other and with the Definition of § 402, and in case of continuity of the derivative all four definitions (Definitions I, II, and III of this section and the Definition of § 402) are consistent.

Example 1. The signum function (Example 1, § 206) has infinite derivatives from the right and from the left at $x = 0$. (Fig. 402.) Its right-hand and left-hand derivatives and the right-hand and left-hand limits of the derivative are all zero.

Example 2. The function defined by $y = x \sin \frac{1}{x}$ if $x \neq 0$ and $y = 0$ if $x = 0$ is everywhere continuous, even at the point $x = 0$ (cf. Ex. 8, § 208), but it has no derivative of any kind at $x = 0$. (Fig. 403.) Relative to the point $a = 0$, the fraction $\frac{\Delta y}{\Delta x}$ oscillates infinitely many times between $+1$ and -1 as $\Delta x \to 0$. For the manner in which Definition III applies to this function, cf. Example 3

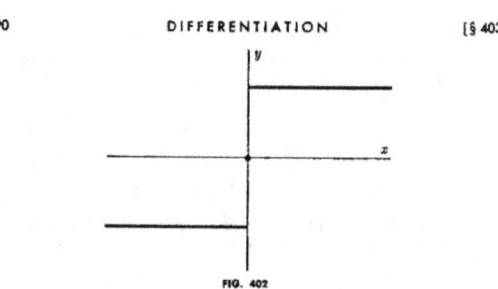

FIG. 402

Example 3. The function defined by $y = x^2 \sin \frac{1}{x}$ if $x \neq 0$ and $y = 0$ if $x = 0$ has a derivative for every value of x. When $x = 0$ this derivative has the value $f'(0) = \lim_{\Delta x \to 0} \frac{\Delta y}{\Delta x} = \lim_{\Delta x \to 0} \frac{\Delta x^2 \sin\left(\frac{1}{\Delta x}\right)}{\Delta x} = \lim_{\Delta x \to 0} \Delta x \sin\left(\frac{1}{\Delta x}\right) = 0$ (cf. Ex. 8, § 208). However, the derivative $f'(x)$ *is not continuous at* $x = 0$ and, in fact, its one-sided limits as $x \to 0$ (Definition III) both fail to exist! To show this, we differentiate $y = x^2 \sin \frac{1}{x}$ according to formula: $\frac{dy}{dx} = 2x \sin \frac{1}{x} - \cos \frac{1}{x}$, when $x \neq 0$. The first term of this expression tends to zero as $x \to 0$, but the second term approaches no limit. Therefore $\frac{dy}{dx}$ can approach no limit. (Why?) Cf. Exs. 47–49, § 408. (See Fig. 404.)

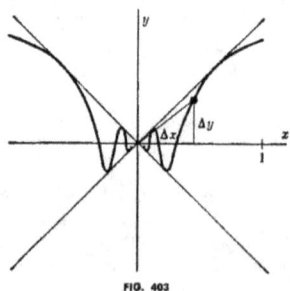

FIG. 403

§ 404]   EXERCISES   91

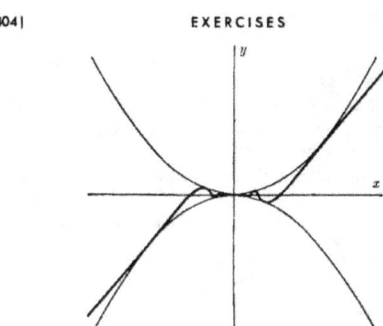

FIG. 404

404. EXERCISES

In Exercises 1-6, differentiate the given function by direct use of the Definition of § 402 (without appeal to any differentiation formulas).

1. $y = x^2 - 4x + 7$.
2. $y = x^3$.
3. $y = \dfrac{1}{x^2}$.
4. $y = \dfrac{3x + 2}{5x - 4}$.
5. $y = \sqrt{x}$.

Hint: Multiply numerator and denominator of $(\sqrt{x + h} - \sqrt{x})/h$ by a quantity which rationalizes the numerator.

6. $y = \sqrt[3]{x}$.

Hint: Multiply numerator and denominator of $(\sqrt[3]{x + h} - \sqrt[3]{x})/h$ by a quantity (consisting of three terms) which rationalizes the numerator.

7. Prove that if $f(x)$ is defined in a neighborhood of $x = \xi$, and if $f'(\xi)$ exists and is positive (negative), then within some neighborhood of ξ, $f(x) < f(\xi)$ for $x < \xi$ and $f(x) > f(\xi)$ for $x > \xi$ ($f(x) > f(\xi)$ for $x < \xi$ and $f(x) < f(\xi)$ for $x > \xi$). *Hint:* If $\lim\limits_{\Delta x \to 0} \dfrac{f(\xi + \Delta x) - f(\xi)}{\Delta x} > 0$, then by Ex. 9, § 208, there exists a deleted neighborhood of 0 such that for any Δx within this deleted neighborhood $\dfrac{f(\xi + \Delta x) - f(\xi)}{\Delta x} > 0$.

8. Derive the formula $\dfrac{d}{dx}(x^n) = nx^{n-1}$, if n is a positive integer.

*9. Derive the formula $\dfrac{d}{dx}(x^n) = nx^{n-1}$, if n is a positive rational number and x is positive (without assuming in the process that x^n is differentiable). *Hint:* With standard notation, $y = x^{p/q}$, $y^q = x^p$, and $(y + \Delta y)^q = (x + \Delta x)^p$, so that

$y^q + qy^{q-1}\Delta y + \cdots = x^p + px^{p-1}\Delta x + \cdots$. Cancel first terms, divide by Δx, and let $\Delta x \to 0$. Since $\Delta y \to 0$ (Ex. 21, § 216), the formula follows by algebraic simplification (cf. Ex. 22, § 216).

★10. Derive the formula $\frac{d}{dx}(x^n) = nx^{n-1}$, if n is any rational number and x is positive. *Hint:* Use Ex. 9 to show that x^n is differentiable. If n is negative, let $m = -n$ and differentiate the quotient $1/x^m$.

NOTE 1. The formula $\frac{d}{dx}(x^n) = nx^{n-1}$ is shown in Exercise 7, § 602, to be valid for all real values of n for $x > 0$.

11. Prove that any rational function (cf. Ex. 20, § 208) of a single variable is differentiable wherever it is defined.

12. Give an example of a function for which $\Delta x \to 0$ does not imply $\Delta y \to 0$. Give an example of a continuous function for which $\Delta y \to 0$ does not imply $\Delta x \to 0$.

13. By mathematical induction extend the chain rule for differentiating a composite function to the case of n functions: $y = f_1(f_2(f_3 \cdots (f_n(x)) \cdots))$.

14. Prove that the derivative of a monotonically increasing (decreasing) differentiable function $f(x)$ satisfies the inequality $f'(x) \geqq 0\ (\leqq 0)$. Does strict increase (decrease) imply a strict inequality?

In Exercises 15-20, discuss differentiability of the given function $f(x)$ (where $f(0) = 0$). Is the derivative continuous wherever it is defined?

15. $x^2 \operatorname{sgn} x$ (cf. Example 1, § 206).   **16.** $x \cos \frac{1}{x}$.

17. $x^2 \cos \frac{1}{x}$.   **18.** $x^2 \sin \frac{1}{x^2}$.

★19. $x^{\frac{3}{2}} \sin \frac{1}{x}$.   **★20.** $x^{\frac{1}{2}} \cos \frac{1}{\sqrt[3]{x}}$.

★21. Discuss differentiability of the function $f(x) = 0$ if $x \leqq 0$, $f(x) = x^n$ if $x > 0$. For what values of n does $f'(x)$ exist for all values of x? For what values of n is $f'(x)$ continuous for all values of x?

★22. If $f(x)$ is the function of Exercise 21, for what values of n does the kth derivative of $f(x)$ exist for all values of x? For what values of n is the kth derivative continuous for all values of x?

★23. If $f(x) = x^n \sin \frac{1}{x}$ for $x > 0$ and $f(0) = 0$, find $f'(x)$. For what values of n does $f'(x)$ exist for all nonnegative values of x? For what values of n is $f'(x)$ continuous for all nonnegative values of x?

★24. If $f(x) = x^n \sin \frac{1}{x}$ for $x > 0$ and $f(0) = 0$, find $f''(x)$. For what values of n does $f''(x)$ exist for all nonnegative values of x? For what values of n is $f''(x)$ continuous for all nonnegative values of x?

NOTE 2. For the function $e^{-\frac{1}{x^2}}$, which behaves in a curious fashion near the origin, see Exercise 52, § 419.

§ 405] ROLLE'S THEOREM 93

★**25.** By mathematical induction establish **Leibnitz's Rule**: *If u and v are functions of x, each of which possesses derivatives of order n, then the product also does and*

$$\frac{d^n}{dx^n}(uv) = \frac{d^n u}{dx^n} \cdot v + \binom{n}{1}\frac{d^{n-1}u}{dx^{n-1}} \cdot \frac{dv}{dx} + \binom{n}{2}\frac{d^{n-2}u}{dx^{n-2}} \cdot \frac{d^2 v}{dx^2} + \cdots + \frac{d^n v}{dx^n},$$

where the coefficients are the binomial coefficients (Ex. 35, § 107).

★**26.** If y is a function of u, and u is a function of x, each possessing derivatives of as high an order as desired, establish the following formulas for higher order derivatives of y with respect to x:

$$\frac{d^2 y}{dx^2} = \frac{d^2 y}{du^2}\left(\frac{du}{dx}\right)^2 + \frac{dy}{du}\frac{d^2 u}{dx^2},$$

$$\frac{d^3 y}{dx^3} = \frac{d^3 y}{du^3}\left(\frac{du}{dx}\right)^3 + 3\frac{d^2 y}{du^2}\frac{du}{dx}\frac{d^2 u}{dx^2} + \frac{dy}{du}\frac{d^3 u}{dx^3},$$

$$\frac{d^4 y}{dx^4} = \frac{d^4 y}{du^4}\left(\frac{du}{dx}\right)^4 + 6\frac{d^3 y}{du^3}\left(\frac{du}{dx}\right)^2\frac{d^2 u}{dx^2} + 3\frac{d^2 y}{du^2}\left(\frac{d^2 u}{dx^2}\right)^2$$
$$+ 4\frac{d^2 y}{du^2}\frac{du}{dx}\frac{d^3 u}{dx^3} + \frac{dy}{du}\frac{d^4 u}{dx^4}.$$

405. ROLLE'S THEOREM AND THE LAW OF THE MEAN

From the fact that a function continuous on a closed interval has a maximum value there (Theorem II, § 213) stem many of the most important propositions of pure and applied mathematics. In this section we initiate a sequence of these theorems.

Theorem I. *If $f(x)$ is continuous on the closed interval $[a, b]$ and differentiable in the open interval (a, b), and if $f(x)$ assumes either its maximum or minimum value for the closed interval $[a, b]$ at an interior point ξ of the interval, then $f'(\xi) = 0$.*

Proof. Assume the hypotheses of the theorem and, for definiteness, let $f(\xi)$ be the *maximum* value of $f(x)$ for the interval $[a, b]$, where $a < \xi < b$. (The student should supply the details for the case where $f(\xi)$ is the *minimum* value of $f(x)$.) Consider the difference quotient

(1) $$\frac{\Delta y}{\Delta x} = \frac{f(\xi + \Delta x) - f(\xi)}{\Delta x}$$

for values of Δx so small numerically that $\xi + \Delta x$ is also in the open interval (a, b). Since $f(\xi)$ is the maximum value of $f(x)$, $f(\xi) \geq f(\xi + \Delta x)$, and $\Delta y \leq 0$. Therefore $\Delta y/\Delta x \geq 0$ for $\Delta x < 0$ and $\Delta y/\Delta x \leq 0$ for $\Delta x > 0$. Hence (in the limit) the derivative from the left at ξ is nonnegative and the derivative from the right at ξ is nonpositive. By hypothesis these one-sided derivatives are equal, and must therefore both equal zero.

Theorem II. Rolle's Theorem. *If $f(x)$ is continuous on the closed interval $[a, b]$, if $f(a) = f(b) = 0$, and if $f(x)$ is differentiable in the open*

94   DIFFERENTIATION   [§ 405

interval (a, b), *then there is some point* ξ *of the open interval* (a, b) *such that* $f'(\xi) = 0$.

Proof. If $f(x)$ vanishes identically the conclusion is trivial. If $f(x)$ is somewhere positive it attains its maximum value at some interior point, and if it is somewhere negative it attains its minimum value at some interior point. In either case, the conclusion follows from Theorem I.

The geometric interpretation of Rolle's Theorem (Fig. 405) is that the graph of the function $f(x)$ has a horizontal tangent for at least one intermediate point.

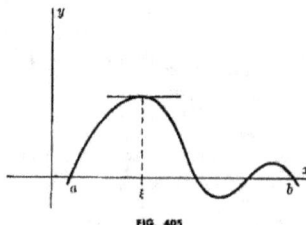

FIG. 405

Theorem III. Law of the Mean (Mean Value Theorem for Derivatives). *If $f(x)$ is continuous on the closed interval $[a, b]$, and if $f(x)$ is differentiable in the open interval (a, b), then there is some point ξ of the open interval (a, b) such that*

$$(2) \qquad f'(\xi) = \frac{f(b) - f(a)}{b - a}.$$

Proof. The geometric interpretation is suggested in Figure 406: the tangent line to the graph of the function $f(x)$, at some appropriate point between a and b, is parallel to the secant line between the points with abscissas a and b. The figure also suggests a proof. Let $K = \dfrac{f(b) - f(a)}{b - a}$ be the slope of the secant line, so that

$$(3) \qquad f(b) = f(a) + K(b - a).$$

The equation of this secant line (the straight line through the point $(b, f(b))$ with slope K) can be written in the form

$$(4) \qquad y = f(b) - K(b - x).$$

For an arbitrary x on the closed interval $[a, b]$, since the curve has the equation $y = f(x)$, the vertical distance from the curve to the secant (ϕ in Figure 406) is given by the expression

§ 405] LAW OF THE MEAN 95

(5) $$\phi(x) = f(b) - f(x) - K(b - x).$$

It is a simple matter to verify that the function $\phi(x)$ defined by (5) satisfies the conditions of Rolle's Theorem (check the details), so that the conclusion is valid. That is, there exists a number ξ of the open interval (a, b)

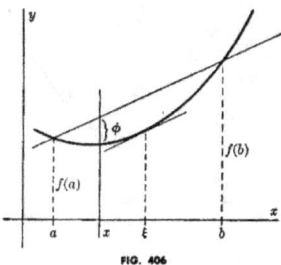

FIG. 406

where the derivative $\phi'(x) = -f'(x) + K$ vanishes, and we have the conclusion sought:

$$f'(\xi) = K = \frac{f(b) - f(a)}{b - a}.$$

A more general form of the Law of the Mean which is useful in evaluating indeterminate forms (§§ 413–416) is given in the following theorem. The proof is requested in Exercise 14, § 408, where hints are given.

Theorem IV. Generalized Law of the Mean (Generalized Mean Value Theorem for Derivatives). *If $f(x)$ and $g(x)$ are continuous on the closed interval $[a, b]$, if $f(x)$ and $g(x)$ are differentiable in the open interval (a, b), and if $g'(x)$ does not vanish in the open interval (a, b), then there is some point ξ of the open interval (a, b) such that*

(6) $$\frac{f'(\xi)}{g'(\xi)} = \frac{f(b) - f(a)}{g(b) - g(a)}.$$

The geometric interpretation is similar to that for the preceding Law of the Mean. The curve this time is given parametrically, where for convenience we relabel the independent variable with the letter t. The coordinates of a point (x, y) on the curve are given in terms of the parameter t by the functions $x = g(t)$ and $y = f(t)$, where $a \leq t \leq b$ (Fig. 407). In this case, by a formula from Calculus (cf. Ex. 46, § 408), the slope of the tangent at the point where $t = \xi$ is the left-hand member of (6), while the slope of the secant joining the points corresponding to $t = a$ and $t = b$ is the right-hand member of (6).

We conclude this section with some Notes, whose proofs are requested in the Exercises of § 408.

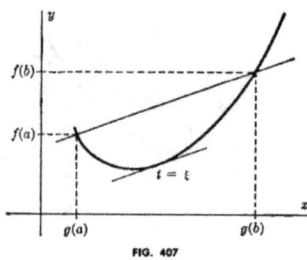

FIG. 407

NOTE 1. Rolle's Theorem and the Law of the Mean remain valid if $f'(x)$ is permitted to be either positively or negatively infinite in the open interval (a, b). (Ex. 11, § 408.)

NOTE 2. Formulas (2) and (6) are valid, under corresponding assumptions, in case $a > b$. (Ex. 12, § 408.)

NOTE 3. If $f(x)$ is differentiable in a neighborhood of $x = a$, then for any x in this neighborhood there exists a point ξ between a and x ($\xi = a$ if $x = a$) such that

(7) $$f(x) = f(a) + f'(\xi)(x - a).$$

(Ex. 13, § 408.)

NOTE 4. If $f(x)$ is differentiable in a neighborhood of $x = a$, and if h is sufficiently small numerically, then there exists a number θ such that $0 < \theta < 1$ and

(8) $$f(a + h) = f(a) + f'(a + \theta h)h.$$

(Ex. 13, § 408.)

Example 1. Use the Law of the Mean to establish the inequality $\sin x < x$, for $x > 0$.

Solution. If $a = 0, h = x$, and $f(x) = \sin x$, the equation $f(a + h) = f(a) + f'(a + \theta h) \cdot h$ becomes $\sin x = \cos (\theta x) \cdot x$. If $0 < x \leq 1$, then $0 < \theta x < 1$ and $\cos (\theta x) < 1$, so that $\sin x < x$. If $x > 1$, $\sin x \leq 1 < x$.

Example 2. Use the Law of the Mean to establish the inequalities

(9) $$\frac{h}{1 + h} < \ln (1 + h) < h,$$

if $h > -1$ and $h \neq 0$.

§ 406] LAW OF THE MEAN 97

Solution. If $a = 1$, and $f(x) = \ln x$, the equation $f(a + h) = f(a) + f'(a + \theta h) \cdot h$ becomes $\ln(1 + h) = \dfrac{h}{1 + \theta h}$. If $h > 0$, the inequalities $0 < \theta < 1$ imply $1 < 1 + \theta h < 1 + h$, and hence $\dfrac{1}{1 + h} < \dfrac{1}{1 + \theta h} < 1$, whence (9) follows. If $-1 < h < 0$, the inequalities $0 < \theta < 1$ imply $1 > 1 + \theta h > 1 + h > 0$, and hence $\dfrac{1}{1 + h} > \dfrac{1}{1 + \theta h} > 1$, whence (9) follows.

406. CONSEQUENCES OF THE LAW OF THE MEAN

It is a trivial fact that a constant function has a derivative that is identically zero. The converse, though less trivial, is also true.

Theorem I. *A function with an identically vanishing derivative throughout an interval must be constant in that interval.*

Proof. If $f(x)$ is differentiable and nonconstant in an interval, there are two points, a and b, of that interval where $f(a) \neq f(b)$. By the Law of the Mean there must be a point ξ between a and b such that

$$f'(\xi) = [f(b) - f(a)]/(b - a) \neq 0,$$

in contradiction to the basic assumption.

Theorem II. *Two differentiable functions whose derivatives are equal throughout an interval must differ by a constant in that interval.*

Proof. This is an immediate consequence of Theorem I, since the difference of the two functions has an identically vanishing derivative and must therefore be a constant.

A direct consequence of the definition of a derivative is that monotonic differentiable functions have derivatives of an appropriate sign (Ex. 14, § 404). A converse is stated in the following theorem:

Theorem III. *If $f(x)$ is continuous over an interval and differentiable in the interior, and if $f'(x) \geq 0$ (≤ 0) there, then $f(x)$ is monotonically increasing (decreasing) on the interval. If furthermore $f'(x) > 0$ (< 0), then $f(x)$ is strictly increasing (decreasing).*

Proof. Let x_1 and x_2 be points of the interval such that $x_1 < x_2$. Then, by the Law of the Mean, there is a number x_3 between x_1 and x_2 such that $f(x_2) - f(x_1) = f'(x_3)(x_2 - x_1)$. The resulting inequalities give the desired conclusions.

An important relation between monotonicity of a function and the nonvanishing of its derivative is contained in the following theorem:

Theorem IV. *A function continuous over an interval and having a nonzero derivative throughout at least the interior of that interval is strictly mono-*

tonic *(over the interval), and its derivative is of constant sign (wherever it is defined in the interval).*

Proof. By Theorem III, it is sufficient to show that the derivative is of constant sign. Accordingly, we seek a contradiction to the assumption that there exist two points, a and b $(a < b)$, of an interval throughout which the function $f(x)$ has a nonzero derivative, and such that $f'(a)$ and $f'(b)$ have *opposite* signs. On the closed interval $[a, b]$ the continuous function $f(x)$ has a maximum value at some point ξ of $[a, b]$ and a minimum value at some point η of $[a, b]$, where $\xi \neq \eta$ (why must ξ and η be distinct?). By Theorem I, § 405, neither ξ nor η can be an interior point of $[a, b]$. Therefore either $\xi = a$ and $\eta = b$ or $\xi = b$ and $\eta = a$. However, if $\xi = a$ and $\eta = b$, $f'(a) \leq 0$ and $f'(b) \leq 0$ (why?), whereas if $\xi = b$ and $\eta = a$, $f'(a) \geq 0$ and $f'(b) \geq 0$. Either conclusion is a contradiction to the assumption that $f'(a)$ and $f'(b)$ have opposite signs.

Corollary. *If a function has a nonzero derivative over an interval, the inverse function exists and is differentiable, and consequently the formula* $\frac{dx}{dy} = 1 / \frac{dy}{dx}$ *is valid whenever its right-hand member exists over an interval.*

407. THE EXTENDED LAW OF THE MEAN

In order to motivate an extension of the Law of the Mean to include higher order derivatives, we consider heuristically the problem of trying to approximate a given function $f(x)$ in a neighborhood of a point $x = a$ by means of a polynomial $p(x)$. The higher the degree of $p(x)$, the better we should expect to be able to approximate $f(x)$ (assuming, as we shall for this introductory discussion, that $f(x)$ not only is continuous but has derivatives of as high an order as we wish to consider, for x in the neighborhood of a). If $p(x)$ is a constant (degree zero), a reasonable value for this constant, if it is to approximate $f(x)$ for x near a, is clearly $f(a)$. If $p(x)$ is a polynomial of (at most) the first degree, it should certainly approximate $f(x)$ if its graph is the line tangent to the graph of $y = f(x)$ at the point $(a, f(a))$, that is, if $p(a) = f(a)$ and $p'(a) = f'(a)$. In this case, $p(x)$ has the form

(1) $$p(x) = f(a) + f'(a)(x - a).$$

More generally, let us approximate $f(x)$ in the neighborhood of $x = a$ by a polynomial $p(x)$ of degree $\leq n$ with the property that $p(a) = f(a)$, $p'(a) = f'(a)$, $\cdots$, $p^{(n)}(a) = f^{(n)}(a)$. We first express $p(x)$ by means of the substitution of $a + (x - a)$ for x, and subsequent expansion, in the form

(2) $$p(x) = p_0 + p_1(x - a) + \cdots + p_n(x - a)^n$$

§ 407] EXTENDED LAW OF THE MEAN

Successive differentiation and substitution in the equations $p(a) = f(a)$, $p'(a) = f'(a), \cdots, p^{(n)}(a) = f^{(n)}(a)$ lead to an evaluation of the coefficients in (2) in terms of the function $f(x)$, and hence to the following expression for $p(x)$ (check the details):

(3) $\quad p(x) = f(a) + f'(a)(x - a) + \dfrac{f''(a)}{2!}(x - a)^2 + \cdots + \dfrac{f^{(n)}(a)}{n!}(x - a)^n.$

The Law of the Mean, in the form

(4) $\quad f(x) = f(a) + f'(\xi)(x - a),$

states that $f(x)$ can be represented by an expression which resembles the approximating polynomial (1) of (at most) the first degree, differing from it only by the substitution of ξ for a in the coefficient of the last term. It is not altogether unreasonable to expect that $f(x)$ can be represented more generally by an expression which resembles the approximating polynomial (3) of degree $\leq n$, differing from it only by the substitution of ξ for a in the coefficient of the last term. Our objective in this section is to show that this is indeed the case.

Theorem. Extended Law of the Mean (Mean Value Theorem). *If $f(x), f'(x), \cdots, f^{(n-1)}(x)$ are continuous on the closed interval $[a, b]$, and if $f^{(n)}(x)$ exists in the open interval (a, b), then there is some point ξ of the open interval (a, b) such that*

(5) $\quad f(b) = f(a) + f'(a)(b - a) + \dfrac{f''(a)}{2!}(b - a)^2 + \cdots$
$\qquad\qquad + \dfrac{f^{(n-1)}(a)}{(n - 1)!}(b - a)^{n-1} + \dfrac{f^{(n)}(\xi)}{n!}(b - a)^n.$

Proof. The methods used in establishing the Law of the Mean in § 405 can be extended to the present theorem. Let the constant K be defined by the equation

(6) $\quad f(b) = f(a) + f'(a)(b - a) + \dfrac{f''(a)}{2!}(b - a)^2 + \cdots$
$\qquad\qquad + \dfrac{f^{(n-1)}(a)}{(n - 1)!}(b - a)^{n-1} + \dfrac{K}{n!}(b - a)^n,$

and define a function $\phi(x)$ by replacing a by x in (6), and rearranging terms:

(7) $\quad \phi(x) \equiv f(b) - f(x) - f'(x)(b - x) - \dfrac{f''(x)}{2!}(b - x)^2 - \cdots$
$\qquad\qquad - \dfrac{f^{(n-1)}(x)}{(n - 1)!}(b - x)^{n-1} - \dfrac{K}{n!}(b - x)^n.$

This function $\phi(x)$, for the interval $[a, b]$, satisfies the conditions of Rolle's Theorem (check this), and therefore its derivative must vanish for some point ξ of the open interval (a, b):

100   DIFFERENTIATION   [§ 408

(8) $$\phi'(\xi) = 0.$$

Routine differentiation of (7) gives

(9) $$\phi'(x) = -f'(x) + f'(x) - f''(x)(b-x) + f''(x)(b-x) - \cdots \\ - \frac{f^{(n)}(x)}{(n-1)!}(b-x)^{n-1} + \frac{K}{(n-1)!}(b-x)^{n-1},$$

in a form where all of the terms except the last two cancel in pairs. Equation (8) becomes, therefore, on simplification:

(10) $$K = f^{(n)}(\xi),$$

and the proof is complete.

Notes similar to those of § 405 apply to this section.

Note 1. The Extended Law of the Mean remains valid if $f^{(n)}(x)$ is permitted to be either positively or negatively infinite in the open interval (a, b). (Ex. 15, § 408.)

Note 2. Formula (5) is valid, under corresponding assumptions, in case $a > b$. (Ex. 16, § 408.)

Note 3. If $f^{(n)}(x)$ exists at every point of an interval I (open, closed, or half-open) that contains the point $x = a$, then for any x belonging to I there exists a point ξ between a and x ($\xi = a$ if $x = a$) such that

(11) $$f(x) = f(a) + f'(a)(x-a) + \frac{f''(a)}{2!}(x-a)^2 + \cdots \\ + \frac{f^{(n-1)}(a)}{(n-1)!}(x-a)^{n-1} + \frac{f^{(n)}(\xi)}{n!}(x-a)^n.$$

(Ex. 17, § 408.)

Note 4. If $f^{(n)}(x)$ exists at every point of an interval I (open, closed, or half-open) that contains the point $x = a$, then for any h such that $a + h$ belongs to I there exists a number θ such that $0 < \theta < 1$ and

(12) $$f(a+h) = f(a) + f'(a)h + \frac{f''(a)}{2!}h^2 + \cdots \\ + \frac{f^{(n-1)}(a)}{(n-1)!}h^{n-1} + \frac{f^{(n)}(a+\theta h)}{n!}h^n.$$

(Ex. 17, § 408.)

408. EXERCISES

In Exercises 1-2, find a value for ξ as prescribed by Rolle's Theorem. Draw a figure.

1. $f(x) = \cos x$, for $\frac{1}{2}\pi \leq x \leq \frac{3}{2}\pi$.
2. $f(x) = x^3 - 6x^2 + 6x - 1$, for $\frac{1}{3}(5 - \sqrt{21}) \leq x \leq 1$.

In Exercises 3-4, find a value for ξ as prescribed by the Law of the Mean, (2), § 405. Draw a figure.

3. $f(x) = \ln x$, for $1 \leq x \leq e$.

4. $f(x) = px^2 + qx + r$, for $a \leq x \leq b$.

In Exercises 5-6, find a value for θ as prescribed by the Law of the Mean, (8), § 405. Draw a figure.

5. $f(x) = \ln x$, for $a = e$, $h = 1 - e$.
6. $f(x) = px^2 + qx + r$, a and h arbitrary.

In Exercises 7-8, find a value for ξ as prescribed by the Generalized Law of the Mean, (6), § 405. Draw a figure.

7. $f(x) = 2x + 5$, $g(x) = x^2$, for $0 < b \leq x \leq a$.
8. $f(x) = x^3$, $g(x) = x^2$, for $1 \leq x \leq 3$.

In Exercises 9-10, find a value for ξ as prescribed by the Extended Law of the Mean, (5), § 407.

9. $f(x) = \dfrac{1}{1-x}$, $a = 0$, arbitrary n, and $b < 1$.
10. $f(x) = \ln x$, $a = 1$, $b = 3$, $n = 3$.

11. Prove Note 1, § 405. Explain why the function $x^{\frac{2}{3}}$ on the closed interval $[-1, 1]$ is excluded.
12. Prove Note 2, § 405.
13. Prove Notes 3 and 4, § 405.
14. Prove Theorem IV, § 405. *Hint*: First show that $g(b) \neq g(a)$, by using Rolle's Theorem with the function $h(x) = g(x) - g(a)$. Then let
$$K = (f(b) - f(a))/(g(b) - g(a))$$
and define the function $\phi(x) = f(x) - f(a) - K[g(x) - g(a)]$. Proceed as with the proof of the Law of the Mean.

15. Prove Note 1, § 407.
16. Prove Note 2, § 407.
17. Prove Notes 3 and 4, § 407.
18. The functions $f(x) = \dfrac{1}{x}$ and $g(x) = \dfrac{1}{x} + \operatorname{sgn} x$ (Example 1, § 206) have identical derivatives, but do not differ by a constant. Explain how this is possible in the presence of Theorem II, § 406.

In Exercises 19-30, use the Law of the Mean to establish the given inequalities. (Assume the standard properties of the transcendental functions. Cf. §§ 428-431.)

19. $\tan x > x$ for $0 < x < \frac{1}{2}\pi$.
20. $|\sin a - \sin b| \leq |a - b|$. (Cf. Ex. 44.)
21. $\dfrac{b-a}{b} < \ln \dfrac{b}{a} < \dfrac{b-a}{a}$, for $0 < a < b$.
22. $\sqrt{1+h} < 1 + \frac{1}{2}h$, for $-1 < h < 0$ or $h > 0$. More generally, for these values of h and $0 < p < 1$, $(1 + h)^p < 1 + ph$. (Cf. Exs. 38-41.)
23. $(1 + h)^p > 1 + ph$, for $-1 < h < 0$ or $h > 0$, and $p > 1$ or $p < 0$. (Cf. Exs. 38-41.)
24. $\dfrac{h}{1+h^2} < \operatorname{Arctan} h < h$, for $h > 0$.

25. $x < \text{Arcsin } x < \dfrac{x}{\sqrt{1-x^2}}$, for $0 < x < 1$.

26. $\text{Arctan}(1+h) \leq \dfrac{\pi}{4} + \dfrac{h}{2}$, for $h > -1$.

27. $\left|\dfrac{\cos ax - \cos bx}{x}\right| \leq |a-b|$, for $x \neq 0$.

28. $\dfrac{\sin px}{x} < p$, for $p > 0$ and $x > 0$.

29. $e^a(b-a) < e^b - e^a < e^b(b-a)$, for $a < b$.

30. $ae^{-ax} < \dfrac{1-e^{-ax}}{x} < a$, for $a > 0$ and $x > 0$.

In Exercises 31-34, use the Extended Law of the Mean to establish the given inequalities.

31. $\cos x \geq 1 - \dfrac{x^2}{2}$.

32. $\cos x > 1 - \dfrac{x^2}{2}$, for $x \neq 0$.

33. $x - \dfrac{x^3}{6} < \sin x < x$, for $x > 0$.

34. $1 + x + \dfrac{x^2}{2} < e^x < 1 + x + \dfrac{x^2}{2}e^x$, for $x > 0$.

★35. Use the trigonometric identity
$$\cos u - \cos v = -2\sin \tfrac{1}{2}(u+v) \sin \tfrac{1}{2}(u-v)$$
to establish the inequality
$$\cos a - \cos b < \dfrac{b^2 - a^2}{2}, \quad \text{for } 0 \leq a < b.$$
Prove that
$$\dfrac{\cos ax - \cos bx}{x^2} < \dfrac{b^2 - a^2}{2}, \quad \text{for } 0 \leq a < b, \; x \neq 0.$$

★36. Prove that $\dfrac{2}{\pi} < \dfrac{\sin x}{x} < 1$, for $0 < x < \dfrac{\pi}{2}$. *Hint:* For the first inequality, show that $\sin x / x$ is a decreasing function.

★37. Prove that $\dfrac{4}{\pi} > \dfrac{\tan x}{x} > 1$, for $0 < x < \dfrac{\pi}{4}$. (Cf. Ex. 36.)

38. Use the Law of the Mean to establish the following inequalities, for the designated ranges of p, assuming in each case that $x > 0$ and $x \neq 1$:
$$p(x-1)x^{p-1} < x^p - 1 < p(x-1), \quad \text{for } 0 < p < 1,$$
$$p(x-1)x^{p-1} > x^p - 1 > p(x-1), \quad \text{for } p < 0 \text{ or } p > 1.$$

★39. By solving the left-hand inequalities of Exercise 38 for x^p, establish the following inequalities for the designated ranges of p and $h = x - 1$:
$$\dfrac{1+h}{1+h-ph} < (1+h)^p < 1 + ph,$$
for $0 < p < 1$ and either $-1 < h < 0$ or $h > 0$.
$$\dfrac{1+h}{1+h-ph} > (1+h)^p > 1 + ph,$$

for $p > 1$ and either $-1 < h < 0$ or $0 < h < \dfrac{1}{p-1}$,

or for $p < 0$ and either $h > 0$ or $\dfrac{-1}{1-p} < h < 0$.

Conclude that for any real number p, the expression $(1 + h)^p$, for sufficiently small $|h|$, is between the two numbers $(1 + h)/(1 + h - ph)$ and $1 + ph$ (being equal to them if they are equal). (Cf. Exs. 22-23.)

★**40.** Show that if n is an integer greater than 1, and if either $h > 0$ or $-1 < h < 0$, then

$$\frac{1+h}{1+h-\dfrac{h}{n}} < \sqrt[n]{1+h} < 1 + \frac{h}{n}.$$

(Cf. Ex. 39.)

★**41.** Show that if n is an integer greater than 1, and if either $h > 0$ or $-\tfrac{1}{2} < h < 0$, then

$$\frac{1+h}{1+h+\dfrac{h}{n}} > \frac{1+h-\dfrac{h}{n}}{1+h} > \frac{1}{\sqrt[n]{1+h}} > \frac{1}{1+\dfrac{h}{n}} > 1 - \frac{h}{n}.$$

(Cf. Exs. 39-40.)

★**42.** Let $f(x)$ be differentiable, with $f'(x) \geqq 0$ ($f'(x) \leqq 0$), on an interval, and assume that on no subinterval does $f'(x)$ vanish identically. Prove that $f(x)$ is strictly increasing (decreasing) on the interval.

★**43.** Establish the inequality $\sin x < x$, for $x > 0$ (Example 1, § 405) by applying Exercise 42 to the function $x - \sin x$. Similarly, establish the inequality $\tan x > x$ for $0 < x < \dfrac{\pi}{2}$ (Ex. 19).

★**44.** Let $f(x)$ be differentiable, with $f'(x) \geqq k$ ($f'(x) \leqq k$), on an interval and assume that on no subinterval does $f'(x)$ equal k identically. Prove that for the graph of $y = f(x)$, on the given interval, every secant line has slope $> k$ ($< k$). As a consequence, show that if $a \neq b$,

$$|\sin a - \sin b| < |a - b|,$$
$$|\cos a - \cos b| < |a - b|.$$

Hint: Let $g(x) = f(x) - kx$, and use Exercise 42.

45. Prove that a function differentiable at every point of an interval is monotonic there if and only if its derivative does not change sign there.

46. Let $x = g(t)$ and $y = f(t)$ be continuous over a closed interval $a \leqq t \leqq b$ and differentiable in the interior, and assume that $g'(t)$ does not vanish there. Prove that y, as a function of x, is continuous over a corresponding interval and differentiable in the interior, and that for any interior point

$$\frac{dy}{dx} = \frac{f'(t)}{g'(t)}.$$

Interpret the results geometrically.

★**47.** It was shown in Example 3, § 403, that although a function may be differentiable for all values of the independent variable, its derivative may not be continuous. In spite of this fact, a derivative shares with continuous functions the intermediate value property of Theorem IV, § 213. Prove the theorem: *If $f(x)$ is the derivative of some function $g(x)$, on an interval, then $f(x)$ assumes (as a value) every number between any two of its values.* Hint: Let c and d be any two distinct values of $f(x)$, let r be any number between c and d, and apply Theorem IV, § 406, to the function $\phi(x) \equiv g(x) - rx$.

★**48.** Show by the example $f(x) \equiv x^2 \sin \frac{1}{x^2}$ ($f(0) \equiv 0$) that derivatives do not always share with continuous functions the property of being bounded on closed intervals (Theorem I, § 213). That is, exhibit a closed interval at every point of which $f(x)$ is differentiable but on which $f'(x)$ is unbounded.

★**49.** Prove that among the discontinuities discussed in § 210 for functions in general, *derivatives* can have discontinuities only of type (iii), where *not both* one-sided limits exist. In other words, show that for derivatives existing throughout an interval the discontinuities of Example 3, § 403, and Exercise 48 are the rule and not the exception. Hint: If $\lim_{x \to c+} f'(x)$ exists and $\neq f'(c)$, there exist positive numbers ϵ and δ such that $c < x < c + \delta$ implies $|f'(x) - f'(c)| \geqq \epsilon$. Use Ex. 47. (Also cf. Ex. 54.)

★**50.** Show by the example $f(x) \equiv x + 2x^2 \sin (1/x)$ ($f(0) \equiv 0$) that the hypotheses (i) $f'(x)$ exists in a neighborhood of the point $x = a$, and (ii) $f'(a) > 0$, do not imply that there exists some neighborhood of $x = a$ throughout which $f(x)$ is increasing.

★**51.** The **Wronskian determinant** of two differentiable functions, $f(x)$ and $g(x)$, is defined:

$$W(f, g) \equiv \begin{vmatrix} f(x) & g(x) \\ f'(x) & g'(x) \end{vmatrix}.$$

Prove that if the Wronskian of f and g never vanishes over an interval, then between any two roots of the equation $f(x) = 0$ ($g(x) = 0$) there must exist at least one root of the equation $g(x) = 0$ ($f(x) = 0$). (Example: $f(x) = \sin x$, $g(x) = \cos x$.) Hint: Observe first that $f(x)$ and $g(x)$ never vanish simultaneously. Let a and b ($a < b$) be two roots of $f(x)$ and assume that $g(x)$ does not vanish for $a < x < b$. Apply Rolle's Theorem to the quotient $f(x)/g(x)$, to obtain a contradiction.

★**52.** Prove that if a function has a bounded derivative in an open interval it is uniformly continuous there.

★**53.** Prove that if $f(x)$ has a bounded derivative on an open interval (a, b), then $f(a+)$ and $f(b-)$ exist and are finite. Hint: Cf. Ex. 52, and Ex. 15, § 308.

★**54.** Prove that if the right-hand (left-hand) limit of the derivative of a function (Definition III, § 403) exists and is finite, then the right-hand (left-hand) derivative (Definition II, § 403) exists and is equal to it. Hint: For the case $x \to a+$, show that there exists an interval of the form $(a, a + h)$, where $h > 0$, in which $f'(x)$ is bounded (cf. Ex. 9, § 208), and use Exercise 53 to infer that $f(a+) \equiv \lim_{x \to a+} f(x)$ exists. Redefine $f(a) \equiv f(a+)$ and use the Law of the Mean in the form $[f(a + h) - f(a+)]/h = f'(a + \theta h)$. (Cf. Ex. 49.)

409. MAXIMA AND MINIMA

We shall assume that the reader is familiar with the standard routine of finding maximum and minimum values of a function $y = f(x)$: (i) differentiate; (ii) set the derivative equal to zero; (iii) solve the equation $f'(x) = 0$ for x; (iv) test the values of x thus obtained, by using either the first or second derivative of the function; and (v) substitute in $f(x)$ the appropriate values of x to find the maximum and minimum values of $y = f(x)$.

Inasmuch as this routine gives only a partial answer to the story of maxima and minima, we present in this section a more complete summary of the pertinent facts and tests.

Let $f(x)$ be a function defined over a set A, and let ξ be a point of A. If the inequality $f(\xi) \geqq f(x)$ ($f(\xi) \leqq f(x)$) holds for every x in A, we say that $f(x)$ has a **maximum** (**minimum**) value on A equal to $f(\xi)$. According to Theorem II, § 213, such maximum and minimum values exist if A is a closed interval and $f(x)$ is continuous on A. If a function has a maximum (minimum) value on its domain of definition, this value is called the **absolute maximum** (**minimum**) value of the function. If ξ is a point where $f(x)$ is defined, and if within some neighborhood of ξ the inequality $f(\xi) \geqq f(x)$ ($f(\xi) \leqq f(x)$) holds whenever $f(x)$ is defined, we say that $f(\xi)$ is a **relative maximum** (**minimum**) value of $f(x)$. By a **critical value** of x for a function $f(x)$ we mean any point c of the domain of definition of $f(x)$ such that either (i) $f'(c)$ does not exist (as a finite quantity) or (ii) $f'(c) = 0$.

Theorem I. *If a function has a maximum or minimum value on an interval at a point ξ of the interval, then ξ is either an end-point of the interval or a critical value for the function.*

Proof. This theorem extends Theorem I, § 405, to an arbitrary (not necessarily closed) interval and drops continuity and differentiability assumptions. The details of the proof, however, are identical.

Some different kinds of maxima for a function continuous on a closed interval are illustrated in Figure 408.

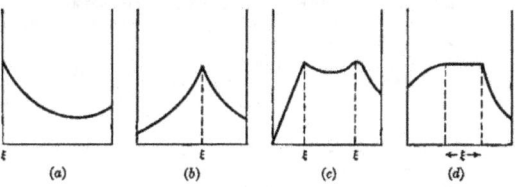

Theorem II. First Derivative Test. *If $f(x)$ is continuous at a point $x = \xi$ and differentiable in a deleted neighborhood of ξ, and if in this deleted neighborhood $f'(x) > 0$ for $x < \xi$ and $f'(x) < 0$ for $x > \xi$ ($f'(x) < 0$ for $x < \xi$ and $f'(x) > 0$ for $x > \xi$), then $f(x)$ has a relative maximum (minimum) value at $x = \xi$. If, on the other hand, $f'(x)$ is of constant sign throughout the deleted neighborhood, $f(x)$ has neither a relative maximum nor a relative minimum value at $x = \xi$.*

Proof. Let x be an arbitrary point in the deleted neighborhood of ξ. By the Law of the Mean, § 405, there exists a point ξ' between ξ and x such that $f(x) - f(\xi) = f'(\xi')(x - \xi)$. Examination of each individual case leads to an appropriate inequality of the form $f(x) > f(\xi)$ or $f(x) < f(\xi)$. (Check the details.)

NOTE 1. The conditions assumed in Theorem II are sufficient but not necessary, even if $f'(x)$ is continuous and $f'(\xi) = 0$ (cf. Exs. 21-22, § 412).

Theorem III. Second Derivative Test. *If $f(x)$ is differentiable in a neighborhood of a critical value ξ, and if $f''(\xi)$ exists and is negative (positive), then $f(x)$ has a relative maximum (minimum) value at $x = \xi$.*

Proof. Assume $f''(\xi) < 0$. Then, by Exercise 7, § 404, within some neighborhood of ξ, $f'(x) > f'(\xi) = 0$ for $x < \xi$ and $f'(x) < f'(\xi) = 0$ for $x > \xi$. By the First Derivative Test, $f(x)$ has a relative maximum value at $x = \xi$. (Supply the corresponding details for the case $f''(\xi) > 0$.)

NOTE 2. No conclusion regarding maximum or minimum of a function can be drawn from the vanishing of the second derivative at a critical value—the function may have a maximum value or a minimum value or neither at such a point (cf. Ex. 9, § 412).

A useful extension of the Second Derivative Test is the following:

Theorem IV. *Let $f(x)$ be a function which, in some neighborhood of the point $x = \xi$, is defined and has derivatives $f'(x), f''(x), \cdots, f^{(n-1)}(x)$, of order $\leq n - 1$, where $n > 1$. If $f'(\xi) = f''(\xi) = \cdots = f^{(n-1)}(\xi) = 0$, and if $f^{(n)}(\xi)$ exists and is different from zero, then (i) if n is even, $f(x)$ has a relative maximum value or a relative minimum value at $x = \xi$ according as $f^{(n)}(\xi)$ is negative or positive, and (ii) if n is odd, $f(x)$ has neither a relative maximum value nor a relative minimum value at $x = \xi$.*

Proof. Owing to the vanishing of the derivatives of order $< n - 1$ at the point ξ, the Extended Law of the Mean provides the formula

$$(1) \qquad f(x) - f(\xi) = \frac{f^{(n-1)}(\xi')}{(n-1)!}(x - \xi)^{n-1},$$

where x is in a suitably restricted deleted neighborhood of ξ, and ξ' is between ξ and x. The proof resolves itself into determining what happens to the sign of the right-hand member of (1), as x changes from $x < \xi$ to

$x > \xi$, according to the various possibilities for the sign of $f^{(n)}(\xi)$ and the parity of n. We give the details for the case $f^{(n)}(\xi) < 0$, and suggest that the student furnish the corresponding details for the case $f^{(n)}(\xi) > 0$. By Exercise 7, § 404 (applied to the function $f^{(n-1)}(x)$), as x changes from $x < \xi$ to $x > \xi$, $f^{(n-1)}(x)$ changes from $+$ to $-$. Therefore if n is even, the right-hand member of (1) is negative whether $x < \xi$ or $x > \xi$, whence $f(x) < f(\xi)$ for x in the deleted neighborhood of ξ, and $f(\xi)$ is a relative maximum value of $f(x)$. If n is odd, the right-hand member of (1) is positive for $x < \xi$ and negative for $x > \xi$, so that $f(\xi)$ is neither a relative maximum value nor a relative minimum value of $f(x)$ at $x = \xi$.

Example. Examine the function $f(x) = \dfrac{x^{\frac{2}{3}}}{x^2 + 8}$ for critical values of x and relative and absolute maxima and minima. Find its maximum and minimum values (when they exist) for the intervals $[1, 3]$, $(-1, 2)$, and $[1, +\infty)$.

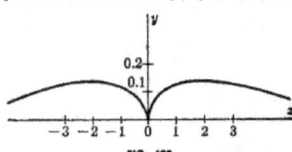

FIG. 409

Solution. The graph of $f(x)$ (Fig. 409) is symmetrical with respect to the y-axis, $\lim\limits_{x \to \infty} f(x) = 0$, and from the definition of one-sided derivatives, the derivative from the right at $x = 0$ is $+\infty$ and the derivative from the left at $x = 0$ is $-\infty$. For $x \neq 0$,

$$f'(x) = \frac{4(4 - x^2)}{3\sqrt[3]{x}\,(x^2 + 8)^2}.$$

The critical values of x are $x = 0, 2$, and -2. At $x = 0$ the function has both a relative and an absolute minimum value of 0. At $x = \pm 2$ the function has a relative and an absolute maximum value of $\tfrac{1}{12}\sqrt[3]{4} = 0.1326$ (approximately, to four decimal places). On the interval $[1, 3]$ the function has a maximum of $f(2)$ and a minimum of $f(1) = \tfrac{1}{9} = 0.1111$ ($f(3) = 0.1224$). On the interval $(-1, 2)$ $f(x)$ has a minimum of $f(0) = 0$, but no maximum. On the interval $[1, +\infty)$ $f(x)$ has a maximum of $f(2)$, but no minimum.

At this point, the student may wish to do a few of the Exercises of § 412. Some that are suitable for performance now are Exercises 1-22.

410. DIFFERENTIALS

The student of Calculus becomes familiar with the differential notation, and learns to appreciate its convenience in the treatment of composite,

108   DIFFERENTIATION   [§ 410

inverse, and implicit functions and functions defined parametrically, and in the technique of integration by substitution. Differentials also lend themselves simply and naturally to such procedures as the solving of differential equations. Such techniques and their legitimacy will not be discussed here. In the present section we restrict ourselves to basic definitions and theoretical facts.

If $y = f(x)$ is a differentiable function of x, we introduce two symbols, dx and dy, devised for the purpose of permitting the derivative symbol to be regarded and manipulated as a fraction.

To this end we let dx denote an arbitrary real number, and define $dy = d(f(x))$ to be a function of the *two* independent variables x and dx, prescribed by the equation

(1) $$dy \equiv f'(x)\,dx.$$

The differentials dx and dy are interpreted geometrically in Figure 410.

Although Calculus was the common invention of Sir Isaac Newton (1642-1727) and Gottfried Wilhelm von Leibnitz (1646-1716), the differential notation is due to Leibnitz. Its principal importance lies ultimately

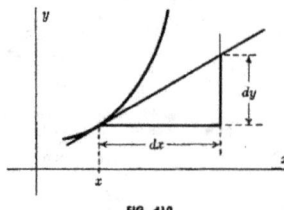

FIG. 410

in the fact that formula (1), initially true under the hypothesis that y is a function of the independent variable x, remains true under any possible reinterpretation of the dependence or independence of the variables x and y. This fact is given explicit formulation in the two theorems that follow. Before proceeding to these theorems we point out a further justification of the differential notation which is of so elementary a character that it is frequently overlooked, but which follows directly from the definition (1): If two variables are related by the identity relation, $y = x$, their differentials are also related by the identity relation, $dy = dx$ (Ex. 23, § 412).

Theorem I. *If $x = \phi(y)$ is a differentiable function of the independent variable y in a certain interval, if $\phi'(y) \neq 0$ in this interval, and if dy and dx denote the differentials of the independent variable y and the dependent variable x, respectively, related by definition by the equation $dx = \phi'(y)\,dy$, then if*

§ 411] APPROXIMATIONS BY DIFFERENTIALS 109

$y = f(x)$ *denotes the inverse function of* $x = \phi(y)$, *these differentials are also related by equation* (1): $dy = f'(x)\,dx$.

Proof. By the Corollary to Theorem IV, § 406, the derivatives dx/dy and dy/dx are reciprocals, so that $\phi'(y) = 1/f'(x)$. Therefore
$$dx = (1/f'(x))\,dy,$$
and $dy = f'(x)\,dx$.

Theorem II. *If* $y = f(x)$ *is a differentiable function of the variable* x, *and if* x *is a differentiable function of the variable* t, *then if* x *and* y *are both regarded as dependent variables, depending on the independent variable* t, *their differentials are related by equation* (1): $dy = f'(x)\,dx$.

Proof. Let $x = \phi(t)$ and $y = \psi(t) = f(\phi(t))$ denote the dependence of x and y on the independent variable t. Then by definition, $dx = \phi'(t)\,dt$ and $dy = \psi'(t)\,dt$. By the Chain Rule (Theorem I, § 402),
$$dy/dt = (dy/dx)(dx/dt),$$
or $\psi'(t) = f'(x)\phi'(t) = f'(\phi(t))\phi'(t)$, so that
$$dy = \psi'(t)\,dt = f'(\phi(t))(\phi'(t)\,dt) = f'(x)\,dx.$$

This completes the proof and shows that Theorem II is, in essence, simply a reformulation of the Chain Rule.

411. APPROXIMATIONS BY DIFFERENTIALS

In order to compare the differentials dx and dy, on the one hand, and the increments Δx and Δy, on the other, we observe that whereas the differentials can be associated with the *tangent* to the curve $y = f(x)$ (Fig. 410), the increments are associated with the *curve* itself. It is often convenient to regard the real number dx as an increment in the variable $x : dx = \Delta x$. In this case the quantities just discussed find their geometric interpretation in Figure 411. Throughout this section we shall identify dx and Δx. It

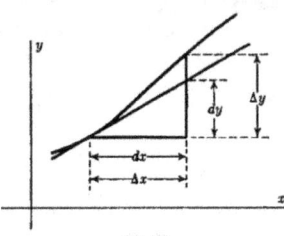

FIG. 411

should be noted, however, that under this assumption dy and Δy are *not* in general the same.

Although in theory the increment Δy is a simpler concept than the differential dy, in practice the differential is usually easier to compute than the increment, and dy is often a useful approximation to Δy. The statement that for a numerically small increment $dx = \Delta x$, the quantity dy is a good approximation to the quantity Δy, has a precise formulation, and means much more than the statement that dy and Δy are approximately equal (which is trivially true since they are both approximately equal to zero). The precise relation (under the assumption that the given function is differentiable at the point in question) is given by equation (4), § 402, rewritten in the form

(1) $$\Delta y = dy + \epsilon \Delta x,$$

where, *for a fixed x*, $\epsilon = \epsilon(\Delta x)$ is an infinitesimal function of Δx ($\epsilon \to 0$ as $\Delta x \to 0$).

Using equation (1) we can formulate the statement that (for small $|dx| = |\Delta x|$) dy is a good approximation to Δy by means of the equation

(2) $$\lim_{\Delta x \to 0} \frac{\Delta y - dy}{\Delta x} = 0,$$

or, in case dy is not identically zero, by the equation

(3) $$\lim_{\Delta x \to 0} \frac{\Delta y}{dy} = 1. \quad (\text{Ex. 24, § 412}).$$

Example 1. If $y = f(x) = x^3$, then $\Delta y = f(x + \Delta x) - f(x) = 3x^2 \Delta x + 3x \Delta x^2 + \Delta x^3$, and $dy = 3x^2 \, dx$. Therefore, in equation (1), the function $\epsilon(\Delta x)$ is $3x \Delta x + \Delta x^2$, equation (2) takes the form $\lim_{\Delta x \to 0}(3x \Delta x + \Delta x^2) = 0$, and equation (3) becomes $\lim_{\Delta x \to 0} \frac{3x^2 + 3x \Delta x + \Delta x^2}{3x^2} = 1.$

Example 2. Prove that for numerically small h the quantity $\sqrt{1 + h}$ is closely approximated by $1 + \frac{1}{2}h$.

Solution. Let $y = f(x) = \sqrt{x}$, and let x change from 1 to $1 + h$. Then $dx = \Delta x = h$, and $dy = \frac{dx}{2\sqrt{x}} = \frac{h}{2}$. Therefore the value of y changes from $\sqrt{1} = 1$ by an amount approximately equal to $dy = \frac{1}{2}h$. The new value of y (which is $\sqrt{1 + h}$) is therefore approximately equal to $1 + \frac{1}{2}h$. (Cf. Ex. 41, § 408.)

The Extended Law of the Mean provides a measure of the accuracy of approximation of dy for Δy. Under the assumption that $f''(x)$ exists in the neighborhood of $x = a$, the Extended Law of the Mean ((12), § 407, for $n = 2$) ensures the existence of a number θ between 0 and 1 such that

(4) $$f(a + h) = f(a) + f'(a)h + \tfrac{1}{2}f''(a + \theta h)h^2,$$

§ 412] EXERCISES 111

for numerically small h. Expressed in terms of differentials, with $\Delta x = h$, $\Delta y = f(a + h) - f(a)$ and $dy = f'(a)h$, (4) becomes

(5) $$\Delta y - dy = \tfrac{1}{2} f''(a + \theta \Delta x) \Delta x^2.$$

If B is a bound for the absolute value of $f''(x)$ (that is, $|f''(x)| \leq B$) for x in a certain neighborhood of $x = a$, and if Δx is so restricted that $a + \Delta x$ (and therefore $a + \theta \Delta x$) is in this neighborhood, then (5) gives the inequality

(6) $$|\Delta y - dy| \leq \tfrac{1}{2} B \Delta x^2.$$

Example 3. Find an estimate for the accuracy of the approximation established in Example 2, if $|h| \leq 0.1$.

Solution. If $f(x) = \sqrt{x}$, then $f'(x) = \dfrac{1}{2\sqrt{x}}$, $f''(x) = -\dfrac{1}{4x\sqrt{x}}$, and formula (5) becomes

(7) $$\Delta y - dy = (\sqrt{1 + h} - 1) - (\tfrac{1}{2} h) = -\dfrac{h^2}{8(1 + \theta h)^{\frac{3}{2}}}.$$

If h is positive (whether $h \leq 0.1$ or $h > 0.1$), $1 + \theta h > 1$ and therefore the third member of (7) is between $-h^2/8$ and 0. We thus have the inequalities

(8) $$1 + \tfrac{1}{2} h - \tfrac{1}{8} h^2 < \sqrt{1 + h} < 1 + \tfrac{1}{2} h, \quad h > 0.$$

(Illustration: $1.058 < \sqrt{1.12} < 1.06$.)

If $-0.1 \leq h < 0$, $1 + \theta h > 0.9$, and the third member of (7) is between $-h^2/6$ and 0. Therefore

(9) $$1 + \tfrac{1}{2} h - \tfrac{1}{6} h^2 < \sqrt{1 + h} < 1 + \tfrac{1}{2} h, \quad -0.1 \leq h < 0.$$

(Illustration: $0.9694 < \sqrt{0.94} < 0.97$.)

NOTE. Formula (7) permits sharper results than those just obtained. For example, if h is between -0.1 and 0, the third member of (7) is between $-h^2/6$ and $-h^2/8$, so that $0.9694 < \sqrt{0.94} < 0.9696$. For computations where a high degree of accuracy is desired, however, the methods of this section are inadequate, and the reader is referred to Chapter 8.

412. EXERCISES

In Exercises 1-4, find the relative and absolute maximum and minimum values of the function, and the intervals within which the function is increasing, or decreasing. Draw a figure.

1. $x^3 - 6x^2 + 9x + 5$.
2. $\dfrac{2x}{1 + x^2}$.
3. $x^{\frac{2}{3}}(1 - x)$.
4. $\sqrt{x(x - 1)^2}$.

In Exercises 5-8, find the maximum and minimum values of the given function (whenever they exist) for the designated interval. Draw a figure.

5. $x + 2x^2 - 4x^3$, $[-\tfrac{1}{2}, 1]$.
6. $\dfrac{x^2 + 100}{x^2 - 25}$, $[-1, 3]$.

112   DIFFERENTIATION   [§ 412

7. $\cos x + \cos 2x$, $(-\infty, +\infty)$.   **8.** xe^x, $(-\infty, 0)$.

9. Show that the three functions x^4, $-x^4$, and x^3 all have the property of possessing a continuous second derivative which vanishes at $x = 0$, whereas at this point the functions have an absolute minimum, an absolute maximum, and neither a maximum nor a minimum, respectively. Thus justify Note 2, § 409.

10. Show that the function $x^m(x-1)^n$, where m and n are positive integers, has a relative minimum regardless of the values of m and n, and that it has an absolute minimum if and only if m and n have the same parity ($m+n$ is even). Draw figures.

★11. Show that the function $|x|^p \cdot |x-1|^q$, where p and q are positive, has a relative maximum value of $p^p q^q / (p+q)^{p+q}$.

★12. Show that the function $\dfrac{ax+b}{cx+d}$ has neither a relative maximum nor a relative minimum unless it is a constant.

★13. If $a_1, a_2, \cdots, a_n$ are given constants, show that the expression $\sum\limits_{i=1}^{n}(a_i - x)^2$ is minimized if and only if x is their arithmetic mean, $x = \dfrac{1}{n}\sum\limits_{i=1}^{n} a_i$.

★14. Show that the function $x^3 + 3px + q$ has either a relative maximum and a relative minimum or neither. Draw figures corresponding to the various cases.

15. A publisher is planning on putting out a new magazine, and an efficiency expert has estimated that if the price per copy is x cents, the profit is given by a formula of the type

$$\text{Profit} = K\left[\frac{x-a}{x^2+25} - b\right],$$

where K, a, and b are positive constants. Determine what price should be set for maximum profit, in each of the following cases:

(a) x is a multiple of 5, $a = 10$, $b = .02$;
(b) x is a multiple of 5, $a = 11$, $b = .02$;
(c) x is an integer, $a = 10$, $b = .02$;
(d) x is an integer, $a = 11$, $b = .03$.

16. A truck has a top speed of 60 miles per hour and, when traveling at the rate of x miles per hour, consumes gasoline at the rate of $\dfrac{1}{200}\left(\dfrac{400}{x} + x\right)$ gallons per mile. This truck is to be taken on a 200 mile trip by a driver who is to be paid at the rate of b dollars per hour plus a commission of c dollars. Since the time required for this trip, at x miles per hour, is $200/x$, the total cost, if gasoline costs a dollars per gallon, is

$$\left(\frac{400}{x}+x\right)a + \frac{200}{x}b + c.$$

Find the most economical possible speed under each of the following sets of conditions:

(a) $b = 0$;
(b) $a = .25$, $b = 1.25$, $c = 5$;
(c) $a = .20$, $b = 4$ (a crew).

17. A problem in maxima and minima can frequently be simplified by such devices as eliminating constant terms and factors, squaring, and taking reciprocals. Suppose, for example, we wish to find the values of x on the interval $1 \leq x \leq 2$ that maximize and minimize the function $17x/50\sqrt{x^4 + 2x^2 + 2}$. The value of x that maximizes (minimizes) this function is the same as that which maximizes (minimizes) the function $x/\sqrt{x^4 + 2x^2 + 2}$ or $x^2/(x^4 + 2x^2 + 2)$, and is the same as that which minimizes (maximizes) the reciprocal $(x^4 + 2x^2 + 2)/x^2$ or, equivalently, $x^2 + 2x^{-2}$. Let us now substitute $t = x^2$, and seek the value of t ($1 \leq t \leq 4$) that minimizes (maximizes) the function $g(t) = t + 2t^{-1}$. Since $g'(t) = 1 - 2t^{-2}$, $g(t)$ is minimized on the interval by $t = \sqrt{2}$, and is maximized by $t = 4$. Therefore the original function is maximized on $1 \leq x \leq 2$ by $x = \sqrt[4]{2}$ and minimized by $x = 2$. Discuss this technique in general for a function $y = f(x)$ defined on a set A, a strictly monotonic function $v = \phi(y)$ over the range of $f(x)$, and a suitably restricted substitution function $t = t(x)$. For the function $\phi(y)$, treat in particular $y + k$, py, y^n, $\sqrt[n]{y}$, and $1/y$, where k, p, and n are constants. (Be careful to explain under what circumstances squaring is legitimate.)

★18. Prove that the function
$$a|x| + b|x - 1|, \text{ where } a \neq b,$$
has a relative minimum value for the interval $(-\infty, +\infty)$ if and only if $a + b \geq 0$, and that if $a + b \geq 0$ its minimum value is the smaller of the two numbers a and b.

★19. Find the value of x that minimizes the function $t\sqrt{x^2 + a^2} + s|b - x|$ on the interval $(-\infty, +\infty)$, where a, b, s, and t are positive constants. Thus solve the following problem: The shore of a lake extends for a considerable distance along the x-axis of a co-ordinate system. The lake lies in the first two quadrants and has an island on the y-axis at the point $A : (0, a)$. A man on the island wishes to go to the point $B : (b, 0)$ on the positive half of the x-axis by rowing straight to the point $P : (x, 0)$ at a rate of s miles an hour, and walking from P to B at a rate of t miles an hour. What course should he steer?

★20. Find the minimum value of the function $|x^2 - ax| + rx^2$ on the interval $[0, +\infty)$, where a is a positive number and r is a real number.

★21. Show that the function $f(x)$ defined to be $x^4 \left(2 + \sin \dfrac{1}{x}\right)$ when $x \neq 0$ and 0 when $x = 0$ has an absolute strict minimum at $x = 0$ ($f(x) > f(0)$ if $x \neq 0$) and a continuous derivative everywhere, but that this derivative does not change sign from $-$ to $+$ as x changes from $x < 0$ to $x > 0$. Thus show that the conditions of the First Derivative Test (§ 409) are only sufficient and not necessary for a maximum (minimum). (Cf. Ex. 22 for a converse of the First Derivative Test.) Sketch a graph of the function given above. (It oscillates infinitely many times between the two curves $y = x^4$ and $y = 3x^4$.)

★22. Prove the following converse of the First Derivative Test: Let $f(x)$ be differentiable in a neighborhood of a point $x = \xi$ at which $f(x)$ has a relative maximum (minimum) value, and assume that in this neighborhood $f'(x)$ vanishes at only a finite number of points. Then there exists a neighborhood of $x = \xi$

within which $f'(x) > 0$ for $x < \xi$ and $f'(x) < 0$ for $x > \xi$ ($f'(x) < 0$ for $x < \xi$ and $f'(x) > 0$ for $x > \xi$). (Cf. Ex. 21.)

23. Prove that if two variables are related by the identity relation, $y = x$, their differentials are also related by the identity relation $dy = dx$.

24. If y is a function of x which is differentiable for a particular value of x, prove the limit statements (2) and (3), § 411: As $\Delta x \to 0$, $\dfrac{\Delta y - dy}{\Delta x} \to 0$, and if furthermore $dy \neq 0$, $\dfrac{\Delta y}{dy} \to 1$.

In Exercises 25-28, express the given relation $x = \phi(y)$ in the form $y = f(x)$, write down both equations $dx = \phi'(y)\, dy$ and $dy = f'(x)\, dx$, and thus verify Theorem I, § 410. Draw figures.

25. $x = y^2$, $y > 0$.
26. $x = y^2 - 4y + 5$, $y < 2$.
27. $x = \ln(y^2 + 1)$, $y > 0$.
28. $x = \cos y$, $0 < y < \pi$.

In Exercises 29-32, use the given relations $x = \phi(t)$ and $y = \psi(t)$ to express y as a function of x, $y = f(x)$, write down the three equations $dx = \phi'(t)\, dt$, $dy = \psi'(t)\, dt$, and $dy = f'(x)\, dx$, and thus verify Theorem II, § 410. Draw figures.

29. $x = t$, $y = 5t^2 - 7t - 6$.
30. $x = t^3$, $y = t^2 - t$, $t < 0$.
31. $x = \ln t$, $y = e^t$, $t > 0$.
32. $x = \cos t$, $y = \sin t$, $0 < t < \tfrac{1}{2}\pi$.

In Exercises 33-35, find Δy, dy, and $\epsilon(\Delta x)$, and verify the limit statements (2) and (3), § 411.

33. $y = x^4 - 5x^2 + 7$.
34. $y = \dfrac{1}{x}$.
35. $y = \sqrt{x}$. *Hint:* Rationalize a numerator (cf. Ex. 5, § 404).

36. Prove that if $y = f(x)$ has a continuous second derivative y'', then
$$y'' = \lim_{\Delta x \to 0} 2 \cdot \frac{\Delta y - dy}{\Delta x^2}.$$
Use this relation to obtain directly (without first finding the first derivative) the second derivative of each of the functions of Exercises 33-35. (Cf. Ex. 11, § 421.)

In Exercises 37-48, use differentials to obtain an approximation to the given number, or the given function for values of the variable near the specified value.

37. $\sqrt{110}$ (use $10.5^2 = 110.25$).
38. $\ln(1 + h)$, h near 0.
39. $\ln(0.94)$ (cf. Ex. 38).
40. $\sin x$, x near 0.
41. $\tan x$, x near 0.

§ 413] L'HOSPITAL'S RULE. INTRODUCTION 115

42. $\cos x$, x near $\frac{\pi}{3}$.

43. Arctan x, x near 0.

44. $\sqrt[n]{1+h}$, h near 0, n an integer >1.

45. $\ln \cos x$, x near 0.

46. $\frac{\ln(1+h)}{1+h}$, h near 0.

47. $e^{\sin x}$, x near 0.

48. $f(x) = x^2 \sin \frac{1}{x}$ ($f(0) = 0$), x near 0.

In Exercises 49-60, verify the given inequalities, which give estimates of the errors in the approximations of Exercises 37-48.

★49. $10.48808 < \sqrt{110} < 10.48810$.

★50. $h - \frac{1}{2}h^2 < \ln(1+h) < h$, $h > 0$,
$\qquad h - \frac{1}{2}h^2 < \ln(1+h) < h$, $-0.1 \leq h < 0$.

★51. $-0.0622 < \ln 0.94 < -0.06$.

★52. $x - \frac{x^3}{6} < \sin x < x$, $x > 0$. *Hint:* Show that
$$\Delta y - dy = -\frac{1}{6} \cos(\theta x)\, x^3,\ 0 < \theta < 1.$$

★53. $x + \frac{1}{3}x^3 < \tan x < x + \frac{1}{2}x^3$, $0 < x < 0.1$. *Hint:* Show that
$\Delta y - dy = \frac{1}{3}(1 + 3t^2)(1 + t^2)x^3$, where $t = \tan(\theta x)$, $0 < \theta < 1$.

★54. $\frac{1}{2} - \frac{1}{2}\sqrt{3}(x - \frac{1}{3}\pi) - \frac{1}{4}(x - \frac{1}{3}\pi)^2 \leq \cos x \leq \frac{1}{2} - \frac{1}{2}\sqrt{3}(x - \frac{1}{3}\pi)$,
$\qquad 0 \leq x \leq \frac{1}{3}\pi$.

★55. $x - \frac{1}{3}x^3 < \text{Arctan } x < x$, $0 < x < \frac{1}{2}\pi$.

★56. $1 + \frac{h}{n} - \frac{(n-1)h^2}{2n^2} < \sqrt[n]{1+h} < 1 + \frac{h}{n}$, $h > 0$;
$\qquad 1 + \frac{h}{n} - \frac{(n-1)h^2}{n^2} < \sqrt[n]{1+h} < 1 + \frac{h}{n}$, $-\frac{1}{2} < h < 0$.

★57. $-x^2 < \ln \cos x < -\frac{1}{2}x^2$, $0 < x < \frac{1}{4}\pi$.

★58. $h - \frac{9}{4}h^2 < \frac{\ln(1+h)}{1+h} < h - h^2$, $0 < |h| < 0.1$.

★59. $1 + x + 0.4x^2 < e^{\sin x} < 1 + x + 0.61x^2$, $0 < |x| < 0.1$.

★60. $x^2 \sin \frac{1}{x} \leq x^2$.

413. L'HOSPITAL'S RULE. INTRODUCTION

In the following three sections we consider the most important types of indeterminate forms, with some proofs given in the text, and others deferred to the following Exercises. The principles established in these sections are called upon in many of the problems in curve tracing that follow. Evaluation of indeterminate expressions by means of infinite series is discussed in § 812.

116 DIFFERENTIATION [§ 414

414. THE INDETERMINATE FORM 0/0

The statement that 0/0 is an indeterminate form means that the fact that two functions have the limit 0, as the independent variable approaches some limit, does not in itself imply anything about the limit of their quotient. The four examples x/x, x^2/x, x/x^2, and $\left(x \sin \frac{1}{x}\right)/x$, show that as $x \to 0+$, the quotient of functions, each tending toward zero, may have a limit 1, or 0, or $+\infty$, or it may have no limit at all, finite or infinite. That the involvement of infinity or the apparent division by zero does not in itself constitute an indeterminacy was seen in Exercises 31, 32, 50, § 208. Furthermore, the fact that 0/0 is an indeterminate form, in the sense explained above, certainly does not mean that a quotient of functions, each tending toward zero, cannot have a limit. Indeed, the simple examples above show this, as does any evaluation of a derivative as the limit of the quotient of two increments.

Frequently limits of quotients of functions, each tending toward zero, can be determined by the device known as l'Hospital's Rule. This is stated first in general form. The behavior of the independent variable is then specified in the separate cases. The letter a represents a real number.

Theorem. L'Hospital's Rule. *If $f(x)$ and $g(x)$ are differentiable functions and $g'(x) \neq 0$, for values of x concerned, if $\lim f(x) = \lim g(x) = 0$, and if*

$$\lim \frac{f'(x)}{g'(x)} = L \text{ (finite, } +\infty, -\infty, \text{ or } \infty\text{)},$$

then

$$\lim \frac{f(x)}{g(x)} = L.$$

Case 1. $x \to a+$. (Proof below.)
Case 2. $x \to a-$. (Ex. 31, § 417.)
Case 3. $x \to a$. (Ex. 32, § 417.)
Case 4. $x \to +\infty$. (Proof below.)
Case 5. $x \to -\infty$. (Ex. 33, § 417.)
Case 6. $x \to \infty$. (Ex. 34, § 417.)

Proof of Case 1. Let $f(a)$ and $g(a)$ be defined (or redefined if necessary) to be zero. Then they are both continuous on some closed interval $[a, a + \epsilon]$, where $\epsilon > 0$. The number ϵ can be chosen so small that $g'(x)$ does not vanish in the open interval $(a, a + \epsilon)$ and the conditions of the Generalized Law of the Mean (Theorem IV, § 405) are fulfilled for any x such that $a < x \leq a + \epsilon$. Thus, for any such x there exists a number ξ

§ 415] THE INDETERMINATE FORM ∞/∞

such that $a < \xi < x$ and
$$\frac{f(x)}{g(x)} = \frac{f'(\xi)}{g'(\xi)}.$$

As $x \to a+$, $\xi \to a+$ and the limit of the right-hand member of this equation exists (finite or infinite) by hypothesis. Therefore the limit of the left-hand member of the equation also exists (finite or infinite) and is equal to it.

Proof of Case 4. Use reciprocals:
$$L \equiv \lim_{x \to +\infty} \frac{f'(x)}{g'(x)} = \lim_{t \to 0+} \frac{f'(1/t)}{g'(1/t)} = \lim_{t \to 0+} \frac{-f'(1/t)t^{-2}}{-g'(1/t)t^{-2}}$$

(multiplying and dividing by $-t^{-2}$)
$$= \lim_{t \to 0+} \frac{\frac{d}{dt}f(1/t)}{\frac{d}{dt}g(1/t)} = \lim_{t \to 0+} \frac{f(1/t)}{g(1/t)} = \lim_{x \to +\infty} \frac{f(x)}{g(x)}.$$

The next-to-the-last equality is true by Case 1. The sequence of equalities implies that the limit under consideration exists and is equal to L.

Example 1. $\lim_{x \to 0} \frac{\sin x}{x} = \lim_{x \to 0} \frac{\cos x}{1} = 1$. (Cf. Ex. 16, § 604.)

Example 2.
$$\lim_{x \to 0} \frac{\sin x - x}{x^3} = \lim_{x \to 0} \frac{\cos x - 1}{3x^2} = \lim_{x \to 0} \frac{-\sin x}{6x} = \lim_{x \to 0} \frac{-\cos x}{6} = -\frac{1}{6}.$$

In this case l'Hospital's Rule is iterated. The existence of each limit implies that of the preceding and their equality.

Example 3. $\lim_{x \to +\infty} \frac{e^{-x}}{\frac{1}{x}} = \lim_{x \to +\infty} \frac{-e^{-x}}{-\frac{1}{x^2}} = \lim_{x \to +\infty} \frac{e^{-x}}{\frac{2}{x^3}}$!

Things are getting worse! See Example 1, § 415.

It is important before applying l'Hospital's Rule to check on the indeterminacy of the expression being treated. The following example illustrates this.

Example 4. A routine and thoughtless application of l'Hospital's Rule may yield an incorrect result as follows:
$$\lim_{x \to 1} \frac{2x^2 - x - 1}{x^2 - x} = \lim_{x \to 1} \frac{4x - 1}{2x - 1} = \lim_{x \to 1} \frac{4}{2} = 2.$$

The first equality is correct, and the answer is obtained by direct substitution of 1 for x in the continuous function $(4x - 1)/(2x - 1)$, to give the correct value of 3.

415. THE INDETERMINATE FORM ∞/∞

The symbol ∞/∞ indicates that a limit is being sought for the quotient of two functions, each of which is becoming infinite (the absolute value

118 **DIFFERENTIATION** [§ 415

approaches $+\infty$) as the independent variable approaches some limit. L'Hospital's Rule is again applicable, but the proof is more difficult.

Theorem. L'Hospital's Rule. *If $f(x)$ and $g(x)$ are differentiable functions and $g'(x) \neq 0$, for values of x concerned, if $\lim f(x) = \lim g(x) = \infty$, and if*

$$\lim \frac{f'(x)}{g'(x)} = L \text{ (finite, } +\infty, -\infty, \text{ or } \infty),$$

then

$$\lim \frac{f(x)}{g(x)} = L.$$

Case 1. $x \to a+$. (Ex. 35, § 417.)
Case 2. $x \to a-$. (Ex. 36, § 417.)
Case 3. $x \to a$. (Ex. 37, § 417.)
Case 4. $x \to +\infty$. (Proof below.)
Case 5. $x \to -\infty$. (Ex. 38, § 417.)
Case 6. $x \to \infty$. (Ex. 39. § 417.)

★ *Proof of Case 4.* Observe that whenever x is sufficiently large to prevent the vanishing of $f(x)$ and $g(x)$, and N_1 is sufficiently large to prevent the vanishing of $g'(\xi)$ for $\xi > N_1$, the generalized mean value theorem guarantees the relation

(1) $$\frac{f(x) - f(N_1)}{g(x) - g(N_1)} = \frac{f(x)}{g(x)} \cdot \frac{1 - f(N_1)/f(x)}{1 - g(N_1)/g(x)} = \frac{f'(\xi)}{g'(\xi)},$$

and therefore,

(2) $$\frac{f(x)}{g(x)} = \frac{f'(\xi)}{g'(\xi)} \frac{1 - g(N_1)/g(x)}{1 - f(N_1)/f(x)},$$

for $x > N_1$ and a suitable ξ between x and N_1. First choose N_1 so large that if $\xi > N_1$, then $f'(\xi)/g'(\xi)$ is within a specified degree of approximation of L. (If L is infinite, the term *approximate* should be interpreted liberally, in accordance with the definitions of infinite limits, § 206). Second, using the hypotheses that $\lim_{x \to +\infty} |f(x)| = \lim_{x \to +\infty} |g(x)| = +\infty$, let N_2 be so large that if $x > N_2$, then the fraction $[1 - g(N_1)/g(x)]/[1 - f(N_1)/f(x)]$ is within a specified degree of approximation of the number 1. In combination, by equation (2), these two approximations guarantee that $f(x)/g(x)$ approximates L. This completes the outline of the proof, but for more complete rigor, we present the "epsilon" details for the case where L is finite. (Cf. Ex. 40, § 417.)

Let L be an arbitrary real number and ϵ an arbitrary positive number. We shall show first that there exists a positive number δ such that $|y - L| < \frac{1}{2}\epsilon$ and $|z - 1| < \delta$, imply $|yz - L| < \epsilon$. To do this we use the

§415] THE INDETERMINATE FORM ∞/∞

triangle inequality to write

$$|yz - L| \leq |yz - y| + |y - L| = |y| \cdot |z - 1| + |y - L|.$$

If $|y - L| < \tfrac{1}{2}\epsilon$, $|y| < |L| + \tfrac{1}{2}\epsilon$, so that if $\delta = \tfrac{1}{2}\epsilon(|L| + \tfrac{1}{2}\epsilon)^{-1}$ the assumed inequalities imply

$$|yz - L| < (|L| + \tfrac{1}{2}\epsilon)\frac{\epsilon}{2(|L| + \tfrac{1}{2}\epsilon)} + \frac{\epsilon}{2} = \epsilon,$$

which is the desired result. The rest is simple. First choose N_1 such that

$$\xi > N_1 \text{ implies } \left|\frac{f'(\xi)}{g'(\xi)} - L\right| < \tfrac{1}{2}\epsilon,$$

and second choose $N_2 > N_1$ such that

$$x > N_2 \text{ implies } \left|\frac{1 - g(N_1)/g(x)}{1 - f(N_1)/f(x)} - 1\right| < \delta.$$

Then, by (2),

$$x > N_2 \text{ implies } \left|\frac{f(x)}{g(x)} - L\right| < \epsilon.$$

Example 1. Show that $\lim\limits_{x \to +\infty} \dfrac{x^a}{e^x} = 0$ for any real a.

Solution. If $a \leq 0$ the expression is not indeterminate. Assume $a > 0$. Then $\lim\limits_{x \to +\infty} \dfrac{x^a}{e^x} = \lim\limits_{x \to +\infty} \dfrac{ax^{a-1}}{e^x}$, and if this process is continued, an exponent for x is ultimately found that is zero or negative. This example shows that e^x increases, as $x \to +\infty$, faster than any power of x, and therefore faster than any polynomial.

Example 2. Show that $\lim\limits_{x \to +\infty} \dfrac{\ln x}{x^a} = 0$ for any $a > 0$.

Solution. $\lim\limits_{x \to +\infty} \dfrac{\ln x}{x^a} = \lim\limits_{x \to +\infty} \dfrac{\frac{1}{x}}{ax^{a-1}} = \lim\limits_{x \to +\infty} \dfrac{1}{ax^a} = 0$. (Cf. Ex. 7, § 602.)
In other words, $\ln x$ increases, as $x \to +\infty$, more slowly than any positive power of x.

Example 3. Show that $\lim\limits_{n \to +\infty} \dfrac{e^n}{n!} = 0$.

Solution. This is an indeterminate form to which l'Hospital's Rule does not apply, since $n!$ (unless the Gamma function is used to define $n!$ for all positive real numbers) cannot be differentiated. We can establish the limit as follows: Let n be greater than 3. Then

$$\frac{e^n}{n!} = \left(\frac{e}{1} \cdot \frac{e}{2} \cdot \frac{e}{3}\right)\left(\frac{e}{4} \cdots \frac{e}{n}\right) < \frac{e^3}{6} \cdot \left(\frac{e}{4}\right)^{n-3}.$$

As $n \to +\infty$, the last factor approaches zero.

Example 4. Criticize: $\lim\limits_{x \to +\infty} \dfrac{\sin x}{x} = \lim\limits_{x \to +\infty} \dfrac{\cos x}{1}$, and therefore does not exist!

120   DIFFERENTIATION   [§ 416

Solution. The given expression is not indeterminate, and the hypotheses of l'Hospital's Rule are invalid. Since $|\sin x/x| \leq 1/x$, $x > 0$, the limit is 0.

416. OTHER INDETERMINATE FORMS

In the sense discussed in § 414, the forms
$$0 \cdot \infty, \quad \infty - \infty, \quad 0^0, \quad \infty^0, \quad \text{and} \quad 1^\infty$$
are indeterminate (Ex. 42, § 417.) The first type can often be evaluated by writing the product $f(x) g(x)$ as a quotient and then using l'Hospital's Rule (Example 1). The second type sometimes lends itself to rearrangement, use of identities, or judicious multiplication by unity (Examples 2-3). The remaining three forms are handled by first taking a logarithm: if $y = f(x)^{g(x)}$, then $\ln y = g(x) \ln (f(x))$, and an indeterminacy of the first type above results. Then, by continuity of the exponential function (cf. § 602), $\lim y = \lim (e^{\ln y}) = e^{\lim (\ln y)}$. (Examples 4-6). Finally, other devices, including separation of determinate from indeterminate expressions and substitution of a reciprocal variable, are possible (Examples 7-9).

Example 1. Find $\lim_{x \to 0+} x^a \ln x$.

Solution. If $a \leq 0$, the expression is not indeterminate (cf. Ex. 7, § 602), and the limit is $-\infty$. If $a > 0$, the limit can be written
$$\lim_{x \to 0+} \frac{\ln x}{x^{-a}} = \lim_{x \to 0+} \frac{1/x}{-ax^{-a-1}} = \lim_{x \to 0+} \frac{x^a}{-a} = 0.$$
In other words, whenever the expression $x^a \ln x$ is indeterminate, the algebraic factor "dominates" the logarithmic factor (cf. Example 2, § 415).

Example 2. $\lim_{x \to \frac{\pi}{2}} (\sec x - \tan x) = \lim_{x \to \frac{\pi}{2}} \frac{1 - \sin x}{\cos x} = \lim_{x \to \frac{\pi}{2}} \frac{-\cos x}{-\sin x} = 0.$

Alternatively,
$$\lim_{x \to \frac{\pi}{2}} (\sec x - \tan x) = \lim_{x \to \frac{\pi}{2}} \frac{\sec^2 x - \tan^2 x}{\sec x + \tan x} = \lim_{x \to \frac{\pi}{2}} \frac{1}{\sec x + \tan x} = 0.$$

Example 3. $\lim_{x \to \infty} [\sqrt{x^2 - a^2} - |x|] = \lim_{x \to \infty} \frac{\sqrt{x^2 - a^2} - |x|}{1} \cdot \frac{\sqrt{x^2 - a^2} + |x|}{\sqrt{x^2 - a^2} + |x|}$
$$= \lim_{x \to \infty} \frac{-a^2}{\sqrt{x^2 - a^2} + x} = 0.$$

Example 4. Find $\lim_{x \to 0+} x^x$.

Solution. Let $y = x^x$. Then $\ln y = x \ln x$ and, by Example 1, $\ln y \to 0$. Therefore, by continuity of the function e^x, $y = e^{\ln y} \to e^0 = 1$.

Example 5. Find $\lim_{x \to +\infty} (1 + ax)^{\frac{1}{x}}, a > 0.$

§ 417] EXERCISES

Solution. If $y = (1 + ax)^{\frac{1}{x}}$, $\ln y = \dfrac{\ln(1 + ax)}{x} \to 0$.
Therefore $y = e^{\ln y} \to e^0 = 1$.

Example 6. Show that $\lim\limits_{x \to 0} (1 + ax)^{\frac{1}{x}} = e^a$.

Solution. If $a \neq 0$ and if $y = (1 + ax)^{\frac{1}{x}}$,
$$\lim_{x \to 0} \ln y = \lim_{x \to 0} \frac{\ln(1 + ax)}{x} = \lim_{x \to 0} \frac{a}{1 + ax} = a,$$
and $y = e^{\ln y} \to e^a$. (Cf. Ex. 10, § 602.)

Example 7. Find $\lim\limits_{x \to 0+} x\, e^{\frac{1}{x}}$.

Solution. If this is written $\lim\limits_{x \to 0+} \dfrac{e^{\frac{1}{x}}}{\frac{1}{x}}$, differentiation leads to the answer.
However, the limit can be written $\lim\limits_{t \to +\infty} \dfrac{e^t}{t} = +\infty$.

Example 8. $\lim\limits_{x \to 0} (\csc^2 x - x \csc^3 x) = \lim\limits_{x \to 0} \dfrac{\sin x - x}{x^3} \cdot \dfrac{x^3}{\sin^3 x}$
$$= \left(-\frac{1}{6}\right) \cdot \left(\lim_{x \to 0} \frac{x}{\sin x}\right)^3 = -\frac{1}{6},$$
by Example 2, § 414, and the continuity of the function x^3 (the limit of the cube is the cube of the limit).

Example 9. $\lim\limits_{x \to 1} \dfrac{12 \sin \frac{\pi}{2x} \ln x}{(x^2 + 5)(x - 1)}$
$$= \lim_{x \to 1} \frac{12 \sin \frac{\pi}{2x}}{x^2 + 5} \cdot \lim_{x \to 1} \frac{\ln x}{x - 1} = \frac{12 \cdot 1}{6} \cdot \lim_{x \to 1} \frac{\frac{1}{x}}{1} = 2.$$

417. EXERCISES

In Exercises 1-30, evaluate the limit.

1. $\lim\limits_{x \to 2} \dfrac{3x^2 + x - 14}{x^2 - x - 2}$.
2. $\lim\limits_{x \to -3} \dfrac{x^3 + x + 30}{4x^3 + 11x^2 + 9}$.

3. $\lim\limits_{x \to 1} \dfrac{\ln x}{x^2 + x - 2}$.
4. $\lim\limits_{x \to 1} \dfrac{\cos \frac{1}{2}\pi x}{x - 1}$.

5. $\lim\limits_{x \to 0} \dfrac{\cos x - 1 + \frac{1}{2}x^2}{x^4}$.
6. $\lim\limits_{x \to 0} \dfrac{\ln(1 + x) - x}{\cos x - 1}$.

7. $\lim\limits_{x \to \pi} \dfrac{\sin^2 x}{\tan^2 4x}$.
8. $\lim\limits_{x \to \infty} \dfrac{\sin(1/x)}{\operatorname{Arc\,tan}(1/x)}$.

9. $\lim_{x\to 0} \dfrac{a^x - 1}{b^x - 1}$.

10. $\lim_{x\to 0} \dfrac{\tan x - x}{\operatorname{Arc\,sin} x - x}$.

11. $\lim_{x\to\infty} \dfrac{8x^3 - 5x^2 + 1}{3x^3 + x}$.

12. $\lim_{x\to\frac{1}{2}\pi} \dfrac{\tan x - 6}{\sec x + 5}$.

13. $\lim_{x\to\frac{1}{2}\pi^-} \dfrac{\ln \sin 2x}{\ln \cos x}$.

14. $\lim_{x\to 0+} \dfrac{\ln \sin x}{\ln \sin 2x}$.

15. $\lim_{x\to +\infty} \dfrac{\cosh x}{e^x}$ (cf. § 607).

16. $\lim_{x\to\frac{1}{2}-} \dfrac{\ln(1 - 2x)}{\tan \pi x}$.

17. $\lim_{x\to\infty} \dfrac{(\ln x)^n}{x}$, $n > 0$.

18. $\lim_{x\to\infty} \dfrac{a^x}{x^b}$, $a > 1$, $b > 0$.

19. $\lim_{x\to 0+} x(\ln x)^n$, $n > 0$.

20. $\lim_{x\to a} (x - a) \tan \dfrac{\pi x}{2a}$.

21. $\lim_{x\to\frac{1}{2}\pi} \left[x \tan x - \dfrac{\pi}{2} \sec x \right]$.

22. $\lim_{x\to 1+} \left[\dfrac{x}{x - 1} - \dfrac{1}{\ln x} \right]$.

23. $\lim_{x\to +\infty} (1 + x^2)^{\frac{1}{x}}$.

24. $\lim_{x\to 0} (1 + 2 \sin x)^{\cot x}$.

25. $\lim_{x\to 0+} x^{\frac{1}{\ln x}}$.

26. $\lim_{x\to 0} (x + e^{2x})^{\frac{1}{x}}$.

27. $\lim_{x\to 0+} x^{x^x}(x^{x^x}\text{ or }x^{(x^x)})$.

28. $\lim_{x\to 0} [\ln(1 + x)]^x$.

29. $\lim_{x\to 0} (\cos 2x)^{\frac{1}{x^2}}$.

30. $\lim_{x\to 0+} \left[\dfrac{\ln x}{(1 + x)^2} - \ln \dfrac{x}{1 + x} \right]$.

31. Prove Case 2 of l'Hospital's Rule, § 414.

32. Prove Case 3 of l'Hospital's Rule, § 414. *Hint:* Make sensible use of Cases 1 and 2.

33. Prove Case 5 of l'Hospital's Rule, § 414.

34. Prove Case 6 of l'Hospital's Rule, § 414.

35. Prove Case 1 of l'Hospital's Rule, § 415. *Hint:* Apply Case 4 (cf. proof of Case 4, § 414).

★**36.** Prove Case 2 of l'Hospital's Rule, § 415.

★**37.** Prove Case 3 of l'Hospital's Rule, § 415.

★**38.** Prove Case 5 of l'Hospital's Rule, § 415.

★**39.** Prove Case 6 of l'Hospital's Rule, § 415.

★**40.** For Case 4 of l'Hospital's Rule, § 415, supply the "epsilon" details for the case $L = +\infty$.

41. Prove that the forms $(+0)^{+\infty}$, $(+\infty)^{+\infty}$, and $a^{+\infty}$ (where $a > 0$ and $a \neq 1$) are determinate. What can you say about $(+0)^{-\infty}$? $(+\infty)^{-\infty}$? $(+0)^\infty$? $(+\infty)^\infty$?

42. Show by examples that all of the forms of § 416 are indeterminate, as stated.

43. Criticize the following alleged "proof" of l'Hospital's Rule for the form $0/0$ as $x \to a+$: By the Law of the Mean, for any $x > a$, there exist ξ_1 and ξ_2 between a and x such that $f(x) - f(a+) = f'(\xi_1)$ and $g(x) - g(a+) = g'(\xi_2)$.

§ 418] CURVE TRACING

Therefore

$$\frac{f(x)}{g(x)} = \frac{f(x) - f(a+)}{g(x) - g(a+)} = \frac{f'(\xi_1)}{g'(\xi_1)} \to \frac{f'(a+)}{g'(a+)} = \lim_{x \to a+} \frac{f'(x)}{g'(x)}.$$

418. CURVE TRACING

It is not our purpose in this section to give an extensive treatment of curve tracing. Rather, we wish to give the reader an opportunity to review in practice such topics from differential calculus as increasing and decreasing functions, maximum and minimum points, symmetry, concavity, and points of inflection. Certain basic principles we do wish to recall explicitly, however.

(*i*) *Composition of ordinates.* The graph of a function represented as the sum of terms can often be obtained most simply by graphing the separate terms, and adding the ordinates visually, as indicated in Figure 412.

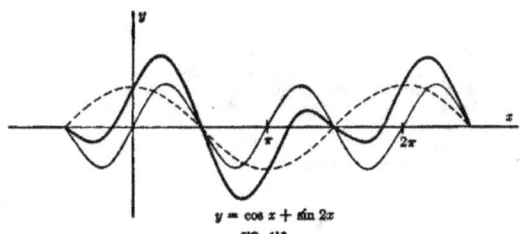

$y = \cos x + \sin 2x$
FIG. 412

(*ii*) *Dominant terms.* If different terms dominate an expression for different values of the independent variable, the general shape of the curve can often be inferred. For example, for positive values of x, the function $x + \frac{1}{x}$ is dominated by the second term if x is small and by the first term if x is large (Fig. 413).

(*iii*) *Vertical and horizontal asymptotes.* A function represented as a quotient $f(x)/g(x)$ of continuous functions has a vertical asymptote at a point a where $g(a) = 0$ and $f(a) \neq 0$. If $\lim f(x)/g(x)$, as x becomes infinite $(+\infty, -\infty, \text{ or } \infty)$, exists and is a finite number b, then $y = b$ is a horizontal asymptote. (Fig. 414.)

(*iv*) *Other asymptotes.* If $f(x) - mx - b$ approaches zero as x becomes infinite $(+\infty, -\infty, \text{ or } \infty)$, then the line $y = mx + b$ is an asymptote for

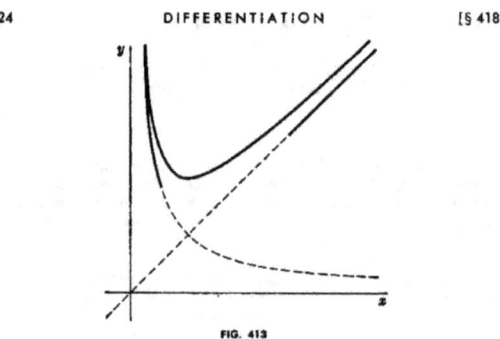

FIG. 413

the graph of $f(x)$. For the function $x + e^x$, for example, the line $y = x$ is an asymptote as $x \to -\infty$. (Fig. 415.)

(v) *Two factors.* Certain principles used for functions represented as sums have their applications to functions represented as products. The functions $e^{-ax} \sin bx$ and $e^{-ax} \cos bx$, useful in electrical theory, are good examples. (Fig. 416.)

Vanishing factors often determine the general shape of a curve in neighborhoods of points where they vanish. For example, the graph of

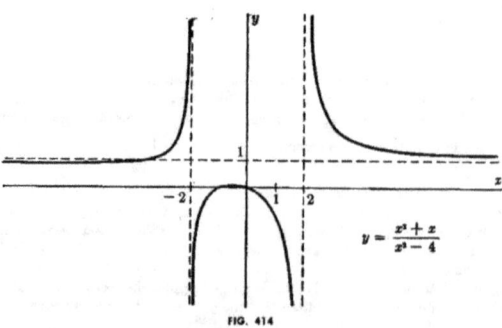

$$y = \frac{x^2 + x}{x^2 - 4}$$

FIG. 414

§ 418] CURVE TRACING 125

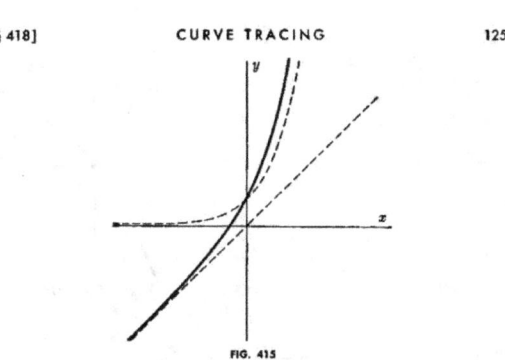

FIG. 415

$y^2 = x^2(2 - x)$ is approximated by that of $y^2 = 2x^2$ for x near 0 and by that of $y^2 = 4(2 - x)$ for x near 2. (Fig. 417.)

A further aid in graphing an equation like that of Figure 417, of the form $y^2 = f(x)$, is graphing the function $f(x)$ itself to determine the values of x for which $f(x)$ is positive, zero, or negative, and hence for which y is double-valued, zero, or imaginary. (Fig. 418.)

(vi) *Parametric equations.* If $x = f(t)$ and $y = g(t)$, we recall two formulas:

(1) $$y' = \frac{dy}{dx} = \frac{g'(t)}{f'(t)};$$

(2) $$y'' = \frac{d^2y}{dx^2} = \frac{\frac{dy'}{dt}}{f'(t)}.$$

The folium of Descartes,

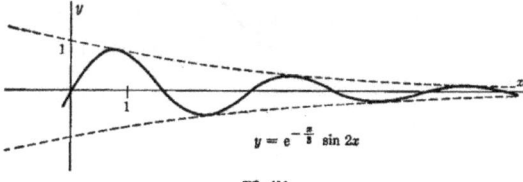

$y = e^{-\frac{x}{3}} \sin 2x$

FIG. 416

126 DIFFERENTIATION [§ 418

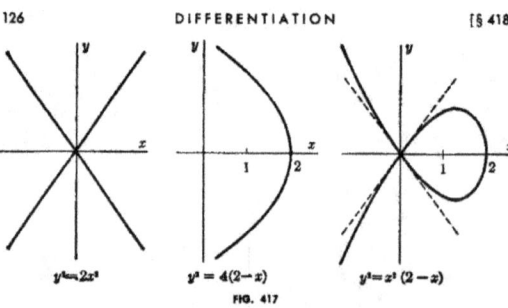

FIG. 417

(3) $$x = \frac{3at}{t^3 + 1}, \quad y = \frac{3at^2}{t^3 + 1},$$

is illustrated in Figure 419. Since

(4) $$\frac{dy}{dx} = \frac{t(t^3 - 2)}{2t^3 - 1},$$

horizontal tangents correspond to the values of t for which the numerator of (4) vanishes: $t = 0$ (the point $(0, 0)$) and $t = \sqrt[3]{2}$ (the point $(a\sqrt[3]{2}, a\sqrt[3]{4})$). Vertical tangents correspond to the values of t for which (4) becomes infinite: $t = \infty$ (the point $(0, 0)$) and $t = \sqrt[3]{\frac{1}{2}}$ (the point $(a\sqrt[3]{4}, a\sqrt[3]{2})$).

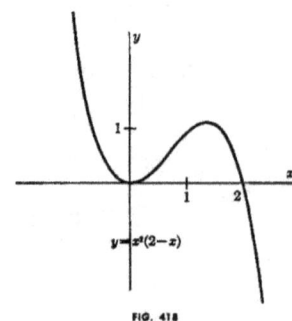

FIG. 418

§ 419] EXERCISES 127

Since $x + y = 3at/(t^2 - t + 1)$, the line $x + y + a = 0$ is an asymptote (let $t \to -1$).

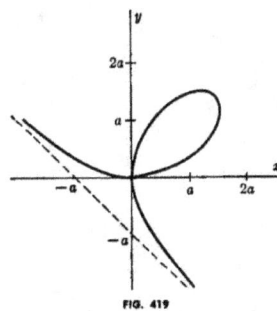

FIG. 419

419. EXERCISES

In Exercises 1-50, graph the equation, showing essential shape and asymptotic behavior.

1. $y = x^2(x^2 - 9)$.
2. $y = \dfrac{2}{x^2 + 1}$.
3. $y = \dfrac{x^2 - 4}{x^2 - 9}$.
4. $y = \dfrac{x^2 - 9}{x^2 - 4}$.
5. $y = x + \dfrac{4}{x}$.
6. $y = x^2 + \dfrac{2}{x}$.
7. $y = x + \dfrac{4}{x^2}$.
8. $y^2 = x^2(x^2 - 9)$.
9. $y^2 = \dfrac{x^2 - 4}{x^2 - 9}$.
10. $y^2 = \dfrac{x^2 - 9}{x^2 - 4}$.
11. $y^2 = (x - 1)(x - 2)^2(x - 3)^3$.
12. $y^2 = (x - 1)^3(x - 2)(x - 3)^2$.
13. $x^2 + xy + y^2 = 4$.
14. $x^2 + 4xy + y^2 = 4$.
15. $y = e^{-x^2}$.
16. $y = e^{1/x}$.
17. $y = xe^x$.
18. $y = x^2 e^{-x}$.
19. $y = xe^{-x^2}$.
20. $y = e^x - x$.
21. $y = \ln |x|$.
22. $y = x \ln x$.
23. $y = \dfrac{\ln x}{x}$.
24. $y = \dfrac{x}{\ln x}$.
25. $y = \dfrac{1}{x \ln x}$.
26. $y = x^2 \ln x$.

27. $y = x + \sin x$.
28. $y = \ln x \cdot e^{-x}$.
29. $y = \ln(1 + x^2)$.
30. $y = \ln \sin x$.
31. $y = e^{-x} \cos 2x$.
32. $y = \frac{g}{k}(1 - e^{-kx})$.
33. The Witch of Agnesi, $x^2 y = 4a^2(2a - y)$.
34. The Cissoid of Diocles, $y^2(2a - x) = x^3$.
35. The Catenary, $y = \frac{1}{2}a(e^{x/a} + e^{-x/a})$.
36. The Folium of Descartes, $x^3 + y^3 = 3axy$.
37. The Parabola, $\pm x^{\frac{1}{2}} \pm y^{\frac{1}{2}} = a^{\frac{1}{2}}$.
38. The Hypocycloid, $x^{\frac{2}{3}} + y^{\frac{2}{3}} = a^{\frac{2}{3}}$.
39. The Lemniscate of Bernoulli, $(x^2 + y^2)^2 = a^2(x^2 - y^2)$.
40. The Ovals of Cassini, $(x^2 + y^2 + a^2)^2 - 4a^2 x^2 = c^4$.
41. $x = \frac{t^2 - 1}{t^2 + 1}, \; y = \frac{2t}{t^2 + 1}$.
42. $x = t + \ln t, \; y = t + e^t$.
43. The Ellipse, $x = a \cos t, \; y = b \sin t$.
44. The Cycloid, $x = a(t - \sin t)$, $y = a(1 - \cos t)$.
45. The Hypocycloid, $x = a \cos^3 t, \; y = a \sin^3 t$.
46. The Cardioid, $x = a(2 \cos t - \cos 2t)$, $y = a(2 \sin t - \sin 2t)$.
47. The Serpentine, $x = a \cot t, \; y = b \sin t \cos t$.
48. The Witch of Agnesi, $x = a \cot t, \; y = a \sin^2 t$.
49. The Hypocycloid of Three Cusps, $x = 2a \cos t + a \cos 2t$, $y = 2a \sin t - a \sin 2t$.
50. The Hyperbolic Spiral, $x = \frac{a}{t} \cos t, \; y = \frac{a}{t} \sin t$.

51. Graph $f(x) = \frac{1}{1 + e^{-1/x}}$, and show that $f(0+) = 1$ and $f(0-) = f'(0+) = f'(0-) = 0$.

52. Graph $f(x) = e^{-1/x^2}$, $x \neq 0$, $f(0) = 0$. Prove that $\lim_{x \to 0} x^{-n} f(x) = 0$ for every positive integer n, and hence show that $f(x)$ has everywhere continuous derivatives of all orders, all of which vanish at $x = 0$. (Cf. Example 2, § 805, for a use of this function as a counterexample.)

★53. Show by a graph that a function $f(x)$ exists having the following properties: (i) $f(x)$ is strictly increasing, (ii) $f'(x)$ exists for every real x, (iii) $f(x)$ is bounded above, (iv) the statement $\lim_{x \to +\infty} f'(x) = 0$ is false.

★54. Show that the function $f(x) = x/(1 + e^{1/x})$, $x \neq 0$, $f(0) = 0$, is everywhere continuous, and has unequal right-hand and left-hand derivatives at $x = 0$.

★55. Graph $(y - x^2)^2 = x^5$, with particular attention to a neighborhood of the origin.

★420. WITHOUT LOSS OF GENERALITY

One of the standard techniques of an analytic proof is to reduce a general proposition to a special case "without loss of generality." This means that it is possible to construct a proof of the general theorem on the assumption that the special form of that theorem is true. Following the establishment of this inference, proof of the special case implies proof of the general proposition. This device was used in two proofs given in § 204: in the proof that the limit of the product of two sequences is the product of their limits (Theorem VIII) we saw that it could be assumed without loss of generality that one of the limits is zero; in the proof that the limit of the quotient of two sequences is the quotient of their limits (Theorem IX) we assumed without loss of generality that the numerators were identically equal to unity. Another instance of the same principle is the proof of the Law of the Mean (§ 405) by reducing it to the special case called Rolle's Theorem.

In the following Exercises are a few problems in showing that the special implies the general.

★421. EXERCISES

In Exercises 1–10, prove the stated proposition on the basis of the assumption given in the braces { }.

★1. If m and n are relatively prime positive integers, there exist positive integers x and y such that $mx - ny = 1$. {If m and n are relatively prime positive integers, there exist positive integers x and y such that $|mx - ny| = 1$.}

★2. If m and n are nonzero integers and if d is their greatest common divisor, then there exist integers x and y such that $mx + ny = d$. {If m and n are relatively prime positive integers, there exist integers x and y such that $mx + ny = 1$.}

★3. If $I_1, I_2, \cdots I_n$ are open intervals and if I is the set of all x such that x is a member of every I_k, $k = 1, 2, \cdots, n$, then I is either empty or an open interval. {$n = 2$.}

★4. The Schwarz inequality (Ex. 43, § 107). {All of the numbers $a_1, \cdots, a_n, b_1, \cdots, b_n$ are positive.}

★5. $\frac{\sin px}{x} < p$, for $p > 0$ and $x > 0$. {$p = 1$.}

★6. $\lim_{x \to 0+} x^p \ln x = 0$, $p > 0$. {$p = 1$.}

★7. Any bounded sequence $\{a_n\}$ that does not converge to a contains a subsequence $\{a_{n_k}\}$ converging to some $b \neq a$. {$|a_n - a| \geq \epsilon > 0$ for all n.}

★8. The trigonometric functions are continuous where they are defined. {$\sin x$ and $\cos x$ are continuous at $x = 0$.}

★9. If $f'(x) \leq g'(x)$ for $a \leq x \leq b$, then $f(x) - f(a) \leq g(x) - g(a)$. {$f(x) = 0$ for $a \leq x \leq b$.}

★11. Show that in proving that if $f(x)$ and $f'(x)$ are defined in a neighborhood of $x = a$ and if $f''(a)$ exists, then

(2) $$f''(a) = \lim_{h \to 0} \frac{f(a + h) + f(a - h) - 2f(a)}{h^2},$$

it may be assumed without loss of generality, that $a = 0$, $f(0) = 0$, $f'(0) = 0$, and $f''(0) = 0$. Hence prove (2). *Hints:* To show that one may assume $f(0) = 0$, define a new function $g(x) = f(x) - f(0)$. To show that one may assume $f'(0) = 0$, define a new function $g(x) = f(x) - f'(0) \cdot x$. (Cf. Ex. 36, § 412.)

5

Integration

501. THE DEFINITE INTEGRAL

It will be assumed that the reader is already familiar with some of the properties and many of the applications of the definite integral. It is our purpose in this section to give a precise definition and a few of the simpler properties of the integral, with analytical proofs that do not depend on the persuasion of a geometrical picture.

We shall be dealing in the main with a fixed closed interval $[a, b]$. On such an interval a finite set of points $a = a_0 < a_1 < a_2 < \cdots < a_n = b$ is called a **net**, and denoted $\mathfrak{N}$. (Cf. Fig. 501.) The closed intervals $[a_{i-1}, a_i], i = 1, 2, \cdots, n$, are called the **subintervals** of $[a, b]$ for the net $\mathfrak{N}$, and their lengths are denoted $\Delta x_i = a_i - a_{i-1}, i = 1, 2, \cdots, n$. The maximum length of the subintervals is called the **norm** of the net $\mathfrak{N}$ and denoted $|\mathfrak{N}| : |\mathfrak{N}| = \max \Delta x_i, i = 1, 2, \cdots, n$.

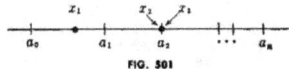

FIG. 501

Let $f(x)$ be defined over the closed interval $[a, b]$, and (for a net $\mathfrak{N}$) let x_i be an arbitrary point of the ith subinterval ($a_{i-1} \leq x_i \leq a_i$), $i = 1, 2, \cdots, n$ (Fig. 501). Consider the sum

(1) $$\sum_{i=1}^{n} f(x_i) \Delta x_i = f(x_1) \Delta x_1 + \cdots + f(x_n) \Delta x_n.$$

The definite integral of the function $f(x)$ is defined as the limit of sums of the form (1). We shall first explain what is meant by such a limit.

Definition I. *The limit statement*

(2) $$\lim_{|\mathfrak{N}| \to 0} \sum_{i=1}^{n} f(x_i) \Delta x_i = I,$$

where I is a (finite) number, means that corresponding to $\epsilon > 0$ there exists $\delta > 0$ such that for any net $\mathfrak{N}$ of norm less than δ and any choice of points x_i such that $a_{i-1} \leq x_i \leq a_i$, $i = 1, 2, \cdots, n$, the inequality

(3) $$\left| \sum_{i=1}^{n} f(x_i) \Delta x_i - I \right| < \epsilon$$

holds.

NOTE 1. Closely though Definition I may resemble the definition of the limit of a single-valued function of a real variable as the independent variable approaches 0, the type of limit just introduced should be recognized as a new concept. Although, for a given net $\mathfrak{N}$ and points $x_1, x_2, \cdots, x_n$, the sum (1) is uniquely determined, the limit (2) is formed with respect to the norm alone as the independent variable. For a given positive number $p < b - a$ there are infinitely many nets $\mathfrak{N}$ of norm $|\mathfrak{N}|$ equal to p, and for each such net there are infinitely many choices of the points $x_1, x_2, \cdots, x_n$. In other words, as a function of the independent variable $|\mathfrak{N}|$, the sum (1) is an infinitely many valued function,† and it is the limit of such a function that is prescribed in Definition I. It should not be forgotten, however, that each sum (1) appearing in the inequality (3) is simply a number obtained by adding together a finite number of terms.

Definition II. *A function $f(x)$, defined on $[a, b]$, is **integrable**‡ there if and only if the limit (2) exists (and is finite). In case the limit exists it is called the **definite integral**‡ of the function, and denoted*

(4) $$\int_a^b f(x)\, dx \equiv \lim_{|\mathfrak{N}| \to 0} \sum_{i=1}^{n} f(x_i) \Delta x_i.$$

Definition III. *If $b < a$,*

(5) $$\int_a^b f(x)\, dx \equiv -\int_b^a f(x)\, dx,$$

in case the latter integral exists. Furthermore,

(6) $$\int_a^a f(x)\, dx \equiv 0.$$

Certain questions naturally come to mind. Does the limit (2) always

---

† The expression "infinitely many valued" should be interpreted here to mean "possibly infinitely many valued" since for any constant function $f(x)$, the sums (1) can have only one value (cf. Theorem VII). It can be shown that for any function $f(x)$ that is not constant on $[a, b]$, and for any positive number δ less than half the length of the interval $[a, b]$, the sum (1), as a function of $|\mathfrak{N}|$, is strictly infinitely many valued for the particular value $|\mathfrak{N}| = \delta$.

‡ The terms *Riemann integrable* and *Riemann integral* are also used, especially if it is important to distinguish the integral defined in this section from some other type, such as the Riemann-Stieltjes integral (§ 517) or the Lebesgue integral (not discussed in this book). When a definite integral is to be distinguished from an improper integral (§§ 511-515), it is customary to call it a *proper integral*.

exist? If it does not always exist, under what circumstances does it exist? When it does exist is it unique?

Let us remark first that the limit (2) *never* exists for unbounded functions. In other words, *every integrable function is bounded.* To see this, let $\mathfrak{N}$ be an arbitrary net and let $f(x)$ be unbounded in the kth subinterval $[a_{k-1}, a_k]$. Then whatever may be the choice of points x_i for $i \neq k$, the point x_k can be picked so that the sum (1) is numerically larger than any preassigned quantity.

On the other hand, not all bounded functions are integrable. For example, the function of Example 6, § 201, which is 1 on the rational numbers from 0 to 1 and 0 on the irrational numbers from 0 to 1, is not integrable on [0, 1]. For, no matter how small the norm of a net $\mathfrak{N}$ may be, every subinterval contains both rational and irrational points ((v), § 112, and Ex. 2, § 113) and the sums (1) can be made to have either the extreme value 1 (if every point x_i is chosen to be rational) or the extreme value 0 (if every point x_i is chosen to be irrational). The limit (2), then, cannot exist for this function (let $\epsilon \equiv \frac{1}{2}$).

One answer to the question of integrability lies in the concept of *continuity.* The function just considered, which is not integrable, is *nowhere* continuous. At the opposite extreme is a function which is *everywhere* continuous (on a closed interval). We shall see that such a function is *always* integrable (Theorem VIII). Between these two extremes are bounded functions which are somewhere but not everywhere continuous. It will be seen that such functions are certainly integrable if they have only *finitely* many discontinuities (Theorem IX), that they *may* be integrable even with *infinitely* many discontinuities (Example 2, § 502), and that a *criterion* for integrability lies in the intriguing concept of *continuity almost everywhere* (Ex. 54, § 503).

We proceed now to the establishment of some of the simpler properties of the definite integral.

Theorem I. *If* $\lim\limits_{|\mathfrak{N}|\to 0} \sum\limits_{i=1}^{n} f(x_i)\,\Delta x_i$ *exists, the limit is unique.*

Proof. Assume that

$$\lim_{|\mathfrak{N}|\to 0} \sum_{i=1}^{n} f(x_i)\,\Delta x_i = I \text{ and } \lim_{|\mathfrak{N}|\to 0} \sum_{i=1}^{n} f(x_i)\,\Delta x_i = J,$$

where $I > J$, and let $\epsilon \equiv \frac{1}{2}(I - J)$. Then there exists a positive number δ so small that for any net $\mathfrak{N}$ of norm less than δ, and for any choice of points $x_1, x_2, \cdots, x_n$, the following inequalities hold simultaneously:

$$I - \epsilon < \sum_{i=1}^{n} f(x_i)\,\Delta x_i < J + \epsilon.$$

But this implies $I - \epsilon < J + \epsilon$, or $\epsilon > \frac{1}{2}(I - J)$. (Contradiction.)

Theorem II. *If $f(x)$ and $g(x)$ are integrable on $[a,b]$, and if $f(x) \leq g(x)$ there, then*

$$\int_a^b f(x)\, dx \leq \int_a^b g(x)\, dx.$$

Proof. Let $I \equiv \int_a^b f(x)\, dx$ and $J \equiv \int_a^b g(x)\, dx$, assume $I > J$, and let $\epsilon \equiv \frac{1}{2}(I - J)$. Then there exists a positive number δ so small that for any net $\mathfrak{N}$ of norm less than δ and any choice of points $x_1, x_2, \cdots, x_n$, the following inequalities hold simultaneously:

$$\sum_{i=1}^n g(x_i)\, \Delta x_i < J + \epsilon = I - \epsilon < \sum_{i=1}^n f(x_i)\, \Delta x_i.$$

But this implies an inequality inconsistent with the assumed inequality $f(x) \leq g(x)$.

Theorem III. *If $f(x)$ is integrable on $[a, b]$ and if k is a constant, then $k\,f(x)$ is integrable on $[a, b]$ and*

(7) $$\int_a^b k\,f(x)\, dx = k \int_a^b f(x)\, dx.$$

Proof. If $k = 0$, the proof is trivial. If $k \neq 0$, and if $\epsilon > 0$ is given, let $\delta > 0$ be such that $|\mathfrak{N}| < \delta$ implies $\left|\sum_{i=1}^n f(x_i)\, \Delta x_i - I\right| < \epsilon/|k|$, and therefore

$$\left|\sum_{i=1}^n k\,f(x_i)\, \Delta x_i - k\,I\right| < |k|\epsilon/|k| = \epsilon.$$

Theorem IV. *If $f(x)$ and $g(x)$ are integrable on $[a, b]$, then so are their sum and difference,† and*

(8) $$\int_a^b [f(x) \pm g(x)]\, dx = \int_a^b f(x)\, dx \pm \int_a^b g(x)\, dx.$$

Proof. Let $I \equiv \int_a^b f(x)\, dx$ and $J \equiv \int_a^b g(x)\, dx$, and if $\epsilon > 0$ is given, let $\delta > 0$ be such that $|\mathfrak{N}| < \delta$ implies simultaneously the inequalities

$$\left|\sum_{i=1}^n f(x_i)\, \Delta x_i - I\right| < \tfrac{1}{2}\epsilon, \quad \left|\sum_{i=1}^n g(x_i)\, \Delta x_i - J\right| < \tfrac{1}{2}\epsilon,$$

and hence, by the triangle inequality,

$$\left|\sum_{i=1}^n [f(x_i) \pm g(x_i)]\, \Delta x_i - (I \pm J)\right|$$

$$\leq \left|\sum_{i=1}^n f(x_i)\, \Delta x_i - I\right| + \left|\sum_{i=1}^n g(x_i)\, \Delta x_i - J\right| < \epsilon.$$

† For products and quotients, see Exercises 25 and 27, § 502.

§ 501] THE DEFINITE INTEGRAL

Theorem V. *If $a < b < c$, and if $f(x)$ is integrable on the intervals $[a, b]$ and $[b, c]$, then it is integrable on the interval $[a, c]$ and*

(9) $$\int_a^c f(x)\, dx = \int_a^b f(x)\, dx + \int_b^c f(x)\, dx.$$

Proof. Let $|f(x)| < K$ for $a \leq x \leq c$, let

$$I \equiv \int_a^b f(x)\, dx \text{ and } J \equiv \int_b^c f(x)\, dx,$$

and let $\epsilon > 0$ be given. Let δ be a positive number less than $\epsilon/4K$ and so small that for any net on $[a, b]$ or $[b, c]$ with norm less than δ any sum of the form (1) differs from I or J, respectively, by less than $\frac{1}{4}\epsilon$. We shall show that for $[a, c]$, $|\mathfrak{N}| < \delta$ implies $\left|\sum_{i=1}^n f(x_i)\, \Delta x_i - (I + J)\right| < \epsilon$. Accordingly, for such a net $\mathfrak{N}$ let $a_{k-1} < b \leq a_k$ (the kth subinterval is the first containing the point b), and write the sum (1) in the form

$$S \equiv \sum_{i=1}^{k-1} f(x_i)\, \Delta x_i + f(x_k)\, \Delta x_k + \sum_{i=k+1}^n f(x_i)\, \Delta x_i.$$

The following sum,

$$S' \equiv \left[\sum_{i=1}^{k-1} f(x_i)\, \Delta x_i + f(b)(b - a_{k-1})\right] + \left[f(b)(a_k - b) + \sum_{i=k+1}^n f(x_i)\, \Delta x_i\right],$$

can be considered as made up of two parts, which approximate I and J, each by less than $\frac{1}{4}\epsilon$. Thus $|S' - (I + J)| < \frac{1}{2}\epsilon$. On the other hand, $|S - S'| = |f(x_k) - f(b)|\, \Delta x_k < 2K(\epsilon/4K) = \frac{1}{2}\epsilon$. Therefore,

$$|S - (I + J)| \leq |S - S'| + |S' - (I + J)| < \tfrac{1}{2}\epsilon + \tfrac{1}{2}\epsilon = \epsilon.$$

NOTE 2. By virtue of Definition III, the relation (9) is universally true whenever the three integrals exist, whatever may be the order relation between the numbers a, b, and c. For example, if $c < a < b$, then

$$\int_c^b f(x)\, dx = \int_c^a f(x)\, dx + \int_a^b f(x)\, dx.$$

Hence

$$\int_a^c f(x)\, dx = -\int_c^a f(x)\, dx = \int_c^b f(x)\, dx - \int_c^a f(x)\, dx = \int_a^b f(x)\, dx + \int_b^c f(x)\, dx.$$

By mathematical induction, the relation (9) can be extended to an arbitrary number of terms:

(10) $$\int_{a_0}^{a_n} f(x)\, dx = \sum_{i=1}^n \int_{a_{i-1}}^{a_i} f(x)\, dx,$$

where $a_0, a_1, \cdots, a_n$ are any $n + 1$ real numbers, and where every integral of (10) is assumed to exist. (The student should satisfy himself regarding (9), by considering other order relations between a, b, and c, including possible equalities of some of these letters, and he should give the details of the proof of (10). (Cf. Ex. 1, § 503.)

Theorem VI. *If the values of a function defined on a closed interval are changed at a finite number of points of the interval, neither the integrability nor the value of the integral is affected.*

Proof. Thanks to mathematical induction, the proof can (and will) be reduced to showing the following: If $f(x)$ is integrable on $[a, b]$, with integral I, and if $g(x)$ is defined on $[a, b]$ and equal to $f(x)$ at every point of $[a, b]$ except for one point c, then $g(x)$ is integrable on $[a, b]$ with integral I. For any net $\mathfrak{R}$, the terms of the sum $\sum_{i=1}^{n} g(x_i) \Delta x_i$ must be identical with the terms of the sum $\sum_{i=1}^{n} f(x_i) \Delta x_i$, with the exception of at most two terms (in case $x_{i-1} = x_i = c$ for some i). Therefore

$$\left| \sum_{i=1}^{n} g(x_i) \Delta x_i - \sum_{i=1}^{n} f(x_i) \Delta x_i \right| \leq 2(|f(c)| + |g(c)|) \cdot |\mathfrak{R}|.$$

Thus, for a given $\epsilon > 0$, let δ be a positive number less than

$$\epsilon/4(|f(c)| + |g(c)|)$$

and so small that $|\mathfrak{R}| < \delta$ implies $\left| \sum_{i=1}^{n} f(x_i) \Delta x_i - I \right| < \frac{1}{2}\epsilon$. Then $|\mathfrak{R}| < \delta$ implies

$$\left| \sum_{i=1}^{n} g(x_i) \Delta x_i - I \right| \leq \left| \sum_{i=1}^{n} g(x_i) \Delta x_i - \sum_{i=1}^{n} f(x_i) \Delta x_i \right| + \left| \sum_{i=1}^{n} f(x_i) \Delta x_i - I \right| < \frac{1}{2}\epsilon + \frac{1}{2}\epsilon = \epsilon.$$

NOTE 3. Theorem VI makes it possible to define integrability and integral for a function which is defined on a closed interval except for a finite number of points. This is done by assigning values to the function at the exceptional points in any manner whatsoever. Theorem VI assures us that the result of applying Definition II is independent of the values assigned. Since the assignment of values does not affect the value of the integral, where it exists, we shall assume that the definition is extended to include such functions even though they remain undefined at the exceptional points.

Theorem VII. *If $f(x)$ is constant, $f(x) \equiv k$, on the interval $[a, b]$, then $f(x)$ is integrable there and*

$$\int_a^b f(x) \, dx = k(b - a).$$

Proof. For any net $\mathfrak{R}$, $\sum_{i=1}^{n} f(x_i) \Delta x_i = k \sum_{i=1}^{n} \Delta x_i = k(b - a)$.

For the sake of convenience and accessibility we state now the three best-known sufficient conditions for integrability (the first being a special case of the second). The proofs are given in § 502 and Exercise 33, § 503.

§ 501] THE DEFINITE INTEGRAL

Theorem VIII. *A function continuous on a closed interval is integrable there.*

Theorem IX. *A function defined and bounded on a closed interval and continuous there except for a finite number of points is integrable there.*

Theorem X. *A function defined and monotonic on a closed interval is integrable there.*

Example. Prove that $\int_0^b x\, dx = \tfrac{1}{2}b^2$ if $b > 0$, and, more generally, that
$$\int_a^b x\, dx = \tfrac{1}{2}(b^2 - a^2),$$
where a and b are any real numbers.

Solution. Let us first observe that since the function $f(x) \equiv x$ is everywhere continuous the integrals exist. The problem is one of *evaluation*. To evaluate $\int_0^b x\, dx$, where $b > 0$, we form a particular simple net by means of the points
$$a_0 = 0,\; a_1 = b/n,\; \cdots,\; a_i = ib/n,\; \cdots,\; a_n = nb/n = b,$$
and choose $x_i = a_i$, $i = 1, 2, \cdots, n$. The sum $\sum_{i=1}^{n} f(x_i)\,\Delta x_i$ becomes
$$\left[\frac{b}{n} + \frac{2b}{n} + \cdots + \frac{nb}{n}\right] \cdot \frac{b}{n} = \frac{b^2}{n^2}[1 + 2 + \cdots + n] = \frac{b^2}{n^2} \cdot \frac{n(n+1)}{2},$$
the last equality having been obtained in Exercise 12, § 107. Therefore $\int_0^b x\, dx = \lim_{n \to +\infty} \tfrac{1}{2}b^2 \frac{n+1}{n} = \tfrac{1}{2}b^2$, as stated. It is left to the student to show, first, that $\int_0^b x\, dx = \tfrac{1}{2}b^2$ if $b \leq 0$ (cf. Exs. 7-8, § 503) and then, by using formula (9) of Theorem V and formula (5) of Definition III, that $\int_a^b x\, dx = \tfrac{1}{2}(b^2 - a^2)$.

Other evaluations of this type are given in Exercises 15-21, § 503.

In conclusion we present a useful basic theorem (give the proof in Ex. 5, § 503).

Theorem XI. First Mean Value Theorem for Integrals. *If $f(x)$ is continuous on $[a, b]$ there exists a point ξ such that $a < \xi < b$ and*

(11) $$\int_a^b f(x)\, dx = f(\xi) \cdot (b - a).$$

For other mean value theorems for integrals, see Exercise 6, § 503, Exercise 14, § 506, and Exercises 27-29, § 518. Also see Exercise 9, § 506.

At this point, the student may wish to do a few of the Exercises of § 503. Some that are suitable for performance now are Exercises 1-30.

138   INTEGRATION   [§ 502

★502. MORE INTEGRATION THEOREMS

Definition. A *step-function* is a function, defined on a closed interval $[a, b]$, that is constant in the interior of each subinterval of some net on $[a, b]$. (Fig. 502.)

Notation

Step-function: $\sigma(x)$ or $\tau(x)$.
Net: $\mathfrak{N} : a = \alpha_0 < \alpha_1 < \cdots < \alpha_m = b$.
Constant values: σ_i or τ_i, $i = 1, \cdots, m$.

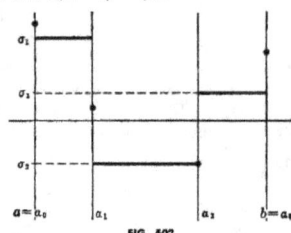

FIG. 502

Theorem I. *Any step-function is integrable and, with the preceding notation,* $\int_a^b \sigma(x)\,dx = \sum_{i=1}^{m} \sigma_i(\alpha_i - \alpha_{i-1})$.

Proof. We need prove this only for the case $m = 1$, since Theorem V, § 501, and mathematical induction will extend this special case to the general result. Accordingly, let $\sigma(x)$ be identically equal to a constant k for $a < x < b$, and redefine $\sigma(x)$, if necessary, at the end-points of the interval so that $\sigma(x) \equiv k$ for $a \leq x \leq b$. Since the new function is integrable with integral $k(b - a)$ (Theorem VII, § 501), the original step-function is also integrable with the same integral (Theorem VI, § 501), and the proof is complete.

The purpose of introducing step-functions in our discussion of the definite integral is to make use of the fact established in the following theorem that a function is integrable if and only if it can be appropriately "squeezed" between two step-functions. (Cf. Fig. 503.)

Theorem II. *A function $f(x)$, defined on a closed interval $[a, b]$, is integrable there if and only if, corresponding to an arbitrary positive number ϵ, there exist step-functions $\sigma(x)$ and $\tau(x)$ such that*

(1) $\qquad\qquad\qquad \sigma(x) \leq f(x) \leq \tau(x)$,

for $a \leq x \leq b$, and

(2) $$\int_a^b [\tau(x) - \sigma(x)] \, dx < \epsilon.$$

Proof. We first establish the "if" part by assuming, for $\epsilon > 0$, the existence of step-functions $\sigma(x)$ and $\tau(x)$ satisfying (1) and (2). Let us observe initially that for *any* step-functions satisfying (1),

$$\int_a^b \sigma(x) \, dx \leq \int_a^b \tau(x) \, dx$$

(Theorem II, § 501), so that the *least upper bound* I of the integrals $\int_a^b \sigma(x) \, dx$, for *all* $\sigma(x) \leq f(x)$, and the *greatest lower bound* J of the integrals $\int_a^b \tau(x) \, dx$, for *all* $\tau(x) \geq f(x)$, both exist and are finite, and $I \leq J$ (supply the details). Because of (2) and the arbitrariness of $\epsilon > 0$, integrals of the form $\int_a^b \sigma(x) \, dx$ and $\int_a^b \tau(x) \, dx$ can be found arbitrarily close to each other, so that I cannot be *less* than J, and therefore $I = J$. Now let $\epsilon > 0$ be given, and choose step-functions $\sigma(x)$ and $\tau(x)$ satisfying (1) and (2) and such that

$$\int_a^b \sigma(x) \, dx > I - \tfrac{1}{3}\epsilon \text{ and } \int_a^b \tau(x) \, dx < J + \tfrac{1}{3}\epsilon.$$

Then choose $\delta > 0$ so that $|\mathfrak{N}| < \delta$ implies

$$\sum_{i=1}^n \sigma(x_i) \, \Delta x_i > \int_a^b \sigma(x) \, dx - \tfrac{1}{3}\epsilon,$$

$$\sum_{i=1}^n \tau(x_i) \, \Delta x_i < \int_a^b \tau(x) \, dx + \tfrac{1}{3}\epsilon.$$

For any such net $\mathfrak{N}$,

$$I - \epsilon < \sum_{i=1}^n \sigma(x_i) \, \Delta x_i \leq \sum_{i=1}^n f(x_i) \, \Delta x_i \leq \sum_{i=1}^n \tau(x_i) \, \Delta x_i < I + \epsilon.$$

Therefore (Definition II, § 501), $f(x)$ is integrable on $[a, b]$, and

$$\int_a^b f(x) \, dx = I = J.$$

We now prove the "only if" part by assuming that $f(x)$ is integrable on $[a, b]$, with integral $I = \int_a^b f(x) \, dx$, and letting $\epsilon > 0$ be given. Choose a net $\mathfrak{N}$, to be held fixed, of such a small norm that

$$I - \tfrac{1}{3}\epsilon < \sum_{i=1}^n f(x_i) \, \Delta x_i < I + \tfrac{1}{3}\epsilon,$$

for all x_i in $[a_{i-1}, a_i]$, $i = 1, 2, \cdots, n$. For each $i = 1, 2, \cdots, n$, let σ_i be the greatest lower bound of $f(x)$ for $a_{i-1} \leq x \leq a_i$ ($f(x)$ is bounded since it is integrable) and let τ_i be the least upper bound of $f(x)$ for $a_{i-1} \leq x \leq a_i$,

140   INTEGRATION   [§ 502

Then (give the details in Ex. 32, § 503)

$$I - \tfrac{1}{3}\epsilon \leq \sum_{i=1}^{n} \sigma_i \Delta x_i \leq \sum_{i=1}^{n} f(x_i) \Delta x_i \leq \sum_{i=1}^{n} \tau_i \Delta x_i \leq I + \tfrac{1}{3}\epsilon,$$

and if the step-functions $\sigma(x)$ and $\tau(x)$ are defined to have the values σ_i and τ_i, respectively, for $a_{i-1} < x < a_i$, $i = 1, 2, \cdots, n$, and the values $f(a_i)$ for $x = a_i$, $i = 0, 1, \cdots, n$, then $\sigma(x) \leq f(x) \leq \tau(x)$ and

$$\int_a^b [\tau(x) - \sigma(x)]\, dx = \sum_{i=1}^{n} (\tau_i - \sigma_i)\, \Delta x_i \leq \tfrac{2}{3}\epsilon < \epsilon.$$

NOTE. The definite integral $\int_a^b f(x)\, dx$ is defined as the limit of sums of the form $\sum_{i=1}^{n} f(x_i) \Delta x_i$. If, for a given net $\mathfrak{N}$ and points x_i from the subinterval $[a_{i-1}, a_i]$, $i = 1, 2, \cdots, n$, a step-function $\sigma(x)$ is defined to have the values $f(x_i)$ for $a_{i-1} < x < a_i$, $i = 1, 2, \cdots, n$, and arbitrary values at the points a_i, $i = 0, 1, \cdots, n$ (cf. Fig. 503), then $\sum_{i=1}^{n} f(x_i) \Delta x_i = \int_a^b \sigma(x)\, dx$, and the definite integral of $f(x)$ can be thought of as the limit of the definite integrals of such "approximating" step-functions:

$$\int_a^b f(x)\, dx = \lim_{|\mathfrak{N}| \to 0} \int_a^b \sigma(x)\, dx.$$

With the aid of the theorem just established it is easy to prove that *any function continuous on a closed interval is integrable there:*

Proof of Theorem VIII, § 501. Let $f(x)$ be continuous on the closed interval $[a, b]$. Then $f(x)$ is uniformly continuous there (Theorem II, § 307), and therefore, if ϵ is any positive number there exists a positive number δ such that $|x' - x''| < \delta$ implies $|f(x') - f(x'')| < \epsilon/(b - a)$. Let $\mathfrak{N}$ be any net of norm less than δ, and let σ_i and τ_i, for $i = 1, 2, \cdots, n$, be defined as the minimum and maximum values, respectively, of $f(x)$ on the subinterval $[a_{i-1}, a_i]$. If x_i' and x_i'' are points of $[a_{i-1}, a_i]$ such that

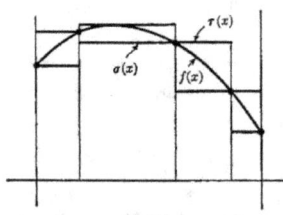

FIG. 503

§ 502] MORE INTEGRATION THEOREMS

$f(x_i') = \sigma_i$ and $f(x_i'') = \tau_i$, $i = 1, 2, \cdots, n$, and if the step-functions $\sigma(x)$ and $\tau(x)$ are defined to have the values σ_i and τ_i, respectively, for $a_{i-1} < x < a_i$, $i = 1, 2, \cdots, n$, and the values $f(a_i)$ for $x = a_i$, $i = 0, 1, \cdots, n$ (see Fig. 503), then $\sigma(x) \leqq f(x) \leqq \tau(x)$ and

$$\int_a^b [\tau(x) - \sigma(x)]\,dx = \sum_{i=1}^n (\tau_i - \sigma_i)\,\Delta x_i$$

$$= \sum_{i=1}^n |f(x_i') - f(x_i'')|\,\Delta x_i < \frac{\epsilon}{b-a}\sum_{i=1}^n \Delta x_i = \epsilon.$$

We generalize the theorem just proved, to permit a finite number of discontinuities:

Proof of Theorem IX, § 501. Thanks to Theorem V, § 501, and mathematical induction we need prove this only for the case of a function $f(x)$ that is defined and bounded on a closed interval and continuous in the interior. Because of the theorem just established, this special case is a simple consequence of the following theorem.

Theorem III. *If $f(x)$ is bounded on $[a, b]$ and integrable on every $[c, d]$, where $a < c < d < b$, then $f(x)$ is integrable on $[a, b]$.*

Proof. If $|f(x)| < K$ for $a \leqq x \leqq b$, and if $\epsilon > 0$, let $0 < \eta < \epsilon/8K$. Construct step-functions $\sigma(x)$ and $\tau(x)$ such that $\sigma(x) = -K$ and $\tau(x) = K$

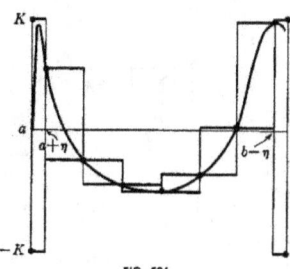

FIG. 504

on the intervals $[a, a + \eta)$ and $(b - \eta, b]$ (see Fig. 504), and such that for the closed interval $[a + \eta, b - \eta]$, $\sigma(x) \leqq f(x) \leqq \tau(x)$ and

$$\int_{a+\eta}^{b-\eta} [\tau(x) - \sigma(x)]\,dx < \tfrac{1}{2}\epsilon$$

(see Theorem II). Then for the interval $[a, b]$, $\sigma(x) \leqq f(x) \leqq \tau(x)$ and

$$\int_a^b [\tau(x) - \sigma(x)]\,dx < 2K\eta + \tfrac{1}{2}\epsilon + 2K\eta < \epsilon.$$

142   INTEGRATION   [§ 502

Example 1. The function $\sin \frac{1}{x}$ (whether or however it is defined at $x = 0$) is integrable on the interval $[0, 1]$.

Example 2. Let $f(x)$ be a function defined on $[0, 1]$ as follows: If x is irrational let $f(x) = 0$; if x is a positive rational number equal to p/q, where p and q are relatively prime positive integers (cf. Ex. 23, § 107), let $f(x) = 1/q$; let $f(0) = 1$. (See Fig. 505.) Prove that $f(x)$ is continuous at every irrational point of $[0, 1]$ and discontinuous with a removable discontinuity at every rational point of $[0, 1]$. Prove that in spite of having infinitely many discontinuities $f(x)$ is integrable on $[0, 1]$. Show, in fact, that $\int_0^1 f(x)\,dx = 0$.

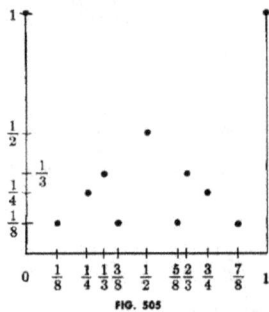

FIG. 505

Solution. If $x_0 = p/q$ we have only to take $\epsilon = 1/q$ to show that $f(x)$ is discontinuous at $x = x_0$, since there are points arbitrarily near x_0 where $f(x) = 0$. Now let x_0 be an irrational number and let $\epsilon > 0$. Since there are only a finite number of positive integers $\leq 1/\epsilon$, there are only a finite number of rational numbers p/q of the interval $[0, 1]$ for which $f(p/q) \geq \epsilon$. If $\delta > 0$ is chosen so small that the interval $(x_0 - \delta, x_0 + \delta)$ excludes all of these finitely many rational points, then $|x - x_0| < \delta$ implies $f(x) < \epsilon$, whether x is rational or irrational, and continuity at x_0 is established. If x_0 is rational and if $f(x_0)$ is redefined to be 0, continuity at x_0 is established by the same argument. Finally, if $\epsilon > 0$, since $f(x) \geq 0$ all that remains to be shown is the existence of a step-function $\tau(x) \geq f(x)$ such that $\int_0^1 \tau(x)\,dx < \epsilon$. Let $x_1, x_2, \cdots, x_n$ be the rational points of $[0, 1]$ such that $f(x_i) \geq \frac{1}{2}\epsilon$, $i = 1, 2, \cdots, n$, and let I_i be a neighborhood of x_i of length $< \epsilon/2n$, $i = 1, 2, \cdots, n$. Define $\tau(x)$ to be 1 on all of the intervals $I_1, I_2, \cdots, I_n$ and $\frac{1}{2}\epsilon$ on the remaining points of $[0, 1]$. Then $\int_0^1 \tau(x)\,dx$ can be split into two parts, each $< \frac{1}{2}\epsilon$. (Cf. Ex. 47, § 1025.)

503. EXERCISES

1. Extend Theorems IV and V, § 501, to an arbitrary finite number of terms, and Theorem V to an arbitrary order relation between a, b, and c. (Cf. Note 2, § 501.)

2. Assuming that $f(x)$ is integrable on $[a, b]$ and that $|f(x)| \leq K$ there, prove that
$$\left| \int_a^b f(x)\, dx \right| \leq K(b - a).$$
It can be proved (cf. Ex. 38) that if $f(x)$ is integrable on $[a, b]$, then so is $|f(x)|$. Assuming that both $f(x)$ and $|f(x)|$ are integrable on $[a, b]$, prove that
$$\left| \int_a^b f(x)\, dx \right| \leq \int_a^b |f(x)|\, dx.$$
Hint: Use Theorem II, § 501, and the inequalities
$$-K \leq f(x) \leq K \quad \text{or} \quad -|f(x)| \leq f(x) \leq |f(x)|.$$

3. Prove that if $f(x)$ is continuous on $[a, b]$ and $f(x) \geq 0$ but not identically zero there, then $\int_a^b f(x)\, dx > 0$. *Hint:* Use Ex. 18, § 212.

4. Prove that if $f(x)$ and $g(x)$ are continuous on $[a, b]$ and $f(x) \leq g(x)$ but $f(x)$ and $g(x)$ are not identical there, then $\int_a^b f(x)\, dx < \int_a^b g(x)\, dx$. (Cf. Ex. 3.)

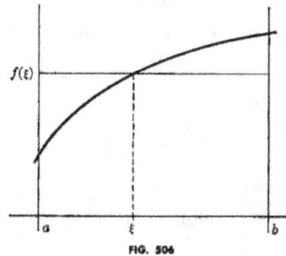

FIG. 506

5. Prove and give a geometric interpretation to the First Mean Value Theorem for Integrals (Theorem XI, § 501). Also consider the case $a > b$. (See Fig. 506.) *Hint:* If m and M are the minimum and maximum values, respectively, of $f(x)$ on $[a, b]$ and if $f(x)$ is not a constant, use Ex. 4 to show that $m < \left[\int_a^b f(x)\, dx \right] / (b - a) < M$. Conclude by applying Theorem IV, § 213.

6. Prove the following generalized form of the First Mean Value Theorem for Integrals (Theorem XI, § 501): *If $f(x)$ and $g(x)$ are continuous on $[a, b]$ and if $g(x)$ never changes sign there, then there exists a point ξ such that $a < \xi < b$ and*

(2) $$\int_a^b f(x) g(x) \, dx = f(\xi) \int_a^b g(x) \, dx.$$

Also consider the case $a > b$. (Cf. Ex. 5.)

7. A function $f(x)$ is said to be **even** if and only if the equality $f(-x) = f(x)$ holds for all x in the domain of definition of the function. (Examples: constants, x^{2n}, $\cos x$, $|x|$.) Prove that if $f(x)$ is even on $[-a, a]$, and integrable on $[0, a]$, where $a > 0$, then $f(x)$ is integrable on $[-a, a]$ and $\int_{-a}^{a} f(x) \, dx = 2 \int_0^a f(x) \, dx$. Prove that if $f(x)$ is even on $(-\infty, +\infty)$ and a and b are any real numbers, then $\int_{-b}^{-a} f(x) \, dx = \int_a^b f(x) \, dx$, whenever the integrals exist.

8. A function $f(x)$ is said to be **odd** if and only if the equality $f(-x) = -f(x)$ holds for all x in the domain of definition of the function. (Examples: $10x$, x^{2n+1}, $\sin x$, the signum function of Example 1, § 206.) Prove that if $f(x)$ is odd on $[-a, a]$, and integrable on $[0, a]$, where $a > 0$, then $f(x)$ is integrable on $[-a, a]$ and $\int_{-a}^{a} f(x) \, dx = 0$. Prove that if $f(x)$ is odd on $(-\infty, +\infty)$ and a and b are any real numbers, then $\int_{-a}^{-b} f(x) \, dx = \int_a^b f(x) \, dx$, whenever the integrals exist.

9. Prove that any sum of even functions is even and any sum of odd functions is odd. Prove that the product of two even functions and the product of two odd functions are even, and that the product of an even function and an odd function is odd. (Cf. Exs. 7-8.)

10. Prove that the only function that is both even and odd is identically zero. (Cf. Exs. 7-9.)

11. Prove that any function whose domain contains the negative of every one of its members is uniquely representable as the sum of an even function and an odd function. (Cf. Exs. 7-10.) *Hint:*
$$f(x) = \tfrac{1}{2}[f(x) + f(-x)] + \tfrac{1}{2}[f(x) - f(-x)].$$

12. Prove that if $f(x)$ is even (odd) and differentiable, then $f'(x)$ is odd (even). (Cf. Exs. 7-11.)

★13. Let $f(x)$ be defined to be $x + 1$ for all real x, and let $g(x)$ be defined to be $x^2 + 1$ for $x > 0$, x^2 for $x < 0$, undefined for $x = 0$. By means of these examples show that the derivative of a function which is neither even nor odd may be either even or odd. On the other hand, show that if the domain of a differentiable function $f(x)$ is an open interval of the form $(-a, a)$ or $(-\infty, +\infty)$ and if $f'(x)$ is odd, then $f(x)$ is even; similarly, that if $f'(x)$ is even, then $f(x)$ plus a suitable constant is odd. (Cf. Exs. 7-12.) *Hint:* $f(x)$ and $f(-x)$ have the same derivative, and must therefore differ by a constant.

14. If $f(x)$ is the bracket function of Example 5, § 201, $f(x) = [x]$, evaluate $\int_0^1 f(x) \, dx$ and $\int_0^3 f(x) \, dx$. More generally, if n is a positive integer, evaluate $\int_0^n f(x) \, dx$.

§ 503]   EXERCISES   145

15. Prove that $\int_a^b x^2\,dx = \frac{1}{3}(b^3 - a^3)$. (Cf. the Example, § 501, and Ex. 13, § 107.)

16. Prove that $\int_a^b x^3\,dx = \frac{1}{4}(b^4 - a^4)$. (Cf. the Example, § 501, and Ex. 14, § 107.)

17. Prove that $\int_a^b x^4\,dx = \frac{1}{5}(b^5 - a^5)$. (Cf. the Example, § 501, and Ex. 15, § 107.)

★18. Prove that if m is a positive integer, then
$$\int_a^b x^m\,dx = \frac{1}{m+1}(b^{m+1} - a^{m+1}).$$
(Cf. the Example, § 501, and Ex. 42, § 107.)

★19. Prove that $\int_a^b \sin x\,dx = \cos a - \cos b$. *Hint:* As in the Example, § 501, let $b > 0$ and $a = 0$, and write
$$\int_0^b \sin x\,dx = \lim_{n \to +\infty} \left(\sin \frac{b}{n} + \cdots + \sin \frac{nb}{n}\right) \cdot \frac{b}{n}.$$
Multiply each term by $2\sin\frac{b}{2n}$, and use the identity
$$2\sin A \sin B = \cos(A - B) - \cos(A + B)$$
to obtain
$$\int_0^b \sin x\,dx = \lim_{n \to +\infty}\left[\left(\cos\frac{b}{2n} - \cos\frac{3b}{2n}\right) + \left(\cos\frac{3b}{2n} - \cos\frac{5b}{2n}\right) + \cdots \right.$$
$$\left. + \left(\cos\frac{(2n-1)b}{2n} - \cos\frac{(2n+1)b}{2n}\right)\right] \cdot \frac{b}{2n \sin\frac{b}{2n}}.$$

★20. Prove that $\int_a^b \cos x\,dx = \sin b - \sin a$. (Cf. Ex. 19.)

★21. Prove that $\int_a^b e^x\,dx = e^b - e^a$.

★22. Prove the **Trapezoidal Rule** for approximating a definite integral: *If $f(x)$ is integrable on $[a, b]$, if $[a, b]$ is subdivided into n equal subintervals of length Δx, and if the values of $f(x)$ at the $n + 1$ points of subdivision, $x_0, x_1, x_2, \cdots, x_n$, are $y_0, y_1, y_2, \cdots, y_n$, respectively, then*
$$\int_a^b f(x)\,dx = \lim_{n \to +\infty} (\tfrac{1}{2}y_0 + y_1 + y_2 + \cdots + y_{n-1} + \tfrac{1}{2}y_n)\,\Delta x.$$
Also prove the following estimate for the error in the trapezoidal formula, assuming existence of $f''(x)$ on $[a, b]$:
$$\int_a^b f(x)\,dx - [\tfrac{1}{2}y_0 + y_1 + \cdots + \tfrac{1}{2}y_n]\,\Delta x = -\frac{b-a}{12}f''(\xi)\,\Delta x^2,$$
where $a < \xi < b$.

Hints: For the first part, write the expression in brackets in the form $(y_1 + \cdots + y_n) + (\tfrac{1}{2}y_0 - \tfrac{1}{2}y_n)$. For the second part, reduce the problem to that of approximating $\int_a^b f(x)\,dx$ by a *single* trapezoid, and assume without loss of generality that the interval $[a, b]$ is $[-h, h]$. (Cf. Ex. 24, § 216, Ex. 47, § 408.) The problem reduces to that of evaluating K, defined by the equation

$$\int_a^b f(x)\,dx = \frac{b-a}{2}[f(b) + f(a)] + K(b-a)^3,$$

or
$$\int_{-h}^{h} f(t)\,dt = h[f(h) + f(-h)] + 8K h^3.$$

Define the function

$$\phi(x) \equiv \int_{-x}^{x} f(t)\,dt - x[f(x) + f(-x)] - 8K x^3.$$

Show that $\phi(h) = \phi(0) = 0$, and hence there must exist, by Rolle's Theorem (§ 407), a number x_1 between 0 and h such that $\phi'(x_1) = 0$. (Cf. Ex. 2, § 506.)

★23. Prove **Simpson's Rule** for approximating a definite integral: *Under the assumptions and notation of Exercise 22, where n is even,*

$$\int_a^b f(x)\,dx = \lim_{n\to+\infty} \tfrac{1}{3}(y_0 + 4y_1 + 2y_2 + 4y_3 + 2y_4 + \cdots + 4y_{n-1} + y_n)\,\Delta x.$$

Also prove the following estimate for the error in Simpson's Rule, assuming existence of $f''''(x)$ on $[a, b]$:

$$\int_a^b f(x)\,dx - \tfrac{1}{3}[y_0 + 4y_1 + \cdots + y_n]\,\Delta x = -\frac{b-a}{180} f''''(\xi)\,\Delta x^4,$$

where $a < \xi < b$. (Cf. Example 5, § 813.)

Hint: Proceed as in Ex. 22, making use of the auxiliary function

$$\phi(x) \equiv \int_{-x}^{x} f(t)\,dt - \frac{x}{3}[f(-x) + 4f(0) + f(x)] - Kx^5.$$

Evaluate $\phi'''(x_1)$. (Cf. Ex. 2, § 506.)

★24. Evaluate $\lim\limits_{n\to+\infty} \left(\dfrac{1}{n+1} + \dfrac{1}{n+2} + \cdots + \dfrac{1}{2n} \right)$. *Hint:* The sum, when rewritten $\left(\dfrac{1}{1+\frac{1}{n}} + \dfrac{1}{1+\frac{2}{n}} + \cdots + \dfrac{1}{1+\frac{n}{n}} \right)\cdot\dfrac{1}{n}$, can be interpreted as one of the approximating sums for $\int_0^1 \dfrac{dx}{1+x}$. (For the specific evaluation of this as $\ln 2$, see § 504.)

★25. Evaluate $\lim\limits_{n\to+\infty} n\cdot\left(\dfrac{1}{n^2+1^2} + \dfrac{1}{n^2+2^2} + \cdots + \dfrac{1}{2n^2} \right)$. (Cf. Ex. 24.)

★26. Evaluate $\lim\limits_{n\to+\infty} \left(\dfrac{1}{\sqrt{4n^2-1^2}} + \dfrac{1}{\sqrt{4n^2-2^2}} + \cdots + \dfrac{1}{\sqrt{3n^2}} \right)$. (Cf. Ex. 24.)

★27. Evaluate

$$\lim_{n\to+\infty} n\cdot\left[\frac{1}{(2n+3)^2 - 1^2} + \frac{1}{(2n+6)^2 - 2^2} + \cdots + \frac{1}{(5n)^2 - n^2} \right].$$

(Cf. Ex. 24.)

★28. It can be proved (cf. Ex. 31) that a function integrable on a closed interval is integrable on any closed subinterval. Assuming that $f(x)$ is integrable

on $[a, b]$ and also on $[c, d]$ for all c and d such that $a \leq c < d \leq b$, prove that
$\int_a^b f(x)\, dx = \lim_{\eta \to 0+} \int_{a+\eta}^b f(x)\, dx = \lim_{\eta \to 0+} \int_a^{b-\eta} f(x)\, dx$. *Hint:* If $|f(x)| \leq K$ for
$a \leq x \leq b$, $\left| \int_{a+\eta}^b f(x)\, dx - \int_a^b f(x)\, dx \right| = \left| \int_a^{a+\eta} f(x)\, dx \right| \leq K\eta$.

★29. It can be proved (cf. Ex. 35) that if $f(x)$ and $g(x)$ are integrable on $[a, b]$, then so is $f(x)\, g(x)$, and hence so are $[f(x)]^2$ and $[g(x)]^2$. Assuming that these functions are all integrable on $[a, b]$, establish the **Schwarz** (or **Cauchy**) inequality (cf. Ex. 43, § 107, Ex. 26, § 711, Ex. 14, § 717, Exs. 38, 40, § 1225):

(3) $$\left[\int_a^b f(x)\, g(x)\, dx \right]^2 \leq \int_a^b [f(x)]^2\, dx \cdot \int_a^b [g(x)]^2\, dx.$$

Hint: First show that $[f(x) + tg(x)]^2$ is integrable for all real t. With the notation $A = \int_a^b [f(x)]^2\, dx$, $B = \int_a^b f(x)\, g(x)\, dx$, and $C = \int_a^b [g(x)]^2\, dx$, show that $A + 2Bt + Ct^2 \geq 0$ for all real t. If $C = 0$, then $B = 0$, and if $C \neq 0$ the discriminant of $A + 2Bt + Ct^2$ must be nonpositive.

★30. Use the Schwarz inequality to establish the **Minkowski inequality** (cf. Ex. 44, § 107, Ex. 14, § 717), assuming that the integrals exist (cf. Ex. 29):

(4) $$\left\{ \int_a^b [f(x) + g(x)]^2\, dx \right\}^{\frac{1}{2}} \leq \left\{ \int_a^b [f(x)]^2\, dx \right\}^{\frac{1}{2}} + \left\{ \int_a^b [g(x)]^2\, dx \right\}^{\frac{1}{2}}.$$

Hint: Use the hint of Ex. 44, § 107. (Also cf. Exs. 39, 40, § 1225.)

★31. Prove that a function integrable on a closed interval is integrable on any closed subinterval.

★32. Supply the details requested in the second part of the proof of Theorem II, § 502. *Hint:* If $\sum \tau_i \Delta x_i > I + \frac{1}{2}\epsilon$, define $\eta = \frac{1}{n}[\sum \tau_i \Delta x_i - (I + \frac{1}{2}\epsilon)] > 0$. For each $i = 1, \cdots, n$, choose x_i such that $f(x_i) > \tau_i - (\eta/\Delta x_i)$, and obtain a contradiction.

★33. Prove Theorem X, § 501. *Hint:* Assume for definiteness that $f(x)$ is monotonically increasing on $[a, b]$, and for a given net $\mathfrak{N}$ define the stepfunctions $\sigma(x) \equiv f(a_{i-1})$ for $a_{i-1} \leq x < a_i$ ($\sigma(b) \equiv f(b)$) and $\tau(x) \equiv f(a_i)$ for $a_{i-1} < x \leq a_i$ ($\tau(a) \equiv f(a)$), $i = 1, 2, \cdots, n$. Then $\sigma(x) \leq f(x) \leq \tau(x)$ and
$\int_a^b [\tau(x) - \sigma(x)]\, dx = \sum_{i=1}^n [f(a_i) - f(a_{i-1})]\, \Delta x_i \leq |\mathfrak{N}| \cdot \sum_{i=1}^n [f(a_i) - f(a_{i-1})] = |\mathfrak{N}| \cdot (f(b) - f(a))$.

★34. Prove that $f(x)$ is integrable on $[a, b]$ if and only if corresponding to $\epsilon > 0$ there exists $\delta > 0$ such that for every net of norm less than δ and every choice of points x_i and x_i' such that $a_{i-1} \leq x_i \leq a_i$ and $a_{i-1} \leq x_i' \leq a_i$, $i = 1, 2, \cdots, n$,

(5) $$\sum_{i=1}^n |f(x_i) - f(x_i')|\, \Delta x_i < \epsilon.$$

Hint: First show that the preceding statement is equivalent to the same statement with (5) replaced by

(6) $$\left| \sum_{i=1}^n f(x_i)\, \Delta x_i - \sum_{i=1}^n f(x_i')\, \Delta x_i \right| < \epsilon.$$

148   INTEGRATION   [§ 503

If $f(x)$ is integrable, with integral equal to I, and if ϵ is a given positive number, let δ be a positive number such that for any net whose norm is less than δ, $|\sum f(x_i) \Delta x_i - I| < \frac{1}{2}\epsilon$. Then (6) follows from

$$|\sum f(x_i) \Delta x_i - \sum f(x_i') \Delta x_i| \leq |\sum f(x_i) \Delta x_i - I| + |\sum f(x_i') \Delta x_i - I|.$$

On the other hand, if the condition stated exists, let $\mathfrak{N}$ be any fixed net for which $\sum |f(x_i) - f(x_i')| \Delta x_i < \frac{1}{2}\epsilon$. Define σ_i and τ_i as in the second part of the proof of Theorem II, § 502, and show that $\sum (\tau_i - \sigma_i) \Delta x_i \leq \frac{1}{2}\epsilon < \epsilon$.

★35. Prove that the product of two integrable functions is integrable, and extend the result to the product of any finite number of integrable functions. (For related theorems, see Exercises 37 and 41.) *Hint:*

$$\sum |f(x_i) g(x_i) - f(x_i') g(x_i')| \cdot \Delta x_i$$
$$= \sum |g(x_i) [f(x_i) - f(x_i')] + f(x_i') [g(x_i) - g(x_i')]| \cdot \Delta x_i$$
$$\leq K \cdot [\sum |f(x_i) - f(x_i')| \Delta x_i + \sum |g(x_i) - g(x_i')| \Delta x_i].$$

(Cf. Ex. 34.)

★36. Prove that the reciprocal of an integrable function is integrable if the given function is bounded from zero. That is, if $f(x)$ is integrable on $[a, b]$ and if $|f(x)| \geq \eta > 0$ on $[a, b]$, then $1/f(x)$ is integrable on $[a, b]$. *Hint:* Use Ex. 34 and

$$\sum_{i=1}^{n} \left| \frac{1}{f(x_i)} - \frac{1}{f(x_i')} \right| \Delta x_i = \sum_{i=1}^{n} \frac{|f(x_i') - f(x_i)|}{|f(x_i) f(x_i')|} \Delta x_i \leq \frac{1}{\eta^2} \sum_{i=1}^{n} |f(x_i) - f(x_i')| \Delta x_i.$$

What happens if we merely assume $|f(x)| > 0$?

★37. Prove that the quotient of two integrable functions is integrable if the second (denominator) function is bounded from zero (cf. Exs. 35, 36).

★38. Prove that if $f(x)$ is integrable on $[a, b]$, then so is $|f(x)|$. Construct an example to show that the converse implication is false. *Hints:*

$$\sum | |f(x_i)| - |f(x_i')| | \Delta x_i \leq \sum |f(x_i) - f(x_i')| \Delta x_i$$

(cf. Ex. 34). Consider a function like that of Example 6, § 201, with values ± 1.

★39. If $f(x)$ and $g(x)$ are defined on a set A, the functions $M(x) \equiv \max [f(x), g(x)]$ and $m(x) \equiv \min [f(x), g(x)]$ are defined, for each x in A, to be the larger and smaller, respectively, of the two numbers $f(x)$ and $g(x)$ (equal to them if they are equal). (See Fig. 507.) Prove that if $f(x)$ and $g(x)$ are integrable on $[a, b]$, then so are $M(x)$ and $m(x)$. *Hint:* $M(x) = \frac{1}{2}[f(x) + g(x)] + \frac{1}{2}|f(x) - g(x)|$, $m(x) = \frac{1}{2}[f(x) + g(x)] - \frac{1}{2}|f(x) - g(x)|$. (Cf. Ex. 38.)

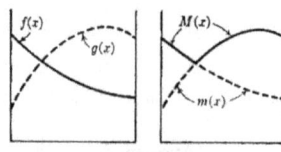

FIG. 507

§ 503]   EXERCISES   149

★40. If $f(x)$ is defined on a set A, the nonnegative functions $f^+(x)$ and $f^-(x)$ are defined: $f^+(x) \equiv \max[f(x), 0]$, $f^-(x) \equiv \max[-f(x), 0]$. (See Fig. 508.) Prove that if $f(x)$ is integrable on $[a, b]$ then so are $f^+(x)$ and $f^-(x)$. Prove that the integrability of any two of the following four functions implies that of all: $f(x)$, $f^+(x)$, $f^-(x)$, $|f(x)|$. Hint: $f(x) = f^+(x) - f^-(x)$, $|f(x)| = f^+(x) + f^-(x)$. (Cf. Ex. 39.)

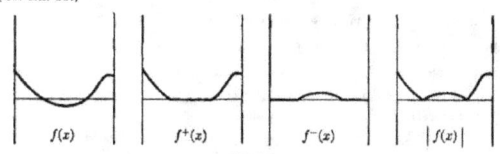

FIG. 508

★41. Prove **Bliss's Theorem**:† If $f(x)$ and $g(x)$ are integrable on $[a, b]$, then the limit $\lim_{|\mathfrak{N}|\to 0} \sum_{i=1}^{n} f(x_i) g(x_i') \Delta x_i$, where $a_{i-1} \leq x_i \leq a_i$ and $a_{i-1} \leq x_i' \leq a_i$, $i = 1, 2, \cdots, n$, the limit being interpreted in the sense of Definition 1, § 501, exists and is equal to $\int_a^b f(x)g(x)\,dx$. Hint: Use the identity

$$\sum f(x_i)\, g(x_i')\, \Delta x_i = \sum f(x_i)\, g(x_i)\, \Delta x_i + \sum f(x_i)\, [g(x_i') - g(x_i)]\, \Delta x_i,$$

the inequality

$$|\sum f(x_i)\, [g(x_i') - g(x_i)]\, \Delta x_i|' \leq K \cdot \sum |g(x_i') - g(x_i)|\, \Delta x_i,$$

and Exs. 34 and 35.

★42. Bliss's Theorem (Ex. 41) is used in such applied problems as work performed by a variable force and total fluid force on a submerged plate. In the latter case, for example, assume that a vertical plane area is submerged in a liquid of constant density ρ, and that the width at depth x is $w(x)$ (cf. Fig. 509). Then (with appropriate continuity assumptions) the element of area, between

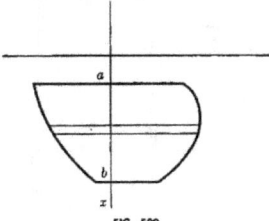

FIG. 509

† Cf. G. A. Bliss, "A Substitute for Duhamel's Theorem," *Annals of Mathematics*, Vol. 16 (1914-15). pp. 45-49.

150   INTEGRATION   [§ 503

x_i and $x_i + \Delta x_i$ is neither $w(x_i) \Delta x_i$ nor $w(x_i + \Delta x_i) \Delta x_i$, but $w(\eta_i) \Delta x_i$ for some appropriate intermediate $\Re_i$ (cf. the Mean Value Theorem for Integrals, and Ex. 5). Furthermore, the total force on this element of area is greater than the area times the pressure at the top of the strip, and less than the area times the pressure at the bottom of the strip. It is therefore equal to product of the area and an intermediate pressure, $\rho \xi_i$. Therefore the total force on the plate is the sum of the individual elements of force: $F = \sum_{i=1}^{n} \rho \xi_i w(\eta_i) \Delta x_i$. According to Bliss's Theorem, the limit of this sum can be expressed as a definite integral, and we have the standard formula, $F = \rho \int_a^b x w(x) \, dx$. Discuss the application of Bliss's Theorem to the work done in pumping the fluid contents of a tank to a certain height above the tank.

★43. It can be proved (cf. Ex. 55) that a bounded continuous function of an integrable function is integrable. The reverse happens to be false: an integrable function of a bounded continuous function may not be integrable. (A counterexample can be constructed with the aid of the "Cantor set," well known in Lebesgue Theory.) By means of the following two functions show that an integrable function of an integrable function may not be integrable: Let $g(x)$ be the function of Example 2, § 502, and let $f(x)$ be the signum function (Example 1, § 206). Then $f(g(x))$ is the function of Example 6, § 201, discussed in a paragraph following Definition III, § 501.

★44. The numbers I and J defined in the first part of the proof of Theorem II, § 502, are called the **lower** and **upper integral**, respectively, of $f(x)$ on $[a, b]$, and are written $\underline{\int_a^b} f(x) \, dx = I$, $\overline{\int_a^b} f(x) \, dx = J$. Prove that for a bounded function $f(x)$ the lower and upper integrals always exist, that $\underline{\int_a^b} f(x) \, dx \leq \overline{\int_a^b} f(x) \, dx$, and that $f(x)$ is integrable if and only if its lower and upper integrals are equal and, in the case of integrability, its integral is equal to their common value.

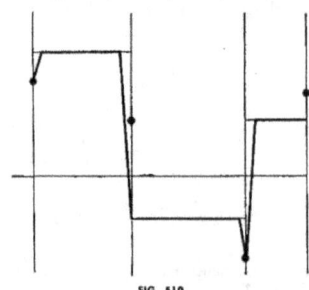

FIG. 510

★45. Prove that if $\sigma(x)$ is any step-function on the interval $[a, b]$, and if $\epsilon > 0$, then there exist continuous functions $\phi(x)$ and $\psi(x)$ on $[a, b]$ such that $\phi(x) \leq \sigma(x) \leq \psi(x)$ and $\int_a^b [\psi(x) - \phi(x)]\, dx < \epsilon$. Hence prove that if $f(x)$ is any bounded function on $[a, b]$, $\underline{\int_a^b} f(x)\, dx = \sup \int_a^b \phi(x)\, dx$, taken for all continuous functions $\phi(x)$ such that $\phi(x) \leq f(x)$, and $\overline{\int_a^b} f(x)\, dx = \inf \int_a^b \psi(x)\, dx$, taken for all continuous functions $\psi(x)$ such that $\psi(x) \geq f(x)$. Prove that Theorem II, § 502, remains true if the word *step-functions* is replaced by the words *continuous functions*. (Cf. Ex. 44.) *Hint:* See Fig. 510, for $\phi(x)$.

★★46. Define a step-function of **positive compact type** to be either the function identically zero or a nonnegative step-function whose nonzero values are assumed as constants on disjoint closed intervals (of positive length). (See Fig. 511.) Prove that if $f(x)$ is a bounded nonnegative function on $[a, b]$, then $\underline{\int_a^b} f(x)\, dx = \sup \int_a^b \sigma(x)\, dx$ taken for all step-functions $\sigma(x)$ of positive compact type such that $(0 \leq) \sigma(x) \leq f(x)$.

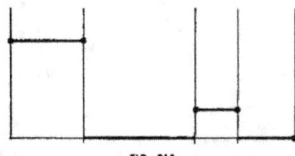

FIG. 511

★★47. Assume $K \geq f(x) \geq g(x) \geq 0$ on $[a, b]$, and let $\sigma(x)$ be a step-function such that $f(x) \geq \sigma(x) \geq 0$ and $\int_a^b \sigma(x)\, dx > \underline{\int_a^b} f(x)\, dx - \epsilon$, where $\epsilon > 0$. Establish the existence of a step-function $\tau(x)$ of positive compact type (Ex. 46) such that $g(x) \geq \tau(x) \, (\geq 0)$, $\sigma(x) \geq \tau(x)$, and $\int_a^b \tau(x)\, dx > \underline{\int_a^b} g(x)\, dx - \epsilon$. *Hint:* Abbreviating $\int_a^b F(x)\, dx$ and $\underline{\int_a^b} F(x)\, dx$ by $\int F\, dx$ and $\underline{\int} F\, dx$, respectively, define $\eta \equiv \int \sigma\, dx - \underline{\int} f\, dx + \epsilon > 0$, and let $\rho(x)$ be a step-function such that $0 \leq \rho(x) \leq g(x)$ and $\int \rho\, dx > \underline{\int} g\, dx - \eta$. If $\phi(x) \equiv \min(\sigma(x), \rho(x))$ and $\psi(x) \equiv \max(\sigma(x), \rho(x))$, then $\phi(x)$ and $\psi(x)$ are step-functions, $\int \psi\, dx \leq \underline{\int} f\, dx$, and $\sigma(x) + \rho(x) = \phi(x) + \psi(x)$. Then $\int \phi\, dx + \int \psi\, dx = \int \sigma\, dx + \int \rho\, dx > \underline{\int} f\, dx + \underline{\int} g\, dx - \epsilon$, and $\int \phi\, dx > \underline{\int} g\, dx - \epsilon$. Finally, use Ex. 46.

★★48. Assume $K \geq f_1(x) \geq f_2(x) \geq \cdots \geq f_n(x) \geq f_{n+1}(x) \geq \cdots$ and
$$\lim_{n \to +\infty} f_n(x) = 0$$
for every x on $[a, b]$. Prove that $\int_a^b f_n(x)\,dx \to 0$. *Hint:* With the notation of Ex. 47, assume $\int f_n\,dx \nrightarrow 0$. Then, since $\int f_n\,dx \downarrow$, $\lim_{n \to +\infty} \int f_n\,dx$ exists and is equal to a positive number 3ϵ. By Ex. 47, there exists a sequence $\{\sigma_n(x)\}$ of step-functions of positive compact type such that $(0 \leq) \sigma_n(x) \leq f_n(x)$, $\sigma_n(x) \geq \sigma_{n+1}(x)$, and $\int \sigma_n\,dx > 2\epsilon$, for $n = 1, 2, \cdots$. If A_n is defined to be the set of points such that $\sigma_n(x) \geq \epsilon/(b-a)$, then $\{A_n\}$ is a compact nonempty set, for $n = 1, 2, \cdots$, and A_n is a decreasing sequence (Ex. 26, § 312). If x_0 is a point common to every set of this sequence (Ex. 26, § 312), $f_n(x_0) \geq \sigma_n(x_0) \geq \epsilon/(b-a)$, and $f_n(x_0) \nrightarrow 0$. (Contradiction.)

★★49. Prove the **Lebesgue Theorem on Bounded Convergence** for Riemann Integrals:† *If $\{f_n(x)\}$ is a sequence of Riemann-integrable functions converging on $[a, b]$ to a Riemann-integrable function $f(x)$, and if $|f_n(x)| \leq K$ for $a \leq x \leq b$ and $n = 1, 2, \cdots$, then $\lim_{n \to +\infty} \int_a^b f_n(x)\,dx = \int_a^b f(x)\,dx = \int_a^b \lim_{n \to +\infty} f_n(x)\,dx$.* *Hint:* Define $g_n(x) \equiv |f_n(x) - f(x)|$, $h_n(x) \equiv \sup(g_n(x), g_{n+1}(x), \cdots)$. Then (Exs. 16-18, § 305), $h_n(x) \downarrow$, $h_n(x) \to 0$, and the result of Ex. 48 shows that $\int h_n\,dx \to 0$. Therefore, since $0 \leq g_n(x) \leq h_n(x)$, $\int g_n\,dx \to 0$, whence $\int f_n\,dx \to \int f\,dx$. (Cf. Ex. 47, § 515.)

★★50. Let $r_1, r_2, r_3, \cdots$ be the rational numbers on the interval $[0, 1]$, and define the function $f_n(x)$ to be 1 if $x = r_1, r_2, \cdots,$ or r_n, and 0 otherwise, $0 \leq x \leq 1$. Show that $\lim_{n \to +\infty} f_n(x) = f(x)$ exists and that the convergence is bounded and monotonic (Ex. 49), but that $f(x)$ is not Riemann-integrable. (Even if all of the functions $f_n(x)$ are continuous on $[0, 1]$ and the convergence is bounded and monotonic, the limit function need not be Riemann-integrable.)

★★51. Let $f(x)$ be defined on a closed interval $[\alpha, \beta]$ and assume that there exists a positive number η such that the oscillation $\omega(x)$ (cf. Ex. 31, § 312) of $f(x)$ at each point x of $[\alpha, \beta]$ is less than η. Prove that there exists a net $\mathfrak{N}$ for $[\alpha, \beta]$ such that on each closed subinterval of $\mathfrak{N}$ the difference between the least upper bound of $f(x)$ and the greatest lower bound of $f(x)$ is less than η (that is, the oscillation of $f(x) < \eta$). *Hint:* Each point of $[\alpha, \beta]$ has a neighborhood I in which the oscillation of $f(x)$ is less than η. Use the Heine-Borel theorem (Ex. 28, § 312) to reduce this covering of $[\alpha, \beta]$ to a finite covering. Finally, obtain the desired net from this finite covering.

† This theorem was stated by Arzelà in 1885, before the invention of the Lebesgue Integral, and given a simplified proof (whose essential ideas we have outlined here) by Hausdorff in 1927. If the integrals concerned are Lebesgue integrals, the integrability of the limit function $f(x)$ is a consequence of the other hypotheses and need not be specifically assumed. (Cf. Ex. 50.) However, the theorem as stated above has useful applications in cases where the limit function is known to be Riemann-integrable.

§ 503]   EXERCISES   153

★★52. A set S of real numbers is said to be of (Lebesgue) **measure zero** if and only if corresponding to $\epsilon > 0$ there exists a (finite or infinite) sequence of open intervals $I_1, I_2, \cdots$ of lengths $L_1, L_2, \cdots$, which cover S (cf. Ex. 28, § 312) and such that for every n,

$$L_1 + L_2 + \cdots + L_n < \epsilon.$$

Prove that any finite set of points is of measure zero. Prove that any denumerable set (cf. Ex. 12, § 113) is of measure zero. (The rational numbers form a set of measure zero.) Prove that any subset of a set of measure zero is of measure zero. *Hint:* For a denumerable set $x_1, x_2, \cdots, x_n, \cdots$, if $\epsilon > 0$, let I_n be a neighborhood of x_n of length $L_n \leq \epsilon/2^n$, $n = 1, 2, \cdots$.

★★53. Let $S_1, S_2, \cdots$ be a (finite or infinite) sequence of sets of measure zero and let S be the set of points x such that x is a number of at least one set S_n of the sequence (S is the "union" of the sets of the sequence). Prove that S is of measure zero. *Hint:* If $\epsilon > 0$, define for each m the sequence $I_1^{(m)}$, $I_2^{(m)}, \cdots$ of open intervals covering S_m, of lengths $L_1^{(m)}, L_2^{(m)}, \cdots$ such that for every n, $L_1^{(m)} + L_2^{(m)} + \cdots + L_n^{(m)} < \epsilon/2^m$. The entire collection of intervals $I_n^{(m)}$ can be arranged as a sequence (cf. Ex. 13, § 113).

★★54. A property is said to hold **almost everywhere** if and only if the set of points where it fails is of measure zero. Prove that a function $f(x)$, defined on a closed interval, is integrable there if and only if it is bounded and almost everywhere continuous. *Hints:* (i) If $f(x)$ is integrable on $[a, b]$ and if $\epsilon > 0$ and $\eta > 0$, there exist step-functions $\sigma(x)$ and $\tau(x)$, over the same net $\mathfrak{M}$, such that $\sigma(x) \leq f(x) \leq \tau(x)$ and $\int_a^b [\tau(x) - \sigma(x)] \, dx < \epsilon\eta$. Then the sum of the lengths of the subintervals of $\mathfrak{M}$ for which $\tau_i - \sigma_i \geq \epsilon$ is less than η. By appropriately enclosing the points of $\mathfrak{M}$ in small neighborhoods, show that the set D_ϵ of points x where $\omega(x) \geq \epsilon$ (cf. Ex. 32, § 312) is contained in a finite set of open intervals the sum of whose lengths $< \eta$, and is therefore a set of measure zero. Conclude by referring to Ex. 33, § 312, and Ex. 53 above. (ii) With the notation of Ex. 33, § 312, assume that D is of measure zero. Then (Ex. 52) D_ϵ, for any $\epsilon > 0$, is of measure zero. If $|f(x)| < K$ and if $\epsilon > 0$, let $I_1, I_2, \cdots$ be a sequence of open intervals covering D_η, where $\eta = \epsilon/2(b - a)$, of such small lengths $L_1, L_2, \cdots$ that for every n, $L_1 + L_2 + \cdots + L_n < \epsilon/4K$. By the Heine-Borel Theorem (Ex. 28, § 312) there exists a finite sequence $I_1, I_2, \cdots, I_n$ covering D_η. On each of the closed subintervals $[\alpha, \beta]$ of $[a, b]$ remaining after removal of $I_1, I_2, \cdots, I_n$, use Ex. 51 to construct step-functions $\sigma(x)$ and $\tau(x)$ such that $\sigma(x) \leq f(x) \leq \tau(x)$ and such that the sum of the integrals $\int [\tau(x) - \sigma(x)] \, dx$ over these intervals $< \eta(b - a) = \frac{1}{2}\epsilon$. Extend $\sigma(x)$ and $\tau(x)$ to the entire interval $[a, b]$ by defining them equal to $-K$ and K, respectively on the intervals $I_1, I_2, \cdots, I_n$. Then $\int_a^b [\tau(x) - \sigma(x)] \, dx < \epsilon$.

★★55. Prove that a bounded continuous function of an integrable function is integrable. In particular, if $f(x)$ is integrable on $[a, b]$ and $p > 0$, prove that $|f(x)|^p$ is integrable on $[a, b]$. *Hint:* The composite function is continuous almost everywhere (cf. Ex. 54).

★★56. Prove that if $f(x)$ is integrable and positive on a closed interval $[a, b]$, then $\int_a^b f(x)\,dx > 0$. *Hint:* By Ex. 54 $f(x)$ is continuous at some point of $[a, b]$.

★★**57.** Prove that in Exercise 56 the word *positive* may be replaced by the expression *positive almost everywhere*. In fact, prove that if $f(x)$ is a nonnegative integrable function, $\int_a^b f(x)\,dx = 0$ if and only if $f(x) = 0$ almost everywhere.

504. THE FUNDAMENTAL THEOREM OF INTEGRAL CALCULUS

Evaluation of a definite integral by actually taking the limit of a sum, as in the Example of § 501 and Exercises 15-21, § 503, is usually extremely arduous. Historically, many of the basic concepts of the definite integral as the limit of a sum were appreciated by the ancient Greeks long before the invention of the Differential Calculus by Newton and Leibnitz. Before the invention of derivatives, an area or a volume could be computed only by such a limiting process as is involved in the definition of the definite integral. The introduction of the notion of a derivative provided a spectacular impetus to the development of mathematics, not only by its immediate application as a rate of change, but also by permitting the evaluation of a definite integral by the process of reversing differention, known as *antidifferention* or *integration*. It is the purpose of this section to study certain basic relations between the two fundamental concepts of Calculus, the derivative and the definite integral.

Theorem I. *Let $f(x)$ be defined and continuous on a closed interval $[a, b]$ or $[b, a]$, and define the function $F(x)$ on this interval:*

(1) $$F(x) \equiv \int_a^x f(t)\,dt.†$$

Then $F(x)$ is differentiable there with derivative $f(x)$:

(2) $$F'(x) = f(x).$$

(Cf. Exs. 22-23, § 506.)

Proof. Let x_0 and $x_0 + \Delta x$, where $\Delta x \neq 0$, belong to the given interval. Then, by Theorem V and Note 2, § 501,

---

† Since the letter x designates the upper limit of integration, it might be confusing to use this same symbol simultaneously in a second role as the variable of integration, and write $\int_a^x f(x)\,dx$. Note that the integral $\int_a^b f(x)\,dx$ is *not* a function of x. Rather, the letter x in this case serves only as a *dummy variable* (cf. Ex. 36, § 107), and any other letter could be substituted, thus: $\int_a^b f(x)\,dx = \int_a^b f(t)\,dt = \int_a^b f(u)\,du$. Therefore, instead of $\int_a^x f(x)\,dx$ we write $\int_a^x f(t)\,dt$, $\int_a^x f(u)\,du$, etc.

§ 504] FUNDAMENTAL THEOREM

$$F(x_0 + \Delta x) - F(x_0) = \int_a^{x_0+\Delta x} f(t)\,dt - \int_a^{x_0} f(t)\,dt = \int_{x_0}^{x_0+\Delta x} f(t)\,dt.$$

By the First Mean Value Theorem for Integrals (Theorem XI, § 501),

$$\frac{F(x_0 + \Delta x) - F(x_0)}{\Delta x} = \frac{1}{\Delta x}\int_{x_0}^{x_0+\Delta x} f(t)\,dt = f(\xi)$$

for some number ξ between x_0 and $x_0 + \Delta x$. Therefore if $\Delta x \to 0$, $\xi \to x_0$ and

$$F'(x_0) = \lim_{\Delta x \to 0} \frac{F(x_0 + \Delta x) - F(x_0)}{\Delta x} = f(x_0),$$

as stated in the theorem.

If a **primitive** or **antiderivative** or **indefinite integral** of a given function is defined to be any function whose derivative is the given function, Theorem I asserts that *any continuous function has a primitive*. The next theorem gives a method for evaluating the definite integral of any continuous function in terms of a given primitive.

Theorem II. Fundamental Theorem of Integral Calculus. *If $f(x)$ is continuous on the closed interval $[a, b]$ or $[b, a]$ and if $F(x)$ is any primitive of $f(x)$ on this interval, then*

(3) $$\int_a^b f(x)\,dx = F(b) - F(a).$$

(Cf. Exs. 19-20, § 506.)

Proof. By Theorem I and the hypotheses of Theorem II, the two functions $\int_a^x f(t)\,dt$ and $F(x)$ have the same derivative on the given interval. Therefore, by Theorem II, § 406, they differ by a constant:

(4) $$\int_a^x f(t)\,dt - F(x) = C.$$

Substitution of $x = a$ gives the value of this constant: $C = -F(a)$. Upon substitution for C in (4) we have $\int_a^x f(t)\,dt = F(x) - F(a)$ which, for the particular value $x = b$, is the desired result.

By virtue of the Fundamental Theorem of Integral Calculus, the integral symbol $\int$, suggested by the letter S (the definite integral is the limit of a *sum*), is appropriate for the *indefinite* integral as well as the *definite* integral. If $F(x)$ is an arbitrary indefinite integral of a function $f(x)$, we write the equation

(5) $$\int f(x)\,dx = F(x) + C,$$

where C is an arbitrary **constant of integration**, and call the symbol $\int f(x)\, dx$ **the indefinite integral** of $f(x)$. The Fundamental Theorem can thus be considered as an expression of the relation between the two kinds of integrals, definite and indefinite.

505. INTEGRATION BY SUBSTITUTION

The question before us is essentially one of appropriateness of notation: In the expression $\int f(x)\, dx$, does the "dx," which looks like a differential, really behave like one? Happily, the answer is in the affirmative:

Theorem. *If $f(v)$ is a continuous function of v and if $v(x)$ is a continuously differentiable function of x, then*

$$(1) \qquad \int f(v(x))\, v'(x)\, dx = \int f(v(x))\, d(v(x)) = \int f(v)\, dv.$$

If $v(a) = c$ and $v(b) = d$, then

$$(2) \qquad \int_a^b f(v(x))\, v'(x)\, dx = \int_c^d f(v)\, dv.$$

Proof. Equation (1) is a statement that a function $F(v)$ whose derivative with respect to v is $f(v)$ has a derivative with respect to x equal to $f(v)\dfrac{dv}{dx}$, and this statement is a reformulation for the rule for differentiating a function of a function. Equation (2) follows from (1) by the Fundamental Theorem of Integral Calculus: $F(v(b)) - F(v(a)) = F(d) - F(c)$.

506. EXERCISES

1. Let the constant a and the variable x belong to an interval I throughout which the function $f(x)$ is continuous. Prove that

$$(1) \qquad \frac{d}{dx}\int_a^x f(t)\, dt = f(x), \quad \frac{d}{dx}\int_x^a f(t)\, dt = -f(x).$$

Hint: Use Theorem V and Note 2, § 501.

2. Let $u(x)$ and $v(x)$ be differentiable functions whose values lie in an interval I throughout which $f(t)$ is continuous. Prove that

$$(2) \qquad \frac{d}{dx}\int_{u(x)}^{v(x)} f(t)\, dt = f(v(x))\, v'(x) - f(u(x))\, u'(x).$$

Hint: Write $\int_{u(x)}^{v(x)} f(t)\, dt = \int_a^{v(x)} f(t)\, dt - \int_a^{u(x)} f(t)\, dt$, and define $F(v) = \int_a^v f(t)\, dt$, with $v = v(x)$.

In Exercises 3-8, find the derivative with respect to x. The letters a and b represent constants. (Cf. Exs. 1-2.)

3. $\int_a^b \sin x^2\, dx.$ 4. $\int_a^x t^2\, dt.$

5. $\int_x^b \sin t^2\, dt.$ 6. $\int_a^{x^2} \sin t^2\, dt.$

7. $\int_{x^2}^{x^4} \sin t^2\, dt.$ 8. $\int_{\sin^3 x}^{\sin x^4} \sin t^2\, dt.$

9. Assume necessary conditions of continuity on $f(x)$ and $f'(x)$ in an interval $[a, b]$ and use the Fundamental Theorem of Integral Calculus to show that the Mean Value Theorems for Integrals and Derivatives are merely alternative expressions of the same fact. *Hint:* Note that $f(x)$ is a primitive for $f'(x)$.

10. Show that the integration by parts formula,

$$(3) \qquad \int u\, dv = uv - \int v\, du,$$

is equivalent to the formula for the differential of the product of two functions: $d(uv) = u\, dv + v\, du$. On occasion integrations by parts can be more readily evaluated by judicious differentiation of products than by use of (3). (Cf. Exs. 11-12.)

11. Obtain the integrals

$$(4) \qquad \int e^{ax} \sin bx\, dx = \frac{e^{ax}}{a^2 + b^2}(a \sin bx - b \cos bx) + C,$$

$$(5) \qquad \int e^{ax} \cos bx\, dx = \frac{e^{ax}}{a^2 + b^2}(b \sin bx + a \cos bx) + C,$$

as follows: (i) Differentiate the products $e^{ax} \sin bx$ and $e^{ax} \cos bx$; (ii) multiply both members of each of the resulting equations by constants so that addition or subtraction eliminates either the terms involving $\cos bx$ or the terms involving $\sin bx$. (Cf. Ex. 10.)

12. Find a primitive of $x^4 e^x$ by the methods suggested in Exercises 10-11. Devise additional examples to illustrate both methods of integrating by parts.

13. Establish the following integration by parts formula, where $u(x)$, $v(x)$, $u'(x)$, and $v'(x)$ are continuous on $[a, b]$:

$$(6) \qquad \int_a^b u(x)\, v'(x)\, dx = u(b)\, v(b) - u(a)\, v(a) - \int_a^b v(x)\, u'(x)\, dx.$$

14. Prove the **Second Mean Value Theorem for Integrals**: *If $f(x)$, $\phi(x)$, and $\phi'(x)$ are continuous on the closed interval $[a, b]$, and if $\phi(x)$ is monotonic there* (equivalently, by Exercise 45, § 408, $\phi'(x)$ does not change sign there), *then there exists a number ξ such that $a < \xi < b$ and*

$$(7) \qquad \int_a^b f(x)\, \phi(x)\, dx = \phi(a) \int_a^\xi f(x)\, dx + \phi(b) \int_\xi^b f(x)\, dx.$$

(For a generalization that eliminates assumptions on $\phi'(x)$ see Ex. 28, § 518; also cf. Ex. 29, § 518.) *Hint:* Use integration by parts, letting $F(x) = \int_a^x f(t)\, dt$:

$$\int_a^b f(x)\, \phi(x)\, dx = \phi(x)\, F(x) \Big]_a^b - \int_a^b F(x)\, \phi'(x)\, dx,$$

and apply the generalized form of the First Mean Value Theorem (Ex. 6, § 503) to this last integral.

15. Show that the function $f(x)$ defined on $[0, 1]$ to be 1 if x is rational and 0 if x is irrational (Example 6, § 201) has neither integral nor primitive there. *Hint:* Cf. Ex. 47, § 408.

16. Show that although $\int_a^x f(t)\, dt$ is always a primitive of a given continuous function $f(x)$, not every primitive of $f(x)$ can be written in the form $\int_a^x f(t)\, dt$. *Hint:* Consider $f(x) = \cos x$.

17. Show that the function $f(x) = 0$ for $0 \leq x \leq 1$ and $f(x) = 1$ for $1 < x \leq 2$ is integrable on the interval $[0, 2]$, although it has no primitive there. (Cf. Ex. 47, § 408.)

18. Show that the function $F(x) = x^2 \sin \frac{1}{x^2}$ ($F(0) = 0$) has an unbounded derivative $f(x)$ on $[0, 1]$, so that a function may have a primitive without being integrable.

19. Prove the following general form for the Fundamental Theorem of Integral Calculus: *If $f(x)$ is integrable on the interval $[a, b]$, and if it has a primitive $F(x)$ there, then $\int_a^b f(x)\, dx = F(b) - F(a)$.* *Hint:* Let $a_0 < a_1 < \cdots < a_n$ be an arbitrary net $\mathfrak{N}$ on $[a, b]$, and write
$$F(b) - F(a) = \sum_{i=1}^{n} \{F(a_i) - F(a_{i-1})\} = \sum_{i=1}^{n} f(x_i)\, \Delta x_i$$
(with the aid of the Law of the Mean, § 405).

★20. Generalize Exercise 19 as follows: *If $F(x)$ is continuous at every point of $[a, b]$, if $F'(x)$ exists at all but a finite number of points of $[a, b]$, and if $f(x) = F'(x)$ is integrable on $[a, b]$, then $\int_a^b f(x)\, dx = F(b) - F(a)$.* (Cf. Ex. 5, § 508.) *Hint:* Use the method suggested in Ex. 19 but require the points of $\mathfrak{N}$ to include the points where $F'(x)$ fails to exist.

★21. Prove the following general form for integration by parts: If $u(x)$ and $v(x)$ are continuous on $[a, b]$, if $u'(x)$ and $v'(x)$ exist for all but a finite number of points of $[a, b]$, and if two (and therefore all three) of the functions $(uv)'$, uv', and vu' are integrable on $[a, b]$, then
$$\int_a^b u(x)\, v'(x)\, dx = u(b)\, v(b) - u(a)\, v(a) - \int_a^b v(x)\, u'(x)\, dx.$$
(Cf. Exs. 13, 20; also Ex. 7, § 508.)

★22. Prove that if $f(x)$ is integrable on $[a, b]$, then $F(x) = \int_a^x f(t)\, dt$ is continuous on $[a, b]$. *Hint:* $|F(x) - F(x_0)| \leq \left| \int_{x_0}^x |f(t)|\, dt \right|$.

★23. Prove that if $f(x)$ is integrable on $[a, b]$, then $F(x) = \int_a^x f(t)\, dt$ is differentiable at every point x at which $f(x)$ is continuous, and at such points of continuity of $f(x)$, $F'(x) = f(x)$. *Hint:*
$$\left| \frac{F(x) - F(x_0)}{x - x_0} - f(x_0) \right| \leq \left| \frac{1}{x - x_0} \int_{x_0}^x |f(t) - f(x_0)|\, dt \right|.$$

*507. SECTIONAL CONTINUITY AND SMOOTHNESS

In many applications certain discontinuous functions play an important role. In such applications it is necessary, however, that the discontinuities be within reason, both as to their nature and as to their number. An extremely useful class of such functions, frequently employed in the study of Fourier Series, is given in the following definition.

Definition I. *A function $f(x)$ is **sectionally continuous** on the closed interval $[a, b]$ if and only if it is continuous there except for at most a finite number of removable or jump discontinuities. That is, the one-sided limits $f(a+)$ and $f(b-)$ exist and are finite and $f(x+)$ and $f(x-)$ exist and are finite for all x between a and b, and $f(x+) = f(x-) = f(x)$ for all x between a and b with at most a finite number of exceptions. For sectional continuity the function $f(x)$ may or may not be defined at a or b or at any of the exceptional points of discontinuity.* (See Fig. 512.)

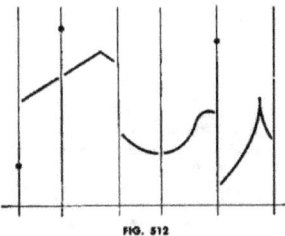

FIG. 512

An example of a sectionally continuous function is the signum function (§ 206) on the interval $[-1, 1]$. Another example is the derivative of $|x|$ (on the same interval), which is identical with the signum function, except that it is not defined at $x = 0$. Neither $1/x$ nor $\sin(1/x)$ is sectionally continuous on $[0, 1]$, or $[-1, 1]$, since there is no finite limit for either, as $x \to 0+$.

One of the most important properties of a sectionally continuous function is its integrability. This statement is made more precise and complete in the following theorem, whose proof is requested in Exercise 4, § 508, with hints.

Theorem. *If $f(x)$ is sectionally continuous on the closed interval $[a, b]$, and is continuous there except possibly at the points*

$$a = a_0 < a_1 < \cdots < a_n = b,$$

160   INTEGRATION   [§ 508

then the integrals $\int_a^b f(x)\,dx$ and $\int_{a_{i-1}}^{a_i} f(x)\,dx$, $i = 1, 2, \cdots, n$, exist and

(1) $$\int_a^b f(x)\,dx = \sum_{i=1}^n \int_{a_{i-1}}^{a_i} f(x)\,dx.$$

The values of the integrals in formula (1) are not affected by any values that might be assigned or reassigned to $f(x)$ at the exceptional points.

Closely related to sectional continuity is the more restrictive property of sectional smoothness:

Definition II. *A function $f(x)$ is **sectionally smooth** on the closed interval $[a, b]$ if and only if both $f(x)$ and $f'(x)$ are sectionally continuous there.* (Cf. Ex. 8, § 508.)

Examples. On the interval $[-1, 1]$ the following functions are sectionally smooth: (a) the signum function (§ 206); (b) $|x|$; (c) $f(x) = x - 1$ if $x > 0$, $f(x) = x + 1$ if $x < 0$. On $[-1, 1]$ the following functions are sectionally continuous but not sectionally smooth: (d) $\sqrt{1 - x^2}$; (e) $\sqrt[3]{x^2}$; (f) $x^2 \sin(1/x)$ (Example 3, § 403).

★508. EXERCISES

★1. Show that the function $[x]$ (Example 5, § 201) is sectionally smooth on any finite closed interval. Draw the graph of its derivative.

★2. Show that $|\sin x|$ is sectionally smooth on any finite interval. Draw the graphs of the function and its derivative.

★3. Prove that a function that is sectionally continuous on a closed interval is bounded there. Must it have a maximum value and a minimum value there? Must it take on all values between any two of its values?

★4. Prove the Theorem of § 507. *Hint:* For the interval $[a_{i-1}, a_i]$, if $f(x)$ is defined or redefined: $f(a_{i-1}) = f(a_{i-1}+)$ and $f(a_i) = f(a_i-)$, then $f(x)$ becomes continuous there. Apply Theorem V, § 501, and induction.

★5. Let $F(x)$ be sectionally smooth on $[a, b]$, with exceptional points a_i, $i = 1, 2, \cdots, n - 1$, and possibly $a_0 = a$ and $a_n = b$, and denote by J_i the jump of $F(x)$ at a_i: $J_i \equiv F(a_i+) - F(a_i-)$, $i = 1, 2, \cdots, n - 1$. Prove that

(1) $$\int_a^b F'(x)\,dx = F(b-) - F(a+) - \sum_{i=1}^{n-1} J_i.$$

(Cf. Ex. 20, § 506.)

★6. Prove that if $f(x)$ and $g(x)$ are sectionally continuous (smooth) then so are $f(x) + g(x)$ and $f(x) g(x)$. Extend by induction to n terms and n factors.

★7. Generalize Exercise 5 to the following integration by parts formula: If $F(x)$ is sectionally smooth on $[a, b]$ and if $G(x)$ has a continuous derivative there, then (with the notation of Ex. 5)

§ 509] REDUCTION FORMULAS 161

(2) $\int_a^b F'(x) \, G(x) \, dx = F(b-) \, G(b) - F(a+) \, G(a)$
$$- \int_a^b F(x) \, G'(x) \, dx - \sum_{i=1}^{n-1} J_i \, G(a_i).$$

(Cf. Ex. 21, § 506.)

★8. Prove that if $f'(x)$ is sectionally continuous on $[a, b]$, then $f(x)$ is sectionally smooth there. Furthermore, prove that $f(x)$ has a right-hand and a left-hand derivative at each point between a and b, and a right-hand derivative at a and a left-hand derivative at b. (Cf. Exs. 53-54, § 408.)

★9. If $f(x)$ is sectionally continuous on the closed interval $[a, b]$, then $F(x) \equiv \int_a^x f(t) \, dt$ is continuous there and $F'(x)$ exists and is equal to $f(x)$ there with at most a finite number of exceptions (Exs. 22-23, § 506). Prove that any function $\Phi(x)$ that is continuous on $[a, b]$ such that $\Phi'(x)$ exists and is equal to $f(x)$ there with at most a finite number of exceptions must differ from $F(x)$ by at most a constant. If $\Phi(x)$ is not everywhere continuous, what can you say?

★509. REDUCTION FORMULAS

It often happens that the routine evaluation of an integral involves repeated applications of integration by parts, all such integrations by parts being of the same tedious type. For example, in evaluating $I = \int \sin^{10} x \, dx$, we might proceed:

$$I = \int \sin^9 x \, d(-\cos x) = -\sin^9 x \cos x + 9 \int \sin^8 x \cos^2 x \, dx$$
$$= -\sin^9 x \cos x + 9 \int \sin^8 x \, dx - 9I,$$

so that

(1) $\quad I = -\tfrac{1}{10} \sin^9 x \cos x + \tfrac{9}{10} \int \sin^8 x \, dx.$

We have succeeded in reducing the exponent from 10 to 8. We could repeat this labor to reduce the new exponent from 8 to 6; then from 6 to 4; etc.

A more satisfactory method is to establish a *single formula* to handle all integrals of a single type. We present a few derivations of such **reduction formulas** in Examples, and ask for more in the Exercises of the following section. Since differentiation is basically a simpler process than integration, we perform our integrations by parts by means of differentiating certain products.

Example 1. Express $\int \sin^m x \cos^n x \, dx$, where $m + n \neq 0$, in terms of an integral with reduced exponent on $\sin x$ (cf. Exs. 1-2, § 510).

Solution. The derivative of the product $\sin^p x \cos^q x$ is

$$p \sin^{p-1} x \cos^{q+1} x - q \sin^{p+1} x \cos^{q-1} x$$
$$= p \sin^{p-1} x \cos^{q-1} x (1 - \sin^2 x) - q \sin^{p+1} x \cos^{q-1} x$$
$$= p \sin^{p-1} x \cos^{q-1} x - (p+q) \sin^{p+1} x \cos^{q-1} x.$$

In other words,

(2) $\quad p \int \sin^{p-1} x \cos^{q-1} x \, dx - (p+q) \int \sin^{p+1} x \cos^{q-1} x \, dx = \sin^p x \cos^q x + C.$

Letting $p = m - 1$ and $q = n + 1$, and absorbing the constant of integration with the $\int$, we have the formula sought:

(3) $\quad \int \sin^m x \cos^n x \, dx = -\frac{\sin^{m-1} x \cos^{n+1} x}{m+n} + \frac{m-1}{m+n} \int \sin^{m-2} x \cos^n x \, dx.$

Equation (1) is a special case of (3), with $m = 10$, $n = 0$.

Often it is desirable to increase a negative exponent.

Example 2. Establish the reduction formula ($m \neq 1$) (cf. Exs. 1-2, § 510):

(4) $\quad \int \frac{\cos^n x \, dx}{\sin^m x} = -\frac{\cos^{n+1} x}{(m-1) \sin^{m-1} x} - \frac{n-m+2}{m-1} \int \frac{\cos^n x \, dx}{\sin^{m-2} x}.$

Solution. In (2), let $p = -m + 1$ and $q = n + 1$.

Example 3. Establish the reduction formula ($n \neq 1$):

(5) $\quad \int \frac{dx}{(a^2 + x^2)^n} = \frac{x}{2(n-1)a^2(a^2 + x^2)^{n-1}} + \frac{2n-3}{2(n-1)a^2} \int \frac{dx}{(a^2 + x^2)^{n-1}}.$

Solution. The derivative of the product $x(a^2 + x^2)^m$ is

$$(a^2 + x^2)^m + 2mx^2(a^2 + x^2)^{m-1}$$
$$= (a^2 + x^2)^m + 2m[(a^2 + x^2) - a^2](a^2 + x^2)^{m-1}$$
$$= (1 + 2m)(a^2 + x^2)^m - 2ma^2(a^2 + x^2)^{m-1}.$$

In other words,

(6) $\quad (1 + 2m) \int (a^2 + x^2)^m \, dx - 2ma^2 \int (a^2 + x^2)^{m-1} \, dx = x(a^2 + x^2)^m + C.$

Now let $m = -n + 1$, from which we obtain (5).

Useful reduction formulas are given in nearly every Table of Integrals.

★510. EXERCISES

In Exercises 1-12, establish the reduction formula.

★1. $\int \sin^m x \cos^n x \, dx = \frac{\sin^{m+1} x \cos^{n-1} x}{m+n} + \frac{n-1}{m+n} \int \sin^m x \cos^{n-2} x \, dx \quad (m+n \neq 0).$

★2. $\int \frac{\sin^m x \, dx}{\cos^n x} = \frac{\sin^{m+1} x}{(n-1) \cos^{n-1} x} - \frac{m-n+2}{n-1} \int \frac{\sin^m x \, dx}{\cos^{n-2} x} \quad (n \neq 1).$

★3. $\int \tan^n x \, dx = \frac{\tan^{n-1} x}{n-1} - \int \tan^{n-2} x \, dx \quad (n \neq 1).$

§ 510]   EXERCISES

★4. $\int \cot^n x \, dx = -\dfrac{\cot^{n-1} x}{n-1} - \int \cot^{n-2} x \, dx \ (n \neq 1).$

★5. $\int \sec^n x \, dx = \dfrac{\sec^{n-2} x \tan x}{n-1} + \dfrac{n-2}{n-1} \int \sec^{n-2} x \, dx \ (n \neq 1).$

★6. $\int \csc^n x \, dx = -\dfrac{\csc^{n-2} x \cot x}{n-1} + \dfrac{n-2}{n-1} \int \csc^{n-2} x \, dx \ (n \neq 1).$

★7. $\int x^n \sin x \, dx = -x^n \cos x + n x^{n-1} \sin x - n(n-1) \int x^{n-2} \sin x \, dx.$

★8. $\int x^n \cos x \, dx = x^n \sin x + n x^{n-1} \cos x - n(n-1) \int x^{n-2} \cos x \, dx.$

★9. $\int x^m (ax+b)^n \, dx = \dfrac{x^m (ax+b)^{n+1}}{a(m+n+1)}$
$\qquad - \dfrac{mb}{a(m+n+1)} \int x^{m-1}(ax+b)^n \, dx \ (m+n+1 \neq 0).$

★10. $\int x^m(ax+b)^n \, dx = \dfrac{x^{m+1}(ax+b)^n}{m+n+1}$
$\qquad + \dfrac{nb}{m+n+1} \int x^m (ax+b)^{n-1} \, dx \ (m+n+1 \neq 0).$

★11. $\int \dfrac{x^n \, dx}{\sqrt{ax^2+bx+c}} = \dfrac{x^{n-1}}{a(n-1)}$
$\qquad - \dfrac{b}{a} \int \dfrac{x^{n-1} \, dx}{\sqrt{ax^2+bx+c}} - \dfrac{c}{a} \int \dfrac{x^{n-2} \, dx}{\sqrt{ax^2+bx+c}} \ (n \neq 1).$

★12. $\int x^n \sqrt{2ax-x^2} \, dx = -\dfrac{x^{n-1}(2ax-x^2)^{\frac{3}{2}}}{n+2}$
$\qquad + \dfrac{a(2n+1)}{n+2} \int x^{n-1} \sqrt{2ax-x^2} \, dx \ (n \neq -2).$

In Exercises 13-20, perform the integration, using reduction formulas above.

★13. $\int \sin^4 x \, dx.$ ★14. $\int \cos^5 x \, dx.$

★15. $\int \cot^5 \dfrac{x}{5} \, dx.$ ★16. $\int \sec^7 x \, dx.$

★17. $\int x^4 \sin 2x \, dx.$ ★18. $\int x^{\frac{3}{2}}(x+4)^{\frac{1}{2}} \, dx.$

★19. $\int \dfrac{x^3 \, dx}{\sqrt{x^2+x+1}}.$ ★20. $\int x^2 \sqrt{6x-x^2} \, dx.$

★★21. Use mathematical induction to verify the formula:
$$\int_a^b (x-a)^m (b-x)^n \, dx = (b-a)^{m+n+1} \dfrac{m! \, n!}{(m+n+1)!} \ (m \text{ and } n \text{ positive integers}).$$

★★22. Establish the reduction formula, for $\alpha, \beta, n \geq 1$ (cf. Ex. 14, § 1538):
$$\int_0^1 x^n x^{\alpha-1}(1-x)^{\beta-1} \, dx = \dfrac{n+\alpha-1}{n+\alpha+\beta-1} \int_0^1 x^{n-1} x^{\alpha-1}(1-x)^{\beta-1} \, dx.$$

511. IMPROPER INTEGRALS, INTRODUCTION

The definite integral $\int_a^b f(x)\,dx$, as defined in § 501, has meaning only if a and b are finite and $f(x)$ is bounded on the interval $[a, b]$. In the following two sections we shall extend the definition so that under certain circumstances the symbol $\int_a^b f(x)\,dx$ shall be meaningful even when the interval of integration is infinite or the function $f(x)$ is unbounded.

For the sake of conciseness, parentheses will be used in some of the definitions to indicate alternative statements. For a discussion of the use of parentheses for alternatives, see the Preface. Also see Chapter 14.

512. IMPROPER INTEGRALS, FINITE INTERVAL

Definition I. *Let $f(x)$ be (Riemann-) integrable on the interval $[a, b - \epsilon]$ $([a + \epsilon, b])$ for every number ϵ such that $0 < \epsilon < b - a$, but not integrable on the interval $[a, b]$, and assume that*

$$\lim_{\epsilon \to 0+} \int_a^{b-\epsilon} f(x)\,dx \quad \left(\lim_{\epsilon \to 0+} \int_{a+\epsilon}^b f(x)\,dx \right)$$

*exists. Under these conditions the **improper integral** $\int_a^b f(x)\,dx$ is defined to be this limit:*

(1) $\qquad \int_a^b f(x)\,dx \equiv \lim_{\epsilon \to 0+} \int_a^{b-\epsilon} f(x)\,dx \quad \left(\lim_{\epsilon \to 0+} \int_{a+\epsilon}^b f(x)\,dx \right).$

If the limit in (1) *is finite the integral $\int_a^b f(x)\,dx$ is **convergent** to this limit and the function $f(x)$ is said to be **improperly integrable** on the half-open interval $[a, b)$ $((a, b])$; if the limit in* (1) *is infinite or does not exist, the integral is **divergent**.*

NOTE 1. A function improperly integrable on a half-open interval is necessarily unbounded there. In fact, it is unbounded in every neighborhood of the end-point of the interval that is not included. (Cf. Theorem III, § 502, and Ex. 28, § 503.)

Definition II. *Let $[a, b]$ be a given finite interval, let $a < c < b$, and let both integrals $\int_a^c f(x)\,dx$ and $\int_c^b f(x)\,dx$ be convergent improper integrals in the sense of Definition I. Then the **improper integral** $\int_a^b f(x)\,dx$ is **convergent** and defined to be:*

(2) $\qquad \int_a^b f(x)\,dx \equiv \int_a^c f(x)\,dx + \int_c^b f(x)\,dx.$

§ 512] IMPROPER INTEGRALS, FINITE INTERVAL 165

If either integral on the right-hand side of (2) *diverges, so does* $\int_a^b f(x)\,dx$.

Note 2. There are four ways in which the integrals (2) may be improper, corresponding to the points in the neighborhoods of which $f(x)$ is unbounded: (i) $c-$ and $c+$, (ii) $a+$ and $c+$, (iii) $c-$ and $b-$, and (iv) $a+$ and $b-$. In this last case (iv), the definition (2) is meaningful only if the value of $\int_a^b f(x)\,dx$ is independent of the interior point c. This independence is indeed a fact (cf. Ex. 11, § 515). As illustration of this independence of the point c, see Example 4, below.

Note 3. In case the function $f(x)$ is (Riemann-) integrable on the interval $[a, b]$ it is often convenient to refer to the integral $\int_a^b f(x)\,dx$ as a **proper** integral, in distinction to the improper integrals defined above, and to call the integral $\int_a^b f(x)\,dx$ **convergent**, even though it is not improper. (Cf. Example 3, below, and Ex. 28, § 503.)

Example 1. Evaluate $\int_4^5 \frac{dx}{\sqrt{x-4}}$.

Solution. The integrand becomes infinite as $x \to 4+$. The given improper integral has the value
$$\lim_{\epsilon \to 0+} \int_{4+\epsilon}^5 \frac{dx}{\sqrt{x-4}} = \lim_{\epsilon \to 0+} \left[2\sqrt{x-4}\right]_{4+\epsilon}^5 = \lim_{\epsilon \to 0+} [2 - 2\sqrt{\epsilon}] = 2.$$

Example 2. Evaluate $\int_{-1}^1 \frac{dx}{x^2}$.

Solution. A thoughtless, brash, and incorrect "evaluation" would yield the ridiculous negative result:
$$\int_{-1}^1 \frac{dx}{x^2} = \left[-\frac{1}{x}\right]_{-1}^1 = [-1] - [1] = -2.$$
Since the integrand becomes infinite as $x \to 0$, a correct evaluation is
$$\int_{-1}^1 \frac{dx}{x^2} = \lim_{\epsilon \to 0+} \int_{-1}^{-\epsilon} \frac{dx}{x^2} + \lim_{\epsilon \to 0+} \int_\epsilon^1 \frac{dx}{x^2}$$
$$= \lim_{\epsilon \to 0+} \left(\frac{1}{\epsilon} - 1\right) + \lim_{\epsilon \to 0+} \left(-1 + \frac{1}{\epsilon}\right) = +\infty,$$
and the integral diverges.

Example 3. For what values of p does $\int_0^1 \frac{dx}{x^p}$ converge?

Solution. If $p \leq 0$, the integral is proper, and therefore converges (Note 3). If $0 < p < 1$,
$$\int_0^1 x^{-p}\,dx = \lim_{\epsilon \to 0+} \left[\frac{x^{1-p}}{1-p}\right]_\epsilon^1 = \frac{1}{1-p},$$
and the integral converges. If $p = 1$, $\lim_{\epsilon \to 0+} \left[\ln x\right]_\epsilon^1 = \lim_{\epsilon \to 0+} (-\ln \epsilon) = +\infty$,

and if $p > 1$, $\lim\limits_{\epsilon \to 0+} \left[\dfrac{x^{1-p}}{1-p} \right]_\epsilon^1 = +\infty$. Therefore the given integral converges if and only if $p < 1$.

Example 4. Evaluate $\displaystyle\int_{-2}^{2} \dfrac{dx}{\sqrt{4-x^2}}$.

Solution. The integrand becomes infinite at both end-points of the interval $[-2, 2]$, and we therefore evaluate according to Definition II, choosing some number c in the interior of this interval. The simplest value of c is 0. We find, then, that

$$\int_{-2}^{0} \dfrac{dx}{\sqrt{4-x^2}} = \lim_{\epsilon \to 0+} \left[\operatorname{Arcsin} \dfrac{x}{2} \right]_{-2+\epsilon}^{0} = -\operatorname{Arcsin}(-1) = \dfrac{\pi}{2},$$

and

$$\int_{0}^{2} \dfrac{dx}{\sqrt{4-x^2}} = \lim_{\epsilon \to 0+} \left[\operatorname{Arcsin} \dfrac{x}{2} \right]_{0}^{2-\epsilon} = \operatorname{Arcsin}(1) = \dfrac{\pi}{2},$$

and the value of the given integral is π. Notice that any other value of c between -2 and 2 could have been used (Note 2, above):

$$\int_{-2}^{c} \dfrac{dx}{\sqrt{4-x^2}} + \int_{c}^{2} \dfrac{dx}{\sqrt{4-x^2}} = \left(\operatorname{Arcsin} \dfrac{c}{2} + \dfrac{\pi}{2} \right) + \left(\dfrac{\pi}{2} - \operatorname{Arcsin} \dfrac{c}{2} \right) = \pi.$$

NOTE 4. Under certain circumstances a student may evaluate an improper integral on a finite interval, with correct result, without forming a limit, or even without recognizing that the given integral is improper. This would be the case with Example 1, above—but not with Example 2. The following theorem justifies such a method and simplifies many evaluations:

Theorem. *If $f(x)$ is continuous in the open interval (a, b), and if there exists a function $F(x)$ which is continuous over the closed interval $[a, b]$ and such that $F'(x) = f(x)$ in the open interval (a, b), then the integral $\int_a^b f(x)\,dx$, whether proper or improper, converges and*

$$\int_a^b f(x)\,dx = F(b) - F(a).$$

(Cf. Ex. 15, § 515.)

Proof. By the Fundamental Theorem of Integral Calculus, for any c between a and b, and sufficiently small positive ϵ and η,

$$\int_{a+\epsilon}^{c} f(x)\,dx + \int_{c}^{b-\eta} f(x)\,dx = F(b-\eta) - F(a+\epsilon),$$

and the result follows from the continuity of $F(x)$ at a and b.

Example 5. In Example 1, above, let $F(x) = 2\sqrt{x-4}$. Then

$$\int_4^5 \dfrac{dx}{\sqrt{x-4}} = \left[2\sqrt{x-4} \right]_4^5 = 2.$$

513. IMPROPER INTEGRALS, INFINITE INTERVAL

Definition I. Let $f(x)$ be (Riemann-) integrable on the interval $[a, u]$ for every number $u > a$, and assume that $\lim_{u \to +\infty} \int_a^u f(x)\, dx$ exists. Under these conditions the **improper integral** $\int_a^{+\infty} f(x)\, dx$ is defined to be this limit:

$$(1) \qquad \int_a^{+\infty} f(x)\, dx = \lim_{u \to +\infty} \int_a^u f(x)\, dx.$$

*If the limit in (1) is finite the improper integral is **convergent** to this limit and the function $f(x)$ is said to be **improperly integrable** on the interval $[a, +\infty)$; if the limit in (1) is infinite or does not exist, the integral is **divergent**.*

A similar definition holds for the improper integral $\int_{-\infty}^{a} f(x)\, dx$.

Definition II. Let $f(x)$ be (Riemann-) integrable on every finite closed interval, and assume that both improper integrals $\int_0^{+\infty} f(x)\, dx$ and $\int_{-\infty}^0 f(x)\, dx$ converge. Then the **improper integral** $\int_{-\infty}^{+\infty} f(x)\, dx$ is **convergent** and defined to be:

$$(2) \qquad \int_{-\infty}^{+\infty} f(x)\, dx = \int_{-\infty}^0 f(x)\, dx + \int_0^{+\infty} f(x)\, dx.$$

If either integral on the right-hand side of (2) diverges, so does $\int_{-\infty}^{+\infty} f(x)\, dx$.

NOTE 1. The improper integral (2) could have been defined unambiguously: $\int_{-\infty}^{+\infty} f(x)\, dx = \int_{-\infty}^{c} f(x)\, dx + \int_{c}^{+\infty} f(x)\, dx$, where c is an arbitrary number. (Cf. Ex. 12, § 515.)

Improper integrals on finite and infinite intervals are often combined:

Definition III. Let $f(x)$ be improperly integrable on the interval $(a, c]$ and on the interval $[c, +\infty)$, where c is any constant greater than a. Then the **improper integral** $\int_a^{+\infty} f(x)\, dx$ is **convergent** and defined to be:

$$(3) \qquad \int_a^{+\infty} f(x)\, dx = \int_a^c f(x)\, dx + \int_c^{+\infty} f(x)\, dx.$$

If either integral on the right-hand side of (3) diverges, so does $\int_a^{+\infty} f(x)\, dx$.

168   INTEGRATION   [§ 514

Similar statements hold for a similarly improper integral $\int_{-\infty}^{c} f(x)\,dx$.

Note 2. The improper integral (3) is independent of c. (Cf. Ex. 12, § 515.)

Example 1. Evaluate $\int_{0}^{+\infty} e^{-ax}\,dx,\ a > 0$.

Solution. $\int_{0}^{+\infty} e^{-ax}\,dx = \lim_{u \to +\infty}\left[\dfrac{e^{-ax}}{-a}\right]_{0}^{u} = \lim_{u \to +\infty}\left[\dfrac{1}{a} - \dfrac{e^{-au}}{a}\right] = \dfrac{1}{a}$.

Example 2. Evaluate $\int_{-\infty}^{+\infty} \dfrac{dx}{a^2 + x^2},\ a > 0$.

Solution.

$$\int_{-\infty}^{+\infty}\frac{dx}{a^2+x^2} = \lim_{u\to-\infty}\left[\frac{1}{a}\operatorname{Arctan}\frac{x}{a}\right]_{u}^{0} + \lim_{v\to+\infty}\left[\frac{1}{a}\operatorname{Arctan}\frac{x}{a}\right]_{0}^{v}$$

$$= \lim_{u\to-\infty}\left[-\frac{1}{a}\operatorname{Arctan}\frac{u}{a}\right] + \lim_{v\to+\infty}\left[\frac{1}{a}\operatorname{Arctan}\frac{v}{a}\right] = \frac{1}{a}\cdot\frac{\pi}{2} + \frac{1}{a}\cdot\frac{\pi}{2} = \frac{\pi}{a}.$$

Example 3. For what values of p does $\int_{1}^{+\infty}\dfrac{dx}{x^p}$ converge?

Solution. If $p \neq 1$, $\int_{1}^{u} x^{-p}\,dx = \left[\dfrac{x^{1-p}}{1-p}\right]_{1}^{u}$, and the integral converges if $p > 1$ and diverges if $p < 1$. Similarly, $\int_{1}^{+\infty}\dfrac{dx}{x} = \lim_{u\to+\infty}\ln u = +\infty$. Therefore the given integral converges if and only if $p > 1$, and its value, for such p, is $1/(p-1)$.

Example 4. For what values of p does $\int_{0}^{+\infty}\dfrac{dx}{x^p}$ converge?

Solution. For this improper integral to converge, both $\int_{0}^{1}$ and $\int_{1}^{+\infty}$ must converge. But they never converge for the same value of p. (Examples 3, §§ 512, 513.) Answer: none.

514. COMPARISON TESTS. DOMINANCE

For a nonnegative function, convergence of an improper integral means (in a sense determined by the definition of the improper integral concerned) that *the function is not too big*: on a finite interval the function does not become infinite too fast, and on an infinite interval the function does not approach 0 too slowly. This means that whenever a nonnegative function $g(x)$ has a convergent improper integral, any well-behaved nonnegative function $f(x)$ less than or equal to $g(x)$ also has a convergent improper integral. We make these ideas precise:

Definition I. *The statement that a function* $g(x)$ **dominates** *a function* $f(x)$ *on a set* A *means that both functions are defined for every member* x *of* A *and that for every such* x, $|f(x)| \leq g(x)$.

§ 514]   COMPARISON TESTS. DOMINANCE   169

Note 1. Any dominating function is automatically nonnegative, although a dominated function may have negative values.

Theorem I. Comparison Test. *If a function $g(x)$ dominates a nonnegative function $f(x)$ on the interval $[a, b)$ $\bigl([a, +\infty)\bigr)$, if both functions are integrable on the interval $[a, c]$ for every c such that $a < c < b$ $(a < c)$, and if the improper integral $\int_a^b g(x)\,dx$ $\left(\int_a^{+\infty} g(x)\,dx\right)$ converges, then so does $\int_a^b f(x)\,dx$ $\left(\int_a^{+\infty} f(x)\,dx\right)$.*

Similar statements apply to other types of improper integrals defined in §§ 512–513.

Proof. Since the two functions are nonnegative, the two integrals $\int_a^c f(x)\,dx$ and $\int_a^c g(x)\,dx$ are monotonically increasing functions of c, and both have limits as $c \to b-$ $(c \to +\infty)$. (Cf. § 215.) The inequalities $0 \leq \int_a^c f(x)\,dx \leq \int_a^c g(x)\,dx$ imply, thanks to Theorem VII, § 207, the inequalities $0 \leq \int_a^b f(x)\,dx \leq \int_a^b g(x)\,dx$ $\left(0 \leq \int_a^{+\infty} f(x)\,dx \leq \int_a^{+\infty} g(x)\,dx\right)$, and the proof is complete.

Example 1. Since, for $x \geq 1$, $\dfrac{1}{x^2 + 5x + 17} < \dfrac{1}{x^2}$, the convergence of $\int_1^{+\infty} \dfrac{dx}{x^2}$ implies that of $\int_1^{+\infty} \dfrac{dx}{x^2 + 5x + 17}$.

Note 2. An interchange of the roles of $f(x)$ and $g(x)$ in Theorem I furnishes a comparison test for divergence.

Example 2. Since, for $0 < x \leq 1$, $\dfrac{1}{x^2 + 5x} \geq \dfrac{1}{6x}$, the divergence of $\int_0^1 \dfrac{dx}{x}$ implies that of $\int_0^1 \dfrac{dx}{x^2 + 5x}$.

A convenient method for establishing dominance, and hence the convergence of an improper integral, can be formulated in terms of the "big O" notation (this is an upper case letter O derived from the expression "order of magnitude"):

Definition II. *The notation $f = O(g)$ (read "f is big O of g"), or equivalently $f(x) = O(g(x))$, as x approaches some limit (finite or infinite), means that within some deleted neighborhood of that limit, $f(x)$ is dominated by some positive constant multiple of $g(x)$: $|f(x)| \leq K \cdot g(x)$. If simultaneously $f = O(g)$ and $g = O(f)$, the two functions $f(x)$ and $g(x)$ are said to be of the **same order of magnitude** as x approaches its limit.*[†]

[†] Similar to the "big O" notation is the "little o" notation: $f = o(g)$ means that within some deleted neighborhood of the limiting value of x, an inequality of the form $|f(x)| \leq K(x)\,g(x)$ holds, where $K(x) \geq 0$ and $K(x) \to 0$. Obviously $f = o(g)$ implies

170 INTEGRATION

Example 3. As $x \to +\infty$, $\sin x = O(1)$ and $\ln x = O(x)$. (Cf. Ex. 26, § 515.)

Big O and order of magnitude relationships are usually established by taking limits:

Theorem II. *Let $f(x)$ and $g(x)$ be positive functions and assume that $\lim \dfrac{f(x)}{g(x)} = L$ exists, in a finite or infinite sense. Then:*

(i) $0 \leq L < +\infty$ implies $f = O(g)$,
(ii) $0 < L \leq +\infty$ implies $g = O(f)$,
(iii) $0 < L < +\infty$ implies that $f(x)$ and $g(x)$ are of the same order of magnitude.

(Give the details of the proof in Ex. 25, § 515.)

Example 4. As $x \to +\infty$, $\dfrac{1}{x+1} - \dfrac{1}{x} = O\left(\dfrac{1}{x^2}\right)$, since $\left[\dfrac{1}{x+1} - \dfrac{1}{x}\right] / \left[\dfrac{1}{x^2}\right] = \dfrac{-x^2}{x^2+x}$, and $\lim\limits_{x \to +\infty} \left|\dfrac{-x^2}{x^2+x}\right| = 1 < +\infty$.

Note 3. The "big O" notation is also used in the following sense: $f(x) = g(x) + O(h(x))$ means that $f(x) - g(x) = O(h(x))$. Thus, as $x \to +\infty$, $\dfrac{1}{x+1} = \dfrac{1}{x} + O\left(\dfrac{1}{x^2}\right)$, by Example 4.

Theorem III. *If $f(x)$ and $g(x)$ are nonnegative on $[a, b]$ and integrable on $[a, c]$ for every c such that $a < c < b$, and if $f = O(g)$ as $x \to b-$, then the convergence of $\int_a^b g(x)\, dx$ implies that of $\int_a^b f(x)\, dx$.*

Similar statements apply to other types of improper integrals defined in §§ 512-513.

Proof. The convergence of $\int_c^b K g(x)\, dx$ implies that of $\int_c^b f(x)\, dx$ by Theorem 1.

Note 4. An interchange of the roles of $f(x)$ and $g(x)$ in Theorem III furnishes a test for divergence. If $f(x)$ and $g(x)$ are of the same order of magnitude, then their two integrals either both converge or both diverge.

Example 5. Test for convergence or divergence: $\int_0^1 \dfrac{dx}{1-x^3}$.

Solution. The integrand, $f(x)$, becomes infinite at $x = 1$. Write
$$f(x) = \dfrac{1}{(1-x)(1+x+x^2)},$$
and compare it with $g(x) = \dfrac{1}{1-x}$. Since
$$\lim_{x \to 1-} \dfrac{f(x)}{g(x)} = \lim_{x \to 1-} \dfrac{1}{1+x+x^2} = \tfrac{1}{3} > 0,$$
$g = O(f)$ as $x \to 1-$, and since $\int_0^1 g(x)\, dx$ diverges, $\int_0^1 f(x)\, dx$ diverges.

Example 6. Test for convergence or divergence: $\int_0^5 \frac{dx}{\sqrt[3]{7x + 2x^4}}$.

Solution. The integrand, $f(x)$, becomes infinite at $x = 0$. Write
$$f(x) = \frac{1}{\sqrt[3]{x}\sqrt[3]{7 + 2x^3}}$$
and compare it with $g(x) = \frac{1}{\sqrt[3]{x}}$. Since
$$\lim_{x \to 0+} \frac{f(x)}{g(x)} = \lim_{x \to 0+} \frac{1}{\sqrt[3]{7 + 2x^3}} = \frac{1}{\sqrt[3]{7}} < +\infty,$$
$f = O(g)$ as $x \to 0+$, and since $\int_0^5 g(x)\, dx$ converges, $\int_0^5 f(x)\, dx$ converges.

Example 7. Test for convergence or divergence: $\int_1^{+\infty} \frac{dx}{\sqrt{x^3 + 2x + 2}}$.

Solution. The integrand, as $x \to +\infty$, is of the same order of magnitude as $x^{\frac{3}{2}}$. Therefore the given integral converges.

Example 8. (The *Beta Function*.) Determine the values of p and q for which
$$B(p, q) = \int_0^1 x^{p-1}(1 - x)^{q-1}\, dx$$
converges. (Cf. § 1414.)

Solution. The integrand $f(x)$ has possible discontinuities at $x = 0$ and $x = 1$. As $x \to 0+$, $f(x)$ is of the same order of magnitude as x^{p-1}, and as $x \to 1-$, $f(x)$ is of the same order of magnitude as $(1-x)^{q-1}$. Therefore convergence of both $\int_0^c f(x)\, dx$ and $\int_c^1 f(x)\, dx$, for $0 < c < 1$, is equivalent to the two inequalities $p - 1 > -1$ and $q - 1 > -1$. Therefore $\int_0^1 f(x)\, dx$ converges if and only if both p and q are positive.

Example 9. (The *Gamma Function*.) Determine the values of α for which
$$\Gamma(\alpha) = \int_0^{+\infty} x^{\alpha-1} e^{-x}\, dx$$
converges. (Cf. § 1413.)

Solution. If $f(x)$ is the integrand, $f(x)$ is of the same order of magnitude as $x^{\alpha-1}$ as $x \to 0+$, and $f(x) = O\left(\frac{1}{x^2}\right)$ as $x \to +\infty$ (this is true by l'Hospital's Rule in case $\alpha > -1$). Since $\int_0^1 x^{\alpha-1}\, dx$ converges if and only if $\alpha > 0$, and $\int_1^{+\infty} \frac{dx}{x^2}$ converges, the given integral converges if and only if α is positive.

515. EXERCISES

In Exercises 1-10, evaluate every convergent improper integral and specify those that diverge.

172   INTEGRATION   [§ 515

1. $\displaystyle\int_0^3 \frac{dx}{\sqrt{9-x^2}}.$ 2. $\displaystyle\int_{-1}^1 \frac{dx}{\sqrt[3]{x}}.$

3. $\displaystyle\int_{-2}^2 \frac{dx}{x^4}.$ 4. $\displaystyle\int_0^{\frac{\pi}{2}} \sqrt{\sin x \tan x}\, dx.$

5. $\displaystyle\int_0^{+\infty} \frac{dx}{\sqrt{e^x}}.$ 6. $\displaystyle\int_0^{+\infty} \sin x\, dx.$

7. $\displaystyle\int_1^{+\infty} \frac{dx}{x\sqrt{x^2-1}}.$ 8. $\displaystyle\int_2^{+\infty} \frac{dx}{x(\ln x)^k}.$

9. $\displaystyle\int_0^{+\infty} \frac{e^{-\sqrt{x}}}{\sqrt{x}}\, dx.$ 10. $\displaystyle\int_{-\infty}^{+\infty} \frac{dx}{1+4x^2}.$

11. Prove that the existence and the value of the improper integral (2), § 512, does not depend on the value of c in case (iv) of Note 2, § 512. *Hint:* Let $a + \epsilon < c < d < b - \eta$. Then

$$\int_{a+\epsilon}^c + \int_c^{b-\eta} = \int_{a+\epsilon}^c + \int_c^d + \int_d^{b-\eta} = \int_{a+\epsilon}^d + \int_d^{b-\eta}.$$

12. Prove the independence of c of the improper integrals referred to in the two Notes of § 513. (Cf. Ex. 11.) Illustrate each by an example.

13. Let $f(v)$ be continuous for $c < v < d$, let $v(x)$ have values between c and d, and a continuous derivative $v'(x)$, for $a < x < b$, and assume that $v(x) \to c$ as $x \to a+$ and $v(x) \to d$ as $x \to b-$. Prove that if the integral $\int_c^d f(x)\, dv$ converges (whether proper or improper) then the left-hand member in the following integration by substitution formula also converges and

(1) $\qquad\displaystyle\int_a^b f(v(x))\, v'(x)\, dx = \int_c^d f(v)\, dv.$

Evaluate the integral of Example 4, § 512, by use of this formula and the substitution $x = 2 \sin v$. Illustrate with other examples. *Hint:* If $F(v)$ is a primitive of $f(v)$, $c < v < d$,

$$\int_{a+\epsilon}^{b-\eta} f(v(x))\, v'(x)\, dx = F(v(b-\eta)) - F(v(a+\epsilon)).$$

14. Discuss the integration by substitution formula (1) for cases involving infinite intervals. Illustrate with examples.

In Exercises 15-24, merely establish convergence or divergence. Do not try to evaluate. (Cf. §§ 1329, 1408, 1410.)

15. $\displaystyle\int_0^{+\infty} \frac{dx}{\sqrt{1+x^4}}.$ 16. $\displaystyle\int_2^{+\infty} \frac{x\, dx}{\sqrt{x^4-1}}.$

17. $\displaystyle\int_{-\infty}^{+\infty} e^{-x^2}\, dx.$ 18. $\displaystyle\int_0^1 \frac{\ln x\, dx}{\sqrt{x}}.$

19. $\displaystyle\int_0^1 \frac{dx}{\sqrt{x}\ln x}.$ 20. $\displaystyle\int_{-\infty}^0 e^x \ln |x|\, dx.$

21. $\displaystyle\int_{-\infty}^{+\infty} \frac{2x\, dx}{e^x - e^{-x}}.$ 22. $\displaystyle\int_0^1 \frac{\ln x\, dx}{1-x}.$

23. $\int_{-1}^{1} e^{\frac{1}{x}} dx.$ **24.** $\int_{0}^{1} \sqrt[3]{x \ln(1/x)}\, dx.$

25. Prove Theorem II, § 514.

26. Prove that if p is an arbitrary positive number, then

(2) $\qquad \ln x = o(x^p), \quad x = o(e^{px}), \quad \text{as } x \to +\infty;$

(3) $\qquad \ln x = o(x^{-p}), \quad \frac{1}{x} = o(e^{\frac{p}{x}}), \quad \text{as } x \to 0+.$

27. Show that the functions in any "big O" relationship can be multiplied or divided by any function whose values are positive: If $f(x) = O(g(x))$ and $h(x) > 0$, then $f(x) h(x) = O(g(x) h(x))$; conversely, if $f(x) h(x) = O(g(x) h(x))$, then $f(x) = O(g(x))$. Is the same thing true for the "little o" relation?

28. Prove that $\int_{3}^{+\infty} \frac{dx}{x(\ln x)^p}$ converges if and only if $p > 1$.

29. Prove that $\int_{3}^{+\infty} \frac{dx}{x(\ln x)(\ln \ln x)^p}$ converges if and only if $p > 1$. More generally, prove that $\int_{a}^{+\infty} \frac{dx}{x(\ln x)(\ln \ln x) \cdots (\ln \ln \cdots \ln x)^p}$, where a is sufficiently large, converges if and only if $p > 1$.

★30. The **Cauchy principal value** of the improper integral $\int_{-\infty}^{+\infty} f(x)\, dx$ is denoted and defined

$$(P)\int_{-\infty}^{+\infty} f(x)\, dx = \lim_{u \to +\infty} \int_{-u}^{u} f(x)\, dx,$$

provided the integral $\int_{-u}^{u} f(x)\, dx$ and its limit exist. Prove that whenever the improper integral $\int_{-\infty}^{+\infty} f(x)\, dx$ converges, its Cauchy principal value exists and is equal to it. Give examples to show that the converse is false. (Cf. Ex. 31.)

★31. For each of the cases (i) and (ii), below, define a Cauchy principal value of the improper integral $\int_{-a}^{a} f(x)\, dx$ assuming $f(x)$ is integrable on $[u, v]$ for every u and v such that

(i) $-a < u < v < a$; (ii) $-a \leq u < v < 0$ and $0 < u < v \leq a$.

Give a sufficient condition for the existence of the integral in each case. (Cf. Ex. 30.)

In Exercises 32-35, establish the given relation.

★32. $\int_{0}^{1} \frac{x^{p-1}}{x+1}\, dx = \int_{1}^{+\infty} \frac{x^{-p}\, dx}{x+1}, \; p > 0.$

★33. $\int_{0}^{+\infty} \frac{x^{p-1}\, dx}{x+1} = \int_{0}^{+\infty} \frac{x^{-p}\, dx}{x+1}, \; 0 < p < 1.$

34. $\int_{0}^{1} x^{p-1}(1-x)^q\, dx = \frac{q}{p} \int_{0}^{1} x^p (1-x)^{q-1}\, dx, \; p > 0, \, q > 0.$

★35. $\int_{0}^{+\infty} \frac{dx}{(x+p)\sqrt{x}} = \frac{\pi}{\sqrt{p}}, \; p > 0.$

In Exercises 36 and 37, use integration by parts and mathematical induction to verify the formulas.

★36. Wallis's Formulas.

$$\int_0^1 \frac{x^n\, dx}{\sqrt{1-x^2}} = \int_0^{\frac{\pi}{2}} \sin^n x\, dx = \int_0^{\frac{\pi}{2}} \cos^n x\, dx$$

$$= \begin{cases} \dfrac{2\cdot 4\cdot 6\cdots(n-1)}{3\cdot 5\cdot 7\cdots n}, & \text{if } n \text{ is an odd integer} > 1; \\ \dfrac{1\cdot 3\cdot 5\cdots(n-1)}{2\cdot 4\cdot 6\cdots n}\cdot \dfrac{\pi}{2}, & \text{if } n \text{ is an even integer} > 0. \end{cases}$$

★37. $\int_0^{+\infty} x^n e^{-ax}\, dx = \dfrac{n!}{a^{n+1}}$ (n a positive integer, $a > 0$).

In Exercise 38–41, state and prove the analogue for improper integrals of the specified theorem of § 501.

★38. Theorem II.   **★39.** Theorem III.
★40. Theorem IV.   **★41.** Theorem V.

★42. Show by an example that a continuous function improperly integrable on $[0, +\infty)$ need not have a zero limit at $+\infty$, in contrast to the fact that the general term of a convergent infinite series must tend toward zero (§ 703). (For another example see Ex. 46.) *Hint:* See Figure 513.

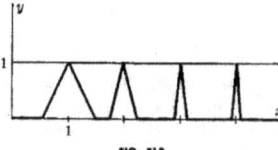

FIG. 513

★43. Prove the **Cauchy Criterion** for convergence of improper integrals: *If $f(x)$ is integrable on $[a, c]$ for every c such that $a < c < b$ ($a < c$), then the improper integral $\int_a^b f(x)\, dx$ $\left(\int_a^{+\infty} f(x)\, dx\right)$ converges if and only if corresponding to $\epsilon > 0$ there exists a number $c = c(\epsilon)$ such that $a < c < b$ ($a < c$) and such that $c < u < v < b$ ($c < u < v$) implies $\left|\int_u^v f(x)\, dx\right| < \epsilon$.*

★44. If $f(x)$ is integrable on $[a, c]$ for every c such that $a < c < b$ ($a < c$), then $\int_a^b f(x)\, dx$ $\left(\int_a^{+\infty} f(x)\, dx\right)$ is said to **converge absolutely** if and only if the improper integral $\int_a^b |f(x)|\, dx$ $\left(\int_a^{+\infty} |f(x)|\, dx\right)$ converges. Prove that an absolutely convergent improper integral is convergent.

★45. An improper integral is said to be **conditionally convergent** if and only

§ 516] **BOUNDED VARIATION**

if it is convergent but not absolutely convergent. Prove that $\int_0^{+\infty} \frac{\sin x}{x}\, dx$ is conditionally convergent.

★46. Prove that $\int_0^{+\infty} \cos x^2\, dx$ converges, by integrating $\int_u^v \cos x^2\, dx$ by parts to obtain $\frac{1}{2v} \sin v^2 - \frac{1}{2u} \sin u^2 + \frac{1}{2} \int_u^v \frac{\sin x^2}{x^2}\, dx$. Also consider $\int_0^{+\infty} x \cos x^4\, dx$ (where the integrand is unbounded). (Cf. Ex. 42.)

★★47. Prove the **Lebesgue Theorem on Dominated Convergence** for Improper Integrals: *If $\{f_n(x)\}$ is a sequence of functions improperly integrable on $[a, b)$ ($[a, +\infty)$) and converging there to an (improperly) integrable function $f(x)$, and if there exists a function $g(x)$ improperly integrable on that interval such that $|f_n(x)| \leq g(x)$ for $a \leq x < b$ ($a \leq x < +\infty$) and $n = 1, 2, \cdots$, then*

$$\lim_{n \to +\infty} \int_a^b f_n(x)\, dx = \int_a^b f(x)\, dx = \int_a^b \lim_{n \to +\infty} f_n(x)\, dx$$

$$\left(\lim_{n \to +\infty} \int_a^{+\infty} f_n(x)\, dx = \int_a^{+\infty} f(x)\, dx = \int_a^{+\infty} \lim_{n \to +\infty} f_n(x)\, dx \right).$$

(*Similar statements apply to other types of improper integrals.*) *Hint:* For the interval $[a, b)$, and any $\epsilon > 0$, let c be a number between a and b such that $\int_c^b g(x)\, dx < \frac{1}{3}\epsilon$. Then for any n,

$$\left| \int_a^b f_n(x)\, dx - \int_a^b f(x)\, dx \right| \leq \left| \int_a^c f_n(x)\, dx - \int_a^c f(x)\, dx \right|$$
$$+ \int_c^b |f_n(x)|\, dx + \int_c^b |f(x)|\, dx \leq \left| \int_a^c f_n(x)\, dx - \int_a^c f(x)\, dx \right| + \tfrac{2}{3}\epsilon.$$

Now use Ex. 49, § 503.

★★48. Prove that $\lim_{n \to +\infty} \sqrt[n]{\frac{n!}{n^n}} = \frac{1}{e}$, by adapting the technique of Exercises 24–27, § 503, to improper integrals. (Cf. Ex. 36, § 711.)

★★**516. BOUNDED VARIATION**

The material of this section is presented primarily for its relation to the Riemann-Stieltjes integral defined in the following section, although it is also related to other topics in analysis, such as length of arc (discussed in Chapter 11). The first part of § 517 and the first exercises of § 518 do not depend on the present section.

Definition I. *A function $f(x)$, defined on a closed interval $[a, b]$, is of bounded variation there if and only if, for all possible nets $\mathfrak{N}$, consisting of points $a = a_0 < a_1 < a_2 < \cdots < a_n = b$, the sums*

(1) $$\sum_{i=1}^{n} |f(a_i) - f(a_{i-1})|$$

*are bounded. The **total variation** of a function $f(x)$ (of bounded variation)*

176 INTEGRATION [§ 516

on the interval $[a, b]$ is defined to be the least upper bound of the set of all possible values of (1), and denoted:

(2) $$V(f, [a, b]) \equiv \sup_{\Re} \sum_{i=1}^{n} |f(a_i) - f(a_{i-1})|.$$

Let us first show that the concept just introduced is far from a vacuous one by presenting two important classes of functions of bounded variation.

Theorem I. *Any function $f(x)$ defined and monotonic on a closed interval $[a, b]$ is of bounded variation there and*
$$V(f, [a, b]) = |f(b) - f(a)|.$$

Proof. Every sum (1) is equal to $|f(b) - f(a)|$.

Theorem II. *Any function $f(x)$ continuous on a closed interval $[a, b]$ and having a bounded derivative in the interior (a, b) is of bounded variation there.*

Proof. Let $|f'(x)| \leq K$ and use the Law of the Mean (§ 405) to obtain
$$\sum_{i=1}^{n} |f(a_i) - f(a_{i-1})| = \sum_{i=1}^{n} |f'(x_i)| \Delta x_i \leq K(b - a).$$

An explicit formulation for the total variation of a function is often possible:

Theorem III. *If $f(x)$ is differentiable on $[a, b]$ and if $|f'(x)|$ is integrable there, then $f(x)$ is of bounded variation there and*

(3) $$V(f, [a, b]) = \int_a^b |f'(x)| \, dx.$$

Proof. By the Law of the Mean (§ 405), the sums (1) have the form

(4) $$\sum_{i=1}^{n} |f(a_i) - f(a_{i-1})| = \sum_{i=1}^{n} |f'(x_i)| \Delta x_i,$$

and their least upper bound must be *at least* as large as $I = \int_a^b |f'(x)| \, dx$. If a sum S of the form (1) *greater* than I were assumed to exist, a contradiction could be obtained as follows: Let $\epsilon \equiv S - I > 0$, and choose $\delta > 0$ so that $|\Re| < \delta$ implies $\left| \sum_{i=1}^{n} |f'(x_i)| \Delta x_i - I \right| < \epsilon$. Then add points to the net $\Re_1$ corresponding to the sum S until a net $\Re$ of norm less than δ is obtained. By the triangle inequality, each time such a point is added to the net, the sum (1) is either increased or unchanged. The final result, for the net $\Re$ of norm less than δ is the set of incompatible inequalities:

$$I + \epsilon = S \leq \sum_{i=1}^{n} |f(a_i) - f(a_{i-1})| = \sum_{i=1}^{n} |f'(x_i)| \Delta x_i < I + \epsilon.$$

Next we give a few statements to be used later, with references to the Exercises of § 518 for proofs:

Theorem IV. *If $f(x)$ is of bounded variation on $[a, b]$ and if*
$$a \leq c < d \leq b,$$
then $f(x)$ is of bounded variation on $[c, d]$ and

(5) $$V(f, [c, d]) \leq V(f, [a, b]).$$

(Cf. Ex. 32, § 518.)

Theorem V. *If $f(x)$ is of bounded variation on $[a, b]$ and on $[b, c]$, then $f(x)$ is of bounded variation on $[a, c]$ and*

(6) $$V(f, [a, c]) = V(f, [a, b]) + V(f, [b, c]).$$

(Cf. Ex. 33, § 518.)

Theorem VI. *If $f(x)$ and $g(x)$ are of bounded variation on $[a, b]$, then so are $f(x) \pm g(x)$, $f(x) g(x)$, and $k f(x)$, where k is a constant, and*

(7) $$V(f \pm g, [a, b]) \leq V(f, [a, b]) + V(g, [a, b]),$$
(8) $$V(kf, [a, b]) = |k| \cdot V(f, [a, b]).$$

(Cf. Ex. 34, § 518.)

We now introduce three new functions associated with a given function $f(x)$ of bounded variation on a given interval $[a, b]$:

Definition II. *The **total variation**, **positive variation**, and **negative variation functions**, and their symbols, are defined by the three equations, respectively:*

(9) $$v(x) \equiv V(f, [a, x]), a < x \leq b; v(a) \equiv 0.$$
(10) $$p(x) \equiv \tfrac{1}{2}[v(x) + f(x) - f(a)], a \leq x \leq b.$$
(11) $$n(x) \equiv \tfrac{1}{2}[v(x) - f(x) + f(a)], a \leq x \leq b.$$

Theorem VII. *The three functions $v(x)$, $p(x)$, and $n(x)$ of Definition II are nonnegative monotonically increasing functions on the interval $[a, b]$, and*

(12) $$v(x) = p(x) + n(x),$$
(13) $$f(x) = p(x) - n(x) + f(a).$$

Proof. Formulas (12) and (13) follow immediately from (10) and (11). The total variation function is nonnegative by definition and monotonically increasing by Theorem IV. Since the negative variation of any function of bounded variation is the positive variation of the negative of that function, we need prove only the two statements regarding $p(x)$. The nonnegativeness of $p(x)$ follows from the inequality

$$V(f, [a, x]) \geq |f(x) - f(a)|,$$

while for $a \leq c \leq d$, the inequality $p(c) \leq p(d)$ reduces to the inequality

178   INTEGRATION   [§ 516]

$$V(f, [c, d]) \geq |f(d) - f(c)|,$$

thanks to Theorem V.

We are now in a position to establish a simple and useful criterion for bounded variation.

Theorem VIII. *A function $f(x)$ defined on an interval $[a, b]$ is of bounded variation there if and only if it can be represented as the difference between two monotonically increasing functions.*

Proof. If $f(x)$ is of bounded variation, equation (12) expresses $f(x)$ as the difference between the two monotonically increasing functions

$$(p(x) + f(a)) \quad \text{and} \quad n(x).$$

On the other hand, the difference between two monotonic functions is the difference between two functions of bounded variations (Theorem I) and is therefore of bounded variation (Theorem VI).

Example 1. The function $\sqrt{x}$ is of bounded variation on $[0, 1]$, by Theorem I.

Example 2. The function $y = x^2 \sin \frac{1}{x}$ ($y = 0$ if $x = 0$) (Example 3, § 403) is of bounded variation on $[0, 1]$, by Theorem II.

Example 3. Prove that the function $y = x \sin \frac{1}{x}$ ($y = 0$ if $x = 0$) (Example 2, § 403) is not of bounded variation on $[0, 1]$ (although it is continuous there).

Solution. Let $a_0 < a_1 < \cdots < a_n$ be the points

$$0, \frac{2}{\pi(2n - 3)}, \frac{2}{\pi(2n - 5)}, \cdots, \frac{2}{5\pi}, \frac{2}{3\pi}, \frac{2}{\pi}, 1.$$

Then

$$\sum_{i=1}^{n} |f(a_i) - f(a_{i-1})| = \left|\sin 1 - \frac{2}{\pi}\right| + \left(\frac{2}{\pi} + \frac{2}{3\pi}\right) + \left(\frac{2}{3\pi} + \frac{2}{5\pi}\right) + \cdots$$
$$+ \left(\frac{2}{\pi(2n - 5)} + \frac{2}{\pi(2n - 3)}\right) + \frac{2}{\pi(n - 3)} > \frac{1}{3} + \frac{1}{5} + \cdots + \frac{1}{2n - 3}.$$

Because of the divergence of the harmonic series $\sum \frac{1}{n}$ (Chapter 7), this expression can be made arbitrarily large. (Cf. Example 1, § 1105.)

Example 4. Find the total, positive, and negative variation functions for $f(x) = \sin x$ on the interval $[0, 2\pi]$.

Solution. By Theorem III, $v(x) = \int_0^x |\cos t| \, dt$, so that for

$$0 \leq x \leq \frac{\pi}{2}, \quad v(x) = \int_0^x \cos t \, dt = \sin x,$$

$$\frac{\pi}{2} \leq x \leq \frac{3\pi}{2}, \quad v(x) = 1 - \int_{\frac{\pi}{2}}^x \cos t \, dt = 2 - \sin x,$$

§ 517]   THE RIEMANN-STIELTJES INTEGRAL

$$\frac{3\pi}{2} \leq x \leq 2\pi, \quad v(x) = 3 + \int_{\frac{\pi}{2}}^{x} \cos t \, dt = 4 + \sin x.$$

Similarly, in the ranges $\left[0, \frac{\pi}{2}\right]$, $\left[\frac{\pi}{2}, \frac{3\pi}{2}\right]$, and $\left[\frac{3\pi}{2}, 2\pi\right]$, respectively, the function $p(x) = \sin x$, 1, and $2 + \sin x$, and the function $n(x) = 0$, $1 - \sin x$, and 2. The graphs are given in Figure 514.

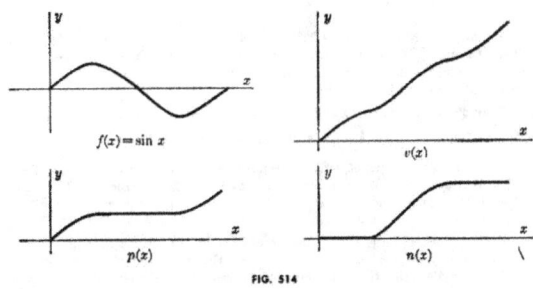

FIG. 514

★517. THE RIEMANN-STIELTJES INTEGRAL

A generalization of the definite integral, with useful applications in many applied fields including Physics and Statistics, is given in the following definition:

Definition I. *Let $f(x)$ and $g(x)$ be defined and bounded on a closed interval $[a, b]$. Then $f(x)$ is **Riemann-Stieltjes integrable** with respect to $g(x)$ on $[a, b]$, with **Riemann-Stieltjes integral** I, if and only if corresponding to $\epsilon > 0$ there exists $\delta > 0$ such that for any net*

$$\mathfrak{N} = \{a = a_0 < a_1 < \cdots < a_n = b\}$$

of norm $|\mathfrak{N}| < \delta$ and any choice of points x_i, $a_{i-1} \leq x_i \leq a_i$, $i = 1, 2, \cdots, n$:

(1) $$\left| \sum_{i=1}^{n} f(x_i)[g(a_i) - g(a_{i-1})] - I \right| < \epsilon.$$

Letting $\Delta g_i = g(a_i) - g(a_{i-1})$ and $\int_a^b f(x) \, dg(x) = I$, we write (1) in limit notation:

(2) $$\int_a^b f(x) \, dg(x) = \lim_{\mathfrak{N} \to 0} \sum_{i=1}^{n} f(x_i) \Delta g_i.$$

180   INTEGRATION   [§ 517

NOTE 1. If $g(x) \equiv x$, the Riemann-Stieltjes integral (2) reduces to the Riemann or definite integral:

(3) $$\int_a^b f(x)\, dg(x) = \int_a^b f(x)\, dx, \quad \text{if} \quad g(x) \equiv x.$$

Example 1. If $f(x)$ is defined and bounded on $[a, b]$ and continuous at $x = c$ $(a < c < b)$, and if $g(x) \equiv 0$ for $a \leq x < c$ and $g(x) \equiv p$ for $c \leq x \leq b$, show that $f(x)$ is Riemann-Stieltjes integrable with respect to $g(x)$ on $[a, b]$, and $\int_a^b f(x)\, dg(x) = pf(c)$.

Solution. For any net $\mathfrak{N}$, let k be the positive integer such that $a_{k-1} < c \leq a_k$. Then the sum $\sum f(x_i)\, \Delta g_i$ reduces to the single term $pf(x_k)$, which (because of continuity of $f(x)$ at $x = c$) approximates $pf(c)$ to any desired degree of accuracy if the norm of $\mathfrak{N}$ is sufficiently small.

Example 2. If $f(x)$ and $g(x)$ are both $\equiv 0$ for $0 \leq x < \frac{1}{2}$ and $\equiv 1$ for $\frac{1}{2} \leq x \leq 1$, show that $\int_0^1 f(x)\, dg(x)$ does not exist.

Solution. For any net $\mathfrak{N}$, let k be the positive integer such that $a_{k-1} < \frac{1}{2} \leq a_k$. Then the sum $\sum f(x_i)\, \Delta g_i$ reduces to the single term $f(x_k)$, which has the value 0 or 1 according as $x_k < \frac{1}{2}$ or $x_k \geq \frac{1}{2}$. Thus, for $\epsilon = \frac{1}{2}$ there is no $\delta > 0$ guaranteeing the inequality (1), whatever the number I may be!

NOTE 2. Example 1, above, illustrates the effect of a jump discontinuity in the function $g(x)$ at a point of continuity of the function $f(x)$ (Ex. 9, § 518), while Example 2 illustrates the general principle that whenever $f(x)$ and $g(x)$ have a common point of discontinuity the integral $\int_a^b f(x)\, dg(x)$ fails to exist (Ex. 10, § 518).

The following theorem establishes the differential nature of the symbol $dg(x)$, and shows how certain Riemann-Stieltjes integrals can be written as standard Riemann integrals.

Theorem I. *If $f(x)$ is defined and $g(x)$ is differentiable at every point of a closed interval $[a, b]$, and if $f(x)$ and $g'(x)$ are integrable there, then $f(x)$ is Riemann-Stieltjes integrable with respect to $g(x)$ there, and*

$$\int_a^b f(x)\, dg(x) = \int_a^b f(x)\, g'(x)\, dx.$$

Proof. If $I = \int_a^b f(x)\, g'(x)\, dx$ (cf. Ex. 35, § 503), and $\epsilon > 0$, we wish to establish an inequality of the form (1), which can be rewritten with the aid of the Law of the Mean in the form

(4) $$\left| \sum_{i=1}^n f(x_i)\, g'(\xi_i)\, \Delta x_i - I \right| < \epsilon.$$

To this end we appeal to Bliss's Theorem (Ex. 41, § 503), and produce the required $\delta > 0$.

§ 517] THE RIEMANN-STIELTJES INTEGRAL 181

The following theorem shows the reciprocal relation between $f(x)$ and $g(x)$ in the definition of Riemann-Stieltjes integrability (cf. Ex. 11, § 518, for hints on a proof):

Theorem II. Integration by Parts. *If $f(x)$ is integrable with respect to $g(x)$ on $[a, b]$, then $g(x)$ is integrable with respect to $f(x)$ there and*

$$(5) \quad \int_a^b f(x)\,dg(x) + \int_a^b g(x)\,df(x) = f(b)\,g(b) - f(a)\,g(a).$$

The question of existence of the integral of a function $f(x)$ with respect to a function $g(x)$ in case the function $g(x)$ is not differentiable, or possibly not even continuous, naturally arises. The simplest useful sufficient condition corresponds to the condition of continuity of $f(x)$ for the existence of the definite integral of $f(x)$:

Theorem III. *If, on the interval $[a, b]$, one of the functions $f(x)$ and $g(x)$ is continuous and the other monotonic, the integral $\int_a^b f(x)\,dg(x)$ exists.*

Proof. Thanks to Theorem II and the similarity in behavior of monotonically increasing and decreasing functions, we shall assume without loss of generality that $f(x)$ is continuous and $g(x)$ monotonically increasing (with $g(b) > g(a)$ since the case $g(b) = g(a)$ is trivial).

Our object is to find a number I which is approximated by all possible sums $\sum f(x_i)(g(a_i) - g(a_{i-1})) = \sum f(x_i)\,\Delta g_i$ associated with nets of sufficiently small norm. For a given net $\mathfrak{N}$, let m_i and M_i be the minimum and maximum values, respectively, of the continuous function $f(x)$ for $a_{i-1} \leq x \leq a_i$, $i = 1, 2, \cdots, n$. For any choice of x_i such that $a_{i-1} \leq x_i \leq a_i$, $i = 1, 2, \cdots, n$, $m_i \leq f(x_i) \leq M_i$, and since $g(x)$ is assumed to be monotonically increasing, $\Delta g_i \geq 0$, $i = 1, 2, \cdots, n$. Therefore

$$(6) \quad \sum_{i=1}^n m_i\,\Delta g_i \leq \sum_{i=1}^n f(x_i)\,\Delta g_i \leq \sum_{i=1}^n M_i\,\Delta g_i.$$

The extreme left-hand and right-hand terms of (6) are called the **lower** and **upper sums** for the net $\mathfrak{N}$, and written $L(\mathfrak{N})$ and $U(\mathfrak{N})$, respectively.

In our quest for the desired number I, we shall define two numbers, I and J, and then prove that they are equal:

$$(7) \quad \begin{cases} I \equiv \sup \sum_{i=1}^n m_i\,\Delta g_i = \sup \text{ (all lower sums)}, \\ J \equiv \inf \sum_{i=1}^n M_i\,\Delta g_i = \inf \text{ (all upper sums)}. \end{cases}$$

In order to establish the equality of the numbers I and J, we first affirm that $I \leq J$ and that this inequality is a consequence of the fact that *every* lower sum is less than or equal to every upper sum: *If $\mathfrak{M}$ and $\mathfrak{N}$ are any two nets on $[a, b]$,*

(8) $$L(\mathfrak{M}) \leqq U(\mathfrak{N}).$$

Reasonable though (8) may appear, it needs proof, and detailed hints are given in Exercise 12, § 518. In Exercise 13, § 518, the student is asked to prove that (8) implies $I \leqq J$.

Finally, with $I \leqq J$ established, we wish to show that $I = J$ and that if $\epsilon > 0$ there exists $\delta > 0$ such that

(9) $$\left| \sum_{i=1}^{n} f(x_i) \Delta g_i - I \right| < \epsilon$$

whenever $|\mathfrak{N}| < \delta$. By use of the inequalities (6) and

(10) $$L(\mathfrak{N}) \leqq I \leqq J \leqq U(\mathfrak{N}),$$

the desired equality $I = J$ and inequality (9) follow from the fact that if δ is chosen so small that (thanks to the uniform continuity of $f(x)$)

$$|x' - x''| < \delta \text{ implies } |f(x') - f(x'')| < \frac{\epsilon}{g(b) - g(a)},$$

then $|\mathfrak{N}| < \delta$ implies

$$U(\mathfrak{N}) - L(\mathfrak{N}) = \sum_{i=1}^{n} (M_i - m_i) \Delta g_i < \frac{\epsilon}{g(b) - g(a)} \sum_{i=1}^{n} \Delta g_i = \epsilon.$$

(Details are requested in Ex. 14, § 518.)

A few facts about the result of adding or subtracting functions or of multiplying by constants are herewith assembled (proofs are requested in Ex. 15, § 518):

Theorem IV. *If $f_i(x)$ is (Riemann-Stieltjes) integrable with respect to $g_j(x)$ for $i, j = 1, 2,$ and if c is any constant, then $f_1(x) + f_2(x)$ is integrable with respect to $g_1(x)$, $f_1(x)$ is integrable with respect to $g_1(x) + g_2(x)$, $c f_1(x)$ is integrable with respect to $g_1(x)$, $f_1(x)$ is integrable with respect to $c\, g_1(x)$, and (with a simplified notation suggested by $\int f\, dg \equiv \int_a^b f(x)\, dg(x)$):*

(11) $$\int (f_1 + f_2)\, dg_1 = \int f_1\, dg_1 + \int f_2\, dg_1,$$

(12) $$\int f_1\, d(g_1 + g_2) = \int f_1\, dg_1 + \int f_1\, dg_2,$$

(13) $$\int c f_1\, dg_1 = \int f_1\, d(cg_1) = c \int f_1\, dg_1.$$

Definition II. *If $b < a$,*

(14) $$\int_a^b f(x)\, dg(x) \equiv - \int_b^a f(x)\, dg(x),$$

THE RIEMANN-STIELTJES INTEGRAL

in case the latter integral exists. Furthermore,

(15) $$\int_a^a f(x)\, dg(x) = 0.$$

For integrals over adjoining intervals we have:

Theorem V. *For any three numbers a, b, and c,*

(16) $$\int_a^c f(x)\, dg(x) = \int_a^b f(x)\, dg(x) + \int_b^c f(x)\, dg(x),$$

provided the three integrals exist.

Proof. We shall assume $a < b < c$ (cf. Exs. 16-17, § 518). For $\epsilon > 0$, let $\delta > 0$ be such that $|\mathfrak{R}| < \delta$ implies that any sum corresponding to any of the three integrals approximates that integral within $\tfrac{1}{3}\epsilon$. Then the sum of two such approximating sums for the two integrals on the right-hand side of (16) must approximate the left-hand side within $\tfrac{2}{3}\epsilon$. Therefore the two sides of (16) are constants differing by less than ϵ, and must consequently be equal.

★★Finally, a simple application of Theorems II-IV and the fact (Theorem VIII, § 516) that any function of bounded variation can be represented as the difference between two monotonically increasing functions yields the following generalization of Theorem III:

★★**Theorem VI.** *If, on the interval $[a, b]$, one of the functions $f(x)$ and $g(x)$ is continuous and the other of bounded variation, the integral $\int_a^b f(x)\, dg(x)$ exists.*

Example 3. Evaluate the Riemann-Stieltjes integral $\int_0^{\frac{\pi}{2}} x\, d\sin x$.

Solution. By Theorem I, this integral is equal to $\int_0^{\frac{\pi}{2}} x \cos x\, dx$. However, it is simpler to use the integration by parts formula (Theorem II) directly:

$$\int_0^{\frac{\pi}{2}} x\, d\sin x = \frac{\pi}{2} \sin \frac{\pi}{2} - 0 \sin 0 - \int_0^{\frac{\pi}{2}} \sin x\, dx = \frac{\pi}{2} - 1.$$

Example 4. Evaluate the Riemann-Stieltjes integral $\int_0^3 e^{2x}\, d\,[x]$, where $[x]$ is the bracket function of § 201.

Solution. The integration by parts formula gives

$$\int_0^3 e^{2x}\, d\,[x] = 3e^6 - 0\cdot e^0 - \int_0^3 [x]\, d\,e^{2x},$$

which, by Theorem I, is equal to

$$3e^6 - 2\int_1^2 e^{2x}\, dx - 4\int_2^3 e^{2x}\, dx = e^2 + e^4 + e^6.$$

184   INTEGRATION   [§ 518

The result could be obtained directly by using the fact that the function $[x]$ in the original form of the integral makes contributions only at its jumps, of amounts determined by the values of e^{2x} and the size of the jumps (cf. Example 1, and Exs. 8-9, § 518.)

★518. EXERCISES

The notation $[x]$ indicates the bracket function of § 201.

In Exercises 1–6, evaluate the Riemann-Stieltjes integral.

★1. $\int_0^1 x \, dx^2$.
★2. $\int_0^2 x \, d\,[x]$.

★3. $\int_0^{\frac{\pi}{2}} \cos x \, d \sin x$.
★4. $\int_0^3 e^x \, d\{x - [x]\}$.

★5. $\int_{-1}^1 e^x \, d\,[x]$.
★6. $\int_0^\pi x \, d\,[\cos x]$.

★7. Prove that if $g(x)$ is defined on $[a, b]$, then $\int_a^b dg(x)$ exists and is equal to $g(b) - g(a)$.

★8. Prove that if $f(x)$ is continuous on $[0, n]$, then $\int_0^n f(x) \, d\,[x] = f(1) + f(2) + \cdots + f(n)$.

★9. For a net $\mathfrak{N}: a = a_0 < a_1 < \cdots < a_n = b$, let $g(x)$ be a step-function constant for $a_{i-1} < x < a_i$, $i = 1, 2, \cdots, n$, and having jumps $J_0 \equiv g(a+) - g(a)$, $J_i \equiv g(a_i+) - g(a_i-)$, $i = 1, 2, \cdots, n-1$, and $J_n \equiv g(b) - g(b-)$. If $f(x)$ is continuous on $[a, b]$, prove that

(1) $$\int_a^b f(x) \, dg(x) = \sum_{i=0}^n J_i f(a_i).$$

Hint: Since the integral exists (Theorem VI, § 517) choose a net of arbitrarily small norm such that each a_i, $i = 1, 2, \cdots, n-1$, is interior to some subinterval. Then let each a_i, $i = 0, 1, \cdots, n$, be a point chosen for evaluating $f(x)$.

★10. Prove that if $f(x)$ and $g(x)$ are both discontinuous at a point c, where $a \leq c \leq b$, then $\int_a^b f(x) \, dg(x)$ does not exist. *Hint*: For $a < c < b$, and c nonremovable for $g(x)$, let $\mathfrak{N}$ be a net not containing c and let $a_{k-1} < c < a_k$. Then $|\Delta g_k|$ can always be made $\geq \eta$, a fixed positive number. Also, for two suitable numbers, x_k and x_k', in $[a_{k-1}, a_k]$, $|f(x_k) - f(x_k')| \geq \xi$, a fixed positive number. Hence, for any $\delta > 0$, there is a net $\mathfrak{N}$ of norm $< \delta$, and two sums $\sum f(x_i) \, \Delta g_i$ which differ numerically by at least $\eta \xi$ (for $i \neq k$, let $x_i = x_i'$).

★11. Prove Theorem II, § 517. *Hint*: Expansion of $\sum_{i=1}^n g(x_i)[f(a_i) - f(a_{i-1})]$ and rearrangement of terms leads to the following identical formulation, where $x_0 \equiv a$ and $x_{n+1} \equiv b$:

$$f(b) \, g(b) - f(a) \, g(a) - \sum_{i=0}^n f(a_i) \, (g(x_{i+1}) - g(x_i)).$$

★12. Prove (8), § 517. *Hint:* If $\mathfrak{M}$ and $\mathfrak{N}$ are arbitrary nets on $[a, b]$, let $\mathcal{P}$ be the net consisting of all points appearing in either $\mathfrak{M}$ or $\mathfrak{N}$ (or both). Then, since $\mathcal{P}$ contains both $\mathfrak{M}$ and $\mathfrak{N}$,
$$L(\mathfrak{M}) \leq L(\mathcal{P}) \leq U(\mathcal{P}) \leq U(\mathfrak{N}).$$

★13. Prove that (8), § 517, implies $I \leq J$.

★14. Supply the details requested at the end of the proof of Theorem III, § 517.

★15. Prove Theorem IV, § 517.

★16. Show that formula (16), Theorem V, § 517, holds whatever the order relation between a, b, and c may be.

★17. Show by the example
$$f(x) = 0 \text{ for } 0 \leq x < 1, \quad f(x) = 1 \text{ for } 1 \leq x \leq 2,$$
$$g(x) = 0 \text{ for } 0 \leq x \leq 1, \quad g(x) = 1 \text{ for } 1 < x \leq 2$$
that the existence of the integrals on the right-hand side of (16), Theorem V, § 517, do not imply the existence of the integral on the left-hand side.

★18. Prove the *integration by substitution* formula for Riemann-Stieltjes integrals: If $\phi(t)$ is continuous and strictly monotonic on $[a, b]$, and if $\phi(a) = c$ and $\phi(b) = d$, then the equality
$$(2) \qquad \int_c^d f(x) \, dg(x) = \int_a^b f(\phi(t)) \, dg(\phi(t))$$
holds, the existence of either integral implying that of the other. (Cf. Exs. 19-21.)

★★19. Prove that if the assumption in Exercise 18 that $\phi(t)$ is strictly monotonic on $[a, b]$ is replaced by the assumption that $\phi(t)$ is monotonic on $[a, b]$, then the existence of the left-hand integral of (2) implies that of the right-hand integral, and their equality. (Cf. Exs. 20-21.)

★★20. Show by means of the following example that under the assumptions of Exercise 19, the existence of the right-hand integral of (2) does not imply that of the left-hand integral:
$f(x) = 0$ for $0 \leq x < \frac{1}{2}$, $f(x) = 1$ for $\frac{1}{2} \leq x \leq 1$;
$g(x) = 0$ for $0 \leq x \leq \frac{1}{2}$, $g(x) = 1$ for $\frac{1}{2} < x \leq 1$;
$\phi(t) = \frac{3}{2}t$ for $0 \leq t \leq \frac{1}{3}$, $\phi(t) = \frac{1}{2}$ for $\frac{1}{3} < t < \frac{2}{3}$, $\phi(t) = \frac{1}{2}(3x - 1)$ for $\frac{2}{3} \leq t \leq 1$.

★★21. Prove that if the assumption in Exercise 18 that $\phi(t)$ is strictly monotonic is dropped, then formula (2) holds whenever both integrals exist. *Hint:* Use Theorem IV, § 213, to show that by introducing appropriate points between those of a given net on $[a, b]$, a net $\mathfrak{N}: a = a_0 < a_1 < \cdots < a_n = b$ can be obtained such that with an appropriate choice of $t_1, t_2, \cdots, t_n$ all terms of
$$\sum_{i=1}^{n} f(\phi(t_i)) \left[g(\phi(t_i)) - g(\phi(t_{i-1})) \right]$$
cancel except those corresponding to a monotonic sequence of points $\{\phi(t_{i_k})\}$.

★22. Prove that if $f(x) \geq 0$ and $g(x)$ monotonically increasing on $[a, b]$, then the integral $\int_a^b f(x) \, dg(x)$, if it exists, is nonnegative. State and prove a generalization of this for which $f_1(x) \leq f_2(x)$ implies $\int_a^b f_1(x) \, dg(x) \leq \int_a^b f_2(x) \, dg(x)$.

★23. Prove that if $f(x)$ is continuous and nonnegative on $[a, b]$ but not identically 0 there, and if $g(x)$ is strictly increasing on $[a, b]$, then

$$\int_a^b f(x)\, dg(x) > 0.$$

State and prove a generalization of this for which $f_1(x) \leq f_2(x)$ implies

$$\int_a^b f_1(x)\, dg(x) < \int_a^b f_2(x)\, dg(x)$$

(Cf. Ex. 22.)

★24. Prove the analogue of Exercise 23 for the case $f(x)$ continuous and positive on $[a, b]$, and $g(x)$ monotonically increasing but nonconstant there. Also generalize to $f_1(x)$ and $f_2(x)$.

★25. Prove the **First Mean Value Theorem** for Riemann-Stieltjes integrals (cf. Exs. 5-6, § 503): *If $f(x)$ is continuous and $g(x)$ is monotonic (strictly monotonic) on $[a, b]$, then there exists a point ξ such that $a \leq \xi \leq b$ $(a < \xi < b)$, and such that*

$$(3) \qquad \int_a^b f(x)\, dg(x) = f(\xi)\, [g(b) - g(a)].$$

Hint: Assume $g(x)$ increasing, and let $M = \sup f(x)$ and $m = \inf f(x)$ for $a \leq x \leq b$. Then

$$m[g(b) - g(a)] \leq \int_a^b f(x)\, dg(x) \leq M\, [g(b) - g(a)].$$

(See Ex. 23 for the strict inequalities.)

★26. Show by an example that equation (3) may not hold for $a < \xi < b$ if $g(x)$ is not *strictly* monotonic. *Hint:* Let $g(x) = [x]$ on $[0, 1]$.

★27. Prove the **Second Mean Value Theorem** for Riemann-Stieltjes integrals (cf. Ex. 14, § 506): *If $f(x)$ is monotonic (strictly monotonic) and $g(x)$ is continuous on $[a, b]$, then there exists a point ξ such that $a \leq \xi \leq b$ $(a < \xi < b)$ and such that*

$$(4) \qquad \int_a^b f(x)\, dg(x) = f(a)\, [g(\xi) - g(a)] + f(b)\, [g(b) - g(\xi)].$$

Hint: Use (3), Ex. 25, and the integration by parts formula (Theorem II, § 517).

★28. Prove the following form of the Second Mean Value Theorem for Riemann integrals (cf. Ex. 14, § 506): *If $f(x)$ is monotonic (strictly monotonic) and if $h(x)$ is continuous on $[a, b]$, then there exists a point ξ such that $a \leq \xi \leq b$ $(a < \xi < b)$ such that*

$$(5) \qquad \int_a^b f(x)\, h(x)\, dx = f(a) \int_a^\xi h(x)\, dx + f(b) \int_\xi^b h(x)\, dx.$$

Hint: Let $g(x)$ be a primitive of $h(x)$, and use (4).

★29. Prove the following Bonnet form of the Second Mean Value Theorem for Riemann integrals: *If $f(x) \geq 0$ and monotonically decreasing (or, alternatively, $f(x) \leq 0$ and monotonically increasing) and if $h(x)$ is continuous on $[a, b]$, then there exists a point ξ of $[a, b]$ such that*

$$(6) \qquad \int_a^b f(x)\, h(x)\, dx = f(a) \int_a^\xi h(x)\, dx.$$

If $f(x) \geq 0$ and monotonically increasing (or, alternatively, $f(x) \leq 0$ and monotonically decreasing) and if $h(x)$ is continuous on $[a, b]$, then there exists a point ξ

of $[a, b]$ such that

(7) $$\int_a^b f(x) h(x) \, dx = f(b) \int_\xi^b h(x) \, dx.$$

Hint: Use (5), redefining $f(x)$ to be 0 at either b or a.

★**30.** It is trivial (is it not?) that if $g(x)$ is a constant then $\int_a^b f(x) \, dg(x) = 0$. Prove the following converse: If $f(x)$ and $g(x)$ are bounded on $[a, b]$, and if $\int_a^b f(x) \, dg(x)$ exists and is equal to 0 for all monotonic functions $f(x)$, then $g(x)$ is a constant. *Hint:* By Ex. 10, $g(x)$ is continuous, and by Theorem II, § 517, $\int_a^b g(x) \, df(x) = f(b) g(b) - f(a) g(a)$. Let $f(x)$ be a step-function with one jump.

★★**31.** Prove that $\sin x^2$ is of bounded variation on every finite interval.

★★**32.** Prove Theorem IV, § 516.

★★**33.** Prove Theorem V, § 516.

★★**34.** Prove Theorem VI, § 516.

In Exercises 35-36, find the total, positive, and negative variation functions for the given function on the given interval.

★★**35.** $x - [x]; [0, 4]$.

★★**36.** $2x - x^2; [-1, 2]$.

★★**37.** Prove that any function of bounded variation on an interval is bounded there.

★★**38.** Prove that a function of bounded variation has only jump or removable discontinuities. *Hint:* Cf. Theorem VIII, § 516, and Ex. 28, § 216.

★★**39.** Prove that the set of points of discontinuity of a function of bounded variation is either finite or denumerable. *Hint:* Cf. Ex. 29, § 216.

★★**40.** Prove that a function of bounded variation on $[a, b]$ is integrable there. *Hint:* Four proofs are available (Theorem VIII, § 516; Theorem VI, § 517; Ex. 34, § 503; Ex. 54, § 503).

★★**41.** Prove that the total, positive, and negative variation functions of a function of bounded variation $f(x)$ are continuous wherever $f(x)$ is continuous. *Hint:* For right-hand continuity of $v(x)$ at $x = c$ use nets on $[c, c + h]$:
$$V(f, [c, c + h]) \leq \sup_{\mathfrak{A}} \{|f(a_1) - f(c)| + V(f, [a_1, c + h])\}.$$

★★**42.** Let $f(x)$ be continuous and $g(x)$ be of bounded variation on $[a, b]$, and let $v(x)$ be the total variation function of $g(x)$. Prove that
$$\left| \int_a^b f(x) \, dg(x) \right| \leq \int_a^b |f(x)| \, dv(x).$$

NOTE. Further facts on functions of bounded variation and Riemann-Stieltjes integrals are given in Theorems I, II, III, § 1526.

6

Some Elementary Functions

★601. THE EXPONENTIAL AND LOGARITHMIC FUNCTIONS

The function x^a has been defined for integral exponents and arbitrary x in Chapter 1, and for positive rational exponents and nonnegative x in Chapter 2. In this section and the Exercises of the next, we study the expression a^b ($a > 0$, b real) and the logarithmic function.

One method of procedure is to define the exponential function a^x, first for integral exponents, then for all rational exponents, and finally by a limiting process, for all real exponents. Having defined the exponential function one can then define and study its inverse, the logarithmic function. For simplicity, we have chosen to reverse the order and define the logarithmic function first, by means of an integral, and then the exponential function as its inverse. We finally show that the exponential expression thus obtained agrees with the special cases previously studied. All of this is presented in the following section, by means of exercises arranged in logical order. (For another method cf. § 912.)

★602. EXERCISES

★1. Prove that the logarithmic function *defined* by the equation

$$(1) \qquad \ln x = \int_1^x \frac{dt}{t}, \quad x > 0,$$

has the following five properties:

(i) $\ln 1 = 0$;
(ii) $\ln x$ is strictly increasing;
(iii) $\ln x$ is continuous and, in fact, differentiable, with derivative $\frac{1}{x}$;
(iv) $\lim_{x \to +\infty} \ln x = +\infty$;
(v) $\lim_{x \to 0+} \ln x = -\infty$.

§ 602] EXERCISES 189

Hints for (iv) *and* (v):

$$\int_1^{2^n} \frac{dt}{t} = \int_1^2 \frac{dt}{t} + \int_2^4 \frac{dt}{t} + \int_4^8 \frac{dt}{t} + \cdots + \int_{2^{n-1}}^{2^n} \frac{dt}{t}$$
$$> \int_1^2 \frac{dt}{2} + \int_2^4 \frac{dt}{4} + \cdots + \int_{2^{n-1}}^{2^n} \frac{dt}{2^n} = \frac{n}{2}.$$

$$\int_{1/n}^1 \frac{dt}{t} = \int_n^1 \frac{d(1/u)}{1/u} = \int_1^n \frac{du}{u}.$$

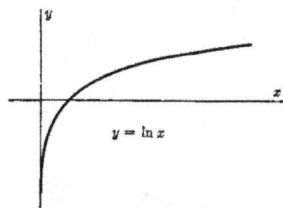

FIG. 601

★2. Prove the following laws for ln x:

(i) $\ln (xy) = \ln x + \ln y$, $x > 0$, $y > 0$;

(ii) $\ln \left(\dfrac{x}{y}\right) = \ln x - \ln y$, $x > 0$, $y > 0$;

(iii) $\ln (x^n) = n \ln x$, $x > 0$, n an integer;

(iv) $\ln (\sqrt[n]{x}) = \dfrac{1}{n} \ln x$, $x > 0$, n a positive integer;

(v) $\ln (x^r) = r \ln x$, $x > 0$, r rational.

(Cf. § 214 and Exs. 21-22, § 216.) It follows from Exercise 6, below, that (v) holds for any real r. *Hints:*

(i): $\int_1^{xy} \dfrac{dt}{t} = \int_1^x \dfrac{dt}{t} + \int_x^{xy} \dfrac{dt}{t} = \int_1^x \dfrac{dt}{t} + \int_x^{xy} \dfrac{d(t/x)}{t/x}$;

(ii): use (i) with $x = y \cdot \dfrac{x}{y}$;

(iv): use (iii) with $x = (\sqrt[n]{x})^n$.

★3. Let the function $e^x = \exp(x)$ be *defined* as the inverse of the logarithmic function:
$$y = e^x = \exp(x) \quad \text{if and only if} \quad x = \ln y.$$
Prove that e^x has the following three properties:

(i) e^x is defined, positive, and strictly increasing for all real x;

(ii) $e^0 = 1$, $\lim\limits_{x \to +\infty} e^x = +\infty$, $\lim\limits_{x \to -\infty} e^x = 0$;

(iii) e^x is continuous and, in fact, differentiable, with derivative e^x.

★4. Prove the following laws for e^x:

(i) $e^x e^y = e^{x+y}$,

(ii) $\dfrac{e^x}{e^y} = e^{x-y}$,

(iii) if r is rational, $(e^x)^r = e^{rx}$.

Hints: (i): let $a \equiv e^x$, $b \equiv e^y$, $c \equiv e^{x+y}$; then $x = \ln a$, $y = \ln b$, $x + y = \ln c = \ln (ab)$; (ii): use (i) with e^{x-y} and e^y.

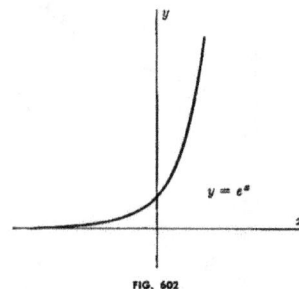

FIG. 602

★5. Define the number e as the value of e^x, for $x = 1$, $e \equiv \exp(1)$, or, equivalently, as the number whose (natural) logarithm is 1. Prove that if x is rational, then the exponential function e^x, defined in Exercise 3, is identical with the function e^x previously defined in Exercise 22, § 216.

★6. For an arbitrary positive base a, define the exponential function:

$$a^x \equiv e^{x \ln a}.$$

Prove the following ten properties of a^x:

(i) a^x is defined and positive for all real x;

(ii) if $a = 1$, $a^x = 1$ for all x;

(iii) if $a = e$, $a^x = e^x$ (Ex. 3), for all x;

(iv) if x is rational, a^x, as defined here, is identical with the function a^x previously defined in Exercise 22, § 216;

(v) if $a > 1$ ($a < 1$), a^x is strictly increasing (decreasing);

(vi) $a^x a^y = a^{x+y}$;

(vii) $\dfrac{a^x}{a^y} = a^{x-y}$;

(viii) $(a^x)^y = a^{xy}$;

(ix) $a^x b^x = (ab)^x$, $a > 0$, $b > 0$;

(x) a^x is continuous and, in fact, differentiable, with derivative $\ln a \cdot a^x$.

★7. The function x^a, for $x > 0$ and arbitrary real a, is defined as in Exercise 6:
$$x^a = e^{a \ln x}.$$
Prove the following properties of x^a:

(i) If $a > 0$ $(a < 0)$, x^a is strictly increasing (decreasing);

(ii) if $a > 0$ $(a < 0)$, $\lim_{x \to +\infty} x^a = +\infty$ (0) and $\lim_{x \to 0+} x^a = 0$ $(+\infty)$;

(iii) x^a is continuous and, in fact, differentiable, with derivative ax^{a-1}.

★8. If $a > 0$, prove that the function x^a, defined as in Exercise 7 for $x > 0$, and defined to be 0 when $x = 0$, has the three properties stated in Exercise 7, except that if $0 < a < 1$, the derivative of x^a at $x = 0$ is $+\infty$.

★9. Prove that $e = \lim_{x \to 0} (1 + x)^{\frac{1}{x}}$. *Hint:* At $x = 1$,
$$\frac{d}{dx} \ln x = \lim_{h \to 0} \frac{\ln(1+h) - \ln(1)}{h} = \lim_{h \to 0} \ln(1+h)^{\frac{1}{h}} = 1.$$

★10. Prove that $\lim_{x \to 0} (1 + ax)^{\frac{1}{x}} = e^a$.

★11. Define $\log_a x$, where $a > 0$ and $a \neq 1$, $x > 0$, as the inverse of the function a^x. That is, $\log_a x$ is that unique number y such that $a^y = x$. Prove the standard laws of logarithms, and the change of base formulas:
$$\log_a x = \log_a b \log_b x = \frac{\log_b x}{\log_b a}.$$
Prove that $a^{\log_a x} = \log_a (a^x) = x$.

★★12. Show that the function $\log_a x$ defined as $\ln x / \ln a$ is the inverse of the function a^x and hence identical with the function of Exercise 11. Derive its other properties from this definition alone.

★603. THE TRIGONOMETRIC FUNCTIONS

In a first course in Trigonometry the six basic trigonometric functions are defined. Their definitions there and their subsequent treatment in Calculus, however, are usually based on geometric arguments and intuitive appeal unfortified by a rigorous analytic background. It is the purpose of this section and the following section of exercises to present purely analytic definitions and discussion of the trigonometric functions. That these definitions correspond to those of the reader's previous experience is easily shown after the concept of arc length is available. (Cf. § 1109.)

The development here is restricted to the sine and cosine functions, since the remaining four trigonometric functions are readily defined in terms of those two. Furthermore, the calculus properties of the other four are immediately obtainable, once they are established for $\sin x$ and $\cos x$. These properties, as well as those of the inverse trigonometric functions, will be assumed without specific formulation here. (For another method cf. § 912.)

★604. EXERCISES

★1. Prove that the function defined by the equation
$$\operatorname{Arcsin} x \equiv \int_0^x \frac{dt}{\sqrt{1-t^2}}, \quad -1 < x < 1,$$
is strictly increasing, continuous, and, in fact, differentiable with derivative $(1 - x^2)^{-\frac{1}{2}}$. (Fig. 603.)

★2. Let the function $\sin x$ be *defined* as the inverse of the function prescribed in Exercise 1:
$$y = \sin x \quad \text{if and only if} \quad x = \operatorname{Arcsin} y, \text{ for } -\operatorname{Arcsin} \tfrac{3}{4} < x < \operatorname{Arcsin} \tfrac{3}{4}.$$

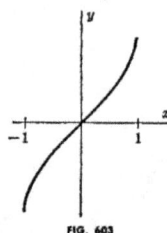

FIG. 603

If the number π is defined: $\pi = 4 \operatorname{Arcsin} \tfrac{1}{2}\sqrt{2}$, and if the number b is defined: $b = \operatorname{Arcsin} \tfrac{3}{4}$, prove that over the interval $(-b, b)$, containing $\tfrac{1}{4}\pi$, $\sin x$ is strictly increasing, continuous, and, in fact, differentiable with derivative $(1 - \sin^2 x)^{\frac{1}{2}} \equiv [1 - (\sin x)^2]^{\frac{1}{2}}$. (Fig. 604.) Also, $\sin 0 = 0$, $\sin \tfrac{1}{4}\pi = \tfrac{1}{2}\sqrt{2}$.

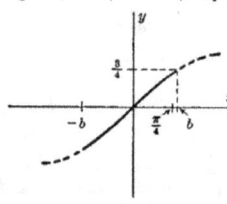

FIG. 604

★3. Let the function $\cos x$ be *defined*, on the interval $(-b, b)$ of Exercise 2, as the positive square root of $1 - \sin^2 x$. Prove that $\cos x$ is differentiable and that the derivatives of $\sin x$ and $\cos x$ are $\cos x$ and $-\sin x$, respectively. Also, $\cos 0 = 1$ and $\cos \tfrac{1}{4}\pi = \tfrac{1}{2}\sqrt{2}$. *Hint:* If $y = \cos x \neq 0$, $y^2 = 1 - \sin^2 x$, $2 \cos x \, dy/dx = -2 \sin x \cos x$.

★4. Let $s(x)$ and $c(x)$ be any two functions defined and differentiable on an open interval $(-a, a)$, $a > 0$, and possessing there the following four properties:

(1) $$(s(x))^2 + (c(x))^2 = 1,$$

(2) $$\frac{d}{dx}(s(x)) = c(x),$$

(3) $$\frac{d}{dx}(c(x)) = -s(x),$$

(4) $$c(0) = 1.$$

Prove that if α and β are any two numbers which, together with their sum $\alpha + \beta$, belong to the interval $(-a, a)$, then the following two identities hold:

(5) $$s(\alpha + \beta) = s(\alpha) c(\beta) + c(\alpha) s(\beta),$$

(6) $$c(\alpha + \beta) = c(\alpha) c(\beta) - s(\alpha) s(\beta).$$

Hence prove that if α and 2α belong to $(-a, a)$, then

(7) $$s(2\alpha) = 2s(\alpha) c(\alpha),$$

(8) $$c(2\alpha) = (c(\alpha))^2 - (s(\alpha))^2.$$

Hint: For (5), let $\gamma = \alpha + \beta$, and prove that the function $s(x) c(\gamma - x) + c(x) s(\gamma - x)$ has an identically vanishing derivative and is therefore a constant. Then let $x = \alpha$ and $x = 0$, in turn.

★5. Let $s(x)$ and $c(x)$ be any two functions satisfying (1)-(4) of Exercise 4, on the open interval $(-a, a)$, $a > 0$. Define two new functions in the open interval $(-2a, 2a)$ by means of the formulas

(9) $$S(x) \equiv 2s(\tfrac{1}{2}x) c(\tfrac{1}{2}x),$$

(10) $$C(x) \equiv (c(\tfrac{1}{2}x))^2 - (s(\tfrac{1}{2}x))^2.$$

Prove that $S(x) = s(x)$ and $C(x) = c(x)$ on $(-a, a)$, and that properties (1)-(4), and therefore also (5)-(8), hold for $S(x)$ and $C(x)$ on $(-2a, 2a)$ (where s and c are replaced by S and C, respectively).

★6. Prove that a repeated application of the extension definitions of Exercise 5 provide two functions, $\sin x$ and $\cos x$, defined and differentiable for all real numbers, possessing properties (1)-(8) (where $s(x)$ and $c(x)$ are replaced by $\sin x$ and $\cos x$, respectively), and agreeing with the originally defined $\sin x$ and $\cos x$ on the interval $(-b, b)$ of Exercise 2.

★7. Prove that $\sin x$ and $\cos x$, as defined for all real numbers x, are odd and even functions, respectively. (Cf. Exs. 7-8, § 503.)

★8. Prove that $\sin \tfrac{1}{2}\pi = 1$, and $\cos \tfrac{1}{2}\pi = 0$, and that $\tfrac{1}{2}\pi$ is the smallest positive number whose sine is 1 and also the smallest positive number whose cosine is 0. Hence show that in the closed interval $[0, \tfrac{1}{2}\pi]$, $\sin x$ and $\cos x$ are strictly increasing and decreasing, respectively. *Hint*: If $2k \leq \tfrac{1}{2}\pi$, and if $\sin 2k = 1$, then $\cos 2k = \cos^2 k - \sin^2 k = 0$, and $\cos k = \sin k = \tfrac{1}{2}\sqrt{2}$, and $k = \tfrac{1}{4}\pi$.

★9. Prove that $\sin \pi = 0$ and $\cos \pi = -1$, and that π is the smallest positive number whose sine is 0 and also the smallest positive number whose cosine is -1.

★10. Prove that $\sin 2\pi = 0$ and $\cos 2\pi = 1$, and that 2π is the smallest positive number whose cosine is 1.

194   SOME ELEMENTARY FUNCTIONS   [§ 605

★11. Prove that $\sin x$ and $\cos x$ are periodic with period 2π. That is, $\sin (x + 2\pi) = \sin x$ and $\cos (x + 2\pi) = \cos x$ for all x, and if either $\sin (x + k) = \sin x$ or $\cos (x + k) = \cos x$ for all x, then $k = 2n\pi$, for some integer n.

★12. Prove that $\sin x$ and $\cos x$ have the familiar signs in the appropriate "quadrants"; namely, $+, +, -, -$, and $+, -, -, +$, respectively, for x in the ranges $(0, \tfrac{1}{2}\pi)$, $(\tfrac{1}{2}\pi, \pi)$, $(\pi, \tfrac{3}{2}\pi)$, $(\tfrac{3}{2}\pi, 2\pi)$.

★13. Prove that if λ and μ are any two real numbers such that $\lambda^2 + \mu^2 = 1$, then there is a unique value of x in the half open interval $[0, 2\pi)$ such that $\sin x = \lambda$ and $\cos x = \mu$. *Hint:* Let x_0 be the unique number in the closed interval $[0, \tfrac{1}{2}\pi]$ whose sine is $|\lambda|$, and choose for x the appropriate number according to Ex. 12: $x_0, \pi - x_0, \pi + x_0, 2\pi - x_0$.

★14. Prove that $-1 \leq \sin x \leq 1$ and $-1 \leq \cos x \leq 1$ for all x.

★15. Prove that $\sin x$ and $\cos x$ have, everywhere, continuous derivatives of all orders.

★16. Prove that $\lim\limits_{x \to 0} \dfrac{\sin x}{x} = 1$.

Hint: Let $f(x) = \sin x$. Then $f'(0) = \lim\limits_{h \to 0} \dfrac{f(0 + h) - f(0)}{h}$.

★★17. Prove that the function $f(x) = \dfrac{\sin x}{x}$, $x \neq 0$, $f(0) = 1$, (i) is everywhere continuous; (ii) is everywhere differentiable; (iii) has everywhere a continuous nth derivative for all positive integers n.

605. SOME INTEGRATION FORMULAS

We start by asking two questions. (i) If $\int \dfrac{dx}{x} = \ln x + C$, and if $\ln x$ is defined only for $x > 0$, how does one evaluate the simple integral $\int_{-3}^{-2} \dfrac{dx}{x}$,

(ii) If $\int \dfrac{dx}{a^2 - x^2} = \dfrac{1}{2a} \ln \dfrac{a + x}{a - x} + C$ and $\int \dfrac{dx}{x^2 - a^2} = \dfrac{1}{2a} \ln \dfrac{x - a}{x + a} + C'$,

where $a > 0$, why cannot each of these integration formulas be obtained from the other by a mere change in sign? In other words, why does their sum, which should be a constant, give formally the result

$$\frac{1}{2a} \ln \left(\frac{a + x}{a - x} \cdot \frac{x - a}{x + a} \right) + C + C' \text{ or } \frac{1}{2a} \ln (-1) + C + C'$$

which does not even exist!

The answers to these questions lie most simply in the use of absolute values. We know that if $x > 0$, then $\dfrac{d}{dx} (\ln (x)) = \dfrac{1}{x}$ and that if $x < 0$, then $\dfrac{d}{dx} (\ln (-x)) = \dfrac{1}{-x} (-1) = \dfrac{1}{x}$. That is, $\ln x$ and $\ln (-x)$ both have the same derivative, formally, but have completely distinct domains of definition, $\ln x$ being defined for $x > 0$ and $\ln (-x)$ for $x < 0$. The func-

§ 605] SOME INTEGRATION FORMULAS

tion $\ln |x|$ encompasses both and is defined for any nonzero x. Thus the single integration formula

(1) $$\int \frac{dx}{x} = \ln |x| + C$$

is applicable under all possible circumstances. The integration of question (i) is therefore simple: $\int_{-3}^{-2} \frac{dx}{x} = \left[\ln |x| \right]_{-3}^{-2} = \ln |-2| - \ln |-3| = \ln \frac{2}{3}$.

For similar reasons the formulas of question (ii) are only apparently incompatible, since they apply to different domains of definition. The first is applicable if $x^2 < a^2$ and the second if $x^2 > a^2$. (Ex. 11, § 606.) However, again a single integration formula is available which is universally applicable, and can be written alternatively in the two forms:

(2) $$\int \frac{dx}{a^2 - x^2} = \frac{1}{2a} \ln \left| \frac{a + x}{a - x} \right| + C;$$

(2') $$\int \frac{dx}{x^2 - a^2} = \frac{1}{2a} \ln \left| \frac{x - a}{x + a} \right| + C.$$

The seeming paradox is resolved by the fact that $|-1| = 1$, whose logarithm certainly exists (although it vanishes).

More generally, any of the standard integration formulas that involve logarithms become universally applicable when absolute values are inserted. The student should establish this fact in detail for the following formulas (Ex. 12, § 606). (In formulas (7)-(14), a represents a positive constant.)

(3) $$\int \tan x \, dx = -\ln |\cos x| + C = \ln |\sec x| + C;$$

(4) $$\int \cot x \, dx = \ln |\sin x| + C = -\ln |\csc x| + C;$$

(5) $$\int \sec x \, dx = \ln |\sec x + \tan x| + C;$$

(6) $$\int \csc x \, dx = \ln |\csc x - \cot x| + C;$$

(7) $$\int \frac{dx}{\sqrt{x^2 \pm a^2}} = \ln |x + \sqrt{x^2 \pm a^2}| + C, |x| > a \text{ for the } - \text{ case};$$

(8) $$\int \sqrt{x^2 \pm a^2} \, dx = \tfrac{1}{2} [x\sqrt{x^2 \pm a^2} \pm a^2 \ln |x + \sqrt{x^2 \pm a^2}|] + C, |x| > a \text{ for the } - \text{ case};$$

(9) $$\int \frac{dx}{x\sqrt{a^2 \pm x^2}} = \frac{1}{a} \ln \left| \frac{\sqrt{a^2 \pm x^2} - a}{x} \right| + C, |x| < a \text{ for the } - \text{ case}.$$

Since the derivative of $\sec x \tan x$ is

$$\sec^3 x + \sec x \tan^2 x = \sec^3 x + \sec x (\sec^2 x - 1) = 2\sec^3 x - \sec x,$$

and since by (5), above, the derivative of $\ln|\sec x + \tan x|$ is $\sec x$, we have, by addition,

(10) $\int \sec^3 x \, dx = \tfrac{1}{2} \sec x \tan x + \tfrac{1}{2} \ln|\sec x + \tan x| + C.$

Finally, we give four more integration formulas, which involve inverse trigonometric functions. These are valid for the principal value ranges specified. The student should draw the graphs of the functions involved and verify the statements just made, as well as show that the formulas that follow are not all valid if the range of the inverse function is unrestricted (Ex. 13, § 606).

(11) $\int \dfrac{dx}{\sqrt{a^2 - x^2}} = \operatorname{Arcsin} \dfrac{x}{a} + C,\ -a < x < a,\ -\dfrac{\pi}{2} \leq \operatorname{Arcsin} \dfrac{x}{a} \leq \dfrac{\pi}{2};$

(12) $\int \dfrac{dx}{a^2 + x^2} = \dfrac{1}{a} \operatorname{Arctan} \dfrac{x}{a} + C,\ -\dfrac{\pi}{2} < \operatorname{Arctan} \dfrac{x}{a} < \dfrac{\pi}{2};$

(13) $\int \sqrt{a^2 - x^2}\, dx = \tfrac{1}{2} x \sqrt{a^2 - x^2} + \tfrac{1}{2} a^2 \operatorname{Arcsin} \dfrac{x}{a} + C,\ |x| < a;$

(14) $\int \dfrac{dx}{x\sqrt{x^2 - a^2}} = \begin{cases} -\dfrac{1}{a} \operatorname{Arcsin}\left(\dfrac{a}{x}\right) + C,\ x > a, \\ \dfrac{1}{a} \operatorname{Arcsin}\left(\dfrac{a}{x}\right) + C,\ x < -a. \end{cases}$

606. EXERCISES

In Exercises 1-10, perform the integration, and specify any limitations on the variable x.

1. $\int \tan 5x \, dx.$
2. $\int \sec 4x \, dx.$
3. $\int \dfrac{dx}{\sqrt{2 - x^2}}.$
4. $\int \sqrt{x^2 + 4x}\, dx.$
5. $\int \dfrac{dx}{\sqrt{4x^2 - 4x + 5}}.$
6. $\int \sqrt{x - 3x^2}\, dx.$
7. $\int \dfrac{dx}{x\sqrt{5 - 2x^2}}.$
8. $\int \dfrac{\sec^3 \sqrt{x}}{\sqrt{x}}\, dx.$
9. $\int \dfrac{dx}{3x^2 + 5x - 7}.$
10. $\int \dfrac{dx}{3x^2 + 5x + 7}.$

11. Show that in formula (2), § 605, the absolute value signs can be removed if $|x| < a$, and similarly for formula (2′), § 605, if $|x| > a$. *Hints:* The quotient of $a + x$ and $a - x$ is positive if and only if their product is positive.

12. Establish formulas (3)-(9), § 605, by direct evaluation rather than mere

[§ 607] HYPERBOLIC FUNCTIONS

differentiation of the right-hand members. *Hints:* For (3)-(6) express the functions in terms of sines and cosines. For (7)-(9) make trigonometric substitutions and, if necessary, use (10), § 605.

13. Establish formulas (11)-(14), § 605, by direct evaluation. (Cf. Ex. 12.)

607. HYPERBOLIC FUNCTIONS

The hyperbolic functions, called *hyperbolic sine, hyperbolic cosine,* etc., are defined:

$$(1) \quad \sinh x \equiv \frac{e^x - e^{-x}}{2}, \quad \coth x \equiv \frac{1}{\tanh x},$$
$$\cosh x \equiv \frac{e^x + e^{-x}}{2}, \quad \text{sech } x \equiv \frac{1}{\cosh x},$$
$$\tanh x \equiv \frac{\sinh x}{\cosh x}, \quad \text{csch } x \equiv \frac{1}{\sinh x}.$$

These six functions bear a close resemblance to the trigonometric functions. For example, $\sinh x$, $\tanh x$, $\coth x$, and $\text{csch } x$ are odd functions and $\cosh x$ and $\text{sech } x$ are even functions. (Ex. 21, § 609.) Furthermore, the hyperbolic functions satisfy identities that are similar to the basic trigonometric identities (verify the details in Ex. 22, § 609):

(2) $\cosh^2 x - \sinh^2 x = 1;$

(3) $1 - \tanh^2 x = \text{sech}^2 x;$

(4) $\coth^2 x - 1 = \text{csch}^2 x;$

(5) $\sinh (x \pm y) = \sinh x \cosh y \pm \cosh x \sinh y;$

(6) $\cosh (x \pm y) = \cosh x \cosh y \pm \sinh x \sinh y;$

(7) $\tanh (x \pm y) = \dfrac{\tanh x \pm \tanh y}{1 \pm \tanh x \tanh y};$

(8) $\sinh 2x = 2 \sinh x \cosh x;$

(9) $\cosh 2x = \cosh^2 x + \sinh^2 x = 2 \cosh^2 x - 1 = 2 \sinh^2 x + 1.$

The differentiation formulas (and therefore the corresponding integration formulas, which are omitted here) also have a familiar appearance (Ex. 23, § 609):

(10) $d(\sinh x)/dx = \cosh x;$

(11) $d(\cosh x)/dx = \sinh x;$

(12) $d(\tanh x)/dx = \text{sech}^2 x;$

(13) $d(\coth x)/dx = -\text{csch}^2 x;$

(14) $d(\text{sech } x)/dx = -\text{sech } x \tanh x;$

(15) $d(\text{csch } x)/dx = -\text{csch } x \coth x.$

198 SOME ELEMENTARY FUNCTIONS [§ 607

The graphs of the first four hyperbolic functions are given in Figure 605.
A set of integration formulas (omitting those that are mere reformulations of the differentiation formulas (10)-(15) follows (Ex. 24, § 609):

(16) $$\int \tanh x \, dx = \ln \cosh x + C;$$

(17) $$\int \coth x \, dx = \ln |\sinh x| + C;$$

(18) $$\int \operatorname{sech} x \, dx = \operatorname{Arctan} (\sinh x) + C;$$

(19) $$\int \operatorname{csch} x \, dx = \ln \left| \tanh \frac{x}{2} \right| + C.$$

NOTE. The trigonometric functions are sometimes called the **circular functions** because of their relation to a circle. For example, the parametric equations of the circle $x^2 + y^2 = a^2$ can be written $x = a \cos \theta$, $y = a \sin \theta$. In analogy with this, the hyperbolic functions are related to a hyperbola. For example, the parametric equations of the rectangular hyperbola $x^2 - y^2 = a^2$ can be written $x = a \cosh \theta$, $y = a \sinh \theta$. In this latter case, however, the parameter θ does not represent the polar coordinate angle for the point (x, y).

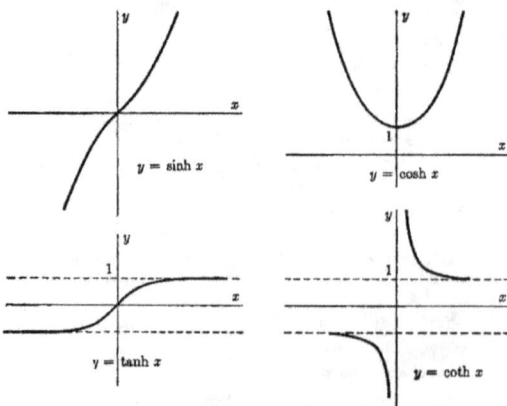

FIG. 605

608. INVERSE HYPERBOLIC FUNCTIONS

The integrals of certain algebraic functions are expressed in terms of inverse trigonometric functions (§ 605). In a similar fashion, the integrals of certain other algebraic functions can be expressed in terms of inverse hyperbolic functions. The four hyperbolic functions whose inverses are the most useful in this connection are the first four, $\sinh x$, $\cosh x$, $\tanh x$, and $\coth x$. The graphs and notation are given in Figure 606. For simplicity, by analogy with the principal value ranges for the inverse trigonometric functions, we choose only the nonnegative values of $\cosh^{-1} x$, and write for these principal values $\mathrm{Cosh}^{-1} x$.

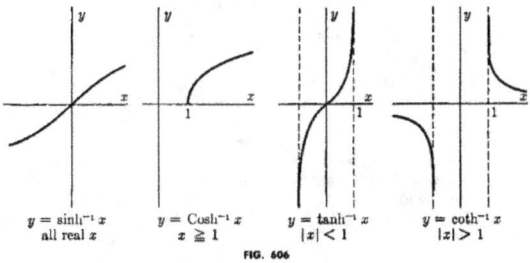

$y = \sinh^{-1} x$ $y = \mathrm{Cosh}^{-1} x$ $y = \tanh^{-1} x$ $y = \coth^{-1} x$
all real x $x \geq 1$ $|x| < 1$ $|x| > 1$

FIG. 606

The inverse hyperbolic functions can be expressed in terms of functions discussed previously:

(1) $\qquad \sinh^{-1} x = \ln(x + \sqrt{x^2 + 1})$, all real x;

(2) $\qquad \mathrm{Cosh}^{-1} x = \ln(x + \sqrt{x^2 - 1})$, $x \geq 1$;

(3) $\qquad \tanh^{-1} x = \tfrac{1}{2} \ln \dfrac{1+x}{1-x}$; $|x| < 1$;

(4) $\qquad \coth^{-1} x = \tfrac{1}{2} \ln \dfrac{x+1}{x-1}$; $|x| > 1$.

We prove (2) and leave the rest for the student (Ex. 25, § 609). If $x = \dfrac{e^y + e^{-y}}{2}$, let the positive quantity e^y be denoted by p. Then $2x = p + \dfrac{1}{p}$, or $p^2 - 2xp + 1 = 0$. Solving this equation we get $p = x \pm \sqrt{x^2 - 1}$. Since y is to be chosen nonnegative, $p = e^y \geq 1$. If

200   SOME ELEMENTARY FUNCTIONS   [§ 609

$x > 1$, $x + \sqrt{x^2 - 1} > 1$, and since the product of this and $x - \sqrt{x^2 - 1}$ is equal to 1, $x - \sqrt{x^2 - 1} < 1$. We therefore reject the minus sign and have $p = x + \sqrt{x^2 - 1}$, or equation (2). Notice that the expressions in (3) and (4) exist for the specified values of x. For example, in (3), $(1 + x)$ and $(1 - x)$ have a positive quotient if and only if they have a positive product, and $1 - x^2 > 0$ if and only if $|x| < 1$.

The derivatives of the four inverse hyperbolic functions considered here can either be obtained from the derivatives of the corresponding hyperbolic functions (Ex. 26, § 609) or from formulas (1)-(4) (Ex. 27, § 609). They are:

(5) $$\frac{d}{dx} \sinh^{-1} x = \frac{1}{\sqrt{1 + x^2}}, \text{ all real } x;$$

(6) $$\frac{d}{dx} \text{Cosh}^{-1} x = \frac{1}{\sqrt{x^2 - 1}}, x > 1;$$

(7) $$\frac{d}{dx} \tanh^{-1} x = \frac{1}{1 - x^2}, |x| < 1;$$

(8) $$\frac{d}{dx} \coth^{-1} x = \frac{1}{1 - x^2}, |x| > 1.$$

The corresponding integration formulas are mere reformulations of formulas (2) and (7) of § 605, with appropriate ranges specified. They can be established independently by differentiations of the right-hand members (Ex. 28, § 609). The letter a represents a positive number throughout.

(9) $$\int \frac{dx}{\sqrt{a^2 + x^2}} = \sinh^{-1} \frac{x}{a} + C, \text{ all real } x;$$

(10) $$\int \frac{dx}{\sqrt{x^2 - a^2}} = \text{Cosh}^{-1} \frac{x}{a} + C, x > a;$$

(11) $$\int \frac{dx}{a^2 - x^2} = \frac{1}{a} \tanh^{-1} \frac{x}{a} + C, |x| < a;$$

(12) $$\int \frac{dx}{x^2 - a^2} = -\frac{1}{a} \coth^{-1} \frac{x}{a} + C, |x| > a.$$

609. EXERCISES

In Exercises 1-10, differentiate the given function.

1. $\cosh 3x$.
2. $\sinh^2 x$.
3. $\tanh (2 - x)$.
4. $x \coth x^2$.
5. $\ln \sinh 2x$.
6. $e^{ax} \cosh bx$.
7. $\sinh^{-1} 4x$.
8. $\text{Cosh}^{-1} e^x$.
9. $\tanh^{-1} x^2$.
10. $\coth^{-1} (\sec x)$.

§ 610] NUMBERS AND FUNCTIONS 201

In Exercises 11-20, perform the indicated integration, expressing your answer in terms of hyperbolic functions or their inverses.

11. $\int \tanh 6x \, dx$.
12. $\int e^x \coth e^x \, dx$.
13. $\int \cosh^2 x \, dx$.
14. $\int \tanh^2 10x \, dx$.
15. $\int \sinh^2 x \, dx$.
16. $\int x^2 \tanh^{-1} x \, dx$.
17. $\int \frac{dx}{\sqrt{x^2 - 2}}$.
18. $\int \frac{dx}{\sqrt{4x^2 - 4x + 5}}$.
19. $\int \frac{dx}{3x^2 + 5x - 7}$,
20. $\int \frac{dx}{7 - 5x - 3x^2}$,
$(3x^2 + 5x - 7 > 0)$.
$(7 - 5x - 3x^2 > 0)$.

21. Prove that $\cosh x$ and $\operatorname{sech} x$ are even functions and that the other four hyperbolic functions are odd (cf. Exs. 7-8, § 503).

22. Establish the identities (2)-(9), § 607.

23. Establish the differentiation formulas (10)-(15), § 607.

24. Establish the integration formulas (16)-(19), § 607. *Hint for* (19): $\int \frac{dx}{\sinh x} = \int \frac{du}{u^2 - 1}$, where $u = \cosh x$, and $\frac{\cosh x - 1}{\cosh x + 1} = \frac{2 \sinh^2 \frac{1}{2}x}{2 \cosh^2 \frac{1}{2}x}$.

25. Establish formulas (1), (3), (4), § 608.

26. Establish formulas (5)-(8), § 608, by use of (10)-(13), § 607. *Hint for* (5): Let $y = \sinh^{-1} x$. Then $x = \sinh y$, $dx/dy = \cosh y$, and $dy/dx = 1/\sqrt{1 + \sinh^2 y}$.

27. Establish formulas (5)-(8), § 608, by use of (1)-(4), § 608.

28. Establish formulas (9)-(12), § 608.

★29. Show that $\int \operatorname{csch} x \, dx = -\coth^{-1}(\cosh x) + C$.

★30. Establish formulas (7)-(9), § 605, by means of hyperbolic substitutions.

★610. CLASSIFICATION OF NUMBERS AND FUNCTIONS

Certain classes of numbers (integers, rational numbers, and irrational numbers) were defined in Chapter I. Another important class of numbers is the algebraic numbers, an **algebraic number** being defined to be a root of a **polynomial equation**

(1) $\qquad a_0 x^n + a_1 x^{n-1} + \cdots + a_n = 0$,

where the coefficients $a_0, a_1, \cdots, a_n$ are integers. Examples of algebraic numbers are $\frac{3}{4}$ (a root of the equation $4x - 3 = 0$) (in fact, *every rational number is algebraic*), $-\sqrt[3]{5}$ (a root of the equation $x^3 + 5 = 0$), and $\sqrt[3]{5} + \sqrt{3}$ (a root of the equation $x^6 - 9x^4 - 10x^3 + 27x^2 - 90x - 2 = 0$). It can be shown† that any number that is the sum, difference, product, or

† Cf. Birkhoff and MacLane, *A Survey of Modern Algebra* (New York, The Macmillan Company, 1944).

quotient of algebraic numbers is algebraic, and that any (real) root of an algebraic number is an algebraic number. Therefore any number (like $\sqrt{3} - \sqrt[3]{8/9}/4\sqrt[4]{61}$) that can be obtained from the integers by a finite sequence of sums, differences, products, quotients, powers, and roots is algebraic. However, it should be appreciated that not every algebraic number is of the type just described. In fact, it is shown in Galois Theory (cf. the Birkhoff and MacLane book just referred to) that the general equation of degree 5 or higher cannot be solved in terms of radicals. In particular, the real root of the equation $x^5 - x - 1 = 0$ (this number is algebraic by definition) cannot be expressed in the finite form described above.

A **transcendental number** is any number that is not algebraic. The most familiar transcendental numbers are e and π. The transcendental character of e was established by Hermite in 1873, and that of π by Lindemann in 1882.† For a discussion of the transcendence of these two numbers see Felix Klein, *Elementary Mathematics from an Advanced Standpoint* (New York, Dover Publications, 1945). The first number to be proved transcendental was neither e nor π, but a number constructed artificially for the purpose by Liouville in 1844. (Cf. the Birkhoff and MacLane reference, page 413.) The existence of a vast infinite supply of transcendental numbers was provided by Cantor in 1874 in his theory of transfinite cardinal numbers. This amounts to showing (Ex. 10, § 612) that the algebraic numbers are denumerable (Ex. 12, § 113) and that the real numbers are not (Ex. 30, § 711). This method, however, merely establishes existence without specifically producing examples.

Functions are classified in a manner similar to that just outlined for numbers. The role played by integers above is now played by polynomials. A **rational function** is any function that can be expressed as the quotient of two polynomials, an **algebraic function** is any function $f(x)$ that satisfies (identically in x) a polynomial equation

(2) $\qquad a_0(x)(f(x))^n + a_1(x)(f(x))^{n-1} + \cdots + a_n(x) = 0,$

where $a_0(x), a_1(x), \cdots, a_n(x)$ are polynomials, and a **transcendental function** is any function that is not algebraic.

Examples of polynomials are $2x - \sqrt{3}$ and $\pi x^5 + \sqrt{e}$. Examples of rational functions that are not polynomials are $1/x$ and $(x + \sqrt{3})/(5x - 6)$. Examples of algebraic functions that are not rational are $\sqrt{x}$ and $1/\sqrt{x^2 + \pi}$. Examples of transcendental functions are e^x, $\ln x$, $\sin x$, and $\cos x$. (Cf. Exs. 11-13, § 612.)

† It is easy to show that e is irrational (Ex. 34, § 811). A proof of the irrationality of π was given in 1761 by Lambert. For an elementary proof, and further discussion of both the irrationality and transcendence of π, see Ivan Niven, *Irrational Numbers* (Carus Mathematical Monographs, 1956).

★611. THE ELEMENTARY FUNCTIONS

Most of the familiar functions which one encounters at the level of elementary calculus, like $\sin 2x$, e^{-x^2}, and $\frac{1}{2}\operatorname{Arctan} \frac{1}{2}x$, are examples of what are called *elementary functions*. In order to define this concept, we describe first the elementary operations on functions. The **elementary operations** on functions $f(x)$ and $g(x)$ are those that yield any of the following: $f(x) \pm g(x)$, $f(x)\,g(x)$, $f(x)/g(x)$, $\{f(x)\}^a$, $a^{f(x)}$, $\log_a f(x)$, and $T(f(x))$, where T is any trigonometric or inverse trigonometric function. The **elementary functions** are those generated by constants and the independent variable by means of a finite sequence of elementary operations. Thus

$$(x^2 + 17\pi)^e \operatorname{Arcsin}\,[(\log_3 \cos x)(e^{\sqrt{x}})]$$

is an elementary function. Its derivative is also an elementary function. In fact (cf. Ex. 9, §612), the derivative of any elementary function is an elementary function. Reasonable questions to be asked at this point are, "Is there ever any occasion to study functions that are *not* elementary? What are some examples?" The answer is that nonelementary functions arise in connection with infinite series, differential equations, integral equations, and equations defining functions implicitly. For example, the differential equation $\frac{dy}{dx} = \frac{\sin x}{x}$ ($=1$ if $x = 0$) has a nonelementary solution

$$(1) \qquad \int_0^x \frac{\sin t}{t}\,dt$$

whose power series expansion (cf. Chapter 8) is

$$x - \frac{x^3}{3 \cdot 3!} + \frac{x^5}{5 \cdot 5!} - \frac{x^7}{7 \cdot 7!} + \cdots.$$

Other examples of nonelementary functions which are integrals of elementary functions are the *probability integral*,

$$(2) \qquad \int_0^x e^{-t^2}\,dt,$$

the *Fresnel sine integral* and *cosine integral*,

$$(3) \qquad \int_0^x \sin t^2\,dt \quad \text{and} \quad \int_0^x \cos t^2\,dt,$$

and the *elliptic integrals of the first and second kind*,

$$(4) \qquad \int_0^x \frac{dt}{\sqrt{1 - k^2 \sin^2 t}} \quad \text{and} \quad \int_0^x \sqrt{1 - k^2 \sin^2 t}\,dt.$$

Extensive tables for these and other nonelementary functions exist.†

† Cf. Eugen Jahnke and Fritz Emde, *Tables of Functions* (New York, Dover Publications, 1943).

Proofs that these functions are not elementary are difficult and will not be given in this book.† For some evaluations, see §§ 1329, 1408, 1410.

The inverse of an elementary function need not be elementary. For example, the inverse of the function $x - a \sin x$ ($a \neq 0$), the function defined implicitly by *Kepler's equation*

$$(5) \qquad y - a \sin y - x = 0,$$

is not elementary.† Again, the proof is omitted.

An important lesson to be extracted from the above considerations is that inability to integrate an elementary function in terms of elementary functions does not mean that the integral fails to exist. In fact, many important nonelementary functions owe their existence to the (Riemann) integration process. This is somewhat similar to the position of the logarithmic, exponential, and trigonometric functions as developed in this chapter, which were defined, ultimately, in terms of Riemann integrals of algebraic functions.

One more final remark. The reader might ask, "Apart from the importance of the functions defined above, if all we seek is an example of a nonelementary function, isn't the bracket function $[x]$ of § 201 one such?" Perhaps the most satisfactory reply is that whereas $[x]$ is elementary in intervals, the functions presented above might be described as "nowhere elementary." A behind-the-scenes principle here is that of analytic continuation, important in the theory of analytic functions of a complex variable. (Cf. § 815.)

★612. EXERCISES

★1. Prove that $\sqrt{5} - \sqrt{2}$ is algebraic by finding a polynomial equation with integral coefficients of which it is a root.

★2. Prove that $1/x$ is not a polynomial. *Hint:* It is not sufficient to remark that it does not *look* like one. Show that $1/x$ has some property that no polynomial can have, such as a certain kind of limit as $x \to \infty$. (Cf. Ex. 4.)

★3. Prove that $\frac{x^2}{x+1}$ is not a polynomial (cf. Ex. 2).

★4. Prove that $\sqrt{x^2 + 1}$ is not a polynomial. *Hint:* Differentiate the function a large number of times.

★5. Prove that $\sqrt{x^2 + 1}$ is not rational. (Cf. Ex. 4.)

★6. Prove that the hyperbolic functions and their inverses are elementary.

★7. Prove that x^x is elementary.

★8. Prove that $\int_0^x t^n \sin t \, dt$, where n is a positive integer, is elementary.

★9. Prove that the derivative of any elementary function is elementary.

† Cf. J. F. Ritt, *Integration in Finite Terms* (New York, Columbia University Press, 1948).

★10. Prove that the algebraic numbers are denumerable (cf. Ex. 12, § 113). *Hint:* A useful device of Cantor was to let $p(x) = a_0 x^n + \cdots + a_n = 0$ be the (unique!) polynomial equation of lowest degree and smallest positive leading coefficient a_0 ($a_0, \cdots, a_n$ being integers) satisfied by a given algebraic number a and to define the **height** of a to be the positive integer $n - 1 + |a_0| + \cdots |a_n|$. Show that there are only finitely many algebraic numbers having a given height.

★11. Prove that e^x is transcendental. *Hint:* Let
$$a_0(x) e^{nx} + a_1(x) e^{(n-1)x} + \cdots + a_{n-1}(x) e^x + a_n(x) = 0,$$
where $a_0(x), \cdots, a_n(x)$ are polynomials and $a_n(x)$ is not identically 0. Form the limit as $x \to -\infty$. Then $\lim_{x \to -\infty} a_n(x) = 0$ (!).

★12. Prove that $\ln x$ is transcendental. *Hint:* Let $y = \ln x$, and assume that y satisfies identically a polynomial equation, arranged in the form
$$b_0(y) x^n + b_1(y) x^{n-1} + \cdots + b_{n-1}(y) x + b_n(y) = 0,$$
where $b_0(y)$ is not identically 0. Divide every term by x^n, and let $x \to +\infty$.

★13. Prove that $\sin x$ and $\cos x$ are transcendental. *Hint for* $\sin x$: Let
$$a_0(x) \sin^n x + a_1(x) \sin^{n-1} x + \cdots + a_{n-1}(x) \sin x + a_n(x) = 0,$$
where $a_0(x), \cdots, a_n(x)$ are polynomials and $a_n(x)$ is not identically 0. Let $x = k\pi$, where k is so large that $a_n(x) \neq 0$.

★14. Prove that the system of algebraic numbers does not satisfy the axiom of completeness (§ 114). (Cf. Ex. 2, § 116.)

7

Infinite Series of Constants

701. BASIC DEFINITIONS

If $a_1, a_2, \cdots, a_n, \cdots$ is a sequence of numbers, the expression

$$(1) \qquad \sum_{n=1}^{+\infty} a_n = a_1 + a_2 + \cdots + a_n + \cdots$$

(also written $\sum a_n$ if there is no possible misinterpretation) is called an **infinite series** or, in brief, a **series**. The numbers $a_1, a_2, \cdots, a_n, \cdots$ are called **terms**, and the number a_n is called the **nth term** or **general term**.

The sequence $\{S_n\}$, where

$$S_1 \equiv a_1, S_2 \equiv a_1 + a_2, \cdots, S_n \equiv a_1 + a_2 + \cdots + a_n,$$

is called the **sequence of partial sums** of the series (1). An infinite series **converges** or **diverges** according as its sequence of partial sums converges or diverges. In case of convergence, the series (1) is said to have a **sum** equal to the limit of the sequence of partial sums, and if $S \equiv \lim_{n \to +\infty} S_n$, we write

$$(2) \qquad \sum a_n = \sum_{n=1}^{+\infty} a_n = a_1 + \cdots + a_n + \cdots = S,$$

and say that the series **converges to** S.

In case $\lim_{n \to +\infty} S_n$ exists and is infinite the series $\sum a_n$, although divergent, is said to have an **infinite sum**:

$$(3) \qquad \sum a_n = \sum_{n=1}^{+\infty} a_n = a_1 + \cdots + a_n + \cdots = +\infty, -\infty, \text{ or } \infty,$$

and the series is said to **diverge to** $+\infty, -\infty,$ or ∞.

Example 1. $\frac{1}{2} + \frac{1}{4} + \cdots + \frac{1}{2^n} + \cdots = 1$, since the sum of the first n terms is

$$S_n = \frac{1}{2} + \frac{1}{4} + \cdots + \frac{1}{2^n} = 1 - \frac{1}{2^n},$$

and $S_n \to 1$ (cf. Example 3, § 202).

Example 2. The series whose general term is $2n$ diverges to $+\infty$:
$$2 + 4 + 6 + \cdots + 2n + \cdots = +\infty.$$

Example 3. The series
$$1 - 1 + 1 - 1 + \cdots + (-1)^{n+1} + \cdots$$
diverges since its sequence of partial sums is
$$1, 0, 1, 0, 1, 0, 1, \cdots.$$
This sequence diverges since, if it converged to S, every subsequence would also converge to S, and S would equal both 0 and 1, in contradiction to the uniqueness of the limit of a sequence. (Cf. § 204; also Example 4, § 202.)

702. THREE ELEMENTARY THEOREMS

Direct consequences of Theorems I, II, and X, § 204, are the following:

Theorem I. *The alteration of a finite number of terms of a series has no effect on convergence or divergence (although it will in general change the sum in case of convergence).*

Theorem II. *If a series converges or has an infinite sum, this sum is unique.*

Theorem III. *Multiplication of the terms of a series by a nonzero constant k does not affect convergence or divergence. If the original series converges, the new series converges to k times the sum of the original, for any constant k:*

$$(1) \qquad \sum_{n=1}^{+\infty} k a_n = k \sum_{n=1}^{+\infty} a_n.$$

(Cf. Ex. 27, § 707.)

703. A NECESSARY CONDITION FOR CONVERGENCE

Theorem. *If a series converges, its general term tends toward zero as n becomes infinite. Equivalently, if the general term of a series does not tend toward zero as n becomes infinite, the series diverges.*

Proof. Assume the series, $a_1 + a_2 + \cdots$, converges to S. Then $S_n = S_{n-1} + a_n$, or $a_n = S_n - S_{n-1}$. Since the sequence of partial sums converges to S, both S_n and S_{n-1} tend toward S as n becomes infinite. Hence (Theorem VII, § 204),

$$\lim_{n \to +\infty} a_n = \lim_{n \to +\infty} (S_n - S_{n-1}) = S - S = 0.$$

Example. Show that the series $2 + 4 + \cdots + 2n + \cdots$ and $0 + 1 + 0 + 1 + \cdots$ diverge.

Solution. In neither case does the general term tend toward zero, although

208 INFINITE SERIES OF CONSTANTS [§ 704

NOTE. The condition of the theorem is necessary but *not sufficient* for convergence. It will be shown later (Theorem IV, § 708) that the *harmonic series* $1 + \frac{1}{2} + \frac{1}{3} + \cdots$ diverges, although the general term, $\frac{1}{n}$, converges to zero.

704. THE GEOMETRIC SERIES

The series
$$(1) \qquad a + ar + ar^2 + \cdots + ar^{n-1} + \cdots$$
is known as a **geometric series**.

Theorem. *The geometric series* (1), *where* $a \neq 0$, *converges if* $|r| < 1$, *with* $\frac{a}{1-r}$ *as its sum, and diverges if* $|r| \geq 1$.

Proof. By use of the formula of algebra for the sum of the terms of a finite geometric progression, the sum of the first n terms of (1) can be written
$$S_n = \frac{a - ar^n}{1 - r} = \frac{a}{1-r} - \frac{a}{1-r} r^n.$$

If $|r| < 1$, $\lim_{n \to +\infty} r^n = 0$ (Theorem XIII, § 204), and therefore
$$\lim_{n \to \infty} S_n = \frac{a}{1-r} - \frac{a}{1-r} \lim_{n \to +\infty} r^n = \frac{a}{1-r}.$$

If $|r| \geq 1$, $|ar^{n-1}| = |a| \cdot |r|^{n-1} \geq |a|$, and the general term of (1) does not tend toward zero. Therefore, by the Theorem of § 703, the series (1) diverges.

Example. Determine for each of the following series whether the series converges or diverges. In case of convergence, find the sum:

(a) $\frac{1}{2} + \frac{1}{4} + \frac{1}{8} + \frac{1}{16} + \cdots$; (b) $2 + 3 + \frac{9}{2} + \frac{27}{4} + \cdots$;
(c) $3 - 2 + \frac{4}{3} - \frac{8}{9} + \cdots$.

Solution. (a) The series converges since $r = \frac{1}{2} < 1$, and the sum is $\frac{\frac{1}{2}}{1-\frac{1}{2}} = 1$.
(b) The series diverges since $r = \frac{3}{2} > 1$. (c) The series converges since $r = -\frac{2}{3}$, and $|r| < 1$. The sum is $\frac{3}{1-(-\frac{2}{3})} = \frac{9}{5}$.

705. POSITIVE SERIES

For a **positive series** (a series whose terms are positive) or, more generally, a **nonnegative series** (a series whose terms are nonnegative), there is a simple criterion for convergence.

Theorem. *The series $\sum_{n=1}^{+\infty} a_n$, where $a_n \geq 0$, converges if and only if the sequence of partial sums is bounded; that is, if and only if there is a number A such that $a_1 + \cdots + a_n \leq A$ for all values of n. If this condition is satisfied, the sum of the series is less than or equal to A. A series of nonnegative terms either converges, or diverges to $+\infty$.*

Proof. Since $S_{n+1} = S_n + a_n$, the condition $a_n \geq 0$ is equivalent to the condition $S_n \uparrow$, that is, that the sequence $\{S_n\}$ is monotonically increasing. The Theorem of this section is therefore a direct consequence of the Note following Theorem XIV, § 204.

706. THE INTEGRAL TEST

A convenient test for establishing convergence or divergence for certain series of positive terms makes use of the technique of integration.

Theorem. Integral Test. *If $\sum a_n$ is a positive series and if $f(x)$ is a positive monotonically decreasing function, defined for $x \geq a$, where a is some real number, and if (for integral values of n) $f(n) = a_n$ for $n > a$, then the improper integral*

$$(1) \qquad \int_a^{+\infty} f(x)\, dx$$

and the infinite series

$$(2) \qquad \sum_{n=1}^{+\infty} a_n$$

either both converge, or both diverge to $+\infty$.

Proof. Let us observe that both (1) and (2) always exist, in the finite or $+\infty$ sense. (Why? Supply details in Ex. 28, § 707.) We wish to show that the finiteness of either implies the finiteness of the other. With the aid of results given in §§ 511, 702, and 705, we have only to show that if K is a fixed positive integer $\geq a$ and if N is an arbitrary integer $> K$, then the boundedness (as a function of N) of either

$$(3) \qquad I(N) \equiv \int_K^N f(x)\, dx$$

or

$$(4) \qquad J(N) \equiv \sum_{n=K}^N a_n$$

implies that of the other. (Cf. Ex. 29, § 707.) Let us look at the geometric representations of (3) and (4) given in Figure 701. Each term of (4) is the area of a rectangle of width 1 and height a_n. This suggests that if (3) is written as the sum

$$I(N) = \int_K^N f(x)\,dx = \sum_{n=K}^{N-1} \int_n^{n+1} f(x)\,dx, \tag{5}$$

then each term of (5) is trapped between successive terms of (4). Indeed, for the interval $n \leq x \leq n+1$, since $f(x)$ is monotonically decreasing,
$$a_n = f(n) \geq f(x) \geq f(n+1) = a_{n+1},$$
so that
$$a_n = \int_n^{n+1} a_n\,dx \geq \int_n^{n+1} f(x)\,dx \geq \int_n^{n+1} a_{n+1}\,dx = a_{n+1}.$$

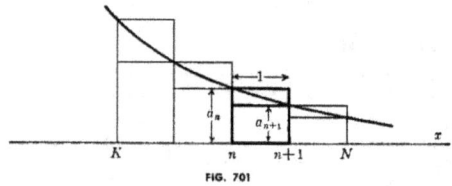

FIG. 701

Therefore
$$\sum_{n=K}^{N-1} a_n \geq \int_K^N f(x)\,dx \geq \sum_{n=K}^{N-1} a_{n+1}.$$
Consequently (in the notation of (3) and (4)), since
$$\sum_{n=K}^{N-1} a_{n+1} = \sum_{n=K+1}^{N} a_n = J(N) - a_K,$$
$$J(N-1) \geq I(N) \geq J(N) - a_K.$$

These two inequalities show that the boundedness of either $J(N)$ or $I(N)$ implies that of the other (Ex. 30, § 707) and the proof is complete.

Example 1. Show that the *p*-series
$$\sum_{n=1}^{+\infty} \frac{1}{n^p}$$
converges if $p > 1$ and diverges if $p \leq 1$.

Solution. This follows immediately from Example 3, § 512, if $p \geq 0$. If $p < 0$, the conclusion is trivial.

Example 2. Show that $\sum_{n=2}^{+\infty} \dfrac{1}{n(\ln n)^p}$ converges if and only if $p > 1$.

Solution. The integral $\displaystyle\int_2^{+\infty} \frac{dx}{x(\ln x)^p}$ converges if and only if $p > 1$. (Ex. 28, § 513.)

707. EXERCISES

In Exercises 1-4, find an expression for the sum of the first n terms of the series. Thus find the sum of the series, if it exists.

1. $\frac{1}{1\cdot 2} + \frac{1}{2\cdot 3} + \frac{1}{3\cdot 4} + \cdots$. *Hint:* $\frac{1}{n(n+1)} = \frac{1}{n} - \frac{1}{n+1}$.

2. $\frac{1}{1\cdot 3} + \frac{1}{3\cdot 5} + \frac{1}{5\cdot 7} + \cdots$.

3. $\frac{1}{3} + \frac{1}{9} + \frac{1}{27} + \frac{1}{81} + \cdots$. 4. $1 + 2 + 3 + 4 + \cdots$.

In Exercises 5-8, write down the first four terms and the general term of the series whose sequence of partial sums is given.

5. $2, 2\frac{1}{2}, 2\frac{3}{4}, 2\frac{7}{8}, \cdots$. 6. $1, 3, 1, 3, 1, 3, \cdots$.

7. $1.3, 0.91, 1.027, 0.9919, \cdots, 1 - (-.3)^n, \cdots$.

8. $2, 1.3, 1.09, 1.027, \cdots, 1 + (.3)^{n-1}, \cdots$.

In Exercises 9-12, determine whether the given geometric series is convergent or not, and find its sum in case of convergence.

9. $12 - 8 + \cdots$. 10. $8 - 12 + \cdots$.

11. $0.27 + 0.027 + \cdots$. 12. $101 - 100 + \cdots$.

In Exercises 13-16, find as a quotient of integers the rational number represented by the repeating decimal, by means of geometric series.

13. $0.5555\cdots$. *Hint:* $0.5555\cdots$ means $0.5 + 0.05 + 0.005 + \cdots$.

14. $3.285555\cdots$. *Hint:* Write the number $3.28 + 0.005555\cdots$.

15. $6.1727272\cdots$. 16. $0.428571428571428571\cdots$.

In Exercises 17-26, establish convergence or divergence by the integral test of § 506.

17. $\sum \frac{1}{2n+1}$. 18. $\sum \frac{n}{n^2+3}$.

19. $\sum \frac{1}{n^2+4}$. 20. $\sum \frac{n}{e^n}$.

21. $\sum_{n=3}^{+\infty} \frac{1}{n^2-4}$. 22. $\sum_{n=4}^{+\infty} \frac{1}{\sqrt{n^2-9}}$.

23. $\sum \frac{\ln n}{n}$. 24. $\sum_{n=3}^{+\infty} \frac{1}{n(\ln n)(\ln \ln n)^2}$.

25. $\sum \frac{1}{(n+1)^3}$. 26. $\sum \frac{1}{n^3+1}$.

27. Establish formula (1) of Theorem III, § 702, for the case of an infinite sum $(+\infty, -\infty, \text{ or } \infty)$.

28. Explain why both (1) and (2), § 706 must exist, as either finite or infinite limits. (Cf. Theorem X, § 501.)

29. For the Theorem of § 706 prove that it is sufficient to show that the boundedness of either (3) or (4) implies that of the other. In particular, explain why we may use K in place of a in (3) and K in place of 1 in (4).

30. Supply the details needed in the last sentence of the proof of § 706.

31. Prove that $\sum_{n=3}^{+\infty} \frac{1}{n(\ln n)(\ln \ln n)^p}$ converges if and only if $p > 1$. More generally, prove that $\sum_{n=N}^{+\infty} \frac{1}{n(\ln n)(\ln \ln n) \cdots (\ln \ln \cdots \ln n)^p}$, where N is sufficiently large for $(\ln \ln \cdots \ln n)$ to be defined, converges if and only if $p > 1$. (Cf. Ex. 29, § 513.)

★32. Indicate by a graph how you could define a function $f(x)$ having the properties: $f(x)$ is positive, continuous, defined for $x \geq 0$, $\int_0^{+\infty} f(x)\,dx$ converges, and $\sum_{n=1}^{+\infty} f(n)$ diverges. (Cf. Ex. 40, § 513.)

★33. Indicate by a graph how you could define a function $f(x)$ having the properties: $f(x)$ is positive, continuous, defined for $x \geq 0$, $\int_0^{+\infty} f(x)\,dx$ diverges, and $\sum_{n=1}^{+\infty} f(n)$ converges.

708. COMPARISON TESTS. DOMINANCE

In establishing the integral test of § 706 we have already made use of the fact that a series of positive terms converges if and only if its sequence of partial sums is *bounded*. In order to prove convergence for a series of positive terms it is necessary to show only that the terms *approach zero fast enough* to keep the partial sums bounded. One of the simplest and most useful ways of doing this is to compare the terms of the given series with those of a series whose behavior is known. The ideas are similar to those relevant to comparison tests for improper integrals (§ 512).

Definition I. *The statement that a series $\sum b_n$ **dominates** a series $\sum a_n$ means that $|a_n| \leq b_n$ for every positive integer n.*

NOTE 1. Any dominating series consists automatically of nonnegative terms, although a dominated series may have negative terms.

Theorem I. Comparison Test. *Any nonnegative series dominated by a convergent series converges. Equivalently, any series that dominates a divergent nonnegative series diverges.*

Proof. We shall prove only the first form of the statement (cf. Ex. 21, § 711). Assume $0 \leq a_n \leq b_n$ and that $\sum b_n$ converges. The convergence of $\sum b_n$ implies the boundedness of the partial sums $\sum_{n=1}^{N} b_n$, which implies the boundedness of the partial sums $\sum_{n=1}^{N} a_n$. By the Theorem of § 705, the proof is complete.

Example 1. Prove that the series
$$1 + \frac{1}{2^2} + \frac{1}{3^3} + \cdots \frac{1}{n^n} + \cdots$$
converges.

Solution. Since $\frac{1}{n^n} \leq \frac{1}{2^n}$, for $n > 1$, the given series is dominated by a convergent geometric series (except for the first term).

Example 2. Prove that the series
$$\frac{1}{2} + \frac{1}{3} + \frac{1}{2^2} + \frac{1}{3^2} + \frac{1}{2^3} + \frac{1}{3^3} + \cdots$$
converges.

Solution. This series is dominated by
$$\frac{1}{2} + \frac{1}{2} + \frac{1}{2^2} + \frac{1}{2^2} + \frac{1}{2^3} + \frac{1}{2^3} + \cdots,$$
whose partial sums are bounded by twice the sum of the convergent geometric series $\sum \frac{1}{2^n}$.

Let us adapt the "big O" notation of § 514 to the present topic:[†]

Definition II. *If $\{a_n\}$ and $\{b_n\}$ are two sequences, the notation $a_n = O(b_n)$ (read "a sub n is big O of b sub n"), means that there exist a positive integer N and a positive number K such that $n > N$ implies $|a_n| \leq K b_n$. If simultaneously $a_n = O(b_n)$ and $b_n = O(a_n)$, the two sequences (and also the two series $\sum a_n$ and $\sum b_n$) are said to be of the **same order of magnitude** at $+\infty$.*

NOTE 2. As in § 512, the "big O" notation is also used in the following sense: $a_n = b_n + O(c_n)$ means $a_n - b_n = O(c_n)$.

Big O and order of magnitude relationships are usually established by taking limits:

Theorem II. *Let $\{a_n\}$ and $\{b_n\}$ be sequences of positive terms and assume that $\lim_{n \to +\infty} \frac{a_n}{b_n} = L$ exists, in a finite or infinite sense. Then:*

(i) $0 \leq L < +\infty$ *implies* $a_n = O(b_n)$,
(ii) $0 < L \leq +\infty$ *implies* $b_n = O(a_n)$,
(iii) $0 < L < +\infty$ *implies that* $\{a_n\}$ *and* $\{b_n\}$ *are of the same order of magnitude at* $+\infty$.

(Give the details of the proof in Ex. 22, § 711. Cf. Theorem II, § 512.)

[†] The "little o" notation of § 512 takes the form for sequences: $a_n = o(b_n)$ means that there exist a positive integer N and a sequence of nonnegative terms K_n such that $K_n \to 0$ and such that $n > N$ implies $|a_n| \leq K_n b_n$. Obviously $a_n = o(b_n)$ implies $a_n = O(b_n)$.

Theorem III. *If $\sum a_n$ and $\sum b_n$ are series of nonnegative terms and if $a_n = O(b_n)$, then the convergence of $\sum b_n$ implies that of $\sum a_n$. Equivalently, the divergence of $\sum a_n$ implies that $\sum b_n$.*

Proof. The convergence of $\sum_{n=N+1}^{+\infty} Kb_n$ implies that of $\sum_{n=N+1}^{+\infty} a_n$ by Theorem I.

Corollary. *Two positive series that have the same order of magnitude either both converge or both diverge.*

Because of the usefulness of the *p*-series, $\sum \frac{1}{n^p}$, given in Example 1, § 706, as a "test series," we restate here the facts established in that Example.

Theorem IV. *The **p-series** $\sum_{n=1}^{+\infty} \frac{1}{n^p}$ converges if $p > 1$ and diverges if $p \leq 1$. As a special case, the **harmonic series**, $\sum_{n=1}^{+\infty} \frac{1}{n}$, diverges.†*

(Cf. Ex. 29, § 707, for a sequence of test series similar to the *p*-series, each converging more slowly than the preceding.)

The technique of the comparison test for positive series, particularly in the form of Theorem III, consists usually of four steps:
 (*i*) get a "feeling" for how rapidly the terms of the given series are approaching zero;
 (*ii*) construct a new test-series of positive terms which dominates, is dominated by, or is of the same order of magnitude as the original series;
 (*iii*) establish the necessary inequalities, or the limit of the quotient of the general terms;
 (*iv*) infer convergence if the dominating series converges, and divergence if the dominated series diverges.

In practice, one frequently postpones applying a comparison test to a given series until after one has tried one of the more automatic tests (like the ratio test) given in subsequent sections.

Example 3. Test for convergence or divergence:
$$\frac{1}{1 \cdot 3} + \frac{2}{5 \cdot 7} + \frac{3}{9 \cdot 11} + \frac{4}{13 \cdot 15} + \cdots.$$

† The function defined by the convergent *p*-series is called the **Riemann Zeta-function**:

(1) $$\zeta(z) \equiv \sum_{n=1}^{+\infty} \frac{1}{n^z}.$$

This function is defined for complex values of z, by (1) if the real part of $z > 1$, and by other means (infinite products, improper integrals) for other values of z. Its values have been tabulated (cf. Eugen Jahnke and Fritz Emde, *Tables of Functions* (New York, Dover Publications, 1943). For further comments, see the Note, § 720.

§ 709] THE RATIO TEST

Solution. The general term is $\dfrac{n}{(4n-3)(4n-1)}$, which evidently is of the same order of magnitude at $+\infty$ as $\dfrac{1}{n}$ (the limit of the quotient is $\tfrac{1}{16}$). Since the test series $\sum \dfrac{1}{n}$ diverges, so does the given series.

Example 4. Prove that if $a_n = O(c_n)$ and $b_n = O(d_n)$, then $a_n + b_n = O(c_n + d_n)$ and $a_n b_n = O(c_n d_n)$. Prove that if $a_n = O(c_n)$ and b_n and d_n are sequences of positive numbers which are of the same order of magnitude at $+\infty$, then $\dfrac{a_n}{b_n} = O\left(\dfrac{c_n}{d_n}\right)$.

Solution. For the first part, if $|a_n| \leq K_1 c_n$ and $|b_n| \leq K_2 d_n$, then $|a_n + b_n| \leq \max(K_1, K_2)(c_n + d_n)$ and $|a_n b_n| \leq K_1 K_2 c_n d_n$. For the second part, since $1/b_n = O(1/d_n)$, the result follows from the product form of the first part.

Example 5. Test for convergence or divergence:
$$\frac{\ln 5}{1 \cdot 2} + \frac{\sqrt{3}\ln 9}{2 \cdot 3} + \frac{\sqrt{5}\ln 13}{3 \cdot 4} + \cdots.$$

Solution. The general term is $\dfrac{\sqrt{2n-1}\ln(4n+1)}{n(n+1)}$. Inasmuch as $\sqrt{2n-1} = O(n^{\frac{1}{2}})$ and $\ln(4n+1) = O(n^p)$ for every $p > 0$ (by l'Hospital's Rule), the numerator $= O(n^q)$ for every $q > \tfrac{1}{2}$ (cf. Example 4). Since the denominator is of the same order of magnitude as n^2 at $+\infty$, the general term $= O(n^r)$ for every $r > -\tfrac{3}{2}$. Finally, since $\sum n^r$ converges for every $r < -1$, we have only to choose r between $-\tfrac{3}{2}$ and -1 to establish the convergence of the given series.

709. THE RATIO TEST

One of the most practical routine tests for convergence of a positive series makes use of the ratio of consecutive terms.

Theorem. *Ratio test.* Let $\sum a_n$ be a positive series, and define the *test ratio*
$$r_n \equiv \frac{a_{n+1}}{a_n}.$$

Assume that the limit of this test ratio exists:
$$\lim_{n \to +\infty} r_n = \rho,$$

where $0 \leq \rho \leq +\infty$. *Then*
  (i) *if* $0 \leq \rho < 1$, $\sum a_n$ *converges*;
  (ii) *if* $1 < \rho \leq +\infty$, $\sum a_n$ *diverges*;
  (iii) *if* $\rho = 1$, $\sum a_n$ *may either converge or diverge, and the test fails.*

Proof of (i). Assume $\rho < 1$, and let r be any number such that $\rho < r < 1$. Since $\lim_{n \to +\infty} r_n = \rho < r$, we may choose a neighborhood of ρ that excludes r.

Since every r_n, for n greater than or equal to some N, lies in this neighborhood of ρ, we have:
$$n \geq N \text{ implies } r_n < r.$$
This gives the following sequence of inequalities:

$$\frac{a_{N+1}}{a_N} < r, \text{ or } a_{N+1} < ra_N,$$

$$\frac{a_{N+2}}{a_{N+1}} < r, \text{ or } a_{N+2} < ra_{N+1} < r^2 a_N,$$

$$\frac{a_{N+3}}{a_{N+2}} < r, \text{ or } a_{N+3} < ra_{N+2} < r^3 a_N,$$
$$\cdots \cdots \cdots \cdots.$$

Thus, each term of the series

(1) $$a_{N+1} + a_{N+2} + a_{N+3} + \cdots$$

is less than the corresponding term of the series

(2) $$ra_N + r^2 a_N + r^3 a_N + \cdots.$$

But the series (2) is a geometric series with common ratio $r < 1$, and therefore converges. Since (2) dominates (1), the latter series also converges. Therefore (Theorem I, § 702), the original series $\sum a_n$ converges.

Proof of (ii). By reasoning analogous to that employed above, we see that since $\lim\limits_{n \to +\infty} r_n > 1$, whether the limit is finite or infinite, there must be a number N such that $r_n \geq 1$ whenever $n \geq N$. In other words,

(3) $$n \geq N \text{ implies } \frac{a_{n+1}}{a_n} \geq 1, \text{ or } a_{n+1} \geq a_n.$$

The inequalities (3) state that beyond the first N terms, each term is at least as large as the preceding term. Since these terms are positive, the limit of the general term cannot be 0 (take $\epsilon = a_N > 0$), and therefore (§ 703) the series $\sum a_n$ diverges.

Proof of (iii). For any p-series the test ratio is $r_n = \frac{a_{n+1}}{a_n} = \left(\frac{n}{n+1}\right)^p$, and since the function x^p is continuous at $x = 1$ (Ex. 7, § 602),

$$\lim_{n \to +\infty} r_n = \lim_{n \to +\infty} \left(\frac{n}{n+1}\right)^p = \left[\lim_{n \to +\infty} \frac{n}{n+1}\right]^p = 1^p = 1.$$

If $p > 1$ the p-series converges and if $p \leq 1$ the p-series diverges, but in either case $\rho = 1$.

NOTE 1. For convergence it is important that the *limit* of the test ratio be less than 1. It is not sufficient that the test ratio itself be always less than 1. This is shown by the harmonic series, which diverges, whereas the test ratio $n/(n+1)$ is always less than 1. However, if an inequality of the form

$\frac{a_{n+1}}{a_n} \leq r < 1$ holds for all sufficiently large n (whether the limit ρ exists or not), the series $\sum a_n$ converges.

NOTE 2. For divergence it is sufficient that the test ratio itself be greater than 1. In fact, if $r_n \geq 1$ for all sufficiently large n, the series $\sum a_n$ diverges, since this inequality is the inequality (3) upon which the proof of (ii) rests.

NOTE 3. The ratio test may fail, not only by the equality of ρ and 1, but by the failure of the limit ρ to exist, finitely or infinitely. For example, the series of Example 2, § 708, converges, although the test ratio $r_n = a_{n+1}/a_n$ has no limit, since $r_{2n-1} = 2^n/3^n \to 0$ and $r_{2n} = 3^n/2^{n+1} \to +\infty$.

NOTE 4. The ratio test provides a simple proof that certain limits are zero. That is, if $\lim_{n \to +\infty} a_{n+1}/a_n$, where $a_n > 0$ for all n, exists and is less than 1, then $a_n \to 0$. This is true because a_n is the general term of a convergent series.

Example 1. Prove that $\lim_{n \to +\infty} \frac{x^n}{n!} = 0$, for every real number x.

Solution. Without loss of generality we can assume $x > 0$ (take absolute values). Then $\frac{x^{n+1}}{(n+1)!} \div \frac{x^n}{n!} = \frac{x}{n+1} \to 0$.

Example 2. Use the ratio test to establish convergence of the series
$$1 + \frac{1}{1!} + \frac{1}{2!} + \frac{1}{3!} + \cdots + \frac{1}{(n-1)!} + \cdots.$$

Solution. Since $a_n = \frac{1}{(n-1)!}$, $a_{n+1} = \frac{1}{n!}$ and
$$r_n = \frac{a_{n+1}}{a_n} = \frac{(n-1)!}{n!} = \frac{1}{n}.$$

Therefore $\rho = \lim_{n \to +\infty} r_n = 0 < 1$.

Example 3. Use the ratio test to establish convergence of the series
$$\frac{1}{2} + \frac{2^2}{2^2} + \frac{3^2}{2^3} + \frac{4^2}{2^4} + \cdots + \frac{n^2}{2^n} + \cdots.$$

Solution. Since $a_n = \frac{n^2}{2^n}$, $a_{n+1} = \frac{(n+1)^2}{2^{n+1}}$ and
$$r_n = \frac{a_{n+1}}{a_n} = \left(\frac{n+1}{n}\right)^2 \cdot \frac{2^n}{2^{n+1}} = \frac{1}{2}\left(\frac{n+1}{n}\right)^2.$$

Therefore $\rho = \lim_{n \to +\infty} r_n = \frac{1}{2} \lim_{n \to +\infty} \left(1 + \frac{1}{n}\right)^2 = \frac{1}{2} < 1$.

Example 4. Test for convergence or divergence:
$$\frac{1}{3} + \frac{2!}{3^2} + \frac{3!}{3^3} + \cdots + \frac{n!}{3^n} + \cdots.$$

Solution. Since $a_n = \frac{n!}{3^n}$, $a_{n+1} = \frac{(n+1)!}{3^{n+1}}$ and
$$r_n = \frac{(n+1)!}{n!} \cdot \frac{3^n}{3^{n+1}} = \frac{n+1}{3}.$$

218 INFINITE SERIES OF CONSTANTS [§ 710]

Therefore $\rho = \lim\limits_{n \to +\infty} \frac{n+1}{3} = +\infty$, and the series diverges.

NOTE 5. Experience teaches us that if the general term of a series involves the index n either exponentially or factorially (as in the preceding Examples) the ratio test can be expected to answer the question of convergence or divergence, while if the index n is involved only algebraically or logarithmically (as in the p-series and Examples 3 and 5, § 708), the ratio test can be expected to fail.

710. THE ROOT TEST

A test somewhat similar to the ratio test is the following, whose proof is requested in Exercise 23, § 711:

Theorem. Root Test. *Let $\sum a_n$ be a nonnegative series, and assume the existence of the following limit:*

$$\lim_{n \to +\infty} \sqrt[n]{a_n} = \sigma,$$

where $0 \leq \sigma \leq +\infty$. Then

(i) *if $0 \leq \sigma < 1$, $\sum a_n$ converges;*
(ii) *if $1 < \sigma \leq +\infty$, $\sum a_n$ diverges;*
(iii) *if $\sigma = 1$, $\sum a_n$ may either converge or diverge, and the test fails.*

NOTE 1. The inequality $\sqrt[n]{a_n} < 1$, for all n, is not sufficient for convergence, although an inequality of the form $\sqrt[n]{a_n} \leq r < 1$, for $n > N$, does guarantee convergence.

NOTE 2. For divergence it is sufficient to have an inequality of the form $\sqrt[n]{a_n} \geq 1$, for $n > N$, since this precludes $\lim\limits_{n \to +\infty} a_n = 0$.

NOTE 3. The ratio test is usually easier to apply than the root test, but the latter is more powerful. (Cf. Exs. 34-35, § 711.)

Example. Use the root test to establish convergence of the series

$$1 + \frac{2}{2^1} + \frac{3}{2^2} + \frac{4}{2^3} + \cdots.$$

Solution. Since $\sqrt[n]{a_n} = \sqrt[n]{\frac{n}{2^{n-1}}} = \frac{\sqrt[n]{n}}{2^{\frac{n-1}{n}}}$, the problem of finding $\lim\limits_{n \to +\infty} \sqrt[n]{a_n}$ can be reduced to that of finding the two limits $\lim\limits_{n \to +\infty} \sqrt[n]{n}$ and $\lim\limits_{n \to +\infty} 2^{1-\frac{1}{n}}$. The second of these is not indeterminate, owing to the continuity of the function 2^x at $x = 1$, and has the value $2^1 = 2$. To evaluate $\lim\limits_{n \to +\infty} \sqrt[n]{n}$, or $\lim\limits_{x \to +\infty} x^{\frac{1}{x}}$, let $y = x^{\frac{1}{x}}$, and take logarithms: (cf. § 416): $\ln y = \frac{\ln x}{x}$. Then, by l'Hospital's

Rule (§ 415). $\lim_{x \to +\infty} \ln y = \lim_{x \to +\infty} \frac{1-x}{1} = 0$, and $y \to e^0 = 1$. Therefore $\lim_{n \to +\infty} \sqrt[n]{a_n} = \frac{1}{2} < 1$, and $\sum a_n$ converges.

711. EXERCISES

In Exercises 1-20, establish convergence or divergence of the given series.

1. $\frac{1}{3} + \frac{\sqrt{2}}{5} + \frac{\sqrt{3}}{7} + \frac{\sqrt{4}}{9} + \cdots$.

2. $\frac{1}{\sqrt{1 \cdot 2}} + \frac{1}{\sqrt{2 \cdot 3}} + \frac{1}{\sqrt{3 \cdot 4}} + \frac{1}{\sqrt{4 \cdot 5}} + \cdots$.

3. $\frac{\sqrt{3}}{2 \cdot 4} + \frac{\sqrt{5}}{4 \cdot 6} + \frac{\sqrt{7}}{6 \cdot 8} + \frac{\sqrt{9}}{8 \cdot 10} + \cdots$.

4. $\frac{\sqrt{2}+1}{3^3-1} + \frac{\sqrt{3}+1}{4^3-1} + \frac{\sqrt{4}+1}{5^3-1} + \frac{\sqrt{5}+1}{6^3-1} + \cdots$.

5. $\frac{1}{3} + \frac{1 \cdot 2}{3 \cdot 5} + \frac{1 \cdot 2 \cdot 3}{3 \cdot 5 \cdot 7} + \frac{1 \cdot 2 \cdot 3 \cdot 4}{3 \cdot 5 \cdot 7 \cdot 9} + \cdots$.

6. $\frac{1!}{2^5} + \frac{2!}{2^6} + \frac{3!}{2^7} + \frac{4!}{2^8} + \cdots$.

7. $\frac{1}{\ln 2} + \frac{1}{\ln 3} + \frac{1}{\ln 4} + \frac{1}{\ln 5} + \cdots$.

8. $\frac{2!}{4!} + \frac{3!}{5!} + \frac{4!}{6!} + \frac{5!}{7!} + \cdots$.

9. $\sum_{n=1}^{+\infty} \frac{\sqrt{n}}{n^2+4}$.

10. $\sum_{n=1}^{+\infty} \frac{\ln n}{n\sqrt{n+1}}$.

11. $\sum_{n=1}^{+\infty} \frac{n^4}{n!}$.

12. $\sum_{n=1}^{+\infty} \frac{3^{2n-1}}{n^2+1}$.

13. $\sum_{n=1}^{+\infty} e^{-n^2}$.

14. $\sum_{n=2}^{+\infty} \frac{1}{(\ln n)^n}$.

15. $\sum_{n=1}^{+\infty} \frac{\sqrt{n+1} - \sqrt{n}}{n^\alpha}$.

16. $\sum_{n=1}^{+\infty} \frac{n!}{n^n}$.

17. $\sum_{n=2}^{+\infty} \frac{1}{(\ln n)^n}$.

18. $\sum_{n=1}^{+\infty} \frac{1}{1+\alpha^n}$, $\alpha > -1$.

19. $\sum_{n=1}^{+\infty} (\sqrt[n]{n} - 1)^n$.

20. $\sum_{n=1}^{+\infty} r^n |\sin n\alpha|$, $\alpha > 0, r > 0$.

21. Prove the comparison test (Theorem I, § 708) for divergence.
22. Prove Theorem II, § 708.
23. Prove the root test, § 710.
24. Prove that if $a_n \geq 0$ and there exists a number $k > 1$ such that $\lim_{n \to +\infty} n^k a_n$ exists and is finite, then $\sum a_n$ converges.

25. Prove that if $a_n \geq 0$ and $\lim_{n \to +\infty} n a_n$ exists and is positive, then $\sum a_n$ diverges.

★26. Prove the *Schwarz* (or *Cauchy*) *inequality* for nonnegative series: If $a_n \geq 0$ and $b_n \geq 0$, $n = 1, 2, \cdots$, then

(1) $$\left(\sum_{n=1}^{+\infty} a_n b_n\right)^2 \leq \sum_{n=1}^{+\infty} a_n^2 \sum_{n=1}^{+\infty} b_n^2,$$

with the following interpretations: (i) if both series on the right-hand side of (1) converge, then the series on the left-hand side also converges and (1) holds; (ii) if either series on the right-hand side of (1) has a zero sum, then so does the series on the left. (Cf. Ex. 43, § 107, Ex. 29, § 503, Ex. 14, § 717.)

★27. Prove that if $\sum a_n^2$ is a convergent series, then $\sum |a_n|/n$ is also convergent. (Cf. Ex. 26.)

★28. Prove that any series of the form $\sum_{n=1}^{+\infty} \frac{d_n}{10^n}$, where $d_n = 0, 1, 2, \cdots, 9$, converges. Hence show that any decimal expansion $0.d_1 d_2 \cdots$ represents some real number r, where $0 \leq r \leq 1$.

★29. Prove the converse of Exercise 28: If $0 \leq r \leq 1$, then there exists a decimal expansion $0.d_1 d_2 \cdots$ representing r. Show that this decimal expansion is unique unless r is positive and representable by a (unique) terminating decimal (cf. Ex. 5, § 113), in which case r is also representable by a (unique) decimal composed, from some point on, of repeating 9's.

★30. Show that Exercise 29 establishes a one-to-one correspondence between the points of the half-open interval $(0, 1]$ and all nonterminating decimals $0.d_1 d_2 \cdots$. Prove that any subset of a denumerable set (cf. Ex. 12, § 113) is either finite or denumerable, and hence in order to establish that the real numbers are **nondenumerable** (neither finite nor denumerable) it is sufficient to show that the set S of nonterminating decimals $0.d_1 d_2 \cdots$ is nondenumerable. Finally, prove that the set S is nondenumerable by assuming the contrary and obtaining a contradiction, as follows: Assume S is made up of the sequence $\{x_n\}$:

$$x_1: \quad .d_{11} \quad d_{12} \quad d_{13} \quad d_{14} \quad \cdots$$
$$x_2: \quad .d_{21} \quad d_{22} \quad d_{23} \quad d_{24} \quad \cdots$$
$$x_3: \quad .d_{31} \quad d_{32} \quad d_{33} \quad d_{34} \quad \cdots$$
$$\cdots \cdots$$

Construct a new sequence $\{a_n\}$, where a_n is one of the digits $1, 2, \cdots, 9$, and $a_n \neq d_{nn}$, for $n = 1, 2, \cdots$. Finally, show that the decimal $0.a_1 a_2 a_3 \cdots$ simultaneously must and cannot belong to S.

★31. Establish the following form of the ratio test: The positive series $\sum a_n$ converges if $\overline{\lim}_{n \to +\infty} \frac{a_{n+1}}{a_n} < 1$, and diverges if $\underline{\lim}_{n \to +\infty} \frac{a_{n+1}}{a_n} > 1$. (Cf. Ex. 16, § 305.)

★32. Establish the following form of the root test: If $\sum a_n$ is a nonnegative series and if $\sigma = \overline{\lim}_{n \to +\infty} \sqrt[n]{a_n}$, then $\sum a_n$ converges if $\sigma < 1$ and diverges if $\sigma > 1$. (Cf. Ex. 31.)

★33. Apply Exercises 31 and 32 to the series $\frac{1}{2} + \frac{1}{3} + \frac{1}{2^2} + \frac{1}{3^2} + \cdots$ of Example 2, § 708, and show that $\overline{\lim}\frac{a_{n+1}}{a_n} = +\infty$, $\underline{\lim}\frac{a_{n+1}}{a_n} = 0$, $\overline{\lim}\sqrt[n]{a_n} = 1/\sqrt{2}$, and $\underline{\lim}\sqrt[n]{a_n} = 1/\sqrt{3}$. Thus show that the ratio test of § 709 and that of Exercise 31 both fail, that the root test of § 710 fails, and that the root test of Exercise 32 succeeds in establishing convergence of the given series.

★34. Prove that if $\lim_{n\to+\infty}\frac{a_{n+1}}{a_n}$ exists, then $\lim_{n\to+\infty}\sqrt[n]{a_n}$ also exists and is equal to it. *Hints:* For the case $\lim_{n\to+\infty}\frac{a_{n+1}}{a_n} = L$, where $0 < L < +\infty$, let α and β be arbitrary numbers such that $0 < \alpha < L < \beta$. Then for $n \geqq$ some N, $\alpha a_n < a_{n+1} < \beta a_n$. Hence

$$\alpha a_N < a_{N+1} < \beta a_N$$
$$\alpha^2 a_N < \alpha a_{N+1} < a_{N+2} < \beta a_{N+1} < \beta^2 a_N$$
$$\cdots$$
$$\alpha^p a_N < a_{N+p} < \beta^p a_N.$$

Thus, for $n > N$,
$$\alpha^n \frac{a_N}{\alpha^N} < a_n < \beta^n \frac{a_N}{\beta^N},$$
and
$$\alpha \leqq \left\{\begin{array}{c}\overline{\lim}_{n\to+\infty}\sqrt[n]{a_n}\\ \underline{\lim}_{n\to+\infty}\sqrt[n]{a_n}\end{array}\right\} \leqq \beta.$$

★35. The example 1, 2, 1, 2, $\cdots$ shows that $\lim_{n\to+\infty}\sqrt[n]{a_n}$ may exist when $\lim_{n\to+\infty}\frac{a_{n+1}}{a_n}$ does not. Find an example of a convergent positive series $\sum a_n$ for which this situation is also true.

★36. Prove that $\lim_{n\to+\infty}\sqrt[n]{\frac{n^n}{n!}} = e$. (Cf. Ex. 34, above; also Ex. 48, § 515.)

★712. MORE REFINED TESTS

The tests discussed in preceding sections are those most commonly used in practice. There are occasions, however, when such a useful test as the ratio test fails, and it is extremely difficult to devise an appropriate test series for the comparison test. We give now some sharper criteria which may sometimes be used in the event that $\lim_{n\to+\infty}\frac{a_{n+1}}{a_n} = 1$.

Theorem I. Kummer's Test. *Let $\sum a_n$ be a positive series, and let $\{p_n\}$ be a sequence of positive constants such that*

$$(1) \qquad \lim_{n\to+\infty}\left[p_n\frac{a_n}{a_{n+1}} - p_{n+1}\right] = L$$

exists and is positive $(0 < L \leqq +\infty)$. *Then $\sum a_n$ converges.*

If the limit (1) exists and is negative ($-\infty \leq L < 0$) (or, more generally, if $p_n \frac{a_n}{a_{n+1}} - p_{n+1} \leq 0$ for $n \geq N$), and if $\sum \frac{1}{p_n}$ diverges, then $\sum a_n$ diverges.

Proof. Let r be any number such that $0 < r < L$. Then (cf. the proof of the ratio test) there must exist a positive integer N such that $n \geq N$ implies $p_n \frac{a_n}{a_{n+1}} - p_{n+1} > r$. For any positive integer m, then, we have the sequence of inequalities:

$$p_N a_N - p_{N+1} a_{N+1} > r a_{N+1},$$
$$p_{N+1} a_{N+1} - p_{N+2} a_{N+2} > r a_{N+2},$$
$$\cdots \cdots$$
$$p_{N+m-1} a_{N+m-1} - p_{N+m} a_{N+m} > r a_{N+m}.$$

Adding on both sides we have, because of cancellations by pairs on the left:

(2) $\qquad p_N a_N - p_{N+m} a_{N+m} > r(a_{N+1} + \cdots + a_{N+m}).$

Using the notation S_n for the partial sum $a_1 + \cdots + a_n$, we can write the sum in parentheses of (2) as $S_{N+m} - S_N$, and obtain by rearrangement of terms:

$$r S_{N+m} < r S_N + p_N a_N - p_{N+m} a_{N+m} < r S_N + p_N a_N.$$

Letting B denote the constant $(r S_N + p_N a_N)/r$, we infer that $S_n < B$ for $n > N$. In other words, the partial sums of $\sum a_n$ are bounded and hence (§ 705) the series converges.

To prove the second part we infer from the inequality $p_n \frac{a_n}{a_{n+1}} - p_{n+1} \leq 0$, which holds for $n \geq N$, the sequence of inequalities

$$p_N a_N \leq p_{N+1} a_{N+1} \leq \cdots \leq p_n a_n,$$

for any $n > N$. Denoting by A the positive constant $p_N a_N$, we conclude from the comparison test, the inequality

$$a_n \geq A \cdot \frac{1}{p_n}, \text{ for } n > N,$$

and the divergence of $\sum \frac{1}{p_n}$, that $\sum a_n$ also diverges.

Theorem II. Raabe's Test. Let $\sum a_n$ be a positive series and assume that

$$\lim_{n \to +\infty} n\left(\frac{a_n}{a_{n+1}} - 1\right) = L$$

exists (finite or infinite). Then
  (i) if $1 < L \leq +\infty$, $\sum a_n$ converges;
  (ii) if $-\infty \leq L < 1$, $\sum a_n$ diverges;
  (iii) if $L = 1$, $\sum a_n$ may either converge or diverge, and the test fails.

Proof. The first two parts are a consequence of Kummer's Test with $p_n \equiv n$. (Cf. Ex. 5, § 713, for further suggestions.)

In case the limit L of Raabe's Test exists and is equal to 1, a refinement is possible:

Theorem III. *Let $\sum a_n$ be a positive series and assume that*
$$\lim_{n \to +\infty} \ln n \left[n \left(\frac{a_n}{a_{n+1}} - 1 \right) - 1 \right] = L$$
exists (finite or infinite). Then

(i) *if $1 < L \leq +\infty$, $\sum a_n$ converges;*
(ii) *if $-\infty \leq L < 1$, $\sum a_n$ diverges;*
(iii) *if $L = 1$, $\sum a_n$ may either converge or diverge, and the test fails.*

Proof. The first two parts are a consequence of Kummer's Test with $p_n \equiv n \ln n$. (Cf. Exs. 9-10, § 713, for further suggestions.)

Example 1. Test for convergence or divergence:
$$\left(\frac{1}{2}\right)^p + \left(\frac{1 \cdot 3}{2 \cdot 4}\right)^p + \left(\frac{1 \cdot 3 \cdot 5}{2 \cdot 4 \cdot 6}\right)^p + \cdots.$$

Solution. Since $\lim_{n \to +\infty} \frac{a_{n+1}}{a_n} = 1$, the ratio test fails, and we turn to Raabe's test. We find
$$n\left(\frac{a_n}{a_{n+1}} - 1\right) = n\left[\left(\frac{2n}{2n-1}\right)^p - 1\right] = \frac{2n}{2n-1} \cdot \frac{(1+x)^p - 1}{2x},$$
where $x = (2n - 1)^{-1}$. The limit of this expression, as $n \to +\infty$, is (by l'Hospital's Rule):
$$\lim_{x \to 0} \frac{(1+x)^p - 1}{2x} = \lim_{x \to 0} \frac{p(1+x)^{p-1}}{2} = \frac{p}{2}.$$
Therefore the given series converges for $p > 2$ and diverges for $p < 2$. For the case $p = 2$, see Example 2.

Example 2. Test for convergence or divergence
$$\left(\frac{1}{2}\right)^2 + \left(\frac{1 \cdot 3}{2 \cdot 4}\right)^2 + \left(\frac{1 \cdot 3 \cdot 5}{2 \cdot 4 \cdot 6}\right)^2 + \cdots.$$

Solution. Both the ratio test and Raabe's test fail (cf. Example 1). Preparing to use Theorem III, we simplify the expression
$$n\left(\frac{a_n}{a_{n+1}} - 1\right) - 1 = n \frac{4n-1}{4n^2 - 4n + 1} - 1 = \frac{3n-1}{(2n-1)^2}.$$
Since $\lim_{n \to +\infty} \ln n \frac{3n-1}{(2n-1)^2} = 0 < 1$, the given series diverges.

★713. EXERCISES

In Exercises 1-4, test for convergence or divergence.

★1. $\sum_{n=1}^{+\infty} \frac{2 \cdot 4 \cdot 6 \cdots 2n}{1 \cdot 3 \cdot 5 \cdots (2n+1)}$. (Cf. Ex. 4.)

★2. $\sum_{n=1}^{+\infty} \frac{1 \cdot 3 \cdots (2n-1)}{2 \cdot 4 \cdots 2n} \cdot \frac{1}{2n+1}$.

★3. $\sum_{n=1}^{+\infty} \frac{1 \cdot 3 \cdots (2n-1)}{2 \cdot 4 \cdots 2n} \cdot \frac{4n+3}{2n+2}$.

★4. $\sum_{n=1}^{+\infty} \left[\frac{2 \cdot 4 \cdot 6 \cdots 2n}{1 \cdot 3 \cdot 5 \cdots (2n+1)} \right]^n$.

★5. Prove Theorem II, § 712. *Hint:* For examples for part (*iii*), define the terms of $\sum_{n=1}^{+\infty} a_n$ inductively, with $a_1 = 1$, and $\frac{a_n}{a_{n+1}} = 1 + \frac{1}{n} + \frac{k}{n \ln n}$. Then use Theorem III, § 712.

★6. Prove that the **hypergeometric series**
$$1 + \frac{\alpha \cdot \beta}{1 \cdot \gamma} + \frac{\alpha(\alpha+1)\beta(\beta+1)}{1 \cdot 2 \cdot \gamma \cdot (\gamma+1)} + \cdots$$
$$+ \frac{\alpha(\alpha+1) \cdots (\alpha+n-1) \beta(\beta+1) \cdots (\beta+n-1)}{n!\, \gamma(\gamma+1) \cdots (\gamma+n-1)} + \cdots$$
converges if and only if $\gamma - \alpha - \beta > 0$ (γ not 0 or a negative integer).

★7. Prove **Gauss's Test:** *If the positive series $\sum a_n$ is such that*
$$\frac{a_n}{a_{n+1}} = 1 + \frac{h}{n} + O\left(\frac{1}{n^2}\right),$$
then $\sum a_n$ converges if $h > 1$, and diverges if $h \leq 1$. Show that $O(1/n^2)$ could be replaced by the weaker condition $O(1/n^\alpha)$, where $\alpha > 1$.

★8. State and prove limit superior and limit inferior forms for the Theorems of § 712. (Cf. Exs. 31-32, § 711.)

★9. Prove the first two parts of Theorem III, § 712. (Cf. Ex. 11.) *Hint:* This is equivalent to showing (using Kummer's test with $p_n = n \ln n$) that
$$\ln n \left[n \left(\frac{a_n}{a_{n+1}} - 1 \right) - 1 \right] - \left\{ n \ln n \frac{a_n}{a_{n+1}} - (n+1) \ln (n+1) \right\} \to 1,$$
or
$$(n+1) \ln (n+1) - [n \ln n + \ln n] = (n+1) \ln \left(\frac{n+1}{n} \right) \to 1.$$

★★10. Prove the following refinement of Theorem III, § 712: Let $\sum a_n$ be a positive series such that
$$\lim_{n \to +\infty} \ln \ln n \left\{ \ln n \left[n \left(\frac{a_n}{a_{n+1}} - 1 \right) - 1 \right] - 1 \right\} = L$$
exists (finite or infinite). Then (i) if $1 < L \leq +\infty$, $\sum a_n$ converges, and (ii) if $-\infty \leq L < 1$, $\sum a_n$ diverges. *Hint:* The problem is to show (cf. Ex. 9) that

(1) $\quad (n+1) \ln (n+1) \ln \ln (n+1)$
$\qquad - \{n \ln n \ln \ln n + \ln n \ln \ln n + \ln n \ln\} \to 1.$

To establish this limit, show first that if $\epsilon_n \to 0$, then $\ln (1 + \epsilon_n) = \epsilon_n + O(\epsilon_n^2)$ (cf. Ex. 21, § 408), and, more generally, if $\{\alpha_n\}$ is a sequence of positive numbers and if $\epsilon_n / \alpha_n \to 0$, then $\ln (\alpha_n + \epsilon_n) = \ln (\alpha_n) + \frac{\epsilon_n}{\alpha_n} + O\left(\frac{\epsilon_n^2}{\alpha_n^2} \right)$. Thus

$$\ln(n+1) = \ln n + \frac{1}{n} + O\left(\frac{1}{n^2}\right)$$

and

$$\ln \ln (n+1) = \ln\left[\ln n + \frac{1}{n} + O\left(\frac{1}{n^2}\right)\right] = \ln \ln n + \frac{1}{\ln n}\left[\frac{1}{n} + O\left(\frac{1}{n^2}\right)\right]$$
$$+ O\left(\left[\frac{1}{n} + O\left(\frac{1}{n^2}\right)\right]^2\right) = \ln \ln n + \frac{1}{n \ln n} + O\left(\frac{1}{n^2}\right).$$

Now form the product $(n+1)\ln(n+1)\ln\ln(n+1)$, subtract the terms in the braces of (1), and take a limit.

★★11. Prove the third part of Theorem III, § 712. *Hint:* Cf. Exs. 5 and 10.

714. SERIES OF ARBITRARY TERMS

If a series has terms of one sign, as we have seen for nonnegative series, there is only one kind of divergence—to infinity—and convergence of the series is equivalent to the boundedness of the partial sums. We wish to turn our attention now to series whose terms may be either positive or negative—or zero. The behavior of such series is markedly different from that of nonnegative series, but we shall find that we can make good use of the latter to clarify the former.

715. ALTERNATING SERIES

An **alternating series** is a series of the form

(1) $$c_1 - c_2 + c_3 - c_4 + \cdots,$$

where $c_n > 0$ for every n.

Theorem. *An alternating series* (1) *whose terms satisfy the two conditions*
  (i) $c_{n+1} < c_n$ *for every* n,
  (ii) $c_n \to 0$ *as* $n \to +\infty$,
converges. If S and S_n denote the sum, and the partial sum of the first n terms, respectively, of the series (1),

(2) $$|S_n - S| < c_{n+1}.$$

Proof. We break the proof into six parts (cf. Fig. 702.):

A. The partial sums S_{2n} (consisting of an even number of terms) form an increasing sequence.

B. The partial sums S_{2n-1} (consisting of an odd number of terms) form a decreasing sequence.

C. For every m and every n, $S_{2m} < S_{2n-1}$.

D. S exists.

E. For every m and every n, $S_{2m} < S < S_{2n-1}$.

F. The inequality (2) holds.

A: Since $S_{2n+2} = S_{2n} + (c_{2n+1} - c_{2n+2})$, and since $c_{2n+2} < c_{2n+1}$, $S_{2n+2} > S_{2n}$.

B: Since $S_{2n+1} = S_{2n-1} - (c_{2n} - c_{2n+1})$, and since $c_{2n+1} < c_{2n}$, $S_{2n+1} < S_{2n-1}$.

C: If $2m < 2n - 1$,
$$S_{2n-1} - S_{2m} = (c_{2m+1} - c_{2m+2}) + \cdots + (c_{2n-3} - c_{2n-2}) + c_{2n-1} > 0,$$
and $S_{2m} < S_{2n-1}$. If $2m > 2n - 1$,
$$S_{2m} - S_{2n-1} = -(c_{2n} - c_{2n+1}) - \cdots - (c_{2m-2} - c_{2m-1}) - c_{2m} < 0,$$
and $S_{2m} < S_{2n-1}$.

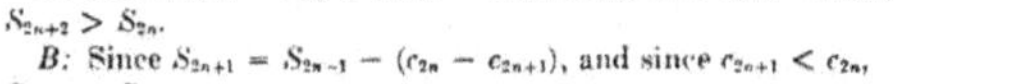

$S_2 \quad S_4 \quad S_6 \quad \cdots \quad S_{2n} \quad S \quad S_{2n-1} \quad \cdots \quad S_5 \quad S_3 \quad S_1$

FIG. 702

D: From *A*, *B*, and *C* it follows that $\{S_{2n}\}$ and $\{S_{2n-1}\}$ are bounded monotonic sequences and therefore converge. We need only show that their limits are equal. But this is true, since
$$\lim_{n \to +\infty} S_{2n+1} - \lim_{n \to +\infty} S_{2n} = \lim_{n \to +\infty} (S_{2n+1} - S_{2n}) = \lim_{n \to +\infty} c_{2n+1} = 0.$$

E: By the fundamental theorem on convergence of monotonic sequences (Theorem XIV, § 204), for an arbitrary fixed n and variable m, $S_{2m} < S_{2n+1}$ implies $S \leq S_{2n+1} < S_{2n-1}$. Similarly, for an arbitrary fixed m and variable n, $S_{2m+2} < S_{2n+1}$ implies $S_{2m} < S_{2m+2} \leq S$.

F: On the one hand,
$$0 < S_{2n-1} - S < S_{2n-1} - S_{2n} = c_{2n},$$
while, on the other hand
$$0 < S - S_{2n} < S_{2n+1} - S_{2n} = c_{2n+1}.$$

Example 1. The alternating harmonic series
$$1 - \tfrac{1}{2} + \tfrac{1}{3} - \tfrac{1}{4} + \cdots$$
converges, since the conditions of the Theorem above are satisfied.

Example 2. Prove that the series $\sum_{n=1}^{+\infty} (-1)^n \dfrac{\ln n}{n}$ converges.

Solution. All of the conditions of the alternating series test are obvious except for the inequality $\dfrac{\ln (n+1)}{n+1} < \dfrac{\ln n}{n}$. The simplest way of establishing this is to show that $f(x) \equiv \dfrac{\ln x}{x}$ is strictly decreasing since its derivative is
$$f'(x) = \frac{1 - \ln x}{x^2} < 0, \quad \text{for} \quad x > e.$$

A: Since $S_{2n+2} = S_{2n} + (c_{2n+1} - c_{2n+2})$, and since $c_{2n+2} < c_{2n+1}$, $S_{2n+2} > S_{2n}$.

B: Since $S_{2n+1} = S_{2n-1} - (c_{2n} - c_{2n+1})$, and since $c_{2n+1} < c_{2n}$, $S_{2n+1} < S_{2n-1}$.

C: If $2m < 2n - 1$,
$$S_{2n-1} - S_{2m} = (c_{2m+1} - c_{2m+2}) + \cdots + (c_{2n-3} - c_{2n-2}) + c_{2n-1} > 0,$$
and $S_{2m} < S_{2n-1}$. If $2m > 2n - 1$,
$$S_{2m} - S_{2n-1} = -(c_{2n} - c_{2n+1}) - \cdots - (c_{2m-2} - c_{2m-1}) - c_{2m} < 0,$$
and $S_{2m} < S_{2n-1}$.

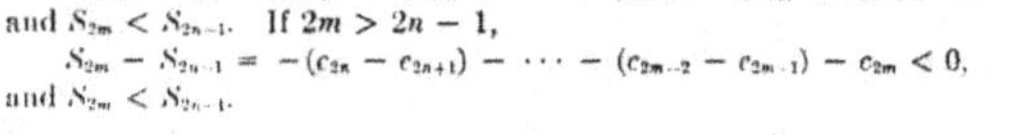

$$S_1 \quad S_3 \quad S_5 \quad \cdots \quad S_{2n} \quad S \quad S_{2n-1} \quad \cdots \quad S_6 \quad S_4 \quad S_2$$

FIG. 702

D: From *A*, *B*, and *C* it follows that $\{S_{2n}\}$ and $\{S_{2n-1}\}$ are bounded monotonic sequences and therefore converge. We need only show that their limits are equal. But this is true, since
$$\lim_{n \to +\infty} S_{2n+1} - \lim_{n \to +\infty} S_{2n} = \lim_{n \to +\infty} (S_{2n+1} - S_{2n}) = \lim_{n \to +\infty} c_{2n+1} = 0.$$

E: By the fundamental theorem on convergence of monotonic sequences (Theorem XIV, § 204), for an arbitrary fixed n and variable m, $S_{2m} < S_{2n+1}$ implies $S \leq S_{2n+1} < S_{2n-1}$. Similarly, for an arbitrary fixed m and variable n, $S_{2m+2} < S_{2n+1}$ implies $S_{2m} < S_{2m+2} \leq S$.

F: On the one hand,
$$0 < S_{2n-1} - S < S_{2n-1} - S_{2n} = c_{2n},$$
while, on the other hand
$$0 < S - S_{2n} < S_{2n+1} - S_{2n} = c_{2n+1}.$$

Example 1. The **alternating harmonic series**
$$1 - \tfrac{1}{2} + \tfrac{1}{3} - \tfrac{1}{4} + \cdots$$
converges, since the conditions of the Theorem above are satisfied.

Example 2. Prove that the series $\sum_{n=1}^{+\infty} (-1)^n \dfrac{\ln n}{n}$ converges.

Solution. All of the conditions of the alternating series test are obvious except for the inequality $\dfrac{\ln(n+1)}{n+1} < \dfrac{\ln n}{n}$. The simplest way of establishing this is to show that $f(x) \equiv \dfrac{\ln x}{x}$ is strictly decreasing since its derivative is
$$f'(x) = \frac{1 - \ln x}{x^2} < 0, \quad \text{for} \quad x > e.$$

716. ABSOLUTE AND CONDITIONAL CONVERGENCE

We introduce some notation. Let $\sum_{n=1}^{+\infty} a_n$ be a given series of arbitrary terms. For every positive integer n we define p_n to be the larger of the two numbers a_n and 0, and q_n to be the larger of the two numbers $-a_n$ and 0 (in case $a_n = 0$, $p_n = q_n = 0$):

(1) $$p_n \equiv \max(a_n, 0), \quad q_n \equiv \max(-a_n, 0).$$

Then p_n and q_n are nonnegative numbers (at least one of them being zero) satisfying the two equations

(2) $$p_n - q_n = a_n, \quad p_n + q_n = |a_n|,$$

and the two inequalities

(3) $$0 \leq p_n \leq |a_n|, \quad 0 \leq q_n \leq |a_n|.$$

The two nonnegative series $\sum p_n$ and $\sum q_n$ are called the **nonnegative** and **nonpositive parts**, respectively, of the series $\sum a_n$. A third nonnegative series related to the original is $\sum |a_n|$, called the **series of absolute values** of $\sum a_n$.

We now assign labels to the partial sums of the series considered:

$$S_n \equiv a_1 + \cdots + a_n, \qquad A_n \equiv |a_1| + \cdots + |a_n|,$$
$$P_n \equiv p_1 + \cdots + p_n, \qquad Q_n \equiv q_1 + \cdots + q_n.$$

From (1) and (2) we deduce:

(4) $$P_n - Q_n = S_n, \quad P_n + Q_n = A_n,$$

and are ready to draw some conclusions about convergence.

From the first equation of (4) we can solve for either P_n or Q_n

$$(P_n = Q_n + S_n \quad \text{and} \quad Q_n = P_n - S_n),$$

and conclude (with the aid of a limit theorem) that the convergence of any two of the three series $\sum a_n$, $\sum p_n$, and $\sum q_n$ implies the convergence of the third. This means that if *both* the nonnegative and nonpositive parts of a series converge the series itself must also converge. It also means that if a series converges, then the nonnegative and nonpositive parts must either *both converge* or *both diverge*. (Why?) The alternating harmonic series of § 715 is an example of a convergent series whose nonnegative and nonpositive parts both diverge.

The inequalities (3) and the second equation of (4) imply (by the comparison test and a limit theorem) that the series of absolute values, $\sum |a_n|$ converges if and only if *both* the nonnegative and nonpositive parts of the series converge. Furthermore, in case the series $\sum |a_n|$ converges, if we define $P \equiv \lim_{n \to +\infty} P_n$, $Q \equiv \lim_{n \to +\infty} Q_n$, $S \equiv \lim_{n \to +\infty} S_n$, and $A \equiv \lim_{n \to +\infty} A_n$, we have from (4), $P - Q = S$, $P + Q = A$.

We give a definition and a theorem embodying some of the results just obtained:

Definition. *A series $\sum a_n$ **converges absolutely** (or is **absolutely convergent**) if and only if the series of absolute values, $\sum |a_n|$ converges. A series **converges conditionally** (or is **conditionally convergent**) if and only if it converges and does not converge absolutely.*

Theorem I. *An absolutely convergent series is convergent. In case of absolute convergence, $|\sum a_n| \leq \sum |a_n|$.*

Note 1. For an absolutely convergent series, the nonnegative and nonpositive parts both converge. For a conditionally convergent series, the nonnegative and nonpositive parts both diverge.

Note 2. An alternative proof of Theorem I is provided by the Cauchy Criterion (cf. Exs. 17-21, § 717 for a discussion).

Note 3. The tests established for convergence of nonnegative series are of course immediately available for absolute convergence of arbitrary series. We state here the ratio test for arbitrary series, of nonzero terms, both because of its practicality and because the conclusion for $\rho > 1$ is not simply that the series fails to converge absolutely but that it *diverges* (cf. Ex. 12, § 717).

Theorem II. Ratio Test. *Let $\sum a_n$ be a series of nonzero terms, and define the test ratio $r_n \equiv a_{n+1}/a_n$. Assume that the limit of the absolute value of this test ratio exists:*
$$\lim_{n \to +\infty} \left| \frac{a_{n+1}}{a_n} \right| = \rho,$$
where $0 \leq \rho \leq +\infty$. Then
  (i) *if $0 \leq \rho < 1$, $\sum a_n$ converges absolutely;*
  (ii) *if $1 < \rho \leq +\infty$, $\sum a_n$ diverges;*
  (iii) *if $\rho = 1$, $\sum a_n$ may converge absolutely, or converge conditionally, or diverge, and the test fails.*

Note 4. The ratio test never establishes convergence in the case of a conditionally convergent series.

Example 1. The alternating p-series,
$$1 - \frac{1}{2^p} + \frac{1}{3^p} - \frac{1}{4^p} + \cdots,$$
converges absolutely if $p > 1$, converges conditionally if $0 < p \leq 1$, and diverges if $p \leq 0$.

★Example 2. Show that the series $\frac{1}{2} - \frac{1 \cdot 3}{2 \cdot 4} + \frac{1 \cdot 3 \cdot 5}{2 \cdot 4 \cdot 6} - \frac{1 \cdot 3 \cdot 5 \cdot 7}{2 \cdot 4 \cdot 6 \cdot 8} + \cdots$
converges conditionally.

Solution. By Example 1, § 712, the series fails to converge absolutely. In order to prove that the series converges, we can use the alternating series test to reduce the problem to showing that the general term tends toward 0 (clearly $c_n \downarrow$). Letting $p = 3$ in Example 1, § 712, we know that $c_n^2 \to 0$. Therefore $c_n \to 0$.

717. EXERCISES

In Exercises 1-10, test for absolute convergence, conditional convergence, or divergence.

1. $\sum_{n=1}^{+\infty} \frac{(-1)^{n-1}}{2n+3}$.
2. $\sum_{n=1}^{+\infty} \frac{(-1)^n n}{n+2}$.
3. $\sum_{n=1}^{+\infty} \frac{(-1)^n \sqrt[3]{n^2+5}}{\sqrt{n^3+n+1}}$.
4. $\sum_{n=1}^{+\infty} (-1)^{n-1} \frac{n^4}{(n+1)!}$.
5. $\sum_{n=1}^{+\infty} (-1)^n \frac{n \ln n}{e^n}$.
6. $\sum_{n=1}^{+\infty} (-1)^n \frac{\cos n\alpha}{n^2}$.
7. $e^{-x}\cos x + e^{-2x}\cos 2x + e^{-3x}\cos 3x + \cdots$.
8. $1 + r\cos\theta + r^2\cos 2\theta + r^3\cos 3\theta + \cdots$.
9. $\frac{1}{2(\ln 2)^p} - \frac{1}{3(\ln 3)^p} + \frac{1}{4(\ln 4)^p} - \cdots$.

★10. $\left(\frac{1}{2}\right)^p - \left(\frac{1\cdot 3}{2\cdot 4}\right)^p + \left(\frac{1\cdot 3\cdot 5}{2\cdot 4\cdot 6}\right)^p - \cdots$.

11. Prove that if the condition (i) of § 715 is replaced by $c_{n+1} \leq c_n$, the conclusion is altered only by replacing (2), § 715, by $|S_n - S| \leq c_{n+1}$.

12. Prove the ratio test (Theorem II) of § 716.

13. Show by three counterexamples that each of the three conditions of the alternating series test, § 715, is needed in the statement of that test (that is, the alternating of signs, the decreasing nature of c_n, and the limit of c_n being 0).

★**14.** Prove the **Schwarz** (or **Cauchy**) and the **Minkowski inequalities** for series:

If $\sum_{n=1}^{+\infty} a_n^2$ and $\sum_{n=1}^{+\infty} b_n^2$ converge, then so do $\sum_{n=1}^{+\infty} a_n b_n$ and $\sum_{n=1}^{+\infty} (a_n + b_n)^2$, and

$$\left[\sum_{n=1}^{+\infty} a_n b_n\right]^2 \leq \sum_{n=1}^{+\infty} a_n^2 \sum_{n=1}^{+\infty} b_n^2,$$

$$\left[\sum_{n=1}^{+\infty} (a_n + b_n)^2\right]^{\frac{1}{2}} \leq \left[\sum_{n=1}^{+\infty} a_n^2\right]^{\frac{1}{2}} + \left[\sum_{n=1}^{+\infty} b_n^2\right]^{\frac{1}{2}}.$$

(Cf. Exs. 43-44, § 107, Exs. 29-30, § 503, Ex. 26, § 711; Exs. 38-40, § 1225.)

★★**15.** Let $\sum_{n=1}^{+\infty} a_n$ be a given series, with partial sums $S_n = a_1 + \cdots + a_n$. Define the sequence of *arithmetic means*

$$\sigma_n = \frac{S_1 + \cdots + S_n}{n}.$$

The series $\sum a_n$ is said to be **summable** *by Cesàro's method of arithmetic means of order* 1 (for short, *summable* $(C, 1)$) if and only if $\lim_{n \to +\infty} \sigma_n$ exists and is finite.

Show that Exercises 16-19, § 205, prove that summability $(C, 1)$ is a generalization of convergence: that any convergent series is summable $(C, 1)$ with $\lim_{n\to+\infty} \sigma_n = \lim_{n\to+\infty} S_n$, that for nonnegative series summability $(C, 1)$ is identical with convergence, and that there are divergent series (whose terms are not of one sign) which are summable $(C, 1)$. *Hint:* Consider $1 - 1 + 1 - 1 + \cdots$

16. A series $\sum a_n$ is *summable by Cesàro's method of arithmetic means of order* 2 (for short, *summable* (C. 2)) if and only if

$$\lim_{n \to +\infty} \frac{\sigma_1 + \cdots + \sigma_n}{n}$$

(in the notation of Exercise 15) exists and is finite. Show that summability $(C, 1)$ implies summability $(C, 2)$, but not conversely. Generalize to summability (C, r).

17. Prove the **Cauchy criterion** for convergence of an infinite series: *An infinite series $\sum_{n=1}^{+\infty} a_n$ converges if and only if corresponding to $\epsilon > 0$ there exists a number N such that $n > m > N$ implies $|a_m + a_{m+1} + \cdots + a_n| < \epsilon$.* (Cf. § 302.)

18. Prove that an infinite series $\sum_{n=1}^{+\infty} a_n$ converges if and only if corresponding to $\epsilon > 0$ there exists a number N such that $n > N$ and $p > 0$ imply $|a_n + a_{n+1} + \cdots + a_{n+p}| < \epsilon$. (Cf. Ex. 17.)

19. Prove that an infinite series $\sum_{n=1}^{+\infty} a_n$ converges if and only if corresponding to $\epsilon > 0$ there exists a positive integer N such that $n > N$ implies $|a_N + a_{N+1} + \cdots + a_n| < \epsilon$. (Cf. Ex. 2, § 305.)

20. Show by an example that the condition of Exercise 18 is not equivalent to the following: $\lim_{n \to +\infty} (a_n + \cdots + a_{n+p}) = 0$ for every $p > 0$. (Cf. Ex. 4, § 305, Ex. 43, § 904.)

21. Use the Cauchy criterion of Exercise 17 to prove that an absolutely convergent series is convergent.

****22.** Prove the following **Abel test**: *If the partial sums of a series $\sum_{n=1}^{+\infty} a_n$ are bounded and if $\{b_n\}$ is a monotonically decreasing sequence of nonnegative numbers whose limit is 0, then $\sum a_n b_n$ converges.* Use this fact to establish the convergence in the alternating series test as phrased in Exercise 11. *Hint:* Let $S_n = a_1 + \cdots + a_n$, and assume $|S_n| < K$ for all n. Then

$$\left| \sum_{i=m}^{n} a_i b_i \right| = \left| \sum_{i=m}^{n} (S_i - S_{i-1}) b_i \right| = \left| \sum_{i=m}^{n-1} S_i (b_i - b_{i+1}) + S_n b_n - S_{m-1} b_m \right|$$

$$\leq K \left[\sum_{i=m}^{n-1} (b_i - b_{i+1}) + b_n + b_m \right] = 2K b_m.$$

****23.** Prove the following **Abel test**: *If $\sum a_n$ converges and if $\{b_n\}$ is a bounded monotonic sequence, then $\sum a_n b_n$ converges.* *Hint:* Assume for definiteness that $b_n \downarrow$, let $b_n \to b$, write $a_n b_n = a_n(b_n - b) + a_n b$, and use Ex. 22.

****24.** Start with the harmonic series, and introduce $+$ and $-$ signs according to the following patterns:

(i) in pairs: $1 + \frac{1}{2} - \frac{1}{3} - \frac{1}{4} + \frac{1}{5} + \frac{1}{6} - - \cdots$;

(ii) in groups of 1, 2, 3, 4, $\cdots$: $1 - \frac{1}{2} - \frac{1}{3} + + + - - - - \cdots$;

(iii) in groups of 1, 2, 4, 8, $\cdots$: $1 - \frac{1}{2} - \frac{1}{3} + + + + \cdots$.

Show that (i) and (ii) converge and (iii) diverges. (Cf. Ex. 22.) *Hint for* (ii): In Abel's test, Ex. 22, let $a_n = \pm \dfrac{1}{\sqrt{n}}$ and $b_n = \dfrac{1}{\sqrt{n}}$.

★★25. If A is an arbitrary (finite or) denumerable set of real numbers, $a_1, a_2, a_3, \cdots$ (which may in particular be discrete like the integers, or dense like the rational numbers), construct a bounded monotonic function whose set of points of discontinuity is precisely A, as follows: Let $\sum_{n=1}^{\infty} p_n$ be a (finite or) convergent infinite series of positive numbers with sum p. Define $f(x)$ to be 0 if $x < \inf(A)$, and otherwise equal to the sum of all terms p_m of $\sum p_n$ such that $a_m \leq x$. Prove the following six properties of $f(x)$: (i) $f(x)$ is monotonically increasing on $(-\infty, +\infty)$; (ii) $\lim\limits_{x \to -\infty} f(x) = 0$; (iii) $\lim\limits_{x \to +\infty} f(x) = p$; (iv) $f(x)$ is everywhere continuous from the right; (v) $f(a_n) - f(a_n-) = p_n$ for every n; (vi) $f(x)$ is continuous at every point not in A. (Cf. Ex. 29, § 216.)

718. GROUPINGS AND REARRANGEMENTS

A series $\sum b_n$ is said to arise from a given series $\sum a_n$ by **grouping of terms** (or by the **introduction of parentheses**) if every b_n is the sum of a finite number of consecutive terms of $\sum a_n$, and every pair of terms a_m and a_n, where $m < n$, appear as terms in a unique pair of terms b_p and b_q, respectively, where $p \leq q$. For example, the grouping

$$(a_1 + a_2) + (a_3) + (a_4 + a_5) + (a_6) + \cdots$$

gives rise to the series $\sum b_n$, where $b_1 = a_1 + a_2$, $b_2 = a_3$, $b_3 = a_4 + a_5$, $b_4 = a_6, \cdots$.

Theorem I. *Any series arising from a convergent series by grouping of terms is convergent, and has the same sum as the original series.*

Proof. The partial sums of the new series form a subsequence of the partial sums of the original.

NOTE. The example

$$(2 - 1\tfrac{1}{2}) + (1\tfrac{1}{3} - 1\tfrac{1}{4}) + (1\tfrac{1}{5} - 1\tfrac{1}{6}) + \cdots$$

shows that grouping of terms may convert a divergent series into a convergent series. Equivalently, removal of parentheses may destroy convergence.

A series $\sum b_n$ is said to arise from a given series $\sum a_n$ by **rearrangement (of terms)** if there exists a one-to-one correspondence between the terms of $\sum a_n$ and those of $\sum b_n$ such that whenever a_m and b_n correspond, $a_m = b_n$. For example, the series

$$\frac{1}{4} + 1 + \frac{1}{16} + \frac{1}{9} + \frac{1}{36} + \frac{1}{25} + \cdots$$

is a rearrangement of the p-series, with $p = 2$.

Theorem II. Dirichlet's Theorem. *Any series arising from an absolutely convergent series by rearrangement of terms is absolutely convergent, and has the same sum as the original series.*

Proof. We prove first that the theorem is true for nonnegative series. Let $\sum a_n$ be a given nonnegative series, convergent with sum A and let $\sum b_n$ be any rearrangement. If B_n is any partial sum of $\sum b_n$, the terms of B_n consist of a finite number of terms of $\sum a_n$, and therefore form a part of some partial sum A_m of $\sum a_n$. Since the terms are assumed to be nonnegative, $B_n \leq A_m$, and hence, $B_n \leq A$. Therefore the partial sums of the nonnegative series $\sum b_n$ are bounded, and $\sum b_n$ converges to a sum $B \leq A$. Since $\sum a_n$ is a rearrangement of $\sum b_n$, the symmetric relation $A \leq B$ also holds, and $A = B$.

If $\sum a_n$ is absolutely convergent, and if $\sum b_n$ is any rearrangement, the nonnegative and nonpositive parts of $\sum b_n$ are rearrangements of the nonnegative and nonpositive parts of $\sum a_n$, respectively. Since these latter both converge, say with sums P and Q, respectively, their rearrangements will also both converge, with sums P and Q, respectively, by the preceding paragraph. Finally, $\sum a_n = P - Q = \sum b_n$, and the proof is complete.

★**Theorem III.** *The terms of any conditionally convergent series can be rearranged to give either a divergent series or a conditionally convergent series whose sum is an arbitrary preassigned number.*

★*Proof.* We prove one case, and leave the rest as an exercise. (Ex. 2, § 721.) Let $\sum a_n$ be a conditionally convergent series, with divergent nonnegative and nonpositive parts $\sum p_n$ and $\sum q_n$, respectively, and let c be an arbitrary real number. Let the rearrangement be determined as follows: first put down terms $p_1 + p_2 + \cdots + p_{m_1}$ until the partial sum first exceeds c. Then attach terms $-q_1 - q_2 - q_3 - \cdots - q_{n_1}$ until the total partial sum first falls short of c. Then attach terms $p_{m_1+1} + \cdots + p_{m_2}$ until the total partial sum first exceeds c. Then terms $-q_{n_1+1} - \cdots - q_{n_2}$, etc. Each of these steps is possible because of the divergence of $\sum p_n$ and $\sum q_n$. The resulting rearrangement of $\sum a_n$ converges to c since $p_n \to 0$ and $q_n \to 0$.

Example. Rearrange the terms of the series obtained by doubling all terms of the alternating harmonic series so that the resulting series is the alternating harmonic series, thus convincing the unwary that $2 = 1$.

Solution. Write the terms

$$2[1 - \tfrac{1}{2} + \tfrac{1}{3} - \tfrac{1}{4} + \tfrac{1}{5} - \tfrac{1}{6} + \tfrac{1}{7} - \tfrac{1}{8} + \tfrac{1}{9} - \cdots]$$
$$= 2 - 1 + \tfrac{2}{3} - \tfrac{1}{2} + \tfrac{2}{5} - \tfrac{1}{3} + \tfrac{2}{7} - \tfrac{1}{4} + \tfrac{2}{9} - \cdots$$
$$= (2 - 1) - \tfrac{1}{2} + (\tfrac{2}{3} - \tfrac{1}{3}) - \tfrac{1}{4} + (\tfrac{2}{5} - \tfrac{1}{5}) - \tfrac{1}{6} + \cdots$$
$$= 1 - \tfrac{1}{2} + \tfrac{1}{3} - \tfrac{1}{4} + \tfrac{1}{5} - \tfrac{1}{6} + \tfrac{1}{7} - \tfrac{1}{8} + \tfrac{1}{9} - \cdots.$$

719. ADDITION, SUBTRACTION, AND MULTIPLICATION OF SERIES

Definition I. *If $\sum a_n$ and $\sum b_n$ are two series, their **sum** $\sum c_n$ and **difference** $\sum d_n$ are series defined by the equations*

(1) $\qquad c_n = a_n + b_n, \quad d_n = a_n - b_n.$

Theorem I. *The sum and difference of two convergent series, $\sum a_n = A$ and $\sum b_n = B$, converge to $A + B$ and $A - B$, respectively. The sum and difference of two absolutely convergent series are absolutely convergent.*

The proof is left as an exercise (Ex. 3, § 721).

The product of two series is a more difficult matter. The definition of a product series is motivated by the form of the product of polynomials or, more generally, power series (treated in Chapter 8):

$$(a_0 + a_1 x + a_2 x^2 + \cdots)(b_0 + b_1 x + b_2 x^2 + \cdots)$$
$$= a_0 b_0 + (a_0 b_1 + a_1 b_0)x + (a_0 b_2 + a_1 b_1 + a_2 b_0)x^2 + \cdots.$$

For convenience we revise slightly our notation for an infinite series, letting the terms have subscripts $0, 1, 2, \cdots$, and write

$$\sum a_n = \sum_{n=0}^{+\infty} a_n \quad \text{and} \quad \sum b_n = \sum_{n=0}^{+\infty} b_n.$$

Definition II. *If $\sum_{n=0}^{+\infty} a_n$ and $\sum_{n=0}^{+\infty} b_n$ are two series, their **product** $\sum_{n=0}^{+\infty} c_n$ is defined:*

(2) $\qquad c_0 = a_0 b_0, \; c_1 = a_0 b_1 + a_1 b_0, \cdots,$

$$c_n = \sum_{k=0}^{n} a_k b_{n-k} = a_0 b_n + a_1 b_{n-1} + \cdots + a_n b_0, \cdots,$$

The basic questions are these: If $\sum a_n$ and $\sum b_n$ converge, with sums A and B, and if $\sum c_n$ is their product series, does $\sum c_n$ converge? If $\sum c_n$ converges to C is $C = AB$? If $\sum c_n$ does not necessarily converge, what conditions on $\sum a_n$ and $\sum b_n$ guarantee convergence of $\sum c_n$?

The answers, in brief, are: The convergence of $\sum a_n$ and $\sum b_n$ does not guarantee convergence of $\sum c_n$ (Ex. 5, § 721). If $\sum c_n$ does converge, then $C = AB$. (This result is due to Abel. Cf. Ex. 20, § 911.) If both $\sum a_n$ and $\sum b_n$ converge, and if *one* of them converges absolutely, then $\sum c_n$ converges (to AB). (This result is due to Mertens. Cf. Ex. 20, § 721.) If *both* $\sum a_n$ and $\sum b_n$ converge absolutely, then $\sum c_n$ converges absolutely (to AB). (This is our next theorem.)

Theorem II. *The product series of two absolutely convergent series is absolutely convergent. Its sum is the product of their sums.*

Proof. Let $\sum_{n=0}^{+\infty} a_n$ and $\sum_{n=0}^{+\infty} b_n$ be the given absolutely convergent series, let $\sum_{n=0}^{+\infty} c_n$ be their product series, and define the series $\sum_{n=0}^{+\infty} d_n$ to be

$$a_0 b_0 + a_0 b_1 + a_1 b_0 + a_0 b_2 + a_1 b_1 + a_2 b_0$$
$$+ a_0 b_3 + a_1 b_2 + a_2 b_1 + a_3 b_0 + a_0 b_4 + \cdots,$$

the terms following along the diagonal lines suggested in Figure 703.

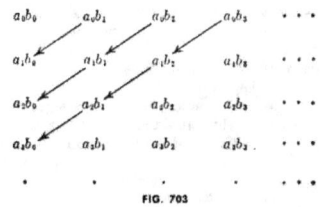

FIG. 703

Furthermore, let

$$A_n = a_0 + a_1 + \cdots + a_n, \qquad B_n = b_0 + b_1 + \cdots + b_n,$$
$$C_n = c_0 + c_1 + \cdots + c_n, \qquad D_n = d_0 + d_1 + \cdots + d_n.$$

Observe that every c_n is obtained by the grouping of $(n+1)$ terms of the series $\sum d_n$, that every C_n is a partial sum of the series $\sum d_n$ with terms occupying a triangle in the upper left-hand corner of Figure 703, and that every $A_n B_n$ is a sum (not a strict "partial sum") of certain terms of the series $\sum d_n$ occupying a square in the upper left-hand corner of Figure 703.

We prove the theorem first for the case of nonnegative series $\sum a_n$ and $\sum b_n$. In this case, since every finite set of terms from $\sum d_n$ is located in some square in the upper left-hand corner of Figure 703, every D_m is less than or equal to *some* $A_n B_n$. That is, if $A = \lim_{n \to +\infty} A_n$ and $B = \lim_{n \to +\infty} B_n$, the inequality $D_m \leq AB$ holds for every m. Therefore the series $\sum d_n$ converges, the limit $D = \lim_{n \to +\infty} D_n$ is finite, and $D \leq AB$. On the other hand, since $A_n B_n$ is a sum of terms of the convergent series $\sum d_n$, the inequality $A_n B_n \leq D$ holds for every n, and hence $AB \leq D$. Thus $D = AB$. Finally, since $\{C_n\}$ is a subsequence of $\{D_n\}$ (resulting from introducing parentheses in the series $\sum d_n$), $C = \lim_{n \to +\infty} C_n = D = AB$.

§ 720] SOME AIDS TO COMPUTATION

If the given series, $\sum a_n$ and $\sum b_n$, have terms of arbitrary sign, the conclusion sought is a consequence of Dirichlet's Theorem (§ 718): Since, by the preceding paragraph, the series $\sum d_n$ converges absolutely, any rearrangement converges absolutely to the same sum. The sequence $\{C_n\}$ is a subsequence of the sequence of partial sums of $\sum d_n$, and the sequence $\{A_n B_n\}$ is a subsequence of the sequence of partial sums of an appropriate rearrangement of $\sum d_n$. Therefore $\lim_{n\to+\infty} C_n = \lim_{n\to+\infty} A_n B_n = \lim_{n\to+\infty} D_n$.

★720. SOME AIDS TO COMPUTATION

The only techniques available from the preceding sections of this chapter for evaluating series are (i) the sum $a/(1-r)$ of a geometric series and (ii) the estimate $|S_n - S| < c_{n+1}$ for an alternating series. We give in this section some further means of estimating the sum of a convergent series, with illustrative examples. These will be available for computation work presented in the next chapter.

If a nonnegative series is dominated by a convergent geometric series, the formula $a/(1-r)$ provides an estimate for the sum. A useful formulation of this method makes use of the test ratio (give the proof in Ex. 6, § 721):

Theorem I. *If* $a_n > 0$, $S_n \equiv a_1 + \cdots + a_n$, $S \equiv \sum_{n=1}^{+\infty} a_n$, $r_n \equiv \dfrac{a_{n+1}}{a_n}$, $r_n \downarrow$, *and* $r_n \to \rho < 1$, *then for any n for which* $r_n < 1$,

(1) $$S_n + \frac{a_{n+1}}{1 - \rho} \leqq S \leqq S_n + \frac{a_{n+1}}{1 - r_{n+1}}.$$

Example 1. Compute the sum of the series
$$1 + \frac{1}{3} + \frac{1}{2!\cdot 5} + \frac{1}{3!\cdot 7} + \cdots + \frac{1}{n!\,(2n+1)} + \cdots,$$
to three decimal places. (The sum is $\int_0^1 e^{x^2}\,dx$. Cf. Example 4, § 810.)

Solution. The sum S_6 is between 1.4625 and 1.4626. Since $r_n \to \rho = 0$, we estimate:
$$\frac{a_7}{1-0} = 0.000107, \quad \text{and} \quad \frac{a_7}{1-r_7} = \frac{0.000107}{1-0.121} < 0.00013.$$

Therefore, from (1), S must lie between 1.4626 and 1.4628. Its value to three decimal places is therefore 1.463.

If a nonnegative series is not dominated by a convergent geometric series $\left(\text{for example, if } \dfrac{a_{n+1}}{a_n} \to 1\right)$, Theorem I cannot be used. However, in this case the process of integration (cf. the integral test, § 706) can sometimes be used, as expressed in the following theorem (give the proof in Ex. 7, § 721):

Theorem II. *If $f(x)$ is a positive monotonically decreasing function, for $x \geq a$, if $\sum_{n=1}^{+\infty} a_n$ is a convergent positive series, with $f(n) = a_n$ for $n > a$, if $S \equiv \sum a_n$, and if $S_n \equiv a_1 + \cdots + a_n$, then for $n > a$,*

$$(6) \quad S_n + \int_{n+1}^{+\infty} f(x)\, dx < S < S_n + \int_{n}^{+\infty} f(x)\, dx.$$

A much sharper estimate is provided by the following theorem (hints for a proof are given in Ex. 21, § 721):

★★**Theorem III.** *Under the hypotheses of Theorem II and the additional assumption that $f''(x)$ is a positive monotonically decreasing function for $x \geq a$, then for $n > a$,*

$$(7) \quad S_n + \frac{a_{n+1}}{2} + \int_{n+1}^{+\infty} f(x)\, dx - \frac{f'(n+2)}{12}$$
$$< S < S_n + \frac{a_{n+1}}{2} + \int_{n+1}^{+\infty} f(x)\, dx - \frac{f'(n)}{12}.$$

Example 2. Estimate the sum of the p-series with $p = 2$, using 10 terms.

Solution. With the aid of a table of reciprocals we find that $S_{10} = 1.549768$. With $f(x) = x^{-2}$, $\int_{11}^{+\infty} f(x)\, dx = \frac{1}{11} = 0.090909$ and $\int_{10}^{+\infty} f(x)\, dx = \frac{1}{10} = 0.100000$. Thus the estimate of Theorem II places the sum S between 1.640 and 1.650, with an accuracy of one digit in the second decimal place.

★★Further computations give $\frac{1}{2} a_{11} = 0.004132$, $-\frac{1}{12} f'(12) = \frac{1}{6 \cdot 12^2} = 0.000096$, and $-\frac{1}{12} f'(10) = \frac{1}{6 \cdot 10^3} = 0.000167$. Thus the estimate of Theorem III places the sum S between 1.6449 and 1.6450, with an accuracy of one digit in the fourth decimal place.

NOTE. The sum in Example 2 can be specifically evaluated by means of the techniques of Fourier series (cf. Example 2, § 1510; Example 1, § 1512; also Example 2, § 1512; Exs. 2-4, 20, 22, § 1515; Ex. 18, § 1528):

$$\frac{\pi^2}{6} = 1 + \frac{1}{2^2} + \frac{1}{3^2} + \frac{1}{4^2} + \cdots.$$

If an alternating series converges slowly, the estimate given in the alternating series test (§ 715) is very crude unless an excessively large number of terms is used. For example, to compute the value of the alternating harmonic series to four decimal places would require at least ten thousand terms! This particular series happens to converge to $\ln 2$ (cf. § 807), and this fortuitous circumstance permits a simpler and more speedy evaluation (cf. Example 4, § 813). However, this series will be used in the following example to illustrate a technique frequently useful in evaluating

§ 721]   EXERCISES   237

slowly converging alternating series:

Example 3. Evaluate the alternating harmonic series,
$$S = 1 - \tfrac{1}{2} + \tfrac{1}{3} - \tfrac{1}{4} + \cdots$$
to four decimal places.

Solution. We start by evaluating the sum of the first 10 terms: $S_{10} = 0.645635$. We wish to estimate the remainder:
$$x = \tfrac{1}{11} - \tfrac{1}{12} + \tfrac{1}{13} - \tfrac{1}{14} + \cdots.$$

If we double, remove parentheses (the student should justify this step), and introduce parentheses, we find
$$2x = (\tfrac{1}{11} + \tfrac{1}{11}) - (\tfrac{1}{12} + \tfrac{1}{12}) + (\tfrac{1}{13} + \tfrac{1}{13}) - \cdots$$
$$= \tfrac{1}{11} + \tfrac{1}{11} - \tfrac{1}{12} - \tfrac{1}{12} + \tfrac{1}{13} + \tfrac{1}{13} - \cdots$$
$$= \tfrac{1}{11} + (\tfrac{1}{11} - \tfrac{1}{12}) - (\tfrac{1}{12} - \tfrac{1}{13}) + (\tfrac{1}{13} - \tfrac{1}{14}) - \cdots$$
$$= \frac{1}{11} + \frac{1}{11 \cdot 12} - \frac{1}{12 \cdot 13} + \frac{1}{13 \cdot 14} - \cdots.$$

Again doubling, and removing and introducing parentheses, we have
$$4x = \frac{2}{11} + \frac{1}{11 \cdot 12} + \frac{2}{11 \cdot 12 \cdot 13} - \frac{2}{12 \cdot 13 \cdot 14} + \cdots,$$
or
$$2x = \frac{25}{11 \cdot 24} + \frac{1}{11 \cdot 12 \cdot 13} - \frac{1}{12 \cdot 13 \cdot 14} + \cdots.$$

Once more:
$$4x = \frac{25}{11 \cdot 12} + \frac{1}{11 \cdot 12 \cdot 13} + \frac{3}{11 \cdot 12 \cdot 13 \cdot 14} - \frac{3}{12 \cdot 13 \cdot 14 \cdot 15} + \cdots.$$

The sum of the first two terms of this series is 0.189977, and the remainder is less than the term $3/11 \cdot 12 \cdot 13 \cdot 14 < 0.000126$. Therefore x is between 0.04749 and 0.04753, and S is between 0.69312 and 0.69317. An estimate to four places is $0.6931+$. (The actual value to five places is 0.69315.)

721. EXERCISES

1. Prove that any series arising from a divergent nonnegative series by grouping of terms is divergent. Equivalently, if the introduction of parentheses into a nonnegative series produces a convergent series, the original series is convergent.

2. Can the terms of a conditionally convergent series be rearranged to give a series whose partial sums (i) tend toward $+\infty$? (ii) tend toward $-\infty$? (iii) tend toward ∞ but neither $+\infty$ nor $-\infty$? (iv) are bounded and have no limit?

3. Prove Theorem I, § 719.

★4. Prove that if $\sum a_n$ and $\sum b_n$ are nonnegative series with sums A and B, respectively, and if $\sum c_n$ is their product series, with sum C, then $C = AB$ under all circumstances of convergence or divergence, with the usual conventions about infinity $((+\infty) \cdot (+\infty) = +\infty$, (positive number) $\cdot (+\infty) = +\infty$) and the additional convention $0 \cdot (+\infty) = 0$.

★5. Prove that if $\sum_{n=0}^{+\infty} a_n$ and $\sum_{n=0}^{+\infty} b_n$ are both the series

$$1 - \frac{1}{\sqrt{2}} + \frac{1}{\sqrt{3}} - \frac{1}{\sqrt{4}} + \cdots$$

and if $\sum_{n=0}^{+\infty} c_n$ is their product series, then $\sum a_n$ and $\sum b_n$ converge, while $\sum c_n$ diverges. *Hint:* Show that $|c_n| \geq 1$.

★6. Prove Theorem I, § 720.

★7. Prove Theorem II, § 720.

★8. Prove the commutative law for product series: The product series of $\sum a_n$ and $\sum b_n$ is the same as the product series of $\sum b_n$ and $\sum a_n$.

★9. Prove the associative law for product series: Let $\sum d_n$ be the product series of $\sum a_n$ and $\sum b_n$, and let $\sum e_n$ be the product series of $\sum b_n$ and $\sum c_n$. Then the product series of $\sum a_n$ and $\sum e_n$ is the same as the product series of $\sum d_n$ and $\sum c_n$.

★10. Prove the distributive law for multiplying and adding series: The product series of $\sum a_n$ and $\sum (b_n + c_n)$ is the sum of the product series of $\sum a_n$ and $\sum b_n$ and the product series of $\sum a_n$ and $\sum c_n$.

★11. Using the evaluation of the series $\sum n^{-2}$ given in the Note, § 720, show that

$$\frac{\pi^2}{8} = 1 + \frac{1}{3^2} + \frac{1}{5^2} + \frac{1}{7^2} + \cdots.$$

★12. Using the evaluation of the series $\sum n^{-2}$ given in the Note, § 720, show that

$$\frac{\pi^2}{12} = 1 - \frac{1}{2^2} + \frac{1}{3^2} - \frac{1}{4^2} + \cdots.$$

In Exercises 13-18, compute to four significant digits.

★13. $(e =)\ 1 + \frac{1}{1!} + \frac{1}{2!} + \frac{1}{3!} + \cdots$.

★14. $(\ln 2 =)\ \frac{1}{2} + \frac{1}{2 \cdot 2^2} + \frac{1}{3 \cdot 2^3} + \frac{1}{4 \cdot 2^4} + \cdots$.

★15. $(\zeta(3) =)\ 1 + \frac{1}{2^3} + \frac{1}{3^3} + \frac{1}{4^3} + \cdots$.

★16. $\frac{5}{2^2 \cdot 3^2} + \frac{9}{4^2 \cdot 5^2} + \frac{13}{6^2 \cdot 7^2} + \frac{17}{8^2 \cdot 9^2} + \cdots$.

★17. $\left(\frac{\pi}{4} =\right)\ 1 - \frac{1}{3} + \frac{1}{5} - \frac{1}{7} + \cdots$.

★18. $\left(\ln \frac{3}{2} =\right)\ \frac{1}{2} - \frac{1}{2 \cdot 2^2} + \frac{1}{3 \cdot 2^3} - \frac{1}{4 \cdot 2^4} + \cdots$.

★★19. Prove that the sequence

$$\{C_n\} = \left\{1 + \frac{1}{2} + \cdots + \frac{1}{n} - \ln n\right\}$$

is decreasing and bounded below. Hence $C = \lim_{n \to +\infty} C_n$ exists. (The number

§ 721] EXERCISES 239

C is known as **Euler's constant**. It is believed to be transcendental, but its transcendence has never been established.) (Cf. Ex. 22. Also, § 1418.)

★★20. Prove the theorem of Mertens: *If $\sum_{n=0}^{+\infty} a_n$ converges absolutely to A and if $\sum_{n=0}^{+\infty} b_n$ converges to B, then the product series $\sum_{n=0}^{+\infty} c_n$ converges to AB.* *Hint*: By virtue of Exs. 8–10, it may be assumed without loss of generality that $\sum a_n$ is nonnegative, $A > 0$, and $B = 0$. Under these assumptions prove that $\sum c_n$ converges to 0: Define $A_n \equiv \sum_{k=0}^{n} a_k$, $B_n \equiv \sum_{k=0}^{n} b_k$, $C_n \equiv \sum_{k=0}^{n} c_k$. Then

$$C_n = a_0 b_0 + (a_0 b_1 + a_1 b_0) + \cdots + (a_0 b_n + \cdots + a_n b_0)$$
$$= a_0 B_n + a_1 B_{n-1} + \cdots + a_n B_0.$$

For a given $\epsilon > 0$, first choose N such that $m > N$ implies $|B_m| < \epsilon/2A$, and therefore
$$|a_0 B_n + \cdots + a_{n-N-1} B_{N+1}| < \tfrac{1}{2}\epsilon.$$
Then choose $N' > N$ such that $n > N'$ implies
$$|a_{n-N}| < \frac{\epsilon}{2(N+1)} \cdot \max(|B_0|, |B_1|, \cdots, |B_N|, 1).$$

★★21. Prove Theorem III, § 720. *Hints*: Let $R_n \equiv a_{n+1} + a_{n+2} + \cdots = \tfrac{1}{2}a_{n+1} + \tfrac{1}{2}(a_{n+1} + a_{n+2}) + \tfrac{1}{2}(a_{n+2} + a_{n+3}) + \cdots \equiv \tfrac{1}{2}a_n + T_{n+1}$. Interpret T_{n+1} as a sum of areas of trapezoids, and use the trapezoidal formula error estimate (Ex. 22, § 503) to write $\tfrac{1}{2}(a_m + a_{m+1}) = \int_{m}^{m+1} f(x)\, dx + \tfrac{1}{12} f''(\xi_m)$, where $m < \xi_m < m+1$. Thus $T_{n+1} - \int_{n+1}^{+\infty} f(x)\, dx = \tfrac{1}{12} \sum_{m=n+1}^{+\infty} f''(\xi_m)$. But

$$\tfrac{1}{12} \int_{n+2}^{+\infty} f''(x)\, dx < \tfrac{1}{12} \sum_{m=n+1}^{+\infty} f''(m) < \sum_{m=n+1}^{+\infty} f''(\xi_m) < \tfrac{1}{12} \sum_{m=n}^{+\infty} f''(m)$$
$$< \tfrac{1}{12} \int_{n}^{+\infty} f''(x)\, dx.$$

★★22. Find an estimate of Euler's constant (Ex. 19), by using the technique of the proof of Theorem III, § 720 (Ex. 21), and $9\tfrac{1}{2}$ terms of the harmonic series:
$$C = 1 + \tfrac{1}{2} + \cdots + \tfrac{1}{9} + \tfrac{1}{2} \cdot \tfrac{1}{10} - \ln 10$$
$$+ \lim_{n \to +\infty} \left\{ \left[\tfrac{1}{2}\left(\tfrac{1}{10} + \tfrac{1}{11}\right) - \int_{10}^{11} \frac{dx}{x} \right] + \left[\tfrac{1}{2}\left(\tfrac{1}{11} + \tfrac{1}{12}\right) - \int_{11}^{12} \frac{dx}{x} \right] + \cdots \right\}.$$

8

Power Series

801. INTERVAL OF CONVERGENCE

A series of the form

$$(1) \quad \sum a_n x^n = \sum_{n=0}^{+\infty} a_n x^n = a_0 + a_1 x + a_2 x^2 + \cdots$$

is called a **power series in** x. A series of the form

$$(2) \quad \sum a_n (x - a)^n = \sum_{n=0}^{+\infty} a_n (x - a)^n = a_0 + a_1(x - a) + \cdots$$

is called a **power series in** $(x - a)$. More generally, a series of the form

$$(3) \quad \sum a_n [u(x)]^n = \sum_{n=0}^{+\infty} a_n [u(x)]^n = a_0 + a_1 u(x) + \cdots,$$

where $u(x)$ is a function of x, is called a **power series in** $u(x)$. It is principally series (1) or (2) that will be of interest in this chapter, and the single expression *power series* will be used to mean either series (1) or series (2).

A power series is an example of a series of *functions*. For any fixed value of x the series becomes a series of *constants*; but convergence or divergence of this series of constants depends, ordinarily, on the value of x. One is frequently interested in the question, "For what values of x does a given power series converge?" The answer is fairly simple: The values of x for which a power series converges always form an interval, which may degenerate to a single point, or encompass all real numbers, or be a finite interval, open, closed, or half-open. To prove this result we formulate it for simplicity for the power series $\sum a_n x^n$, first establishing a lemma:

Lemma. *If a power series $\sum a_n x^n$ converges for $x = x_1$ and if $|x_2| < |x_1|$, then the series converges absolutely for $x = x_2$.*

Proof. Assume $\sum a_n x_1^n$ converges. Then $\lim_{n \to +\infty} a_n x_1^n = 0$. Therefore the sequence $\{a_n x_1^n\}$, being convergent, is bounded. Let $|a_n x_1^n| < K$ for all n. If $|x_2| < |x_1|$, we may write

(4) $\qquad |a_n x_2{}^n| = |a_n x_1{}^n| \cdot \left|\dfrac{x_2}{x_1}\right|^n < Kr^n,$ where $0 \leq r < 1.$

The series $\sum Kr^n$ is a convergent geometric series. Therefore, by comparison, $\sum |a_n x_2{}^n|$ converges, and $\sum a_n x_2{}^n$ converges absolutely.

What this lemma says, in part, is that *any* point of convergence of the power series $\sum a_n x^n$ is at least as close to 0 as *any* point of divergence.

Theorem I. *Let S be the set of points x for which a power series $\sum a_n x^n$ converges. Then either (i) S consists only of the point $x = 0$, or (ii) S consists of all real numbers, or (iii) S is an interval of one of the following forms: $(-R, R), [-R, R], (-R, R],$ or $[-R, R),$ where R is a positive real number.*

Proof. The simplest proof rests on the concept of *least upper bound* (§ 114). (For a proof based on convergent sequences, cf. Ex. 17, § 802.) If neither (i) nor (ii) holds, then (by the preceding lemma) the series $\sum a_n x^n$ must converge for some positive number x_C and diverge for some larger positive number x_D. That is, S contains some positive numbers and is bounded above. Let R be defined to be the least upper bound of S. Then (iii) is a consequence of the Lemma. (Why?)

NOTE 1. The statements just established for the power series (1) apply to the power series (2), the only change being that the interval of convergence has the point $x = a$ instead of the point $x = 0$ as midpoint.

The set S of points x for which a power series $\sum a_n(x - a)^n$ converges is called the **interval of convergence,** and the number R of Theorem I (and Note 1) is called the **radius of convergence**† of $\sum a_n(x - a)^n$. In case (i) of Theorem I we define $R \equiv 0$, and in case (ii) we write $R \equiv +\infty$, so that in general $0 \leq R \leq +\infty$.

A further consequence of the Lemma is the Theorem:

Theorem II. *At any point interior to the interval of convergence of a power series the convergence is absolute.*

NOTE 2. All of the eventualities stated in Theorem I exist (Ex. 13, § 802).

NOTE 3. At an end-point of the interval of convergence a power series may diverge, converge conditionally, or converge absolutely. At the two end-points all combinations are possible, except that if a power series converges absolutely at one end-point it must also converge absolutely at the other end-point. (Cf. Ex. 14, § 802.)

The usual procedure in determining the interval of convergence of a power series is to start with the ratio test, although this may fail (cf. Ex-

---

† R is called the *radius of convergence* because of the analogous situation with complex numbers, where the *interval* of convergence is replaced by a *circle* of convergence whose radius is R.

ample 3). The success of the ratio test depends on the existence of the limit of the ratio of successive coefficients. We specify this relationship:

Theorem III. *The radius of convergence of the power series* $\sum a_n(x-a)^n$ *is*

(5) $$R = \lim_{n \to +\infty} \left|\frac{a_n}{a_{n+1}}\right|,$$

where $0 \leq R \leq +\infty$, *provided this limit exists.*

Proof. We give the details for $0 < R < +\infty$. (Cf. Ex. 15, § 802.) The test ratio for the series $\sum a_n(x-a)^n$ is $\frac{a_{n+1}(x-a)}{a_n}$, whose absolute value has the limit $\lim_{n \to +\infty} \left|\frac{a_{n+1}}{a_n}\right| \cdot |x-a| = \frac{|x-a|}{R}$. Therefore, by the ratio test, $\sum a_n(x-a)^n$ converges absolutely if $|x-a| < R$ and diverges if $|x-a| > R$.

After the radius of convergence has been found, the end-points of the interval of convergence should be tested. For such points the ratio test cannot give any information, for if $\frac{a_{n+1}}{a_n} R$ had a limit for $R > 0$, the limit (5) would exist, and the absolute value of the test ratio must have the limit 1. To test the end-points one is forced to use some type of comparison test, or a refined test of the type discussed in § 712. (The root test also fails at the end-points—cf. Ex. 16, § 802.)

★**Note 4.** A universally valid formula for the radius of convergence of a power series $\sum a_n(x-a)^n$ makes use of *limit superior* (Ex. 16, § 305):

$$R = \frac{1}{\overline{\lim}_{n \to +\infty} \sqrt[n]{a_n}},$$

with the conventions $1/0 = +\infty$, $1/+\infty = 0$. (Give the proof in Ex. 18, § 802, and cf. Ex. 16, § 802.)

Example 1. Determine the interval of convergence for the series
$$1 + x^2 + \frac{x^4}{2!} + \frac{x^6}{3!} + \cdots.$$
(Cf. § 807.)

Solution. This should be treated as a power series in powers of x^2. The radius of convergence is
$$R = \lim_{n \to +\infty} \frac{(n+1)!}{n!} = \lim_{n \to +\infty} (n+1) = +\infty.$$
Therefore the interval of convergence is $(-\infty, +\infty)$.

Example 2. Determine the interval of convergence for the series
$$(x-1) - \frac{(x-1)^2}{2} + \frac{(x-1)^3}{3} - \frac{(x-1)^4}{4} + \cdots.$$
(Cf. § 807.)

§ 802]   EXERCISES   243

Solution. The radius of convergence is $R = \lim_{n \to +\infty} \frac{n+1}{n} = 1$, and the midpoint of the interval of convergence is $x = 1$. We test for convergence at the end-points of the interval, $x = 2$ and $x = 0$. The value $x = 2$ gives the convergent alternating harmonic series and $x = 0$ gives minus the divergent harmonic series. The interval of convergence is $(0, 2]$.

Example 3. Determine the interval of convergence for the series
$$\frac{1}{2} + \frac{x}{3} + \frac{x^2}{2^2} + \frac{x^3}{3^2} + \frac{x^4}{2^3} + \frac{x^5}{3^3} + \frac{x^6}{2^4} + \frac{x^7}{3^4} + \cdots.$$

(Cf. Example 2, § 708.)

Solution. The limit (5) does not exist. However, the intervals of convergence of the two series
$$\frac{1}{2} + \frac{x^2}{2^2} + \frac{x^4}{2^3} + \cdots \quad \text{and} \quad \frac{x}{3} + \frac{x^3}{3^2} + \frac{x^5}{3^3} + \cdots$$

are $(-\sqrt{2}, \sqrt{2})$ and $(-\sqrt{3}, \sqrt{3})$, respectively. Therefore the given series converges absolutely for $|x| < \sqrt{2}$ and diverges for $|x| \geqq \sqrt{2}$. The interval of convergence is $(-\sqrt{2}, \sqrt{2})$.

Example 4. Determine the values of x for which the series
$$xe^{-x} + 2x^2e^{-2x} + 3x^3e^{-3x} + 4x^4e^{-4x} + \cdots$$

converges.

Solution. Either the ratio test or the root test shows that the series converges absolutely for $|x\,e^{-x}| < 1$ (otherwise it diverges). The inequality $|x\,e^{-x}| < 1$ is equivalent to $|x| < e^x$, and is satisfied for all nonnegative x. To determine the negative values of x which satisfy this inequality, we let $\alpha = -x$, and solve the equation $\alpha = e^{-\alpha}$ ($\alpha = 0.567$, approximately). Then the given series converges if and only if $x > -\alpha$, and the convergence is absolute.

802. EXERCISES

In Exercises 1-10, determine the interval of convergence, and specify the nature of any convergence at each end-point of the interval of convergence.

1. $1 - \frac{2x}{1!} + \frac{(2x)^2}{2!} - \frac{(3x)^3}{3!} + \cdots$.

2. $x - \frac{x^3}{3} + \frac{x^5}{5} - \frac{x^7}{7} + \cdots$.

3. $1 + \frac{x^2}{2!} + \frac{x^4}{4!} + \frac{x^6}{6!} + \cdots$.

4. $1 + x + 2!x^2 + 3!x^3 + 4!x^4 + \cdots$.

5. $(x+1) - \frac{(x+1)^2}{4} + \frac{(x+1)^3}{9} - \frac{(x+1)^4}{16} + \cdots$.

6. $(x-2) + \frac{(x-2)^3}{3!} + \frac{(x-2)^5}{5!} + \frac{(x-2)^7}{7!} + \cdots$.

7. $\frac{(\ln 2)(x-5)}{\sqrt{2}} + \frac{(\ln 3)(x-5)^2}{\sqrt{3}} + \frac{(\ln 4)(x-5)^3}{\sqrt{4}} + \cdots$.

8. $(x-1) + \frac{(x-1)^3}{3} + \frac{(x-1)^5}{5} + \frac{(x-1)^7}{7} + \cdots$.

★9. $1 - \frac{1}{2}x + \frac{1\cdot 3}{2\cdot 4}x^2 - \frac{1\cdot 3\cdot 5}{2\cdot 4\cdot 6}x^3 + \frac{1\cdot 3\cdot 5\cdot 7}{2\cdot 4\cdot 6\cdot 8}x^4 - \cdots$.

★10. $x + \frac{x^3}{6} + \frac{1\cdot 3}{2\cdot 4}\cdot\frac{x^5}{5} + \frac{1\cdot 3\cdot 5}{2\cdot 4\cdot 6}\frac{x^7}{7} + \cdots$.

11. Determine the values of x for which the series
$$\frac{1}{x-3} + \frac{1}{2(x-3)^2} + \frac{1}{3(x-3)^3} + \frac{1}{4(x-3)^4} + \cdots$$
converges, and specify the type of convergence.

12. Determine the values of x for which the series
$$\sin x - \frac{\sin^3 x}{3} + \frac{\sin^5 x}{5} - \frac{\sin^7 x}{7} + \cdots.$$
converges, and specify the type of convergence.

13. Prove Note 2, § 801.
14. Prove Note 3, § 801.
15. Prove Theorem III, § 801, for the cases $R = 0$ and $R = +\infty$.
16. Prove Note 4, § 801, for the case where $\lim_{n\to+\infty} \sqrt[n]{a_n}$ exists (replacing $\overline{\lim}$ by lim).

★17. Give a proof of Theorem I, § 801, based on convergent sequences, as follows*: Assume that neither condition (i) nor (ii) of that theorem holds. Show first that there exist positive points of convergence and of divergence which are arbitrarily close. Then construct two sequences $\{c_n\}$ and $\{d_n\}$ of points of convergence and divergence, respectively, where $0 \leq c_n \leq c_{n+1} < d_{n+1} \leq d_n$, for $n = 1, 2, \cdots$, and define R to be their common limit. Prove that R is the radius of convergence of the given series.

★18. Prove Note 4, § 801. (Cf. Ex. 32, § 711.)

★19. Apply Note 4, § 801 to Example 3, § 801. (Cf. Ex. 33, § 711.)

★20. Show by examples that Theorem III, § 801, cannot be generalized in the manner of Note 4, § 801, by the use of limits superior or inferior. (Cf. Exs. 31-33, § 711.)

803. TAYLOR SERIES

We propose to discuss in this section some formal procedures, nearly all of which need justification and will be discussed in future sections. The purpose of this discussion is to motivate an important formula, and raise some questions.

Let us suppose that a power series $\sum a_n(x-a)^n$ has a positive radius of convergence ($0 < R \leq +\infty$), and let $f(x)$ be the function defined by this series wherever it converges. That is,

(1) $\qquad f(x) \equiv a_0 + a_1(x-a) + a_2(x-a)^2 + \cdots$.

We now differentiate term-by-term, as if the infinite series were simply a finite sum:
$$f'(x) = a_1 + 2a_2(x - a) + 3a_3(x - a)^2 + \cdots.$$
Again:
$$f''(x) = 2a_2 + 2 \cdot 3a_3(x - a) + 3 \cdot 4a_4(x - a)^2 + \cdots.$$
And so forth:
$$f'''(x) = 3! \, a_3 + 2 \cdot 3 \cdot 4a_4(x - a) + 3 \cdot 4 \cdot 5a_5(x - a)^2 + \cdots,$$
$$\cdots\cdots\cdots\cdots\cdots\cdots.$$

Upon substitution of $x = a$, we have:
$$f(a) = a_0, f'(a) = a_1, f''(a) = 2! \, a_2, f'''(a) = 3! \, a_3, \cdots,$$
or, if we solve for the coefficients a_n:

(2) $\qquad a_0 = f(a), \; a_1 = f'(a), \; a_2 = \dfrac{f''(a)}{2!}, \; \cdots, \; a_n = \dfrac{f^{(n)}(a)}{n!}, \; \cdots.$

This suggests that if a power series $\sum a_n(x - a)^n$ converges to a function $f(x)$, then the coefficients of the power series should be determined by the values of that function and its successive derivatives according to equations (2). In other words, we should expect:

(3) $\quad f(x) = f(a) + f'(a)(x - a) + \dfrac{f''(a)}{2!}(x - a)^2 + \cdots$
$$+ \dfrac{f^{(n)}(a)}{n!}(x - a)^n + \cdots,$$

and, in particular, for $a = 0$:

(4) $\qquad f(x) = f(0) + f'(0)x + \cdots + \dfrac{f^{(n)}(0)}{n!} x^n + \cdots.$

Now suppose $f(x)$ has derivatives of all orders, at least in a neighborhood of the point $x = a$. Then $f^{(n)}(a)$ is defined for every n, and the series (3) exists. Regardless of any question of convergence or (in case of convergence) equality in (3), we *define* the series $\sum_{n=0}^{+\infty} \dfrac{f^{(n)}(a)}{n!}(x - a)^n$† on the right-hand side of (3) as the **Taylor series for the function** $f(x)$ **at** $x = a$. The particular case $\sum_{n=0}^{+\infty} \dfrac{f^{(n)}(0)}{n!} x^n$ given in (4) is known as the **Maclaurin series for** $f(x)$.

We ask two questions:

(i) For a given function $f(x)$, is the Taylor series expansion (3) universally valid in some neighborhood of $x = a$, and if not, what criteria are there for the relation (3) to be true?

† For convenience in notation we define $f^{(0)}(x) = f(x)$, and recall that $0! = 1$. Although for $x = a$ and $n = 0$ an indeterminacy 0^0 develops, let us agree that *in this instance* 0^0 shall be defined to be 1.

The answer to this question is a major concern of the remaining sections of this chapter.

(ii) For a given power series $\sum a_n(x - a)^n$, converging to a function $f(x)$ in an interval with positive radius of convergence, is the Taylor series equation (3) true?

We answer this second question affirmatively, now, but defer the proof to the next chapter (Ex. 13, § 911):

Theorem I. *If $\sum_{n=0}^{+\infty} a_n(x - a)^n$ has a positive radius of convergence, and if $f(x) \equiv \sum a_n(x - a)^n$ in the interval of convergence of the series, then throughout that interval $f(x)$ is continuous; throughout the interior of that interval $f(x)$ has (continuous) derivatives of all orders, and relations (2) hold: $a_n = f^{(n)}(a) / n!$, for $n = 0, 1, 2, \cdots$. The given series is the Taylor series for the function $f(x)$ at $x = a$.*

One immediate consequence of this theorem is the uniqueness of a power series $\sum a_n(x - a)^n$ converging to a given function:

Theorem II. Uniqueness Theorem. *If a function $f(x)$ is equal to the sum of a power series $\sum a_n(x - a)^n$ in a neighborhood of $x = a$, and if $f(x)$ is also equal to the sum of a power series $\sum b_n(x - a)^n$ in a neighborhood of $x = a$, then these two power series are identical, coefficient by coefficient: $a_n = b_n$, $n = 0, 1, 2, \cdots$.*

Proof. Each power series is the Taylor series for $f(x)$ at $x = a$.

804. TAYLOR'S FORMULA WITH A REMAINDER

In § 407 we obtained from the Extended Law of the Mean a formulation for expanding a function $f(x)$ in terms closely related to the Taylor series discussed in the preceding section. Let us repeat the formula of Note 3, § 407, with a slight change in notation, in the following Definition and Theorem I:

Definition. *If $f^{(n)}(x)$ exists at every point of an interval I containing the point $x = a$, then the **Taylor's Formula with a Remainder** for the function $f(x)$, for any point x of I, is*

$$(1) \quad f(x) = f(a) + f'(a)(x - a) + \frac{f''(a)}{2!}(x - a)^2 + \cdots + \frac{f^{(n-1)}(a)}{(n - 1)!}(x - a)^{n-1} + R_n(x).$$

*The quantity $R_n(x)$ is called the **remainder after n terms**.*

The principal substance of Note 3, § 407, is an explicit evaluation of the

§ 804] TAYLOR'S FORMULA WITH A REMAINDER   247

Theorem I. Lagrange Form of the Remainder. *Let n be a fixed positive integer. If $f^{(n)}(x)$ exists at every point of an interval I (open, closed, or half-open) containing $x = a$, and if x is any point of I, then there exists a point ξ_n between a and x ($\xi_n = a$ if $x = a$) such that the remainder after n terms, in Taylor's Formula with a Remainder, is*

$$(2) \qquad R_n(x) = \frac{f^{(n)}(\xi_n)}{n!}(x-a)^n.$$

The principal purpose of this section is to obtain two other forms of the remainder $R_n(x)$. We first establish an integral form of the remainder (Theorem II), from which both the Lagrange form and the Cauchy form (Theorem III) can be derived immediately (cf. Ex. 1, § 806).

Assuming continuity of all of the derivatives involved, we start with the obvious identity $\int_0^{x-a} f'(x-t)\,dt = f(x) - f(a)$, and integrate by parts, repeatedly:

$$f(x) - f(a) = \int_0^{x-a} f'(x-t)\,dt = \Big[t f'(x-t)\Big]_0^{x-a} + \int_0^{x-a} t f''(x-t)\,dt$$

$$= f'(a)(x-a) + \int_0^{x-a} f''(x-t)\,d\left(\frac{t^2}{2!}\right)$$

$$= f'(a)(x-a) + \left[\frac{t^2}{2!} f''(x-t)\right]_0^{x-a} + \int_0^{x-a} \frac{t^2}{2!} f'''(x-t)\,dt$$

$$= f'(a)(x-a) + \frac{f''(a)}{2!}(x-a)^2 + \int_0^{x-a} f'''(x-t)\,d\left(\frac{t^3}{3!}\right)$$

$$= \cdots\cdots.$$

Iteration of this process an appropriate number of times leads to the formula written out in the following theorem:

Theorem II. Integral Form of the Remainder. *Let n be a fixed positive integer. If $f^{(n)}(x)$ exists and is continuous throughout an interval I containing the point $x = a$, and if x is any point of I, then the remainder after n terms, in Taylor's Formula with a Remainder, can be written*

$$(3) \qquad R_n(x) = \frac{1}{(n-1)!}\int_0^{x-a} t^{n-1} f^{(n)}(x-t)\,dt.$$

Suppose for the moment that $x > a$. Then $x - a > 0$ and, by the First Mean Value Theorem for Integrals (Theorem XI, § 501) there exists a number η_n such that $0 < \eta_n < x - a$ and, from (3),

$$R_n(x) = \frac{1}{(n-1)!}\eta_n^{n-1} f^{(n)}(x - \eta_n)\cdot(x-a).$$

In other words, there exists a number $\xi_n = x - \eta_n$ between a and x such that

$$R_n(x) = \frac{1}{(n-1)!}(x - \xi_n)^{n-1} f^{(n)}(\xi_n) \cdot (x - a).$$

On the other hand, if $x \leq a$, we obtain the same result by reversing the inequalities or replacing them by equalities. We have the conclusion:

Theorem III. Cauchy Form of the Remainder. *Let n be a fixed positive integer. If $f^{(n)}(x)$ exists and is continuous throughout an interval I containing the point $x = a$, and if x is any point of I, then there exists a point ξ_n between a and x ($\xi_n = a$ if $x = a$) such that the remainder after n terms, in Taylor's Formula with a Remainder is*

(4) $$R_n(x) = \frac{(x-a)(x-\xi_n)^{n-1}}{(n-1)!} f^{(n)}(\xi_n).$$

805. EXPANSIONS OF FUNCTIONS

Definition. *A series of functions $\sum u_n(x)$ represents a function $f(x)$ on a certain set A if and only if for every point x of the set A the series $\sum u_n(x)$ converges to the value of the function $f(x)$ at that point.*

Immediately after the question of *convergence* of the Taylor series of a function comes the question, "Does the Taylor series of a given function *represent* the function throughout the interval of convergence?" The clue to the answer lies in Taylor's Formula with a Remainder. To clarify the situation we introduce the notation of partial sums (which now depend on x):

(1) $$S_n(x) \equiv f(a) + f'(a)(x-a) + \cdots + \frac{f^{(n-1)}(a)}{(n-1)!}(x-a)^{n-1}.$$

Taylor's Formula with a Remainder now assumes the form

(2) $$f(x) = S_n(x) + R_n(x),$$

or

(3) $$R_n(x) = f(x) - S_n(x).$$

Immediately from (3) and the definition of the sum of a series, we have the theorem:

Theorem. *If I is an interval containing $x = a$ at each point of which $f(x)$ and all of its derivatives exist, and if $x = x_0$ is a point of I, then the Taylor series for $f(x)$ at $x = a$ represents $f(x)$ at the point x_0 if and only if*

$$\lim_{n \to +\infty} R_n(x_0) = 0.$$

Determining for a particular function whether its Taylor series at $x = a$ represents the function for some particular $x = x_0$, reduces, then, to deter-

§ 806] EXERCISES 249

mining whether $R_n(x_0) \to 0$. Techniques for doing this vary with the function concerned. In the following section we shall make use of different forms of the remainder, in order to show that certain specific expansions represent the given functions. To make any general statement specifying conditions under which a given function is represented by its Taylor series is extremely difficult. We can say that some functions, such as polynomials, e^x, and $\sin x$ (cf. Example 1, below, Examples 5 and 6, § 808, and § 815) are *always* represented by *all* of their Taylor series, and that other functions, such as $\ln x$ and $\tan x$, are represented by Taylor series for only *parts* of their domains. Example 2, below, gives an extreme case of a function possessing a Taylor series which converges everywhere but represents the function only at one point!

Example 1. Prove that every polynomial is everywhere represented by all of its Taylor series.

Solution. If $f(x)$ is a polynomial, $f^{(n)}(x)$ exists for all n and x, and $f^{(n)}(x) = 0$ if n is greater than the degree of the polynomial. Therefore, if the Lagrange form of the remainder is used, $R_n(x)$ is identically zero for sufficiently large n, and $\lim_{n \to +\infty} R_n(x) = 0$.

Example 2. Show that the function $f(x)$ defined to be 0 when $x = 0$, and otherwise $f(x) = e^{-1/x^2}$, is not represented by its Maclaurin series, although the function has derivatives of all orders everywhere.

Solution. The function has (continuous) derivatives of all orders at every point except possibly $x = 0$. At $x = 0$, $f^{(n)}(x) = 0$ for every $n = 0, 1, 2, \cdots$ (cf. Ex. 52, § 419), so that $f^{(n)}(x)$ exists and is continuous for every $n = 0, 1, 2, \cdots$ and every x. The Maclaurin series for $f(x)$ is thus

$$0 + 0 \cdot x + 0 \cdot x^2 + 0 \cdot x^3 + \cdots$$

which represents the function identically 0 everywhere, but the function $f(x)$ only at $x = 0$.

806. EXERCISES

1. Derive the Lagrange form of the remainder in Taylor's formula (Theorem I, § 804) from the integral form (Theorem II, § 804), using the additional hypothesis of continuity of $f^{(n)}(x)$. *Hint:* Use the generalized form of the First Mean Value Theorem for Integrals (Ex. 6, § 503).

2. Show that $|x|$, $\ln x$, $\sqrt{x}$, and $\cot x$ have no Maclaurin series. Show that x^p has a Maclaurin series if and only if p is a nonnegative integer.

3. Show that if the Taylor series for $f(x)$ at $x = a$ represents $f(x)$ near a, it can be written in increment and differential notation:

$$\Delta y = dy + \frac{f''(a)}{2!} dx^2 + \frac{f'''(a)}{3!} dx^3 + \cdots.$$

Hence show that approximations by differentials (cf. § 411) are those provided by the partial sum through terms of the first degree of the Taylor series of

807. SOME MACLAURIN SERIES

In this section we derive the Maclaurin series for the five functions e^x, $\sin x$, $\cos x$, $\ln(1+x)$, and $(1+x)^m$, and (sometimes with the aid of future exercises) show that in each case the function is represented by its Maclaurin series throughout the interval of convergence.

I. The exponential function, $f(x) = e^x$. Since, for $n = 0, 1, 2, \cdots$, $f^{(n)}(x) = e^x$, $f^{(n)}(0) = 1$, and the Maclaurin series is

$$(1) \qquad 1 + x + \frac{x^2}{2!} + \frac{x^3}{3!} + \cdots + \frac{x^n}{n!} + \cdots.$$

The test ratio is $\frac{x}{n}$, whose limit is 0. Therefore the series (1) converges absolutely for all x, and the radius of convergence is infinite: $R = +\infty$.

To show that *the series (1) represents the function e^x for all real x*, we choose an arbitrary $x \neq 0$, and use the Lagrange form of the remainder:

$$(2) \qquad R_n(x) = \frac{e^{\xi_n}}{n!} x^n,$$

where ξ_n is between 0 and x. We observe first that for a fixed x, ξ_n (although it depends on n and is not constant) satisfies the inequality $\xi_n < |x|$, and therefore e^{ξ_n} is bounded above by the constant $e^{|x|}$. Thus the problem has been reduced to showing that $\lim_{n \to +\infty} \frac{x^n}{n!} = 0$. But $\frac{x^n}{n!}$ is the general term of (1), and since (1) has already been shown to converge for all x, the general term must tend toward 0.

II. The sine function, $f(x) = \sin x$. The sequence $\{f^{(n)}(x)\}$ is $\sin x$, $\cos x$, $-\sin x$, $-\cos x$, $\sin x$, $\cdots$, and the sequence $\{f^{(n)}(0)\}$ is $0, 1, 0, -1, 0, 1, \cdots$. Therefore the Maclaurin series is

$$(3) \qquad x - \frac{x^3}{3!} + \frac{x^5}{5!} - \frac{x^7}{7!} + \cdots + (-1)^{n-1} \frac{x^{2n-1}}{(2n-1)!} + \cdots.$$

(Here we have dropped all zero terms, and used n to indicate the sequence of remaining nonzero terms, instead of the original exponent for the series $\sum \frac{f^{(n)}(0)}{n!} x^n$.) The test ratio of (3) is

$$-\frac{x^2}{2n(2n+1)},$$

whose limit is 0. Therefore (3) converges absolutely for all x, and the radius of convergence is infinite: $R = +\infty$.

To show that *the series (3) represents the function $\sin x$ for all real x*, we choose an arbitrary $x \neq 0$, and again use the Lagrange form of the remainder:

(4) $$R_n(x) = \frac{g_n(\xi_n)}{n!} x^n,$$

where ξ_n is between 0 and x, $g_n(x)$ is $\pm \sin x$ or $\pm \cos x$, and n is now used to indicate the number of terms in the original form $\sum \frac{f^{(n)}(0)}{n!} x^n$ of the Maclaurin series. The proof that $\lim_{n \to +\infty} R_n(x) = 0$ follows the same lines as the proof for e^x, with the aid of the inequality $|g_n(\xi_n)| \leq 1$.

III. The cosine function, $f(x) = \cos x$. The details of the analysis are similar to those for $\sin x$. The Maclaurin series is

(5) $$1 - \frac{x^2}{2!} + \frac{x^4}{4!} - \frac{x^6}{6!} + \cdots + (-1)^{n-1} \frac{x^{2n}}{(2n)!},$$

which converges absolutely for all real x ($R = +\infty$). *The series (5) represents the function $\cos x$ for all real x.*

IV. The natural logarithm, $f(x) = \ln(1 + x)$. The sequence $\{f^{(n)}(x)\}$ is

$\ln(1+x)$, $(1+x)^{-1}$, $-(1+x)^{-2}$, $2!(1+x)^{-3}$, $-3!(1+x)^{-4}$,
$$\cdots, (-1)^{n-1}(n-1)!(1+x)^{-n}, \cdots,$$

and hence the sequence $\{f^{(n)}(0)\}$ is

$$0, 1, -1, 2!, -3!, \cdots, (-1)^{n-1}(n-1)!, \cdots.$$

Therefore the Maclaurin series is

(6) $$x - \frac{x^2}{2} + \frac{x^3}{3} - \frac{x^4}{4} + \cdots + (-1)^{n-1} \frac{x^n}{n} + \cdots.$$

Theorem III, § 801, gives the radius of convergence: $R = 1$. If $x = 1$, the series (6) is the conditionally convergent alternating harmonic series, and if $x = -1$, the series (6) is the divergent series $\sum -\frac{1}{n}$. The interval of convergence is therefore $-1 < x \leq 1$.

To show that *the series (6) represents the function $\ln(1+x)$ throughout the interval of convergence*, we shall derive a form of the remainder appropriate to $\ln(1+x)$ alone. Using a formula from College Algebra for the sum of a geometric progression, we have, for any $t \neq -1$:

(7) $$1 - t + t^2 - t^3 + \cdots + (-t)^{n-2} = \frac{1 - (-t)^{n-1}}{1+t}.$$

Therefore, if we solve for $\frac{1}{1+t}$, and integrate from 0 to x, where $-1 < x \leq 1$, we have:

$$\int_0^x \frac{dt}{1+t} = \int_0^x [1 - t + \cdots + (-1)^n t^{n-2}] dt + (-1)^{n-1} \int_0^x \frac{t^{n-1} dt}{1+t},$$

or

(8) $$\ln(1+x) = x - \frac{x^2}{2} + \frac{x^3}{3} - \cdots + (-1)^n \frac{x^{n-1}}{n-1} + R_n(x),$$

where

(9) $$R_n(x) = (-1)^{n-1} \int_0^x \frac{t^{n-1} dt}{1+t}.$$

If $0 \leq x \leq 1$, $|R_n(x)| \leq \int_0^x t^{n-1} dt = \frac{x^n}{n} \leq \frac{1}{n}$, and $\lim_{n \to +\infty} R_n(x) = 0$. If $-1 < x < 0$,

$$|R_n(x)| \leq \left| \int_0^x \left| \frac{t^{n-1}}{1+t} \right| dt \right| \leq \frac{1}{1+x} \left| \int_0^x |t^{n-1}| dt \right|$$
$$= \frac{1}{1+x} \left| \int_0^x t^{n-1} dt \right| = \frac{|x|^n}{n(1+x)} < \frac{1}{n(1+x)},$$

and $\lim_{n \to +\infty} R_n(x) = 0$. Therefore the series (6) converges to $\ln(1+x)$ for $-1 < x \leq 1$. In particular, if $x = 1$, an interesting special case results:

(10) $$\ln 2 = 1 - \frac{1}{2} + \frac{1}{3} - \frac{1}{4} + \cdots.$$

V. The binomial function $f(x) = (1+x)^m$, where m is any real number. The sequence $\{f^{(n)}(x)\}$ is

$$(1+x)^m, m(1+x)^{m-1}, m(m-1)(1+x)^{m-2}, \cdots,$$

and the sequence $\{f^{(n)}(0)\}$ is $1, m, m(m-1), \cdots$. Therefore the Maclaurin series is the **binomial series**

(11) $$1 + mx + \frac{m(m-1)}{2!} x^2 + \cdots + \binom{m}{n} x^n + \cdots,$$

where

(12) $$\binom{m}{n} = \frac{m(m-1)\cdots(m-n+1)}{n!},$$

and is called the **binomial coefficient** of x^n. If m is a nonnegative integer the binomial series (11) has only a finite number of nonzero terms and hence converges to $f(x) = (1+x)^m$ for all real x. Assume now that m is not a nonnegative integer. Since $\binom{m}{n} / \binom{m}{n+1} = \frac{n+1}{m-n} \to -1$, we know from Theorem III, § 801, that the radius of convergence is 1. The behavior of (11) at the endpoints of the interval of convergence depends on the value of m. We state the facts here, but defer the proofs to the Exercises (Ex. 36, § 811):

(i) $m \geq 0$: (11) converges absolutely for $x = \pm 1$.
(ii) $m \leq -1$: (11) diverges for $x = \pm 1$.
(iii) $-1 < m < 0$: (11) converges conditionally for $x = 1$, and diverges for $x = -1$.

The binomial series (11) represents the binomial function $(1+x)^m$ *throughout the interval of convergence.* In proving this we shall call upon both the Lagrange and the Cauchy forms of the remainder, depending on whether $0 < x \leq 1$ or $-1 \leq x < 0$:

Assume $0 < x \leq 1$. Then the Lagrange form of the remainder is

(13) $$R_n(x) = \binom{m}{n}(1 + \xi_n)^{m-n} x^n,$$

where $0 < \xi_n < x$. Since, for $n > m$, the inequality $1 + \xi_n > 1$ implies $(1 + \xi_n)^{m-n} < 1$, $|R_n(x)| \leq \left|\binom{m}{n} x^n\right|$ for sufficiently large n. Since $\binom{m}{n} x^n$ is the general term of the binomial series (11), it must tend toward zero whenever that series converges. Therefore $\lim_{n \to +\infty} R_n(x) = 0$ for $0 < x \leq 1$ whenever (11) converges, and we conclude that the binomial series represents the binomial function throughout the interval $0 \leq x < 1$, and also at the point $x = 1$ whenever the series converges there (that is, for $m > -1$).

Assume $-1 \leq x < 0$. Then the Cauchy form of the remainder is

(14) $$R_n(x) = nx(x - \xi_n)^{n-1}\binom{m}{n}(1 + \xi_n)^{m-n},$$

where $x < \xi_n < 0$. We rewrite (14):

(15) $$R_n(x) = n\binom{m}{n} x^n (1 + \xi_n)^{m-1} \left(\frac{1 - \frac{\xi_n}{x}}{1 + \xi_n}\right)^{n-1}.$$

For $-1 \leq x < \xi_n < 0$, $0 < 1 - \frac{\xi_n}{x} \leq 1 + \xi_n$ (check this), so that the last factor of (15) cannot exceed 1 for $n > 1$. Also, if $m > 1$, the inequality $1 + \xi_n < 1$ implies $(1 + \xi_n)^{m-1} < 1$, while if $m \leq 1$ the inequality $1 + \xi_n > 1 + x$ implies $(1 + \xi_n)^{m-1} < (1 + x)^{m-1}$. In any case, then, the last two factors of (15) remain bounded as $n \to +\infty$. The problem before us has been simplified, then, to showing (under the appropriate conditions) that

(16) $$\lim_{n \to +\infty} n\binom{m}{n} x^n = 0.$$

In case $-1 < x < 0$, the relation (16) is easily established by the ratio test (check the details in Ex. 27, § 811). Finally, if $x = -1$, the last item remaining in the proof is to show that $n\binom{m}{n} \to 0$ for $m > 0$. A technique for doing this is suggested in Exercise 37, § 811. We conclude that the binomial series represents the binomial function throughout the interval of convergence.

808. ELEMENTARY OPERATIONS WITH POWER SERIES

From results obtained in Chapter 7 for series of constants (§§ 702, 719) we have the theorem for power series (expressed here for simplicity in terms of powers of x, although similar formulations are valid for power series in powers of $(x - a)$):

Theorem. Addition, Subtraction, and Multiplication. *Let* $\sum_{n=0}^{+\infty} a_n x^n$ *and* $\sum_{n=0}^{+\infty} b_n x^n$ *be two power series representing the functions* $f_1(x)$ *and* $f_2(x)$, *respectively, within their intervals of convergence, and let* γ *be an arbitrary constant. Then* (i) *the power series* $\sum \gamma a_n x^n$ *represents the function* $\gamma f_1(x)$ *throughout the interval of convergence of* $\sum a_n x^n$; (ii) *the power series* $\sum (a_n \pm b_n) x^n$ *represents the function* $f_1(x) \pm f_2(x)$ *for all points common to the intervals of convergence of the two given power series; and* (iii) *if* $c_n \equiv \sum_{k=0}^{n} a_k b_{n-k}$, $n = 0$, 1, 2, $\cdots$, *then the power series* $\sum_{n=0}^{+\infty} c_n x^n$ *represents the function* $f_1(x) f_2(x)$ *for all points interior to both intervals of convergence of the two given power series* (cf. § 909).

In finding the Maclaurin or Taylor series for a given function, it is well to bear in mind the import of the uniqueness theorem (Theorem II, § 803) for power series. This means that the Maclaurin or Taylor series need not be obtained by direct substitution in the formulas defining those series. Any means that produces an appropriate power series representing the function automatically produces the Maclaurin or Taylor series.

Example 1. Since the series for e^x is $1 + x + \frac{x^2}{2!} + \cdots$, the series for e^{-x} is found by substituting $-x$ for x:
$$e^{-x} = 1 - x + \frac{x^2}{2!} - \frac{x^3}{3!} + \cdots.$$
This series expansion is valid for all real x, and is therefore the Maclaurin series for e^{-x}.

Example 2. The Maclaurin series for $\cos 2x$ is
$$\cos 2x = 1 - \frac{(2x)^2}{2!} + \frac{(2x)^4}{4!} - \frac{(2x)^6}{6!} + \cdots,$$
and is valid for all real x.

Example 3. The Maclaurin series for $\sinh x = \frac{1}{2}(e^x - e^{-x})$ is
$$\frac{1}{2}\left(1 + x + \frac{x^2}{2!} + \cdots\right) - \frac{1}{2}\left(1 - x + \frac{x^2}{2!} - \cdots\right) = x + \frac{x^3}{3!} + \frac{x^5}{5!} + \cdots,$$
and is valid for all real x.

Example 4. Find the Maclaurin series for $\sin\left(\frac{\pi}{6} + x\right)$.

Solution. Instead of proceeding in a routine manner, we expand $\sin\left(\frac{\pi}{6} + x\right)$ = $\sin\frac{\pi}{6} \cos x + \cos\frac{\pi}{6} \sin x$, and obtain:

$$\tfrac{1}{2}\left[1 - \tfrac{x^2}{2!} + \tfrac{x^4}{4!} - \cdots\right] + \tfrac{\sqrt{3}}{2}\left[x - \tfrac{x^3}{3!} + \tfrac{x^5}{5!} - \cdots\right]$$
$$= \tfrac{1}{2} + \tfrac{\sqrt{3}}{2}x - \tfrac{1}{2}\tfrac{x^2}{2!} - \tfrac{\sqrt{3}}{2}\tfrac{x^3}{3!} + \tfrac{1}{2}\tfrac{x^4}{4!} + \tfrac{\sqrt{3}}{2}\tfrac{x^5}{5!} - \cdots.$$

Example 5. The Taylor series for e^x at $x = a$ is most easily obtained by writing $e^x = e^a e^{x-a}$ and expanding the second factor by means of the Maclaurin series already established:

$$e^x = e^a\left[1 + (x-a) + \frac{(x-a)^2}{2!} + \frac{(x-a)^3}{3!} + \cdots\right].$$

Example 6. The Taylor series for $\sin x$ at $x = a$ can be found by writing
$$\sin x = \sin[a + (x-a)] = \sin a \cos(x-a) + \cos a \sin(x-a)$$
$$= \sin a\left[1 - \frac{(x-a)^2}{2!} + \frac{(x-a)^4}{4!} - \cdots\right]$$
$$+ \cos a\left[(x-a) - \frac{(x-a)^3}{3!} + \frac{(x-a)^5}{5!} - \cdots\right].$$

If $a = \frac{\pi}{6}$, the coefficients are those of Example 4.

Example 7. The Taylor series for $\ln x$ at $x = a > 0$ can be found by writing
$$\ln x = \ln[a + (x-a)] = \ln a + \ln\left[1 + \frac{x-a}{a}\right]$$
$$= \ln a + \frac{x-a}{a} - \frac{(x-a)^2}{2a^2} + \frac{(x-a)^3}{3a^3} - \cdots.$$

This is valid for $0 < x \leq 2a$.

Example 8. The Taylor series for x^m at $x = a > 0$ can be found by writing
$$x^m = [a + (x-a)]^m = a^m\left[1 + \left(\frac{x-a}{a}\right)\right]^m$$
$$= a^m\left[1 + m\left(\frac{x-a}{a}\right) + \frac{m(m-1)}{2!}\left(\frac{x-a}{a}\right)^2 + \cdots\right].$$

This is valid for $0 < x < 2a$; also at 0 for $m > 0$ and at $2a$ for $m > -1$.

Example 9. The Maclaurin series for $e^x \sin ax$ is found by multiplying the series:

$$\left[1 + x + \frac{x^2}{2!} + \frac{x^3}{3!} + \cdots\right]\left[ax - \frac{a^3 x^3}{3!} + \frac{a^5 x^5}{5!} - \cdots\right].$$

If we wish the terms of degree ≤ 5, we have
$$\left(1 + x + \frac{x^2}{2} + \frac{x^3}{6} + \frac{x^4}{24} + \cdots\right)\left(ax - \frac{a^3 x^3}{6} + \frac{a^5 x^5}{120} + \cdots\right)$$
$$= ax + ax^2 + \frac{a}{6}(3 - a^2)x^3 + \frac{a}{6}(1 - a^2)x^4 + \frac{a}{120}(5 - 10a^2 + a^4)x^5 + \cdots.$$

809. SUBSTITUTION OF POWER SERIES

Sometimes it is important to obtain a power series for a composite function, where each of the constituent functions has a known power series.

The most useful special case of a general theorem for such substitutions is the principal theorem of this section. We begin with a lemma:

★**Lemma.** *If the terms of the doubly infinite array*

(1)
$$c_{11}, c_{12}, c_{13}, \cdots$$
$$c_{21}, c_{22}, c_{23}, \cdots$$
$$c_{31}, c_{32}, c_{33}, \cdots$$
$$\cdots\cdots\cdots,$$

when arranged in any manner to form an infinite series, give an absolutely convergent series whose sum is C, then every row series $c_{m1} + c_{m2} + \cdots$ converges absolutely, and if $c_m \equiv \sum_{n=1}^{+\infty} c_{mn}$, then $\sum_{m=1}^{+\infty} c_m$ converges absolutely, with sum C.

★*Proof.* We first assume that every $c_{mn} \geqq 0$. Then any partial sum of terms in any row is bounded by C, so that each row series converges. Furthermore, the sum of the terms in the rectangle made up of the elements of the first M rows and the first n columns is bounded by C, so that when $n \to +\infty$ we have in the limit $\sum_{m=1}^{M} c_m \leqq C$, and hence $\sum_{m=1}^{+\infty} c_m \leqq C$. On the other hand, if $\epsilon > 0$, there exists a finite sequence of terms of (1) whose sum exceeds $C - \epsilon$. If M is the largest index of the rows from which these terms are selected, then $\sum_{m=1}^{M} c_m$ must also exceed $C - \epsilon$. Therefore $\sum_{m=1}^{+\infty} c_m \geqq C - \epsilon$ and, since ϵ is arbitrarily small, $\sum_{m=1}^{+\infty} c_m \geqq C$. In combination with a preceding inequality this gives $\sum_{m=1}^{+\infty} c_m = C$.

We now remove the assumption that $c_{mn} \geqq 0$, and (by splitting the entire array into nonnegative and nonpositive parts in the manner of § 716) immediately draw every conclusion stated in the lemma, except the equality $\sum_{m=1}^{+\infty} c_m = C$. But this equality follows from the fact that for any $\epsilon > 0$ there exists a number M such that the sum of the absolute values of all terms of (1) appearing below the Mth row is less than ϵ (check the details of this carefully in Ex. 28, § 811). This completes the proof.

Theorem I. Substitution. *Let*

(1) $$y = f(u) \equiv a_0 + a_1 u + a_2 u^2 + \cdots, \text{ and}$$
(2) $$u = g(x) \equiv b_1 x + b_2 x^2 + \cdots,$$

where both power series have positive radii of convergence. Then the composite function $h(x) \equiv f(g(x))$ is represented by a power series having a positive radius of convergence, obtained by substituting the entire series (2) for the quantity u in (1), expanding, and collecting terms:

§ 809] SUBSTITUTION OF POWER SERIES

(3) $\quad y = h(x) = a_0 + a_1(b_1x + b_2x^2 + \cdots) + a_2(b_1x + \cdots)^2 + \cdots$
$= a_0 + a_1b_1x + (a_1b_2 + a_2b_1^2) x^2$
$+ (a_1b_3 + 2a_2b_1b_2 + a_3b_1^3) x^3$
$+ (a_1b_4 + 2a_2b_1b_3 + a_2b_2^2 + 3a_3b_1^2b_2) x^4 + \cdots$

★*Proof.* We first exploit the continuity of $g(x)$ at $x = 0$ (Theorem I, § 803) and observe that if x belongs to a sufficiently small neighborhood of $x = 0$ and if u is defined in terms of x by (2), then u belongs to the interior of the interval of convergence of (1), and the expansions indicated by (1) and the first line of (3) are valid. It now remains to justify the removal of parentheses and the subsequent rearrangement in (3). For this purpose we shall insist on restricting x to so small an interval about $x = 0$ that $v = \sum_{n=1}^{+\infty} |b_n x^n|$ is inside the interval of convergence of (1). Then, as a consequence of the absolute convergence of all series concerned, we can apply the preceding lemma to the double array

(4)
$\quad a_0, \quad 0, \quad 0, \quad 0, \quad \cdots$
$\quad 0, \quad a_1b_1x, \quad a_1b_2x^2, \quad a_1b_3x^3, \quad \cdots$
$\quad 0, \quad 0, \quad a_2b_1^2x^2, \quad 2a_2b_1b_2x^3, \quad \cdots$
$\quad \cdots \cdots \cdots \cdots \cdots$

With a final appeal to § 718 the proof is complete. (Give precise details, particularly for the last step of the proof, in Ex. 29, § 811.)

Example 1. Find the terms of the Maclaurin series for $e^{\sin x}$, through terms of degree 5.

Solution. The series (1) and (2) are

$$y = f(u) = e^u = 1 + u + \frac{u^2}{2!} + \frac{u^3}{3!} + \frac{u^4}{4!} + \cdots,$$

$$u = g(x) = \sin x = x - \frac{x^3}{3!} + \frac{x^5}{5!} - \cdots.$$

The double array (4) becomes

$\quad 1, \quad 0, \quad 0, \quad 0, \quad 0, \quad 0, \quad \cdots$
$\quad 0, \quad x, \quad 0, \quad -\frac{x^3}{6}, \quad 0, \quad \frac{x^5}{120}, \quad \cdots$
$\quad 0, \quad 0, \quad \frac{x^2}{2}, \quad 0, \quad -\frac{x^4}{6}, \quad 0, \quad \cdots$
$\quad 0, \quad 0, \quad 0, \quad \frac{x^3}{6}, \quad 0, \quad -\frac{x^5}{12}, \quad \cdots$
$\quad 0, \quad 0, \quad 0, \quad 0, \quad \frac{x^4}{24}, \quad 0, \quad \cdots$
$\quad 0, \quad 0, \quad 0, \quad 0, \quad 0, \quad \frac{x^5}{120}, \quad \cdots$
$\quad \cdots \cdots \cdots \cdots \cdots$

Therefore the series sought is

$$1 + x + \frac{x^2}{2} - \frac{x^4}{8} - \frac{x^5}{15} + \cdots.$$

★The radius of convergence is infinite, since each basic series converges absolutely, everywhere.

Example 2. Find the terms of the Maclaurin series for $e^{\cos x}$ through terms of degree 6.

Solution. The Maclaurin series for $\cos x$ has a nonzero constant term. Therefore we write

$$e^{\cos x} = e^{1+g(x)} = e \cdot e^{g(x)},$$

where $g(x) = -\frac{x^2}{2} + \frac{x^4}{24} - \frac{x^6}{720} + \cdots$. We proceed as before, obtaining

$$e^{\cos x} = e\left\{1 + \left[-\frac{x^2}{2} + \frac{x^4}{24} - \cdots\right] + \frac{1}{2}\left[-\frac{x^2}{2} + \frac{x^4}{24} - \cdots\right]^2 + \cdots\right\}$$

$$= e\left\{1 + \left[-\frac{x^2}{2} + \frac{x^4}{24} - \frac{x^6}{720}\right] + \frac{1}{2}\left[\frac{x^4}{4} - \frac{x^6}{24}\right] + \frac{1}{6}\left[-\frac{x^6}{8}\right] + \cdots\right\}$$

$$= e\left(1 - \frac{x^2}{2} + \frac{x^4}{6} - \frac{31x^6}{720} + \cdots\right).$$

Before presenting more examples, let us record for future use a convenient device, which sometimes simplifies the work connected with the method of undetermined coefficients, illustrated in the second solution of Example 3, below:

Theorem II. *If* $\sum_{n=0}^{+\infty} a_n x^n$ *represents a function* $f(x)$ *in a neighborhood* I *of* $x = 0$, *then* (i) *if* $f(x)$ *is an even function in* I *the power series* $\sum a_n x^n$ *consists of only even degree terms, and* (ii) *if* $f(x)$ *is an odd function in* I *the power series* $\sum a_n x^n$ *consists of only odd degree terms.*

Proof. We shall give the details only for part (i) (cf. Ex. 32, § 811, for (ii)). Since

$$f(x) = a_0 + a_1 x + a_2 x^2 + a_3 x^3 + \cdots,$$
$$f(-x) = a_0 - a_1 x + a_2 x^2 - a_3 x^3 + \cdots,$$

and therefore

$$f(x) - f(-x) = 2a_1 x + 2a_3 x^3 + 2a_5 x^5 + \cdots.$$

If $f(x)$ is even, the function $f(x) - f(-x)$ is identically 0 in I, and every coefficient in its power series expansion must vanish, by the uniqueness theorem (Theorem II, § 803).

Example 3. Find the terms of the Maclaurin series for $\sec x$ through terms of degree 8, and ★ determine an interval within which the series converges.

First Solution. Since $\sec x = \frac{1}{\cos x} = \frac{1}{1 - g(x)}$, where $g(x) = \frac{x^2}{2!} - \frac{x^4}{4!} +$

§ 810]   INTEGRATION AND DIFFERENTIATION

···, the Maclaurin series for sec x is found by substituting the power series for $g(x)$ in the series
$$\frac{1}{1-u} = 1 + u + u^2 + u^3 + \cdots.$$
We therefore collect terms from
$$1 + \left[\frac{x^2}{2!} - \frac{x^4}{4!} + \frac{x^6}{6!} - \frac{x^8}{8!}\right] + \left[\frac{x^2}{2!} - \frac{x^4}{4!} + \frac{x^6}{6!}\right]^2 + \left[\frac{x^2}{2!} - \frac{x^4}{4!}\right]^3 + \left[\frac{x^2}{2!}\right]^4,$$
and get
$$\sec x = 1 + \frac{x^2}{2} + \frac{5x^4}{24} + \frac{61x^6}{720} + \frac{277x^8}{8064} + \cdots.$$

★The above procedures have been validated for any interval such that $\frac{x^2}{2!} + \frac{x^4}{4!} + \frac{x^6}{6!} + \cdots < 1$, or $\cosh x < 2$. Therefore the series found for sec x converges to sec x within (at least) the interval $(-1.3, 1.3)$. Actually, as is easily shown by the theory of analytic functions of a complex variable, the interval of convergence is $\left(-\frac{\pi}{2}, \frac{\pi}{2}\right)$.

Second Solution. Since sec x is an even function and is represented by its Maclaurin series (cf. the first solution), its Maclaurin series must have the form
$$\sec x = a_0 + a_2 x^2 + a_4 x^4 + a_6 x^6 + \cdots,$$
all coefficients of odd degree terms being 0. We form the product of this series and that of cos x and have
$$1 = (a_0 + a_2 x^2 + a_4 x^4 + \cdots)\left(1 - \frac{x^2}{2!} + \frac{x^4}{4!} - \cdots\right)$$
$$= a_0 + \left(-\frac{a_0}{2!} + a_2\right)x^2 + \left(\frac{a_0}{4!} - \frac{a_2}{2!} + a_4\right)x^4$$
$$+ \left(-\frac{a_0}{6!} + \frac{a_2}{4!} - \frac{a_4}{2!} + a_6\right)x^6 + \left(\frac{a_0}{8!} - \frac{a_2}{6!} + \frac{a_4}{4!} - \frac{a_6}{2!} + a_8\right)x^8 + \cdots.$$
Equating corresponding coefficients, we have the recursion formulas $a_0 = 1$, $a_2 = \frac{a_0}{2!}$, $a_4 = \frac{a_2}{2!} - \frac{a_0}{4!}$, $\cdots$, from which we can evaluate the coefficients, one after the other. The result is the same as that of the first solution.

810. INTEGRATION AND DIFFERENTIATION OF POWER SERIES

As useful adjuncts to the methods of the two preceding sections, we state three theorems, the second and third of which are proved in the next chapter:

Theorem I. *A power series* $\sum_{n=0}^{+\infty} a_n (x - a)^n$ *and its derived series* $\sum_{n=1}^{+\infty} n a_n (x - a)^{n-1}$ *have the same radius of convergence.*

Proof. For simplicity of notation, we assume that $a = 0$. In the first place, since $|a_n| \leq n |a_n|$, for $n = 1, 2, \cdots$, if the derived series converges absolutely for some particular $x = x_0$, then $x_0(\sum na_n x_0^{n-1}) = \sum na_n x_0^n$ converges absolutely and $\sum a_n x_0^n$ also converges absolutely. That is, the radius of convergence of the derived series can be no larger than that of the original series. On the other hand, it can be no smaller, for let $0 < \alpha < \beta$, and assume that the original series converges absolutely for $x = \beta$. We shall show that the derived series converges absolutely for $x = \alpha$. This will conclude the proof (why?). Our contention follows from the fact that $n\, a_n \alpha^n = O(|a_n \beta^n|)$ for the nonzero terms of the series, and this fact in turn follows by means of the ratio test from the limit:

$$\lim_{n \to +\infty} \frac{n\, a_n \alpha^n}{a_n \beta^n} = \lim_{n \to +\infty} n\left(\frac{\alpha}{\beta}\right)^n = \lim_{n \to +\infty} \frac{n}{e^{n \ln \frac{\beta}{\alpha}}} = 0.$$

For an alternative proof, see Ex. 38, § 811.

Theorem II. *If $f(x)$ is represented by a power series $\sum a_n(x - a)^n$ in its interval of convergence I, and if α and β are any two points interior to I, then the series can be integrated term by term:*

$$\int_\alpha^\beta f(x)\, dx = \sum_{n=0}^{+\infty} a_n \int_\alpha^\beta (x - a)^n\, dx = \sum_{n=0}^{+\infty} \frac{a_n}{n+1} [(\beta - a)^{n+1} - (\alpha - a)^{n+1}].$$

(Cf. Ex. 11, § 911.)

Theorem III. *A function $f(x)$ represented by a power series $\sum a_n(x - a)^n$ in the interior of its interval of convergence is differentiable there, and its derivative is represented there by the derived series:*

$$f'(x) = \sum_{n=1}^{+\infty} n\, a_n (x - a)^{n-1}.$$

(Cf. Ex. 12, § 911.)

Example 1. The series

$$\frac{1}{1+t} = 1 - t + t^2 - t^3 + \cdots$$

has radius of convergence $R = 1$. Therefore, for $|x| < 1$, integration from 0 to x gives

$$\ln(1+x) = x - \frac{x^2}{2} + \frac{x^3}{3} - \cdots.$$

Example 2. The series

$$(1+t)^{-\frac{1}{2}} = 1 - \frac{1}{2}t + \frac{1\cdot 3}{2\cdot 4}t^2 - \frac{1\cdot 3\cdot 5}{2\cdot 4\cdot 6}t^3 + \cdots$$

has radius of convergence $R = 1$. Therefore the same is true for

$$(1-t^2)^{-\frac{1}{2}} = 1 + \frac{1}{2}t^2 + \frac{1\cdot 3}{2\cdot 4}t^4 + \frac{1\cdot 3\cdot 5}{2\cdot 4\cdot 6}t^6 + \cdots.$$

From this, by integrating from 0 to x, where $|x| < 1$, we find

$$\operatorname{Arcsin} x = x + \frac{1}{2}\frac{x^3}{3} + \frac{1\cdot 3}{2\cdot 4}\frac{x^5}{5} + \frac{1\cdot 3\cdot 5}{2\cdot 4\cdot 6}\frac{x^7}{7} + \cdots.$$

Example 3. Find the terms of the Maclaurin series for $\tan x$ through terms of degree 9. (Cf. Ex. 9, § 816.)

First Solution. Multiply the power series for $\sin x$ and $\sec x$ (Example 3, § 809):

$$\tan x = \left(x - \frac{x^3}{3!} + \frac{x^5}{5!} - \cdots\right)\left(1 + \frac{x^2}{2} + \frac{5x^4}{24} + \cdots\right)$$

$$= x + \frac{x^3}{3} + \frac{2x^5}{15} + \frac{17x^7}{315} + \frac{62x^9}{2835} + \cdots.$$

Second Solution. Since $\tan x$ is an odd function its Maclaurin series (which exists and represents the function, by the first solution) has the form

$$\tan x = a_1 x + a_3 x^3 + a_5 x^5 + \cdots.$$

Furthermore, since $\cos x \tan x = \sin x$, the coefficients of the product series for $\cos x \tan x$ must be identically equal to those of the sine series:

$$a_1 = 1, \quad -\frac{a_1}{2} + a_3 = -\frac{1}{6}, \quad \frac{a_1}{24} - \frac{a_3}{2} + a_5 = \frac{1}{120}, \cdots.$$

The result of solving these recursion formulas is the same as that found in the first solution.

Third Solution. As in the second solution, let

$$\tan x = a_1 x + a_3 x^3 + a_5 x^5 + a_7 x^7 + \cdots.$$

Differentiation and use of the identity $\sec^2 x = 1 + \tan^2 x$ give

$$a_1 + 3a_3 x^2 + 5a_5 x^4 + 7a_7 x^6 + 9a_9 x^8 + \cdots$$
$$= 1 + [a_1 x + a_3 x^3 + a_5 x^5 + a_7 x^7 + \cdots]^2$$
$$= 1 + a_1^2 x^2 + 2a_1 a_3 x^4 + (2a_1 a_5 + a_3^2)x^6 + \cdots.$$

Equating corresponding coefficients produces the recursion formulas $a_1 = 1$, $3a_3 = a_1^2$, $5a_5 = 2a_1 a_3$, $\cdots$, and the result of the first solution. This is by far the shortest of the three methods.

Example 4. Find the exact sum of the series

$$\frac{1}{1!\cdot 3} + \frac{1}{2!\cdot 4} + \frac{1}{3!\cdot 5} + \cdots + \frac{1}{n!\,(n+2)} + \cdots.$$

Solution. Start with the series

$$e^x = 1 + x + \frac{x^2}{2!} + \frac{x^3}{3!} + \cdots.$$

Multiply by x and integrate from 0 to 1:

$$\int_0^1 x e^x\, dx = \frac{1}{2} + \frac{1}{3} + \frac{1}{2!\cdot 4} + \frac{1}{3!\cdot 5} + \cdots.$$

Since the value of the integral is 1, the original series converges to $\frac{1}{2}$.

811. EXERCISES

In Exercises 1-12, find the Maclaurin series for the given function.

1. $\cosh x = \dfrac{e^x - e^{-x}}{2}$.
2. $\dfrac{1}{1+x^2}$.

3. $\operatorname{Arctan} x = \displaystyle\int_0^x \frac{dt}{1+t^2}$.
4. $\sqrt{e^x}$.

5. $\cos x^2$.
6. $\ln(2 + 3x)$.
7. $(1 - x^4)^{-\frac{1}{2}}$.
8. $\sqrt{4 + x}$.
9. $\ln \dfrac{1 + x}{1 - x}$.
10. $\int_0^x \dfrac{\sin t}{t}\, dt$.
11. $\int_0^x e^{-t^2}\, dt$.
12. $\int_0^x \sin t^2\, dt$.

In Exercises 13-18, find the terms of the Maclaurin series for the given function through terms of the specified degree.

13. $\dfrac{1}{1 + e^x}$; 5.
14. $e^x \cos x$; 5.
15. $\tanh x$; 7.
16. $e^{\tan x}$; 5.
17. $\ln \dfrac{\sin x}{x}$; 6.
18. $\ln \cos x$; 8.

In Exercises 19-22, find the Taylor series for the given function at the specified value of $x = a$.

19. $\cos x$; a.
20. $\cos x$; $\dfrac{\pi}{4}$.
21. $\ln x$; e.
22. x^m; 1.

In Exercises 23-26, find the exact sum of the infinite series.

23. $1 + 2x + 3x^2 + 4x^3 + \cdots + (n + 1)x^n + \cdots$.
24. $\dfrac{x^3}{1 \cdot 3} - \dfrac{x^5}{3 \cdot 5} + \dfrac{x^7}{5 \cdot 7} - \dfrac{x^9}{7 \cdot 9} + \cdots$.
25. $1 - 2x^2 + 3x^4 - 4x^6 + \cdots$.
26. $\dfrac{1}{1 \cdot 2 \cdot 2} - \dfrac{1}{2 \cdot 3 \cdot 2^2} + \dfrac{1}{3 \cdot 4 \cdot 2^3} - \dfrac{1}{4 \cdot 5 \cdot 2^4} + \cdots$.

27. Prove relation (16), § 807, for $-1 < x < 0$.
★28. Check the final details of the proof of the lemma, § 809.
★29. Give the details requested at the end of the proof of Theorem I, § 809.
30. Prove part (ii) of Theorem II, § 809.
★31. By reasoning analogous to that used in part IV, § 807, for the function $\ln(1 + x)$, show that
$$\operatorname{Arctan} x = x - \frac{x^3}{3} + \frac{x^5}{5} - \frac{x^7}{7} + \cdots,$$
for $|x| \leq 1$. In particular, derive the formula
$$\frac{\pi}{4} = 1 - \frac{1}{3} + \frac{1}{5} - \frac{1}{7} + \cdots.$$

★32. Prove that if
$$f(x) = a_k x^k + a_{k+1} x^{k+1} + \cdots,$$
where k is a nonnegative integer and $a_k \neq 0$, within some neighborhood of $x = 0$, then $f(x)$ and x^k are of the same order of magnitude as $x \to 0$. What is the order of magnitude of $1/f(x)$?

★33. Prove that if
$$f(x) = \frac{a_0 + a_1x + a_2x^2 + \cdots}{x^k},$$
where k is a nonnegative integer and $a_0 \neq 0$, within some deleted neighborhood of $x = 0$, then $f(x)$ and x^{-k} are of the same order of magnitude as $x \to 0$. What is the order of magnitude of $1/f(x)$?

★34. Prove that e is irrational. *Hint:* Assume $e = p/q$, where p and q are positive integers, and write $e = S + R$, where
$$S = 1 + \frac{1}{1!} + \frac{1}{2!} + \cdots + \frac{1}{q!},$$
$$R = \frac{1}{(q+1)!} + \frac{1}{(q+2)!} + \cdots,$$
and show that the integer $eq!$ has the form $Sq! + Rq!$, where $Sq!$ is an integer and $Rq!$ is not.

★35. The numbers B_n in the Maclaurin series
$$(1) \qquad \frac{x}{e^x - 1} = \sum_{n=0}^{+\infty} \frac{B_n x^n}{n!},$$
called **Bernoulli numbers**, play an important role in the theory of infinite series (for instance, the coefficients in the Maclaurin series for $\tan x$ are expressible in terms of the Bernoulli numbers). (For a few series expansions involving Bernoulli numbers, see Ex. 9, §816.) By equating the function x and the product of the series (1) and the Maclaurin series for $e^x - 1$, obtain the recursive formula ($n \geq 2$):
$$(2) \qquad \binom{n}{0}B_0 + \binom{n}{1}B_1 + \binom{n}{2}B_2 + \cdots + \binom{n}{n-1}B_{n-1} = 0.$$
Evaluate $B_0, B_1, \cdots, B_{10}$. Prove that $B_{2n+1} = 0$ for $n = 1, 2, \cdots$. *Hint:* What kind of a function is $\frac{x}{e^x - 1} + \frac{1}{2}x$?

★36. Prove the facts given in Part V, §807, regarding convergence at the endpoints of the interval of convergence of the binomial series. *Hints:* Consider only values of $n > m$. Then $\left|\frac{a_n}{a_{n+1}}\right| = \frac{n+1}{n-m}$. If $m > 0$ use Raabe's test. If $m \leq -1$, $\left|\frac{a_n}{a_{n+1}}\right| \leq 1$ and $a_n \not\to 0$. For $-1 < m < 0$ and $x = -1$, use Raabe's test, the terms being ultimately of one sign. For $-1 < m < 0$ and $x = 1$, the series is alternating, $|a_{n+1}| < |a_n|$, and the problem is to show that $a_n \to 0$. To do this, let k be a positive integer greater than $1/(m+1)$, and show that $a_n{}^k \to 0$, as $n \to +\infty$, by establishing the convergence of $\sum a_n{}^k$ with the aid of Raabe's test.

★37. Prove that $\lim_{n \to +\infty} n\binom{m}{n} = 0$, for $m > 0$. *Hint:* Use the method suggested in the hints of Ex. 36, k being a positive integer $> 1/m$.

★★38. Prove Theorem I, §810, by using limits superior and the formula of Note 4, §801.

812. INDETERMINATE EXPRESSIONS

It is frequently possible to find a simple evaluation of an indeterminate expression by means of Maclaurin or Taylor series. This can often be expedited by the "big O" or "order of magnitude" concepts. (Cf. Ex. 10, § 713.)

Example 1. Find $\lim\limits_{x \to 0} \dfrac{1 - \cos x}{x^2}$.

Solution. Since
$$\cos x = 1 - \frac{x^2}{2!} + \frac{x^4}{4!} - \cdots = 1 - \frac{x^2}{2} + O(x^4), \quad \frac{1 - \cos x}{x^2} = \frac{1}{2} + O(x^2) \to \frac{1}{2}.$$

Example 2. Find $\lim\limits_{n \to +\infty} n\{\ln(n + 1) - \ln n\}$.

Solution. We write the expression
$$\ln(n + 1) - \ln n = \ln\left(1 + \frac{1}{n}\right) = \frac{1}{n} + O\left(\frac{1}{n^2}\right).$$

Therefore $n\{\ln(n + 1) - \ln n\} = 1 + O\left(\dfrac{1}{n}\right) \to 1$.

Example 3. Find $\lim\limits_{x \to 0} \dfrac{\ln(1 + x)}{e^{2x} - 1}$.

Solution. $\dfrac{\ln(1 + x)}{e^{2x} - 1} = \dfrac{x + O(x^2)}{2x + O(x^2)} = \dfrac{1 + O(x)}{2 + O(x)}$
$$= [1 + O(x)][\tfrac{1}{2} + O(x)] = \tfrac{1}{2} + O(x) \to \tfrac{1}{2}.$$

813. COMPUTATIONS

The principal techniques most commonly used for computations by means of power series have already been established. In any such computation one wishes to obtain some sort of definite range within which the sum of a series must lie. The usual tools for this are (*i*) an estimate provided by a dominating series (Theorem I, § 720), (*ii*) an estimate provided by the integral test (Theorems II and III, § 720), (*iii*) the alternating series estimate (Theorem, § 715), and (*iv*) some form of the remainder in Taylor's Formula (§ 804).

Another device is to seek a different series to represent a given quantity. For instance, some logarithms can be computed more efficiently with the series

(1) $\qquad \ln\dfrac{1 + x}{1 - x} = 2\left[x + \dfrac{x^3}{3} + \dfrac{x^5}{5} + \dfrac{x^7}{7} + \cdots\right]$

(cf. Ex. 9, § 811) than with the Maclaurin series for $\ln(1 + x)$.

Finally, one must not forget that infinite series are not the only means for computation. We include in Example 5 one illustration of the use of an approximation (Simpson's Rule) to a definite integral.

§ 813] COMPUTATIONS 265

We illustrate some of these techniques in the following examples:

Example 1. Compute the sine of one radian to five decimal places.

Solution. $\sin 1 = 1 - \frac{1}{3!} + \frac{1}{5!} - \frac{1}{7!} = 0.84147$ with an error of less than $1/9! < 0.000003$.

Example 2. Compute $\sqrt{89}$ to six decimal places.

Solution. We approximate first by 9.4, whose square is 88.36, and write
$$\sqrt{89} = \sqrt{88.36 + .64} = 9.4\sqrt{1 + 0.64/88.36}.$$
Computing $(1 + x)^{\frac{1}{2}}$, where $x = 0.64/88.36$, we have, from the binomial series,
$$1 + \frac{1}{2}x - \frac{1 \cdot 1}{2 \cdot 4}x^2 = 1.00361991,$$
with an error of $R_2(x) = \frac{1 \cdot 1 \cdot 3}{2 \cdot 4 \cdot 6}\xi^3$, from (13), § 807. This error is between 0 and 0.000000003, so that $\sqrt{1 + x}$ is between 1.00361991 and 1.00361992, and $\sqrt{89}$ is between 9.4340271 and 9.4340273. To six decimal places, $\sqrt{89} = 9.434027$.

Example 3. Compute e to five decimal places.

Solution. Substitution of $x = 1$ in the Maclaurin series for e^x gives
$$e = 1 + \frac{1}{2!} + \frac{1}{3!} + \cdots + \frac{1}{9!} + R_{10}(1), = 2.718281 + \frac{e^\xi}{10!}$$
where $R_{10}(1)$ is between 0 and $\frac{e}{10!} < 0.000001$. Therefore e lies between 2.718281 and 2.718282, and is equal to 2.71828, to five decimal places.

Example 4. Compute ln 2 by use of series (1), to five decimal places.

Solution. Solving $\frac{1 + x}{1 - x} = 2$ for x, we have $x = \frac{1}{3}$, and
$$\ln 2 = 2[\tfrac{1}{3} + \tfrac{1}{3}(\tfrac{1}{3})^3 + \tfrac{1}{5}(\tfrac{1}{3})^5 + \cdots] = 0.6931468 + [\tfrac{2}{13}(\tfrac{1}{3})^{13} + \cdots].$$
The remainder, in brackets, by the fact that it is dominated by a convergent geometric series with ratio 1/9, is less than $[\tfrac{2}{13}(\tfrac{1}{3})^{13}] \div [1 - \tfrac{1}{9}]$ (cf. Theorem I, § 720). This quantity, in turn, is less than 0.0000001. Therefore ln 2 is between 0.693146 and 0.693148. To five places, ln 2 = 0.69315. (Cf. Example 3, § 720.)

Example 5. Compute π to five significant digits.

First Solution. One method is to evaluate $\frac{\pi}{4}$ by use of the series for Arctan x (cf. Exs. 3 and 31, § 811) with $x = 1$:
$$\frac{\pi}{4} = 1 - \frac{1}{3} + \frac{1}{5} - \frac{1}{7} + \cdots.$$
Although this series converges very slowly, its terms can be combined in the manner of Example 3, § 720, to produce a manageable series for computation.

POWER SERIES [§ 814]

However, if use is made of such a trigonometric identity as

$$\text{Arctan } 1 = \text{Arctan } \tfrac{1}{2} + \text{Arctan } \tfrac{1}{3},$$

two much more rapidly converging series are obtained:

$$\frac{\pi}{4} = \left[\frac{1}{2} - \frac{1}{3}\left(\frac{1}{2}\right)^3 + \frac{1}{5}\left(\frac{1}{2}\right)^5 - \cdots\right] + \left[\frac{1}{3} - \frac{1}{3}\left(\frac{1}{3}\right)^3 + \frac{1}{5}\left(\frac{1}{3}\right)^5 - \cdots\right].$$

(Complete the details in Ex. 21, § 814.)

The student may be interested in finding other trigonometric identities which give even more rapidly converging series. Two such identities are Arctan $1 =$ 2 Arctan $\tfrac{1}{3}$ + Arctan $\tfrac{1}{7}$ and Arctan $1 = 4$ Arctan $\tfrac{1}{5}$ − Arctan $\tfrac{1}{239}$.

Second Solution. Since

$$\frac{\pi}{4} = \int_0^1 \frac{dx}{1 + x^2} = \text{Arctan } 1,$$

an estimate of π is given by use of Simpson's Rule (Ex. 23, § 503). For simplicity, we take $n = 10$ and complete the table:

x_k	$1 + x_k^2$	y_k	$y_k, 2y_k, 4y_k$
0	1	1.000000	1.000000
0.1	1.01	0.990099	3.960396
0.2	1.04	0.961538	1.923076
0.3	1.09	0.917431	3.669724
0.4	1.16	0.862069	1.724138
0.5	1.25	0.800000	3.200000
0.6	1.36	0.735294	1.470588
0.7	1.49	0.671141	2.684564
0.8	1.64	0.609756	1.219512
0.9	1.81	0.552486	2.209944
1.0	2.00	0.500000	0.500000
			23.561942

Therefore $\dfrac{\pi}{4} = \dfrac{23.561942}{30} = 0.785398$ (rounding off to six places to allow for previous round-off errors). Since the fourth derivative of $(1 + x^2)^{-1}$ is numerically less than 24 for $0 < x < 1$ (cf. Ex. 22, § 1409), the error estimate is less than $(24/180)(.1)^4 < 0.000013$. Thus, to four decimal places, $\pi = 3.1416$.

814. EXERCISES

In Exercises 1-10, evaluate the limit by use of Maclaurin or Taylor series.

1. $\displaystyle\lim_{x \to 0} \frac{1 - \cos x}{\tan^2 x}.$

2. $\displaystyle\lim_{x \to 0} \frac{\sin x - x}{\tan x - x}.$

3. $\displaystyle\lim_{x \to 0} \frac{\text{Arcsin } x - x - \dfrac{x^3}{6}}{\text{Arctan } x - x + \dfrac{x^3}{3}}$

4. $\displaystyle\lim_{x \to 0} \frac{2 - x - 2\sqrt{1 - x}}{x^2}.$

§ 815] ANALYTIC FUNCTIONS 267

5. $\lim_{x\to 0}\left[\dfrac{\sin x}{x^3}-\dfrac{1}{x^4}+\dfrac{1}{6x^4}-\dfrac{1}{120x^2}\right]$.
6. $\lim_{x\to 1}\dfrac{(x-1)\ln x}{\sin^2(x-1)}$.

7. $\lim_{x\to 0}\dfrac{(1+x)^{\frac{5}{3}}-(1-x)^{\frac{2}{3}}}{x}$.
8. $\lim_{x\to 0}\dfrac{x\tan x}{\sqrt{1-x^2}-1}$.

9. $\lim_{x\to 1}\dfrac{e^{2x}-e^2}{\ln x}$.

10. $\lim_{n\to +\infty}\ln n\left[(n+1)\ln\dfrac{n+1}{n}-1\right]$.

In Exercises 11–20, compute the given quantity to the specified number of decimal places, by use of Maclaurin series.

11. e^2, 5.
12. $\sqrt{e}$, 5.
13. $\ln 3$, 4.
14. $\cos\frac{1}{2}$, 4.
15. $\sqrt[5]{10}$, 5.
16. $\tan 0.1$, 5.
17. $\int_0^{0.5}\dfrac{\sin t}{t}\,dt$, 4.
18. $\int_0^{0.1}\dfrac{\ln(1+x)}{x}\,dx$, 4.
19. $\int_0^{0.5}\sqrt{1-x^3}\,dx$, 4.
20. $\int_0^1 e^{-x^2}\,dx$, 4.

21. Complete the computation details in the first solution for Example 5, § 813.

★815. ANALYTIC FUNCTIONS

The property that a function may have of being represented by a Taylor series is of such basic importance in analysis that it is given a special name, *analyticity*, as specified in the definition:

Definition. *A function $f(x)$ is **analytic** at $x = a$ if and only if it has a Taylor series at $x = a$ which represents the function in some neighborhood of $x = a$.*†

We state here some of the more important theorems concerning analytic functions, leaving the proofs to the reader (Exs. 3–6, § 816):

Theorem I. *An analytic function of an analytic function is analytic.*

Theorem II. *The sum, difference, product, and quotient of analytic functions are analytic if division by 0 is not involved.*

Theorem III. *A function represented by a power series is analytic in the interior of the interval of convergence.*

Corollary. *If a function is analytic at a point, it is automatically analytic in a neighborhood of that point.*

† For functions of a complex variable this is only one of several definitions of analyticity, but for functions of a real variable this is the only practical definition in common use.

Theorem IV. *If two functions analytic on an interval I are identical on any subinterval, then they are identical throughout I. If a function analytic on an interval can be extended analytically to a larger interval (that is, if the domain of definition can be extended to a larger interval containing the original), then this extension is unique.*

NOTE 1. Theorem I, § 803 can be reworded: An analytic function has derivatives of all orders. Example 2, § 805, shows that the converse is not true. The function of that Example is not analytic at $x = 0$.

NOTE 2. Example 1, § 805, shows that polynomials are everywhere analytic.

NOTE 3. Theorems III and IV imply that if a function $f(x)$ is analytic throughout the interior I of the interval of convergence of a power series $\sum$, and if $\sum$ represents $f(x)$ in any subinterval (however small), it must represent $f(x)$ throughout I. They imply, then, that any identity involving analytic functions, such as $\sin x = \cos(\frac{1}{2}\pi - x)$ (cf. Example 2, below) is universally valid as soon as it has been established for any given interval. They also imply (for example) that a catenary (Ex. 35, § 419) and a parabola cannot coincide, even over an extremely short interval.

Example 1. Prove that the following functions are analytic for the indicated values of x:

$$e^x, \sin x, \cos x, \text{ for all real } x;$$
$$\ln x, x^m, \text{ for all positive } x.$$

Solution. For the functions e^x, $\sin x$, $\ln x$, and x^m, the result is implicit in Examples 5-8, § 808. Cf. Ex. 19, § 811, for $\cos x$.

Example 2. Establish the equation

(1) $$(1 + x)^m = 1 + mx + \frac{m(m-1)}{2!}x^2 + \cdots,$$

for $-1 < x < 1$ by use of the Lagrange form of the remainder only.

Solution. The Lagrange form of the remainder (§ 807) is

(2) $$R_n(x) = \binom{m}{n}(1 + \xi_n)^{m-n}x^n.$$

The ratio test shows that $R_n(x)$ tends toward 0 as a limit for any x such that $0 < x < 1$ (cf. § 807). By Example 1, the left-hand member of (1) is analytic for $x > -1$, and by Theorem III, the right-hand member of (1) is analytic for $-1 < x < 1$. Therefore, since the two members of (1) are identical for $0 < x < 1$, they must be identical for $-1 < x < 1$.

★816. EXERCISES

★1. Prove that $\tan x$, $\cot x$, $\sec x$, and $\csc x$ are analytic where defined.

★2. Prove that if I is a given interval, then there exists no polynomial identical with e^x throughout I.

★3. Prove Theorem I, § 815. *Hint:* Let $f(u)$ be analytic at $u = a$ and let $g(x)$ be analytic at $x = b$, where $a = g(b)$. Let $y = f(u) = a_0 + a_1(u - a) +$

$a_2(u-a)^2 + \cdots$ and $u = g(x) = a + b_1(x-b) + b_2(x-b)^2 + \cdots$. The proposition is a consequence of the substitution theorem (Theorem I) of § 809, with a translation of axes and a change in notation.

★4. Prove Theorem II, § 815.

★5. Prove Theorem III, § 815. *Hint:* Assume without loss of generality that the point under consideration is $x = 0$, lying within the interval of convergence of $f(x) = \sum a_n(x-a)^n$. The series $\sum |a_n| \cdot |x-a|^n$ converges in a neighborhood I of $x = 0$. Let $\epsilon > 0$ belong to I. Then $\sum |a_n| \{|x| + \epsilon\}^n$ converges. Dirichlet's rearrangement theorem (Theorem II) of § 718 shows that
$$\{|a_0| + |a_1|\epsilon + |a_2|\epsilon^2 + \cdots\} + \{|a_1| + 2|a_2|\epsilon + 3|a_3|\epsilon^2 + \cdots\}x + \cdots$$
converges. Hence $\{a_0 + \cdots\} + \{a_1 + \cdots\}x + \cdots$ converges. Show that it converges to $f(x)$ by use of the Lemma, § 809.

★6. Prove Theorem IV, § 815.

★★7. A theorem regarding inverse functions, easily proved in the theory of analytic functions of a complex variable, but whose proof by use of techniques of real variable theory only is much more difficult (cf. K. Knopp, *Theory and Application of Infinite Series* (London, Blackie, 1928)) follows: *If $y = f(x)$ is analytic and has a nonzero derivative at $x = a$, then in some neighborhood of $b = f(a)$, a single-valued inverse function $x = \phi(y)$ is defined and is analytic.* This means that if $y = b + a_1(x-a) + a_2(x-a)^2 + \cdots$ in a neighborhood of a, and $a_1 \neq 0$, then it is possible to solve this equation for x in terms of a power series in powers of $(y-b)$. Prove that in a proof of this theorem it may be assumed without loss of generality that $a = b = 0$ and $a_1 = 1$. (Do not attempt to prove the theorem.)

★★8. Obtain a set of recursion formulas for the coefficients b_n, where
$$x = \phi(y) = b_1 y + b_2 y^2 + b_3 y^3 + \cdots,$$
if $\phi(y)$ is the inverse function of
$$y = f(x) = x + a_2 x^2 + a_3 x^3 + \cdots.$$
(Cf. Ex. 7.) Use this method to obtain the first few terms of the Maclaurin series for $x = \tan y$, the inverse function of $y = \text{Arctan } x = x - \frac{1}{3}x^3 + \frac{1}{5}x^5 - \frac{1}{7}x^7 + \cdots$. (Cf. Example 3, § 810, Ex. 3, § 811.) *Hint:* Substitute the known series for y in the unknown series (or conversely) and equate coefficients in $x = \phi(f(x))$ (or $y = f(\phi(y))$). (Cf. Ex. 9.)

★★9. Establish the series expansions, for x near 0 (cf. Ex. 35, § 811):

$x \coth x = \sum_{n=0}^{+\infty} \frac{B_{2n}}{(2n)!} (2x)^{2n}$, $\qquad x \cot x = \sum_{n=0}^{+\infty} \frac{(-1)^n B_{2n}}{(2n)!} (2x)^{2n}$,

$\tanh x = \sum_{n=1}^{+\infty} \frac{B_{2n}}{(2n)!} 2^{2n}(2^{2n}-1) x^{2n-1}$, $\quad \tan x = \sum_{n=1}^{+\infty} \frac{(-1)^{n+1} B_{2n}}{(2n)!} 2^{2n}(2^{2n}-1) x^{2n-1}$,

$x \operatorname{csch} x = \sum_{n=0}^{+\infty} \frac{B_{2n}}{(2n)!} (2-2^{2n}) x^{2n}$, $\qquad x \csc x = \sum_{n=0}^{+\infty} \frac{(-1)^n B_{2n}}{(2n)!} (2-2^{2n}) x^{2n}$.

Hints: $x \coth x = x + 2x(e^{2x}-1)^{-1}$, $\tanh x = 2 \coth 2x - \coth x$, $\operatorname{csch} x = \frac{1}{2} \coth \frac{1}{2}x - \frac{1}{2} \tanh \frac{1}{2}x$, $\tan x = \cot x - 2 \cot 2x$, and $\csc x = \frac{1}{2} \tan \frac{1}{2}x + \frac{1}{2} \cot \frac{1}{2}x$. Exploit the relationship between the Maclaurin series for $\sinh x$ and $\sin x$, and for $\cosh x$ and $\cos x$.

9

★Uniform Convergence

★901. UNIFORM CONVERGENCE OF SEQUENCES

Let

(1) $$S_1(x), S_2(x), \cdots, S_n(x), \cdots$$

be a sequence of functions defined on a set A. We say that this sequence of functions **converges on A** in case, for every fixed x of A, the sequence of constants $\{S_n(x)\}$ converges. Assume that (1) converges on A, and define

(2) $$S(x) \equiv \lim_{n \to +\infty} S_n(x).$$

Then the rapidity with which $S_n(x)$ approaches $S(x)$ can be expected to depend (rather heavily) on the value of x.

Let us write down explicitly the analytic formulation of (2):

Corresponding to any point x in A and any $\epsilon > 0$, there exists a number $N = N(x, \epsilon)$ (dependent on both x and ϵ) such that $n > N$ implies

$$|S_n(x) - S(x)| < \epsilon.$$

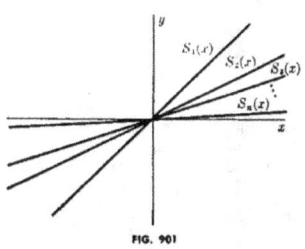

FIG. 901

§901] UNIFORM CONVERGENCE OF SEQUENCES

Example 1. If $S_n(x) \equiv \frac{x}{n}$, and if $A = (-\infty, +\infty)$, then $\lim_{n \to +\infty} S_n(x)$ exists and is equal to 0 for every x in A. (Cf. Fig. 901.) If $\epsilon > 0$, the inequality $|S_n(x) - S(x)| < \epsilon$ is equivalent to $n > |x|/\epsilon$. Therefore $N = N(x, \epsilon)$ can be defined to be $N(x, \epsilon) \equiv |x|/\epsilon$. We can see how N must depend on x. For instance, if we were asked, "How large must n be in order that $|S_n(x) - S(x)| = \frac{x}{n} < 0.001$?" we should be entitled to reply, "Tell us first how large x is." If $x = 0.001$, the second term of the sequence provides the desired degree of approximation, if $x = 1$, we must proceed at least past the 1000th term, and if $x = 1000$, we must choose $n > 1{,}000{,}000$.

If it is possible to find N, in the definition of (2), as a function of ϵ alone, independent of x, then a particularly powerful type of convergence occurs. This is prescribed in the definition:

Definition. *A sequence of functions $\{S_n(x)\}$, defined on a set A, **converges uniformly on A** to a function $S(x)$ defined on A, this being written*

$$S_n(x) \rightrightarrows S(x),$$

if and only if corresponding to $\epsilon > 0$ there exists a number $N = N(\epsilon)$, dependent on ϵ alone and not on the point x, such that $n > N$ implies

$$|S_n(x) - S(x)| < \epsilon$$

for every x in A. (Fig. 902.)

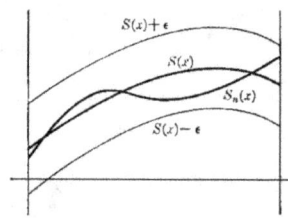

FIG. 902

Let us contrast the definitions of *convergence* and *uniform convergence*. (Cf. § 307.) The most obvious distinction is that convergence is defined *at a point*, whereas uniform convergence is defined *on a set*. These concepts are also distinguished by the order in which things happen. In the case of convergence we have (*i*) the point x, (*ii*) the number $\epsilon > 0$, and (*iii*) the number N, which depends on both x and ϵ. In the case of uniform convergence we have (*i*) the number $\epsilon > 0$, (*ii*) the number N, which depends only on ϵ, and (*iii*) the point x.

Example 2. If $S_n(x) = \frac{x}{n}$, and if A is the interval $(-1000, 1000)$, then the sequence $\{S_n(x)\}$ converges uniformly to 0 on the set A. The function $N(\epsilon)$ can be chosen: $N(\epsilon) = 1000/\epsilon$. Then
$$n > N(\epsilon) \text{ and } |x| \leq 1000$$
imply
$$|S_n(x) - S(x)| = \frac{|x|}{n} < \frac{1000}{1000/\epsilon} = \epsilon.$$

A quick and superficial reading of Example 1 might lead one to believe, since N seems to depend so thoroughly on x, that even when the set A is restricted as it is in Example 2 the convergence could not be uniform. We have seen, however, that it is. Are we really *sure*, now, that the convergence in Example 1 fails to be uniform? In order to show this, we formulate the negation of uniform convergence (give the proof in Ex. 25, § 904):

Negation of Uniform Convergence. *A sequence of functions $\{S_n(x)\}$, defined on a set A, fails to converge uniformly on A to a function $S(x)$ defined on A if and only if there exists a positive number ϵ having the property that for any number N there exist a positive integer $n > N$ and a point x of A such that $|S_n(x) - S(x)| \geq \epsilon$.*

Example 3. Show that the convergence of Example 1 is not uniform on the interval $(-\infty, +\infty)$.

Solution. We can take $\epsilon = 1$. If N is an arbitrary number, let us choose any $n > N$, and hold this n fixed. Next we pick $x > n$. Then (for this pair n and x),
$$S_n(x) - S(x) = \frac{x}{n} > \frac{n}{n} = 1 = \epsilon.$$

NOTE. Any convergent sequence of constants (constant functions) converges uniformly on any set.

Example 4. Show that the sequence $\left\{\frac{x}{e^{nx}}\right\}$ converges uniformly on $[0, +\infty)$.

Solution. For a fixed value of n, the nonnegative function $f(x) = xe^{-nx}$ has a maximum value on $[0, +\infty)$ given by $x = 1/n$, and equal to $1/ne$. Therefore the sequence approaches 0 uniformly on $[0, +\infty)$.

Example 5. Show that the sequence $\left\{\frac{nx}{e^{nx}}\right\}$ converges uniformly on $[\alpha, +\infty)$, for any $\alpha > 0$, but not uniformly on $(0, +\infty)$.

Solution. For $x \geq \alpha$, and a fixed $n > 1/\alpha$, the function $f(x) = nxe^{-nx}$ is monotonically decreasing, with a maximum value of $n\alpha e^{-n\alpha}$. This quantity is independent of x and approaches 0 as $n \to +\infty$. On the entire interval $(0, +\infty)$, the function $f(x) = nxe^{-nx}$ has a maximum value given by $x = 1/n$ and equal to $1/e$. The convergence is therefore not uniform.

★902. UNIFORM CONVERGENCE OF SERIES

Let
$$(1) \quad u_1(x) + u_2(x) + \cdots + u_n(x) + \cdots$$
be a series of functions defined on a set A, and let
$$S_n(x) = u_1(x) + \cdots + u_n(x).$$

We say that this series of functions **converges on A** in case the sequence $\{S_n(x)\}$ converges on A. The series (1) **converges uniformly on A** if and only if the sequence $\{S_n(x)\}$ converges uniformly on A.

NOTE. Any convergent series of constants (constant functions) converges uniformly on any set.

Corresponding to the condition $a_n \to 0$ which is necessary for the convergence of the series of constants $\sum a_n$, we have:

Theorem. *If the series $\sum u_n(x)$ converges uniformly on a set A, then the general term $u_n(x)$ converges to 0 uniformly on A.*

Proof. By the triangle inequality, if $S_n(x) = u_1(x) + \cdots + u_n(x)$ and $S(x) = \sum_{n=1}^{+\infty} u_n(x)$,
$$|u_n(x)| = |S_n(x) - S_{n-1}(x)| = |[S_n(x) - S(x)] + [S(x) - S_{n-1}(x)]|$$
$$\leq |S_n(x) - S(x)| + |S_{n-1}(x) - S(x)|.$$

Let $\epsilon > 0$ be given. If N is chosen such that $n > N - 1$ implies
$$|S_n(x) - S(x)| < \tfrac{1}{2}\epsilon$$
for all x in A, then $n > N$ implies $|u_n(x)| < \epsilon$ for all x in A.

Example. Show that the Maclaurin series for e^x converges uniformly on a set A if and only if A is bounded.

Solution. If the set A is bounded, it is contained in some interval of the form $[-\alpha, \alpha]$. Using the Lagrange form of the remainder in Taylor's formula for e^x at $x = a = 0$, we have (with the standard notation) for any x,
$$|S_n(x) - S(x)| = |R_n(x)| = \frac{e^{\xi x}}{n!}|x|^n \leq \frac{e^\alpha}{n!}\alpha^n.$$

Since $\lim_{n \to +\infty} \frac{e^\alpha}{n!}\alpha^n = 0$ and $\frac{e^\alpha}{n!}\alpha^n$ is independent of x, the uniform convergence on A is established.

If the set A is unbounded, we can show that $\sum_{n=0}^{+\infty} \frac{x^n}{n!}$ fails to converge uniformly on A by showing that the general term does not approach 0 uniformly on A. This we do with the aid of the Negation of Uniform Convergence formulated in § 901: letting ϵ be 1 and n be any fixed positive integer, we can find an x in A such that $|x|^n > n!$

★903. DOMINANCE AND THE WEIERSTRASS M-TEST

The role of dominance in uniform convergence of series of functions is similar to that of dominance in convergence of series of constants (§ 708).

Definition. *The statement that a series of functions $\sum v_n(x)$ **dominates** a series of functions $\sum u_n(x)$ on a set A means that all terms are defined on A and that for any x in A $|u_n(x)| \leq v_n(x)$ for every positive integer n.*

Theorem I. Comparison Test. *Any series of functions $\sum u_n(x)$ dominated on a set A by a series of functions $\sum v_n(x)$ which is uniformly convergent on A is uniformly convergent on A.*

Proof. From previous results for series of constants, we know that the series $\sum u_n(x)$ converges for every x in A. If $u(x) \equiv \sum u_n(x)$ and $v(x) \equiv \sum v_n(x)$, we have (cf. Theorem I, § 716):

$$|[u_1(x) + u_2(x) + \cdots + u_n(x)] - u(x)| = |u_{n+1}(x) + u_{n+2}(x) + \cdots|$$
$$\leq v_{n+1}(x) + v_{n+2}(x) + \cdots = |[v_1(x) + v_2(x) + \cdots + v_n(x)] - v(x)|.$$

If $\epsilon > 0$ and if $N = N(\epsilon)$ is such that for $n > N$,

$$|[v_1(x) + \cdots + v_n(x)] - v(x)| < \epsilon$$

for all x in A, then $|[u_1(x) + \cdots + u_n(x)] - u(x)| < \epsilon$ for all x in A.

Corollary. *A series of functions converges uniformly on a set whenever its series of absolute values converges uniformly on that set.*

Example 1. Since the Maclaurin series for $\sin x$ and $\cos x$ are dominated on any set by the series obtained by substituting $|x|$ for x in the Maclaurin series for e^x, these series for $\sin x$ and $\cos x$ converge uniformly on any bounded set. Neither converges uniformly on an unbounded set. (Cf. the Example, § 902.)

Since any convergent series of constants converges uniformly on any set, we have as a special case of Theorem I the extremely useful test for uniform convergence due to Weierstrass:

Theorem II. Weierstrass M-Test. *If $\sum u_n(x)$ is a series of functions defined on a set A, if $\sum M_n$ is a convergent series of nonnegative constants, and if for every x of A,*

$$(1) \qquad |u_n(x)| \leq M_n, n = 1, 2, \cdots,$$

then $\sum u_n(x)$ converges uniformly on A.

Example 2. Prove that the series $1 + e^{-x} \cos x + e^{-2x} \cos 2x + \cdots$ converges uniformly on any set that is bounded below by a positive constant.

Solution. If $\alpha > 0$ is a lower bound of the set A, then for any x in A, $|e^{-nx} \cos nx| \leq e^{-nx} \leq e^{-n\alpha}$. By the Weierstrass M-test, with $M_n = e^{-n\alpha}$, the given series converges uniformly on A (the series $\sum e^{-n\alpha}$ is a geometric series with common ratio $e^{-\alpha} < 1$). (Cf. Ex. 26, § 904.)

★904. EXERCISES

In Exercises 1-10, use the Weierstrass M-test to show that the given series converges uniformly on the given set.

★1. $\sum n^2 x^n$; $[-\frac{1}{3}, \frac{1}{3}]$.
★2. $\sum \frac{x^n}{n^2}$; $[-1, 1]$.

★3. $\sum \frac{x^{2n-1}}{(2n-1)!}$; $(-1000, 2000)$.
★4. $\sum \frac{x^n}{n^n}$; $x = 1, 2, \cdots, K$.

★5. $\sum \frac{\sin nx}{n^2 + 1}$; $(-\infty, +\infty)$.
★6. $\sum \frac{\sin nx}{e^n}$; $(-\infty, +\infty)$.

★7. $\sum ne^{-nx}$; $[\alpha, +\infty)$, $\alpha > 0$.
★8. $\sum \frac{e^{nx}}{5^n}$; $(-\infty, \alpha]$, $\alpha < \ln 5$.

★9. $\sum \left(\frac{\ln x}{x}\right)^n$; $[1, +\infty)$.
★10. $\sum (x \ln x)^n$; $(0, 1]$.

In Exercises 11-20, show that the sequence whose general term is given converges uniformly on the first of the two given intervals, and that it converges but not uniformly on the second of the two given intervals. The letter η denotes an arbitrarily small positive number.

★11. $\sin^n x$; $[0, \frac{1}{2}\pi - \eta]$; $[0, \frac{1}{2}\pi]$.
★12. $\sqrt[n]{\sin x}$; $[\eta, \frac{1}{2}\pi]$; $(0, \frac{1}{2}\pi]$.

★13. $\frac{x}{x+n}$; $[0, b]$; $[0, +\infty)$.
★14. $\frac{x+n}{n}$; $[a, b]$; $(-\infty, +\infty)$.

★15. $\frac{nx}{1 + nx}$; $[\eta, +\infty)$; $(0, +\infty)$.
★16. $\frac{nx}{1 + n^2 x^2}$; $[\eta, +\infty)$; $(0, +\infty)$.

★17. $\frac{\ln(1 + nx)}{n}$; $[0, b]$; $[0, +\infty)$.
★18. $n^2 x^2 e^{-nx}$; $[\eta, +\infty)$; $(0, +\infty)$.

★19. $\frac{x^n}{1 + x^n}$; $[0, 1 - \eta]$; $[0, 1)$.
★20. $\frac{\sin nx}{1 + nx}$; $[\eta, +\infty)$; $(0, +\infty)$.

In Exercises 21-24, show that the given series converges uniformly on the first of the two given intervals, and that it converges but not uniformly on the second of the two given intervals. The letter η denotes an arbitrarily small positive number.

★21. $\sum x^n$; $[0, 1 - \eta]$; $[0, 1)$.
★22. $\sum \frac{x^n}{n}$; $[0, 1 - \eta]$; $[0, 1)$.

★23. $\sum \frac{1}{n^x}$; $[1 + \eta, +\infty)$; $(1, +\infty)$.
★24. $\sum \frac{nx}{e^{nx}}$; $[\eta, +\infty)$; $(0, +\infty)$.

★25. Prove the Negation of Uniform Convergence, § 901.

★26. Prove that the convergence of the series in Example 2, § 903, is not uniform on $(0, +\infty)$.

★27. Prove that if a sequence of functions converges uniformly on a set, then any subsequence converges uniformly on that set. Show by an example that the converse is false; that is, show that a sequence which converges nonuniformly on a set may have a uniformly convergent subsequence. (Cf. Ex. 28.)

★28. Prove that if a monotonic sequence of functions ($S_{n+1}(x) \leq S_n(x)$ for each x) converges on a set and contains a uniformly convergent subsequence, then the convergence of the original sequence is uniform. (Cf. Ex. 27.)

★29. Prove that if a sequence converges uniformly on a set A, then it converges uniformly on any set contained in A.

★30. Prove that if a sequence converges uniformly on a set A and on a set B, then it converges uniformly on the combined set made up of points that belong either to A or to B (or to both). Show that any convergent sequence converges uniformly on every finite set. Show that if a sequence converges uniformly on an open interval (a, b) and converges at each endpoint, then the sequence converges uniformly on the closed interval $[a, b]$.

★31. Prove that if $S_n(x) \rightrightarrows S(x)$ on a set A, and if $S(x)$ is bounded on A, then the functions $S_n(x)$ ultimately uniformly bounded on A: there exist constants K and N such that $|S_n(x)| \leq K$ for all $n > N$ and all x in A.

★32. Prove that if $S_n(x) \rightrightarrows S(x)$ on a set A, and if each $S_n(x)$ is bounded on A (there exists a sequence of constants $\{K_n\}$ such that $|S_n(x)| \leq K_n$ for each n and all x in A), then the functions $S_n(x)$ are uniformly bounded on A. (Cf. Ex. 31.)

★33. If $f(x)$ is defined on $[\frac{1}{2}, 1]$ and continuous at 1, prove that $\{x^n f(x)\}$ converges for every x of $[\frac{1}{2}, 1]$, and that this convergence is uniform if and only if $f(x)$ is bounded and $f(1) = 0$.

★34. If two sequences $\{f_n(x)\}$ and $\{g_n(x)\}$ converge uniformly on a set A, prove that their sum $\{f_n(x) + g_n(x)\}$ also converges uniformly on A. Show by an example that their product $\{f_n(x)g_n(x)\}$ need not converge uniformly on A. Prove, however, that if both original sequences are uniformly bounded (cf. Ex. 31), then the sequence $\{f_n(x)g_n(x)\}$ converges uniformly. What happens if only one of the sequences is uniformly bounded?

★35. Construct a series $\sum u_n(x)$ such that (i) $\sum u_n(x)$ converges on a set A, (ii) $u_n(x) \rightrightarrows 0$ on A, and (iii) $\sum u_n(x)$ does not converge uniformly on A.

★36. Prove the ratio test for uniform convergence: If $u_n(x)$ are bounded nonvanishing functions on a set A, and if there exist constants N and ρ, where $\rho < 1$, such that $\left|\frac{u_{n+1}(x)}{u_n(x)}\right| \leq \rho$ for all $n > N$ and all x in A, then $\sum u_n(x)$ converges uniformly on A. Use this test to prove that for any fixed x such that $|x| < 1$, the binomial series ((11), § 807), considered as a series of functions of m, converges uniformly for $|m| \leq M$, where M is fixed.

In Exercises 37-42, find an appropriate function $N(\epsilon)$, as prescribed in the Definition of uniform convergence, § 901, for the sequence of the given Exercise (and the set specified in that exercise).

★37. Ex. 11.   ★38. Ex. 13.   ★39. Ex. 15.

★40. Ex. 17. *Hint:* The inequality $\ln(1 + nb) < \epsilon n$ is equivalent to $1 + nb < e^{\epsilon n}$; and $1 + \epsilon n + \frac{\epsilon^2 n^2}{2} < e^{\epsilon n}$. Thus require $1 + nb < 1 + \frac{\epsilon^2 n^2}{2}$.

★41. Ex. 18. *Hint:* $n^2 x^2 e^{-nx}$ is a decreasing function of x for $x > \frac{2}{n}$. First require $n > \frac{2}{\eta}$. Then guarantee $n^2 \eta^2 < \epsilon e^{n\eta}$ (cf. Ex. 40).

★42. Ex. 19.

★43. Prove that the Cauchy condition for convergence of an infinite series $\sum a_n$ (Ex. 17, § 717) can be expressed in the form: $a_n + a_{n+1} + \cdots + a_{n+p} \rightrightarrows 0$, *uniformly in* $p > 0$. (Cf. Exs. 18, 20, § 717.)

★★44. Explain what you would mean by saying that a series of functions is **uniformly summable** $(C, 1)$ on a given set. Prove that a uniformly convergent series is uniformly summable, but not conversely. Generalize to uniform summability (C, r). Give examples. (Cf. Exs. 15-16, § 717.)

★45. Prove the **Cauchy criterion** for uniform convergence of a sequence of functions: *A sequence* $\{S_n(x)\}$ *converges uniformly on a set* A *if and only if corresponding to* $\epsilon > 0$ *there exists* $N = N(\epsilon)$ *such that* $m > N$ *and* $n > N$ *imply* $|S_m(x) - S_n(x)| < \epsilon$ *for every* x *in* A. *Hint:* By § 302, $S(x) \equiv \lim_{n \to +\infty} S_n(x)$ exists. Assume that the convergence is not uniform, use the Negation of § 901, N as prescribed above for $\frac{1}{2}\epsilon$, and the triangle inequality, for fixed $n > N$ and arbitrary $m > N$:
$$|S_n(x) - S(x)| \leq |S_n(x) - S_m(x)| + |S_m(x) - S(x)|.$$

★46. State and prove the Cauchy criterion for uniform convergence of a series of functions. (Cf. Ex. 45; above, Exs. 17-19, § 717.)

★★47. Show that $\sum_{n=1}^{+\infty} \frac{x}{1 + n^2 x^2}$ converges for all x, that it converges uniformly for $x \geq \eta > 0$, and that it does not converge uniformly in any neighborhood of $x = 0$. *Hint:* Use the Cauchy criterion (Ex. 46), together with the ideas used in establishing the integral test (§ 706) to show that uniform convergence of the given series is equivalent to showing that
$$\int_m^n \frac{x\, dt}{1 + x^2 t^2} = \text{Arctan } mx - \text{Arctan } nx$$
is uniformly small for x in a given set, and large m and n.

★★48. Show that $\sum_{n=1}^{+\infty} \frac{x}{n + n^2 x^2}$ converges uniformly for all x. *Hint:* See Ex. 47, and write $x \ln [(mnx^2 + m)/(mnx^2 + n)] = x \ln [1 + 1/nx^2] - x \ln [1 + 1/mx^2]$. Show that $x \ln [1 + 1/nx^2] < \frac{1}{2} \ln 5n^{-\frac{1}{2}}$.

★★49. Prove the **Abel test** for uniform convergence: *If the partial sums of a series* $\sum_{n=1}^{+\infty} u_n(x)$ *of functions are uniformly bounded on a set* A (Ex. 31), *and if* $\{v_n(x)\}$ *is a monotonically decreasing sequence of nonnegative functions converging uniformly to* 0 *on* A, *then the series* $\sum_{n=1}^{+\infty} u_n(x) v_n(x)$ *converges uniformly on* A. (Cf. Ex. 22, § 717.)

★★50. Adapt and prove the Abel test of Exercise 23, § 717, for uniform convergence.

★★51. Prove the alternating series test for uniform convergence: If $\{v_n(x)\}$ is a monotonically decreasing sequence of nonnegative functions converging uniformly to 0 on a set A, then $\sum_{n=1}^{+\infty} (-1)^{n-1} v_n(x)$ converges uniformly on A. (Cf.

278   UNIFORM CONVERGENCE   [§ 905

★★52. Show that $\sum_{n=1}^{+\infty} \frac{(-1)^{n-1}}{n+x}$ converges uniformly for $x \geqq 0$, but does not converge absolutely for any x. (Cf. Ex. 51.)

★★53. Show that $\sum_{n=1}^{+\infty} \frac{(-1)^n x^{2n}}{1+x^{2n}}$ converges uniformly on $[-1+\eta, 1-\eta], \eta > 0$. (Cf. Ex. 51.)

★★54. Show that $\sum_{n=1}^{+\infty} \frac{(-1)^n}{n^x}$ converges uniformly for $x \geqq \eta > 0$, but not uniformly for $x > 0$. (Cf. Ex. 51.)

★★55. Show that $\sum_{n=1}^{+\infty} (-1)^n \frac{x^2+nx}{n^2}$ converges uniformly on every bounded set. Where does it converge absolutely? (Cf. Ex. 51.)

★★56. Show that $\sum_{n=1}^{+\infty} (-1)^n \frac{x}{n} e^{-\frac{x^2}{n^2}}$ converges uniformly on every bounded set, but absolutely only for $x = 0$. (Cf. Ex. 51.)

★905. UNIFORM CONVERGENCE AND CONTINUITY

The example $S_n(x) \equiv x^{\frac{1}{2n-1}}$, $-1 \leqq x \leqq 1$ (Fig. 903), shows that the limit of a sequence of continuous functions need not be continuous. The limit in this case is the signum function (Example 1, § 206), which is discontinuous at $x = 0$. As we shall see, this is possible because the convergence is not uniform. For example, let us set $\epsilon = \frac{1}{2}$. Then however large n may be there will exist a positive number x so small that the inequality

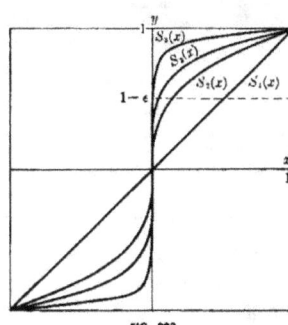

FIG. 903

§ 906]   CONVERGENCE AND INTEGRATION   279

$|S_n(x) - S(x)| < \epsilon$, which is equivalent to $1 - x^{\frac{1}{2n-1}} < \frac{1}{2}$, or $x^{\frac{1}{2n-1}} > \frac{1}{2}$, fails.

In case of uniform convergence we have the basic theorem:

Theorem. *If $S_n(x)$ is continuous at every point of a set A, $n = 1, 2, \cdots$, and if $S_n(x) \rightrightarrows S(x)$ on A, then $S(x)$ is continuous at every point of A.*

Proof. Let a be any point of A, and let $\epsilon > 0$. We first choose a positive integer N such that $|S_N(x) - S(x)| < \frac{1}{3}\epsilon$ for every x of A. Holding N fixed, and using the continuity of $S_N(x)$ at $x = a$, we can find a $\delta > 0$ such that $|x - a| < \delta$ implies $|S_N(x) - S_N(a)| < \frac{1}{3}\epsilon$. (The number δ apparently depends on *both* N and ϵ, but since N is determined by ϵ, δ is a function of ϵ alone—for the fixed value $x = a$.) Now we use the triangle inequality:

$$|S(x) - S(a)| \leq |S(x) - S_N(x)| + |S_N(x) - S_N(a)| + |S_N(a) - S(a)|.$$

Then $|x - a| < \delta$ implies

$$|S(x) - S(a)| < \tfrac{1}{3}\epsilon + \tfrac{1}{3}\epsilon + \tfrac{1}{3}\epsilon = \epsilon,$$

and the proof is complete.

Corollary. *If $f(x) = \sum u_n(x)$, where the series converges uniformly on a set A, and if every term of the series is continuous on A, then $f(x)$ is continuous on A.* (Ex. 19, § 908.)

Example. Show that the function $f(x)$ defined by the series $\sum_{n=0}^{+\infty} e^{-nx} \cos nx$ of Example 2, § 903, is continuous on the set $(0, +\infty)$—that is, for positive x.

Solution. Let $x = a > 0$ be given, and let $\alpha = \frac{1}{2}a$. Then the given series converges uniformly on $[\frac{1}{2}a, +\infty)$, by Example 2, § 903. Therefore $f(x)$ is continuous on $[\frac{1}{2}a, +\infty)$ and, in particular, at $x = a$.

★906. UNIFORM CONVERGENCE AND INTEGRATION

The example illustrated in Figure 904 shows that the limit of the integral (of the general term of a convergent sequence of functions) need not equal the integral of the limit (function). The function $S_n(x)$ is defined to be $2n^2x$ for $0 \leq x \leq 1/2n$, $2n(1 - nx)$ for $1/2n \leq x \leq 1/n$, and 0 for $1/n \leq x \leq 1$. The limit function $S(x)$ is identically 0 for $0 \leq x \leq 1$. For every n, $\int_0^1 S_n(x)\,dx = \frac{1}{2}$ (the integral is the area of a triangle of altitude n and base $1/n$), but $\int_0^1 S(x)\,dx = 0$. Again, the reason that this kind of misbehavior is possible, is that the convergence is not uniform (cf. Ex. 22, § 908). (For another example of the same character, where the functions $S_n(x)$ are defined by single analytic formulas, see Ex. 25, § 908.)

280   UNIFORM CONVERGENCE   [§ 906]

In case of uniform convergence we have the theorem:

Theorem. *If $S_n(x)$ is integrable on $[a, b]$ for $n = 1, 2, \cdots$, and if $S_n(x) \rightrightarrows S(x)$ on $[a, b]$, then $S(x)$ is integrable on $[a, b]$ and*

(1) $$\lim_{n \to +\infty} \int_a^b S_n(x)\, dx = \int_a^b \lim_{n \to +\infty} S_n(x)\, dx = \int_a^b S(x)\, dx.$$

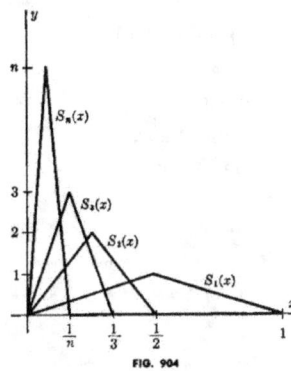

FIG. 904

Proof. We shall first prove that $S(x)$ is integrable on $[a, b]$. The idea is to approximate $S(x)$ by a particular $S_N(x)$, then to squeeze $S_N(x)$ between two step-functions (§ 502), and finally to construct two new step-functions that squeeze $S(x)$. Accordingly, for a given $\epsilon > 0$, we find an index N such that $|S_N(x) - S(x)| < \dfrac{\epsilon}{4(b-a)}$ for $a \leq x \leq b$. Since $S_N(x)$ is integrable on $[a, b]$, there must exist step-functions $\sigma_1(x)$ and $\tau_1(x)$ such that $\sigma_1(x) \leq S_N(x) \leq \tau_1(x)$ on $[a, b]$, and $\int_a^b [\tau_1(x) - \sigma_1(x)]\, dx < \tfrac{1}{2}\epsilon$. Define the new step-functions:

$$\sigma(x) \equiv \sigma_1(x) - \frac{\epsilon}{4(b-a)},\ \tau(x) \equiv \tau_1(x) + \frac{\epsilon}{4(b-a)}.$$

Then, for $a \leq x \leq b$,

$$\sigma(x) < \sigma_1(x) + [S(x) - S_N(x)] \leq S(x),$$
$$\tau(x) > \tau_1(x) + [S(x) - S_N(x)] \geq S(x),$$

§ 907]   CONVERGENCE AND DIFFERENTIATION   281

and
$$\int_a^b [\tau(x) - \sigma(x)] \, dx = \int_a^b \left\{ [\tau_1(x) - \sigma_1(x)] + \frac{\epsilon}{2(b-a)} \right\} dx < \tfrac{1}{2}\epsilon + \tfrac{1}{2}\epsilon = \epsilon.$$

By Theorem II, § 502, $S(x)$ is integrable on $[a, b]$.

We now wish to establish the limit (1). Since the difference between the integrals of two integrable functions is the integral of their difference, (1) is equivalent to

(2) $$\int_a^b [S_n(x) - S(x)] \, dx \to 0.$$

Finally, by Exercises 2 and 38, § 503,
$$\left| \int_a^b [S_n(x) - S(x)] \, dx \right| \leq \int_a^b |S_n(x) - S(x)| \, dx,$$

so that if n is chosen so large that $|S_n(x) - S(x)| < \epsilon/(b-a)$ on $[a, b]$, then $\left| \int_a^b [S_n(x) - S(x)] \, dx \right| < [\epsilon/(b-a)](b-a) = \epsilon$, and the proof is complete.

Corollary. *If $f(x) \equiv \sum u_n(x)$, where the series converges uniformly on $[a, b]$, and if every term of the series is integrable on $[a, b]$, then $f(x)$ is integrable on $[a, b]$, and the series can be integrated term by term:*
$$\int_a^b f(x) \, dx = \int_a^b \sum u_n(x) \, dx = \sum \int_a^b u_n(x) \, dx.$$

(Ex. 20, § 908.)

Example. If $f(x)$ is the function defined by the series $\sum_{n=0}^{+\infty} e^{-nx} \cos nx$ of Example 2, § 903, and if $0 < a < b$, then
$$\int_a^b f(x) \, dx = \sum_{n=0}^{+\infty} \int_a^b e^{-nx} \cos nx \, dx.$$

★907. UNIFORM CONVERGENCE AND DIFFERENTIATION

The example $S_n(x) \equiv \dfrac{x}{1 + nx^2}$, $-1 \leq x \leq 1$ (Fig. 905), shows that even with uniform convergence of differentiable functions to a differentiable function, the limit of the derivatives may not equal the derivative of the limit. Since the maximum and minimum points of $S_n(x)$ are $(n^{-\frac{1}{2}}, \tfrac{1}{2}n^{-\frac{1}{2}})$ and $(-n^{-\frac{1}{2}}, -\tfrac{1}{2}n^{-\frac{1}{2}})$, respectively, $\{S_n(x)\}$ converges uniformly to the function $S(x)$ that is identically 0 on $[-1, 1]$. However,
$$\lim_{n \to +\infty} S_n'(x) = \lim_{n \to +\infty} \frac{1 - nx^2}{(1 + nx^2)^2} = \begin{cases} 1 & \text{if } x = 0, \\ 0 & \text{if } x \neq 0, \end{cases}$$

whereas $S'(x)$ is identically 0.

282   UNIFORM CONVERGENCE   [§ 907

The clue to the problem lies in the uniform convergence of the sequence of *derivatives*, $\{S_n'(x)\}$. We state the basic theorem, and give a proof under an additional assumption (that of continuity of the derivatives), leaving to Exercise 38, § 908, a proof of the general theorem, with hints.

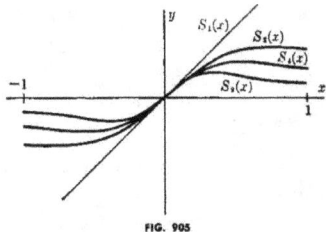

FIG. 905

Theorem. *Assume:*

(i) $S_n(x)$ is differentiable on $[a, b]$, for $n = 1, 2, \cdots$;
(ii) $\{S_n(x)\}$ converges for some point x_0 of $[a, b]$;
(iii) $\{S_n'(x)\}$ converges uniformly on $[a, b]$.

Then

(iv) $\{S_n(x)\}$ converges uniformly on $[a, b]$;
(v) if $S(x) \equiv \lim_{n \to +\infty} S_n(x)$, $S'(x)$ exists on $[a, b]$;
(vi) $S'(x) = \lim_{n \to +\infty} S_n'(x)$.

Proof if every $S_n'(x)$ is continuous. Under the assumption of continuity of $S_n'(x)$ we have, from the Fundamental Theorem of Integral Calculus (§ 504), $S_n(x) - S_n(x_0) = \int_{x_0}^{x} S_n'(t)\, dt$. Therefore, by the Theorem of § 906, the sequence $\{S_n(x) - S_n(x_0)\}$ converges for every x of $[a, b]$. Therefore the limit $\lim_{n \to +\infty} S_n(x)$ exists for every x of $[a, b]$ (why?), and if we define $S(x) \equiv \lim_{n \to +\infty} S_n(x)$ and $T(x) \equiv \lim_{n \to +\infty} S_n'(x)$, we have (§ 906):

(1) $S(x) = \int_{x_0}^{x} T(t)\, dt - S(x_0).$

Since $T(x)$ is continuous on $[a, b]$ (§ 905), the function $S(x)$ is differentiable with derivative $S'(x) = T(x)$ (Theorem I, § 504). Finally, since

$$|S_n(x) - S(x)| = \left| \int_{x_0}^{x} [S_n'(t) - S'(t)]\, dt + [S_n(x_0) - S(x_0)] \right|$$

$$\leq \left| \int_{x_0}^{x} |S_n'(t) - S'(t)|\, dt \right| + |S_n(x_0) - S(x_0)|,$$

the convergence of $\{S_n(x)\}$ is uniform (cf. Ex. 23, § 908).

Corollary. *If $\sum u_n(x)$ is a series of differentiable functions on $[a, b]$, convergent at one point of $[a, b]$, and if the derived series $\sum u_n'(x)$ converges uniformly on $[a, b]$, then the original series converges uniformly on $[a, b]$ to a differentiable function whose derivative is represented on $[a, b]$ by the derived series.* (Ex. 21, § 908.)

Example. Show that if $f(x)$ is the function defined by the series $\sum_{n=0}^{+\infty} e^{-nx} \cos nx$ of Example 2, § 903, then

$$f'(x) = -\sum_{n=0}^{+\infty} n e^{-nx} (\cos nx + \sin nx),$$

for every $x > 0$.

Solution. Let x be a given positive number, and choose a and b such that $0 < a < x < b$. Since the original series has already been shown to converge (uniformly) on $[a, b]$, it remains only to show that the derived series converges uniformly there. This is easily done by the Weierstrass M-test, with $M_n = 2n e^{-na}$ (check the details).

★908. EXERCISES

In Exercises 1-6, show that the convergence fails to be uniform by showing that the limit function is not continuous.

★1. $\lim_{n \to +\infty} \sin^n x$, for $0 \leq x \leq \pi$. ★2. $\lim_{n \to +\infty} e^{-nx^2}$, for $|x| \leq 1$.

★3. $\lim_{n \to +\infty} \dfrac{x^n}{1 + x^n}$, for $0 \leq x \leq 2$.

★4. $\lim_{n \to +\infty} S_n(x)$, where $S_n(x) \equiv \dfrac{\sin nx}{nx}$ for $0 < x \leq \pi$, and $S_n(0) = 1$.

★5. $(1-x) + x(1-x) + x^2(1-x) + \cdots$, for $0 \leq x \leq 1$.

★6. $x^2 + \dfrac{x^2}{1+x^2} + \dfrac{x^2}{(1+x^2)^2} + \cdots$, for $|x| \leq 1$.

In Exercises 7-12, show that the equation is true.

★7. $\lim_{n \to +\infty} \int_{4\pi}^{\pi} \dfrac{\sin nx}{nx}\, dx = 0$. ★8. $\lim_{n \to +\infty} \int_{1}^{2} e^{-nx^2}\, dx = 0$.

★9. $\int_{0}^{\pi} \sum_{n=1}^{+\infty} \dfrac{\sin nx}{n^2}\, dx = \sum_{n=1}^{+\infty} \dfrac{2}{(2n-1)^3}$.

★10. $\int_1^2 \sum_{n=1}^{+\infty} \frac{\ln nx}{n^3} dx = \sum_{n=1}^{+\infty} \frac{\ln 4n - 1}{n^3}$.

★11. $\int_0^\pi \sum_{n=1}^{+\infty} \frac{n \sin nx}{e^n} dx = \frac{2e}{e^2 - 1}$.

★12. $\int_1^2 \sum_{n=1}^{+\infty} ne^{-nx} dx = \frac{e}{e^2 - 1}$.

In Exercises 13-18, show that the equation is true.

★13. $\frac{d}{dx}\left[\sum_{n=1}^{+\infty} \frac{x^n}{n(n+1)}\right] = \sum_{n=0}^{+\infty} \frac{x^n}{n+2}$, for $|x| < 1$.

★14. $\frac{d}{dx}\left[\sum_{n=1}^{+\infty} \frac{n}{x^n}\right] = -\sum_{n=1}^{+\infty} \frac{n^2}{x^{n+1}}$, for $|x| > 1$.

★15. $\frac{d}{dx}\left[\sum_{n=1}^{+\infty} \frac{\sin nx}{n^3}\right] = \sum_{n=1}^{+\infty} \frac{\cos nx}{n^2}$, for any x.

★16. $\frac{d}{dx}\left[\sum_{n=1}^{+\infty} \frac{\sin nx}{n^2 x}\right] = \sum_{n=1}^{+\infty} \left[\frac{\cos nx}{n^2 x} - \frac{\sin nx}{n^3 x^2}\right]$, for $x \neq 0$.

★17. $\frac{d}{dx}\left[\sum_{n=1}^{+\infty} \frac{1}{n^2(1 + nx^2)}\right] = -2x \sum_{n=1}^{+\infty} \frac{1}{n^2(1 + nx^2)^2}$, for any x.

★18. $\frac{d}{dx}\left[\sum_{n=1}^{+\infty} e^{-nx} \sin knx\right] = \sum_{n=1}^{+\infty} ne^{-nx}[k \cos knx - \sin knx]$, for $x > 0$.

★19. Prove the Corollary to the Theorem of § 905.
★20. Prove the Corollary to the Theorem of § 906.
★21. Prove the Corollary to the Theorem of § 907.
★22. Prove that the convergence of the sequence illustrated in Figure 904, § 906, is not uniform.
★23. Complete the final details of the proof of § 907.
★24. Let $S_n(x) = \frac{1}{n} e^{-nx}$, $S(x) = \lim_{n \to +\infty} S_n(x)$. Show that $\lim_{n \to +\infty} S_n'(0) \neq S'(0)$.
★25. Let $S_n(x) = nxe^{-nx^2}$, $S(x) = \lim_{n \to +\infty} S_n(x)$. Show that
$$\lim_{n \to +\infty} \int_0^1 S_n(x) \, dx \neq \int_0^1 S(x) \, dx.$$

★26. Construct an example to show that it is possible to have a sequence of continuous functions converge nonuniformly to a continuous function, on a closed interval $[a, b]$.

★27. Construct an example to show that it is possible to have a sequence $\{S_n(x)\}$ of continuous functions converging to a continuous function $S(x)$ nonuniformly on a closed interval $[a, b]$, but still have $\int_a^b S_n(x) \, dx \to \int_a^b S(x) \, dx$.
Hint: Construct functions like those illustrated in Figure 904, § 906, but having uniform maximum values.

★28. Construct an example to show that it is possible to have $S_n(x) \rightrightarrows S(x)$ and $S_n'(x) \to S'(x)$ nonuniformly on an interval. *Hint:* Work backwards from the result of Ex. 27, letting $\{S_n'(x)\}$ be the sequence of that exercise, and $S_n(x) \equiv \int_0^x S_n'(t)\,dt$.

★**29.** Show that $S_n(x) \equiv \dfrac{n^2 x^2}{1 + n^3 x^3}$ furnishes an example of the type requested in Exercise **27**, for the interval $[-1, 1]$.

★**30.** Show that $S_n(x) \equiv \dfrac{\sin nx}{n} \rightrightarrows 0$ for all x, but that $\{S_n'(x)\}$ converges only for integral multiples of 2π.

★★**31.** Prove that
$$1 - m + \frac{m(m-1)}{2!} - \frac{m(m-1)(m-2)}{3!} + \cdots$$
converges, but not uniformly, for $0 \leq m \leq 1$. *Hint:* The limit function is discontinuous.

★★**32.** Show that although $nx(1-x)^n \to 0$ nonuniformly for $0 \leq x \leq 1$, $\int_0^1 nx(1-x)^n \to 0$. *Hint:* The functions are uniformly bounded (Ex. 31, § 904). Cf. Ex. 49, § 503, or proceed from first principles.

★★**33.** Show that the series $\sum_{n=1}^{+\infty} \dfrac{1}{n}\left(\dfrac{x}{x-1}\right)^n$ converges uniformly for $a \leq x \leq \frac{1}{2}$ and that the derived series converges uniformly for $a \leq x \leq b < \frac{1}{2}$.

★★**34.** Prove that if $u_n(x)$ is improperly integrable on $[a, +\infty)$, for $n = 1, 2, \cdots$, and if $\sum x^k u_n(x)$, where $k > 1$, is dominated by a convergent series of constants, then the series $\sum u_n(x)$ can be integrated term by term, from a to $+\infty$:
$$\int_a^{+\infty} \sum_{n=1}^{+\infty} u_n(x)\,dx = \sum_{n=1}^{+\infty} \int_a^{+\infty} u_n(x)\,dx.$$
Hint: Use Ex. 47, § 515. The hint given in that Exercise, together with the Weierstrass M-test provides a proof of the present proposition without the necessity of using Ex. 49, § 503.

★★**35.** Show by an example that uniform convergence of a sequence of functions on an infinite interval is not sufficient to guarantee that the integral of the limit is the limit of the integral. (Cf. Ex. 40, § 515.) *Hint:* Consider $S_n(x) \equiv 1/n$ for $0 \leq x \leq n$, and $S_n(x) \equiv 0$ for $x > n$.

★★**36.** Prove that any monotonic convergence of continuous functions on a closed interval (more generally, on any compact set) to a continuous function there is uniform. *Hint:* Assume without loss of generality that for each x belonging to the closed interval A, $S_n(x) \downarrow$ and $S_n(x) \to 0$. If the convergence were *not* uniform there would exist a sequence $\{x_k\}$ of points A such that $S_{n_k}(x_k) \geq \epsilon > 0$. Assume without loss of generality that $\{x_k\}$ converges: $x_k \to \bar{x}$. Show that for every n, $S_n(\bar{x}) \geq \epsilon$. (Alternatively, use Ex. 28, § 312.)

★★**37.** Prove the **Moore-Osgood Theorem**: *If $S_n(x)$ is a sequence of functions defined for x in a deleted neighborhood J of $x = c$, and if*

(i) $S_n \equiv \lim_{x \to c} S_n(x)$ *exists and is finite for every* $n = 1, 2, \cdots$;

(ii) $f(x) \equiv \lim_{n \to +\infty} S_n(x)$ *exists and is finite for every x in $J(x \neq c)$;*

(iii) *the convergence in (ii) is uniform in J;*

then

(iv) $\lim_{n \to +\infty} S_n$ *exists and is finite;*

(v) $\lim_{x \to c} f(x)$ *exists and is finite;*

(vi) *the limits in (iv) and (v) are equal.*

Similar results hold for $x \to c+$, $c-$, $+\infty$, $-\infty$, and ∞. (Cf. § 1014.)

Hints: For (iv), write
$$|S_m - S_n| \leq |S_m - S_m(x)| + |S_m(x) - S_n(x)| + |S_n(x) - S_n|,$$
let N be such that $m > N$ and $n > N$ imply $|S_m(x) - S_n(x)| < \epsilon/3$, for all x in J, and let $x \to c$. For (v), write
$$|f(x') - f(x'')| \leq |f(x') - S_n(x')| + |S_n(x') - S_n(x'')| + |S_n(x'') - f(x'')|,$$
let N be such that $n > N$ implies $|S_n(x) - f(x)| < \epsilon/3$ for all x in J, and use (i).

★★38. Prove the Theorem of § 907, without the assumption of continuity for the derivatives. *Hints:* First write
$$S_m(x) - S_n(x) = \{[S_m(x) - S_n(x)] - [S_m(x_0) - S_n(x_0)]\} + \{S_m(x_0) - S_n(x_0)\}$$
$$= [S_m'(\xi) - S_n'(\xi)](x - x_0) + \{S_m(x_0) - S_n(x_0)\},$$
to establish $S_n(x) \rightrightarrows S(x)$. Then use the Moore-Osgood Theorem (Ex. 37) to obtain
$$\lim_{x \to c} \lim_{n \to +\infty} \frac{S_n(x) - S_n(c)}{x - c} = \lim_{n \to +\infty} \lim_{x \to c} \frac{S_n(x) - S_n(c)}{x - c},$$
by writing
$$\frac{S_m(x) - S_m(c)}{x - c} - \frac{S_n(x) - S_n(c)}{x - c} = \frac{S_m(x) - S_n(x)}{x - c} - \frac{S_m(c) - S_n(c)}{x - c}$$
$$= S_m'(\xi) - S_n'(\xi),$$
(Cf. Ex. 45, § 904.)

★★39. Prove that if a sequence of differentiable functions with uniformly bounded derivatives (Ex. 31, § 904) converges on a closed interval, then the convergence is uniform. *Hints:* Assume $S_n(x) \to S(x)$ and $|S_n'(x)| \leq K$ on $[a, b]$. First use $|S_n(x') - S_n(x'')| \leq K \cdot |x' - x''|$ and
$$|S(x') - S(x'')| \leq |S(x') - S_n(x')| + |S_n(x') - S_n(x'')| + |S_n(x'') - S(x'')|$$
to show that $S(x)$ is continuous. Assume the convergence is not uniform, and let $x_k \to \bar{x}$ such that $|S_{n_k}(x_k) - S(x_k)| \geq \epsilon > 0$. But
$$|S_{n_k}(x_k) - S(x_k)| \leq |S_{n_k}(x_k) - S_{n_k}(\bar{x})| + |S_{n_k}(\bar{x}) - S(\bar{x})| + |S(\bar{x}) - S(x_k)|.$$

★★40. Let $f_n(x)$ and $f(x)$ be Riemann-Stieltjes integrable with respect to $g(x)$ on $[a, b]$, and assume $f_n(x) \rightrightarrows f(x)$ on $[a, b]$. Prove that
$$\int_a^b f_n(x)\,dg(x) \to \int_a^b f(x)\,dg(x).$$

★★41. Let $f_1(x)$, $f_2(x)$, $\cdots$ be a sequence of "sawtooth" functions (Fig. 906), defined for all real x, where the graph of $f_n(x)$ is made up of line segments of slope ± 1, such that $f_n(x) = 0$ for $x = \pm m \cdot 4^{-n}$, $m = 0, 1, 2, \cdots$, and $f_n(x) = \frac{1}{2} \cdot 4^{-n}$ for $x = \frac{1}{2} \cdot 4^{-n} + m \cdot 4^{-n}$, $m = 0, 1, 2, \cdots$. Let $f(x) = \sum_{n=0}^{+\infty} f_n(x)$. Prove that $f(x)$ is everywhere continuous and nowhere differentiable.† *Hint for non-*

† This example is modeled after one due to Van der Waerden. Cf. E. C. Titchmarsh, *Theory of Functions* (Oxford, Oxford University Press, 1932).

differentiability: If a is any fixed point, show that for any $n = 1, 2, \cdots$ a number h_n can be chosen as one of the numbers 4^{-n-1} or -4^{-n-1} such that $f_n(a + h_n) - f_n(a) = \pm h_n$. Then $f_m(a + h_n) - f_m(a)$ has the value $\pm h_n$ for $m \leq n$, and otherwise vanishes. Hence the difference quotient $[f(a + h_n) - f(a)]/h_n$ is an integer of the same parity as n (even if n is even and odd if n is odd). Therefore its limit as $n \to +\infty$ cannot exist as a finite quantity.

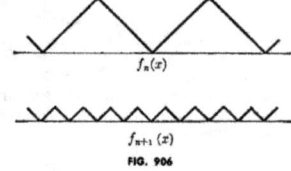

FIG. 906

★★**42.** Modify the construction of Exercise 41 to prove the following generalization (cf. the Note below): *Let $\rho(t)$ be a positive-valued function defined for $t > 0$ such that $\lim\limits_{t \to 0+} \rho(t) = 0$. Then there exists a function $f(x)$, defined and continuous for all real x, having the property that corresponding to any real number a there is a sequence of numbers $\{h_n\}$ such that $h_n \to 0$ and $|f(a + h_n) - f(a)|/\rho(|h_n|) \to +\infty$.* *Hint:* Let $f_n(x)$ be a sawtooth function, somewhat like that of Ex. 41, having maximum value $\frac{1}{2} \cdot 4^{-n}$, and minimum value 0 occurring for $x =$ all integral multiples of a number α_n defined inductively to be a sufficiently high integral power of $\frac{1}{4}$ to ensure the following inequality, where $A_k = 4^{-k}\alpha_k^{-1}$ is the absolute value of the slopes of the segments in the graph of $f_k(x)$, $k = 1, 2, \cdots, n - 1$, (α_1 being defined to be $\frac{1}{4}$):

$$A_n > A_1 + A_2 + \cdots + A_{n-1} + n[\rho(\alpha_n/4)]/\alpha_n, \quad n > 1.$$

Then choose $h_n = \pm\alpha_n/4$ such that $|f_n(a + h_n) - f_n(a)|/|h_n| = A_n$. Define $f(x) \equiv \sum\limits_{n=1}^{+\infty} f_n(x)$.

★★**NOTE.** The statement of Exercise 42 is of interest in connection with the concept of **modulus of continuity** of a function $f(x)$ on an interval I, which is a function of a positive independent variable δ, denoted $\omega(\delta)$ and defined:

$$\omega(\delta) \equiv \sup |f(x_1) - f(x_2)|,$$

formed for all x_1 and x_2 of I such that $|x_1 - x_2| < \delta$. The function $\omega(\delta)$ is monotonic, and hence approaches a limit ω, as $\delta \to 0+$. The statement $\omega = 0$ is clearly equivalent to the statement that $f(x)$ is uniformly continuous on I. The example of Exercise 42 shows that for functions continuous (and hence uniformly continuous) over a closed interval, there is no bound to the slowness with which the modulus of continuity can approach 0 as $\delta \to 0+$. (With the notation of Ex. 42, $\omega(\delta_n) > \rho(\delta_n)$ for a sequence $\{\delta_n\}$ approaching 0.) The method of proof suggested in Exercise 42 is due to F. Koehler. Also cf. W. S. Loud, "Functions with Prescribed Lipschitz condition," *Proc. A. M. S.*, Vol. 2, No. 3 (June 1951), pp. 358-360.

★909. POWER SERIES. ABEL'S THEOREM

Chapter 8 contains three important theorems on power series (Theorem I, § 803, Theorems II and III, § 810), whose proofs have been deferred. These theorems (having to do with continuity, integration, and differentiation of power series within the interval of convergence) are simple corollaries (cf. Exs. 10-13, § 911) of the following basic theorem:

Theorem I. *A power series converges uniformly on any interval whose end-points lie in the interior of its interval of convergence.*

Proof. For simplicity of notation we shall assume that the series has the form $\sum_{n=0}^{+\infty} a_n x^n$ (cf. Ex. 9, § 911). Let $R > 0$ be the radius of convergence of $\sum a_n x^n$, and let I be an interval with endpoints α and β, where $\max(|\alpha|, |\beta|) = r < R$. Choose a fixed γ such that $r < \gamma < R$, and define $M_n \equiv |a_n \gamma^n|$, $n = 0, 1, \cdots$. Since γ is interior to the interval of convergence of $\sum a_n x^n$, $\sum a_n \gamma^n$ converges absolutely. Therefore the convergent series of constants $\sum M_n$ dominates the series $\sum a_n x^n$ throughout I, and by the Weierstrass M-test, the uniform convergence desired is established.

Theorem I and its corollaries in Chapter 8 have to do only with points in the *interior* of the interval of convergence of a power series. Behavior at the end-points of the interval usually involves more subtle and delicate questions. One of the most useful and elegant tools for treating convergence at and near the end-points is due to Abel. In this section we present the statement of Abel's Theorem, together with two corollaries (whose proofs are requested in Exercises 14 and 15, § 911). The proof of Abel's Theorem is given in the following section.

Theorem II. Abel's Theorem. *If* $\sum_{n=0}^{+\infty} a_n$ *is a convergent series of constants, then the power series* $\sum_{n=0}^{+\infty} a_n x^n$ *converges uniformly for* $0 \leq x \leq 1$.

Corollary I. *If a power series converges at an end-point P of its interval of convergence I, it converges uniformly on any closed interval that has P as one of its end-points and any interior point of I as its other end-point. If a power series converges at both end-points of its interval of convergence it converges uniformly throughout that interval.*

Corollary II. *If a function continuous throughout the interval of convergence of a power series is represented by that power series in the interior of that interval, then it is represented by that power series at any end-point of that interval at which the series converges.*

Example. Show that
$$\ln(1+x) = x - \frac{x^2}{2} + \frac{x^3}{3} - \cdots, \tag{1}$$
for $-1 < x \leq 1$, by integrating $\sum_{n=0}^{+\infty} (-1)^n x^n$.

Solution. The geometric series $\sum_{n=0}^{+\infty} (-1)^n x^n$ converges for $|x| < 1$. Therefore the relation (1) is valid for $|x| < 1$ (Theorem II, § 810). Since $\ln(1+x)$ is continuous at $x = 1$, the relation (1) is also true for $x = 1$, by Corollary II.

★★910. PROOF OF ABEL'S THEOREM

Notation. If $u_0, u_1, u_2, \cdots, u_n, \cdots$ is a sequence of real numbers, we denote by
$$\max {}_m^n \big| u_m + u_{m+1} + \cdots + u_k \big| \tag{1}$$
the maximum of the $n - m + 1$ numbers
$$|u_m|, |u_m + u_{m+1}|, \cdots, |u_m + u_{m+1} + \cdots + u_n|.$$

Lemma. *If $\{a_n\}$ is any sequence of real numbers, and if $\{b_n\}$ is a monotonically decreasing sequence of nonnegative numbers ($b_n \downarrow$ and $b_n \geq 0$), then*
$$\left| \sum_{k=0}^{n} a_k b_k \right| \leq b_0 \cdot \max {}_0^n \big| a_0 + a_1 + \cdots + a_k \big| \tag{2}$$
and
$$\left| \sum_{k=m}^{n} a_k b_k \right| \leq b_m \cdot \max {}_m^n \big| a_m + a_{m+1} + \cdots + a_k \big|. \tag{3}$$

Proof. The two statements (2) and (3) are identical, except for notation. For convenience we shall prove (2), and then apply (3) in the proof of Abel's Theorem. Letting $A_n \equiv a_0 + a_1 + \cdots + a_n$, $n = 0, 1, 2, \cdots$, we can write the left-hand member of (2) in the form
$$|A_0 b_0 + (A_1 - A_0) b_1 + (A_2 - A_1) b_2 + \cdots + (A_n - A_{n-1}) b_n|$$
$$= |A_0 (b_0 - b_1) + A_1 (b_1 - b_2) + \cdots + A_{n-1}(b_{n-1} - b_n) + A_n b_n|.$$
By the triangle inequality and the assumptions on $\{b_n\}$, this quantity is less than or equal to
$$|A_0|(b_0 - b_1) + |A_1|(b_1 - b_2) + \cdots + |A_{n-1}|(b_{n-1} - b_n) + |A_n| b_n$$
$$\leq \{\text{maximum of } |A_0|, |A_1|, \cdots, |A_n|\} \cdot \{(b_0 - b_1) + \cdots + b_n\}$$
$$= b_0 \cdot \max {}_0^n \big| a_0 + \cdots + a_k \big|.$$

Proof of Abel's Theorem. By the Lemma, if $b_n \equiv x^n$,
$$\left| \sum_{k=m}^{n} a_k x^k \right| \leq 1 \cdot \max {}_m^n \big| a_m + \cdots + a_n \big|.$$

The convergence of $\sum a_n$ implies that for any $\epsilon > 0$ there exists a number N such that $n \geq m > N$ implies $|a_m + \cdots + a_n| < \epsilon$. Therefore, by the Cauchy criterion for uniform convergence (Ex. 45, § 904), the proof is complete.

★911. EXERCISES

In Exercises 1-6, obtain the given expansions by integration, and justify the inclusion of the end-points specified.

★1. $\ln(1 - x) = -x - \dfrac{x^2}{2} - \dfrac{x^3}{3} - \cdots ; \ -1 \leq x < 1$.

★2. $\operatorname{Arctan} x = x - \dfrac{x^3}{3} + \dfrac{x^5}{5} - \cdots ; \ |x| \leq 1$.

★3. $\operatorname{Arcsin} x = x + \dfrac{1}{2}\dfrac{x^3}{3} + \dfrac{1\cdot 3}{2\cdot 4}\dfrac{x^5}{5} + \dfrac{1\cdot 3\cdot 5}{2\cdot 4\cdot 6}\dfrac{x^7}{7} + \cdots ; \ |x| \leq 1$.

★4. $\ln(x + \sqrt{1+x^2}) = x - \dfrac{1}{2}\dfrac{x^3}{3} + \dfrac{1\cdot 3}{2\cdot 4}\dfrac{x^5}{5} - \dfrac{1\cdot 3\cdot 5}{2\cdot 4\cdot 6}\dfrac{x^7}{7} + \cdots , \ |x| \leq 1$.

★★5. $\tfrac{1}{2}[x\sqrt{1 - x^2} + \operatorname{Arcsin} x]$
$\qquad = x - \dfrac{1}{2}\dfrac{x^3}{3} - \dfrac{1\cdot 1}{2\cdot 4}\dfrac{x^5}{5} - \dfrac{1\cdot 1\cdot 3}{2\cdot 4\cdot 6}\dfrac{x^7}{7} - \cdots ; \ |x| \leq 1$.

★★6. $\tfrac{1}{2}[x\sqrt{1 + x^2} + \ln(x + \sqrt{1+x^2})]$
$\qquad = x + \dfrac{1}{2}\dfrac{x^3}{3} - \dfrac{1\cdot 1}{2\cdot 4}\dfrac{x^5}{5} + \dfrac{1\cdot 1\cdot 3}{2\cdot 4\cdot 6}\dfrac{x^7}{7} - \cdots ; \ |x| \leq 1$.

★7. Show that
$$\int_0^x \ln(1+t)\, dt = \dfrac{x^2}{1\cdot 2} - \dfrac{x^3}{2\cdot 3} + \dfrac{x^4}{3\cdot 4} - \cdots .$$
Hence evaluate $\dfrac{1}{1\cdot 2} - \dfrac{1}{2\cdot 3} + \dfrac{1}{3\cdot 4} - \cdots$.

★8. Show that
$$\int_0^x \operatorname{Arctan} t\, dt = \dfrac{x^2}{1\cdot 2} - \dfrac{x^4}{3\cdot 4} + \dfrac{x^6}{5\cdot 6} - \cdots .$$
Hence evaluate $1 - \tfrac{1}{2} - \tfrac{1}{3} + \tfrac{1}{4} + \tfrac{1}{5} - \tfrac{1}{6} - \tfrac{1}{7} + + - - \cdots$.

★9. Prove Theorem I, § 909, for $\sum a_n(x - a)^n$.

★10. Prove that a function defined by a power series is continuous throughout the interval of convergence.

★11. Prove Theorem II, § 810.

★12. Prove Theorem III, § 810.

★13. Prove Theorem I, § 803.

★14. Prove Corollary I of Theorem II, § 909.

★15. Prove Corollary II of Theorem II, § 909.

★16. Show that
$$\int_0^1 \dfrac{\operatorname{Arctan} x}{x}\, dx = \int_0^1 \dfrac{|\ln x|}{1+x^2}\, dx = \sum_{n=0}^{+\infty} \dfrac{(-1)^n}{(2n+1)^2}.$$

★17. Show that
$$\int_0^1 \frac{\text{Arcsin } x}{x} dx = \int_0^1 \frac{|\ln x|}{\sqrt{1-x^2}} dx = \sum_{n=0}^{+\infty} \frac{1 \cdot 3 \cdots (2n-1)}{2 \cdot 4 \cdots 2n} \cdot \frac{1}{(2n+1)^2}.$$

★18. From the validity of the Maclaurin series for the binomial function $(1 + x)^m$ for $|x| < 1$, infer the validity of the expansion for any end-point of the interval of convergence at which the series converges.

★19. Show that
$$\int_0^1 \frac{x^m}{1+x} dx = \frac{1}{m+1} - \frac{1}{m+2} + \frac{1}{m+3} - \cdots,$$
for $m > -1$.

★20. Prove that if $\sum a_n$ and $\sum b_n$ are convergent series, with sums A and B, respectively, if $\sum c_n$ is their product series, and if $\sum c_n$ converges, with sum C, then $C = AB$. *Hint:* Define the three functions $f(x) \equiv \sum a_n x^n$, $g(x) \equiv \sum b_n x^n$, and $h(x) \equiv \sum c_n x^n$, for $0 \leq x \leq 1$. Examine the continuity of these functions for $0 \leq x \leq 1$, and the equation $h(x) = f(x) g(x)$ for $0 \leq x < 1$.

★★21. Verify the expansion of the following elliptic integral (cf. § 611):
$$\int_0^{\frac{\pi}{2}} \frac{dt}{\sqrt{1 - k^2 \sin^2 t}} = \frac{\pi}{2}\left[1 + \left(\frac{1}{2}\right)^2 k^2 + \left(\frac{1\cdot 3}{2\cdot 4}\right)^2 k^4 + \left(\frac{1\cdot 3\cdot 5}{2\cdot 4\cdot 6}\right)^2 k^6 + \cdots \right],$$
for $|k| \leq 1$. *Hint:* Expand by the binomial series:
$$(1 - k^2 \sin^2 t)^{-\frac{1}{2}} = 1 + \frac{1}{2} k^2 \sin^2 t + \frac{1\cdot 3}{2\cdot 4} k^4 \sin^4 t + \frac{1\cdot 3\cdot 5}{2\cdot 4\cdot 6} k^6 \sin^6 t + \cdots,$$
and use Wallis's formulas (Ex. 36, § 515).

★★22. Prove the **Weierstrass Uniform Approximation Theorem:** *If $f(x)$ is continuous on $[a, b]$, and if $\epsilon > 0$, then there exists a polynomial $P(x)$ such that $|f(x) - P(x)| < \epsilon$ for all x in $[a, b]$.*† *Suggested outline:* (i) The binomial series for $\sqrt{1+x}$ converges uniformly to $\sqrt{1+x}$ for $-1 \leq x \leq 0$. (ii) The corresponding series for $\sqrt{1 + (x^2 - 1)}$ converges uniformly to $|x|$ for $|x| \leq 1$. (iii) $|x|$ can be uniformly approximated by polynomials for $|x| \leq 1$. (iv) The theorem is true for any function of the form $m|x - c|$. (v) The theorem is true for any continuous function identically 0 for $a \leq x \leq c$ ($c \leq x \leq b$) and linear for $c \leq x \leq b$ ($a \leq x \leq b$). (vi) The theorem is true for any continuous function with a polygonal (broken-line) graph. (vii) Any continuous function on $[a, b]$ can be uniformly approximated on $[a, b]$ by a continuous function with a polygonal graph. (Cf. § 1523.)

★★23. Prove that if $\int_a^b x^n f(x) dx = 0$ for $n = 0, 1, 2, \cdots$, and if $f(x)$ is continuous on $[a, b]$, then $f(x)$ is identically 0 on $[a, b]$. *Hint:* Use the Weierstrass Uniform Approximation Theorem (Ex. 22) to show that since $\int_a^b P(x) f(x) dx = 0$ for any polynomial $P(x)$, $\int_a^b [f(x)]^2 dx = 0$.

---

† Several proofs of this theorem have been given, the original by Weierstrass in 1885. The proof outlined here was given by Lebesgue in 1898. For references and additional discussion see Hobson, *Theory of Functions* (Washington, D. C., Harren Press, 1950).

★912. FUNCTIONS DEFINED BY POWER SERIES. EXERCISES.

★1. Let the function $e^x = \exp(x)$ be *defined* by the series (1), § 807, and let the function $y = \ln x$ be *defined* by the equation $x = e^y$. Prove the properties of the exponential and logarithmic functions enumerated in § 602. *Hint:* Begin by proving (i), (ii), and (iii) of Ex. 4, § 602, making use of the special case of (i): $e^x e^{-x} = 1$.

★2. Let the functions $\sin x$ and $\cos x$ be *defined* by the series (3) and (5), respectively, of § 807. Prove the properties of the trigonometric functions enumerated in Exercises 4-17, § 604. *Hints:* Start with Ex. 4, § 604. After showing that $\cos 2 < 0$, *define* $\frac{1}{2}\pi < 2$ to be the smallest positive number whose cosine is 0. Then proceed with Ex. 8, § 604.

★★3. Prove that $\ln x$ can be uniquely described as the analytic function with domain $(0, +\infty)$ whose values for $0 < x < 2$ are given by the series

$$(x - 1) - \frac{(x-1)^2}{2} + \frac{(x-1)^3}{3} - \frac{(x-1)^4}{4} + \cdots.$$

(Cf. Theorem IV, § 815.)

★★4. Prove that the number a^b, where a is positive and b real, can be uniquely defined as a particular value of the analytic function with domain $(0, +\infty)$ whose values for $0 < x < 2$ are given by the series

$$1 + b(x-1) + \frac{b(b-1)}{2!}(x-1)^2 + \cdots + \binom{b}{n}(x-1)^n + \cdots.$$

10

Functions of Several Variables

1001. INTRODUCTION

In the first nine chapters of this book attention has been focused on functions of *one real variable*. We wish now to extend our consideration to functions of *several real variables*—where the word *several* means *at least one*. It will be seen, particularly in this chapter, that many concepts, theorems, and proofs generalize from one to several variables almost without change. This is especially true when the one-dimensional formulations are framed with higher dimensional extensions in mind, as they are in Chapter 3. On the other hand, such a topic as integration (Chapter 13), especially when it comes to a change of variables, requires radical reorganization.

The unstarred sections of this chapter, together with §§ 1022-1025, have only Chapters 1 and 2 as prerequisite. With the exception of §§ 1012-1015 (which depend on Chapter 9) the remainder of the chapter requires in addition only Chapter 3, or portions thereof.

We start with some definitions. Some of these include concepts that were starred in Chapter 3, but that call for more generous treatment in spaces of more than one dimension.

1002. NEIGHBORHOODS IN THE EUCLIDEAN PLANE

In the Euclidean space E_1 of one dimension—that is, the space of real numbers—a *neighborhood* of a point a is defined (Definition III, § 110) as an open interval of the form $(a - \epsilon, a + \epsilon)$. Equally well, we can describe this neighborhood as the set of all points x whose distance from a is less than ϵ:

(1) $$|x - a| < \epsilon.$$

In the Euclidean plane E_2 we shall speak of *two* kinds of neighborhoods, circular and square. A **circular neighborhood** of a point (a, b) is the set of all points (x, y) whose distance from (a, b) is less than some fixed positive number ϵ:

294 FUNCTIONS OF SEVERAL VARIABLES [§ 1003

$$(2) \quad \sqrt{(x-a)^2 + (y-b)^2} < \epsilon.$$

It consists of all points inside a circle of radius ϵ and center (a, b). A **square neighborhood** of a point (a, b) is the set of all points (x, y) whose coordinates satisfy two inequalities of the form

$$(3) \quad |x - a| < \epsilon, \quad |y - b| < \epsilon.$$

It consists of all points in a square whose sides are parallel to the coordinate axes and of length 2ϵ, and whose center is (a, b).

As indicated in Figure 1001, any circular neighborhood of a point contains a square neighborhood of the point (for a smaller ϵ, of course), and

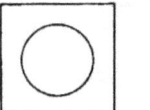 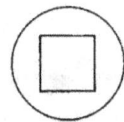

FIG. 1001

any square neighborhood of a point contains a circular neighborhood of the point. This fact will be referred to as the **equivalence** of the two kinds of neighborhoods. (Cf. the Note, below.)

By a **neighborhood** of a point we shall mean either a circular neighborhood or a square neighborhood of the point. By a **deleted neighborhood** of a point we shall mean any neighborhood of that point with the point itself removed.

NOTE. In nearly all the following sections, only the single word *neighborhood* is used. The reader should observe that if either word *circular* or *square* is introduced, the meaning is unaffected. (Cf. Definitions II and III, § 1003.) This is the real significance of the term *equivalence*, as applied to the two types of neighborhoods.

1003. POINT SETS IN THE EUCLIDEAN PLANE

We present now some definitions that are of basic importance in spaces of more than one dimension. Most of these were introduced in Chapter 3, as topics of theoretical interest in the one-dimensional space of real numbers. (Cf. §§ 309–311.) The formulations here are those for the Euclidean plane E_2. (Cf. § 1004.)

Definition I. *If A is a set, the **complement** of A, written A', is the set of all points in the Euclidean plane that are not members of A.*

§ 1003] POINT SETS IN THE EUCLIDEAN PLANE

Example 1. Let A be the set of points (x, y) belonging to a circular neighborhood of some point (a, b): $(x - a)^2 + (y - b)^2 < r^2$, where $r > 0$. Such a set is called an **open disk** (Fig. 1002). Then A' consists of all points (x, y) such that $(x - a)^2 + (y - b)^2 \geq r^2$, and is made up of the circumference of A together with

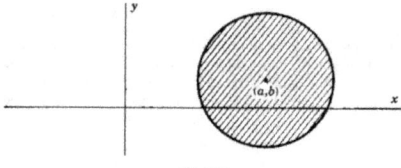

FIG. 1002

the points completely outside the circle. The open disk with center $(0, 0)$ and radius 1 is called the **open unit disk**.

Example 2. If A is a **closed disk** defined by an inequality of the form $(x - a)^2 + (y - b)^2 \leq r^2$, where $r > 0$, then A' consists of the points completely outside the circle. All points of the circumference belong to the set A. The closed disk with center $(0, 0)$ and radius 1 is called the **closed unit disk**.

Definition II. *A set A is **open** if and only if every member of A has some neighborhood contained entirely in A.*

Example 3. Let A be the set of points (x, y) such that both x and y are positive. This set is called the **open first quadrant** (Fig. 1003). Then A is an open set.

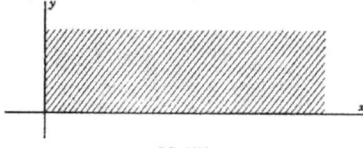

FIG. 1003

To prove this, we choose an arbitrary member (x, y) of A. Then the smaller of the two numbers x and y, $\epsilon \equiv \min(x, y)$, is positive. Furthermore, the corresponding neighborhood, whether circular ((2), § 1002) or square ((3), § 1002), lies entirely within the set A.

Discussion of earlier examples. Any open disk (Example 1) is open (give the details of the proof). A closed disk (Example 2) is *not* open. The point $(1, 0)$, for instance, belongs to the closed unit disk A, but every neighborhood of this point extends partly outside the circle A. In other words, *no* neighborhood of $(1, 0)$

lies entirely in A. Thus, not *every* member of A has a neighborhood contained entirely in A.

Definition III. *A point p is a **limit point** of a set A if and only if every deleted neighborhood of p contains at least one point of A.*

Discussion of earlier examples. If A is either an open disk (Example 1) or a closed disk (Example 2), the set B of all limit points of A is the corresponding closed disk.

Example 4. If A is the open first quadrant of Example 3, then the set B of all limit points of A is the **closed first quadrant** consisting of all points (x, y) such that both x and y are nonnegative.

Note 1. If p is a limit point of a set A we say that the set A *has* p as a limit point, whether p is a member of A or not. For example, both the open unit disk and the closed unit disk (Examples 1 and 2) have the point $(1, 0)$ as a limit point, but only the latter contains $(1, 0)$ as a member.

Note 2. If p is a limit point of a set A, every neighborhood of p contains infinitely many points of A (Ex. 8, § 1005).

Definition IV. *A set is **closed** if and only if it contains all of its limit points.*

Discussion of earlier examples. Any closed disk (Example 2) and the closed first quadrant of Example 4 are closed sets. Neither any open disk (Example 1) nor the open first quadrant of Example 3 is closed (the point $(1, 0)$ is a limit point of the open unit disk and the open first quadrant, but is a member of neither).

Example 5. Let A be the set of all **integral points** of the plane, that is, the set of all points (x, y) such that *both* x and y are integers. This set has *no* limit points. (One way to see this is to observe that *no* neighborhood of *any* point can contain infinitely many points of A.) It follows that the set A *contains* all of its limit points, since otherwise there must exist a limit point of the set A—in fact, a limit point of the set that is not a member of the set.† Thus A is closed.

Example 6. The reasoning used in Example 5 establishes the fact that *any finite set is closed*.

Example 7. The circumference of a disk, defined by the equation $(x - a)^2 + (y - b)^2 = r^2$, is closed. (Why?) It is not open. (Why?)

Example 8. The set made up of all points of the x-axis together with all points of the y-axis is closed. It is not open. (Why?)

Example 9. The positive half of the x-axis, that is, the points (x, y) such that $x > 0$ and $y = 0$, is neither open nor closed. (Why?)

Example 10. Let A be the set of all **rational points** of the plane, that is, the set of all points (x, y) such that *both* x and y are rational. This set is neither open nor closed. (Why?)

---

† It may be held by some that the fact that a set with no limit points contains all of them is completely obvious. It is well, however, to be ready with the machinery to convince the skeptic who claims that the alleged fact is *not* obvious to *him*. The persuasive machinery in this case is to follow to a logical consequence the *denial* of the original assertion and to obtain a contradiction that does not rest on any individual state of mind.

§ 1003] POINT SETS IN THE EUCLIDEAN PLANE 297

NOTE 3. A basic relation between open and closed sets is the following (cf. Theorem I, § 309, Ex. 9, § 1005): *A set is open if and only if its complement is closed.* Equivalently, *a set is closed if and only if its complement is open.* It is of interest that in the Euclidean plane the only sets that are *both* open and closed are the empty set ∅ (cf. Note 3, § 309) and the entire plane. (For a proof of this remarkable fact, see Ex. 34, § 1025. Also cf. Ex. 11, § 312.)

Definition V. *A set is **bounded** if and only if it is contained in some circle, or, equivalently, in some square. A set is **compact** if and only if it is closed and bounded.* (Cf. the footnote, p. 76, Def. II, § 1026, Def., § 1027.)

Discussion of earlier examples. The open and closed unit disks (Examples 1 and 2) are both bounded, and the latter is compact. Neither the open nor the closed first quadrant (Examples 3 and 4) is bounded, and therefore neither is compact. The set of integral points of the plane (Example 5) is unbounded. Since any finite set (Example 6) is bounded and closed, it is compact. The circumference of the unit circle (Example 7) is compact. Not one of Examples 8-10 is bounded, and therefore not one is compact.

Definition VI. *A point p, belonging to a set A, is an **interior point** of A if and only if it has a neighborhood lying entirely in A. The set of all interior points of a set is called its **interior**. A point p, whether it belongs to a set A or not, is a **frontier point** of A if and only if every neighborhood of p contains points of A and of its complement A'. The set of all frontier points of a set is called its **frontier**.* The frontier of a set A is therefore made up of points that are either members of A and limit points of A' or members of A' and limit points of A.†

Example 11. Let A be a set made up of an open disk (Example 1) together with *some* but *not all* of the points of the circumference. This set lies *between* the open disk and the corresponding closed disk (Example 2). It is neither open nor closed. (Why?) Its interior (as also is the interior of either the open or the closed disk is the open disk. (Why?) Its frontier (as also is the frontier of either the open or the closed disk is the circumference of the disk. (Why?)

Discussion of earlier examples. The interior of either the open or the closed first quadrant (Examples 3 and 4) is the open first quadrant, and its frontier is the nonnegative x-axis together with the nonnegative y-axis. The interior of the set A of all integral points (Example 5) is empty (that is, there are no interior points), and its frontier is the set A itself. Similar statements hold for any finite set (Example 6). The interior of the circumference A of any disk (Example 7) is empty, and the frontier of A is the set A itself. The interior of the set A of all rational points (Example 10) is empty, and its frontier is the entire plane. The detailed verification of the preceding statements is left as an exercise for the reader (Ex. 10, § 1005).

† The frontier of a set is sometimes also called its *boundary*. However, since the word *boundary* has other meanings in combinatorial topology and in the theory of surfaces (cf. Exs. 4, 11, 15, § 1331), we shall avoid using this latter term here.

298   FUNCTIONS OF SEVERAL VARIABLES   [§ 1004

NOTE 4. An interesting relation between the concepts of *open set* and *interior of a set* is that if A is an arbitrary set, its interior is always open. (Give a proof in Ex. 11, § 1005.)

Definition VII. *The* **closure** *$\bar{A}$ of a set A consists of all points of A together with all limit points of A.*

Discussion of earlier examples. If A is either an open or closed disk of Example 1 or 2, or a disk of Example 11, its closure $\bar{A}$ is the corresponding closed disk. The closure of either the open or closed first quadrant (Examples 3 and 4) is the closed first quadrant. The closure of the set of all integral points (Example 5) is the set itself. A similar statement holds for any finite set (Example 6). The closure of the set of all rational points (Example 10) is the entire plane. The reader should supply the details of the verification of the preceding statements (Ex. 12, § 1005).

NOTE 5. The term *closure* is appropriate because of the fact that for *any* set A, *its closure $\bar{A}$ is always a closed set*. (Cf. Ex. 13, § 1005.) In a definite sense, $\bar{A}$ is the *smallest* closed set containing A (Ex. 15, § 1005).

Definition VIII. *A* **region** *is a nonempty open set R any two of whose points can be connected by a broken line segment lying entirely in R.* (Cf.

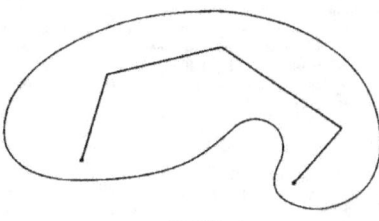

FIG. 1004

Fig. 1004.) *A* **closed region** *is the closure of a region. A* **compact region** *is a bounded closed region.*

NOTE 6. The single word *region* will be reserved for the meaning defined above, and will always refer to an open set. A closed region, therefore, is a region only if it is the entire plane (cf. Note 3), and a compact region is never a region.

1004. SETS IN HIGHER DIMENSIONAL EUCLIDEAN SPACES

In three-dimensional Euclidean space E_3 the distance between two points $p : (x_1, y_1, z_1)$ and $q : (x_2, y_2, z_2)$ is given by the formula

§ 1005]   EXERCISES   299

(1) $$d(p, q) = \sqrt{(x_2 - x_1)^2 + (y_2 - y_1)^2 + (z_2 - z_1)^2}.$$

More generally, in n-dimensional Euclidean space E_n, where a point is *defined* to be an ordered n-tuple of real numbers, the distance between two points $p : (x_1, x_2, \cdots, x_n)$ and $q : (y_1, y_2, \cdots, y_n)$ is *defined* by means of a formula corresponding to (1):

(2) $$d(p, q) = \sqrt{(y_1 - x_1)^2 + (y_2 - x_2)^2 + \cdots + (y_n - x_n)^2}.$$

In a manner analogous to the treatment of neighborhoods in the Euclidean plane, we speak of two kinds of neighborhoods in E_n. A **spherical neighborhood** of a point $p_0 : (a_1, a_2, \cdots, a_n)$ is the set of all points $p : (x_1, x_2, \cdots, x_n)$ whose distance from p_0 is less than some fixed number ϵ:

(3) $$d(p, p_0) < \epsilon.$$

A **cubical neighborhood** of a point $p_0 : (a_1, a_2, \cdots, a_n)$ is the set of all points $p : (x_1, x_2, \cdots, x_n)$ whose coordinates satisfy the n inequalities

(4) $$|x_1 - a_1| < \epsilon, \quad |x_2 - a_2| < \epsilon, \quad \cdots, \quad |x_n - a_n| < \epsilon.$$

The Definitions and Notes of § 1003 are immediately adaptable to the space E_n, virtually the only change being the replacement of the expression *Euclidean plane* by the expression *n-dimensional Euclidean space*. These details, as well as the construction of examples corresponding to those of § 1003, are left to the reader. Perhaps the greatest difficulty in the transition from two- to n-dimensional space lies in the concept of *broken line segment*, in Definition VIII, § 1003, of a region. This depends in turn on the question of *segment between two points*. For a brief discussion of this, see Exercise 21, § 1005.

1005. EXERCISES

In Exercises 1–6, state whether the given set is (a) open, (b) closed, (c) bounded, (d) compact, (e) a region, (f) a closed region, (g) a compact region. Give its (h) set of limit points, (i) closure, (j) interior, and (k) frontier. The space under consideration in each case is the Euclidean plane.

1. All (x, y) such that $a \leq x < b, c \leq y < d$, $a, b, c,$ and d fixed constants such that $a < b$ and $c < d$.

2. All (x, y) such that $xy \neq 0$.

3. All (ρ, θ) such that $\rho^2 \leq \cos 2\theta$, ρ and θ being polar coordinates.

4. All (x, y) such that $y = \sin \frac{1}{x}, x \neq 0$.

★5. All (x, y) such that $y < [x]$ (cf. Example 5, § 201).

★6. All (x, y) such that x and y have the form $x = m + \frac{1}{p}$ and $y = n + \frac{1}{q}$, m and n integers, p and q integers > 1.

In Exercises 7–20, assume that the space under consideration is the Euclidean plane. More generally prove the corresponding result for the space E_n.

7. Prove that a point p is a limit point of a set A if and only if every open set containing p contains at least one point of A different from p.

8. Prove the statement of Note 2, § 1003.

9. Prove the statement of Note 3, § 1003, that a set is open if and only if its complement is closed, without referring to the proof of Theorem I, § 309.

10. Verify the details called for in the *Discussion of earlier examples*, following Example 11, § 1003.

11. Prove the statement of Note 4, § 1003, that if A is an arbitrary set, its interior is always an open set. *Hint:* Let B denote the interior of A, and let p be an arbitrary point of B. Since p is an interior point of A, it has a neighborhood N_p lying entirely in A. Show that N_p lies entirely in B.

12. Verify the details called for in the *Discussion of earlier examples*, following Definition VII, § 1003.

13. Prove the statement of Note 5, § 1003, that if A is an arbitrary set, its closure $\bar{A}$ is always a closed set. *Hint:* Assume that $\bar{A}$ is *not* closed, and let p be a limit point of $\bar{A}$ that is not a member of $\bar{A}$. Since p is neither a member of A nor a limit point of A, it has a neighborhood N_p having no points in common with A. Show that this fact is inconsistent with the assumption that p is a limit point of $\bar{A}$, since N_p must contain some point q belonging to $\bar{A}$.

14. Prove that a set is closed if and only if it is equal to its closure.

★15. Prove that $\bar{A}$ is the smallest closed set containing A in the sense that (i) $\bar{A}$ is a closed set containing A, and (ii) if B is any closed set containing A, then B contains $\bar{A}$. (Cf. Ex. 14.)

★16. Prove that the set of all limit points of any set is always closed.

★17. Prove that the frontier of any set is always closed.

★18. Prove that the interior and the frontier of a set never have any points in common, and together constitute the closure of the set.

★19. Prove that the interior of the complement of any set is equal to the complement of the closure of the set.

★20. Construct an open set that is not the interior of its closure. (Cf. Theorem I, § 1327.)

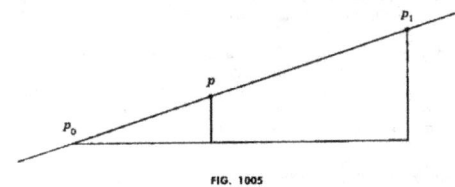

FIG. 1005

★21. If $p_0 : (a, b)$ and $p_1 : (c, d)$ are two distinct points in the Euclidean plane, and if $p : (x, y)$ is an arbitrary point on the line $p_0 p_1$ (cf. Fig. 1005), show that

(1) $\qquad (x - a) : (c - a) = (y - b) : (d - b),$

with a suitable interpretation in case any quantity vanishes. Letting t represent the common value of the ratios in (1), show that the line p_0p_1 can be represented parametrically:

(2) $\qquad x = a + t(c - a), \quad y = b + t(d - b),$

with consideration of the special case of a vertical or horizontal line. Observe that the segment p_0p_1 corresponds to the values $0 \leq t \leq 1$. Generalize these considerations to the space E_n, both for a line through two points $p_0 : (a_1, a_2, \cdots, a_n)$ and $p_1 : (b_1, b_2, \cdots, b_n)$ and the segment p_0p_1:

(3) $\qquad (a_1 + t(b_1 - a_1), \quad a_2 + t(b_2 - a_2), \quad \cdots, \quad a_n + t(b_n - a_n)).$

★**22.** Establish the **triangle inequality** in E_n: if p, q, and r are any three points in E_n, then

(4) $\qquad d(p, r) \leq d(p, q) + d(q, r).$

Hint: Use the Minkowski inequality of Ex. 44, § 107, with $a_i = y_i - x_i$ and $b_i = z_i - y_i$.

1006. FUNCTIONS AND LIMITS

The definition of *function* given in § 201 is sufficiently general to include functions of several independent variables. A function $f(x, y)$ of the two real variables x and y, for example, can be regarded as a function $f(p)$ of the variable point $p : (x, y)$ whose coordinates are x and y. The domain of definition, in this instance, would be a portion of the Euclidean plane; and the range, for a real-valued function, would be a portion of the real number system. In a similar way, a function of n real variables can be conveniently regarded as a function of a point in n-dimensional Euclidean space.

In this section and in following sections, unless explicit statement to the contrary is made, the word *function* will refer to a single-valued real-valued function of a variable point in n-dimensional Euclidean space, defined in a neighborhood of the particular point concerned (or possibly just for points in a set having the particular point as a limit point). For convenience, most definitions and theorems will be formulated for functions of two real variables. The generalizations to n variables are mostly obvious, and will usually be omitted.

We start with the basic definition of a finite limit as the independent variables approach finite limits, phrased in the convenient language of neighborhoods, followed by explicit formulations in terms of deltas and epsilons for both circular and square neighborhoods.

Definition I. *The function $f(x, y)$ has the limit L as the point (x, y) approaches the point (a, b) written*

$$\lim_{(x,\, y) \to (a,\, b)} f(x, y) = L, \quad \lim_{\substack{x \to a \\ y \to b}} f(x, y) = L, \text{ or } f(x, y) \to L \text{ as } (x, y) \to (a, b),$$

if and only if corresponding to an arbitrary neighborhood N_L of L, in the space E_1 of real numbers, there exists a deleted neighborhood $N_{(a,\,b)}$ of the point (a, b) in the Euclidean plane E_2, such that for every point (x, y) of $N_{(a, b)}$, for which $f(x, y)$ is defined, $f(x, y)$ belongs to N_L.

NOTE 1. In terms of circular neighborhoods, Definition I assumes the form: $\lim_{(x, y) \to (a, b)} f(x, y) = L$ if and only if corresponding to an arbitrary positive number ϵ there exists a positive number $\delta = \delta(\epsilon)$ such that $0 < \sqrt{(x - a)^2 + (y - b)^2} < \delta$ implies $|f(x, y) - L| < \epsilon$, for points (x, y) for which $f(x, y)$ is defined. For square neighborhoods the last implication is replaced by $|x - a| < \delta$, $|y - b| < \delta$, and $0 < |x - a| + |y - b|$ imply $|f(x, y) - L| < \epsilon$, for points (x, y) for which $f(x, y)$ is defined.

NOTE 2. As with the limit of a function of a single variable, it is immaterial what the value of the function $f(x, y)$ is at the point (a, b), or whether it is defined there at all. This fact appears in the preceding definition in the use of *deleted* neighborhoods.

Limits where either the independent variables or the dependent variable may become infinite can be formulated in a manner similar to definitions given in Chapter 2. A few samples are given here, others being requested in Exercises 12-13, § 1011. Notice that the Euclidean plane, like the space of real numbers, can be considered to have associated with it different kinds of "points at infinity."

Definition II.
$$\lim_{\substack{x, y \to +\infty}} f(x, y) = \lim_{\substack{x \to +\infty \\ y \to +\infty}} f(x, y) = L,$$

where L is a real number, if and only if corresponding to $\epsilon > 0$ there exists a number $N = N(\epsilon)$ such that $x > N$ and $y > N$ imply $|f(x, y) - L| < \epsilon$, for points (x, y) for which $f(x, y)$ is defined.†

Definition III.
$$\lim_{(x, y) \to \infty} f(x, y) = L,$$

where L is a real number, if and only if corresponding to $\epsilon > 0$ there exists a number $N = N(\epsilon)$ such that $\sqrt{x^2 + y^2} > N$ implies $|f(x, y) - L| < \epsilon$, for points (x, y) for which $f(x, y)$ is defined.

Definition IV.
$$\lim_{(x, y) \to (a, b)} f(x, y) = -\infty$$

if and only if corresponding to any number B there exists a positive number $\delta = \delta(B)$ such that $0 < \sqrt{(x - a)^2 + (y - b)^2} < \delta$ implies $f(x, y) < B$, for points (x, y) for which $f(x, y)$ is defined.

It will be seen that limits for functions of several variables are more varied and interesting than limits for functions of a single real variable.

† Compare the definition of Cauchy sequence, § 302.

One reason for this is that the variable point $p : (x_1, x_2, \cdots, x_n)$ may be considered as approaching a limit point in a much richer variety of ways. Another reason is that whenever the number of dimensions exceeds one, there is the related question of *iterated limits*, considered in the next section, where first one, and then the other, independent variable approaches its limit.

Postponing to later sections (§§ 1014, 1018, Ex. 19, § 1021) the study of *criteria* for the existence of a limit, we give an important *necessary condition* for such existence, with a consequent useful method for establishing *nonexistence* of a limit, depending on the idea of *path of approach*:

Definition V. *Let $\phi(t)$ and $\psi(t)$ be functions that approach finite limits a and b, respectively, but are never simultaneously equal to these limits, as t approaches some limit. Then the point (x, y) is said to **approach the point** (a, b) **as a limit**, along the path $x = \phi(t)$, $y = \psi(t)$. If a function $f(x, y)$ is defined along a path $x = \phi(t)$, $y = \psi(t)$, and if $f(\phi(t), \psi(t))$ approaches a limit L, finite or infinite, as t approaches its limit, the function $f(x, y)$ is said to **approach the limit** L **along the path** $x = \phi(t)$, $y = \psi(t)$.*

A similar definition holds for the case where the point (x, y) approaches an infinite limit. (Cf. Ex. 14, § 1011.)

Theorem I. *If a function $f(x, y)$ has a limit L, finite or infinite, as (x, y) approaches a point (a, b), then the function $f(x, y)$ approaches the limit L as (x, y) approaches the point (a, b) along any path.*

A similar theorem holds for the case where the point (x, y) approaches an infinite limit.

Proof. For definiteness, assume that L is finite (cf. Ex. 15, § 1011), let (x, y) approach (a, b) along the path $x = \phi(t)$, $y = \psi(t)$, and let N_L be an arbitrary neighborhood of L (in E_1). By assumption, there exists a deleted neighborhood $N_{(a, b)}$ of the point (a, b) (in E_2) such that for any point (x, y) in $N_{(a, b)}$ for which $f(x, y)$ is defined, $f(x, y)$ is in N_L. Since (x, y) is assumed to approach (a, b) along the path $x = \phi(t)$, $y = \psi(t)$, it follows that if t is required to be close enough to *its* limit, the point $(\phi(t), \psi(t))$ must be in $N_{(a, b)}$, and hence the number $f(\phi(t), \psi(t))$ must be in N_L. But by definition, this is what we mean by saying that $f(x, y)$ approaches L as (x, y) approaches (a, b) along the given path.

As a corollary (give the proof in Ex. 16, § 1011) we have

Theorem II. *If a function $f(x, y)$ has distinct limits as (x, y) approaches a point (a, b) along two distinct paths, the limit $\lim_{(x, y) \to (a, b)} f(x, y)$ does not exist.*

A similar theorem holds for the case where the point (x, y) approaches an infinite limit.

We consider two examples.

Example 1. Show that $\lim_{(x,y)\to(0,0)} \frac{xy}{x^2+y^2}$ does not exist.

Solution. Let $f(x,y) \equiv xy/(x^2+y^2)$ for $x^2+y^2 \neq 0$, $f(0,0) \equiv 0$. Then $f(x,y)$ is identically zero on each coordinate axis, and therefore its limit, as (x,y) approaches the origin along either axis is 0. On the other hand, let (x,y) approach the origin along the straight line $y = mx$, where m is a nonzero real number. Then
$$f(x, mx) = \frac{mx^2}{x^2 + m^2x^2} = \frac{m}{1+m^2} \to \frac{m}{1+m^2}.$$
Since this limit depends on the path of approach, the limit given initially cannot exist. The surface $z = f(x,y)$ can be visualized by considering it as generated by a line through and perpendicular to the z-axis.

Example 2. Consider the limit of the function
$$f(x,y) \equiv \frac{x^2 y}{x^4 + y^2}, \quad \text{when } x^4 + y^2 \neq 0, \ f(0,0) \equiv 0,$$
by letting (x,y) approach $(0,0)$ along different paths.

Solution. As in Example 1, $f(x,y)$ approaches the limit 0 along each coordinate axis. Along the straight line $y = mx$ ($m \neq 0$), the function has the limit
$$\lim_{x\to 0} f(x, mx) = \lim_{x\to 0} \frac{mx^3}{x^4 + m^2x^2} = \lim_{x\to 0} \frac{mx}{x^2 + m^2} = 0.$$
Is it correct to assume that, since $f(x,y)$ has the limit 0 as (x,y) approaches the origin along an arbitrary straight-line path, the limit $\lim_{(x,y)\to(0,0)} f(x,y)$ exists and is equal to 0? The possibly unexpected answer of "No" is obtained by considering the parabolic path $y = x^2$:
$$\lim_{x\to 0} f(x, x^2) = \lim_{x\to 0} \frac{x^4}{x^4 + x^4} = \frac{1}{2} \neq 0.$$
The surface $z = f(x,y)$, with its parabolic escarpment, is an interesting challenge to visualize.

1007. ITERATED LIMITS

As was mentioned in § 1006, an iterated limit associated with a limit of a function $f(x,y)$, as the point (x,y) approaches a limit, finite or infinite, is obtained by first allowing one independent variable, and then the other, to approach its limit. Thus, corresponding to the limit $\lim_{(x,y)\to(a,b)} f(x,y)$ are the two iterated limits $\lim_{x\to a} \lim_{y\to b} f(x,y)$ and $\lim_{y\to b} \lim_{x\to a} f(x,y)$. Similarly, associated with a limit of a function $f(x,y,z)$ of three variables are six iterated limits and, more generally, for a function of n independent variables there are $n!$ iterated limits.

The principal relation between limits and iterated limits is stated in the Theorem, below. A more difficult and subtle relation is the Moore-Osgood theorem given in § 1014. Other facts are revealed in the Examples that follow below, and summarized in the Notes. (Also cf. Ex. 32, § 1011.)

Theorem. *Of the three limits*

(1) $$\lim_{(x,\,y)\to(a,\,b)} f(x, y),$$

(2) $$\lim_{x\to a}\lim_{y\to b} f(x, y),$$

and

(3) $$\lim_{y\to b}\lim_{x\to a} f(x, y),$$

the existence, finite or infinite, of (1) *and either of the other two implies their equality. Consequently, if all three exist, they are all equal.*

Proof. Assume for definiteness that the limit (1) exists and is finite (cf. Ex. 17, § 1011), and denote this limit by L. We shall show that if the limit (2) exists, it must also be equal to L. Accordingly, let ϵ be an arbitrary positive number and choose a deleted square neighborhood $N_{(a,\,b)}$ of (a, b) such that for all points (x, y) in $N_{(a,\,b)}$ for which $f(x, y)$ is defined, $|f(x, y) - L| < \epsilon$. If δ denotes one-half the length of each side of $N_{(a,\,b)}$, and if x is a number such that $0 < |x - a| < \delta$ and for which the limit function $g(x) \equiv \lim_{y\to b} f(x, y)$ exists, then, since for y sufficiently near b the point (x, y) must lie in $N_{(a,\,b)}$, it follows that $|g(x) - L| \leq \epsilon$. (Why *does it follow?*) But this is precisely what we mean by the statement $\lim_{x\to a} g(x) = L$.

Note 1. The preceding proof has established more than was requested in the statement of the theorem. We know, in fact, that in the presence of the existence of (1), the existence of either *partial limit* $\lim_{y\to b} f(x, y)$ or $\lim_{x\to a} f(x, y)$ implies the existence of the corresponding limit (2) or (3), and hence equality with (1). (Cf. Ex. 20, § 1011.)

Example 1. As shown in Example 1, § 1006, if $f(x, y) \equiv xy/(x^2 + y^2)$ if $x^2 + y^2 \neq 0$, then $\lim_{(x,\,y)\to(0,\,0)} f(x, y)$ does not exist. However, both iterated limits exist and are equal to 0. For instance, $\lim_{x\to 0}\lim_{y\to 0} xy/(x^2 + y^2) = \lim_{x\to 0} 0 = 0$. (An identical statement applies to Example 2, § 1006.)

Example 2. Let $f(x, y) \equiv x + y \sin \frac{1}{x}$ if $x \neq 0$, and let $f(0, y)$ be undefined. Then, since $|f(x, y)| \leq |x| + |y|$, where defined, $\lim_{(x,\,y)\to(0,\,0)} f(x, y) = 0$ (use square neighborhoods and let $\delta = \delta(\epsilon) = \frac{1}{2}\epsilon$ in Definition I, § 1006, as interpreted in Note 1, § 1006). Furthermore, $\lim_{x\to 0}\lim_{y\to 0}\left(x + y\sin\frac{1}{x}\right) = \lim_{x\to 0} x = 0$. On the other hand, $\lim_{y\to 0}\lim_{x\to 0}\left(x + y\sin\frac{1}{x}\right)$ does not exist, since for $y \neq 0$, $\lim_{x\to 0}\left(x + y\sin\frac{1}{x}\right)$ does not exist.

Note 2. The preceding Examples demonstrate that for no two of the three limits (1), (2), (3) do existence and equality imply the existence of the third.

§ 1009] LIMIT AND CONTINUITY THEOREMS 307

Note 1. In terms of circular neighborhoods, Definition II assumes the form: $f(x, y)$ *is continuous at* (a, b) *if and only if corresponding to an arbitrary positive number* ϵ *there exists a positive number* $\delta = \delta(\epsilon)$ *such that* $\sqrt{(x - a)^2 + (y - b)^2} < \delta$ *implies* $|f(x, y) - f(a, b)| < \epsilon$, *for points* (x, y) *for which $f(x, y)$ is defined.* For square neighborhoods the last implication is replaced by $|x - a| < \delta$ *and* $|y - b| < \delta$ *imply* $|f(x, y) - f(a, b)| < \epsilon$, *for points* (x, y) *for which $f(x, y)$ is defined.*

Note 2. Any formulation of continuity employs *full neighborhoods* in distinction to the *deleted neighborhoods* used for limits.

Note 3. Definition II is applicable even when the point (a, b) is not a limit point of the domain of definition of $f(x, y)$. In this case (a, b) is called an *isolated point* of the domain of definition, and $f(x, y)$ is automatically continuous there, even though $\lim_{(x, y) \to (a, b)} f(x, y)$ has no meaning.

The Theorem of § 209, which states that for a continuous function *the limit of the function is the function of the limit*, generalizes to continuous functions of any number of variables (cf. Exs. 22, 25-29, § 1011):

Theorem. *If $f(x, y)$ is continuous at (a, b) and if $\phi(t)$ and $\psi(t)$ are functions such that* $\phi(t) \to a$ *and* $\psi(t) \to b$, *as t approaches some limit, then* $f(\phi(t), \psi(t)) \to f(a, b)$:*

$$\lim f(\phi(t), \psi(t)) = f(\lim \phi(t), \lim \psi(t)).$$

In particular, if the point (x, y) approaches (a, b) along a path $x = \phi(t)$, $y = \psi(t)$, then $f(x, y)$ approaches $f(a, b)$ along the path.

The proof is left as an exercise (Ex. 21, § 1011).

The definitions of *continuity on a set* and *continuity*, given in § 209, remain unchanged, as does the definition of a *removable discontinuity*, in § 210.

1009. LIMIT AND CONTINUITY THEOREMS

The limit theorems of § 207 and the continuity theorems of § 211, and their proofs, are altered only in minor details when applied to functions of several variables. We shall feel free, therefore, to make use of them without specific delineation here. (Cf. Exs. 22, 25-29, § 1011.)

There is, however, an essentially new problem before us now, and that is the relation between a function $f(x, y)$ of two variables (or more generally, n variables) and this same function $f(x, y)$, considered as a function of just one of the independent variables, the other being held fixed. The two most basic theorems in this connection follow, formulated for simplicity for a function of two variables.

Theorem I. *If $f(x, y)$ is continuous at (a, b), then $f(x, b)$ is a continuous function of x at $x = a$, and $f(a, y)$ is a continuous function of y at $y = b$.*

That is, *if a function of two variables is continuous in both variables together, it is continuous in each variable separately.*

Proof. Let $\epsilon > 0$ be given, and let $\delta > 0$ be such that
$$\sqrt{(x-a)^2 + (y-b)^2} < \delta$$
implies $|f(x,y) - f(a,b)| < \epsilon$. Then the inequality $|x - a| < \delta$, since it is equivalent to $\sqrt{(x-a)^2 + (b-b)^2} < \delta$, implies $|f(x,b) - f(a,b)| < \epsilon$. This proves continuity of $f(x,b)$. The other half of the proof is similar.

It will be shown by Example 1, below, that the converse of Theorem I is false: *It is possible for a function of two variables to be continuous in each separately without being continuous in both together.* A partial converse, however, is contained in the following theorem.

Theorem II. *Let $g(x)$ be defined at $x = a$, and let $f(x,y) \equiv g(x)$ for all values of x for which $g(x)$ is defined, and for all values of y. If b is an arbitrary real number, $f(x,y)$ is continuous at (a,b) if and only if $g(x)$ is continuous at a.*

Proof. Half (the "only if" implication) is true by Theorem I. For the other half, assume $g(x)$ is continuous at a, and determine, for a given $\epsilon > 0$, the number $\delta > 0$ so that $|x - a| < \delta$ implies $|g(x) - g(a)| < \epsilon$. Then $\sqrt{(x-a)^2 + (y-b)^2} < \delta$ implies $|x - a| < \delta$, and therefore also $|g(x) - g(a)| = |f(x,y) - f(a,b)| < \epsilon$.

As a consequence of the theorems mentioned at the beginning of this section, and of Theorem II, functions of several variables that are made up, in an obviously simple way, of continuous functions of single variables are continuous. In particular, any polynomial (a polynomial in x and y is a sum of terms of the form $ax^m y^n$, where a is a constant and m and n are nonnegative integers) is continuous, and any rational function (a quotient of polynomials) is continuous wherever defined. As further examples, $e^x \cos y$ and $\cos(nx - y \sin x)$ are continuous functions of x and y together. (Cf. Ex. 22, § 1011.)

Examples. The two examples of § 1006 are continuous everywhere, except at the origin, where they are discontinuous. However, each is a continuous function of each variable for every value of the other! (Verify this.) Example 3, § 1007, is continuous in each open quadrant, and at the origin, but at each point of either axis, different from the origin, the function is discontinuous. At such points continuity is hopeless, since the limit fails to exist. That is, there are no removable discontinuities.

1010. MORE THEOREMS ON CONTINUOUS FUNCTIONS

The theorems of § 213 have their generalizations to n-dimensional space. In these theorems *intervals* are replaced by higher dimensional sets, of which open and closed intervals are special one-dimensional cases. The

§ 1011] EXERCISES

key notion for the first two theorems is *compactness*, and that for the last two is *region*. (Cf. § 1019, and the Theorem, § 1020.) Proofs of the theorems are given in § 1020. The functions are assumed to be real-valued.

Theorem I. *A function continuous on a nonempty compact set is bounded there.* That is, if $f(p)$ is continuous on a nonempty compact set A, there exists a number B such that $|f(p)| \leq B$ for all points p belonging to the set A.

Theorem II. *A function continuous on a nonempty compact set has a maximum and a minimum there.* That is, if $f(p)$ is continuous on a nonempty compact set A, there exist points p_1 and p_2 in A such that $f(p_1) \leq f(p) \leq f(p_2)$ for all points p belonging to A.

Theorem III. *If $f(p)$ is continuous on a region or closed region R, and if p_1 and p_2 are two points of R such that $f(p_1)$ and $f(p_2)$ have opposite signs, then there is a point p_3 in the interior of R for which $f(p_3) = 0$.*

Theorem IV. *A function continuous on a region or closed region assumes in its interior all values between any distinct two of its values.*

1011. EXERCISES

In Exercises 1–10, determine whether the indicated limit exists, and if it does, find it. Give reasons.

1. $\lim\limits_{(x,\, y) \to (0,\, 0)} \dfrac{xy}{x^4 + y^4}$.

2. $\lim\limits_{(x,\, y) \to (0,\, 0)} \dfrac{x + ye^{-x^2}}{1 + y^2}$.

3. $\lim\limits_{(x,\, y) \to (0,\, 0)} \dfrac{x^2 y^2}{x^2 + y^2}$.

4. $\lim\limits_{(x,\, y) \to (0,\, 0)} \dfrac{x^2 y^3}{x^6 + y^4}$.

5. $\lim\limits_{\substack{m \to +\infty \\ n \to +\infty}} \dfrac{m}{m + n}$, m and n positive integers.

6. $\lim\limits_{\substack{m \to +\infty \\ n \to +\infty}} \dfrac{mn}{m^2 + n^2}$, m and n positive integers.

7. $\lim\limits_{\substack{m \to +\infty \\ n \to +\infty}} \dfrac{m + n}{m^2 + n^2}$, m and n positive integers.

8. $\lim\limits_{\substack{m \to +\infty \\ n \to +\infty}} \dfrac{n}{m^2 e^{n/m}}$, m and n positive integers.

9. $\lim\limits_{(x,\, y) \to (0,\, 0)} \dfrac{xy^2}{x^4 + y^2}$.

10. $\lim\limits_{(x,\, y) \to (0,\, 0)} xy \ln (x^2 + y^2)$.

11. Generalize to a function of n independent real variables the formulations of limit given in Definition I and Note 1, § 1006.

12. Give an explicit definition for $\lim\limits_{\substack{x \to +\infty \\ y \to +\infty}} f(x, y) = +\infty$.

13. Give an explicit definition for $\lim\limits_{(x,\, y) \to \infty} f(x, y) = \infty$.

14. Reformulate Definition V, § 1006, for the cases (a) $x \to +\infty$, $y \to +\infty$; (b) $(x, y) \to \infty$.

310   FUNCTIONS OF SEVERAL VARIABLES   [§ 1011

15. Prove Theorem I, § 1006, for the cases (a) $(x, y) \to (a, b)$, $L = \infty$; (b) $(x, y) \to \infty$, $L = +\infty$.

16. Prove Theorem II, § 1006.

17. Prove the Theorem, § 1007, for the case $L = -\infty$.

18. Construct an example illustrating the same principle as that of Example 3, § 1007, but where $x \to +\infty$, $y \to +\infty$.

19. Construct an example illustrating the same principle as that of Example 5, § 1007, but where the independent variables approach finite limits.

20. Prove that if $\lim_{(x,y)\to(a,b)} f(x, y)$ and $\lim_{y\to b} f(x, y)$ both exist, for all x and y near a and b, respectively, then the iterated limit $\lim_{x\to a}\lim_{y\to b} f(x, y)$ exists and is equal to $\lim_{(x,y)\to(a,b)} f(x, y)$. Show that this result is still valid if x and y are restricted by the added condition $x \neq a$, $y \neq b$. (Cf. Note 1, § 1007.)

21. Prove the Theorem of § 1008.

22. Prove the following form of extension of Theorem VI, § 211 ("a continuous function of a continuous function is continuous"): If $f(x, y)$ is continuous at $x = a$, $y = b$, and if $g(r)$ is continuous at $r = f(a, b)$, then $h(x, y) = g(f(x, y))$ is continuous at (a, b). State and prove the corresponding extension of the Theorem of § 209 ("the limit of the function is the function of the limit").

23. Show that $\ln(1 + x^2 + y^2) + e^x \sin y$ is everywhere continuous. (Cf. Ex. 22.)

24. Discuss continuity of the function $[x + y]$, where the brackets indicate the bracket function of Example 5, § 201.

25. Prove the following form of extension of Theorem VI, § 211 (cf. Ex. 22): If $\phi(t)$ and $\psi(t)$ are continuous at $t = t_0$, and if $f(x, y)$ is continuous at $x = a = \phi(t_0)$, $y = b = \psi(t_0)$, then $h(t) = f(\phi(t), \psi(t))$ is continuous at $t = t_0$. (Cf. the Theorem, § 1008.)

In Exercises 26-29, state and prove extensions of the Theorem of § 209 and Theorem VI, § 211, for a composite function of the specified form (cf. Exs. 22, 25).

26. $f(\phi(s), \psi(t))$.   **27.** $f(\phi(s, t), \psi(s, t))$.
28. $f(\phi(s, t), \psi(u, v))$.   **29.** $f(\phi(x), \psi(y, z))$.

30. Formulate and prove the Negation of Continuity of a function of n variables (cf. Ex. 14, § 212).

★31. Let $f(x, y)$ be defined on a closed rectangle R: $a \leq x \leq b$, $c \leq y \leq d$, and assume that for each x, $f(x, y)$ is a continuous function of y, and that for all y, $f(x, y)$ is a *uniformly* continuous function of x in the sense that corresponding to $\epsilon > 0$ there exists $\delta > 0$ such that $|x_1 - x_2| < \delta$ implies $|f(x_1, y) - f(x_2, y)| < \epsilon$ for all x_1 and x_2 of $[a, b]$ and all y of $[c, d]$. Prove that $f(x, y)$ is continuous on R.

★32. Define the multiple-valued function $\overline{\lim}_{y\to b} f(x, y)$ to be the function $g(x)$ having *all* values satisfying the inequalities $\underline{\lim}_{y\to b} f(x, y) \leq g(x) \leq \overline{\lim}_{y\to b} f(x, y)$, and define $\overline{\lim}_{x\to a} f(x, y)$ similarly. Discuss the **generalized iterated limits** $\lim_{x\to a} \overline{\lim}_{y\to b} f(x, y)$ and $\lim_{y\to b} \overline{\lim}_{x\to a} f(x, y)$. Prove that if the limit $\lim_{(x,y)\to(a,b)} f(x, y)$ exists, then both generalized iterated limits exist and are equal to it.

★1012. UNIFORM LIMITS

A *sequence* $\{S_n(x)\}$ of functions of one real variable can be regarded as a *single* function of the *two* real variables x and n:

(1) $$S_n(x) = f(x, n).$$

For example, if x is restricted to the closed interval $[0, 1]$, the function has as its domain of definition the set of all points (x, n) in the Euclidean plane such that $0 \leq x \leq 1$ and $n =$ a positive integer; this domain, therefore, consists of infinitely many horizontal line segments.

Viewed in this way, sequences of functions fall heir to the ideas of limits and iterated limits of §§ 1006 and 1007. Let us confess immediately, however, that whatever interest resides in this fact is principally conceptual. What is more important is to go in the other direction, and carry over the concept of *uniform convergence* to more general functions of two variables.

In order to focus our attention on a specific objective, we shall formulate things in this section in terms of a function $f(x, y)$ and a finite limit c for the variable y. Cases where y has an infinite or a one-sided limit are similar, and some are included in Exercise 1, § 1015. For many properties of uniform limits, the point x may be replaced by a point in E_n (cf. Ex. 2, § 1015).

Definition. *Let $f(x, y)$ be defined for all points x in a set A and all points y in a set B having $y = c$ as a limit point. Then $f(x, y)$ converges to a limit $f(x)$ **uniformly** for x in A, as $y \to c$, written*

(2) $$f(x, y) \rightrightarrows f(x) \text{ on } A, \quad \text{as } y \to c,$$

if and only if corresponding to $\epsilon > 0$ there exists a positive number $\delta = \delta(\epsilon)$, dependent on ϵ alone and not on the point x, such that

(3) $$0 < |y - c| < \delta \quad \text{implies} \quad |f(x, y) - f(x)| < \epsilon$$

*for every x in A. The function $f(x)$ is called a **uniform limit** of $f(x, y)$.*

We formulate for future use:

Negation of Uniform Convergence. *Let $f(x, y)$ have its domain as specified in the preceding Definition. Then $f(x, y)$ fails to converge uniformly to a limit $f(x)$, for x in A, as $y \to c$ if and only if there exists a positive number ϵ having the property that for any positive number δ there exist a point x in A and a point y in B such that*

(4) $$0 < |y - c| < \delta \quad \text{and} \quad |f(x, y) - f(x)| \geq \epsilon.$$

In order to facilitate proofs of the three fundamental theorems on uniform limits, given in the next section, we establish the following relation between uniform limits in general and uniform limits of sequences:

Theorem. *With the assumptions and notation of the preceding Definition,*

(5) $\qquad f(x, y) \rightrightarrows \text{some function } f(x), \quad \text{as} \quad y \to c,$

if and only if for every sequence $\{y_n\}$ of points of B converging to c, and never equal to c,

(6) $\qquad \{f(x, y_n)\} \text{ converges uniformly.}$

In case the uniform limit of (6) exists for every $\{y_n\}$, it is the same for every $\{y_n\}$, and equal to the uniform limit $f(x)$ of (5).

Proof. The "only if" part of the proof should be trivial to the reader at this stage, but we advise his supplying the details.

As a first step in the "if" part of the proof, we observe that if the uniform limit of $\{f(x, y_n)\}$ exists for every $\{y_n\}$, it must be the *same* limit for every $\{y_n\}$ (cf. Ex. 8, § 305). We define $f(x)$ to be this limit, and now seek a contradiction to the assumption that $f(x, y)$ does *not* approach $f(x)$ uniformly, as $y \to c$. According to the Negation of Uniform Convergence, for some positive number ϵ we can choose for every positive integer n the positive number $\delta = 1/n$, and construct sequences of points x_n of A and y_n of B such that

$$0 < |y_n - c| < 1/n \quad \text{and} \quad |f(x_n, y_n) - f(x_n)| \geq \epsilon.$$

The first of these two inequalities guarantees the convergence of $\{y_n\}$ to c, while the other prohibits the uniform convergence of the sequence $\{f(x, y_n)\}$ to the function $f(x)$.

★1013. THREE THEOREMS ON UNIFORM LIMITS

The three basic theorems of §§ 905-907 apply to uniform limits in general. We quote them now for the case where y has a finite limit c. (Cf. Ex. 9, § 1015.) We prove the first, and leave the other two to the exercises (Ex. 10, § 1015).

Theorem I. *Let $f(x, y)$ be defined for all points x in a set A and all points y in a set B having $y = c$ as a limit point. If, for each point y of B, $f(x, y)$ is a continuous function of x for every point x of A, and if $f(x, y) \rightrightarrows f(x)$ on A, as $y \to c$, then $f(x)$ is continuous on A.*

Proof. Under the assumptions of the theorem, if $\{y_n\}$ is a sequence of points of B, none equal to c, converging to c, then, by the Theorem of § 1012, $\{f(x, y_n)\}$ is a sequence of functions of x converging uniformly on A to the function $f(x)$. Therefore, by the Theorem of § 905, $f(x)$ is continuous on A.

Theorem II. *Let $f(x, y)$ be defined for $a \leq x \leq b$, and all points y in a set B having $y = c$ as a limit point. If, for each point y of B, $f(x, y)$ is integrable on $[a, b]$, and if $f(x, y) \rightrightarrows f(x)$ on $[a, b]$, as $y \to c$, then $f(x)$ is integrable on $[a, b]$ and*

$$\lim_{y\to c}\int_a^b f(x,y)dx = \int_a^b \lim_{y\to c} f(x,y)dx = \int_a^b f(x)dx.$$

Theorem III. *Let $f(x,y)$ be defined as in Theorem II, and assume furthermore:*

(i) for each point y of B, $f(x,y)$ as a function of x has a derivative

$$f_1(x,y) = \frac{d}{dx} f(x,y)$$

at every point of $[a,b]$;

(ii) $\lim_{y\to c} f(x_0,y)$ exists for some point x_0 of $[a,b]$;

(iii) $f_1(x,y) \rightrightarrows$ a function $\phi(x)$ on $[a,b]$, as $y \to c$.

Then

(iv) $f(x,y) \rightrightarrows$ a function $f(x)$ on $[a,b]$, as $y \to c$;

(v) $f(x)$ is differentiable on $[a,b]$;

(vi) $f'(x) = \phi(x) = \lim_{y\to c} f_1(x,y)$.

★★1014. THE MOORE-OSGOOD THEOREM

In § 1007 the three limits

(1) $$\lim_{(x,y)\to(a,b)} f(x,y),$$

(2) $$\lim_{x\to a}\lim_{y\to b} f(x,y),$$

and

(3) $$\lim_{y\to b}\lim_{x\to a} f(x,y),$$

were discussed, and it was found that if (1) and either of the two iterated limits exist they must be equal. A stronger relation, given in Note 1, § 1007, is that if (1) exists and if either **partial limit** $\lim_{y\to b} f(x,y)$ or $\lim_{x\to a} f(x,y)$ exists, the corresponding limit (2) or (3) exists (and is equal to (1)). The property of *uniformity* permits a different implication, essentially a converse of the preceding statement, which says in brief that if the *two* partial limits exist, and if *one* of them is uniform, then all three of the limits (1), (2), and (3) exist (and, of course, are equal).

The precise statement follows (cf. Ex. 37, § 908, for another form of this theorem):

Theorem. Moore-Osgood Theorem.† *Let a and b be real numbers, and A and B sets of real numbers, such that a is a limit point but not a member of A and b is a limit point but not a member of B, and let $f(x,y)$ be a real-*

---

† E. H. Moore presented the essential content of this theorem in lectures at the University of Chicago in 1900. W. F. Osgood, in 1906, included the formulation for the case $x \to +\infty$, $y \to +\infty$ through integral values, in the first edition of his *Funktionentheorie*. Cf. W. F. Osgood, *Funktionentheorie* I (Leipzig, B. G. Teubner, 1928), p. 619.

314   FUNCTIONS OF SEVERAL VARIABLES   [§ 1015

valued function whose domain of definition D consists of all points (x, y) such that x and y belong to A and B, respectively. Assume, furthermore:

(i) $g(x) \equiv \lim_{y \to b} f(x, y)$ exists and is finite for every x in A;

(ii) $h(y) \equiv \lim_{x \to a} f(x, y)$ exists and is finite for every y in B;

(iii) either (i) is uniform for x in A or (ii) is uniform for y in B.

Then all three limits (1), (2), and (3) exist and are finite and equal.

Similar statements hold if x or y or both have infinite limits.

Proof. Let us assume for definiteness that $f(x, y) \rightrightarrows g(x)$ on A as $y \to b$. One procedure now would be to let $y_n \to b$ and use Exercise 37, § 908. However, the analytical techniques are sufficiently important to merit an independent proof. For this we rely on the Cauchy criterion for functions of a real variable (§ 304).

We start by proving that the limit (2), $\lim_{x \to a} g(x)$, exists and is finite. We write

(4) $|g(x') - g(x'')| \leq |g(x') - f(x', y)| + |f(x', y) - f(x'', y)|$
$$+ |f(x'', y) - g(x'')|.$$

If $\epsilon > 0$, choose y in B such that $|f(x, y) - g(x)| < \epsilon/3$ for all x in A, and hold this y fixed. Then choose $\delta > 0$ such that for $0 < |x' - a| < \delta$ and $0 < |x'' - a| < \delta$, the middle term on the right in (4) is less than $\epsilon/3$. Then $|g(x') - g(x'')| < \epsilon$, and by the Cauchy criterion, $\lim_{x \to a} g(x)$ exists and is finite. Denote this limit by L.

Now further restrict $\delta > 0$ so that $0 < |y - b| < \delta$ and y in B imply $|f(x, y) - g(x)| < \epsilon/2$ for all x in A and $0 < |x - a| < \delta$ and x in A imply $|g(x) - L| < \epsilon/2$. Then for all (x, y) in D for which $0 < |x - a| < \delta$ and $0 < |y - b| < \delta$,

$$|f(x, y) - L| \leq |f(x, y) - g(x)| + |g(x) - L| < \epsilon,$$

so that the limit (1) exists and is equal to L.

By Note 1, § 1007, the proof is complete.

★1015. EXERCISES

★1. Formulate the definition of a uniform limit of a function $f(x, y)$ as $y \to +\infty$; as $y \to \infty$; as $y \to c+$; as $y \to c-$.

★2. Adapt and prove the Theorem of § 1012, and Theorem I, § 1013, when x is replaced by a point in E_n.

In Exercises 3–4, formulate and establish the Negation of the given statement.

★3. $f(x, y) \rightrightarrows f(x)$ on A, as $y \to c+$.

★4. $S_{mn} \rightrightarrows A_m$ for $m \geq M$, as $n \to +\infty$.

In Exercises 5–8, show that the specified limit is uniform on the first of the two

§ 1016]   MORE GENERAL FUNCTIONS.   MAPPINGS   315

given sets, and that the limit exists but is not uniform on the second of the two given sets. The letter η denotes an arbitrarily small positive number.

★5. $\lim_{y\to 0} \dfrac{xy}{x^2 + y^2}$; $[\eta, +\infty)$; $(0, +\infty)$.

★6. $\lim_{y\to 0+} \dfrac{xy}{1 + xy}$; $[0, b]$; $[0, +\infty)$.

★7. $\lim_{x\to 0} nx + 1$; $a \le n \le b$; $-\infty < n < +\infty$.

★8. $\lim_{n\to\infty} \dfrac{mn}{m^2 + n^2}$; $|m| \le K$; $|m| < +\infty$.

★9. State the three theorems of § 1013 for the case $y \to +\infty$; for the case $y \to c+$.

★10. Prove Theorems II and III, § 1013, for the case $y \to c$; for the case $y \to \infty$.

In Exercises 11-14, find $\delta(\epsilon)$ or $N(\epsilon)$ explicitly, according to the Definition, § 1012, or Exercise 1, above, for the specified interval.

★11. $\lim_{y\to 0} \dfrac{xy}{x^2 + y^2}$; $[\eta, +\infty)$.   ★12. $\lim_{y\to 0+} \dfrac{xy}{1 + xy}$; $[0, b]$.

★13. $\lim_{x\to 0} nx + 1$; $a \le n \le b$.   ★14. $\lim_{n\to\infty} \dfrac{mn}{m^2 + n^2}$; $|m| \le K$.

★15. Verify that $\lim_{y\to 0} \int_1^2 \dfrac{xy}{x^2 + y^2}\, dx = \int_1^2 \lim_{y\to 0} \dfrac{xy}{x^2 + y^2}\, dx$ (cf. Ex. 5).

★16. Verify that $\lim_{y\to 0+} \int_0^b \dfrac{xy}{1 + xy}\, dx = \int_0^b \lim_{y\to 0+} \dfrac{xy}{1 + xy}\, dx$ (cf. Ex. 6).

★17. Discuss the equation $\lim_{y\to 0} \dfrac{d}{dx} \ln(x+y) = \dfrac{d}{dx}\left[\lim_{y\to 0} \ln(x+y)\right]$.

★18. Discuss the equation $\lim_{y\to 0} \dfrac{d}{dx} \dfrac{xy^2}{x^2 + y^2} = \dfrac{d}{dx}\left[\lim_{y\to 0} \dfrac{xy^2}{x^2 + y^2}\right]$, with particular attention to the point $x = 0$.

★19. Prove the following **Cauchy criterion** for a function $f(x, y)$: For x in a set A, $f(x, y) \rightrightarrows f(x)$, as $y \to c$, if and only if corresponding to $\epsilon > 0$ there exists $\delta = \delta(\epsilon) > 0$ such that $0 < |y' - c| < \delta$ and $0 < |y'' - c| < \delta$ imply $|f(x, y') - f(x, y'')| < \epsilon$ for every x in A. (Cf. Ex. 4δ, § 904.)

★20. Assume that $f(x)$ has a continuous derivative $f'(x)$ on a closed interval I. Prove that the difference quotient $[f(x + \Delta x) - f(x)]/\Delta x$ approaches $f'(x)$ uniformly on I as $\Delta x \to 0$.

1016. MORE GENERAL FUNCTIONS. MAPPINGS

It was remarked in § 1006 that the original definition of function (§ 201) permits consideration of real-valued functions of several real variables. This was accomplished by letting the *domain of definition* of the function belong to the n-dimensional space E_n. In similar fashion it is possible to allow the *range of values* more latitude by letting it form part of a Euclidean space E_m. For distinguishing purposes we shall use the term **mapping** or **transformation** to mean a single-valued function whose values are not

316 FUNCTIONS OF SEVERAL VARIABLES [§ 1016

assumed to be real numbers, preserving the word *function* (unless otherwise qualified) to mean "real-valued function." One reason for the use of a distinguishing term is that mappings into higher-dimensional spaces are usually not given by single formulas, but rather by *systems* of real-valued functions:

$$(1) \quad \begin{cases} y_1 = f_1(x_1, x_2, \cdots, x_n), \\ y_2 = f_2(x_1, x_2, \cdots, x_n), \\ \cdots\cdots\cdots\cdots\cdots \\ y_m = f_m(x_1, x_2, \cdots, x_n). \end{cases}$$

It is convenient for many purposes to use functional notation for mappings and write, for instance, $y = f(x)$, $y = T(x)$, $Y = f(X)$, or $q = T(p)$, where x, y, X, Y, p, and q represent points in appropriate spaces. A slight modification of this notation that is particularly suitable for mappings of a space into itself is suggested by the example $f((x, y)) = (2x, 3y)$, which transforms an arbitrary point of the Euclidean plane into the point with twice the abscissa and three times the ordinate. Still another method uses arrows, and would indicate the mapping just considered by the symbols $(x, y) \to (2x, 3y)$.

Definitions of limit and continuity, together with many of their simpler properties, are produced by properly interpreting the word *neighborhood* for the appropriate Euclidean space. We shall feel free to use any of these concepts or properties without boring the reader with details that

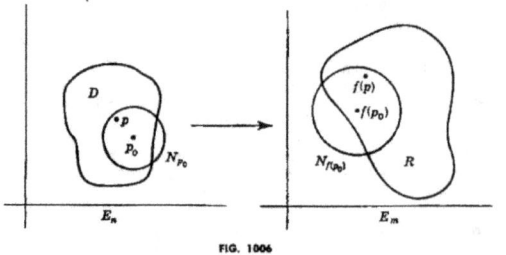

FIG. 1006

are merely tedious repetitions of work already done. One typical definition should suffice (cf. Fig. 1006):

Definition. *A mapping $f(p)$, whose domain of definition D and range of values R belong to E_n and E_m, respectively, is* **continuous** *at the point p_0 in D if and only if corresponding to an arbitrary neighborhood $N_{f(p_0)}$ of $f(p_0)$, in*

§ 1016] MORE GENERAL FUNCTIONS. MAPPINGS 317

E_m, there exists a neighborhood N_{p_0} of p_0, in E_n, such that for every point p of N_{p_0} that belongs to D, $f(p)$ belongs to $N_{f(p_0)}$.

How can one determine whether a mapping given by a system of functions is continuous or not? The answer is simple:

Theorem I. *The mapping* (1) *is continuous at a given point if and only if every* f_k, $k = 1, 2, \cdots, m$, *is continuous at that point.*

Proof. Let the given point be $(a_1, a_2, \cdots, a_n)$ and define
$$b_k \equiv f_k(a_1, a_2, \cdots, a_n), k = 1, 2, \cdots, m.$$
The result follows from the inequalities, true for each $k = 1, 2, \cdots, m$:
$$|y_k - b_k| \leq \sqrt{(y_1 - b_1)^2 + \cdots + (y_m - b_m)^2} \leq |y_1 - b_1| + \cdots + |y_m - b_m|.$$
(Give the details.)

A mapping f with domain D and range R is said to **map D onto R.** If a set A contains R (possibly being identical with R), f is said to **map D into A.** The range R is called the **image** of the domain D, and D is called the **preimage** or **inverse image** of R, under the mapping. If f is continuous on D, R is called a **continuous image** of D. A convenient symbolism for "onto" and "into" mappings is the following:

$$D \xrightarrow[\text{onto}]{f} R \quad \text{and} \quad D \xrightarrow[\text{into}]{f} A.$$

If a mapping $y = f(x)$ of D onto R is *one-to-one*, the **inverse mapping** $x = f^{-1}(y)$ of R onto D exists, and the question of its continuity arises (cf. § 215 on monotonic functions and their inverses). The simplest satisfactory statement on this matter follows (cf. § 1019 for a proof):

Theorem II. *Any continuous one-to-one mapping whose domain is compact has a continuous inverse.*

NOTE. A mapping that is continuous and whose inverse is also continuous, is called **bicontinuous** or **topological.** A topological mapping is also called a **homeomorphism.** Two sets are called **topologically equivalent,** or **homeomorphic,** and either set is called a **homeomorph** of the other, in case there exists a homeomorphism having one of the sets as domain and the other as range. (Cf. §§ 1024-1028 for a brief discussion of other topological ideas.)

The question of *existence* of inverse mappings is considered in Chapter 12.

Example 1. The pair of functions
$$(2) \quad \begin{cases} u = x + y, \\ v = x - y, \end{cases}$$
defines a mapping of the space E_2 whose points are designated (x, y) into the space E_2 whose points are designated (u, v). It is possible to solve explicitly for x and y in terms of u and v:

318   FUNCTIONS OF SEVERAL VARIABLES   [§ 1017

$$(3) \quad \begin{cases} x = \tfrac{1}{2}(u + v), \\ y = \tfrac{1}{2}(u - v). \end{cases}$$

From this fact, and the form of the functions involved, it is clear that the given mapping maps E_2 *onto* E_2, is one-to-one, and has a continuous inverse. Rewriting (2) in the form

$$(4) \quad \begin{cases} u = \sqrt{2}\left(x \cos \tfrac{\pi}{4} + y \sin \tfrac{\pi}{4}\right), \\ v = -\sqrt{2}\left(-x \sin \tfrac{\pi}{4} + y \cos \tfrac{\pi}{4}\right), \end{cases}$$

shows that the mapping is the result of a rotation through 45° together with a change of scale of magnitude $\sqrt{2}$, and a reflection.

Example 2. The mapping

$$(5) \quad \begin{cases} u = x^2 - y^2, \\ v = 2xy, \end{cases}$$

is continuous everywhere in E_2. Since the points (x, y) and $(-x, -y)$ are both mapped into the same point, it is clear that the mapping is one-to-one only at the origin. To study this mapping more thoroughly we write the coordinates in polar form, with $x = r \cos \theta$, $y = r \sin \theta$, $u = \rho \cos \phi$, $v = \rho \sin \phi$, and find:

$$(6) \quad \begin{cases} \rho = r^2, \\ \phi = 2\theta \end{cases}$$

(neglecting multiples of 2π). A point in the xy-plane, therefore, is mapped into a point in the uv-plane having its distance from the origin squared, and its polar angle doubled. Therefore, if the domain of definition of the mapping is sufficiently restricted the mapping becomes one-to-one. For example, the open first quadrant is mapped onto the open upper half-plane in a one-to-one manner. With such a restriction, the inverse mapping is always continuous at any point q other than the origin for the following reason: choose a neighborhood N_p of the corresponding point p in the xy-plane such that N_p subtends an angle at the origin of less than $\pi/4$ (say). Then the image, A, of the closure of N_p contains q in its interior. By Theorem II, the mapping with domain restricted to N_p has its inverse continuous on A, and therefore at q.

Example 3. The mapping

$$(7) \quad (x, y) \to (x, 0),$$

which transforms the point (x, y) into the point on the x-axis having the same abscissa as that of the original point, is everywhere continuous in E_2, but it has no inverse unless the domain is restricted to such an extent that no vertical line meets it in more than one point. It is called a **projection**.

★1017. UNIFORM CONTINUITY

The concept of uniform continuity for mappings in general is virtually the same as that for functions of a single variable:

Definition. *A mapping $f(p)$, defined on a set A, is **uniformly continuous** on A if and only if corresponding to an arbitrary positive number ϵ there exists* a positive number $\delta = \delta(\epsilon)$ such that any two points, p_1 and p_2, of A such that $d(p_1, p_2) < \delta$ map into points $f(p_1)$ and $f(p_2)$ such that $d(f(p_1), f(p_2)) < \epsilon$.

The fact that a real-valued function continuous on a closed interval is uniformly continuous there generalizes to the following (cf. Theorem IV, § 311; cf. § 1019 for a proof):

Theorem. *Any mapping continuous on a compact set is uniformly continuous there.*

★1018. SOME THEORETICAL CONSIDERATIONS

In order to treat adequately the unproved theorems of the preceding sections, we turn our attention to the propositions of Chapter 3, many of which extend to real-valued functions of n variables and, more generally, to mappings with domains in a Euclidean space E_n and ranges in a Euclidean space E_m.

We start by defining convergence of a sequence of points in the space E_m. Let us remark in passing that this is merely an explicit formulation of a limit of a mapping whose domain is the positive integers (a set in the space E_1) and whose range is in the space E_m.

Definition I. *In the Euclidean space E_m, a sequence of points $\{p_n\}$ **converges** to a point p, written*

$$\lim_{n \to +\infty} p_n = p \quad \text{or} \quad \lim_{n \to \infty} p_n = p \quad \text{or} \quad p_n \to p,$$

if and only if corresponding to an arbitrary positive number ϵ there exists a number $N = N(\epsilon)$ such that $n > N$ implies $d(p_n, p) < \epsilon$.

NOTE. The last part of this definition can be phrased: $\cdots$ *if and only if every neighborhood of p contains all but a finite number of terms of the sequence.*

Our first task with sequences will be to establish a simple relation between convergence of sequences of points in E_m and convergence of sequences of real numbers. For notational convenience we formulate the statement and proof for the special case $m = 2$, (Cf. Ex. 9, § 1021.)

Theorem I. *A sequence of points $\{p_n\} = \{(x_n, y_n)\}$ in E_2 converges if and only if both sequences of real numbers $\{x_n\}$ and $\{y_n\}$ converge. More precisely, (i) if $p_n \to p = (a, b)$, then $x_n \to a$ and $y_n \to b$; and (ii) if $x_n \to a$ and $y_n \to b$, then $(x_n, y_n) \to (a, b)$.*

Proof. The theorem is a consequence of the inequalities

$$\left.\begin{array}{l}|x_n - a|\\|y_n - b|\end{array}\right\} \leq \sqrt{(x_n - a)^2 + (y_n - b)^2} \leq |x_n - a| + |y_n - b|.$$

(Cf. Theorem I, § 1016.)

The fundamental theorem on bounded sequences (§ 301) applies to

320 FUNCTIONS OF SEVERAL VARIABLES [§ 1019

Euclidean spaces in general (a **bounded sequence** being one all of whose terms are contained in some neighborhood):

Theorem II. Fundamental Theorem on Bounded Sequences. *Every bounded sequence in E_m contains a convergent subsequence.*

Proof for two dimensions. If $\{x_n, y_n\}$ is bounded, so are the sequences $\{x_n\}$ and $\{y_n\}$. By Theorem I, § 301, the sequence $\{x_n\}$ contains a convergence subsequence. The *corresponding* subsequence of $\{y_n\}$ is again bounded, and contains a convergent subsubsequence, $y_{n_1}, y_{n_2}, y_{n_3}, \cdots$. Then (Theorem IV, § 204) $x_{n_1}, x_{n_2}, x_{n_3}, \cdots$ converges. Therefore the subsequence $\{x_{n_i}, y_{n_i}\}$ of the original sequence converges, by Theorem I.

The Cauchy criterion for convergence of a sequence of points in E_m has formulation and proof almost identical with those of § 302, the only change being the use of the distance $d(p, q)$ between two points (in place of the absolute value of their difference, in the one-dimensional case). We repeat the statements, but omit the proof (Ex. 11, § 1021).

Definition II. *A **Cauchy sequence** of points in E_k† is a sequence $\{p_n\}$ such that*

$$\lim_{m, n \to +\infty} d(p_m, p_n) = 0.$$

Theorem III. Cauchy Criterion. *A sequence of points in E_k converges if and only if it is a Cauchy sequence.*

As a final step in setting the stage for two of the most important fundamental theorems in the theory of mappings, and consequent proofs of earlier theorems, we have the *sequential criterion for continuity of mappings* (give a proof in Ex. 12, § 1021):

Theorem IV. *A necessary and sufficient condition for a mapping $f(p)$ to be continuous at a point $p = p_0$ is that whenever $\{p_n\}$ is a sequence of points that converges to p_0 (and for which $f(p)$ is defined), then $\{f(p_n)\}$ is a sequence of points converging to $f(p_0)$; in short, that $p_n \to p_0$ implies $f(p_n) \to f(p_0)$.*

★★1019. TWO FUNDAMENTAL THEOREMS ON MAPPINGS

We rely heavily, once more, on techniques and results of Chapter 3. In particular, we shall need the concepts of separated and connected sets (§ 309), and the theorems of §§ 310 and 311. As before, recently, we omit proofs which are mere paraphrases of ones given earlier (cf. Exs. 20-21, § 1021). Theorems I and II, § 311, become, for mappings:

Theorem I. *Any continuous image of a compact set is compact.*

Theorem II. *Any continuous image of a connected set is connected.*

† The subscript k is used, in place of m, to permit the customary use of the letter m in this formulation.

§ 1020] THEOREMS ON CONTINUOUS FUNCTIONS

The third and fourth theorems of § 311 become Theorem II, § 1016, and the Theorem of § 1017, respectively.

★★1020. PROOFS OF SOME THEOREMS ON CONTINUOUS FUNCTIONS

The two fundamental theorems of § 1019 provide us with proofs of the theorems of § 1010.

For Theorems I and II of that section, we know that the range of a real-valued function defined and continuous on any nonempty compact set must be a compact set of real numbers. Therefore it must be bounded and, furthermore, it must have a greatest member and a least member (Theorem III, § 309).

Theorem III, § 1010, is a special case of Theorem IV, § 1010, and we shall consider only the latter.

To prove Theorem IV, for the case of a region, we establish a sequence of five lemmas and a theorem:

Lemma 1. *If a connected set A is contained in a set made up of two separated parts, B and C, then A must lie entirely in B or entirely in C.*

Proof. If A lies partly in B and partly in C, let A_1 and A_2 designate the parts of A in B and C, respectively. Then, since B and C are separated, so are A_1 and A_2. (Contradiction.)

Lemma 2. *Let A be a set having the property that for any two points p and q belonging to A there exists a connected set B containing p and q and lying in A. Then A is connected.*

Proof. If A can be split into two separated parts, A_1 and A_2, choose p in A_1 and q in A_2. Then by Lemma 1, there can be no connected set B containing both p and q and lying in A. (Contradiction.)

Lemma 3. *Any line segment is connected.*

Proof. By means of the functions (of a parameter) that define the line segment, it is a continuous image of an interval of real numbers, which is connected (Theorem IV, § 309).

Lemma 4. *Any broken line segment is connected.*

Proof. Assume that a broken line segment A can be split into two parts, B and C. Then, by Lemmas 1 and 3, the first segment must lie entirely in B or entirely in C. Let us say it lies in B. Similarly, the second segment must lie entirely in B or entirely in C, but since it has a point in common with the first segment it must also lie entirely in B. In this way, we can prove by induction that *all* segments must lie in B, so that C must be empty. (Contradiction.)

Lemma 5. *Any region is connected.* (Cf. Ex. 35, § 1025.)

322   FUNCTIONS OF SEVERAL VARIABLES   [§ 1021

Proof. This follows from Lemmas 2 and 4.

Theorem. *A real-valued function continuous on a connected set assumes there all values between any two of its values.*

Proof. Since the range of a real-valued function defined and continuous on any connected set must be a connected set of real numbers, it must be an interval (Theorem IV, § 309). Therefore it contains all numbers between any two of its members.

Proof of Theorem IV, § 1010. For a region, the proof follows from Lemma 5 and the preceding Theorem. To prove Theorem IV, § 1010, for a closed region A, let p_1 and p_2 be points of A and let A be the closure of a region R. Assume $f(p_1) < c < f(p_2)$. Then if p_1 is not a member of R, it must be a limit point of R, and by continuity of the function $f(p)$ at the point p_1, there must exist a point p_1' belonging to R and such that $f(p_1') < c$ (details?). Similarly, there must exist a point p_2' of R such that $f(p_2') > c$. The conclusion follows by applying the case of this theorem already proved for regions.†

1021. EXERCISES

1. Show that the mappings
$$f((x, y)) = (ax, by), ab \neq 0,$$
$$g((x, y)) = (x + y, y),$$
continuously map E_2 onto E_2, and have continuous inverses. Show that the first is produced by changes of scale and possible reflections, and that the second can be thought of as a kind of *shearing* transformation.

2. Show that the mappings
$$f((x, y)) = (e^x, e^y),$$
$$g((x, y)) = (e^x, \text{Arc tan } y),$$
$$h((x, y)) = (\text{Arc tan } x, \text{Arc tan } y),$$
continuously map E_2 onto the open first quadrant, an open half-infinite strip, and an open square, respectively, and have continuous inverses.

3. Show that $(x, y) \to \left(\dfrac{x}{1 + \sqrt{x^2 + y^2}}, \dfrac{y}{1 + \sqrt{x^2 + y^2}} \right)$ continuously maps E_2 onto the open unit circle, and has a continuous inverse.

4. Show that $(x, y) \to \left(\dfrac{x}{x^2 + y^2}, \dfrac{y}{x^2 + y^2} \right), x^2 + y^2 \neq 0$, is continuous and equal to its own inverse.

5. Show that the mapping
$$(x, y) \to (e^x \cos y, e^x \sin y)$$

---

† The simple example of the open unit disk with a radial slit removed shows that since the closure of this region R is the closed unit disk C (as it would have been if the slit had *not* been removed), not *every* interior point of C is a point of R. However, the important relation needed in the preceding proof is that *every* point of R is an interior point of A. (Proof?)

is continuous on E_7, but not one-to-one there. By restricting the domain obtain a single-valued continuous inverse.

6. Show that the mapping
$$f(t) = (\cos t, \sin t),$$
with domain $0 \leq t < 2\pi$ in E_1, and range the circumference of the unit circle in E_2, is continuous and one-to-one, but that the inverse mapping is not continuous.

7. Formulate the *Negation of Continuity* for mappings, corresponding to the Definition of § 1016.

★8. Formulate the *Negation of Uniform Continuity* for mappings, corresponding to the Definition of § 1017.

★9. Formulate the statement and proof of Theorem I, § 1018, for the space E_m. *Hint:* Use superscripts to denote terms of a sequence: $p_n = (x_1^n, x_2^n, \cdots, x_m^n)$.

★10. Prove Theorem II, § 1018. (Cf. Ex. 9.)

★11. Prove Theorem III, § 1018. (Cf. Exs. 9-10.)

★12. Prove Theorem IV, § 1018.

In Exercises 13-14, find $\delta(\epsilon)$ explicitly, according to the Definition, § 1017, of uniform continuity.

★13. $x^2 + y^2$, for $x^2 + y^2 \leq 1$.

★14. $e^x \sin y$, for $x \leq a$.

★15. By means of the example $y = \sin x$ in E_1, show that a continuous image of an open interval may be open, or closed, or neither.

★16. By means of the example $y = e^x$ in E_1, show that a continuous image of a closed set may be closed, or open, or neither.

★17. Define the limit superior and limit inferior of a real-valued function $f(p)$ at a point p_0, where the domain is in E_n:
$$\varlimsup_{p \to p_0} f(p) = \limsup_{p \to p_0} f(p) = \inf_{\delta > 0} \left[\sup_{0 < d(p, p_0) < \delta} f(p) \right],$$
$$\varliminf_{p \to p_0} f(p) = \liminf_{p \to p_0} f(p) = \sup_{\delta > 0} \left[\inf_{0 < d(p, p_0) < \delta} f(p) \right],$$
and prove that they always exist in a finite or infinite sense. (Cf. Exs. 23-24, § 305.)

★18. State and prove results similar to those of Exercise 25, § 305, for functions of n variables. (Cf. Ex. 17.)

★19. Prove that for any real-valued function $f(p)$, where p is a point in E_n, $\varliminf_{p \to p_0} f(p) \leq \varlimsup_{p \to p_0} f(p)$. Prove that $\lim_{p \to p_0} f(p)$ exists if and only if $\varliminf_{p \to p_0} f(p) = \varlimsup_{p \to p_0} f(p)$, and in case of equality,
$$\lim_{p \to p_0} f(p) = \varliminf_{p \to p_0} f(p) = \varlimsup_{p \to p_0} f(p).$$
(Cf. Ex. 26, § 305.)

★★20. Reformulate and prove for E_n the four Theorems of § 310.

★★21. Prove the two Theorems and the statements of the last paragraph of § 1019.

★★22. Prove the **Bolzano-Weierstrass Theorem** for E_n: *Any infinite bounded set in E_n has a limit point.* (Cf. Ex. 13, § 312.)

★★23. Prove that if $\{A_n\}$ is a monotonically decreasing sequence of nonempty compact sets there exists a point common to every set of the sequence. (Cf. Ex. 26, § 312.)

★★24. Prove the **Heine-Borel Theorem** for E_n: *Any open covering of a compact set is reducible to a finite covering*. (Cf. Ex. 28, § 312.) *Hint for E_2*: Make repeated subdivisions of closed rectangles into quarters, and follow the ideas of Ex. 28, § 312.

★★25. Prove, for E_n, the **Converse of the Heine-Borel Theorem** of Exercise 24: *If every open covering of a set is reducible to a finite covering, the set is compact*. *Hint*: Assume that the set A contains a sequence $\{p_n\}$ of points containing no subsequence converging to a point of A. Let C_n be the closure of the set consisting of the points $p_{n+1}, p_{n+2}, \cdots$, and let U_n be the complement of C_n. Then the family of all U_n, $n = 1, 2, \cdots$, is an open covering of A that is not reducible to a finite covering.

★★26. Discuss the concepts of *distance between a point and a nonempty set* (Ex. 19, § 312) and *distance between two nonempty sets* (Ex. 21, § 312), for E_n, proving the corresponding statements of Exercises 19-22, § 312. Show that a hyperbola and its asymptotes provide an example in E_2 playing the role of that of Exercise 21, § 312.

★★27. Prove that the inverse of a homeomorphism (cf. the Note, § 1016) is a homeomorphism.

★★28. Prove that the relation of being homeomorphic is an equivalence relation. That is, if $A \sim B$ means that A is homeomorphic to B, show that $(i) \sim$ is *reflexive*: $A \sim A$ for every A; $(ii) \sim$ is *symmetric*: $A \sim B$ implies $B \sim A$; $(iii) \sim$ is *transitive*: $A \sim B \sim C$ implies $A \sim C$.

★★29. Assume that for each point p of a compact set A in E_n, $f(p, y)$ is a monotonic function of y, and for each y, $f(p, y)$ is continuous on A. Prove that if $f(p) \equiv \lim_{y \to c+} f(p, y)$ is continuous on A, the convergence must be uniform (cf. Ex. 36, § 908, Ex. 2, § 1015).

★★30. Define the **graph** G of a real-valued function of a real variable, $y = f(x)$, to be the set of all points (x, y) in E_2 such that x belongs to the domain D of $f(x)$ and $y = f(x)$. Prove that the mapping that carries the point x of D (in E_1) into the point $(x, f(x))$ of G (in E_2) is continuous on D if and only if $f(x)$ is continuous on D. Generalize.

★★31. Prove that the graph of a continuous real-valued function on a closed interval is compact and connected (cf. Ex. 30).

★★32. A set property that is preserved by all homeomorphisms is called an **intrinsic property**. Prove that compactness and connectedness are intrinsic, while openness and closedness are not. *Hints*: Consider the open interval $(0, 1)$ in the space E_1 and on the x-axis of E_2, and the mapping $y = \tan \frac{\pi}{2} x$ between the intervals $(-1, 1)$, $(-\infty, +\infty)$.

★1022. POINT SET ALGEBRA

A useful tool in analysis is the *algebra of sets*, where sets of "points" are combined in various ways to yield other sets.

The principal operations are *complementation*, *union*, and *intersection*, and the principal relation is *inclusion*.

The complement of a set A in a Euclidean space S was defined in § 1003. In more general theory the word **space** merely means a collection of

[§ 1022]   POINT SET ALGEBRA

"points" or objects thought of as a "universe of discourse," and every set A under consideration is made up of points of S (this includes the empty set $\emptyset$ (cf. Note 3, § 309)). The **complement** of a set A, denoted A', is the set of all points of S that are not members of A.

A collection of sets is commonly designated in either of two ways. If the collection consists of two or three or a suitably small number of sets it is usually simplest to indicate them by distinct letters; thus, A and B, or A, B, and C. If we have a more general finite or infinite collection it is more suitable to use subscript notation; thus, $\{A_k\}$, $k = 1, 2, \cdots, n$, or $\{A_n\}$, $n = 1, 2, \cdots$, or $\{A_\alpha\}$, where α is a member of a general class called an **index set** (which might, for instance, consist of the numbers 1 and 2, or all positive integers, or all real numbers).

The **union** of a collection of sets is the set of all points p having the property that p belongs to *at least one* set of the collection. Notations for union, corresponding to collections mentioned in the preceding paragraph, are

$$A \cup B, \quad A \cup B \cup C, \quad \bigcup_{k=1}^{n} A_k, \quad \bigcup_{n=1}^{+\infty} A_n, \quad \bigcup_{\alpha} A_\alpha.$$

Some finite cases are illustrated in the **Venn diagrams** of Figure 1007.

The **intersection** of a collection of sets is the set of all points p having the property that p belongs to *every* set of the collection. Notations for intersection, corresponding to collections mentioned above, are

$$A \cap B, \quad A \cap B \cap C, \quad \bigcap_{k=1}^{n} A_k, \quad \bigcap_{n=1}^{+\infty} A_n, \quad \bigcap_{\alpha} A_\alpha.$$

Some finite cases are illustrated in Figure 1007.

The statements "A is included in B," "A is contained in B," "A is a subset of B," and "B contains A," denoted

$$A \subset B \quad \text{or} \quad B \supset A,$$

all mean that every point of A is a point of B. Thus, $\emptyset \subset A$, $A \subset A$, $A \subset S$, and $A \cap B \subset A \cup B$ are identities, true of all sets A and B.

Equality between two sets, A and B, written $A = B$, means that A and B are the same set, and is equivalent to the two inclusions $A \subset B$ and $B \subset A$.

Two sets, A and B, are called **disjoint** in case they have no points in common; that is, $A \cap B = \emptyset$.

Many identical relationships follow immediately from the preceding definitions. For example, the **commutative laws**

$$A \cup B = B \cup A \quad \text{and} \quad A \cap B = B \cap A$$

are expressions of the symmetric roles of A and B in the definitions of union and intersection.

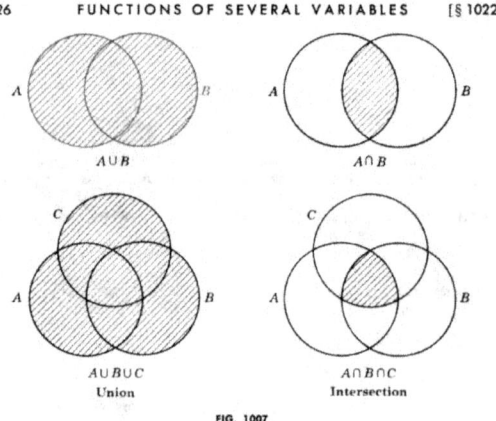

FIG. 1007

A less obvious relation is the following **distributive law**

(1) $$A \cap (B \cup C) = (A \cap B) \cup (A \cap C)$$

(cf. Ex. 3, § 1023), which corresponds to the familiar law of real-number algebra, where $\cap$ replaces multiplication and $\cup$ replaces addition. We give two proofs.

First proof. In order to establish (1) we show first that every point of the set on the left is a member of the set on the right. Assume, then, that p belongs to the left-hand side of (1). Then p is a member of A and also a member of $B \cup C$. Therefore *either* p belongs to both A and B or p belongs to both A and C. In other words, p belongs to $A \cap B$ or to $A \cap C$ (or both), and hence to $(A \cap B) \cup (A \cap C)$, the right-hand side of (1). On the other hand, if we assume that p is a member of the right-hand side of (1), we can retrace the steps of the argument to conclude that p belongs to the left-hand side. Therefore (1) is true.

Second proof. As with any equation (or relation) involving three sets, we can make use of the Venn diagram, illustrated in Figure 1008, shading in the portions that correspond to either side of the given equation. If the two results are identical, the equation is proved. If not, it is disproved. In this case it is proved.

§ 1023] EXERCISES 327

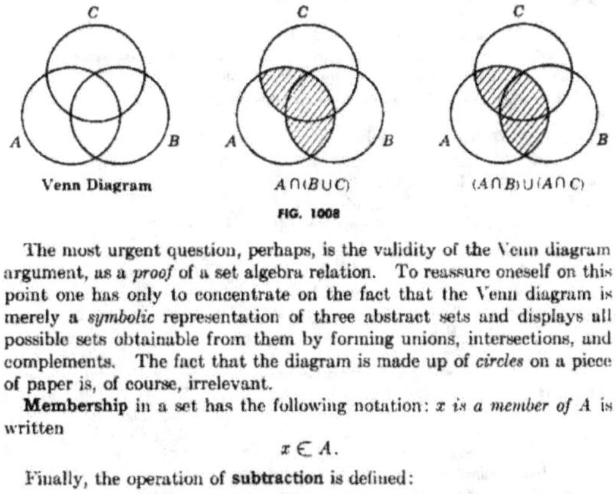

Venn Diagram $A \cap (B \cup C)$ $(A \cap B) \cup (A \cap C)$

FIG. 1008

The most urgent question, perhaps, is the validity of the Venn diagram argument, as a *proof* of a set algebra relation. To reassure oneself on this point one has only to concentrate on the fact that the Venn diagram is merely a *symbolic* representation of three abstract sets and displays all possible sets obtainable from them by forming unions, intersections, and complements. The fact that the diagram is made up of *circles* on a piece of paper is, of course, irrelevant.

Membership in a set has the following notation: *x is a member of A* is written
$$x \in A.$$

Finally, the operation of **subtraction** is defined:
$$A - B \equiv A \cap B'.$$

A number of relationships for the reader to establish, or disprove by counterexample, are given in the following Exercises.

★**1023. EXERCISES**

In Exercises 1-10, prove the given identity.

★1. The **associative law** for unions,
$$A \cup (B \cup C) = (A \cup B) \cup C = A \cup B \cup C.$$
Generalize to n sets.

★2. The **associative law** for intersections,
$$A \cap (B \cap C) = (A \cap B) \cap C = A \cap B \cap C.$$
Generalize to n sets.

★3. The following **distributive law**:
$$A \cup (B \cap C) = (A \cup B) \cap (A \cup C).$$

★4. $A \cup A = A \cap A = A.$ ★5. $\emptyset' = S, \; S' = \emptyset.$
★6. $A \cap A' = \emptyset, \; A \cup A' = S.$ ★7. $(A')' = A.$
★8. $(A \cup B)' = A' \cap B', \; (A \cap B)' = A' \cup B'.$ (Cf. the Note, below.)
★9. The following distributive laws:
$$B \cap (\bigcup_\alpha A_\alpha) = \bigcup_\alpha (B \cap A_\alpha),$$
$$B \cup (\bigcap_\alpha A_\alpha) = \bigcap_\alpha (B \cup A_\alpha).$$

★10. $(\bigcup_\alpha A_\alpha)' = \bigcap_\alpha A_\alpha'$, $(\bigcap_\alpha A_\alpha)' = \bigcup_\alpha A_\alpha'$.

Note. The laws of Exercises 8 and 10, known as **De Morgan's laws**, can be expressed: *The complement of the union is the intersection of the complements and the complement of the intersection is the union of the complements.*

In Exercises 11–14, find a counterexample, and thus disprove the given statement.

★11. $A \cup (B \cap C) = (A \cup B) \cap C$.
★12. $A \cap B' \cap (A \cap B)' = A$.
★13. $A \cup (B \cap C)' \subset (A \cup B') \cap C'$.
★14. $(A \cup B) \cap (C \cup D) \subset (A \cap B) \cup (A \cap C) \cup (A \cap D)$.

In Exercises 15–20, prove or disprove the given statement.

★15. $A \cup (B - C) = (A \cup B) - C$.
★16. $A - (B - C) = (A - B) \cup C$.
★17. $A \cap (B - C) = (A \cap B) - (A \cap C)$.
★18. $A - (A - B) = A \cap B$.
★19. $(A - B)' = B - A$.
★20. $(A - C) \cup (B - C) = (A \cup B) - C$.

★★21. The **limit superior** and **limit inferior** of a sequence $\{A_n\}$ of sets are defined, respectively:

$$\overline{\lim_{n \to +\infty}} A_n = \bigcap_{n=1}^{+\infty} \left[\bigcup_{m=n}^{+\infty} A_m\right], \quad \underline{\lim_{n \to +\infty}} A_n = \bigcup_{n=1}^{+\infty} \left[\bigcap_{m=n}^{+\infty} A_m\right].$$

Prove that the limit superior of a sequence of sets is the set of points p such that p is a member of infinitely many sets of the sequence, and that the limit inferior of a sequence of sets is the set of points p such that p is a member of all but a finite number of sets of the sequence. Conclude that the limit inferior is always a subset of the limit superior. (Cf. Ex. 16, § 305.)

★★22. State and prove theorems concerning limits superior and inferior of sequences of sets corresponding to Exercises 35–37, § 305.

★★23. A sequence $\{A_n\}$ of sets is said to **converge** if and only if its limit superior is equal to its limit inferior and, in case of convergence, the **limit** of the sequence is defined to be this common set. State and prove theorems on limits of sequences of sets, corresponding to Theorems of § 204.

★1024. SET ALGEBRA AND OPERATIONS INVOLVING NEIGHBORHOODS

Many properties of such concepts as *limit point, open set, closed set, interior, frontier,* and *closure,* previously postponed because of awkwardness of formulation, can now be given with relative ease with the help of the notation of point set algebra (§ 1022).

We start by introducing notation for concepts defined in § 1003:

Notation. The set of all limit points of a set A, called its **derived set**, is denoted $D(A)$. The interior and frontier of a set A are denoted $I(A)$ and $F(A)$, respectively.

§ 1024] OPERATIONS INVOLVING NEIGHBORHOODS 329

We give some sample proofs and discussion, and ask for more in the Exercises of § 1025. The underlying space throughout this section and the next will be assumed to be a Euclidean space.

Example 1. Prove the following properties of open sets:

(i) *Arbitrary unions of open sets are open;* that is, if every set A_α of an arbitrary collection $\{A_\alpha\}$ is open, then so is the union $\bigcup_\alpha A_\alpha$.

(ii) *Finite intersections of open sets are open;* that is, if A_k is open for $k = 1, 2, \cdots, n$, then so is $\bigcap_{k=1}^{n} A_k$.

(iii) *The empty set $\emptyset$ is open.*

(iv) *The space S is open.*

Solution. (i): Let p be an arbitrary point of $A = \bigcup_\alpha A_\alpha$. We wish to find a neighborhood N_p of p contained in A. By definition of a union of sets, p must belong to at least one of the sets A_β. But since A_β is assumed to be open, there exists a neighborhood N_p of p contained in A_β. Finally, $N_p \subset A_\beta \subset A$.

(ii): Thanks to mathematical induction, we need prove only that if A and B are open, so is $A \cap B$. Accordingly, let $p \in A \cap B$. Since $p \in A$ there exists a neighborhood $N_p^{(1)}$ of p contained in A. Similarly, there exists a neighborhood $N_p^{(2)}$ of p contained in B. Let N_p be a (smaller) neighborhood of p contained in both $N_p^{(1)}$ and $N_p^{(2)}$. Then $N_p \subset A$ and $N_p \subset B$, so that $N_p \subset A \cap B$.

(iii) and (iv): These are trivial (cf. Note 3, § 1003).

Example 2. Prove that

(1) $$D(A \cup B) \subset D(A) \cup D(B).$$

That is, prove that every limit point of the union of two sets must be a limit point of at least one of the sets. (The two sides are actually equal. Cf. Ex. 6, § 1025.)

Solution. We give first a completely fallacious "proof," in order to show how easy it is to fall victim to self-deception when the argument becomes slightly abstract. With this warning we invite the reader to spot the error in the following step-by-step argument:

1. Assume p is a limit point of $A \cup B$.
2. Then *every* deleted neighborhood N_p of p must contain at least one point q of $A \cup B$.
3. Since $q \in A \cup B$, either $q \in A$ or $q \in B$.
4. If $q \in A$ then $p \in D(A)$.
5. If $q \in B$ then $p \in D(B)$.
6. Therefore $p \in D(A) \cup D(B)$.

What was wrong?

If you know what is wrong, how do you *prove* that the argument is invalid, when the conclusion happens to be correct?

Perhaps the most satisfactory way of disproving a line of incorrect reasoning like that given above is to construct a new situation, which may be without useful substance, but where the *form* of argumentation is the same and a *false* conclusion is reached.

Accordingly, we construct the following **mock space** S: The points of S are the real numbers. A **neighborhood** of a point p of S is defined to be any half-open interval of the form $(p, p + \epsilon)$ or of the form $(p - \epsilon, p]$, where $\epsilon > 0$. Limit points are defined as before: p is a **limit point** of a set if and only if every deleted neighborhood of p contains at least one point of the set. Let A be the set of all negative numbers and let B be the set of all positive numbers. Then the number (or point) 0 is a limit point of $A \cup B$ (the nonzero numbers), but it is neither a limit point of A nor a limit point of B (why?). The six-step "proof" given above must therefore break down, and the place where it breaks down is at steps 4 and 5.† (Cf. Ex. 22, § 1028.)

It is time to give a correct proof of (1). Assume the conclusion to be false. That is, assume that p is neither a limit point of A nor a limit point of B. Then there exist deleted neighborhoods $N_p^{(1)}$ and $N_p^{(2)}$ of p such that $N_p^{(1)}$ contains no points of A and $N_p^{(2)}$ contains no points of B. Let N_p be a (smaller) deleted neighborhood of p contained in *both* $N_p^{(1)}$ and $N_p^{(2)}$. Then N_p contains no points of A and no points of B, and therefore no points of $A \cup B$. Hence p is *not* a limit point of $A \cup B$. (Contradiction.)

Example 3. Prove the following properties of closed sets:

(i) *The closure of the union of two sets is the union of their closures,*
$$\overline{A \cup B} = \bar{A} \cup \bar{B}.$$
(ii) *The closure of the closure of a set is its closure,*
$$\bar{\bar{A}} = \bar{A}.$$
(iii) $\bar{\emptyset} = \emptyset.$

(iv) $\bar{S} = S.$

Solution. Once we know that the closure of any set is closed and that the closure of any closed set is that set (cf. Note 5, § 1003, and Exs. 13-14, § 1005), properties (ii), (iii), and (iv) become trivial (cf. Note 3, § 1003).

To prove (i) we first assume that p is a point belonging to $\bar{A} \cup \bar{B}$, and prove that $p \in \overline{A \cup B}$. Since p must belong to either $\bar{A}$ or $\bar{B}$, let us say for definiteness that $p \in \bar{A}$. If $p \in A$, then $p \in A \cup B \subset \overline{A \cup B}$. On the other hand, if p is a limit point of A, it must certainly be a limit point of the (possibly) larger set $A \cup B$, so that $p \in \overline{A \cup B}$. Thus $\bar{A} \cup \bar{B} \subset \overline{A \cup B}$.

For the final implication we assume that $p \in \overline{A \cup B}$ and prove that $p \in \bar{A} \cup \bar{B}$. If $p \in A \cup B$, then certainly $p \in \bar{A} \cup \bar{B}$ (which is larger if anything). This leads us to the final possibility: if p is a limit point of $A \cup B$, then, by Example 2, $p \in D(A) \cup D(B) \subset \bar{A} \cup \bar{B}$.

★1025. EXERCISES

In this section it is assumed that the space S is a Euclidean space. When a counterexample is sought, the reader is urged to keep in mind the two extreme sets in E_1, one consisting of the rational numbers and the other, its complement, consisting of the irrational numbers.

† From the hypothesis that Jim either walks or rides to school we cannot conclude that he does either exclusively.

§ 1025] EXERCISES

In Exercises 1-14, prove the given statement.

★1. Finite unions of closed sets are closed.
★2. Arbitrary intersections of closed sets are closed.
★3. A point p is a limit point of a set A if and only if every open set containing p contains at least one point q of A different from p.
★4. $A \subset B$ implies $\bar{A} \subset \bar{B}$.   ★5. $A \subset B$ implies $I(A) \subset I(B)$.
★6. $D(A \cup B) = D(A) \cup D(B)$.
★7. $F(A) = F(A') = \bar{A} \cap \overline{A'}$. Therefore $F(A)$ is always closed.
★8. $I(A) \cap F(A) = \emptyset$, $I(A) \cup F(A) = \bar{A}$.
★9. $I(A \cap B) = I(A) \cap I(B)$.   ★10. $A - F(A) = I(A)$.
★11. If A is connected, and if $A \subset B \subset \bar{A}$, then B is connected. Therefore the closure of any connected set is connected. (Cf. Lemma 1, § 1020.)
★12. $\bar{A}$ is the intersection of all closed sets containing A.
★13. $I(A)$ is the union of all open sets contained in A.
★14. $A = F(A)$ if and only if A is closed and without interior points.

In Exercises 15-24, disprove the statement by means of a counterexample in E_1 or E_2.

★15. Arbitrary intersections of open sets are open.
★16. Arbitrary unions of closed sets are closed.
★17. $\bar{A} \subset \bar{B}$ implies $A \subset B$.   ★18. $\bar{A} = \bar{B}$ implies $A = B$.
★19. $A = I(A) \cup F(A)$.   ★20. $A - I(A) = F(A)$.
★21. $A \subset B$ implies $F(A) \subset F(B)$.   ★22. $F(F(A)) = F(A)$.
★23. $\overline{A \cap B} = (\bar{A} \cap B) \cup (A \cap \bar{B})$.   ★24. If A is open, $F(A) = F(\bar{A})$.

In Exercises 25-32, establish the inclusion, and show by a counterexample in E_1 or E_2 that equality is false.

★25. $\overline{A \cap B} \subset \bar{A} \cap \bar{B}$.   ★26. $I(A \cup B) \supset I(A) \cup I(B)$.
★27. $\overline{\cup A_\alpha} \supset \cup \bar{A}_\alpha$.   ★28. $I(\cap A_\alpha) \subset \cap I(A_\alpha)$.
★29. $F(A \cup B) \subset \overline{A \cap B} \cup F(B)$.   ★30. $F(A \cap B) \subset F(A) \cup F(B)$.
★31. $A \cap \bar{B} \subset \overline{A \cap B}$ if A is open, but not in general.
★32. $\bar{A} \cap \bar{B} - \overline{A \cap B} \subset F(A) \cap F(B)$.
★33. The **diameter** of a nonempty bounded set A, denoted $\delta(A)$, is defined to be the supremum of all distances $d(p, q)$ for p and q in A: $\delta(A) = \sup_{p,q \in A} d(p, q)$. Prove that the diameter of a nonempty compact set is actually attained as a maximum distance between two of its points. *Hint:* For every $n = 1, 2, \cdots$, let p_n and q_n be points of A such that $d(p_n, q_n) > \delta(A) - \frac{1}{n}$. Let $p_{n_k} \to p$, and choose a convergent subsequence of $\{q_{n_k}\}$. (Cf. Theorem III, § 309, Ex. 22, § 312.)
★★34. Prove that the only sets of a Euclidean space S that are both open and closed are $\emptyset$ and S. (Cf. Ex. 11, § 312, Ex. 10, § 1028.)
★★35. Prove that every connected open set of E_2 (or, more generally, of E_n) is a region. *Hint:* Let p and q be any two distinct points of the connected open set A, let B be the set of all points of A that *can* be reached by broken line segments lying in A, and let C be the set of all points of A that *cannot* be so reached. Show that B and C are separated. (Cf. Lemma 5, § 1020.)

★★36. Show that if A consists of all points (x, y) in E_2 such that $x \neq 0$ and $y = \sin(1/x)$, and if B is the y-axis in E_2, then $A \cup B$ is connected.

★★37. Let A be an arbitrary nonempty set in a Euclidean space S. A subset C of A is called a **component** of A if and only if it is a maximal connected subset of A, that is, if and only if $C \subset B \subset A$ and B and C connected imply $B = C$. Prove that the components of A constitute a *partitioning* of A, that is, that every point of A belongs to exactly one component of A. *Hint:* Form the union of all connected subsets of A that contain a given point.

★38. A set A in a Euclidean space is **convex** if and only if for every pair of points p and q of A the straight line segment joining p and q is a subset of A. (Cf. Ex. 21, § 1005.) Prove that every open or closed spherical neighborhood, $d(p, p_0) < \epsilon$ and $d(p, p_0) \leq \epsilon$, in E_n, is convex.

★★39. Prove the following theorems on convex sets (cf. Ex. 38):
(a) The closure of any convex set is a convex set.
(b) The intersection of any family of convex sets is a convex set.
(c) Any set is contained in a smallest convex set, called the **convex hull** of the given set, which is the intersection of all convex sets containing the given set.
(d) Any set is contained in a smallest closed convex set, called the **convex closure** of the given set, which is the intersection of all closed convex sets containing the given set.
(e) The closure of the convex hull of any set is the convex closure of that set.

★★40. A set is called an F_σ set (a G_δ set) if and only if it is the union (intersection) of a finite or denumerable family of closed (open) sets. Prove that a set is an F_σ set (a G_δ set) if and only if its complement is a G_δ set (an F_σ set).

★★41. Prove that the set of points of discontinuity of a real-valued function of a real variable is characterized by the property of being an F_σ set. That is, prove that every such set of points of discontinuity is an F_σ set and, conversely, every F_σ set of real numbers is the set of points of discontinuity of some real-valued function of a real variable. Generalize to higher dimensions. (Cf. Ex. 40.) *Hints:* Cf. Ex. 33, § 312, Example 2, § 502.

★★42. A set is **nowhere dense** if and only if its closure has no interior points. Prove that any subset of a nowhere dense set is nowhere dense, and that the union of any finite number of nowhere dense sets is nowhere dense. *Hint for the union:* Assume without loss of generality that there are two closed nowhere dense sets.

★★43. A set is of the **first category** if and only if it is the union of a finite or denumerable family of nowhere dense sets. A set is of the **second category** if and only if it is not of the first category. Prove that any subset of a set of the first category is of the first category, and that the union of any finite or denumerable family of sets of the first category is of the first category. (Cf. Ex. 42.)

★★★44. Prove that any set with nonempty interior in a Euclidean space is of the second category. *Hint:* Assume the contrary, choose a neighborhood N whose closure C lies in the given open set, and let C be the union of the closed nowhere dense sets $F_1, F_2, F_3, \cdots$. First choose a neighborhood N_1 whose closure C_1 lies in N and has no points in common with F_1. Then choose a neighborhood N_2 whose closure C_2 lies in N_1 and has no points in common with F_2. Continue, and form the intersection of $C_1, C_2, C_3, \cdots$. (Cf. Ex. 43.)

★★45. Show that the set of rational numbers in E_1 is of the first category, and the set of irrational numbers is of the second category. (Cf. Exs. 43, 44.)

★★46. Show that the set of rational numbers in E_1 is an F_σ set and not a G_δ set, and that the set of irrational numbers in E_1 is a G_δ set and not an F_σ set. (Cf. Ex. 40.) *Hint:* If the set of irrational numbers were an F_σ set it would be of the first category (cf. Ex. 45).

★★47. Show that there is no real-valued function of a real variable x that is continuous when x is rational and discontinuous when x is irrational. (Cf. Exs. 41, 46.)

★★1026. METRIC SPACES

Much of the work of the preceding portions of this chapter rests heavily on the use of the distance $d(p, q)$ between two points p and q of some Euclidean space. Actually, only a few properties of this distance function are needed. This fact has led mathematicians to extract the properties of distance truly essential to the type of discussion recently pursued here, and to formulate in terms of them an abstract concept known as a *metric space*. As is readily seen, the Euclidean spaces are particular examples of metric spaces, so that a metric space can be considered as a *generalization* of the Euclidean spaces. The principal reasons for studying this type of generalization are, first, that results obtained for such generalized concepts can often be applied in many directions without separate detailed proofs for each special case and, second, that the structure of a proposition or theory can often be seen more clearly and treated more intelligently when it is stripped of its nonessentials and reduced to its basic core.

Definition I. *A **metric space** is a set S together with a real-valued distance function $d(p, q)$, defined for every ordered pair (p, q) of points or members of S, subject to the three laws:*

(i) $d(p, q)$ *is **strictly positive**, that is, for any points p and q of S,*

$$d(p, q) \geq 0, \text{ and}$$
$$d(p, q) = 0 \text{ if and only if } p = q;$$

(ii) $d(p, q)$ *is **symmetric**, that is, for any points p and q of S,*

$$d(p, q) = d(q, p);$$

(iii) $d(p, q)$ *satisfies the **triangle inequality**, that is, for any points p, q, and r of S,*

$$d(p, r) \leq d(p, q) + d(q, r).$$

Such concepts as *neighborhood*, *limit point*, *open set*, *closed set*, *convergence of a sequence* $\{p_n\}$, and *continuity of a mapping* whose domain lies in a metric space and whose range lies in a metric space are immediately available. There is one important notion, however, that needs a new formulation. We take Theorem IV, § 310, as our point of departure in giving one of many possible equivalent definitions of compactness suitable for metric spaces in general (cf. the footnote on page 76, and the Definition of § 1027).

Definition II. *A set A in a metric space S is **compact** if and only if every sequence $\{p_n\}$ of points of A contains a subsequence converging to a point of A.*

For Euclidean spaces, this definition is equivalent to that given before, which identified compactness with the two properties of being closed and bounded (cf. Ex. 20, § 1021). It is not difficult to see that for any metric space a compact set must be both closed and bounded (Ex. 12, § 1028), but Example 1, below, shows that the reverse implication is false.

Many of the most important theorems of analysis rest ultimately on the Cauchy criterion for sequences of real numbers, which stems, by means of the fundamental fact that any bounded sequence of real numbers contains a convergence subsequence, from the *completeness axiom* of the real number system. In metric spaces in general a Cauchy sequence (cf. Definition II, § 1018) may or may not converge, as is shown by the (metric) space of rational numbers, which does not satisfy the property of completeness (cf. the Note, § 302). In order to exploit the power of the Cauchy criterion in abstract metric spaces it is necessary to find some substitute for the axiom of completeness enjoyed by the real number system. It is hopeless to try to formulate a substitute in terms of bounded sequences. The solution is to go directly to the heart of the problem and frame a definition in terms of Cauchy sequences themselves:

Definition III. *A metric space is **complete** if and only if every Cauchy sequence of points of the space converges to a point of the space.*

Example 1. Let S be an arbitrary set, and let the function $d(p, q)$, where p and q are arbitrary points of S, be defined:

$$d(p, q) = \begin{cases} 0 \text{ if } p = q, \\ 1 \text{ if } p \neq q. \end{cases}$$

It is a routine matter to verify the three properties of Definition I, for $d(p, q)$. Therefore S is a metric space. Since, for $\epsilon = \frac{1}{2}$, the set of all points p such that $d(p, p_0) < \epsilon$ consists of the single point p_0, every point is isolated, and there are no limit points. Hence *every* set in S is both open and closed. Furthermore, a sequence $\{p_n\}$ can converge only by being constant from some point on: $p_n = p_0$ for $n > N$. It follows that a set in S is compact if and only if it is finite (that is, consists of a finite number of points). Therefore, if S is infinite, and if A is any infinite set in S, A fails to be compact, whereas A is both closed and bounded (it is contained in any neighborhood of "radius" 2, for example). In a similar way we see, by taking a sequence of distinct points of S that not every bounded sequence contains a convergent subsequence. The space S is trivially complete.

Example 2. Let S be the set of all real-valued functions $f(x)$, defined and continuous on the closed interval $0 \leq x \leq 1$, and define the distance between the points $f(x)$ and $g(x)$:

$$d(f, g) = \left\{ \int_0^1 [f(x) - g(x)]^2 \, dx \right\}^{\frac{1}{2}}.$$

The first two properties of Definition I are easily verified (cf. Ex. 3, § 503). The third follows from the Minkowski inequality (Ex. 30, § 503; cf. Ex. 22, § 1005). The example

$$f_n(x) = \begin{cases} 1, & 0 \leq x \leq \frac{1}{2}, \\ \frac{n+2}{2} - nx, & \frac{1}{2} < x < \frac{1}{2} + \frac{1}{n}, \\ 0, & \frac{1}{2} + \frac{1}{n} \leq x \leq 1, \end{cases}$$

$n = 2, 3, \cdots$, shows that the space S is not complete. The reason, in brief, is that if $f_n \to f$ in this space, the limit function $f(x)$ must be identically 1 for $0 \leq x < \frac{1}{2}$, and identically 0 for $\frac{1}{2} < x \leq 1$. But no such function exists in this space of continuous functions. (Give the details.)

Example 3. Let S be the same set as that of Example 2, and define
$$d(f, g) = \max |f(x) - g(x)|,$$
for $0 \leq x \leq 1$. This space is complete, by the Cauchy criterion for uniform convergence (Ex. 45, § 904) and the fact that a uniform limit of continuous functions is continuous. (Give the details.)

Example 4. Prove that in any metric space every neighborhood is open.

Solution. Let N_p be a neighborhood of a point p, of radius $\epsilon > 0$, consisting of all points whose distance from p is less than ϵ, and let q be a point of N_p. We wish to find a neighborhood N_q of q such that $N_q \subset N_p$. Denote by δ the nonnegative distance $\delta = d(p, q)$. We can now define N_q to be the neighborhood of q having radius $\eta = \epsilon - \delta > 0$. Then, by the triangle inequality, if r is a point of N_q, since $d(r, q) < \eta$, $d(r, p) \leq d(r, q) + d(q, p) < \eta + \delta = \epsilon$, and hence $r \in N_p$. That is, $N_q \subset N_p$.

★★1027. TOPOLOGICAL SPACES

An abstraction that goes beyond the concept a metric space is that of a *topological space*, in which the kind of properties discussed in § 1024 are treated without regard to *distance*. The structure of such a space can be provided in a number of ways. Two of the most common are achieved either by specifying those sets of the space that are to be called *open* and subjecting them to a certain minimal list of requirements or *axioms*, or by describing a *closure operation* that states for every set A of the space what the "closure" $\bar{A}$ of that set is, and subjecting this closure operation to a set of axioms.

The *open set axioms* and the *closure axioms* for a topological space are given in Examples 1 and 3, § 1024, respectively. The arguments used in those two Examples show that according to either set of axioms any metric space is a topological space (check the details). In the present section we discuss very briefly the open set axioms.

In a topological space given by the open set axioms, a **neighborhood** of a point p is defined to be any open set containing p. In terms of neighborhoods, other concepts, like *limit point, closed set, closure, interior,* and *frontier,* have definitions identical with those already given. One concept alone, among those discussed in the preceding sections, requires a new definition, and that is *compactness*. It turns out that sequences do not play the central role in general topological spaces that they do in metric spaces, so that it would be inappropriate to use the same formulation as that of Definition II, § 1026. Instead, we start from the Heine-Borel theorem and formulate the following definition, which can be shown to be equivalent, for metric spaces, to Definition II, § 1026 (cf. the footnote reference, p. 477, Exs. 24-25, § 1021, Ex. 11, § 1028, and the reference immediately below):

Definition. *A set A is compact if and only if every open covering of A can be reduced to a finite covering.*

No attempt will be made to define *Cauchy sequence* or *completeness* for a general topological space.

The student interested in pursuing the subject of topology is referred to J. L. Kelley, *General Topology* (New York, D. Van Nostrand, 1955) and the bibliography contained therein.

Example 1. Let S be an arbitrary set, and let the open sets of S be *only* the empty set $\emptyset$ and the entire space S. This is a topological space, with what is sometimes called the **trivial topology.** The open set axioms are trivially satisfied. In this space every point is a limit point of every nonempty set consisting of more than one point, and every sequence converges to every point of the space. We notice that the limit of a sequence is *not* unique if the space contains more than one point. Furthermore, in this space, *a compact set need not be closed,* since every subset of the space is compact (cf. Exs. 15, 29, § 1028).

Example 2. Let S be an arbitrary set, and let *every* set of S be open. Since the open set axioms are satisfied, S is a topological space. It has what is called the **discrete topology.** Since every set of S is both open and closed, it is the same space topologically as that discussed as a metric space in Example 1, § 1026.

Example 3. Prove that in any topological space a set is closed if and only if its complement is open.

Solution. Assume first that A is an open set. We wish to prove that its complement A' is closed. Assume the contrary. Then there exists a point p of A that is a limit point of A'. Therefore every open set containing p must contain at least one point of A'. But A is an open set containing p and *no* point of A'. (Contradiction.)

Now assume that A' is closed. We wish to prove that A is open. Let p be an arbitrary point of A. Since p is not a limit point of A' (A' contains all of its limit points) there exists an open set A_p containing p and no point of A': $p \in A_p \subset A$. Now form the union $\bigcup_p A_p$ for all points p of A. Then this union is open and equal to A (why?). Hence A is open.

§ 1028] EXERCISES 337

★★1028. EXERCISES

In Exercises 1-8, prove for metric spaces the statements referred to, with appropriate changes in wording.

★★**1.** Theorems I-V, § 204.　　★★**2.** Theorems I-III, § 310.
★★**3.** Theorems I-IV, § 311.　　★★**4.** Notes 2, 4, and 5, § 1003.
★★**5.** Exs. 14-16, § 1005.　　　★★**6.** Ex. 23, § 1021.
★★**7.** Example 2, § 1024.　　　★★**8.** Exs. 1-14, 25-33, § 1025.

★★**9.** Denote by $O_p{}^\epsilon$ the "open sphere" consisting of all points q such that $d(p, q) < \epsilon$, and by $C_p{}^\epsilon$ the "closed sphere" consisting of all points q such that $d(p, q) \leq \epsilon$. Example 4, § 1026, shows that $O_p{}^\epsilon$ is open. Prove that $C_p{}^\epsilon$ is closed. Show by counterexample that $\overline{O_p{}^\epsilon}$ is not always equal to $C_p{}^\epsilon$. *Hint:* Consider Example 1, § 1026, with $\epsilon = 1$.

★★**10.** Prove that a metric space S is connected if and only if the only sets of S both open and closed are $\emptyset$ and S. (Cf. Ex. 34, § 1025.)

★★**11.** Prove the Heine-Borel theorem and its converse for metric spaces: *A set is compact if and only if every open covering is reducible to a finite covering.* (Cf. Exs. 24, 25, § 1021.) *Hints:* For the 'if' part, follow the hint of Ex. 25, § 1021. For the "only if" part, assume A is an infinite compact set and Φ an open covering of A not reducible to a finite covering. Show that for any $\epsilon > 0$ there is a finite subset of A such that any point of A is at a distance $< \epsilon$ from some point of this subset. Thus obtain a denumerable dense subset B of A. Then arrange those neighborhoods of points of B with rational radii that are subsets of sets of Φ in a sequence $N_1, N_2, \cdots$. Finally, show that for some n, $A \subset N_1 \cup N_2 \cup \cdots \cup N_n$.

★★**12.** Prove that any compact set in a metric space is closed and bounded.

★★**13.** Prove that any closed subset of a compact set in a metric space is compact.

★★**14.** Prove that a set in a metric space is compact if and only if every infinite subset has a limit point in the set.

★★**15.** A **Hausdorff space** is a topological space having the property that if p and q are any two distinct points there exist open sets A and B such that $p \in A$, $q \in B$, and $A \cap B = \emptyset$. Prove that every metric space is a Hausdorff space. Prove Theorem II, § 204, for Hausdorff spaces. (Cf. Ex. 24.)

★★**16.** Let S_1 and S_2 be topological spaces, and let f be a mapping of S_1 into S_2. Define the **inverse image** $f^{-1}(B)$ of a set B in S_2 to be the set of all points p in S_1 such that $f(p) \in B$. Prove that f is continuous on S_1 if and only if the inverse image of every open set in S_2 is an open set in S_1.

In Exercises 17-23, prove for topological spaces the statements referred to, with appropriate changes in wording.

★★**17.** Theorems I, III, IV, § 204.

★★**18.** Theorems I and II, § 311, where the domain is considered to be the entire space. *Hint:* Use Ex. 16.

★★**19.** Notes 4 and 5, § 1003.

★★**20.** Exs. 14 and 15, § 1005.

★★**21.** Ex. 23, § 1021, the single word *compact* being replaced by the two words *closed compact*.

★★**22.** Example 2, § 1024. Show that the "mock space" is not a topological space.

★★**23.** Exs. 1-14, 25-32, § 1025.

In Exercises 24–28, show by a counterexample that the statement referred to, although true for metric spaces, is not true for topological spaces in general.

★★24. Theorem II, § 204. (Cf. Ex. 15.)

★★25. Theorems I and II, § 310.

★★26. Theorem III, § 311. *Hint:* Map a space of seventeen points with the discrete topology onto a space of seventeen points with the trivial topology. Cf. Exs. 16, 30.

★★27. Note 2, § 1003.

★★28. Ex. 16, § 1005. *Hint:* Consider a one-point set in a space with the trivial topology.

★★29. Prove that in a Hausdorff space (Ex. 15) every compact set is closed. *Hint:* Assume that A is compact but not closed, and let q be a limit point of A that is not a point of A. For each point p of A let N_p and N_q be disjoint neighborhoods of p and q. Let G_1 be the union of all such N_p, reduced to a finite covering, and let G_2 be the finite intersection of the corresponding N_q.

★★30. Prove that in a Hausdorff space (Ex. 15) Theorem III, § 311, holds. *Hint:* Show that every open set is mapped onto an open set by using Ex. 29 to show that every closed set is mapped onto a closed set. (Cf. Exs. 13, 16, 18.)

★★31. Adapt and prove for topological spaces the statement of Exercise 37 § 1025. Prove that every component of a *space* is closed. Show by an example that the components of a space need not be open.

★★32. Adapt and prove for topological spaces the statements of Exercises 40, 42, 43, § 1025.

★★33. Prove that every complete metric space is of the second category. (Cf. Ex. 32.) *Hint:* In the hint of Ex. 44, § 1025, require radius $(N_n) \to 0$.

11

Arcs and Curves

1101. DUHAMEL'S PRINCIPLE FOR INTEGRALS

The definite integral is defined as the limit of sums of a certain type. For the product of two functions $f(x)$ and $g(x)$ these sums take the form $\sum_{i=1}^{n} f(x_i) g(x_i) \Delta x_i$. In practice one is occasionally faced with the problem of evaluating a limit of sums of the form $\sum_{i=1}^{n} f(x_i) g(x_i') \Delta x_i$ which resemble the standard approximating sums of a definite integral but which differ from the latter in that the two given functions are evaluated at possibly distinct values of the independent variable in the ith subinterval. The fact that this limit exists and is equal to the definite integral $\int_a^b f(x) g(x) \, dx$ is known as **Bliss's Theorem**. A precise statement, together with hints for a proof, is given in Exercise 41, § 503.

In this section we state a useful theorem of which Bliss's Theorem is a simple special case (where $\phi(x, y) = xy$). In the following section a proof is given under the more restrictive assumption that $f(t)$ and $g(t)$ are continuous, while the general proof is deferred to Exercise 28, § 1110, where hints are supplied. For extensions to more than two variables see Exercises 20, 21, and 30, § 1110.

Theorem. Duhamel's Principle for Integrals. *Let $f(t)$ and $g(t)$ be integrable on $[a, b]$ and let $\phi(x, y)$ be everywhere continuous. Then, in the sense of § 501, the limit of the sum $\sum_{i=1}^{n} \phi(f(t_i), g(t_i')) \Delta t_i$, where $a_{i-1} \leq t_i \leq a_i$ and $a_{i-1} \leq t_i' \leq a_i$, as the norm of the net $\mathfrak{N} : a = a_0, a_1, \cdots, a_n = b$ tends toward zero, exists and is equal to the definite integral $\int_a^b \phi(f(t), g(t)) \, dt$, which also exists:*

$$(1) \qquad \lim_{|\mathfrak{N}| \to 0} \sum_{i=1}^{n} \phi(f(t_i), g(t_i')) \Delta t_i = \int_a^b \phi(f(t), g(t)) \, dt.$$

★1102. A PROOF WITH CONTINUITY HYPOTHESES

We give here a proof of Duhamel's Principle under the additional assumption that $f(t)$ and $g(t)$ are continuous for $a \leq t \leq b$ (cf. Ex. 28, § 1110). Since $f(t)$ and $g(t)$ are continuous on a closed interval they are bounded there (Theorem I, § 213), so that there exists a number K such that for $a \leq t \leq b$, $|f(t)| \leq K$ and $|g(t)| \leq K$. Since $\phi(x, y)$ is assumed to be everywhere continuous it is, in particular, continuous on the closed square $A: -K \leq x \leq K, -K \leq y \leq K$. Consequently, by the Theorem of § 1017, $\phi(x, y)$ is uniformly continuous on A. Corresponding to a preassigned positive number ϵ, we can therefore find a positive number η such that $\sqrt{(x_2 - x_1)^2 + (y_2 - y_1)^2} < \eta$ implies $|\phi(x_2, y_2) - \phi(x_1, y_1)| < \epsilon/2(b - a)$ (in which case, of course, $|y_2 - y_1| < \eta$ implies $|\phi(x_1, y_2) - \phi(x_1, y_1)| < \epsilon/2(b - a)$). We now require the positive number δ to be so small that whenever the norm of a net on the closed interval $[a, b]$ is less than δ, then any sum of the form $\sum_{i=1}^{n} \phi(f(t_i), g(t_i)) \Delta t_i$ approximates the integral $\int_a^b \phi(f(t), g(t)) dt$ within $\epsilon/2$. (The integral exists since the integrand is a continuous function of the variable t. Cf. Ex. 25, § 1011.) Then, since $g(t)$ is uniformly continuous on $[a, b]$ (cf. § 307), we can (and do) further require that δ be so small that when the norm of a net is less than δ, and $a_{i-1} \leq t_i \leq a_i$ and $a_{i-1} \leq t_i' \leq a_i$, then $|g(t_i) - g(t_i')| < \eta$. Owing to the properties of the numbers η and δ, whenever the norm of the net is less than δ we have

(1) $$\left|\sum \phi(f(t_i), g(t_i)) \Delta t_i - \int_a^b \phi(f(t), g(t)) dt\right| < \frac{\epsilon}{2},$$

and

(2) $$|\sum \phi(f(t_i), g(t_i')) \Delta t_i - \sum \phi(f(t_i), g(t_i)) \Delta t_i|$$
$$\leq \sum |\phi(f(t_i), g(t_i')) - \phi(f(t_i), g(t_i))| \Delta t_i < \sum \frac{\epsilon}{2(b - a)} \Delta t_i = \frac{\epsilon}{2}.$$

By the triangle inequality,

$$\left|\sum \phi(f(t_i), g(t_i')) \Delta t_i - \int_a^b \phi(f(t), g(t)) dt\right| < \epsilon.$$

Since this statement is the meaning of (1), § 1101, the proof is complete.

1103. ARCS AND CURVES

Definition. *An arc is a continuous image of a closed interval. For example, an arc in E_2 is the set of all points (x, y) whose coordinates are given by two continuous functions of a real variable,*

(1) $$x = f(t), \quad y = g(t),$$

§ 1103] ARCS AND CURVES 341

where $a \leq t \leq b$. The variable t is called a **parameter**, the representation (1) is a **parametrization**, and the functions $f(t)$ and $g(t)$ are **parametrization functions**. A **closed curve** is an arc for which the images of the endpoints, a and b, of the defining interval are identical; in E_2, $f(a) = f(b)$, $g(a) = g(b)$. A **simple arc** is a continuous one-to-one image of a closed interval. A **simple closed curve** is a closed curve for which the end-points of the defining interval are the only two distinct points that map into the same point. (Cf. Fig. 1101.)

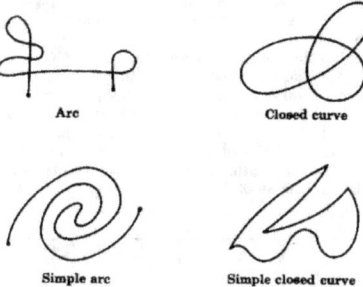

FIG. 1101

A similar definition applies to arcs and curves in higher dimensional spaces.

The single word **curve** will be used to mean the same as *arc* except that the defining interval need not be closed or bounded.

An arc can be thought of as the path traced out by a particle in motion during a closed interval of time. In a closed curve the particle returns to its initial position. In a simple arc it never repeats an earlier position, and in a simple closed curve it repeats only its initial position.

NOTE 1. Since a closed interval is compact, the mapping of the defining interval onto a simple arc has a continuous inverse. Thus, *a simple arc could be defined as a homeomorph of a closed interval.* (Cf. Theorem II and the Note, § 1016.) In a similar way, since in the definition of a closed curve the end-points of the defining interval can be considered as being brought together and identified, *a simple closed curve could be defined as a homeomorph of the circumference of a circle.*

NOTE 2. Since any two closed intervals are homeomorphic (Note, § 1016), it is immaterial what the defining interval in the preceding Definition is. For simplicity, it can be taken to be the closed unit interval $[0, 1]$.

NOTE 3. It is easy to see that the parametrization functions f and g of the above Definition are not unique for a given arc or curve. An example in E_2 is the straight line segment from $(0, 0)$ to $(1, 1)$, given by the three distinct parametrizations $x = t, y = t; x = t^2, y = t^2;$ and $x = 1 - t, y = 1 - t; 0 \leq t \leq 1$.

★★1104. SPACE-FILLING ARCS

In 1890 Peano (cf. § 101) made public a remarkable discovery about certain arcs and curves. He had found that, whatever one's intuitions might say to the contrary, it is possible for an arc or curve to "fill space"—that is, to pass at least once through every point of a higher dimensional set, like a closed square, or a "cube" in a Euclidean space of any number of dimensions.

We present in this section an outline discussion of an arc, described in 1891 by the German mathematician D. Hilbert (1862-1943), that completely fills the closed unit square $A: 0 \leq x \leq 1, 0 \leq y \leq 1$ in E_2. The idea is to subdivide the square A into four parts (each a closed square), then into sixteen, then sixty-four, etc., while the defining unit interval I is correspondingly subdivided (into closed subintervals). The manner of subdivision and setting up a correspondence between the subsquares of A and the subintervals of I is tedious to write analytically, and will be indicated only by a figure (Fig. 1102).

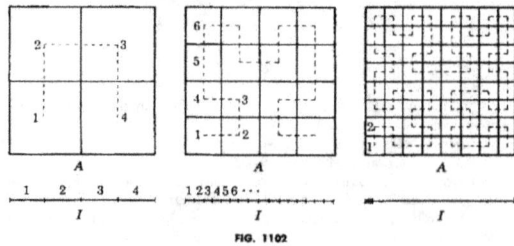

FIG. 1102

We now define the point of A that is to be the image of a given point x of I as follows: At *each* stage of the subdivision suggested in Figure 1102, x belongs to one of the subintervals of I (if x belongs to two such subintervals, select either one). In this way a decreasing sequence of subintervals of I is obtained, whose intersection consists of the single point x. The *corresponding* sequence of subsquares of A is also decreasing, and has exactly one point p in common (cf. Ex. 23, § 1021). This point p is *defined* to be the image of x.

The interested reader can show that the mapping defined in this way does indeed give a space-filling arc, by verifying the following three steps:

1. *The image of x is unique.* That is, if a *different* sequence of subintervals of I were chosen, for a given x, the *same* point p would result. (This is because adjacent subintervals of I correspond to adjacent subsquares of A.)

2. *Every* point p of A is the image of *some* point x of I. (Choose any decreasing sequence of subsquares of A whose intersection is p, and intersect the corresponding subintervals of I.)

3. *The mapping is continuous.* (If an arbitrary neighborhood N_p of p is given, it is possible to choose a sufficiently small neighborhood N_x of x that all subintervals of I contained within N_x correspond to subsquares of A lying inside N_p.)

NOTE 1. The mapping of I onto A, defined above, although single-valued, is *many-to-one* in places; that is, more than one value of x may map into the same point p. For example, the three numbers $\frac{1}{4}$, $\frac{1}{2}$, and $\frac{5}{8}$ are all mapped into the point at the center of A. This fact seems in itself to violate our preconceptions since it could be interpreted as stating that *there are more points in I than there are in A!* On the other hand, any continuous mapping of I onto A *must* be many-to-one in places. This is explained in Note 2.

NOTE 2. The square A is not a *simple* arc. In other words, a mapping of the type discussed above *cannot* be one-to-one. A simple way to see this is to observe that the removal of a single interior point of I "disconnects" I, while no such removal disconnects A. Therefore I and A are not homeomorphic—which they would be, by Theorem II, § 1016, if either could be mapped continuously in a one-to-one manner on the other. A more general expression of this principle is that neighborhoods (or their closures) in Euclidean spaces of different dimensions are never homeomorphic. (Cf. W. Hurewicz and H. Wallman, *Dimension Theory* (Princeton University Press, 1941).)

1105. ARC LENGTH

The length of an arc or curve is defined in terms of lengths of approximating inscribed polygons obtained by joining a finite sequence of points, each to the next, by straight line segments. As will be disclosed in succeeding discussion, such a definition necessitates a formulation based on a particular parametrization. Otherwise, for a given finite set of points on an arc, it would be meaningless to speak of joining one of these points to the "next." The parametrization tells us, for instance, which way to proceed from a point of intersection of a self-crossing arc.

Definition. *Let C be an arc or closed curve in E_2 with parametrization* $\mathfrak{R}: x = f(t)$, $y = g(t)$, $a \leq t \leq b$, *and let* $\mathfrak{N}: a_0 < a_1 < \cdots < a_n = b$ *be a net on $[a, b]$, the corresponding points of C being $p_0, p_1, \cdots, p_n$. Let $L_\mathfrak{N}$ denote the length* $d(p_0, p_1) + d(p_1, p_2) + \cdots + d(p_{n-1}, p_n)$ *of the* **inscribed polygon** *$(p_0p_1 \cdots p_n)$. Then the* **length** *L of the arc or closed curve C is the limit (in the sense of § 501):*

$$(1) \quad L \equiv \lim_{|\mathfrak{N}|\to 0} L_{\mathfrak{N}} = \lim_{|\mathfrak{N}|\to 0} [d(p_0, p_1) + \cdots + d(p_{n-1}, p_n)],$$

provided this limit exists and is finite. In case (1) exists and is finite, C is called **rectifiable**. Otherwise it is **nonrectifiable** and has **infinite length**.

A similar definition applies to higher dimensional spaces.

NOTE. It is shown in § 1106 that an arc or curve is rectifiable if and only if the lengths $L_{\mathfrak{N}}$ are bounded for all possible nets $\mathfrak{N}$.

As illustrated in Example 2, below, an arc may have infinitely many lengths, depending on the parametrization.† However, for *simple* arcs and *simple* closed curves, as will be demonstrated in the next section, *arc length is independent of the parametrization.*

Example 1. Show that the simple arc

$$x = t, \, y = t \sin \frac{1}{t} \text{ for } 0 < t \leq 1, \, x = 0, \, y = 0 \text{ for } t = 0,$$

is not rectifiable. (Cf. Fig. 403.)

Solution. For any given positive integer n, choose for t the values

$$t_0 = 0, \, t_1 = \frac{2}{(2n-1)\pi}, \, t_2 = \frac{2}{(2n-3)\pi}, \, \cdots, \, t_{n-2} = \frac{2}{3\pi}, \, t_{n-1} = \frac{2}{\pi}, \, t_n = 1.$$

The intermediate points of the corresponding inscribed polygon are points of tangency of the graph of $y = x \sin \frac{1}{x}$ with the straight lines $y = \pm x$ (cf. Fig. 403). The length L of the polygon $[p_0 p_1 \cdots p_n]$ is certainly greater than the sum of the absolute values of the ordinates at the points $p_1, p_2, \cdots, p_{n-1}$, or

$$\frac{2}{\pi} + \frac{2}{3\pi} + \frac{2}{5\pi} + \cdots + \frac{2}{(2n-1)\pi}.$$

By the divergence of the harmonic series, these sums are unbounded. Therefore the lengths L are unbounded. (Cf. Example 3, § 516. Also cf. Exs. 24, 27, § 1110.)

★**Example 2.** Find the length of the arc C in E_2 whose points (x, y) satisfy the relations $0 \leq x \leq 1$, $y = 0$, given by the parametrization functions $x = f(t)$, $y = 0$, where the graph of $f(t)$, as shown in Figure 1103, is made up of straight line segments. The interval $0 \leq t \leq 1$ is broken up into $k + 2$ equal parts, where k is a positive odd integer, and α is an arbitrary number such that $0 \leq \alpha \leq 1$. The arc C, in other words, is traced out by a particle moving as follows: first across the arc; then a retracing of amount α; then to the right-hand end of the arc; then back and forth across the arc an arbitrary number of times.

Solution. It is not hard to see that the arc length of C, according to the Definition, is equal to the total straight-line distance covered by the moving point. In other words, it is equal to $1 + 2\alpha + 2 + 2 + \cdots + 2 = k + 2\alpha$. Therefore the length of the arc C can be any real number L such that $L \geq 1$. Clearly the smallest of all possible lengths given by *all* parametrizations is $L = 1$.

† It is possible to formulate a concept of arc length independent of parametrization by defining it to be the greatest lower bound of the lengths L given by all possible parametrizations.

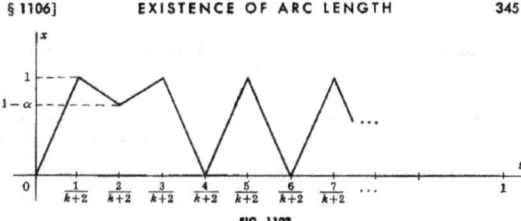

FIG. 1103

★1106. EXISTENCE OF ARC LENGTH

We shall establish in this section a simple condition for the existence of arc length, formulated for convenience for the Euclidean plane, with notation of § 1105:

Theorem. *Let C be an arc or closed curve in E_2, with parametrization $x = f(t)$, $y = g(t)$, $a \leq t \leq b$. Then the limit*

(1) $\qquad L = \lim_{|\mathfrak{N}| \to 0} L_\mathfrak{N} = \lim_{|\mathfrak{N}| \to 0} [d(p_0, p_1) + \cdots + d(p_{n-1}, p_n)]$

always exists, in the finite or infinite sense. It is finite if and only if the polygonal lengths $L_\mathfrak{N}$ are bounded for all nets $\mathfrak{N}$ on $[a, b]$. In any case,

$$L = \sup L_\mathfrak{N}, \text{ for all nets } \mathfrak{N} \text{ on } [a, b].$$

In proving this theorem we shall find it convenient to use the inclusion relation $\mathfrak{N}_1 \subset \mathfrak{N}_2$ for nets $\mathfrak{N}_1$ and $\mathfrak{N}_2$, where every point of $\mathfrak{N}_1$ is a point of $\mathfrak{N}_2$. In this case we call $\mathfrak{N}_2$ **a refinement** of $\mathfrak{N}_1$. We start with a lemma:

Lemma. $\mathfrak{N}_1 \subset \mathfrak{N}_2$ *implies* $L_{\mathfrak{N}_1} \leq L_{\mathfrak{N}_2}$.

Proof of lemma. Adding points to a net adds corners to the inscribed polygon. This means that a single term $d(p_{i-1}, p_i)$ in the length $L_{\mathfrak{N}_1}$ is replaced by several terms whose sum represents a "broken-line distance" between the points p_{i-1} and p_i. By the triangle inequality (generalized to an arbitrary number of terms), this new polygonal length is *at least* as great as the original "straight-line distance" $d(p_{i-1}, p_i)$. Upon addition of all such inequalities we obtain the desired result.

Proof of Theorem. In order to prove the theorem we define L to be the least upper bound of the numbers $L_\mathfrak{N}$, and let B be an arbitrary number less than L. (B is a real number, but L may equal $+\infty$.) The next step is to let $\mathfrak{N}_0$ be a net, to be held fixed, such that

$$L_{\mathfrak{N}_0} > B.$$

We denote by k the number of subintervals of $\mathfrak{N}_0$, and by ϵ the positive number $L_{\mathfrak{N}_0} - B$. We shall have completed the proof when we show the existence of a positive number $\delta = \delta(\epsilon)$ such that

(2) $\qquad\qquad |\mathfrak{N}| < \delta$ implies $L_{\mathfrak{N}} > L_{\mathfrak{N}_0} - \epsilon = B.$

At this point, let us observe that, if $\mathfrak{N}$ is an arbitrary net, the two nets $\mathfrak{N}$ and $\mathfrak{N}_0$ have a common refinement $\mathfrak{N}_1 \equiv \mathfrak{N} \cup \mathfrak{N}_0$, obtained by adding (at most) $k-1$ points to $\mathfrak{N}$. Since $\mathfrak{N}_0 \subset \mathfrak{N}_1$, $L_{\mathfrak{N}_0} \leq L_{\mathfrak{N}_1}$, and we frame our goal as that of finding δ so that

(3) $\qquad\qquad |\mathfrak{N}| < \delta$ implies $L_{\mathfrak{N}} > L_{\mathfrak{N}_1} - \epsilon.$

This can be clarified as the problem of finding δ so that the addition of fewer than k points to any $\mathfrak{N}$ whose norm is less than δ increases the length $L_{\mathfrak{N}}$ by less than ϵ. Still simpler: let us find δ so that the addition of *one* point to $\mathfrak{N}$ increases the length $L_{\mathfrak{N}}$ by less than ϵ/k.

In final form, then, we seek $\delta = \delta(\epsilon) > 0$ such that if α and β are any two points of $[a, b]$ such that $|\beta - \alpha| < \delta$, and if γ is any point between them, then

$$d(p(\alpha), p(\gamma)) + d(p(\gamma), p(\beta)) < d(p(\alpha), p(\beta)) + \epsilon/k,$$

where $p(\alpha)$, $p(\beta)$, and $p(\gamma)$ are the points of C corresponding to α, β, and γ, respectively. The existence of δ for this last inequality is guaranteed by the uniform continuity of the function

$$\begin{aligned}F(r, s, t) &= d(p(r), p(t)) + d(p(t), p(s))\\ &= \sqrt{[f(t) - f(r)]^2 + [g(t) - g(r)]^2}\\ &\quad + \sqrt{[f(t) - f(s)]^2 + [g(t) - g(s)]^2}\end{aligned}$$

over the cube $a \leq r \leq b,\ a \leq s \leq b,\ a \leq t \leq b$, when it is evaluated at the two points (α, β, γ) and (α, β, α), which are separated by a distance $|\gamma - \alpha| < \delta$.

★1107. INDEPENDENCE OF PARAMETRIZATION

In case an arc or closed curve is simple, its length (as defined in § 1105) *is independent of the parametrization.*

In a proof of this fact the first step is to observe that since the length of an arc C, for a given parametrization, is the least upper bound of the lengths $L_{\mathfrak{N}}$, for all nets $\mathfrak{N}$ on the defining interval, it is sufficient to show that *for any two parametrizations every polygonal length $L_{\mathfrak{N}}$ in each parametrization is equal to some polygonal length $L_{\mathfrak{N}}$ in the other parametrization.*

The problem, then, is to show that if Π_1 and Π_2 are any two parametrizations of C, with defining intervals $[a, b]$ and $[c, d]$, respectively, and if $[p_0, p_1, \cdots p_n]$ is an inscribed polygon given by a net $\mathfrak{N}_1$ on $[a, b]$, for the parametrization Π_1, then there exists a net $\mathfrak{N}_2$ on $[c, d]$ that gives the *same*

§ 1108] INTEGRAL FORM FOR ARC LENGTH 347

inscribed polygon for the parametrization Π_2. The attack on this problem is clear: Let $p(t)$ and $P'(t)$ be the mappings given by Π_1 and Π_2, respectively, and let $t(p)$ and $T(p)$ denote their inverses, which are continuous by Theorem II, § 1016. Then the net $\mathfrak{N}_1$, consisting of the points $a = t_0 < t_1 < \cdots < t_n = b$, determines the points $p_0 = p(t_0)$, $p_1 = p(t_1)$, $\cdots$, $p_n = p(t_n)$ on C. The net $\mathfrak{N}_2$ for the parametrization Π_2 must be determined by the inverse mapping $T(p) : s_0 = T(p_0)$, $s_1 = T(p_1)$, $\cdots$, $s_n = T(p_n)$. The real question, and the only question, now, is whether this collection $s_0, s_1, \cdots, s_n$, or its reverse, $s_n, s_{n-1}, \cdots, s_0$, really *is* a net on $[c, d]$. The affirmative answer is given by the fact that the numbers $s_0, s_1, \cdots, s_n$ must align themselves on the interval $[c, d]$ either in an increasing or in a decreasing order. Finally, the fact that this ordering of the points $s_0, s_1, \cdots, s_n$ must be monotonic is seen by considering the composite function

$$\phi(t) \equiv T(p(t)),$$

which maps the interval $[a, b]$ onto the interval $[c, d]$ in a one-to-one continuous manner. Such a function, that is, a real-valued function of a real variable that maps a closed interval onto a closed interval in a one-to-one continuous manner, must be strictly monotonic (cf. Ex. 31, § 216), and the proof is complete.

1108. INTEGRAL FORM FOR ARC LENGTH

It is our purpose in this section to show that if enough more than mere continuity is demanded of the parametrization functions, the length of a rectifiable arc can be expressed as a definite integral. The conditions of the following theorem are not the most general possible (cf. Exs. 23, 29, § 1110). They have been chosen for their simplicity, and are sufficiently general for most purposes. The theorem is formulated for an arc in the Euclidean plane, but an analogous statement, with an analogous proof, applies to higher dimensional spaces (cf. § 1112, and Ex. 22, § 1110).

Theorem I. *Let C be an arc whose parametrization functions $x = f(t)$, $y = g(t)$ and their first derivatives are continuous on the closed interval $[a, b]$. Then C is rectifiable and its length L is given by the formula*

(1) $$L = \int_a^b \sqrt{(f'(t))^2 + (g'(t))^2}\, dt.$$

Proof. Since the integrand is continuous, the integral (1) exists. Let its value be I. If $\epsilon > 0$ we wish to find $\delta > 0$ such that

$$|\mathfrak{N}| < \delta \text{ implies } |L_\mathfrak{N} - I| < \epsilon.$$

Such a number δ is provided by Duhamel's Principle (§ 1101) applied to the function $\phi(x, y) \equiv \sqrt{x^2 + y^2}$, since

$$L_{\mathfrak{N}} = \sum_{i=1}^{n} d(p_{i-1}, p_i) = \sum_{i=1}^{n} \sqrt{[f(a_i) - f(a_{i-1})]^2 + [g(a_i) - g(a_{i-1})]^2}$$
$$= \sum_{i=1}^{n} \sqrt{[f'(t_i)]^2 + [g'(t_i')]^2} \, \Delta t_i.$$

The final expression is given by the Law of the Mean (§ 405) applied to the two functions $f(t)$ and $g(t)$ on the interval $[a_{i-1}, a_i]$.

If a variable upper limit t (in place of b) is used, we shall denote the variable arc length from the fixed point $p(a)$ to the variable point $p(t)$ by the letter s:

(2) $$s = s(t) \equiv \int_a^t \sqrt{(f'(u))^2 + (g'(u))^2} \, du.$$

Here, of course, a new letter is used for the variable of integration.

The arc length s thus becomes a differentiable function of the parameter t and, by Theorem I, § 504, $\dfrac{ds}{dt} = \sqrt{(f'(t))^2 + (g'(t))^2}$, or

(3) $$\left(\frac{ds}{dt}\right)^2 = \left(\frac{dx}{dt}\right)^2 + \left(\frac{dy}{dt}\right)^2.$$

In terms of differentials this takes the form

(4) $$ds^2 = dx^2 + dy^2,$$

a relation which shows how ds is associated with the tangent to the curve. (Cf. Figure 1104.)

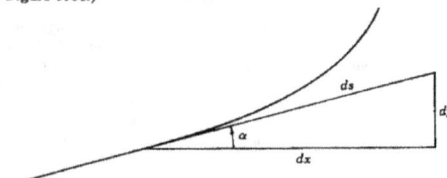

FIG. 1104

A point on a curve $x = f(t)$, $y = g(t)$ at which $f'(t) = g'(t) = 0$ is called a **singular point**. Any other point of the curve is a **regular point**. For example, for the semicubical parabola $x = t^2$, $y = t^3$, the origin is a singular point and all other points are regular.

NOTE 1. The examples $x = t$, $y = t$ and $x = t^3$, $y = t^3$, $-1 \leq t \leq 1$, show that an arc may have a singular point for one parametrization and fail to have one for another.

NOTE 2. A point where a curve crosses itself should not be confused with a singular point. For example, the folium of Descartes has no singular points in the parametrization given in (3), § 418.

§ 1108] INTEGRAL FORM FOR ARC LENGTH

Under the assumption that a curve C has no singular points, the function $s(t)$ is seen to be a differentiable function of t, with a positive derivative so that (§ 406) s is a strictly increasing differentiable function of t, and t is a strictly increasing differentiable function of s. This permits the use of the arc length itself as a parameter:

(5) $$x = F(s), \quad y = G(s).$$

In this case, formula (3) shows that

(6) $$\left(\frac{dx}{ds}\right)^2 + \left(\frac{dy}{ds}\right)^2 = 1.$$

In fact, if α is the inclination of the tangent line to C, it is clear from Figure 1104 that $\frac{dx}{ds} = \cos \alpha$ and $\frac{dy}{ds} = \sin \alpha$, so that equation (6) becomes an expression of a familiar trigonometric identity.

A curve with continuously differentiable parametrization functions and without singular points is called a **smooth curve**. For such a curve the continuity of the functions $\frac{dx}{dt}, \frac{dy}{dt}, \frac{ds}{dt}$, and $\frac{dt}{ds}$, and hence of the functions $\frac{dx}{ds}$ and $\frac{dy}{ds}$, imply the continuity of the inclination α as a function of s. In other words, *a smooth curve C has a continuously turning tangent.*

If the parameter is the variable x, formula (1) for arc length becomes the familiar one of a first course in calculus,

(7) $$L = \int_a^b \sqrt{1 + \left(\frac{dy}{dx}\right)^2}\, dx,$$

a similar formula holding in case the parameter is the variable y.

If a curve is given by polar coordinates, results similar to those just given can be obtained. Suppose ρ and θ are continuous functions of a parameter t, and have continuous derivatives with respect to t. Then the rectangular coordinates x and y, by virtue of the relations

(8) $$x = \rho \cos \theta, \quad y = \rho \sin \theta,$$

are also continuously differentiable as functions of t:

(9) $$\frac{dx}{dt} = -\rho \sin \theta \frac{d\theta}{dt} + \cos \theta \frac{d\rho}{dt}, \quad \frac{dy}{dt} = \rho \cos \theta \frac{d\theta}{dt} + \sin \theta \frac{d\rho}{dt}.$$

Squaring the various members of (9), adding, and combining terms, we have

(10) $$\left(\frac{dx}{dt}\right)^2 + \left(\frac{dy}{dt}\right)^2 = \left(\frac{d\rho}{dt}\right)^2 + \rho^2 \left(\frac{d\theta}{dt}\right)^2.$$

We now substitute in (1) to obtain

(11) $$L = \int_a^b \sqrt{\left(\frac{d\rho}{dt}\right)^2 + \rho^2 \left(\frac{d\theta}{dt}\right)^2}\, dt$$

350   ARCS AND CURVES   [§ 1108

and, if the parameter is θ,

(12) $$L = \int_{\theta_1}^{\theta_2} \sqrt{\rho^2 + \left(\frac{d\rho}{d\theta}\right)^2}\, d\theta.$$

In differential form the relation between arc length and polar coordinates is

(13) $$ds^2 = (d\rho)^2 + (\rho d\theta)^2.$$

Notice that, as with rectangular coordinates, the differential of arc length can be represented as the hypotenuse of a right triangle, this hypotenuse being tangent to the arc. (Cf. Fig. 1105.)

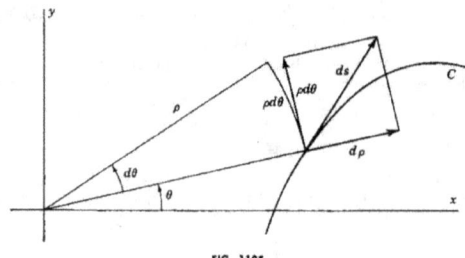

FIG. 1105

An important fact about smooth curves is given in the following theorem:

Theorem II. *If p is fixed point on a smooth curve C and q is a variable point on C that tends toward p as a limit, the quotient of the length of the chord $\overline{pq}$ to the length of the arc $\overset{\frown}{pq}$ tends toward 1 as a limit:*

$$\lim_{q \to p} \frac{\overset{\frown}{pq}}{\overline{pq}} = 1.$$

(See Figure 1106.)

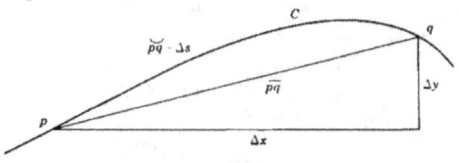

FIG. 1106

§ 1109] THE TRIGONOMETRIC FUNCTIONS 351

Proof. Since $\widetilde{pq} = \Delta s$ and $\overline{pq}^2 = \Delta x^2 + \Delta y^2$, we wish to show that

$$\lim_{\Delta s \to 0} \left(\frac{\overline{pq}}{\widetilde{pq}}\right)^2 = \lim_{\Delta s \to 0} \frac{\Delta x^2 + \Delta y^2}{\Delta s^2} = 1.$$

But this reduces immediately to (6).

All of the results of this section have their three-dimensional (indeed, n-dimensional) analogues (cf. § 1112), which we shall feel free to use, as if they were formally established.

Example. Express the coordinates of the points of the semi-cubical parabola

$$x = t^2, \quad y = t^3,$$

for $t \geq 0$, in terms of arc length as parameter.

First Solution. Taking the origin ($t = 0$) as the starting point, we use formula (2):

$$s = \int_0^t \sqrt{(2u)^2 + (3u^2)^2}\, du = \int_0^t u\sqrt{4 + 9u^2}\, du$$

$$= \left[\frac{(4 + 9u^2)^{\frac{3}{2}}}{27}\right]_0^t = \frac{(4 + 9t^2)^{\frac{3}{2}} - 8}{27}.$$

Solving for t in terms of s gives

$$t = \left[\frac{(27s + 8)^{\frac{2}{3}} - 4}{9}\right]^{\frac{1}{2}},$$

whence

$$x = \frac{(27s + 8)^{\frac{2}{3}} - 4}{9}, \quad y = \frac{[(27s + 8)^{\frac{2}{3}} - 4]^{\frac{3}{2}}}{27}.$$

Second Solution. Elimination of the parameter t gives the equation $y = x^{\frac{3}{2}}$, so that (7) becomes

$$s = \int_0^x \sqrt{1 + \tfrac{9}{4}u}\, du = \frac{8(1 + \tfrac{9}{4}x)^{\frac{3}{2}} - 8}{27}.$$

Solving for x in terms of s gives the previous result.

★1109. REMARK CONCERNING THE TRIGONOMETRIC FUNCTIONS

In §§ 603 and 604 the basic trigonometric functions, $\sin t$ and $\cos t$, were defined, and their principal elementary properties were established. Among these properties is the parametrization of the unit circle (Ex. 13, § 604), $x^2 + y^2 = 1$, by means of the equations $x = \cos t$, $y = \sin t$. We are now in a position to show that the trigonometric functions defined analytically in Chapter 6 are identical with those with which the reader is already familiar on an intuitive basis. For this purpose we *define* the measure α of an angle in standard position to be the arc length of that portion of the unit circle enclosed by it, account being taken of sign and multiples of 2π. Complete consistency between the intuitive and analytic definitions is obtained by demonstrating that the arc length of the curve

$x = \cos t, y = \sin t$, for t between 0 and α, is α. But this is merely a simple application of the integral form for arc length:

$$\int_0^\alpha \sqrt{(-\sin t)^2 + (\cos t)^2}\, dt = \int_0^\alpha 1 \cdot dt = \alpha.$$

The reconciliation of the two definitions is now completed as follows: In the first place, the traditional method of defining the sine and cosine functions is equivalent to considering the point where the unit circle cuts the terminal side of the angle α in standard position, the coordinates of this point being *by definition* $\cos \alpha$ and $\sin \alpha$. The analytic method defines these quantities in a different way, but the point whose coordinates are the quantities $\cos \alpha$ and $\sin \alpha$ of § 604 has been shown to have the property of being the point on the unit circle whose radius vector is the terminal side of the angle α in standard position.

1110. EXERCISES

In Exercises 1-10, find the length of the given curve between the specified points or values of the parameter. The constant a is positive.

1. $y = x^2$; $(0, 0)$ and $(2, 8)$.
2. $y = \sqrt{x}$; $(0, 0)$ and $(4, 2)$.
3. $\rho = a \cos \theta$; $\theta = -\pi/2$ and $\pi/2$.
4. $\rho = k\theta$, $k > 0$; $(0, 0)$ and $(\alpha, k\alpha)$.
5. $x = 5 \cos t, y = 5 \sin t$; $t = 0$ and $2n\pi$.
6. $x = a \cos^3 t, y = a \sin^3 t$; $t = 0$ and 2π.
7. $x^{\frac{2}{3}} + y^{\frac{2}{3}} = a^{\frac{2}{3}}$; as a simple closed curve.
8. $y = a \cosh \frac{x}{a}$; $x = 0$ and $x = b > 0$.
9. $\rho = 2/(1 + \cos \theta)$; $\theta = 0$ and $\pi/2$.
10. $x = e^t \cos t, y = e^t \sin t$; $t = 0$ and $\pi/2$.

In Exercises 11-14, express x and y, or ρ and θ, in terms of arc length as parameter, distances being measured from the specified point or value of the parameter.

11. $x = r \cos \theta, y = r \sin \theta$, r a positive constant, $\theta \geq 0$; $\theta = 0$.
12. $x = \theta - \sin \theta, y = 1 - \cos \theta$, $0 \leq \theta \leq 2\pi$; $\theta = 0$.
13. $x = \cos^3 t, y = \sin^3 t$, $0 \leq t \leq \pi/2$; $t = 0$.
14. $\rho = e^\theta$, $\theta \geq 0$; $\theta = 0$.

15. Criticize the following derivation of the arc length of the cardioid, $\rho = 1 + \cos \theta$: Since $\rho' = -\sin \theta$, $\rho^2 + \rho'^2 = 2 + 2 \cos \theta = 4 \cos^2(\theta/2)$. Hence formula (12) gives $\int_0^{2\pi} 2 \cos (\theta/2)\, d\theta = 4 \sin (\theta/2) \Big]_0^{2\pi} = 0$.

16. Prove that every polygon is an arc.

17. Let A be a set in E_2 shaped like a letter E, consisting of the segments joining the following pairs of points: $(0, 4)$ and $(3, 4)$; $(0, 2)$ and $(2, 2)$; $(0, 0)$ and $(3, 0)$; and $(0, 0)$ and $(0, 4)$. Find a specific parametrization that yields A as an arc. Find its arc length under this parametrization. (Cf. Exs. 18, 26.)

18. Find a specific parametrization that yields the set A of Exercise 17 as a

§ 1111] CYLINDRICAL, SPHERICAL COORDINATES 353

closed curve. Find its arc length under this parametrization. (Cf. Exs. 17, 26.)

19. Prove that any set that can be parametrized as an arc can also be parametrized as a closed curve, and conversely.

20. State Duhamel's Principle for a function of n variables.

★**21.** Prove Duhamel's Principle for a function of n variables each of which is a continuous function of the parameter.

★**22.** Derive the integral form

$$L = \int_a^b \sqrt{\sum_{i=1}^{n} (f_i'(t))^2} \, dt$$

for an arc $x_i = f_i(t)$, $i = 1, 2, \cdots, n$, in E_n, where the derivatives are continuous.

★**23.** Prove the following form of Theorem I, § 1108: *Let C be the arc* $y = f(x)$, *where* $f(x)$ *is continuous and* $f'(x)$ *is integrable on* $[a, b]$. *Then C is rectifiable and its length L is given by the formula*

$$L = \int_a^b \sqrt{1 + (f'(x))^2} \, dx.$$

(Cf. Ex. 29.) *Hints:* First show that Duhamel's Principle is not needed. Next, if $\psi(x) \equiv \sqrt{1 + (f'(x))^2}$, show that $|\psi(x_i) - \psi(x_i')| < |f'(x_i) - f'(x_i')|$. Then use Ex. 34, § 503, and techniques from the proof of Theorem I, § 1108.

★**24.** Show that the arc $y = x^2 \sin(1/x)$ ($y = 0$ when $x = 0$) is rectifiable on $[0, 1]$. (Cf. Ex. 23. Also cf. Example 1, § 1105, Example 2, § 516, and Ex. 27, below.)

★**25.** Prove that the circumference of a circle of radius r is $2\pi r$, and hence justify the definition of π given in § 604. Explain how, if a circle is not considered as a *simple* closed curve, its circumference can be any number $L \geq 2\pi r$.

★★**26.** Show that the set A of Exercise 17 cannot be parametrized as a *simple* arc.

★★**27.** Prove that an arc $x = f(t)$, $y = g(t)$, $a \leq t \leq b$, is rectifiable if and only if both functions $f(t)$ and $g(t)$ are of bounded variation on $[a, b]$ (cf. § 516). Generalize to E_n.

★★**28.** Prove Duhamel's Principle (§ 1101). *Hints:* Use the principle of Ex. 55, § 503, to prove that $\phi(f(t), g(t))$ is integrable. If $\epsilon > 0$, let $\eta > 0$ be such that $\sqrt{(x_2 - x_1)^2 + (y_2 - y_1)^2} < \eta$ implies $|\phi(x_2, y_2) - \phi(x_1, y_1)| < \epsilon/4(b - a)$, and $\delta > 0$ such that $|\Re| < \delta$ implies both (1), § 1102, and $\sum |g(t_i) - g(t_i')| \Delta t_i < \eta \epsilon/8B$, where $|\phi(x, y)| < B$ on A. Split the sums of (2), § 1102, into two parts, one being for the subintervals where the inequality $|g(t_i) - g(t_i')| \geq \eta$ is possible and (consequently) the sum of whose lengths must be $< \epsilon/8B$.

★★**29.** Prove Theorem I, § 1108, under the relaxed conditions that $f(t)$ and $g(t)$ are continuous for $a \leq t \leq b$ and that their derivatives exist for $a < t < b$ and are integrable on $[a, b]$. *Hints:* Use Ex. 28, including the hint showing the integrability of $\phi(f'(t), g'(t)) = \sqrt{(f'(t))^2 + (g'(t))^2}$, and the proof of Theorem I, § 1108.

★★**30.** State and prove Duhamel's Principle (§ 1101) for a function of n variables.

★★**31.** Define and discuss "sectionally smooth" curves.

1111. CYLINDRICAL AND SPHERICAL COORDINATES

Many problems in three dimensions can be more easily attacked with the aid of coordinate systems that are not rectangular. We introduce in this section, two of the most important, and discuss arc length in terms of them. In later sections we shall put them to additional service. We refer each of the new systems of coordinates to a basic rectangular system.

I. Cylindrical coordinates

In this system a point is specified by the polar coordinates (ρ, θ) of its projection on the xy-plane and its directed distance z from the xy-plane. (Cf. Fig. 1107.) With the restrictions

$$\rho \geqq 0,$$
$$0 \leqq \theta < 2\pi,$$

any point not on the z-axis has a unique representation (ρ, θ, z).

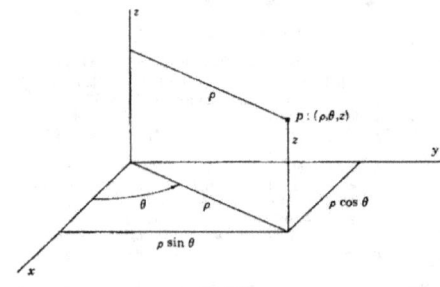

FIG. 1107

If the cylindrical coordinates of a point are given, its rectangular coordinates can be obtained by means of the relations

$$(1) \quad \begin{cases} x = \rho \cos \theta, \\ y = \rho \sin \theta, \\ z = z. \end{cases}$$

Conversely, the cylindrical coordinates of a point are given in terms of the rectangular coordinates by the relations

$$(2) \quad \begin{cases} \rho^2 = x^2 + y^2, \\ \tan \theta = y/x, \\ z = z, \end{cases}$$

the quadrant of θ being determined by the signs of x and y.

These relations can also be used to transform an equation of a surface

§1111] CYLINDRICAL, SPHERICAL COORDINATES 355

in one coordinate system into an equation in the other coordinate system. For example, by means of relations (2), the equation $z = \rho^2$ becomes $z = x^2 + y^2$.

II. Spherical coordinates

The spherical coordinates (r, ϕ, θ) of a point p, as indicated in Figure 1108, are (i) its distance r from the origin O, (ii) the angle ϕ between its

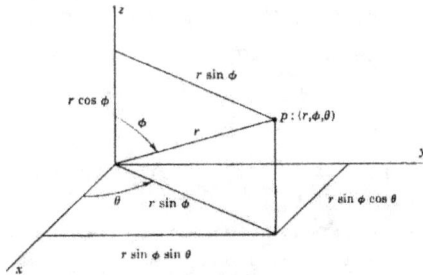

FIG. 1108

radius vector and the z-axis, and (iii) the angular polar coordinate θ of its projection on the xy-plane, subject to the restrictions

$$r \geqq 0,$$
$$0 \leqq \phi \leqq \pi,$$
$$0 \leqq \theta < 2\pi.$$

Again the coordinates of any point not on the z-axis are uniquely determined.

The equations of transformation are

$$(3) \quad \begin{cases} x = r \sin \phi \cos \theta, & r^2 = x^2 + y^2 + z^2, \\ y = r \sin \phi \sin \theta, & \cos \phi = \dfrac{z}{\sqrt{x^2 + y^2 + z^2}}, \\ z = r \cos \phi, & \tan \theta = \dfrac{y}{x} \end{cases}$$

As an example, the equation

$$x^2 + y^2 + z^2 - 2z = 0$$

becomes $r^2 - 2r \cos \phi = 0$, or $r = 2 \cos \phi$.

1112. ARC LENGTH IN RECTANGULAR, CYLINDRICAL, AND SPHERICAL COORDINATES

If the treatment of arc length for a plane curve in rectangular coordinates is paralleled for three dimensions, formulas (1)-(6), § 1108, have the following analogues, where $x = f(t)$, $y = g(t)$, $z = h(t)$ are the parametric equations of the given arc:

(1) $$L = \int_a^b \sqrt{(f'(t))^2 + (g'(t))^2 + (h'(t))^2}\, dt,$$

(2) $$s = s(t) = \int_a^t \sqrt{(f'(u))^2 + (g'(u))^2 + (h'(u))^2}\, du,$$

(3) $$\left(\frac{ds}{dt}\right)^2 = \left(\frac{dx}{dt}\right)^2 + \left(\frac{dy}{dt}\right)^2 + \left(\frac{dz}{dt}\right)^2,$$

(4) $$ds^2 = dx^2 + dy^2 + dz^2,$$

(5) $$x = F(s),\quad y = G(s),\quad z = H(s),$$

(6) $$\left(\frac{dx}{ds}\right)^2 + \left(\frac{dy}{ds}\right)^2 + \left(\frac{dz}{ds}\right)^2 = 1.$$

Details of these derivations are requested in Exercise 11, § 1113.

If the procedure used in § 1108 for deriving a formula for arc length in terms of polar coordinates is followed here, we have only to consider formulas (1) and (3), § 1111, as defining relationships among coordinates all of which are considered as functions of a parameter t. We differentiate those two sets of formulas with respect to t, square, and add, obtaining for the case of cylindrical coordinates

(7) $$\left(\frac{ds}{dt}\right)^2 = \left(\frac{d\rho}{dt}\right)^2 + \rho^2\left(\frac{d\theta}{dt}\right)^2 + \left(\frac{dz}{dt}\right)^2,$$

and for the case of spherical coordinates

(8) $$\left(\frac{ds}{dt}\right)^2 = \left(\frac{dr}{dt}\right)^2 + r^2\left(\frac{d\phi}{dt}\right)^2 + r^2\sin^2\phi\left(\frac{d\theta}{dt}\right)^2.$$

(Cf. Ex. 11, § 1113.) The corresponding formulas for arc length are

(9) $$L = \int_a^b \sqrt{\left(\frac{d\rho}{dt}\right)^2 + \rho^2\left(\frac{d\theta}{dt}\right)^2 + \left(\frac{dz}{dt}\right)^2}\, dt,$$

(10) $$L = \int_a^b \sqrt{\left(\frac{dr}{dt}\right)^2 + r^2\left(\frac{d\phi}{dt}\right)^2 + r^2\sin^2\phi\left(\frac{d\theta}{dt}\right)^2}\, dt.$$

In Exercise 12, § 1113, the student is asked to discuss formulas (7) and (8) in differential form.

§ 1114] CURVATURE IN TWO DIMENSIONS 357

1113. EXERCISES

 1. Write the rectangular coordinates of the points whose cylindrical coordinates are $(3, 0, 2)$ and $(6, \pi/3, -1)$.
 2. Write the cylindrical coordinates of the points whose rectangular coordinates are $(-4, 0, 1)$ and $(6, 2\sqrt{3}, 0)$.
 3. Write the rectangular coordinates of the points whose spherical coordinates are $(7, \pi/2, \pi)$ and $(12, 5\pi/6, 2\pi/3)$.
 4. Write the spherical coordinates of the points whose rectangular coordinates are $(0, -5, 0)$ and $(1, 1, 1)$.
 5. Describe the surfaces $\rho = $ constant, $\theta = $ constant, and $\rho = \sin\theta$.
 6. Describe the surfaces $r = $ constant, $\phi = $ constant, and $f(r, \phi) = 0$.
 7. Transform from cylindrical to rectangular coordinates: $z = \pm\rho$.
 8. Transform from rectangular to cylindrical coordinates: $x^2 + y^2 = z$.
 9. Transform from spherical to rectangular coordinates: $r = 3$.
 10. Transform from rectangular to spherical coordinates: $x^2 + y^2 - 3z^2 = 0$.
 11. Give the details in the derivation of formulas (1)-(10), § 1112.
 12. Write formulas (7) and (8), § 1112, in differential form, and discuss them intuitively. *Hint:* See Figures 1311 and 1312.

In Exercises 13-16, find the length of the given arc between the specified points or values of the parameter. All constants are positive.

 13. $x = a\cos\lambda t, y = a\sin\lambda t, z = \mu t; t = t_1$ and t_2.
 14. $\rho = a, \theta = \lambda t, z = \mu t; t = t_1$ and t_2.
 15. $x = at\cos\lambda t, y = at\sin\lambda t, z = \mu t; t = 0$ and $t > 0$.
 16. $r = \nu t, \phi = \alpha, \theta = \lambda t; t = 0$ and $t > 0$.
 ★17. State and prove the analogue of Theorem II, § 1108, for three dimensions.

1114. CURVATURE AND RADIUS OF CURVATURE IN TWO DIMENSIONS

A measure of the rapidity with which a curve, or its tangent line, is turning at a particular point on the curve is given by the rate of change of the angle made by the tangent line with some fixed direction (which can be taken for convenience to be that of a coordinate axis) with respect to the arc length, measured from some fixed point on the curve. The absolute value of this rate of change (if it exists) is called the **curvature** of the curve at the particular point. That is, if α is the inclination of the tangent line to a curve C (Fig. 1109), and if s is its arc length, the curvature K of C at any particular point is defined to be

$$(1) \qquad K = \left|\frac{d\alpha}{ds}\right|,$$

evaluated at that point.

Assume that the curve C is given parametrically by functions $x = x(t)$ and $y = y(t)$ possessing second derivatives, and let differentiation with respect to the parameter t be indicated by primes. Then, by standard

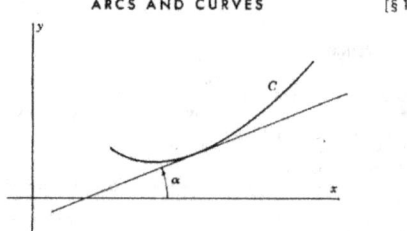

FIG. 1109

differentiation formulas of calculus, and (3), § 1108, at any regular point where C has a finite slope:

$$\frac{d\alpha}{ds} = \frac{d}{dt}\left(\arctan \frac{dy}{dx}\right)\frac{dt}{ds} = \frac{\left(\arctan \frac{y'}{x'}\right)'}{s'}$$
$$= \frac{1}{1 + (y'/x')^2} \cdot \frac{x'y'' - y'x''}{x'^2} \cdot \frac{1}{\sqrt{x'^2 + y'^2}},$$

so that the curvature is given by the formula

$$(2) \qquad K = \frac{|x'y'' - y'x''|}{(x'^2 + y'^2)^{\frac{3}{2}}}.$$

Owing to the symmetry between x and y in formula (2), we know that at a point with a vertical tangent line, where $x' = 0$, an interchange of the roles of the coordinate axes yields the same formula (2). If the student is interested he can carry through the derivation by considering $\alpha = \text{arccot}\,\frac{dx}{dy}$. (Cf. Ex. 25, § 1117.)

If the parameter is arc length, then, by virtue of (6), § 1108,

$$(3) \qquad K = |x'y'' - y'x''|.$$

If the curve is given by an equation of the form $y = f(x)$, so that x is the parameter, then (since $x' = 1$ and $x'' = 0$),

$$(4) \qquad K = \frac{|y''|}{(1 + y'^2)^{\frac{3}{2}}},$$

a similar formula holding in case the parameter is y.

If the curve is given in polar coordinates, and if θ is the parameter, then substitution of $x = \rho \cos \theta$ and $y = \rho \sin \theta$ in (2) gives (check the details):

$$(5) \qquad K = \frac{|\rho^2 + 2\rho'^2 - \rho\rho''|}{(\rho^2 + \rho'^2)^{\frac{3}{2}}}.$$

§ 1115] CIRCLE OF CURVATURE 359

Note 1. The formulas above presuppose that the point in question is not a singular point; that is, that x' and y' are not both zero there (§ 1108), and that the derivatives involved exist. Throughout the remainder of this chapter the existence and continuity of all derivatives used will be assumed without specific statement. In particular *only smooth curves will be considered*.

The **radius of curvature** R of a curve at a given point is defined to be the reciprocal of the curvature at that point,

$$(6) \qquad R = \frac{1}{K}.$$

(If the curvature is zero, the radius of curvature is said to be *infinite*.) Formulas for radius of curvature are given by taking reciprocals of (2)-(5).

Justification for the terminology introduced in this section resides in the two facts that *the curvature of a straight line is zero* and *the radius of curvature of a circle is its radius*. (Cf. Exs. 9-10, § 1117.)

1115. CIRCLE OF CURVATURE

The **center of curvature** of a curve at a given point p on the curve is the point q on the normal to the curve at p, on the concave side of the curve,† whose distance from p is the radius of curvature R of the curve at p. (Cf. Figure 1110.) The **circle of curvature** of the curve at p is the circle with

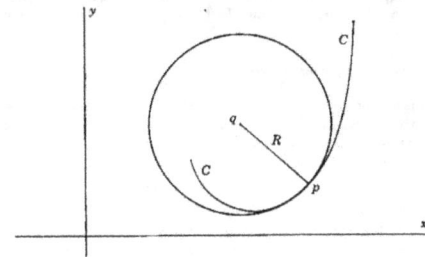

FIG. 1110

center q and radius R. Its tangent at p is the same as that of the given curve at p, and its curvature is that of the given curve at p.

In the next section we shall derive formulas for the coordinates of the center of curvature, but we illustrate by an example a standard method of obtaining the center of curvature without any further use of formulas.

† Concavity is determined by the sign of $d\alpha/ds$ or, in other terms, of $x'y'' - y'x''$, together with the direction of increasing arc length.

360   ARCS AND CURVES   [§ 1116]

Example. Find the center of curvature of the curve $y = x^3$ at the point $(1, 1)$.

Solution. At the point $p: (1, 1)$, $y' = 3$ and $y'' = 6$, and therefore $R = \frac{5}{3}\sqrt{10}$. Since the slope of the curve at p is 3, the slope of the normal there is $-\frac{1}{3}$, and (cf. Fig. 1111) a triangle similar to the one whose hypotenuse is the radius of the circle

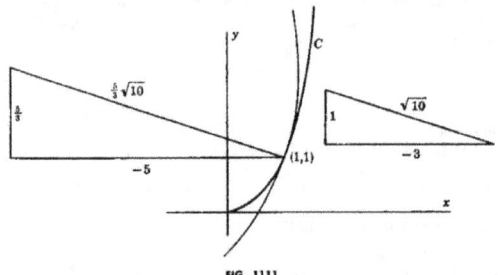

FIG. 1111

of curvature would have sides 1 and -3, and hypotenuse $\sqrt{10}$. Since the triangle desired is larger in the ratio of $\frac{5}{3}$, the x-coordinate of the center must be 5 units to left of 1, and the y-coordinate must be $\frac{5}{3}$ units above 1. That is, the center of curvature is $(-4, 2\frac{2}{3})$.

★1116. EVOLUTES AND INVOLUTES

In this section we shall obtain formulas for the coordinates of the center of curvature of a smooth curve $C: x = x(t)$, $y = y(t)$, at a point where the curvature exists and is nonzero.

If the circle of curvature Γ has center (α, β) and radius R, its equation is

(1) $$(x - \alpha)^2 + (y - \beta)^2 = R^2.$$

Let x and y, for this circle Γ, be functions of a parameter (undesignated), and denote differentiation with respect to this parameter by primes. Since (1) becomes an identity in this parameter, we obtain by differentiation the following two equations:

(2) $$\begin{cases} x'(x - \alpha) + y'(y - \beta) = 0, \\ x''(x - \alpha) + y''(y - \beta) = -(x'^2 + y'^2), \end{cases}$$

true for all points of Γ.

A rather curious and most significant fact is the following: Let x and y be considered as the *original* functions $x(t)$ and $y(t)$ defining the curve C, let differentiation *with respect to the parameter t* be designated by primes, and let (α, β) be the center of curvature and R the radius of curvature at the point p at which x and y and their derivatives are evaluated. *Then equations* (1) *and* (2) *hold*.

Equation (1) is obvious. We shall establish (2) by first proving it for the special case where the parameter for the curve C is x. This can always be done unless C has a vertical tangent at (x, y). (For this latter special case the parameter can be chosen to be y, and the derivation is entirely similar to the one that follows below; the student is asked in Exercise 25, § 1117 to carry through the details.) With x as the parameter, then, equations (2) take the form

$$(3) \quad \begin{cases} (x - \alpha) + \dfrac{dy}{dx}(y - \beta) = 0, \\ \dfrac{d^2y}{dx^2}(y - \beta) = -\left[1 + \left(\dfrac{dy}{dx}\right)^2\right]. \end{cases}$$

The reasons that equations (3) hold are threefold: In the first place, they hold in case y as a function of x is defined by the circle of curvature Γ. In the second place, since C and Γ have a common tangent at the point (x, y), the derivative $\dfrac{dy}{dx}$, when evaluated at that point, is the same for C as it is for Γ. In the third place, since C and Γ have the same curvature and the same concavity at (x, y), the second derivative $\dfrac{d^2y}{dx^2}$, evaluated at that point, is the same for C as it is for Γ (cf. (4), § 1114). Therefore equations (3) hold for C.

Finally, if we substitute

$$\frac{dy}{dx} = \frac{y'}{x'} \quad \text{and} \quad \frac{d^2y}{dx^2} = \frac{1}{x'}\left(\frac{y'}{x'}\right)' = \frac{x'y'' - y'x''}{x'^3}$$

in (3) we obtain equations (2), and the proof is complete.

Solving (2) as a system of two linear equations in the two unknowns $(x - \alpha)$ and $(y - \beta)$, and rearranging terms, we find for the coordinates of the center of curvature:

$$(4) \qquad \alpha = x - \frac{y'(x'^2 + y'^2)}{x'y'' - y'x''}, \quad \beta = y + \frac{x'(x'^2 + y'^2)}{x'y'' - y'x''}.$$

If the point p of the curve, just considered, is regarded as variable and moving along the curve, the circle of curvature can be thought of as rolling along the curve (with variable radius) and its center becomes a variable point tracing out a curve whose equations are given parametrically by (1). Such a curve is called an *evolute*, according to the following definition:

Definition. The locus of centers of curvature of the points of a given curve is called the **evolute** of that curve. If a curve C_1 is the evolute of a curve C_2, then C_2 is called an **involute** of C_1.

NOTE. The evolute of a curve is, by definition, unique. However, as is shown in Exercise 28, § 1117, a curve may have infinitely many involutes.

It has been stated that equations (1) give the evolute of a given curve parametrically. In some cases the parameter can be eliminated to give a single relation between α and β as the equation of the evolute. Often, elimination of the parameter involves using the original equation of the curve.

Example. Find the evolute of the ellipse $\frac{x^2}{a^2} + \frac{y^2}{b^2} = 1$.

First Solution. Letting x be the parameter, we have by implicit differentiation,
$$\frac{dy}{dx} = -\frac{b^2 x}{a^2 y}, \quad \frac{d^2 y}{dx^2} = -\frac{b^2(b^2 x^2 + a^2 y^2)}{a^4 y^3} = -\frac{b^4}{a^2 y^3},$$
so that equations (4) become
$$\alpha = \frac{(a^2 - b^2)x^3}{a^4}, \quad \beta = -\frac{(a^2 - b^2)y^3}{b^4},$$
and elimination of x and y, by solving here and substituting in the original equation, gives $(a\alpha)^{\frac{2}{3}} + (b\beta)^{\frac{2}{3}} = (a^2 - b^2)^{\frac{2}{3}}$. (Cf. Fig. 1112.)

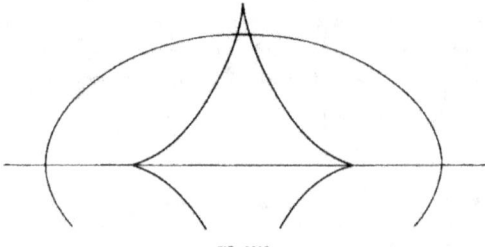

FIG. 1112

Second Solution. Transform the equation to parametric form: $x = a \cos t$, $y = b \sin t$. Then substitution in (4) gives
$$\alpha = a \cos t - \frac{b \cos t(a^2 \sin^2 t + b^2 \cos^2 t)}{ab} = \frac{(a^2 - b^2)}{a} \cos^3 t,$$
$$\beta = b \sin t + \frac{-a \sin t(a^2 \sin^2 t + b^2 \cos^2 t)}{ab} = \frac{(a^2 - b^2)}{-b} \sin^3 t.$$

Elimination of the parameter t by means of a well-known trigonometric identity gives the answer above.

1117. EXERCISES

In Exercises 1-8, find the curvature, radius of curvature, and center of curvature of the given curve at the specified point. Draw a figure and show the circle of curvature.

1. $y = x^2$; $(0, 0)$.
2. $y = x^3$; $(1, 1)$.
3. $y^2 = x^3$; $(1, 1)$.
4. $y^2 = x^3 + 8$; $(1, 3)$.
5. $x = 3t^2, y = 3t - t^3$; $t = 1$.
6. $x = t - \sin t, y = 1 - \cos t$; $t = \pi$.
7. $\rho = 1 - \cos \theta$; $\theta = \pi$.
8. $\rho^2 = \cos 2\theta$; $\theta = 0$.
9. Prove that the curvature of any straight line is identically zero. State and prove a converse. Include the case of a vertical line.
10. Prove that the radius of curvature of any circle is its radius. *Hint:* Write its equations in parametric form $x = a + r \cos \theta, y = b + r \sin \theta$.
11. Show that the point of maximum curvature of a parabola $y = ax^2 + bx + c$ occurs at the vertex.
12. Show that the points of maximum curvature of $y = x^3 - 3x$ do not occur at the maximum and minimum points.
13. Show that the radii of curvature at the ends of the axes of the ellipse $b^2x^2 + a^2y^2 = a^2b^2$ are b^2/a and a^2/b.
14. Find the minimum value of the radius of curvature of the curve $y = e^x$. Check your answer by working the same problem for the inverse function $y = \ln x$.
15. Show that the "circle of curvature" of $y = x^4$ at the origin is the x-axis.
16. The bowl of a goblet is a part of the surface obtained by revolving about the y-axis the curve $y^2 = x^4$. If a spherical pellet is dropped into the goblet, show that this pellet cannot touch the bottom of the container.
★17. Show that the evolute of the cycloid, $x = a(\theta - \sin \theta), y = a(1 - \cos \theta)$ is again a cycloid, which can be obtained by translating the original cycloid.
★18. Demonstrate the unreliability of one's intuitions in the following "reasoning": A wheel is rolling on a straight line. At any instant the point at the top of the wheel is rotating instantaneously about the point at the bottom of the wheel. "Therefore" the cycloid of Exercise 17 has radius of curvature at its maximum point equal to the diameter of the generating circle. (Show that this radius of curvature is actually equal to *twice* the diameter.)

In Exercises 19-24, find the center of curvature at a general point. Draw a figure showing the evolute of the given curve.

★19. $y^2 = 2px$.
★20. $x^2 - y^2 = 1$.
★21. $x = t^2, y = t^3$.
★22. $x = 3t^2, y = 3t - t^3$.
★23. The parabola $\sqrt{x} + \sqrt{y} = 1, 0 \leq x \leq 1, 0 \leq y \leq 1$.
★24. The hypocycloid $4x = 3 \cos \theta + \cos 3\theta, 4y = 3 \sin \theta - \sin 3\theta$.
★25. Prove the validity of formulas (2), §§ 1114 and 1116, in the case of a vertical tangent ($x' = 0$).

364   ARCS AND CURVES   [§ 1117]

★26. Show that the center of curvature of a curve at a point p on the curve is the limiting position of the point of intersection of the normal lines to the curve at p and at a neighboring point q on the curve, as q approaches p along the curve. *Hint:* Choose a rectangular coordinate system so that p is the origin and the positive half of the y-axis is normal to the curve on its concave side. (Cf. Fig. 1113.) The radius of curvature at p is $1/y''$, evaluated at $x = 0$. The y-intercept

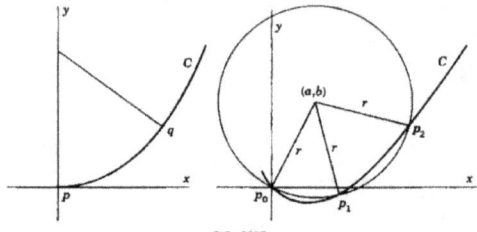

FIG. 1113

of the normal at q is $y + x/y'$ which, by l'Hospital's Rule, tends toward $1/y''$, as x approaches zero.

★27. Let a circle be drawn through three points p_0, p_1, and p_2 of a curve, and let p_1 and p_2 approach p_0 along the curve. The circle in the limiting position of this circle (if it exists) is called the **osculating circle** of the curve at the point p_0. Show that the osculating circle is identical with the circle of curvature. *Hint:* Choose a coordinate system as in Ex. 26, let p_i have coordinates (x_i, y_i), $i = 0, 1, 2$, $(x_0 = y_0 = 0)$ and let $y = f(x)$ be the equation of the curve near the origin. (Cf. Fig. 1113.) Let the circle through p_0, p_1, and p_2 have center (a, b) and radius r, and let $F(x) \equiv (x - a)^2 + (y - b)^2 - r^2$, where $y = f(x)$. Then $F(0) = F(x_1) = F(x_2) = 0$. Therefore, by Rolle's Theorem there exist x_3 between the smallest and the middle one of $0, x_1$, and x_2 and x_4 between the middle one and the largest of $0, x_1$, and x_2 such that $F'(x_3) = F'(x_4) = 0$, and, again, there exists x_5 between x_3 and x_4 such that $F''(x_5) = 0$. Show that $F''(x) = 2 + 2y'^2 + 2(y - b)y''$, and, since $x_5 \to 0$, the facts that $y \to 0$ and $y' \to 0$ imply $b \to 1/y''$, evaluated at $x = 0$. Then use the form of $F'(x)$ to show that $a \to 0$.

★28. *The unwinding string.* Suppose a string is thought of as being first laid out along a curve C and then unwound, as indicated in Figure 1114. What is the curve Γ described by a point on the string? Show that the given curve is the evolute of this new curve, and therefore that, although any curve has by definition, only one evolute, it may have (since there are infinitely many points on the string) infinitely many involutes. *Hint:* Let the given curve C have parametric equations $x = f(s)$, $y = g(s)$, where s is the arc length from a fixed point p_0 on C. Since a tangent line segment of unit length at p, in the appropriate direction, has horizontal and vertical projections equal to $-f'$ and $-g'$, respectively, and the string, in being unwound from tangency at p_0 to tangency at p, increases its length from

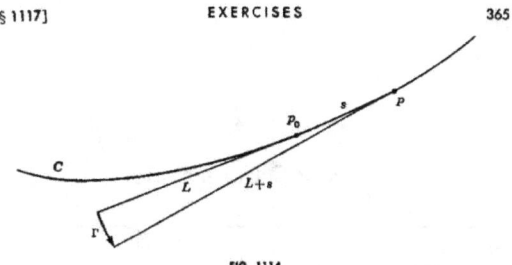

FIG. 1114

an initial length L by an amount equal to the arc length s between p_0 and p to a new length $L + s$, the parametric equations of Γ are $x = f - (L + s)f'$, $y = g - (L + s)g'$. Equations (4), § 1116, applied to Γ, become $\alpha = f$, $\beta = g$, as requested. (Good use must be made of the identity $f'^2 + g'^2 = 1$.)

★29. Find the equations of the spiral involute obtained by "unwinding a string" (cf. Ex. 28) from the unit circle, starting from the point $(1, 0)$. Then verify that the unit circle is the evolute of this involute.

★30. With suitable differentiability assumptions prove that the normal to a curve at any point is tangent to the evolute at the corresponding point. *Hint*: Differentiate formulas (4), § 1116, and show that $\beta' : \alpha' = -x' : y'$.

★31. Prove that if arc length s is the parameter, then
$$K = \sqrt{x''^2 + y''^2}.$$

12

Partial Differentiation

1201. PARTIAL DERIVATIVES

When a function of several real variables is considered as dependent on just one of these variables, the others being held constant, it may possess a derivative with respect to this one variable. Such a derivative (if it exists) is called a *partial derivative*, according to the following definition, which is formulated for simplicity in terms of a function of two variables:

Definition. *Let $u = f(x, y)$ be defined in a region R of the Euclidean plane. The **partial derivatives** of u with respect to x and of u with respect to y, written*

(1)
$$\begin{cases} \dfrac{\partial u}{\partial x} = \dfrac{\partial f}{\partial x} = u_x = f_x = f_x(x, y) = f_1(x, y), \\ \dfrac{\partial u}{\partial y} = \dfrac{\partial f}{\partial y} = u_y = f_y = f_y(x, y) = f_2(x, y), \end{cases}$$

respectively, are the limits of the difference quotients:

(2)
$$\begin{cases} f_1(x, y) = \lim_{h \to 0} \dfrac{f(x + h, y) - f(x, y)}{h}, \\ f_2(x, y) = \lim_{k \to 0} \dfrac{f(x, y + k) - f(x, y)}{k}. \end{cases}$$

The derivatives of the preceding Definition, in distinction to the derivatives *of* derivatives treated in the following section, are called **partial derivatives of the first order** or, for simplicity, **first partial derivatives**.

NOTE. We shall discuss later (cf. § 1218) relative merits and weaknesses of the notations of (1). Let us point out at this time, however, that one advantage of the numerical subscript notation f_1, f_2 is that it stresses the *form* of the function rather than the particular letters used. For example, whether we write $f(x, y, z) = xy^2z^3$ or $f(\xi, \eta, \zeta) = \xi\eta^2\zeta^3$, $f_1(5, 3, 2) = 72$.

Example 1. The partial derivatives of $u = \sin(x + 2y - 3z)$ are $u_x = \cos(x + 2y - 3z)$, $u_y = 2\cos(x + 2y - 3z)$, and $u_z = -3\cos(x + 2y - 3z)$.

Example 2. Direct substitution shows that $u = \dfrac{Ax^2 + By^2}{Cx^2 + Dy^2}$ is a solution of the *partial differential equation* $x \dfrac{\partial u}{\partial x} + y \dfrac{\partial u}{\partial y} = 0$.

1202. PARTIAL DERIVATIVES OF HIGHER ORDER

If a function of several variables has first partial derivatives, these derivatives (which are again functions of these same several variables) may in turn have first partial derivatives. These are called **partial derivatives of the second order**, or **second partial derivatives**, of the original function. Notation for the case of a function $u = f(x, y, z)$ of three variables, and differentiation with respect to x or y is:

$$\frac{\partial}{\partial x}\left(\frac{\partial u}{\partial x}\right) = \frac{\partial^2 u}{\partial x^2} = \frac{\partial^2 f}{\partial x^2} = u_{xx} = f_{xx}(x, y, z) = f_{11}(x, y, z),$$

$$\frac{\partial}{\partial y}\left(\frac{\partial u}{\partial x}\right) = \frac{\partial^2 u}{\partial y \partial x} = \frac{\partial^2 f}{\partial y \partial x} = u_{yx} = f_{yx}(x, y, z) = f_{21}(x, y, z),$$

$$\frac{\partial}{\partial x}\left(\frac{\partial u}{\partial y}\right) = \frac{\partial^2 u}{\partial x \partial y} = \frac{\partial^2 f}{\partial x \partial y} = u_{xy} = f_{xy}(x, y, z) = f_{12}(x, y, z),$$

$$\frac{\partial}{\partial y}\left(\frac{\partial u}{\partial y}\right) = \frac{\partial^2 u}{\partial y^2} = \frac{\partial^2 f}{\partial y^2} = u_{yy} = f_{yy}(x, y, z) = f_{22}(x, y, z).$$

Partial derivatives of order higher than the second, and for functions of any number of variables, are defined and denoted similarly. For example,

$$\frac{\partial}{\partial x}\left(\frac{\partial^2 u}{\partial x^2}\right) = \frac{\partial^3 u}{\partial x^3}, \quad \frac{\partial}{\partial y}\left(\frac{\partial^2 u}{\partial y \partial x}\right) = \frac{\partial^3 u}{\partial y^2 \partial x}, \quad \frac{\partial}{\partial y}\left(\frac{\partial^2 u}{\partial x \partial z}\right) = \frac{\partial^3 u}{\partial y \partial x \partial z}.$$

Partial derivatives in which more than one of the independent variables is a variable of differentiation are called **mixed partial derivatives**. Thus $\dfrac{\partial^2 u}{\partial x \partial y}$ is a mixed partial derivative of second order.

It might appear at first that for a function f of two variables, x and y, there are two mixed partial derivatives of the second order, f_{xy} and f_{yx}, and six of the third order, $f_{xxy}, f_{xyx}, f_{yxx}, f_{xyy}, f_{yxy},$ and f_{yyx}, with a corresponding increase in the number as the order increases, or as more independent variables are present. However, as is shown in the following section, in case the mixed partial derivatives involved are *continuous*, certain equality relations exist among them. For example, for a function f of x and y, $f_{xy} = f_{yx}, f_{xxy} = f_{xyx} = f_{yxx}$, and $f_{xyy} = f_{yxy} = f_{yyx}$, so that there are only one mixed partial derivative of the second order and two of the third order. In general, in the presence of continuity, two higher order mixed partial derivatives are equal whenever the number of times each independent variable is a variable of differentiation is the same for the two. For example, $f_{xyzxx} = f_{xxyzx}$. This general fact is a consequence of the simpler fact that $f_{xy} = f_{yx}$, since this latter equality states that two adjacent subscript letters can be interchanged, and any permutation of subscripts can be obtained by a sequence of adjacent interchanges. Furthermore, in a proof of the relation $f_{xy} = f_{yx}$ the presence of variables other than x and y is irrelevant since they are held fixed. Thus everything depends on the basic fact that $f_{xy} = f_{yx}$ for functions of *two* variables. In the following section this relation is stated with precision, and proved in a slightly more general form than that indicated above.

★1203. EQUALITY OF MIXED PARTIAL DERIVATIVES

Theorem. *Assume that (i) the function f of the two real variables x and y, and its two first partial derivatives f_x and f_y, exist in a region R, and (ii) the mixed partial derivative f_{xy} exists in R and is continuous at the point (x_0, y_0) of R. Then the mixed partial derivative f_{yx} exists at (x_0, y_0) and is equal to f_{xy} at that point.*

Proof. Choose a small square neighborhood of (x_0, y_0) lying in R, and choose h and k so that their absolute values are less than half the length of a side of this neighborhood. Let $A(h, k)$ and $\psi(y)$ be defined by the equations

(1) $\quad A(h, k) \equiv f(x_0 + h, y_0 + k) - f(x_0 + h, y_0) - f(x_0, y_0 + k) + f(x_0, y_0),$

(2) $\quad\quad\quad\quad \psi(y) \equiv f(x_0 + h, y) - f(x_0, y),$

with y restricted to lie appropriately near y_0. (The function ψ also depends on h, but it is its dependence on y that concerns us here.) Then $A(h, k) = \psi(y_0 + k) - \psi(y_0)$, and since $f_y(x, y)$ exists in the chosen neighborhood of (x_0, y_0), $\psi(y)$ is differentiable for y near y_0 (and is therefore also continuous for such y). Thus, by the Law of the Mean for functions of one variable (§ 405),

(3) $\quad\quad\quad A(h, k) = \psi(y_0 + k) - \psi(y_0) = k\psi'(y_0 + \theta_1 k)$

for some θ_1 such that $0 < \theta_1 < 1$ (θ_1 depending on h and k).

This equation can be written

(4) $\quad\quad A(h, k) = k[f_y(x_0 + h, y_0 + \theta_1 k) - f_y(x_0, y_0 + \theta_1 k)].$

Again, by the Law of the Mean, applied to $f_y(x, y_0 + \theta_1 k)$ considered as a function of the single variable x, the quantity in brackets, on the right in (4), can be written $hf_{xy}(x_0 + \theta_2 h, y_0 + \theta_1 k)$, for some θ_2 such that $0 < \theta_2 < 1$, so that

(5) $\quad\quad\quad\quad A(h, k) = hk f_{xy}(x_0 + \theta_2 h, y_0 + \theta_1 k).$

Therefore, for $hk \neq 0$,

(6) $\quad\quad\quad\quad\quad \dfrac{A}{hk} = f_{xy}(x_0 + \theta_2 h, y_0 + \theta_1 k).$

fact that $f_{xy} = f_{yx}$, since this latter equality states that two adjacent subscript letters can be interchanged, and any permutation of subscripts can be obtained by a sequence of adjacent interchanges. Furthermore, in a proof of the relation $f_{xy} = f_{yx}$ the presence of variables other than x and y is irrelevant since they are held fixed. Thus everything depends on the basic fact that $f_{xy} = f_{yx}$ for functions of *two* variables. In the following section this relation is stated with precision, and proved in a slightly more general form than that indicated above.

★1203. EQUALITY OF MIXED PARTIAL DERIVATIVES

Theorem. *Assume that (i) the function f of the two real variables x and y, and its two first partial derivatives f_x and f_y, exist in a region R, and (ii) the mixed partial derivative f_{xy} exists in R and is continuous at the point (x_0, y_0) of R. Then the mixed partial derivative f_{yx} exists at (x_0, y_0) and is equal to f_{xy} at that point.*

Proof. Choose a small square neighborhood of (x_0, y_0) lying in R, and choose h and k so that their absolute values are less than half the length of a side of this neighborhood. Let $A(h, k)$ and $\psi(y)$ be defined by the equations

$$(1) \quad A(h, k) \equiv f(x_0 + h, y_0 + k) - f(x_0 + h, y_0) - f(x_0, y_0 + k) + f(x_0, y_0),$$

$$(2) \quad \psi(y) \equiv f(x_0 + h, y) - f(x_0, y),$$

with y restricted to lie appropriately near y_0. (The function ψ also depends on h, but it is its dependence on y that concerns us here.) Then $A(h, k) = \psi(y_0 + k) - \psi(y_0)$, and since $f_y(x, y)$ exists in the chosen neighborhood of (x_0, y_0), $\psi(y)$ is differentiable for y near y_0 (and is therefore also continuous for such y). Thus, by the Law of the Mean for functions of one variable (§ 405),

$$(3) \quad A(h, k) = \psi(y_0 + k) - \psi(y_0) = k\psi'(y_0 + \theta_1 k)$$

for some θ_1 such that $0 < \theta_1 < 1$ (θ_1 depending on h and k).

This equation can be written

$$(4) \quad A(h, k) = k[f_y(x_0 + h, y_0 + \theta_1 k) - f_y(x_0, y_0 + \theta_1 k)].$$

Again, by the Law of the Mean, applied to $f_y(x, y_0 + \theta_1 k)$ considered as a function of the single variable x, the quantity in brackets, on the right in (4), can be written $hf_{xy}(x_0 + \theta_2 h, y_0 + \theta_1 k)$, for some θ_2 such that $0 < \theta_2 < 1$, so that

$$(5) \quad A(h, k) = hk f_{xy}(x_0 + \theta_2 h, y_0 + \theta_1 k).$$

Therefore, for $hk \neq 0$,

$$(6) \quad \frac{A}{hk} = f_{xy}(x_0 + \theta_2 h, y_0 + \theta_1 k).$$

We now start afresh with the function

(7) $$\phi(x) \equiv f(x, y_0 + k) - f(x, y_0),$$

for x near x_0. For $hk \neq 0$,

(8) $$\frac{A}{hk} = \frac{1}{k} \frac{\phi(x_0 + h) - \phi(x_0)}{h},$$

and since $\phi'(x_0)$ exists, it follows that for $k \neq 0$, $\lim_{h \to 0} \frac{A}{hk}$ exists and is equal to $\frac{1}{k} \phi'(x_0)$, or

(9) $$\lim_{h \to 0} \frac{A}{hk} = \frac{1}{k} [f_x(x_0, y_0 + k) - f_x(x_0, y_0)].$$

By (6), and the continuity of f_{xy} at (x_0, y_0), $\lim_{(h,k) \to (0,0)} \frac{A}{hk}$ ($hk \neq 0$) exists and is equal to $f_{xy}(x_0, y_0)$. Therefore (Note 1, § 1107)

(10) $$\lim_{k \to 0} \lim_{h \to 0} \frac{A}{hk}$$

exists and is equal to $f_{xy}(x_0, y_0)$. Finally, by (9), this iterated limit is, by definition, $f_{yx}(x_0, y_0)$.

1204. EXERCISES

In Exercises 1-4, find the first partial derivatives of the given function with respect to each of the independent variables.

1. $u = (x - y) \sin (x + y)$.
2. $u = xe^y + ye^x$.
3. $f(x, y, z) = \sqrt{x^2 + y^2 + z^2}$.
4. $x = r \sin \phi \cos \theta$.

In Exercises 5-8, find all second order partial derivatives of each function, and verify the appropriate equalities among the mixed partial derivatives.

5. $f(x, y) = Ax^2 + 2Bxy + Cy^2$.
6. $z = \frac{x}{x + y}$.
7. $u = xe^y + y \sin x$.
8. $u = x \operatorname{Arctan} \frac{y}{x}$.

In Exercises 9-16, show that the function u satisfies the given differential equation.

9. $u = \frac{xy}{x + y}$; $x \frac{\partial u}{\partial x} + y \frac{\partial u}{\partial y} = u$.
10. $u = Ax^3 + Bx^2y + Cxy^2 + Dy^3$; $xu_x + yu_y + zu_z = 3u$.
11. $u = (x^2 + y^2 + z^2)^{-\frac{1}{2}}$; $u_x{}^2 + u_y{}^2 + u_z{}^2 = u^4$.
12. $u = \operatorname{Arctan} \frac{y}{x}$; $xu_x + yu_y = 0$.
13. $u = \operatorname{Arctan} \frac{y}{x}$; $xu_y - yu_x = 1$.

14. $u = \dfrac{xy}{x+y}$; $x^2 \dfrac{\partial^2 u}{\partial x^2} + 2xy \dfrac{\partial^2 u}{\partial x \partial y} + y^2 \dfrac{\partial^2 u}{\partial y^2} = 0$.

15. $u = ce^{-n^2 t} \sin nx$; $\dfrac{\partial^2 u}{\partial x^2} = \dfrac{\partial u}{\partial t}$.

16. $u = c \sin akx \cos kt$; $\dfrac{\partial^2 u}{\partial x^2} = a^2 \dfrac{\partial^2 u}{\partial t^2}$.

17. Interpret the first partial derivatives of $z = f(x, y)$ as slopes of curves formed by cutting the surface $z = f(x, y)$ by planes of the form $x = a$ and $y = b$. (Cf. Fig. 1201.)

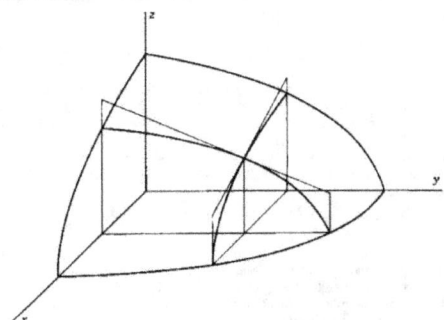

FIG. 1201

18. As a consequence of the Theorem of § 1203, assuming that $f(x, y, z)$ and all of its first and second order derivatives are continuous (in a given region) prove that $f_{xxy} = f_{yxx}$ and $f_{yxx} = f_{xxy}$.

A function $u(x, y)$ is called **harmonic** if and only if it satisfies **Laplace's equation**

$$(1) \qquad \dfrac{\partial^2 u}{\partial x^2} + \dfrac{\partial^2 u}{\partial y^2} = 0.$$

In Exercises 19-22, show that u is harmonic.

19. $u = x^3 - 3xy^2$.   **20.** $u = e^x \sin y$.

21. $u = \ln(x^2 + y^2)$.   **22.** $u = \text{Arctan} \dfrac{y}{x}$.

23. Assuming existence and continuity of derivatives wherever desired, show that if two functions, u and v, satisfy the **Cauchy-Riemann differential equations**

$$(2) \qquad \dfrac{\partial u}{\partial x} = \dfrac{\partial v}{\partial y}, \quad \dfrac{\partial u}{\partial y} = -\dfrac{\partial v}{\partial x},$$

then both functions are harmonic. (Cf. Exs. 19-22.)

★24. If $f(x, y)$ is a given function, define $g(x, y) \equiv f(y, x)$. Let the notation

§ 1205] THE FUNDAMENTAL INCREMENT FORMULA   371

$\phi_u(v, w)$, where the letters u, v, and w represent some arrangement of the variables x and y, mean $\frac{\partial}{\partial u}\phi(v, w)$. Show that

$$g_x(x, y) = f_x(y, x), \quad \text{or} \quad g_1(x, y) = f_2(y, x),$$
$$g_y(x, y) = f_y(y, x), \quad \text{or} \quad g_2(x, y) = f_1(y, x).$$

Illustrate by simple (but not completely trivial) examples.

★25. Let $f(x, y) = xy\frac{x^2 - y^2}{x^2 + y^2}$, for $x^2 + y^2 \neq 0$, $f(0, 0) = 0$. Show that $f_{xy}(0, 0) = 1$ and $f_{yx}(0, 0) = -1$. Why does this not invalidate the Theorem of § 1203?

★26. Show that the example

$$f(x, y) = \frac{xy}{x^2 + y^2}, \quad x^2 + y^2 \neq 0, \quad f(0, 0) = 0,$$

considered in §§ 1006-1008, has first partial derivatives everywhere, although it is not everywhere continuous. In other words, show that although the existence of a derivative implies continuity of a function of one variable (§ 402), this fact does not extend to functions of more than one variable. (Cf. Exs. 27-28.)

★27. Prove that if $f(x, y)$ has *bounded* first partial derivatives in a region it must be continuous there. (Cf. Exs. 26, 28.) *Hints:* Apply the Law of the Mean (§ 405) to each bracketed part of

$$f(x_0 + h, y_0 + k) - f(x_0, y_0) = [f(x_0 + h, y_0 + k) - f(x_0, y_0 + k)]$$
$$+ [f(x_0, y_0 + k) - f(x_0, y_0)].$$

★28. Assume that $f(x, y)$ possesses first partial derivatives $f_1(x, y)$ and $f_2(x, y)$ in a region R and that f_1 and f_2 are continuous at a point p of R. Prove that f is continuous in some neighborhood of p. (Cf. Ex. 27.)

★29. Give an example of a function $f(x, y)$ for which $f_{xy}(x, y)$ is identically 0 but for which $f_{xy}(x, y)$ never exists. *Hint:* Let $f(x, y)$ be a nondifferentiable function of x alone.

1205. THE FUNDAMENTAL INCREMENT FORMULA

If $u = f(x)$ is a differentiable function of the single variable x, equation (4), § 402:

(1) $$\Delta y = \frac{dy}{dx}\Delta x + \epsilon\,\Delta x,$$

where $\epsilon \to 0$ as $\Delta x \to 0$, expresses an increment Δy of the dependent variable, in terms of the derivative, with a correction term involving an infinitesimal. It is the purpose of this section to obtain a corresponding formula for the increment of a dependent variable that is a function of several independent variables. For simplicity the details are presented for the case of a function of two independent variables. The case of three or more independent variables is given as an exercise (Ex. 11, § 1209).

Assume that $u = f(x, y)$ possesses first partial derivatives $f_1(x, y)$ and $f_2(x, y)$ in a region R, and that f_1 and f_2 are continuous at some point (x_0, y_0) of R. Furthermore, take a square neighborhood of (x_0, y_0) that lies in R

and let Δx and Δy be numerically less than half the length of a side of this neighborhood, so that the point $(x_0 + \Delta x, y_0 + \Delta y)$ lies in R. Then the increment Δu of the dependent variable can be written as follows:

$$\Delta u = f(x_0 + \Delta x, y_0 + \Delta y) - f(x_0, y_0)$$
(2)
$$= [f(x_0 + \Delta x, y_0 + \Delta y) - f(x_0, y_0 + \Delta y)]$$
$$+ [f(x_0, y_0 + \Delta y) - f(x_0, y_0)].$$

We now apply the Law of the Mean (§ 405) to each of the bracketed quantities in (2) and obtain

(3) $$\Delta u = f_1(x_0 + \theta_1 \Delta x, y_0 + \Delta y) \Delta x + f_2(x_0, y_0 + \theta_2 \Delta y) \Delta y,$$

where θ_1 and θ_2 are quantities between 0 and 1 that depend on Δx and Δy.

Continuity of the first partial derivatives is now used. Since $(x_0 + \theta_1 \Delta x, y_0 + \Delta y) \to (x_0, y_0)$ and $(x_0, y_0 + \theta_2 \Delta y) \to (x_0, y_0)$, as $\Delta x \to 0$ and $\Delta y \to 0$, $f_1(x_0 + \theta_1 \Delta x, y_0 + \Delta y) \to f_1(x_0, y_0)$ and $f_2(x_0, y_0 + \theta_2 \Delta y) \to f_2(x_0, y_0)$. In other words,

(4) $$\begin{cases} f_1(x_0 + \theta_1 \Delta x, y_0 + \Delta y) = f_1(x_0, y_0) + \epsilon_1, \\ f_2(x_0, y_0 + \theta_2 \Delta y) = f_2(x_0, y_0) + \epsilon_2, \end{cases}$$

where ϵ_1 and ϵ_2 are infinitesimals (tending toward zero as Δx and Δy tend toward zero). Substitution of (4) in (3) gives the Fundamental Increment Formula:

Theorem. Fundamental Increment Formula. *If $u = f(x, y)$ possesses first partial derivatives in a region R, and if these first partial derivatives are continuous at some point (x_0, y_0) of R, then (for increments Δx and Δy sufficiently small numerically) the increment Δu of the independent variable u can be written*

(5) $$\Delta u = \frac{\partial u}{\partial x} \Delta x + \frac{\partial u}{\partial y} \Delta y + \epsilon_1 \Delta x + \epsilon_2 \Delta y,$$

where the partial derivatives are evaluated at the point (x_0, y_0) and ϵ_1 and ϵ_2 are infinitesimals, tending toward zero with Δx and Δy.

The Fundamental Increment Formula for a function u of several variables $x, y, z, \cdots$ takes the form (cf. Ex. 11, § 1209):

(6) $$\Delta u = \frac{\partial u}{\partial x} \Delta x + \frac{\partial u}{\partial y} \Delta y + \frac{\partial u}{\partial z} \Delta z + \cdots + \epsilon_1 \Delta x + \epsilon_2 \Delta y + \epsilon_3 \Delta z + \cdots,$$

where $\epsilon_1, \epsilon_2, \epsilon_3, \cdots$ are infinitesimal functions of the increments $\Delta x, \Delta y, \Delta z, \cdots$.

Whenever the increment of a function u of several variables can be expressed in the form (6), the function u is said **to be differentiable** at the particular point at which the first partial derivatives are evaluated. If a function possesses first partial derivatives in a neighborhood of a point and if these first partial derivatives are continuous at the point, the function is said to be **continuously differentiable** at the point. The preceding Theorem can now be restated: *continuous differentiability implies differentiability*.

NOTE 1. From the form of (6) we can infer that *differentiability of a function at a point implies its continuity there*. (Cf. Ex. 27, § 1204.)

NOTE 2. Mere *existence* of the first partial derivatives of a function does not imply its differentiability. (Cf. Ex. 26, § 1204, and Note 1.)

NOTE 3. *Differentiability does not imply continuous differentiability*. (Cf. the Example, below.)

Example. Define $\phi(x) = x^2 \sin(1/x)$, $x \neq 0$, $\phi(0) = 0$ (Example 3, § 403) and let $f(x, y) = \phi(x) + \phi(y)$. Then $f_1(x, y) = \phi'(x)$ and $f_2(x, y) = \phi'(y)$ are both discontinuous at the origin. On the other hand, f is differentiable there, since formula (5) becomes $\Delta u = 0 \cdot \Delta x + 0 \cdot \Delta y + [\Delta x \sin(1/\Delta x)] \Delta x + [\Delta y \sin(1/\Delta y)] \Delta y$, and the quantities in brackets are infinitesimals.

1206. DIFFERENTIALS

By analogy with the case of a function of one variable (§ 410), we define the **differential**, or **total differential**, of a differentiable function $u = f(x, y)$ of two variables:

$$(1) \qquad du = \frac{\partial u}{\partial x} dx + \frac{\partial u}{\partial y} dy,$$

where dx and dy are two new independent variables. The differential du, then, is a function of *four* independent variables: x and y (the coordinates of the point at which the partial derivatives are evaluated) and dx and dy. These latter quantities, dx and dy, are also called **differentials**. Again (cf. § 411) it is convenient, for most purposes, to identify these differentials with actual increments:

$$(2) \qquad dx = \Delta x, \quad dy = \Delta y.$$

For a function $u = f(x, y, z, \cdots)$ of any number of variables, formulas similar to (1) and (2) hold by definition:

$$(3) \qquad du = \frac{\partial u}{\partial x} dx + \frac{\partial u}{\partial y} dy + \frac{\partial u}{\partial z} dz + \cdots,$$

$$(4) \qquad dx = \Delta x, \quad dy = \Delta y, \quad dz = \Delta z, \cdots.$$

The independence of the variables $dx, dy, dz, \cdots$ in (3) has the following consequence: Whenever the differential of a function u of several variables $x, y, z, \cdots$ is represented as a *linear combination* (cf. § 1502) of the differentials $dx, dy, dz, \cdots$,

$$(5) \qquad du = P\,dx + Q\,dy + R\,dz + \cdots$$

the coefficients must be the first partial derivatives of u:

(6) $\qquad P = \dfrac{\partial u}{\partial x}, \quad Q = \dfrac{\partial u}{\partial y}, \quad R = \dfrac{\partial u}{\partial z}, \cdots.$

This can be seen by equating the right-hand members of (3) and (5), setting first $dx = 1, dy = dz = \cdots = 0$, then $dx = 0, dy = 1, dz = \cdots = 0$, and so forth.

With the notation of differentials the Fundamental Increment Formula (§ 1205) becomes

(7) $\qquad \Delta u = du + \epsilon_1 \Delta x + \epsilon_2 \Delta y + \cdots.$

(Cf. (1), § 411: $\Delta y = dy + \epsilon \Delta x$.) It follows that a differential is a "good approximation" to an increment, for a function of several variables, in the same sense that it is for a function of one variable. (Cf. § 1213.)

Example. For the function $u = x^3 y^3$, find $\Delta u, du$, and the infinitesimals ϵ_1 and ϵ_2 of (7).

Solution. The differential is $du = 3x^2 y^3 dx + 2x^3 y dy$. The increment is $\Delta u = (x + \Delta x)^3 (y + \Delta y)^2 - x^3 y^2$, which can be written

(8) $\qquad \begin{aligned} \Delta u = du &+ (3xy^2 \Delta x^2 + 6x^2 y \Delta x \Delta y + x^3 \Delta y^2) \\ &+ (y^2 \Delta x^3 + 6xy \Delta x^2 \Delta y + 3x^2 \Delta x \Delta y^2) \\ &+ (2y \Delta x^3 \Delta y + 3x \Delta x^2 \Delta y^2) + \Delta x^3 \Delta y^2. \end{aligned}$

The infinitesimals ϵ_1 and ϵ_2 are not uniquely determined, since many distinct groupings of terms on the right in (8) can be used. According to one such grouping,

$\epsilon_1 = \epsilon_1(\Delta x, \Delta y) = 3xy^2 \Delta x + 6x^2 y \Delta y + y^2 \Delta x^2 + 2y \Delta x^2 \Delta y + \Delta x^2 y^2,$
$\epsilon_2 = \epsilon_2(\Delta x, \Delta y) = x^3 \Delta y + 6xy \Delta x^2 + 3x^2 \Delta x \Delta y + 3x \Delta x^2 \Delta y.$

NOTE. It can be shown (cf. Ex. 12, § 1211) that the differential dz of a function $z = f(x, y)$ of two variables is associated with the tangent plane to the surface $z = f(x, y)$ in much the same way that the differential dy of a function $y = f(x)$ of one variable is associated with the tangent line to the curve $y = f(x)$.

1207. CHANGE OF VARIABLES. THE CHAIN RULE

It is our aim in this section to discuss the following question: *If u is a differentiable function of several variables $x, y, z, \cdots$, each of which is a differentiable function of several variables $r, s, t, \cdots$, how can the first partial derivatives of u with respect to the new variables be expressed in terms of the partial derivatives of the given functions?*

The simplest case was discussed in § 1021 If u is a differentiable function of a single variable x which in turn is a differentiable function of a single variable t, then u becomes a differentiable function of t, and $du/dt = (du/dx)(dx/dt)$. The partial derivatives referred to above are generalizations of the ordinary derivatives of this special case.

We proceed to the next simplest case. Let u be a differentiable func-

§ 1207] CHANGE OF VARIABLES. THE CHAIN RULE 375

tion of several variables. For simplicity let us assume that u is a function $f(x, y)$ of *two* independent variables. Furthermore, assume that each of these two variables is a differentiable function of a single variable t. Allowing t to change by an increment Δt ($\neq 0$), we express, with the aid of the fundamental increment formula (5), § 1205, the difference quotient as follows:

(1) $$\frac{\Delta u}{\Delta t} = \frac{\partial u}{\partial x}\frac{\Delta x}{\Delta t} + \frac{\partial u}{\partial y}\frac{\Delta y}{\Delta t} + \epsilon_1 \frac{\Delta x}{\Delta t} + \epsilon_2 \frac{\Delta y}{\Delta t}.$$

Considering u as a function of the single variable t, we let Δt tend toward zero, and take limits of both members of (1). Since x and y are differentiable functions of t, and since ϵ_1 and ϵ_2 are both infinitesimals (as Δt tends toward zero so do Δx and Δy, and consequently so do ϵ_1 and ϵ_2), we conclude by taking the limit of the right-hand member of (1) that u is a differentiable function of t and its derivative is given by the formula

(2) $$\frac{du}{dt} = \frac{\partial u}{\partial x}\frac{dx}{dt} + \frac{\partial u}{\partial y}\frac{dy}{dt}.$$

A similar result (for similar reasons) is valid for a function u of more than two variables:

(3) $$\frac{du}{dt} = \frac{\partial u}{\partial x}\frac{dx}{dt} + \frac{\partial u}{\partial y}\frac{dy}{dt} + \frac{\partial u}{\partial z}\frac{dz}{dt} + \cdots.$$

The form of (3) shows that if u as a function of $x, y, z, \cdots$ and $x, y, z, \cdots$ as functions of t are all *continuously* differentiable, then u as a function of t is too.

Still more generally (give a proof in Exercise 12, § 1209), if $u = f(x, y, z, \cdots)$ is a differentiable function of several variables each of which is a differentiable function of several variables $r, s, t, \cdots$, then u, as a function of these new independent variables, is differentiable and, since formula (3) can be applied to each of the independent variables separately, the following equations, known as the general **chain rule** are valid:

(4) $$\begin{cases} \frac{\partial u}{\partial r} = \frac{\partial u}{\partial x}\frac{\partial x}{\partial r} + \frac{\partial u}{\partial y}\frac{\partial y}{\partial r} + \frac{\partial u}{\partial z}\frac{\partial z}{\partial r} + \cdots, \\ \frac{\partial u}{\partial s} = \frac{\partial u}{\partial x}\frac{\partial x}{\partial s} + \frac{\partial u}{\partial y}\frac{\partial y}{\partial s} + \frac{\partial u}{\partial z}\frac{\partial z}{\partial s} + \cdots, \\ \frac{\partial u}{\partial t} = \frac{\partial u}{\partial x}\frac{\partial x}{\partial t} + \frac{\partial u}{\partial y}\frac{\partial y}{\partial t} + \frac{\partial u}{\partial z}\frac{\partial z}{\partial t} + \cdots, \\ \cdots \end{cases}$$

In case $u = f(x)$ is a differentiable function of a single variable which is a differentiable function of several variables $r, s, t, \cdots$, formulas (4) take the form

$$\begin{cases} \dfrac{\partial u}{\partial r} = \dfrac{du}{dx}\dfrac{\partial x}{\partial r}, \\ \dfrac{\partial u}{\partial s} = \dfrac{du}{dx}\dfrac{\partial x}{\partial s}, \\ \cdots. \end{cases} \quad (5)$$

As before, if the given functions are *continuously* differentiable, so is the resulting composite function.

In § 410 it was pointed out that one consequence of the chain rule is that the formula $dy = (dy/dx)\, dx$ is valid whether x is the independent variable or x is dependent on a third variable. A similar statement is true for functions of several variables.

Theorem. *If u is a differentiable function $f(x, y, z, \cdots)$ of several variables, then*

$$(6) \qquad du = \frac{\partial u}{\partial x} dx + \frac{\partial u}{\partial y} dy + \frac{\partial u}{\partial z} dz + \cdots,$$

whether the variables $x, y, z, \cdots$ are considered as independent variables or as differentiable functions of other variables $r, s, t, \cdots$.

Proof. The problem is to show that the right-hand member of (6), when expressed in terms of the new variables, reduces to $\dfrac{\partial u}{\partial r} dr + \dfrac{\partial u}{\partial s} ds + \cdots$, which we know to be the value of the left-hand member when u is considered as a function of the new variables. To achieve this result we substitute for $dx, dy, \cdots$, in (6), the expressions $\dfrac{\partial x}{\partial r} dr + \dfrac{\partial x}{\partial s} ds + \cdots$, $\dfrac{\partial y}{\partial r} dr + \dfrac{\partial y}{\partial s} ds + \cdots$, $\cdots$, expand the products, and then rearrange terms so that those involving dr, those involving ds, $\cdots$, are separately assembled. (It would be well to check the mechanical details!) If dr is factored from those terms involving it, its coefficient (better check this!) will be found to be $\dfrac{\partial u}{\partial x}\dfrac{\partial x}{\partial r} + \dfrac{\partial u}{\partial y}\dfrac{\partial y}{\partial r} + \cdots$, or $\dfrac{\partial u}{\partial r}$ (formulas (4)). Since the coefficients of the remaining differentials, $ds, dt, \cdots$ are found similarly to be $\dfrac{\partial u}{\partial s}, \dfrac{\partial u}{\partial t}, \cdots$, the proof is complete.

NOTE. The student should be on guard against treating a partial symbol ∂x, or the like, as if it had a meaning of its own similar to that of a total differential dx. For example, in the first equation of (4), to "cancel" the symbols $\partial x, \partial y, \partial z, \cdots$ and conclude that

$$\frac{\partial u}{\partial r} = \frac{\partial u}{\partial r} + \frac{\partial u}{\partial r} + \frac{\partial u}{\partial r} + \cdots$$

§ 1208] HOMOGENEOUS FUNCTIONS 377

is in the same category as stating that $\sin 2\theta/\sin\theta = \sin 2/\sin$ or as proposing the following as a proof of the Remainder Theorem of College Algebra:

$$\begin{array}{r}f\\x-r\overline{\smash{\big)}f(x)}\\\underline{f(x)-f(r)}\\f(r).\end{array}$$

Example 1. If $u = f(x, y) = xy$, then $du = x\,dy + y\,dx$, whether x and y are independent variables, differentiable functions of a new independent variable (this is the differential form for the formula for differentiating a product of two functions), or differentiable functions of several new variables.

Example 2. If $u = e^{xy}$ and $x = r + s$, $y = r - 2s$, then $\frac{\partial u}{\partial x} = ye^{xy}$, $\frac{\partial u}{\partial y} = xe^{xy}$, $\frac{\partial x}{\partial r} = \frac{\partial x}{\partial s} = \frac{\partial y}{\partial r} = 1$, $\frac{\partial y}{\partial s} = -2$. Therefore $\frac{\partial u}{\partial r} = ye^{xy}\cdot 1 + xe^{xy}\cdot 1 = e^{r^2-rs-2s^2}\cdot(2r-s)$, and $\frac{\partial u}{\partial s} = ye^{xy}\cdot 1 + xe^{xy}(-2) = e^{r^2-rs-2s^2}\cdot(-r-4s)$. This result is easily checked by direct substitution before differentiation.

Example 3. Prove that if u is a function of x and y in which x and y occur only in the combination xy, then $x\frac{\partial u}{\partial x} = y\frac{\partial u}{\partial y}$ (assuming differentiability).

Solution. Let $u = f(t)$, where $t = xy$. Then, by (5), $\frac{\partial u}{\partial x} = \frac{du}{dt}\frac{\partial t}{\partial x} = y\frac{du}{dt}$; $\frac{\partial u}{\partial y} = \frac{du}{dt}\frac{\partial t}{\partial y} = x\frac{du}{dt}$. Hence $x\frac{\partial u}{\partial x} = xy\frac{du}{dt} = y\frac{\partial u}{\partial y}$.

★**1208. HOMOGENEOUS FUNCTIONS. EULER'S THEOREM**

For simplicity, statements in this section will be made in terms of functions of two variables. Extensions to several variables are immediate.

Definition. *A function $f(x, y)$ is **homogeneous** of degree n in a region R if and only if for all x, y, and positive λ† such that both (x, y) and $(\lambda x, \lambda y)$ are in R,*

(1) $$f(\lambda x, \lambda y) = \lambda^n f(x, y).$$

Example 1. $f(x, y) = 2x^3 - 8xy^2$ is homogeneous of degree 3, since $f(\lambda x, \lambda y) = 2\lambda^3 x^3 - 8\lambda^3 xy^2 = \lambda^3 f(x, y)$. The region R may be taken to be the entire plane.

Example 2. $f(x, y) = (x^2 + 4y^2)^{-\frac{1}{2}}$ is homogeneous of degree $-\frac{1}{2}$ in the plane (except at the origin).

Example 3. $f(x, y) = x^3 + xy$ is not homogeneous of any degree, since $f(\lambda x, \lambda y)/f(x, y)$ is not independent of x and y.

† Restricting λ to positive values means that $f(x, y)$ is being studied at points (x, y) and $(\lambda x, \lambda y)$ on the *same* side of the origin. It permits inclusion of such functions as $\sqrt{x^2 + y^2}$.

Theorem. Euler's Theorem. *If $f(x, y)$ is continuously differentiable and homogeneous of degree n in a region R, then in R*

(2) $$xf_1(x, y) + yf_2(x, y) = nf(x, y).$$

Proof. Since R is open, it follows that for any fixed point (x_0, y_0) in R, $(\lambda x_0, \lambda y_0)$ is in R for all λ sufficiently near 1 (why?). For such λ we define a function $g(\lambda)$, making use of (1):

(3) $$g(\lambda) \equiv f(\lambda x_0, \lambda y_0) \equiv \lambda^n f(x_0, y_0).$$

Applying the chain rule to (3), where the intermediate variables are the two functions of λ, λx_0 and λy_0, we have

$$g'(\lambda) = x_0 f_1(\lambda x_0, \lambda y_0) + y_0 f_2(\lambda x_0, \lambda y_0) = n\lambda^{n-1} f(x_0, y_0).$$

With $\lambda = 1$, equation (2) is obtained for the fixed but arbitrary point (x_0, y_0) of R.

For a partial converse of Euler's theorem see Exercise 31, § 1209.

1209. EXERCISES

1. Find $\dfrac{du}{dt}$ by the chain rule. Check by substitution before differentiation.

$$u = \ln(1 + x^2 + y^2), \quad x = \cos t, \quad y = \sin t;$$
$$u = e^x \sin y, \quad x = \ln t, \quad y = \operatorname{Arcsin} 3t.$$

2. Find $\dfrac{\partial u}{\partial r}$ and $\dfrac{\partial u}{\partial s}$ by the chain rule. Check by substitution before differentiation.

$$u = x^2, \quad x = re^s;$$
$$u = x^2 + y^2 + z^2, \quad x = r\cos s, \quad y = r\sin s, \quad z = r.$$

In Exercises 3-6, show that the given partial differential equation must be satisfied by a differentiable function of the variables x and y, if these variables occur only in the combination specified.

3. $x + y; \dfrac{\partial u}{\partial x} - \dfrac{\partial u}{\partial y} = 0.$ **4.** $x - y; \dfrac{\partial u}{\partial x} + \dfrac{\partial u}{\partial y} = 0.$

5. $y/x; x\dfrac{\partial u}{\partial x} + y\dfrac{\partial u}{\partial y} = 0.$ **6.** $x^2 + y^2; y\dfrac{\partial u}{\partial x} - x\dfrac{\partial u}{\partial y} = 0.$

In Exercises 7-8, show that the given partial differential equation must be satisfied by a differentiable function of the variables x, y, and z, if these variables occur only in the combinations specified.

7. $x + y + z; \dfrac{\partial u}{\partial x} = \dfrac{\partial u}{\partial y} = \dfrac{\partial u}{\partial z}.$

8. $x - y, y - z, z - x; \dfrac{\partial u}{\partial x} + \dfrac{\partial u}{\partial y} + \dfrac{\partial u}{\partial z} = 0.$

In Exercises 9-10, express in terms of the individual functions and their derivatives.

9. $\dfrac{d}{dx} f(g(x) + h(x)).$ **10.** $\dfrac{d}{dx} f(g(x)h(x)).$

11. Derive the Fundamental Increment Formula (6), § 1205, for a function of several variables, with a careful statement of hypotheses.

12. Prove the statement preceding (4), § 1207, that if u is a differentiable function of $x, y, z, \cdots$ each of which is a differentiable function of $r, s, t, \cdots$, then u is a differentiable function of $r, s, t, \cdots$. *Hint:* Letting $\epsilon_i, \zeta_j, \eta_k, \cdots$ denote infinitesimals, write

$$\Delta u = \frac{\partial u}{\partial x}\Delta x + \frac{\partial u}{\partial y}\Delta y + \cdots + \epsilon_1 \Delta x + \epsilon_2 \Delta y + \cdots$$

$$= \frac{\partial u}{\partial x}\left[\frac{\partial x}{\partial r}\Delta r + \frac{\partial x}{\partial s}\Delta s + \cdots + \zeta_1 \Delta r + \zeta_2 \Delta s + \cdots\right]$$

$$+ \frac{\partial u}{\partial y}\left[\frac{\partial y}{\partial r}\Delta r + \frac{\partial y}{\partial s}\Delta s + \cdots + \eta_1 \Delta r + \eta_2 \Delta s + \cdots\right]$$

$$+ \cdots$$

and reassemble terms in the manner suggested in the proof of the Theorem of § 1207.

13. Use the chain rule to derive the formula for differentiating u^v, where u and v are differentiable functions of x. Hence differentiate x^x and $x^{x^x} = x^{(x^x)}$. *Hint:* Use logarithms.

14. Show that $d(x^{y^z}) = zy^{z-1} y^z\, dx + (\ln x) x^{y^z} y^{z-1} z \, dy + (\ln x)(\ln y) x^{y^z} y^z \, dz$. (Cf. Ex. 13.)

15. Express the derivative of the determinant

$$\begin{vmatrix} a(x) & b(x) & c(x) \\ d(y) & e(y) & f(y) \\ g(z) & h(z) & k(z) \end{vmatrix},$$

where x, y, and z are functions of a variable t, as a sum of three determinants obtained by differentiating a row at a time. Generalize to nth order determinants, and thus obtain a formula for differentiating a determinant all of whose elements a_{ij} are functions of a single variable t: $a_{ij} = a_{ij}(t)$.

16. Prove that if $u = f(x, y)$ is harmonic (cf. Exs. 19-22, § 1204), then so is

$$\phi(x, y) = f\left(\frac{x}{x^2 + y^2}, \frac{y}{x^2 + y^2}\right).$$

★17. Assuming appropriate continuity conditions, prove that if $u = f(x, y)$, $x = \phi(r, s)$, and $y = \psi(r, s)$, then

$$\frac{\partial^2 u}{\partial r^2} = f_1 \phi_{11} + f_2 \psi_{11} + \phi_1^2 f_{11} + 2\phi_1 \psi_1 f_{12} + \psi_1^2 f_{22},$$

$$\frac{\partial^2 u}{\partial r \partial s} = f_1 \phi_{12} + f_2 \psi_{12} + \phi_1 \phi_2 f_{11} + (\phi_1 \psi_2 + \phi_2 \psi_1) f_{12} + \psi_1 \psi_2 f_{22},$$

$$\frac{\partial^2 u}{\partial s^2} = f_1 \phi_{22} + f_2 \psi_{22} + \phi_2^2 f_{11} + 2\phi_2 \psi_2 f_{12} + \psi_2^2 f_{22}.$$

Make up two examples suitable for verifying these formulas.

★18. Obtain formulas similar to those of Exercise 17 for $u = f(x)$, $x = \phi(r, s)$, and verify them for two examples.

★19. Prove that if $f(x, y, z) = g(u)$, where $u = x^2 + y^2 + z^2$, then $f_{xx} + f_{yy} + f_{zz}$ is a function of u only.

In Exercises 20-23, show that the given function is homogeneous. Specify the region and the degree. Verify Euler's theorem in each case.

★20. $\dfrac{xy}{x^2+y^2}$. ★21. $\sqrt{x^2-xy}$.

★22. $\operatorname{Arctan}\dfrac{y}{x}$. ★23. $z\ln\left(1+\dfrac{y}{x}\right)$.

In Exercises 24-25, use Euler's theorem to show that the given function is not homogeneous in any region.

★24. e^{xy}. ★25. $\ln(1+x+y)$.

★26. Prove that if $f(x, y)$ is continuously differentiable and homogeneous of degree n in a region R in the right half-plane, then $f(x, y) = x^n f(1, y/x)$. In particular, show that if $n = 0$, the value of $f(x, y)$ depends only on the ratio of the independent variables—that is, only on the polar angle θ.

★27. Define homogeneity, and establish Euler's theorem, for a function of several variables.

★28. Prove that the first partial derivatives of a homogeneous function of degree n are homogeneous functions of degree $n-1$.

★29. Using appropriate continuity assumptions, establish for a homogeneous function of degree n the formula
$$x^2 f_{11}(x, y) + 2xy f_{12}(x, y) + y^2 f_{22}(x, y) = n(n-1)f(x, y).$$
(Cf. Ex. 28.)

★30. Generalize Exercise 29, both as to the number of variables and as to the order of the derivatives.

★31. Define a function $f(x, y)$ to be **locally homogeneous** in a region R in case, for every point of R, the Definition, § 1208, applies for all values of λ in a suitable neighborhood of $\lambda = 1$. Prove the following extension and converse of Euler's theorem: *If $f(x, y)$ is continuously differentiable in a region R, then it is locally homogeneous of degree n there if and only if $xf_1(x, y) + yf_2(x, y) = nf(x, y)$ there.* *Hint:* Assume the preceding equation holds throughout R, let (x, y) be any point in R, and let I be an open interval containing the number 1 and such that $\lambda \in I$ implies $(\lambda x, \lambda y) \in R$. With x and y fixed, define $g(\lambda) = f(\lambda x, \lambda y)$. Show that $g(\lambda)$ satisfies the equation $\lambda g'(\lambda) = ng(\lambda)$, so that $g(\lambda)$ has the form $g(\lambda) = c\lambda^n$. Substitute $\lambda = 1$ to evaluate c.

★32. Specify conditions on a region so that the condition $xf_1(x, y) + yf_2(x, y) = nf(x, y)$ is equivalent to homogeneity of degree n there. (Cf. Exs. 31, 33.)

★33. Let R be the region consisting of all points (x, y) such that $x > 0$, with the exception of the half-line $x = 1, y \geq 0$. Define $f(x, y) = 0$ except when $x > 1$ and $y > 0$. Define $f(x, y) = y^4/x^4$ if $x > 1$ and $y > 0$. Prove that f is locally homogeneous of degree 0 in R, but not homogeneous there. (Cf. Exs. 26, 31, 32.)

★1210. DIRECTIONAL DERIVATIVES. TANGENTS AND NORMALS

We shall consider now a few applications of the preceding sections to three-dimensional Euclidean space. These same ideas find their analogues in the plane and in higher dimensions (cf. Exs. 9, 11, § 1211).

It will be assumed for the immediate objectives that the student is familiar with such topics from Solid Analytic Geometry as direction numbers and direction cosines (including the formula for the cosine of the angle between two directed lines) and equations of planes and lines. We shall also find it convenient to use the concept of **vector**, meaning the combination of a nonnegative *magnitude* and a *direction*, and represented by a directed line segment (two directed line segments representing the *same* vector whenever one is a *translate* of the other with preservation of sense). A **unit vector** is one of unit magnitude or length. The **components** of a vector are its projections on the coordinate axes. The **zero vector** is the vector of zero magnitude all of whose components vanish.

Let a smooth space curve C (§ 1108) be given. That is, C is prescribed by parametrization functions $x(t)$, $y(t)$, $z(t)$ having continuous first derivatives not all of which vanish at the same point (C has no singular points). Then, since Δx, Δy, Δz represent direction numbers of a secant line of C, so do the quotients $\Delta x/\Delta t$, $\Delta y/\Delta t$, $\Delta z/\Delta t$. Taking the limits of these quotients we see that the derivatives $x'(t)$, $y'(t)$, $z'(t)$ are direction numbers of the tangent line to the curve C. In particular, if the parameter is taken to be arc length s, measured from some fixed point of the curve, the derivatives dx/ds, dy/ds, dz/ds become direction *cosines* of the tangent line, oriented in the direction of increasing arc length (formula (6), § 1112):

$$(1) \qquad \frac{dx}{ds} = \cos \alpha, \quad \frac{dy}{ds} = \cos \beta, \quad \frac{dz}{ds} = \cos \gamma.$$

In other words, if C is given by functions $x(s)$, $y(s)$, $z(s)$, the quantities (1) are the components of the *unit tangent vector*.

We now address ourselves to the following questions: If $u = f(x, y, z)$ is a differentiable function defined in a region R of space, how does u change as the point (x, y, z) moves in a certain direction? In which direction does u change most rapidly? How fast? Just what do these questions really mean?

In the first place, by "rate of change" of u we shall mean *rate of change with respect to distance*. Accordingly, we shall consider the behavior of the function $u = f(x, y, z)$ as the point (x, y, z) is constrained to move along a smooth curve C lying in R, thinking of u as a function of arc length s:

$$(2) \qquad u(s) \equiv f(x(s), y(s), z(s)).$$

The rate of change of u with respect to s takes the form (from the chain rule):

$$(3) \qquad \frac{du}{ds} = \frac{df}{ds} = f_1(x, y, z)\frac{dx}{ds} + f_2(x, y, z)\frac{dy}{ds} + f_3(x, y, z)\frac{dz}{ds}$$

$$= \frac{\partial u}{\partial x}\cos \alpha + \frac{\partial u}{\partial y}\cos \beta + \frac{\partial u}{\partial z}\cos \gamma.$$

It is now evident that the rate of change of u depends on two vectors. The first, called the **gradient** of the function f, has components

$$(4) \qquad f_1(x, y, z), \quad f_2(x, y, z), \quad f_3(x, y, z),$$

and depends only on the function f and the point (x, y, z). The second is the unit tangent vector (1) of the curve C and depends only on the *direction* of the curve at the specified point. Because of this last fact, the derivative (3) is called the **directional derivative** of u in the given direction.

If we examine the form (3) in which the gradient and unit tangent vectors are combined, recalling that a set of direction numbers can be converted to a set of direction cosines by dividing each by the sum of their squares, and appeal to the formula $\cos \theta = \cos \alpha_1 \cos \alpha_2 + \cos \beta_1 \cos \beta_2 + \cos \gamma_1 \cos \gamma_2$ (where θ is the angle between the unit vectors with components $\cos \alpha_i, \cos \beta_i,$ and $\cos \gamma_i, i = 1, 2$), we have the important result:

Theorem I. *The directional derivative of a function $f(x, y, z)$ at a given point and in a given direction is equal to the projection of the gradient of that function, evaluated at the given point, on a directed line having the given direction.*

Corollary. *The maximum and minimum values of the directional derivative of a function $f(x, y, z)$, at a given point, are the magnitude of its gradient and the negative of this magnitude, evaluated at that point, respectively. The function $f(x, y, z)$ has its greatest and least rates of change in the direction of the gradient and in the opposite direction, respectively.*

If the smooth curve C through the given point is chosen to be a straight line parallel to a coordinate axis, and having the same orientation, Δs becomes $\Delta x, \Delta y,$ or Δz, and the directional derivative of $u = f(x, y, z)$ reduces to one of the first partial derivatives of f. In other words, *the directional derivative of a differentiable function is a generalization of a first partial derivative*.

For present purposes we shall define a **surface** as the set of all points (x, y, z) whose coordinates satisfy an equation of the form $f(x, y, z) = 0$, where f is differentiable. A **smooth surface** is a surface for which f has continuous first partial derivatives not all of which vanish identically. Any point of a surface at which these three first partial derivatives all vanish is called a **singular point**. Any other point of the surface is a **regular point**.

Consider now a surface S with equation $f(x, y, z) = 0$, let C be any smooth curve lying wholly in the surface, and let $p : (x_0, y_0, z_0)$ be any point of C that is a regular point of S (it is also a regular point of C). Then $u(s) = f(x(s), y(s), z(s))$ vanishes identically, so that the directional derivative of f also vanishes identically and, in particular, at the point p:

$$(5) \qquad \frac{du}{ds} = f_1 \frac{dx}{ds} + f_2 \frac{dy}{ds} + f_3 \frac{dz}{ds} = 0,$$

§ 1210] DIRECTIONAL DERIVATIVES 383

all derivatives being evaluated at p. By means of the perpendicularity condition for two lines ($\cos \theta = 0$) we conclude that the gradient, evaluated at p, is perpendicular or normal to *all* smooth curves lying in the surface and passing through p. For this reason we define the **normal direction** to a surface $f(x, y, z) = 0$ at a regular point to be that of the gradient of f evaluated at the point. In other words, *the line normal to a surface $f(x, y, z) = 0$ at a regular point has as a set of direction numbers the three first partial derivatives of f evaluated at the point*.

If $f(x, y, z)$ is a differentiable function and if $p : (x_0, y_0, z_0)$ is a point at which f is defined, we can infer immediately, by considering the surface whose equation is $F(x, y, z) \equiv f(x, y, z) - f(x_0, y_0, z_0) = 0$, and agreeing that the zero vector is normal to all vectors and to all surfaces:

Theorem II. *The gradient of a differentiable function, evaluated at any point (x_0, y_0, z_0), is normal to the surface $f(x, y, z) = f(x_0, y_0, z_0)$ at that point.*

Motivated by this theorem and the Corollary to Theorem I we call the maximum directional derivative of a differentiable function $u = f(x, y, z)$ at a point p the **normal derivative** of f at p, and denote it $\dfrac{du}{dn}$ or $\dfrac{df}{dn}$. We have, then,

(6) $$\frac{du}{dn} = \frac{df}{dn} = \sqrt{f_1^2 + f_2^2 + f_3^2}.$$

In conclusion, we assemble some equations arising from the foregoing discussion, pertaining to a curve $C: (x(t), y(t), z(t))$ at a regular point $p : (x_0, y_0, z_0) = (x(t_0), y(t_0), z(t_0))$ and a surface $S: f(x, y, z) = 0$ at a regular point $p : (x_0, y_0, z_0)$:

Normal plane and tangent line to the curve C:

(7) $$x'(t_0)(x - x_0) + y'(t_0)(y - y_0) + z'(t_0)(z - z_0) = 0$$

(8) $$\frac{x - x_0}{x'(t_0)} = \frac{y - y_0}{y'(t_0)} = \frac{z - z_0}{z'(t_0)}.$$

Tangent plane and normal line to the surface S:

(9) $$f_1(x_0, y_0, z_0)(x - x_0) + f_2(x_0, y_0, z_0)(y - y_0) + f_3(x_0, y_0, z_0)(z - z_0) = 0,$$

(10) $$\frac{x - x_0}{f_1(x_0, y_0, z_0)} = \frac{y - y_0}{f_2(x_0, y_0, z_0)} = \frac{z - z_0}{f_3(x_0, y_0, z_0)}.$$

Example. Discuss the gradient of the function $u = x^2 + y^2 + z^2$. In particular, find its normal derivative, and its directional derivative in a direction normal to the paraboloid $z = 3x^2 + y^2$ at the point $(1, 1, 4)$, directed inward.

Solution. Let $f(x, y, z) = x^2 + y^2 + z^2$. The gradient of f has components $2x$, $2y$, $2z$. At any point, then, it has the same direction as the radius vector of that point, and twice the magnitude. It always points directly away from the origin, and is normal to the sphere with center at the origin passing through the point. The function f is increasing most rapidly in that direction, and has a normal derivative of $2\sqrt{x^2 + y^2 + z^2}$. Normals to the paraboloid have direction numbers $6x, 2y, -1$, which become at the given point $6, 2, -1$. Converting these to direction cosines we have $-6/\sqrt{41}, -2/\sqrt{41}, 1/\sqrt{41}$ and, by equation (3) for the directional derivative,

$$\frac{du}{ds} = \left[2\left(-\frac{6}{\sqrt{41}}\right) + 2\left(-\frac{2}{\sqrt{41}}\right) + 8\left(\frac{1}{\sqrt{41}}\right)\right] = -\frac{8}{\sqrt{41}} = -1.249.$$

The normal derivative at this same point is $2\sqrt{18} = 6\sqrt{2} = 8.485$.

★1211. EXERCISES

In Exercises 1-2, find the equations of the tangent line and normal plane to the given curve at the specified point.

★1. $x = a \cos t, y = b \sin t, z = ct, t = \pi/2$.
★2. $x = at \cos t, y = bt \sin t, z = ct, t = \pi/2$.

In Exercises 3-4, find the equations of the normal line and tangent plane to the given surface at the specified point.

★3. $z = e^x \sin y$; $(0, 0, 0)$.
★4. $Ax^2 + By^2 + Cz^2 + 2Dyz + 2Exz + 2Fxy = G$; (x_1, y_1, z_1).

In Exercises 5-6, prove that the surfaces intersect orthogonally (that is, their normals are perpendicular).

★5. $x^2 + y^2 + z^2 - 2x + 4y - 8z = 0$,
$\quad 3x^2 - y^2 + 2z^2 - 6x - 4y - 16z + 31 = 0$.

★6. $\dfrac{x^2}{a^2 - r^2} + \dfrac{y^2}{b^2 - r^2} + \dfrac{z^2}{c^2 - r^2} = 1$,

$\quad \dfrac{x^2}{a^2 - s^2} + \dfrac{y^2}{b^2 - s^2} + \dfrac{z^2}{c^2 - s^2} = 1, \quad \dfrac{x^2}{a^2 - t^2} + \dfrac{y^2}{b^2 - t^2} + \dfrac{z^2}{c^2 - t^2} = 1$,

where $0 < r < a < s < b < t < c$.

In Exercises 7-8, discuss the gradient of the given function. Find the normal derivative, and the directional derivative at the given point in the given direction.

★7. $u = xyz$; $(4, -2, 3)$; $\cos \alpha : \cos \beta : \cos \gamma = 2 : -1 : -2$ directed upward.
★8. $u = x^2 + yz$; $(-3, 1, -2)$; direction making equal acute angles with the positive coordinate axes.

★9. Discuss the gradient and directional derivative of a function $f(x, y)$ of two variables. Show that the gradient of $f(x, y)$ at the point (x_0, y_0) is normal to the curve $f(x, y) = f(x_0, y_0)$. Derive the formulas $\dfrac{du}{ds} = \dfrac{\partial u}{\partial x} \cos \alpha + \dfrac{\partial u}{\partial y} \cos \beta$ and

$\dfrac{du}{dn} = \sqrt{\left(\dfrac{\partial u}{\partial x}\right)^2 + \left(\dfrac{\partial u}{\partial y}\right)^2}.$

★10. Find the normal derivative of the function $u = x^2 - y^2$ at the point $(2, 1)$, and the value of the directional derivative of this function at the same point in the direction of a line making an angle of $120°$ with the positive x-axis (directed upward to the left). (Cf. Ex. 9.)

★11. Discuss the gradient and directional derivative of a function f of n variables.
★12. Show that for a linear function z of two variables x and y, the differential

1212. THE LAW OF THE MEAN

The Law of the Mean for a function of one variable (§ 405) can be expressed by the formula

(1) $$f(a + h) = f(a) + f'(a + \theta h)h,$$

where $0 < \theta < 1$, the assumptions on the function $f(x)$ being continuity on the closed interval $[a, a + h]$ (or $[a + h, a]$ if $h < 0$), and differentiability for points between a and $a + h$.

For a function of several variables the Law of the Mean takes the following form (which we state for simplicity for a function of *two* variables):

Theorem I. Law of the Mean. *Let $f(x, y)$ be continuous in a region or closed region R and differentiable in the interior of R, and let $p : (a, b)$ and $q : (a + h, b + k)$ be any two points of R such that all points $r : (a + \theta h, b + \theta k)$, where $0 < \theta < 1$, of the straight line segment l joining p and q (r not equal to p or q) belong to the interior of R (cf. Fig. 1202). Then there exists a number θ such that $0 < \theta < 1$ and*

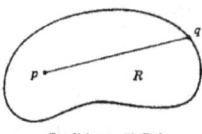 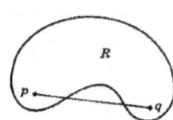

Conditions satisfied   Conditions not satisfied

FIG. 1202

(2) $$f(a + h, b + k) = f(a, b) + f_1(a + \theta h, b + \theta k)h + f_2(a + \theta h, b + \theta k)k.$$

That is, there exists a point r in the interior of l such that for the function $u = f(x, y)$ the increment Δu for the points p and q is equal to the differential du evaluated at r, with $dx = h$ and $dy = k$.

Proof. Let the function $\phi(t)$ of a single real variable t be defined on the unit interval $0 \leq t \leq 1$:

(3) $$\phi(t) \equiv f(a + th, b + tk).$$

Since $\phi(t)$ is continuous on $[0, 1]$ and differentiable in $(0, 1)$, the Law of the Mean (1) can be applied (with $a = 0$, $h = 1$):

(4) $$\phi(1) = \phi(0) + \phi'(\theta).$$

By the chain rule, (2) is the same as (4).

An important consequence of the Law of the Mean is the following (cf. Theorem I, § 406):

Theorem II. *If a function is continuous in a region or closed region R and differentiable in the interior, and if its first partial derivatives vanish throughout the interior, then the function is constant in R.*

Proof. Assume the contrary. That is, assume that the function f assumes distinct values at two points of R. Then, by Theorem IV, § 1010, there are two points in the interior of R at which f has distinct values. Join these two points by a broken line segment lying in R. Then there must be consecutive vertices p and q of this broken line segment such that $f(p) \neq f(q)$. But this is a contradiction to the Law of the Mean (2), since the differential of f vanishes identically between p and q.

If we apply the Extended Law of the Mean (§ 407) to the function $\phi(t) \equiv f(a + th, b + tk)$ (assuming appropriate continuity and differentiability), we obtain for the case $n = 2$:

$$(5) \qquad \phi(1) = \phi(0) + \phi'(0) + \frac{\phi''(\theta)}{2!}, \quad 0 < \theta < 1,$$

or, in terms of the function $f(x, y)$:

$$(6) \quad f(a + h, b + k) = f(a, b) + f_1(a, b)h + f_2(a, b)k$$
$$+ \tfrac{1}{2}[f_{11}(a + \theta h, b + \theta k)h^2$$
$$+ 2f_{12}(a + \theta h, b + \theta k)hk + f_{22}(a + \theta h, b + \theta k)k^2].$$

It is helpful at this stage to use a more compressed notation, showing the binomial behavior of this expansion, with the substituted values of x and y indicated by subscripts:

$$(7) \quad f(a + h, b + k) = f(a, b) + \left[\left(h\frac{\partial}{\partial x} + k\frac{\partial}{\partial y}\right)f(x, y)\right]_{\substack{x=a\\y=b}}$$
$$+ \frac{1}{2}\left[\left(h\frac{\partial}{\partial x} + k\frac{\partial}{\partial y}\right)^2 f(x, y)\right]_{\substack{x=a+\theta h\\y=b+\theta k}}$$

The formula for the **Extended Law of the Mean** in general, for a function of two variables (under suitable continuity and differentiability conditions), can be written:

$$(8) \quad f(a + h, b + k) = f(a, b) + \sum_{i=1}^{n-1} \frac{1}{i!}\left[\left(h\frac{\partial}{\partial x} + k\frac{\partial}{\partial y}\right)^i f(x, y)\right]_{\substack{x=a\\y=b}} + R_n,$$

where $\qquad R_n = \dfrac{1}{n!}\left[\left(h\dfrac{\partial}{\partial x} + k\dfrac{\partial}{\partial y}\right)^n f(x, y)\right]_{\substack{x=a+\theta h\\y=b+\theta k}}, \quad 0 < \theta < 1.$

(The student is asked to establish (8) in Exercise 13, § 1216, and to extend the law of the mean to several variables in Exercises 14, 16, § 1216.) The Law of the Mean (8) is also called **Taylor's Formula with a Remainder.**

1213. APPROXIMATIONS BY DIFFERENTIALS

As in the case of a single variable (§ 411), the Extended Law of the Mean ($n = 2$) provides a measure of the accuracy with which a differential approximates an increment. Let us first recall the formula for a function of one variable ((5), § 411):

(1) $$\Delta u - du = \tfrac{1}{2} f''(a + \theta h) h^2.$$

For a function $u = f(x, y)$ this takes the form ((6), § 1212):

(2) $$\Delta u - du = \tfrac{1}{2} [f_{11} h^2 + 2 f_{12} hk + f_{22} k^2]_{\substack{x = a + \theta h \\ y = b + \theta k}},$$

the differential du being evaluated at $x = a$, $y = b$, with $dx = h$, $dy = k$.

If B is a bound for the second partial derivatives of f, for points near (a, b), that is, if $|f_{ij}| \leq B$, $i, j = 1, 2$, for $|x - a| + |y - b| \leq \delta$, then $|h| + |k| \leq \delta$ implies

$$|f_{11} h^2 + 2 f_{12} hk + f_{22} k^2| \leq B(|h| + |k|)^2 \leq B\delta^2,$$

so that

(3) $$|\Delta u - du| \leq \tfrac{1}{2} B\delta^2.$$

Example 1. A racing car is clocked over a 15-mile course at 5 minutes. However, there is a possible error in the measurement of the course length of as much as 0.03 miles and an error in the measurement of the time of as much as 0.02 minutes. Approximately how accurate is the estimate of 3 miles per minute, or 180 miles per hour, for the average speed?

Solution. Let the course length, measured time, and average speed be denoted by x, t, and v, respectively, with x in miles and t in minutes. Then $v = x/t$ and

$$dv = \frac{1}{t} dx - \frac{x}{t^2} dt.$$

Upon making use of the values $x = 15$ and $t = 5$ and the estimates $|dx| \leq 0.03$ and $|dt| \leq 0.02$, we have $|dv| \leq (0.2)(0.03) + (0.6)(0.02) = 0.018$. Therefore the computed average speed should be accurate to within 0.018 miles per minute, or approximately one mile per hour.

The inequality (3) states that the above error estimate is accurate to within 0.02 miles per hour.

Example 2. A water tank is formed by the bottom half of a right circular cylinder with a horizontal axis, closed at the ends by (vertical) plane semicircles. The radius of this cylinder is to be determined by measuring the length l of the tank and the volume V that it contains. If the quantities l and V are measured with errors up to 1% and 3%, respectively, determine approximately, by means of differentials, the maximum relative error in the computed radius.

Solution. Since $V = \tfrac{1}{2}\pi r^2 l$, $r = (V/\tfrac{1}{2}\pi l)^{\tfrac{1}{2}}$ and

$$\ln r = \tfrac{1}{2} \ln V - \tfrac{1}{2} \ln \tfrac{1}{2}\pi - \tfrac{1}{2} \ln l.$$

Hence $\dfrac{dr}{r} = \dfrac{dV}{2V} - \dfrac{dl}{2l}$ and, with the data above,

388   PARTIAL DIFFERENTIATION   [§1214

$$\left|\frac{dr}{r}\right| \leq \frac{1}{2}\left|\frac{dV}{V}\right| + \frac{1}{2}\left|\frac{dl}{l}\right| \leq \frac{1}{2}(0.01 + 0.03) = 0.02.$$

In other words, the maximum relative error in the determination of the radius is approximately 2%.

We shall not make any attempt to use the inequality (3).

Example 3. Estimate $N = 1.99^{2.98}$.

Solution. It is possible to estimate the given number by considering $u = x^y$, and letting x and y change from 2 and 3, respectively, by amounts $dx = -0.01$ and $dy = -0.02$. However, this method is unrewarding in terms of effort expended. A more practical method is to compute $\ln N = 2.98 \ln 1.99$. Since $\ln 1.99 = \ln [2(1 - 0.005)] = \ln 2 + \ln (1 - 0.005)$, an accurate computation of $\ln 1.99$ is possible by infinite series (cf. § 720). Finally, $N = e^{\ln N}$ is also adaptable to series computation. We omit details. In fact, we reject the problem as inappropriate to this section.

1214. MAXIMA AND MINIMA

If $f(x, y)$ is differentiable in a neighborhood of a point (a, b) at which it has a relative **extremum** (that is, a relative maximum or minimum value), it is clear from elementary considerations (cf. § 409) that the following conditions are *necessary*:

(1) $$f_1(a, b) = 0, \quad f_2(a, b) = 0.$$

It is our purpose in this section to establish a set of conditions that are *sufficient* for a maximum or minimum.† A simple example, like $z = xy$ or $z = x^2 - y^2$ each of which has as graph a hyperbolic paraboloid with a saddle-like shape near the origin, shows that conditions (1) are not sufficient for an extremum. A slightly subtler example is $z = x^2 + 4xy + y^2$. For this surface the two traces $z = x^2$ and $z = y^2$ in the coordinate planes $y = 0$ and $x = 0$ are parabolas opening upward, so that one might expect the origin to be a minimum point. However, for values of x and y arbitrarily near 0 and such that $y = -x$, z is negative, and the origin is *not* an extreme point. This surface is again a hyperbolic paraboloid, owing to the character of the sections $z = x^2 + 4xy + y^2 = $ constant. The fact that these sections are hyperbolas is disclosed by the negativeness of the discriminant

(2) $$AC - B^2 = \begin{vmatrix} A & B \\ B & C \end{vmatrix}$$

of the quadratic expression $Ax^2 + 2Bxy + Cy^2$, where $A = C = 1$, $B = 2$.

The discussion of the preceding paragraph contains the nucleus of a method for approaching the problem of extrema with the aid of the Law of the Mean. Assume that $u = f(x, y)$ has continuous second partial derivatives in a neighborhood of (a, b), and that the first partial derivatives

---

† The method of *Lagrange multipliers* is presented in Exercises 28–37, § 1225.

vanish at that point. Then the Extended Law of the Mean for $n = 2$ ((6), § 1212) can be written

(3) $\quad \Delta u = f(a + h, b + k) - f(a, b)$
$$= f_{11}(a + \theta h, b + \theta k)h^2 + 2f_{12}(a + \theta h, b + \theta k)hk$$
$$+ f_{22}(a + \theta h, b + \theta k)k^2,$$

where $0 < \theta < 1$.

There are three possibilities that present themselves, corresponding to inequalities involving Δu for values of h and k near 0:

(4) $\quad \begin{cases} f \text{ has a maximum value at } (a, b): \Delta u \leq 0, \\ f \text{ has a minimum value at } (a, b): \Delta u \geq 0, \\ f \text{ has neither}: \Delta u \text{ changes sign.} \end{cases}$

In order to study the signs of Δu, we write

(5) $\quad \begin{cases} f_{11}\Delta u = (f_{11}h + f_{12}k)^2 + (f_{11}f_{22} - f_{12}{}^2)k^2, \\ f_{22}\Delta u = (f_{11}f_{22} - f_{12}{}^2)h^2 + (f_{12}h + f_{22}k)^2, \end{cases}$

where the partial derivatives are evaluated at the point $x = a + \theta h$, $y = b + \theta k$. We are now ready to formulate a set of sufficient conditions for maxima and minima:

Theorem. *Let $f(x, y)$ be a function with continuous second partial derivatives in a neighborhood of the point (a, b), and assume that at that point the first partial derivatives vanish: $f_1(a, b) = f_2(a, b) = 0$. Let*

(6) $\quad D(x, y) \equiv f_{11}(x, y)f_{22}(x, y) - [f_{12}(x, y)]^2, \quad D \equiv D(a, b).$

Then

(i) *if $D > 0$, f has an extreme value at (a, b);*
(ii) *if $D > 0$, this extreme value is a maximum or a minimum according as $f_{11}(a, b)$ and $f_{22}(a, b)$ are both negative or both positive;*
(iii) *if $D < 0$, f has neither a maximum nor a minimum at (a, b);*
(iv) *if $D = 0$, any of the preceding alternatives are possible.*

Proof. We observe first that if $D > 0$, $f_{11}(a, b)$ and $f_{22}(a, b)$ must both be nonzero and of the same sign. We infer from (5), and continuity, that for values of h and k sufficiently near 0, the three quantities Δu, f_{11}, and f_{22} are all of the same sign, and conclusions (i) and (ii) follow from (4).

If $D < 0$ the conclusion is less immediate. If $f_{11}(a, b) \neq 0$ we can use (5) to find pairs of points arbitrarily near (a, b) at which $f_{11}\Delta u$, and hence Δu itself, have opposite signs. For example, if $k = 0$ and h is small, $f_{11}\Delta u = (f_{11}h)^2$ is positive; and if $h = -[f_{12}(a, b)/f_{11}(a, b)]k$ and k is small, $f_{11}\Delta u$ is negative (check the details). A similar argument applies if $f_{22}(a, b) \neq 0$. If $D < 0$ and $f_{11}(a, b) = f_{22}(a, b) = 0$, then let $h = k$ and $h = -k$, in turn (how does this show (iii)?).

Finally, the examples $\pm(x^4 + y^4)$ and $x^4 - y^4$ show that if $D = 0$ no conclusion can be drawn.

NOTE 1. The routine procedure in solving an extremum problem is: (i) find the **critical points** (a, b) such that $f_1(a, b) = f_2(a, b) = 0$; (ii) test the function for each critical point. In practice it is often simpler to analyze a given function directly instead of its second derivatives. Techniques such as completing squares, squaring variables, etc., may suggest themselves. (Cf. Ex. 17, § 412. Also Exs. 28–37, § 1225.)

NOTE 2. For functions of more than two variables the attack on the problem of sufficiency conditions for extrema is again the extended law of the mean. However, the analysis of the signs of Δu becomes much more difficult, and depends on the subject of *positive definite quadratic forms*. For a reference, see G. Birkhoff and S. MacLane, *A Survey of Modern Algebra* (New York, the Macmillan Company, 1944).

Example 1. Find the minimum value of the function
$$f(x, y) = x^2 + 5y^2 - 6x + 10y + 6.$$

First Solution. Completion of squares gives
$$f(x, y) = (x - 3)^2 + 5(y + 1)^2 - 8,$$
with a minimum value of -8 at $x = 3, y = -1$.

Second Solution. By (1), the only critical point is $(3, -1)$. The corresponding value of f is -8. Since $f_{11} = 2, f_{12} = 0, f_{22} = 10, D > 0$ and $(3, -1)$ is a minimum point.

Example 2. Test $f(x, y) = x^2 + xy$ for extrema.

First Solution. The only critical point is $(0, 0)$. Since $f(\epsilon, 0) = \epsilon^2 > 0$ and $f(\epsilon, -2\epsilon) = -\epsilon^2 < 0$, the function has no extrema.

Second Solution. At the origin $D = -1 < 0$, so that (as before) the function has no extrema.

Example 3. Find the dimensions and the volume of the largest rectangular parallelepiped that has faces in the coordinate planes and one vertex in the ellipsoid

(7) $$\frac{x^2}{a^2} + \frac{y^2}{b^2} + \frac{z^2}{c^2} = 1.$$

Solution. The problem is to maximize the function $V = xyz$ subject to the *constraint* (7). This is equivalent to finding the values of x, y, z subject to (7) that maximize $x^2y^2z^2$. Still more simply, if we let $u = x^2/a^2, v = y^2/b^2$, and $w = z^2/c^2$, we seek positive numbers u, v, w subject to the constraint

(8) $$u + v + w = 1$$

that maximize $a^2b^2c^2uvw$ or, equivalently, that maximize uvw. Symmetry considerations suggest the solution $u = v = w = \frac{1}{3}$. In analytic terms, we wish to maximize the function of u and v:
$$\phi(u, v) = uv(1 - u - v) = uv - u^2v - uv^2.$$

The critical points for $\phi(u, v)$ are $(0, 0)$ and $(\frac{1}{3}, \frac{1}{3})$, the first clearly inappropriate. At the point $(\frac{1}{3}, \frac{1}{3})$, $\phi_{11}\phi_{22} - \phi_{12}^2 = \frac{1}{3} > 0$ and $\phi_{11} < 0$. Therefore ϕ has a maximum value there. The original parallelepiped requested therefore corresponds to

$u = v = w = \frac{1}{3}$, and has dimensions $x = a/\sqrt{3}$, $y = b/\sqrt{3}$, $z = c/\sqrt{3}$ and volume $abc/3\sqrt{3}$. (Cf. Ex. 32, § 1225.)

NOTE 3. If extreme values are sought for a function on a restricted set that is not open, it is necessary to examine its values at all points of its domain of definition that are not interior points. For example, on the set $x^2 + y^2 \leqq 1$ the function xy has a maximum value of $\frac{1}{2}$ on the frontier given by $x = \cos\theta$, $y = \sin\theta$, for $\theta = \frac{1}{4}\pi$.

1215. TAYLOR SERIES

As with a function of one variable, it is possible that a function of two or more variables has an infinite series expansion. Assuming that the function $f(x, y)$ has continuous partial derivatives of all orders, the form of this expansion, called the **Taylor series of the function** and obtained from Taylor's Formula with a Remainder (formula (8), § 1212), is

$$(1) \quad f(a + h, b + k) = f(a, b) + \sum_{i=1}^{+\infty} \frac{1}{i!}\left[\left(h\frac{\partial}{\partial x} + k\frac{\partial}{\partial y}\right)^i f(x, y)\right]_{\substack{x=a \\ y=b}},$$

where h and k represent the increments $x - a$ and $y - b$, respectively. The condition that the series on the right in (1) converges to the function $f(x, y)$ is that the remainder after n terms,

$$(2) \quad R_n = \left[\left(h\frac{\partial}{\partial x} + k\frac{\partial}{\partial y}\right)^n f(x, y)\right]_{\substack{x=a+\theta h \\ y=b+\theta k}}, \quad 0 < \theta < 1,$$

is an infinitesimal, as a function of n.

A similar formulation holds for a function of more than two variables.

In practice, most Taylor series are most easily obtained by combining Taylor series of functions of a single variable.

Example 1. Find the Taylor series for the function $f(x, y) = e^x \sin y$, through terms of degree 3, for $a = b = 0$.

First Solution. Evaluate:
$$f(0, 0) = e^0 \sin 0 = 0,$$
$$\left[\left(h\frac{\partial}{\partial x} + k\frac{\partial}{\partial y}\right)f(x, y)\right]_{\substack{x=0 \\ y=0}} = h \cdot 0 + k \cdot 1 = k = y - b = y,$$
$$\frac{1}{2}\left[\left(h\frac{\partial}{\partial x} + k\frac{\partial}{\partial y}\right)^2 f(x, y)\right]_{\substack{x=0 \\ y=0}} = \frac{1}{2}[h^2 \cdot 0 + 2hk \cdot 1 + k^2 \cdot 0] = hk = xy,$$
$$\frac{1}{6}\left[\left(h\frac{\partial}{\partial x} + k\frac{\partial}{\partial y}\right)^3 f(x, y)\right]_{\substack{x=0 \\ y=0}} = \frac{1}{6}[3h^2k - k^3] = \frac{1}{6}(3x^2y - y^3).$$

Therefore
$$e^x \sin y = y + xy + \frac{1}{6}(3x^2y - y^3) + \cdots.$$

Second Solution. Form the product:
$$\left(1 + x + \frac{x^2}{2} + \cdots\right)\left(y - \frac{y^3}{6} + \cdots\right) = y + xy + \frac{1}{2}x^2y - \frac{1}{6}y^3 + \cdots.$$

The second solution shows that the Taylor series represents the given function for all real x and y.

Example 2. Find the Taylor series for the function $f(x, y) = \dfrac{1}{xy}$, through terms of degree 2, for $a = 2, b = -1$.

First Solution. Evaluate:

$$f(a, b) = \frac{1}{2(-1)} = -\frac{1}{2},$$

$$\left[\left(h\frac{\partial}{\partial x} + k\frac{\partial}{\partial y}\right)f(x, y)\right]_{\substack{x=a \\ y=b}} = -\frac{h}{2^2(-1)} - \frac{k}{2(-1)^2} = \frac{1}{4}(x-2) - \frac{1}{2}(y+1),$$

$$\frac{1}{2}\left[\left(h\frac{\partial}{\partial x} + k\frac{\partial}{\partial y}\right)^2 f(x, y)\right]_{\substack{x=a \\ y=b}} = \frac{1}{2}\left[\frac{2h^2}{2^3(-1)} + \frac{2hk}{2^2(-1)^2} + \frac{2k^2}{2(-1)^3}\right]$$

$$= -\tfrac{1}{8}(x-2)^2 + \tfrac{1}{4}(x-2)(y+1) - \tfrac{1}{2}(y+1)^2.$$

Therefore

$$\frac{1}{xy} = -\frac{1}{2} + \frac{x-2}{4} - \frac{y+1}{2} - \frac{(x-2)^2}{8} + \frac{(x-2)(y+1)}{4} - \frac{(y+1)^2}{2} + \cdots.$$

Second Solution. Form the product of the two series:

$$\frac{1}{x} = \frac{1}{a+h} = \frac{1}{a}\frac{1}{1+\frac{h}{a}} = \frac{1}{2}\left(1 - \frac{x-2}{2} + \frac{(x-2)^2}{4} - \cdots\right),$$

$$\frac{1}{y} = \frac{1}{b+k} = -\frac{1}{1-k} = -(1 + (y+1) + (y+1)^2 + \cdots).$$

The second solution shows that the Taylor series represents the given function for $|x - 2| < 2$ and $|y + 1| < 1$.

1216. EXERCISES

1. Find a value of θ between 0 and 1 satisfying (2), § 1212, for the function $x^2 + xy$ with $a = 3, b = -5, h = -1, k = 2$.

2. Write out the equation (2), § 1212, for the function $e^x \sin y$ with $a = 0$, $b = 0, h = 1, k = \pi/6$.

3. Find a value of θ between 0 and 1 satisfying (6), § 1212, for the function $x^2 y$ with $a = 1, b = 1, h = 1, k = 2$.

4. Write out equation (6), § 1212, for the function $x^2 \ln y$, with $a = 0, b = 1$, $h = 1, k = e - 1$.

5. The legs of a right triangle are measured to be 20 and 30 feet, with maximum errors of 1 inch. Find approximately the maximum possible error in calculating the area of the triangle. Estimate the accuracy of the preceding computation.

6. Do Exercise 5 for the hypotenuse, instead of the area.

7. Discuss for extrema, considering all possible values of a: $z = x^3 - 3axy + y^3$.

8. Find the maximum and minimum values of $(ax^2 + by^2)e^{-(x^2+y^2)}$ for $0 \leq |a| \leq b$. Consider all possible cases.

9. Find the Taylor series for the function $f(x, y) = e^{x+y}$, through terms of degree 3, for $a = 0, b = 0$, using three methods: by formula (1), § 1215, by expanding e^r and substituting $r = x + y$, and by multiplying two series.

10. Find the Taylor series for the function $f(x, y) = \ln(xy)$, through terms of degree 2, for $a = 1, b = 1$, using two methods.

11. Verify the quadratic approximation formulas for values of x and y near 0:

(a) $e^x \ln(1+y) = y + xy - \frac{y^2}{2}$;

(b) $\cos x \cos y = 1 - \frac{1}{2}(x^2 + y^2)$.

12. Obtain linear approximation formulas for values of x and y near 0:

(a) $\sqrt{\frac{1+x}{1+y}}$; (b) $\text{Arctan}\,\frac{y}{1+x}$.

13. Establish formula (8), § 1212.

14. Generalize the Law of the Mean (Theorem I, § 1212) to a function of several variables.

15. Write out the generalization of equation (2), § 1212, to several variables for the function xyz with $a = 0, b = 0, c = 0, h = 1, k = 2, l = 3$. (Cf. Ex. 14.)

★**16.** Generalize the Extended Law of the Mean (formula (8), § 1212) to a function of several variables.

★**17.** Show that the approximation to an increment of a function given by a differential is the same as that given by the Taylor series of the function through the terms of the first degree. (Cf. Ex. 3, § 806.)

★**18.** Obtain for a function of 3 variables the formula corresponding to (2), § 1213.

★**19.** Show by the following example that a function $f(x, y)$ may be continuously differentiable throughout a region R and such that its first partial derivative $f_1(x, y)$ vanishes identically there, yet not depend solely on y. Hence show the inadequacy of the following reasoning in proving Theorem II, § 1212: "If $f_1(x, y) = 0$, f does not depend on x, and if $f_2(x, y) = 0$, f does not depend on y; therefore f is a constant." Show that if R is a region in E_2 such that every line parallel to the x-axis meets R in an interval, finite or infinite, then the existence and identical vanishing of the partial derivative f_1 of a function f implies that f is a function of y alone. *Example*: Let R consist of the plane with the points (x, y) such that $x = 0, y \geq 0$ deleted, and let $f(x, y) = 0$ except in the first quadrant, where $f(x, y) = y^4$.

★**20.** Show that the Law of the Mean (2), § 1212, if the condition $(a + \theta h, b + \theta k) \in I(R)$ is omitted, fails for the function of Exercise 19, by taking $a = -1, b = 1, h = 2, k = 0$.

★**21.** Say what you can about a function $f(x, y)$ all of whose second partial derivatives vanish identically in a region R. *Hint*: Show that $f_1(x, y) = a$ and $f_2(x, y) = b$, identically, and consider the function $f(x, y) - ax - by$.

★**22.** Let R be a region in E_2 such that every line parallel to a coordinate axis meets R in an interval, finite or infinite. Say what you can about a function $f(x, y)$ having the following property:

(a) $f_{12}(x, y) = f_{21}(x, y) = 0$ identically in R;

(b) $f_{11}(x, y) = f_{22}(x, y) = 0$ identically in R;

(c) $f_1(x, y) = f_2(x, y) = x$ identically in R.

(Cf. Ex. 19.)

1217. DIFFERENTIATION OF AN IMPLICIT FUNCTION

It is often desirable or necessary to treat functions that are not given explicitly, but instead are defined by or involved implicitly in some functional relation. For example, it may be convenient to study properties of the function $y = \sqrt{1 - x^2}$ by means of the equation $x^2 + y^2 = 1$. On the other hand, the inverse function of $x + e^x$ is of necessity not written explicitly, but is *defined* implicitly by the equation $y + e^y = x$. We shall be faced, then, with the study of certain properties of a function $y = \phi(x)$ determined by an equation in the two variables x and y, which can be written in the form $f(x, y) = 0$. More generally, we shall investigate functions of several variables which are implicitly specified. The question of *existence* of a function satisfying a given equation is treated in § 1228, and will not be pursued here. Our concern in this section is to establish and use certain relationships between the derivative (or partial derivatives if there is more than one independent variable) of a function implicitly defined and the partial derivatives of the functions appearing in the defining equation.

A basic fact of the greatest importance is that when a variable is defined, by a given equation, as a function of the remaining variables the given equation, considered as a relation in these remaining variables, is an *identity*. For example, when $\sqrt{1 - x^2}$ is substituted for y in the equation $x^2 + y^2 = 1$, this equation reduces to the identity $x^2 + (1 - x^2) = 1$. Similarly, if $y = \phi(x)$ is the function defined by the equation $y + e^y = x$, then the equation $\phi(x) + e^{\phi(x)} = x$ is true for all x by definition. In this section we shall use the expression "is defined by" to mean "satisfies identically."

We are now ready to derive the basic differentiation formulas for functions assumed to exist (cf. § 1228 for their existence). The ranges of the variables are unspecified for reasons of expediency.

Theorem I. *Let $y = \phi(x)$ be a differentiable function defined by the equation $f(x, y) = 0$, where $f(x, y)$ is differentiable and $f_y(x, y)$ does not vanish. Then*

$$(1) \quad \frac{dy}{dx} = -\frac{f_x}{f_y} = -\frac{f_1(x, \phi(x))}{f_2(x, \phi(x))}.$$

Proof. Let $u = f(x, y)$, and let x and y be considered as functions of the new independent variable t: $x = t$, $y = \phi(t)$. Then u, as a function of t, is differentiable and, by the chain rule,

$$(2) \quad \frac{du}{dt} = \frac{\partial f}{\partial x}\frac{dx}{dt} + \frac{\partial f}{\partial y}\frac{dy}{dt}.$$

As a function of t alone, however, u is identically zero, so that the left-

§ 1217]   IMPLICIT FUNCTION   395

hand member of (2) vanishes identically. Using the fact that $dx/dt = 1$, solving (2) for dy/dt, and relabeling the variable t with the letter x, we obtain (1).

NOTE. A mnemonic device for formula (1) is the following: Write the quotient $-f_x/f_y$ in the form $-(\partial f/\partial x)/(\partial f/\partial y)$, "cancel" ∂f, and obtain $-\partial y/\partial x$. This, of course, is nonsense (cf. the Note, § 1207), but the point is that it tells us where the subscripts x and y belong. (The minus sign must be remembered separately.) This device is a little like one that serves to recall the change of base formula for logarithms (Ex. 11, § 602), $\log_a x = (\log_a b)(\log_b x)$: "cancel" the b's and shift the subscript a.

The proof of the next theorem is nearly identical in method with that of Theorem I, and is requested in Exercise 10, § 1219.

Theorem II. *Let $u = \phi(x, y, z, \cdots)$ be a differentiable function of the k variables $x, y, z, \cdots$ defined by the equation $f(x, y, z, \cdots, u) = 0$, where the function $f(x, y, z, \cdots, u)$ of the $k + 1$ variables $x, y, z, \cdots, u$ is differentiable and $f_u(x, y, z, \cdots, u)$ does not vanish. Then*

$$(3) \quad \begin{cases} \dfrac{\partial u}{\partial x} = -\dfrac{f_x}{f_u} = -\dfrac{f_1(x, y, z, \cdots, \phi(x, y, z, \cdots))}{f_{k+1}(x, y, z, \cdots, \phi(x, y, z, \cdots))}, \\ \dfrac{\partial u}{\partial y} = -\dfrac{f_y}{f_u} = -\dfrac{f_2(x, y, z, \cdots, \phi(x, y, z, \cdots))}{f_{k+1}(x, y, z, \cdots, \phi(x, y, z, \cdots))}, \\ \cdots \end{cases}$$

Example 1. Find the slope of the hyperbola

$$x^2 - 4xy - 3y^2 = 9$$

at the point $(2, -1)$.

First Solution. If $f(x, y) = x^2 - 4xy - 3y^2 - 9$, y is defined as a function of x by the equation $f(x, y) = 0$. By (1),

$$\frac{dy}{dx} = -\frac{f_x}{f_y} = -\frac{2x - 4y}{-4x - 6y} = \frac{x - 2y}{2x + 3y} = \frac{4}{1} = 4.$$

Second Solution. Differentiate with respect to x:

$$2x - 4y - 4x\frac{dy}{dx} - 6y\frac{dy}{dx} = 0,$$

and solve for dy/dx.

Example 2. Find $\dfrac{\partial z}{\partial y}$ if $x^2y - 8xyz = yz + z^3$.

First Solution. Let $f(x, y, z) = 8xyz + yz + z^3 - x^2y$. Then, by (3):

$$\frac{\partial z}{\partial y} = -\frac{\partial f/\partial y}{\partial f/\partial z} = -\frac{8xz + z - x^2}{8xy + y + 3z^2}.$$

Second Solution. Solve

$$x^2 - 8xz - 8xy\frac{\partial z}{\partial y} = z + y\frac{\partial z}{\partial y} + 3z^2\frac{\partial z}{\partial y}.$$

1218. SOME NOTATIONAL PITFALLS

Suppose u is a given function of three variables x, y, and z, and that z is a given function of the two variables x and y. What does $\frac{\partial u}{\partial x}$ mean? This meaning depends on whether u is being considered as a function of all three variables or as a function of only the two variables x and y. The notation $\frac{\partial u}{\partial x}$ does not specify which. This ambiguity becomes particularly confusing if we recklessly apply the chain rule, using the symbol $\frac{\partial u}{\partial x}$ twice:

$$(1) \qquad \frac{\partial u}{\partial x} = \frac{\partial u}{\partial x} + \frac{\partial u}{\partial z}\frac{\partial z}{\partial x},$$

where in the left-hand member u is a function of x and y, and in the right-hand member u is a function of x, y, and z. Such a muddled equation as (1) should be avoided. One of the simplest means of getting away from it is to use functional notation and to be more generous in the use of letters. To be precise, for the question above let $u = f(x, y, z)$ be the given function of three variables. Following the notation of § 1207 and observing that a change of variables is taking place, we can let r and s denote the new variables and represent the old variables in terms of the new by the formulas

$$(2) \qquad x = r, \quad y = s, \quad z = \phi(r, s).$$

Finally, if we write $g(r, s) = f(r, s, \phi(r, s))$, the chain rule gives $\frac{\partial g}{\partial r} = \frac{\partial f}{\partial x}\frac{\partial x}{\partial r} + \frac{\partial f}{\partial y}\frac{\partial y}{\partial r} + \frac{\partial f}{\partial z}\frac{\partial z}{\partial r}$ which, by (2), reduces to $\frac{\partial g}{\partial r} = \frac{\partial f}{\partial x} + \frac{\partial f}{\partial z}\frac{\partial z}{\partial r}$. Finally, upon restoration of the notation x and y for the new variables, equation (1) becomes

$$(3) \qquad \frac{\partial g}{\partial x} = \frac{\partial f}{\partial x} + \frac{\partial f}{\partial z}\frac{\partial z}{\partial x}.$$

In this example the purpose of using two new letters r and s is to clarify the possibly disturbing fact that in the original formulation of the problem the two new variables x and y are identical with two of the original variables. With practice and sophistication the student will learn to dispense with such auxiliary variables as r and s, and proceed immediately to equation (3).

The numerical subscript notation for partial derivatives (§ 1201) is also suitable for resolving such notational problems as that just discussed. Equation (3) becomes

$$(4) \qquad g_1(x, y) = f_1(x, y, \phi(x, y)) + f_3(x, y, \phi(x, y))\phi_1(x, y).$$

A similar problem is the following: Suppose z is defined as a function of x and y by means of the equation

§ 1219] EXERCISES 397

(4) $$z = f(x, y, z).$$

What is $\frac{\partial z}{\partial x}$? It should be clear, in spite of the deceptive equality of (4), that this is *not* the same as $\frac{\partial f}{\partial x}$. One method of answering the question is the following: Define the function $F(x, y, z)$:

(5) $$F(x, y, z) \equiv f(x, y, z) - z,$$

and write (4) in the form

(6) $$F(x, y, z) = 0.$$

Then, by (3), § 1217, $\frac{\partial z}{\partial x} = -\frac{F_x}{F_z} = -\frac{f_x}{f_z - 1} = \frac{f_x}{1 - f_z}$. More simply, perhaps, considering z as a function of x and y, we can differentiate (4) directly:

(7) $$\frac{\partial z}{\partial x} = \frac{\partial f}{\partial x} + \frac{\partial f}{\partial z}\frac{\partial z}{\partial x};$$

and solve for $\frac{\partial z}{\partial x}$, with the same result as before.

Example. If $u = x^2y + y^2z + z^2x$ and if z is defined implicitly as a function of x and y by the equation $x^3 + yz + z^3 = 0$, find $\frac{\partial u}{\partial x}$, where u is considered as a function of x and y alone.

Solution. Let $f(x, y, z) = x^2y + y^2z + z^2x$, $g(x, y, z) = x^3 + yz + z^3$, $z = \phi(x, y)$ be the function determined by $g(x, y, z) = 0$, and $\psi(x, y) = f(x, y, \phi(x, y))$. Then $\psi_1(x, y) = f_1 + f_3\phi_1 = f_1 + f_3\left(-\frac{g_1}{g_3}\right) = 2xy + z^2 - \frac{2x(y^2 + 2zx)}{y + 3z^2} = \frac{6xyz^2 - 4x^2z + yz^3 + 3z^4}{y + 3z^2}$.

1219. EXERCISES

In these Exercises assume differentiability wherever appropriate.

In Exercises 1-4, find $\frac{dy}{dx}$.

1. $x^2 + xy + 3y^2 = 17$. 2. $x^{\frac{2}{3}} + y^{\frac{2}{3}} = a^{\frac{2}{3}}$.
3. $x \sin y + y \sin x = b$. 4. $x = ay + be^{-cy}$.

In Exercises 5-8, find $\frac{\partial z}{\partial x}$ and $\frac{\partial z}{\partial y}$.

5. $Ayz + Bzx + Cxy = D$. 6. $x^3 + y^3 + z^3 = 3axyz$.
7. $x = ze^{yz}$. 8. $xe^z = y \sin z$.

9. If the equation $f(x, y, z) = 0$ defines each variable as a differentiable function of the other two establish the formulas

$$\frac{\partial z}{\partial x} = -\frac{f_1}{f_3}, \quad \frac{\partial z}{\partial y} = -\frac{f_2}{f_3}, \quad \frac{\partial y}{\partial x} = -\frac{f_1}{f_2}, \quad \frac{\partial y}{\partial z} = -\frac{f_3}{f_2}, \quad \frac{\partial x}{\partial y} = -\frac{f_2}{f_1}, \quad \frac{\partial x}{\partial z} = -\frac{f_3}{f_1}.$$

398 PARTIAL DIFFERENTIATION [§1220

where the variable being differentiated is considered as a function of the other two. In what sense are $\frac{\partial x}{\partial z}$ and $\frac{\partial z}{\partial x}$ reciprocals? Is it true that $\frac{\partial x}{\partial y} \frac{\partial y}{\partial z} \frac{\partial z}{\partial x} = -1$? Generalize to the case of n variables.

10. Prove Theorem II, § 1217.

In Exercises 11–14, u is defined as a function of x, y, and z by the first equation, and z is defined as a function of x and y by the second equation. Considering u, as a consequence, as a function of x and y, find $\frac{\partial u}{\partial x}$ and $\frac{\partial u}{\partial y}$.

11. $x^2 + y^3 + z^4 + u^5 = 1, x + y^2 + z^3 = 1$.
12. $u = x^2 + y^2 + z^2 + u^3, z = x^2 + y^2 + z^2$.
13. $u = x + y + z + e^u, z = x + y + \sin z$.
14. $f(x, y, z, u) = 0, g(x, y, z) = 0$.
15. Examine Theorems 1 and 2, § 1217, in the linear case, $f(x, y) = ax + by$, $f(x, y, \cdots, u) = ax + by + \cdots + qu$. Indicate the significance of the nonvanishing of the appropriate derivative.
16. If $z = f(x, y, z)$, find $\frac{\partial x}{\partial y}, \frac{\partial y}{\partial z}, \frac{\partial z}{\partial x}$.
17. If $f(x, y) = g(y, z)$, find $\frac{\partial x}{\partial y}, \frac{\partial y}{\partial z}, \frac{\partial z}{\partial x}$.
18. If $x^2 + y^2 = f(x, y, z)$, find $\frac{\partial x}{\partial y}, \frac{\partial y}{\partial z}, \frac{\partial z}{\partial x}$.
19. If $z = f(x + y + z, xyz)$, find $\frac{\partial x}{\partial y}, \frac{\partial y}{\partial z}, \frac{\partial z}{\partial x}$.
20. If $g(x) = f(x, y, g(y))$, find $\frac{dy}{dx}$.
21. Show that if y is defined as a function of x by the equation $f(x, y) = 0$, then

$$\frac{d^2y}{dx^2} = -\frac{f_1^2 f_{22} - 2f_1 f_2 f_{12} + f_2^2 f_{11}}{f_2^3}.$$

22. Show that if z is defined as a function of x and y by the equation $f(x, y, z) = 0$, then

$$\frac{\partial^2 z}{\partial x \partial y} = -\frac{f_1 f_2 f_{33} - f_1 f_3 f_{23} - f_2 f_3 f_{13} + f_3^2 f_{12}}{f_3^3}.$$

1220. ENVELOPE OF A FAMILY OF PLANE CURVES

Definition. *An **envelope** of a family of plane curves is a curve that has the following two properties: (i) at every one of its points it is tangent to at least one curve of the family; (ii) it is tangent to every curve of the family at at least one point.* ***The envelope*** *of a family of curves is the totality of its envelopes.*

Example 1. The family of curves $y = (x - t)^2$, where t is a parameter, is the set of parabolas obtained by translating the parabola $y = x^2$ in the direction of the x-axis. The x-axis clearly satisfies the definition above, and is therefore an envelope. It appears obvious (intuitively) that it is *the* envelope, since there is no

§ 1220] ENVELOPE OF A FAMILY 399

other curve that is an envelope. The point on the envelope that corresponds to the curve $y = (x - t)^2$, for any fixed t, is given by $x = t, y = 0$.

Example 2. The circles of the family $(x - t)^2 + y^2 = 1$, have centers on the x-axis and radii equal to 1. The line $y = 1$ is an envelope, as is the line $y = -1$. *The* envelope consists of the two lines $y^2 = 1$. The points on the envelope that correspond to the curve $(x - t)^2 + y^2 = 1$, for any fixed t, are given by $x = t$, $y = \pm 1$.

It is our purpose in this section to find necessary conditions which a well-behaved envelope of a well-behaved family of curves must satisfy.

Theorem. *Let*
(1) $$f(x, y, t) = 0$$
be the equation of a family F of curves, where t is a parameter, and let
(2) $$x = \phi(t), \quad y = \psi(t)$$
be the parametric equations of an envelope E of F. Assume furthermore that f, ϕ, and ψ are continuously differentiable and that for each value of t the curve (1) and E have a common tangent at the point $(\phi(t), \psi(t))$. Then the coordinates (2) of each point of E must satisfy, in addition to equation (1), the equation
(3) $$f_3(x, y, t) = 0.$$

Proof. Since (1) and E have a common tangent at the point $(\phi(t), \psi(t))$, this point certainly lies on the curve (1), for each t. Therefore the equation $f(\phi(t), \psi(t), t) = 0$ is an *identity* in t and, by the chain rule,
(4) $$f_1(\phi(t), \psi(t), t)\phi'(t) + f_2(\phi(t), \psi(t), t)\psi'(t) + f_3(\phi(t), \psi(t), t) = 0.$$

Since our objective is to establish the vanishing of the third term of (4), it is sufficient to show that the sum of the first two terms of (4) is zero. This is trivial unless at least one of these terms is nonzero. We assume for definiteness that both $f_2(\phi(t), \psi(t), t)$ and $\psi'(t)$ are nonzero, for the particular value of t considered, and use the fact that (1) and E have a common tangent at $(\phi(t), \psi(t))$. In the first place (§ 1217), the slope of (1) at $(\phi(t), \psi(t))$ is equal to $-f_1(\phi(t), \psi(t), t)/f_2(\phi(t), \psi(t), t)$, which is finite. Therefore the tangent to E at $(\phi(t), \psi(t))$ is not vertical, and $\phi'(t) \neq 0$. Its slope is therefore $\psi'(t)/\phi'(t)$. The equality $-f_1/f_2 = \psi'/\phi'$ is equivalent to the desired vanishing of $f_1\phi' + f_2\psi'$.

NOTE. In simple cases, since equations (1) and (3) are both satisfied by the coordinates of points of the envelope, the envelope can often be found by eliminating t from that pair of equations.

We now consider the two examples given above, and others, in the light of the Theorem and Note.

Example 1 (again). Let $f(x, y, t) = (x - t)^2 - y$. Then $f_3(x, y, t) = -2(x - t)$. The coordinates of points on an envelope must satisfy identically the two equations
$$x - t = 0, \quad (x - t)^2 = y.$$

In other words, $y = 0$, and only points on the x-axis are on the envelope. Therefore (with the aid of the preceding discussion) we know that the x-axis is *the* envelope.

Example 2 (again). Let $f(x, y, t) = (x - t)^2 + y^2 - 1$. Then $f_t(x, y, t) = -2(x - t)$. As in Example 1, the two equations (1) and (3) lead to the envelope. In this case its equation is $y^2 = 1$.

Example 3. Find the envelope of the family of parabolas obtained by arbitrarily translating the parabola $y = x^2$ so that the vertex lies on the line $y = 2x$.

Solution. The family has the equation $f(x, y, t) = (x - t)^2 - (y - 2t) = 0$. Then $f_t(x, y, t) = -2x + 2t + 2$. Elimination of t between equations (1) and (3) leads to the equation $y = 2x - 1$. This line must therefore contain the envelope. Direct verification shows that it *is* the envelope.

Example 4. Discuss the envelope of the family of curves $f(x, y, t) = (x - t)^2 - y^3 = 0$, obtained by translating parallel to the x-axis the semicubical parabola $y^3 = x^2$.

Solution. The procedure used in Example 1 yields the same result: $y = 0$. However, the x-axis is clearly *not* an envelope. In fact, there is *no* envelope. This example shows the importance of checking the result of eliminating t from equations (1) and (3).

Example 5. Discuss the envelope of the family of curves $\sqrt[3]{y} = x - t$.

Solution. If $f(x, y, t) = \sqrt[3]{y} - x + t$, equation (3) becomes $1 = 0$. This contradiction might lead us to believe that there is *no* envelope. However, the conditions of the preceding Theorem are not satisfied: $f(x, y, t)$ is not a differentiable function, since $f_y(x, y, t) = (\frac{1}{3})y^{-\frac{2}{3}}$ becomes infinite on the x-axis.

There are two courses open to us now: (*i*) examine the points where $f(x, y, t)$ fails to be differentiable, and (*ii*) rewrite the original equation. If we choose the latter alternative, and define $f(x, y, t) = y - (x - t)^3 = 0$, we have $f_t(x, y, t) = 3(x - t)^2$. The result, as in preceding examples, is that the x-axis is the envelope.

1221. EXERCISES

In Exercises 1-6, find the envelope of the given family of curves. Draw a figure.

1. The lines $x \cos \phi + y \sin \phi = 3$.
2. The lines $y = tx + t^2$.
3. The circles $(x - t)^2 + y^2 = 2t^2$.
4. The circles $(x - t)^2 + y^2 = t$.
5. The parabolas $x^2 = t(y - t)$.
6. The parabolas $x^2 = t^2(y - t)$.
7. Explain the absence of envelope for the family of hyperbolas $4(x - 3t)^2 - 9(y - 2t)^2 = 36$.
8. Find the envelope of the family obtained by translating parallel to the x-axis the curve $y^2 = x^2(2 - x)$ (cf. Fig. 417, p. 126). Check your answer by maximizing y.
9. Translate the parabola $y = x^2$ by moving its vertex along the parabola $y = x^2$. Find the envelope of the resulting family.
10. A variable circle moves so that it is always tangent to the x-axis and its center remains on the parabola $y = x^2$. Find the envelope.
11. A line segment of fixed length a moves so that its end-points are on the two

§ 1222] SEVERAL FUNCTIONS DEFINED IMPLICITLY 401

coordinate axes. Find the envelope. *Hint:* Denote the intercepts $a \cos \theta$ and $a \sin \theta$.

12. A line moves so that its intercepts a and b have a constant positive sum, $a + b = c > 0$. Find the envelope.

13. The ellipse $\frac{x^2}{a^2} + \frac{y^2}{b^2} = 1$ varies so that its area is constant. Find the envelope.

14. The ellipse $\frac{x^2}{a^2} + \frac{y^2}{b^2} = 1$ varies so that the sum of its semiaxes is a constant c. Find the envelope.

15. The ellipse $\frac{x^2}{a^2} + \frac{y^2}{b^2} = 1$ varies so that the distance $c = \sqrt{a^2 + b^2}$ between the ends of the axes is constant. Find the envelope.

16. A variable circle moves so that it always passes through the origin and its center lies on the hyperbola $x^2 - y^2 = 1$. Find the envelope. *Hint:* Parametrize the hyperbola: $x = \cosh t, y = \sinh t$.

17. Prove that any smooth curve (§ 1108) $x = \phi(t), y = \psi(t)$ is an envelope of its tangents. Prove furthermore that if the curvature of the given curve (§ 1114) exists and is not zero, the curve is *the* envelope of its tangents. *Hint:* Equations (1) and (3), § 1220, become $(y - \psi)\phi' = (x - \phi)\psi'$ and $(y - \psi)\phi'' = (x - \phi)\psi''$, respectively. Show that these together with $\phi'\psi'' \neq \phi''\psi'$ imply $x = \phi, y = \psi$, for each t.

In Exercises 18-20, verify the statements of Exercise 17 for the given curve.

18. $x = a \cos \theta, y = b \sin \theta$.
19. $x^2 + y^2 = 4$.   **20.** $y = x^3$.

★**21.** Let $x = \phi(t), y = \psi(t)$ be a smooth curve with nonzero curvature (§§ 1108, 1114). Prove that the envelope of its normals is its evolute (§ 1116).

In Exercises 22-24, find the evolute requested in the specified Exercise of § 1117, as the envelope of the normals of the given curve. (Cf. Ex. 21.)

★**22.** Ex. 19.   ★**23.** Ex. 21.   ★**24.** Ex. 22.

1222. SEVERAL FUNCTIONS DEFINED IMPLICITLY. JACOBIANS

In § 1217 we considered some properties of a function $u = \phi(x, y, z, \cdots)$ defined implicitly by an equation of the form $f(x, y, z, \cdots, u) = 0$. The problem of an implicit function becomes particularly simple in the linear case (cf. Ex. 15, § 1219). As with linear expressions, the case of more general functions defined implicitly extends to the study of functions defined by *systems* of equations.

Let us review some important facts regarding systems of linear equations.† The most important single theorem is Cramer's Rule, which deals

---
† For a treatment of systems of linear equations see M. Bôcher, *Introduction to Higher Algebra* (New York, The Macmillan Company, 1935) or N. Conkwright, *Introduction to the Theory of Equations* (New York, Ginn and Company, 1941).

with a system of n linear equations in n unknowns. If the determinant D of such a system is nonzero, Cramer's Rule assures us that there is precisely one solution and specifies the form of this solution by means of quotients of determinants. If the determinant D is zero or if the number m of equations and the number n of unknowns are unequal it is possible that the system may have infinitely many solutions or none at all. It is reasonable to expect, however (in the absence of other information), that if $m < n$ there are infinitely many solutions and if $m > n$ there are no solutions. Under any circumstances regarding m and n, whenever a system of linear equations has at least one solution it is possible to obtain all solutions by suitably selecting k of the equations and k of the unknowns ($k \leq m, k \leq n$), and solving these k equations for these k unknowns uniquely in terms of the remaining $n - k$ unknowns.

Because of considerations such as those of the preceding paragraph we shall restrict our discussion of systems of equations to those for which the number of equations and the number of variables considered to be defined by the system as dependent on the remaining variables are equal.

We now proceed to the study of a system of equations of the form

$$(1) \quad \begin{cases} f^{(1)}(x_1, x_2, \cdots, x_m, u_1, u_2, \cdots, u_n) = 0, \\ f^{(2)}(x_1, x_2, \cdots, x_m, u_1, u_2, \cdots, u_n) = 0, \\ \quad \vdots \\ f^{(n)}(x_1, x_2, \cdots, x_m, u_1, u_2, \cdots, u_n) = 0, \end{cases}$$

where the functions are specified by superscripts instead of subscripts to avoid confusion with subscript notation for partial derivatives. The variables $u_1, \cdots, u_n$ are considered to be defined by (1) as functions of $x_1, \cdots, x_m$. The question of *existence* of such solutions is treated in § 1228. In this section we shall deduce certain consequences of their existence, under appropriate conditions. For simplicity the region under consideration will not be specified here, and the details of the discussion will be carried through for the case $n = 2$. Since the method is essentially the same for all n, no loss of generality is suffered.

Accordingly, we consider a system of two equations

$$(2) \quad \begin{cases} f(x, y, \cdots, u, v) = 0, \\ g(x, y, \cdots, u, v) = 0, \end{cases}$$

where f and g are differentiable. Let $u = \phi(x, y, \cdots)$ and $v = \psi(x, y, \cdots)$ be two differentiable functions of the remaining variables that satisfy the two equations (2) identically. Our objective is to express the first partial derivatives of ϕ and ψ in terms of those of f and g. We apply the chain rule to (2), differentiating first with respect to x:

§ 1222] SEVERAL FUNCTIONS DEFINED IMPLICITLY

(3)
$$\begin{cases} \dfrac{\partial f}{\partial x} + \dfrac{\partial f}{\partial u}\dfrac{\partial u}{\partial x} + \dfrac{\partial f}{\partial v}\dfrac{\partial v}{\partial x} = 0, \\ \dfrac{\partial g}{\partial x} + \dfrac{\partial g}{\partial u}\dfrac{\partial u}{\partial x} + \dfrac{\partial g}{\partial v}\dfrac{\partial v}{\partial x} = 0, \end{cases}$$

where a compressed notation has been adopted. The second term of the first equation of (3), for example, could have been written more completely

$$f_u(x, y, \cdots, \phi(x, y, \cdots),\ \psi(x, y, \cdots))\cdot \phi_x(x, y, \cdots).$$

Since system (3) may be regarded as a system of two linear equations in the two unknowns $\dfrac{\partial u}{\partial x}$ and $\dfrac{\partial v}{\partial x}$, we may solve for these two unknowns by Cramer's Rule:

(4)
$$\dfrac{\partial u}{\partial x} = -\dfrac{\begin{vmatrix}\dfrac{\partial f}{\partial x} & \dfrac{\partial f}{\partial v}\\ \dfrac{\partial g}{\partial x} & \dfrac{\partial g}{\partial v}\end{vmatrix}}{\begin{vmatrix}\dfrac{\partial f}{\partial u} & \dfrac{\partial f}{\partial v}\\ \dfrac{\partial g}{\partial u} & \dfrac{\partial g}{\partial v}\end{vmatrix}},\quad \dfrac{\partial v}{\partial x} = -\dfrac{\begin{vmatrix}\dfrac{\partial f}{\partial u} & \dfrac{\partial f}{\partial x}\\ \dfrac{\partial g}{\partial u} & \dfrac{\partial g}{\partial x}\end{vmatrix}}{\begin{vmatrix}\dfrac{\partial f}{\partial u} & \dfrac{\partial f}{\partial v}\\ \dfrac{\partial g}{\partial u} & \dfrac{\partial g}{\partial v}\end{vmatrix}},$$

where, to be sure, the denominator determinant must be different from zero.

Determinants of the type appearing in (4) are of great importance in pure and applied mathematics. They are called *functional determinants* or *Jacobians*, after the German mathematician C. G. J. Jacobi (1804-1851). Their precise definition and a more concise notation for them follow.

Definition. *If the variables $u_1, \cdots, u_n$ are differentiable functions of the variables $x_1, \cdots, x_n$ (and possibly of more variables $x_{n+1}, \cdots, x_m$ as well):*

(5)
$$\begin{cases} u_1 = f^{(1)}(x_1, x_2, \cdots, x_n, x_{n+1}, \cdots, x_m), \\ u_2 = f^{(2)}(x_1, x_2, \cdots, x_n, x_{n+1}, \cdots, x_m), \\ \quad\vdots \\ u_n = f^{(n)}(x_1, x_2, \cdots, x_n, x_{n+1}, \cdots, x_m), \end{cases}$$

the Jacobian or functional determinant of $u_1, u_2, \cdots, u_n$ with respect to $x_1, x_2, \cdots, x_n$ is defined as an nth order determinant, and denoted, as follows:

(6)
$$\dfrac{\partial(u_1, \cdots, u_n)}{\partial(x_1, \cdots, x_n)} = \dfrac{\partial(f^{(1)}, \cdots, f^{(n)})}{\partial(x_1, \cdots, x_n)} \equiv \begin{vmatrix}\dfrac{\partial f^{(1)}}{\partial x_1} & \cdots & \dfrac{\partial f^{(1)}}{\partial x_n}\\ \cdots & \cdots & \cdots \\ \dfrac{\partial f^{(n)}}{\partial x_1} & \cdots & \dfrac{\partial f^{(n)}}{\partial x_n}\end{vmatrix}$$

With this notation equation (4) takes the form

$$(7) \quad \frac{\partial u}{\partial x} = -\frac{\frac{\partial(f, g)}{\partial(x, v)}}{\frac{\partial(f, g)}{\partial(u, v)}}, \quad \frac{\partial v}{\partial x} = -\frac{\frac{\partial(f, g)}{\partial(u, x)}}{\frac{\partial(f, g)}{\partial(u, v)}}.$$

In an entirely similar way, again under the assumption that the Jacobian $\frac{\partial(f, g)}{\partial(u, v)}$ is nonzero, we have for the partial derivatives with respect to y

$$(8) \quad \frac{\partial u}{\partial y} = -\frac{\frac{\partial(f, g)}{\partial(y, v)}}{\frac{\partial(f, g)}{\partial(u, v)}}, \quad \frac{\partial v}{\partial y} = -\frac{\frac{\partial(f, g)}{\partial(u, y)}}{\frac{\partial(f, g)}{\partial(u, v)}},$$

with analogous formulas holding for the partial derivatives with respect to any of the other variables of which u and v are considered to be functions.

The student should observe the resemblance between formulas (7) and (8), and formula (1) § 1217, where $n = 1$. In the simplest case $n = 1$, the Jacobian becomes a partial derivative, and therefore may be regarded as a generalization of a partial derivative. Other similarities will become evident in the sequel.

Notational pitfall. The notation $\frac{\partial u}{\partial x}$, appearing in (7) is ambiguous, in that it tells only that u is considered as a dependent variable, but it does *not* indicate the other. Suppose y, instead of v, were the other dependent variable. Then the formula for $\frac{\partial u}{\partial x}$ would be $-\frac{\partial(f, g)}{\partial(x, y)} \Big/ \frac{\partial(f, g)}{\partial(u, y)}$. This is a different function, even though the notation $\partial u/\partial x$ is the same! One standard notational device to clarify this ambiguity is to specify the other *independent* variables (*the ones being held fixed*) by subscripts. For example, if the only variables involved in (2) were $x, y, u,$ and v, then the partial derivatives $\partial u/\partial x$ just considered would be denoted in formula (7), $\left(\frac{\partial u}{\partial x}\right)_v$; in the instance immediately above, $\left(\frac{\partial u}{\partial x}\right)_y$. A pertinent question in this connection is this: Under what circumstances are $\partial u/\partial x$ and $\partial x/\partial u$ reciprocals? The answer is simple: When the *variables held constant* are the same for both derivatives. (Cf. Ex. 9, § 1219, Ex. 13, § 1225.) With the subscript notation just introduced and where the only variables involved are $x, y, u,$ and v we can say, then, that $(\partial u/\partial x)_v$ and $(\partial x/\partial u)_v$ are reciprocals; so are $(\partial u/\partial x)_y$ and $(\partial x/\partial u)_y$.

We formalize the more general case with a theorem (whose proof is requested in Exercise 14, § 1225):

Theorem. *Let the functions* $f^{(i)}(x_1, \cdots, x_m, u_1, \cdots, u_n), i = 1, \cdots, n,$ *of system* (1) *be differentiable, let* $u_j = u_j(x_1, \cdots, x_m), j = 1, \cdots, n,$ *be* *differentiable functions that satisfy system* (1) *identically in the variables* $x_1, \cdots, x_m$, *and assume that the Jacobian* $\dfrac{\partial(f^{(1)}, \cdots, f^{(n)})}{\partial(u_1, \cdots, u_n)}$, *called the **Jacobian of the system**, be nonzero for all values of the variables concerned. Then the partial derivatives of the functions* u_j *are given by the formula*

$$(9) \qquad \frac{\partial u_j}{\partial x_k} = -\frac{\dfrac{\partial(f^{(1)}, f^{(2)}, \ldots, f^{(j)}, \ldots, f^{(n)})}{\partial(u_1, u_2, \ldots, x_k, \ldots, u_n)}}{\dfrac{\partial(f^{(1)}, f^{(2)}, \ldots, f^{(j)}, \ldots, f^{(n)})}{\partial(u_1, u_2, \ldots, u_j, \ldots, u_n)}},$$

where the numerator Jacobian is obtained from the denominator Jacobian by formally replacing u_j *by* x_k.

1223. COORDINATE TRANSFORMATIONS. INVERSE TRANSFORMATIONS

An algebraic fact which we shall need in this section comes from the theory of determinants (see the references given in § 1222). This states that the product of two nth order determinants can be written as an nth order determinant obtained as follows: Let A and B be nth order determinants whose general elements (ith row and jth column) are a_{ij} and b_{ij}, respectively, and define C to be the nth order determinant whose general element is

$$(1) \qquad c_{ij} = \sum_{k=1}^{n} a_{ik} b_{kj} = a_{i1} b_{1j} + a_{i2} b_{2j} + \cdots + a_{in} b_{nj}.$$

Then $AB = C$.

We omit the proof, but illustrate with a simple example, which the student should check:

Example 1. $\begin{vmatrix} 3 & -1 \\ 5 & 6 \end{vmatrix} \cdot \begin{vmatrix} 1 & 4 \\ -2 & 7 \end{vmatrix} = \begin{vmatrix} 5 & 5 \\ -7 & 62 \end{vmatrix}.$ (This says that the product of 23 and 15 is 345.)

One simple consequence of this fact about the multiplication of determinants should be noted: If both factors, A and B, are nonzero, so is their product AB.

We consider now a system of the form

$$(2) \qquad \begin{cases} x = x(u, v), \\ y = y(u, v), \end{cases}$$

where $x(u, v)$ and $y(u, v)$ are continuously differentiable functions of the variables u and v. The system (2) can be thought of as defining a *point transformation* T, carrying certain points from the uv-plane into points of the xy-plane (cf. § 1016). Under suitably restrictive conditions on the one-to-one *correspondence* between two sets, and the *inverse transformation* T^{-1}, which carries the points in the xy-plane back into points in the uv-plane, will be defined. (Such sufficiency conditions are treated in § 1229.) In the case of a one-to-one correspondence, the system (2) can be considered alternatively as reassigning to a point (u, v) (or (x, y)) a new pair of coordinates (x, y) (or (u, v)). When so considered the transformation is called a **coordinate transformation**. In either case the Jacobian $\frac{\partial(x, y)}{\partial(u, v)}$ is called the **Jacobian of the transformation** (2).

NOTE. Since a coordinate transformation is commonly used for transforming an equation in x and y (or $x, y, \cdots$) into an equation in other variables, and since such a transformation of an equation is usually achieved by substituting for $x, y, \cdots$ their expressions in terms of the new variables, it is customary to represent a coordinate transformation in the form (2), instead of writing the new variables in terms of $x, y, \cdots$, as was done in § 1016.

A simple and familiar example may help to clarify the situation:

Example 2. Consider the relation between rectangular and polar coordinates in the plane, and write

(3) $\qquad \begin{cases} x = u \cos v, \\ y = u \sin v, \end{cases}$

where u and v represent the polar coordinates ρ and θ, respectively. The student is already familiar with (3) as a coordinate transformation. Let us consider it as a point transformation. Figure 1203 illustrates a one-to-one correspondence be-

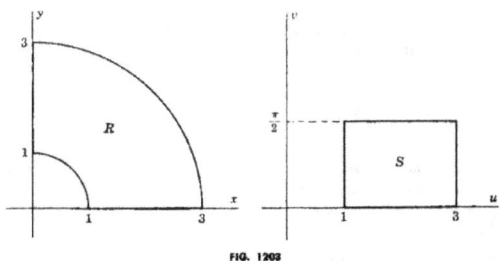

FIG. 1203

tween two regions, S in the uv-plane and R in the xy-plane. If the region S were extended upward until v became greater than 2π, the correspondence would no longer be one-to-one, since the partial annulus R would extend around the origin and overlap itself. Similarly, if u were allowed to be zero, the v-axis in the uv-plane would correspond to the origin in the xy-plane, and the transformation (3) would

§ 1223] COORDINATE TRANSFORMATIONS

be many-to-one instead of one-to-one. The Jacobian of the transformation is
$\begin{vmatrix} \cos v & -u \sin v \\ \sin v & u \cos v \end{vmatrix} = u$, which vanishes only on the v-axis in the uv-plane (the origin in the xy-plane).

Now suppose we have *two* transformations,

(4) $\qquad \begin{cases} x = x(u, v), \\ y = y(u, v), \end{cases} \quad \begin{cases} u = u(r, s), \\ v = v(r, s), \end{cases}$

where the functions are differentiable, so that upon applying the transformations in succession, we have x and y expressed in terms of r and s:

(5) $\qquad \begin{cases} x = \phi(r, s) \equiv x(u(r, s), v(r, s)), \\ y = \psi(r, s) \equiv y(u(r, s), v(r, s)). \end{cases}$†

Since this *composition of transformations* has such a close similarity to the *composition of functions* (we can speak of a transformation of a transformation instead of a function of a function) it might be expected that we should have some sort of chain rule involving Jacobians, corresponding to the familiar formula $\dfrac{dy}{dx} = \dfrac{dy}{du}\dfrac{du}{dx}$ for functions. Let us see.

Applying the chain rule for partial differentiation to (5), we have:

(6) $\qquad \begin{cases} \dfrac{\partial x}{\partial r} = \dfrac{\partial x}{\partial u}\dfrac{\partial u}{\partial r} + \dfrac{\partial x}{\partial v}\dfrac{\partial v}{\partial r}, & \dfrac{\partial x}{\partial s} = \dfrac{\partial x}{\partial u}\dfrac{\partial u}{\partial s} + \dfrac{\partial x}{\partial v}\dfrac{\partial v}{\partial s}, \\ \dfrac{\partial y}{\partial r} = \dfrac{\partial y}{\partial u}\dfrac{\partial u}{\partial r} + \dfrac{\partial y}{\partial v}\dfrac{\partial v}{\partial r}, & \dfrac{\partial y}{\partial s} = \dfrac{\partial y}{\partial u}\dfrac{\partial u}{\partial s} + \dfrac{\partial y}{\partial v}\dfrac{\partial v}{\partial s}. \end{cases}$

This is the spot where we use the multiplication theorem for determinants. We observe, from (6):

$$\begin{vmatrix} \dfrac{\partial x}{\partial u} & \dfrac{\partial x}{\partial v} \\ \dfrac{\partial y}{\partial u} & \dfrac{\partial y}{\partial v} \end{vmatrix} \cdot \begin{vmatrix} \dfrac{\partial u}{\partial r} & \dfrac{\partial u}{\partial s} \\ \dfrac{\partial v}{\partial r} & \dfrac{\partial v}{\partial s} \end{vmatrix} = \begin{vmatrix} \dfrac{\partial x}{\partial r} & \dfrac{\partial x}{\partial s} \\ \dfrac{\partial y}{\partial r} & \dfrac{\partial y}{\partial s} \end{vmatrix},$$

a relation which we shall call the **chain rule for transformations** and which can be written more concisely:

(7) $\qquad \dfrac{\partial(x, y)}{\partial(r, s)} = \dfrac{\partial(x, y)}{\partial(u, v)} \dfrac{\partial(u, v)}{\partial(r, s)}.$

In the form (7), the chain rule bears a very close resemblance to the chain rule for functions of one variable, and can be easily remembered by virtue of the apparent "cancellation" of the symbol $\partial(u, v)$ (which, of course, is meaningless by itself). Observe that we can conclude from (7) that if both

† Observe that the notation of (4) is functional notation, and that since x is *not* the

Jacobians of the transformations (4) that are combined by composition are nonzero, then the Jacobian of the resultant transformation (5) is also nonzero.

An important special case of (7) is that in which the second of the two transformations (4) is the inverse of the first:

(8) $$\begin{cases} x = x(u, v), \\ y = y(u, v), \end{cases} \quad \begin{cases} u = u(x, y), \\ v = v(x, y), \end{cases}$$

where the system (5) reduces to $x = \phi(r, s) \equiv r$ and $y = \psi(r, s) \equiv s$. The resultant Jacobian is then the Jacobian of the **identity transformation** that leaves every point invariant:

$$\frac{\partial(x, y)}{\partial(r, s)} = \begin{vmatrix} 1 & 0 \\ 0 & 1 \end{vmatrix} = 1.$$

The chain rule (7) then reduces to

(9) $$\frac{\partial(x, y)}{\partial(u, v)} \frac{\partial(u, v)}{\partial(x, y)} = 1,$$

stating that *the Jacobians of a transformation and its inverse are reciprocals*. This corresponds to the familiar fact for functions of one variable that dy/dx and dx/dy are reciprocals. From (9) we infer that *whenever a transformation and its inverse exist and are prescribed by differentiable functions, both the transformation and its inverse have nonzero Jacobians*.

Since the functions $u(x, y)$ and $v(x, y)$ of the inverse transformation are defined implicitly by a system of equations

$$f(x, y, u, v) \equiv x - x(u, v) = 0 \quad \text{and} \quad g(x, y, u, v) \equiv y - y(u, v) = 0,$$

their first partial derivatives can be obtained by formulas (7) and (8), § 1222:

$$\frac{\partial u}{\partial x} = -\frac{\frac{\partial(f, g)}{\partial(x, v)}}{\frac{\partial(f, g)}{\partial(u, v)}} = \frac{\frac{\partial y}{\partial v}}{\frac{\partial(x, y)}{\partial(u, v)}}, \quad \frac{\partial v}{\partial x} = -\frac{\frac{\partial(f, g)}{\partial(u, x)}}{\frac{\partial(f, g)}{\partial(u, v)}} = \frac{-\frac{\partial y}{\partial u}}{\frac{\partial(x, y)}{\partial(u, v)}},$$

$$\frac{\partial u}{\partial y} = -\frac{\frac{\partial(f, g)}{\partial(y, v)}}{\frac{\partial(f, g)}{\partial(u, v)}} = \frac{-\frac{\partial x}{\partial v}}{\frac{\partial(x, y)}{\partial(u, v)}}, \quad \frac{\partial v}{\partial y} = -\frac{\frac{\partial(f, g)}{\partial(u, y)}}{\frac{\partial(f, g)}{\partial(u, v)}} = \frac{\frac{\partial x}{\partial u}}{\frac{\partial(x, y)}{\partial(u, v)}}.$$

For the general case of the chain rule for transformations from one n-dimensional space to another (that is, where there are n dependent and n independent variables), the details are completely equivalent to those given above, and will be omitted here (cf. Ex. 15, § 1225). We merely note in passing the particular forms and values of the Jacobians for a change of coordinates in three-dimensional space from rectangular coordinates to cylindrical and spherical coordinates (§ 1111):

(10) Cylindrical: $\dfrac{\partial(x, y, z)}{\partial(\rho, \theta, z)} = \begin{vmatrix} \cos\theta & -\rho\sin\theta & 0 \\ \sin\theta & \rho\cos\theta & 0 \\ 0 & 0 & 1 \end{vmatrix} = \rho.$

(11) Spherical:

$$\dfrac{\partial(x, y, z)}{\partial(r, \phi, \theta)} = \begin{vmatrix} \sin\phi\cos\theta & r\cos\phi\cos\theta & -r\sin\phi\sin\theta \\ \sin\phi\sin\theta & r\cos\phi\sin\theta & r\sin\phi\cos\theta \\ \cos\phi & -r\sin\phi & 0 \end{vmatrix} = r^2\sin\phi.$$

Example 3. Write Laplace's equation $\dfrac{\partial^2 u}{\partial x^2} + \dfrac{\partial^2 u}{\partial y^2} = 0$ (cf. Exs. 19-22, § 1204) in polar coordinates.

Solution. Since $x = \rho\cos\theta$ and $y = \rho\sin\theta$,

(12) $\begin{cases} \dfrac{\partial u}{\partial \rho} = \dfrac{\partial u}{\partial x}\cos\theta + \dfrac{\partial u}{\partial y}\sin\theta, \\ \dfrac{\partial u}{\partial \theta} = -\dfrac{\partial u}{\partial x}\rho\sin\theta + \dfrac{\partial u}{\partial y}\rho\cos\theta. \end{cases}$

Solving (12) for $\dfrac{\partial u}{\partial x}$ and $\dfrac{\partial u}{\partial y}$, we have

(13) $\begin{cases} \dfrac{\partial u}{\partial x} = \dfrac{\partial u}{\partial \rho}\cos\theta - \dfrac{\partial u}{\partial \theta}\dfrac{\sin\theta}{\rho}, \\ \dfrac{\partial u}{\partial y} = \dfrac{\partial u}{\partial \rho}\sin\theta + \dfrac{\partial u}{\partial \theta}\dfrac{\cos\theta}{\rho}. \end{cases}$

Putting each equation of (13) to double duty we get

$\dfrac{\partial^2 u}{\partial x^2} = \dfrac{\partial}{\partial x}\left[\dfrac{\partial u}{\partial x}\right] = \dfrac{\partial}{\partial \rho}\left[\dfrac{\partial u}{\partial \rho}\cos\theta - \dfrac{\partial u}{\partial \theta}\dfrac{\sin\theta}{\rho}\right]\cos\theta - \dfrac{\partial}{\partial \theta}\left[\dfrac{\partial u}{\partial \rho}\cos\theta - \dfrac{\partial u}{\partial \theta}\dfrac{\sin\theta}{\rho}\right]\dfrac{\sin\theta}{\rho}$

$= \left[\dfrac{\partial^2 u}{\partial \rho^2}\cos\theta + \dfrac{\partial u}{\partial \theta}\dfrac{\sin\theta}{\rho^2} - \dfrac{\partial^2 u}{\partial \rho\partial\theta}\dfrac{\sin\theta}{\rho}\right]\cos\theta$

$\qquad - \left[-\dfrac{\partial u}{\partial \rho}\sin\theta + \dfrac{\partial^2 u}{\partial\theta\partial\rho}\cos\theta - \dfrac{\partial u}{\partial \theta}\dfrac{\cos\theta}{\rho} - \dfrac{\partial^2 u}{\partial \theta^2}\dfrac{\sin\theta}{\rho}\right]\dfrac{\sin\theta}{\rho}.$

Similarly,

$\dfrac{\partial^2 u}{\partial y^2} = \left[\dfrac{\partial^2 u}{\partial \rho^2}\sin\theta - \dfrac{\partial u}{\partial \theta}\dfrac{\cos\theta}{\rho^2} + \dfrac{\partial^2 u}{\partial \rho\partial\theta}\dfrac{\cos\theta}{\rho}\right]\sin\theta$

$\qquad + \left[\dfrac{\partial u}{\partial \rho}\cos\theta + \dfrac{\partial^2 u}{\partial\theta\partial\rho}\sin\theta - \dfrac{\partial u}{\partial \theta}\dfrac{\sin\theta}{\rho} + \dfrac{\partial^2 u}{\partial \theta^2}\dfrac{\cos\theta}{\rho}\right]\dfrac{\cos\theta}{\rho}.$

Addition of these expressions gives

(14) $\qquad \dfrac{\partial^2 u}{\partial x^2} + \dfrac{\partial^2 u}{\partial y^2} = \dfrac{\partial^2 u}{\partial \rho^2} + \dfrac{1}{\rho^2}\dfrac{\partial^2 u}{\partial \theta^2} + \dfrac{1}{\rho}\dfrac{\partial u}{\partial \rho} = 0.$

Alternatively, Laplace's equation can be written

(15) $\qquad \rho\dfrac{\partial}{\partial \rho}\left(\rho\dfrac{\partial u}{\partial \rho}\right) + \dfrac{\partial^2 u}{\partial \theta^2} = 0.$

1224. FUNCTIONAL DEPENDENCE

Definition. *If $u = u(x, y, \cdots)$, $v = v(x, y, \cdots)$, $\cdots$ are m differentiable functions of the n variables $x, y, \cdots$, and if these m functions satisfy, identically in the variables $x, y, \cdots$ in some region R, an equation of the form*

$$(1) \qquad F(u, v, \cdots) = 0,$$

*where F is differentiable for the appropriate values of the m variables $u, v, \cdots$ and where not all of its m first partial derivatives vanish simultaneously, then the functions $u, v, \cdots$ are said to be **functionally dependent** in R.*

Our principal objective in this section will be to establish an important *necessary* condition for functional dependence. In considering such a *consequence* of functional dependence we shall concern ourselves only with the case $m \leq n$ because, under ordinary circumstances, if $m > n$ functional dependence normally holds. For example, the two functions of one variable $u = e^x$ and $v = \sin x$ are related by the equation $\sin \ln u - v = 0$. Similarly, if $m = 3$ and $n = 2$ we can ordinarily solve the first two of the equations $u = f(x, y)$, $v = g(x, y)$, $w = h(x, y)$ for x and y in terms of u and v and by substituting the results in the third, obtain a relation between $u, v,$ and w.

Theorem. *If m functions of n variables, where $m \leq n$, are functionally dependent in a region R, then every mth order Jacobian of these m functions with respect to m of the n variables vanishes identically in R.*

Proof. For definiteness and simplicity we give the details for the case of *two* functions $u = u(x, y, \cdots)$ and $v = v(x, y, \cdots)$ of the n variables $x, y, \cdots$, where $n \geq 2$. The method is general (cf. Ex. 16, § 1225). We select an arbitrary pair of independent variables, r and s, and establish the identical vanishing of the Jacobian $\partial(u, v)/\partial(r, s)$. To achieve this we use the chain rule for the function $F(u, v)$ which by hypothesis vanishes identically in the variables $x, y, \cdots$ and whose two first partial derivatives do not vanish simultaneously. The chain rule, then, for the variables r and s, gives the two equations:

$$(2) \qquad \begin{cases} \dfrac{\partial F}{\partial u}\dfrac{\partial u}{\partial r} + \dfrac{\partial F}{\partial v}\dfrac{\partial v}{\partial r} = 0, \\ \dfrac{\partial F}{\partial u}\dfrac{\partial u}{\partial s} + \dfrac{\partial F}{\partial v}\dfrac{\partial v}{\partial s} = 0. \end{cases}$$

If the Jacobian $\partial(u, v)/\partial(r, s)$ were different from zero, at some point of R, Cramer's Rule would state that the only values of $\partial F/\partial u$ and $\partial F/\partial v$ that satisfy system (2) are zero. This contradicts the assumption that $\partial F/\partial u$ and $\partial F/\partial v$ never vanish simultaneously. Therefore the Jacobian $\partial(u, v)/\partial(r, s)$ must vanish identically.

The theorem just established provides a means of establishing the functional *independence* of a set of functions: if the Jacobians specified in the theorem are not all identically zero, then the functions are *not* functionally related. However, the substantial and difficult theorem is the converse of this one. In § 1230 such a converse will be established. We should observe, however, that Theorem II, § 1212, is a simple special case of the converse: If there is just *one* function u, the Jacobians of our theorem reduce to the first partial derivatives of that function. If these all vanish, the function is dependent (by itself) since it is a constant and satisfies an equation of the form $F(u) \equiv u - c = 0$.

NOTE. Functional dependence should not be confused with *linear dependence* (which means that one of the functions is a linear combination $c_1 f_1 + c_2 f_2 + \cdots$ of the others). For example, $\sin x$ and $\cos x$ are functionally dependent ($\sin^2 x + \cos^2 x = 1$) and linearly independent. (Cf. § 1533.)

1225. EXERCISES

In these Exercises assume differentiability wherever appropriate.

In Exercises 1-2, if the variables u and v are defined implicitly as functions of the remaining variables, find their first (partial) derivatives.

1. $x^2 + 2uv = 1, x^2 - u^2 + v^2 = 1$.
2. $u + e^v = x + y, e^u + v = x - y$.

In Exercises 3-4, if the variables u, v, and w are defined implicitly as functions of the remaining variables, find their first (partial) derivatives.

3. $u + v + w = x, uv + uw + vw = x^2, uvw = 1$.
4. $u + \sin v = x, v + \sin w = y, w + \sin u = 1$.

In Exercises 5-6, find the indicated first partial derivatives, where the notation is that of the Notational Pitfall, § 1222.

5. If $x + y + z + u = 1, x^2 + y^2 + z^2 + u^2 = 1$, find $\left(\dfrac{\partial u}{\partial x}\right)_y, \left(\dfrac{\partial u}{\partial x}\right)_z, \left(\dfrac{\partial x}{\partial u}\right)_y, \left(\dfrac{\partial x}{\partial u}\right)_z$.

6. If $x + y + z + u + v = 1, x^2 + y^2 + z^2 + u^2 + v^2 = 1$, find $\left(\dfrac{\partial u}{\partial x}\right)_{yz}, \left(\dfrac{\partial u}{\partial x}\right)_{yv}, \left(\dfrac{\partial u}{\partial x}\right)_{zv}$.

In Exercises 7-9, verify that the Jacobians of the direct and inverse transformations are reciprocals, for the designated Exercise of § 1021.

7. Ex. 1. 8. Ex. 2. 9. Ex. 5.

10. Find the Jacobian of the transformation of Exercise 4, § 1021. Explain why, if the transformation is equal to its own inverse, its Jacobian is not equal to its own reciprocal and therefore have the value ± 1.

In Exercises 11-12, show that the given sets of functions are not functionally dependent.

11. $u = \sin x + \sin y, v = \sin (x + y)$.

12. $u = e^x + \ln y + xyz$, $v = \ln x + e^y + xyz$.

13. Prove the statement in the Notational Pitfall, § 1222, about the reciprocal nature of $\partial u/\partial x$ and $\partial x/\partial u$.

14. Prove the Theorem of § 1222.

15. State and prove the chain rule for transformations in n-dimensional spaces (§ 1223).

16. Prove the Theorem, § 1224, for arbitrary m and n.

★17. Let a space curve be given as the intersection of two surfaces $f(x, y, z) = g(x, y, z) = 0$. Show that the three Jacobians

$$(1) \qquad \frac{\partial(f, g)}{\partial(y, z)}, \quad \frac{\partial(f, g)}{\partial(z, x)}, \quad \frac{\partial(f, g)}{\partial(x, y)}$$

(assuming they are not all zero) are a set of direction numbers for the tangent to the curve. *Hint:* The tangent line is "tangent" to both surfaces.

★18. Let a surface $f(x, y, z) = 0$ have its coordinates prescribed parametrically: $x(u, v)$, $y(u, v)$, $z(u, v)$. Show that the three Jacobians

$$(2) \qquad \frac{\partial(y, z)}{\partial(u, v)}, \quad \frac{\partial(z, x)}{\partial(u, v)}, \quad \frac{\partial(x, y)}{\partial(u, v)}$$

(assuming they are not all zero) are a set of direction numbers for the normal to the surface. *Hint:* As an equation in u and v, $f(x, y, z) = 0$ is an identity. Show that the Jacobians (2) are proportional to f_1, f_2, f_3.

★19. If $u = f(x, y, z)$, and if z is defined as a function of x and y by the equation $g(x, y, z) = 0$, show that u as a function of x and y has first partials

$$\frac{\partial u}{\partial x} = \frac{\frac{\partial(f, g)}{\partial(x, z)}}{\frac{\partial g}{\partial z}}, \quad \frac{\partial u}{\partial y} = \frac{\frac{\partial(f, g)}{\partial(y, z)}}{\frac{\partial g}{\partial z}}.$$

★20. If $w = f(x, y, z, u, v)$, and if u and v are defined as functions of x, y, z by the equations $g(x, y, z, u, v) = h(x, y, z, u, v) = 0$, show that w as a function of x, y, z has first partials

$$\frac{\partial w}{\partial x} = \frac{\frac{\partial(f, g, h)}{\partial(x, u, v)}}{\frac{\partial(g, h)}{\partial(u, v)}}, \quad \frac{\partial w}{\partial y} = \frac{\frac{\partial(f, g, h)}{\partial(y, u, v)}}{\frac{\partial(g, h)}{\partial(u, v)}}, \quad \frac{\partial w}{\partial z} = \frac{\frac{\partial(f, g, h)}{\partial(z, u, v)}}{\frac{\partial(g, h)}{\partial(u, v)}}.$$

★21. Let u and v be functions of a single variable x which is a function of two variables r and s. Show that the Jacobian of u and v with respect to r and s vanishes identically.

★22. Let u and v be functions of three variables x, y, z, each of which is a function of the two variables r and s. Establish the following chain rule:

$$\frac{\partial(u, v)}{\partial(r, s)} = \frac{\partial(u, v)}{\partial(y, z)}\frac{\partial(y, z)}{\partial(r, s)} + \frac{\partial(u, v)}{\partial(z, x)}\frac{\partial(z, x)}{\partial(r, s)} + \frac{\partial(u, v)}{\partial(x, y)}\frac{\partial(x, y)}{\partial(r, s)}.$$

Hint: The complete expansion of the left-hand side consists of 18 terms, while that on the right contains 24 terms. However, 6 of these cancel out and the rest match up.

★23. If $u = f(x, y)$, and if x and y are defined as functions of y and z and of x and z, respectively, by the equation $g(x, y, z) = 0$, show that (cf. the Notational Pitfall, § 1222):

$$\left(\frac{\partial u}{\partial x}\right)_z - \left(\frac{\partial u}{\partial z}\right)_y = \left(\frac{\partial u}{\partial y}\right)_z \left(\frac{\partial y}{\partial x}\right)_z = -\left(\frac{\partial u}{\partial x}\right)_z \left(\frac{\partial x}{\partial z}\right)_y.$$

★**24.** If $x = \rho \cos \theta$, $y = \rho \sin \theta$, show that
$$\left(\frac{\partial u}{\partial x}\right)^2 + \left(\frac{\partial u}{\partial y}\right)^2 = \left(\frac{\partial u}{\partial \rho}\right)^2 + \frac{1}{\rho^2}\left(\frac{\partial u}{\partial \theta}\right)^2.$$

★**25.** Show that the Cauchy-Riemann differential equations
$$\frac{\partial u}{\partial x} = \frac{\partial v}{\partial y}, \quad \frac{\partial u}{\partial y} = -\frac{\partial v}{\partial x},$$
(cf. Ex. 23, § 1204) can be expressed as follows in polar coordinates:
$$\frac{\partial u}{\partial \rho} = \frac{1}{\rho}\frac{\partial v}{\partial \theta}, \quad \frac{\partial v}{\partial \rho} = -\frac{1}{\rho}\frac{\partial u}{\partial \theta}.$$

★**26.** Show that Laplace's equation in three rectangular coordinates,
$$\frac{\partial^2 u}{\partial x^2} + \frac{\partial^2 u}{\partial y^2} + \frac{\partial^2 u}{\partial z^2} = 0,$$
can be written as follows in cylindrical coordinates:
$$\frac{\partial^2 u}{\partial \rho^2} + \frac{1}{\rho^2}\frac{\partial^2 u}{\partial \theta^2} + \frac{1}{\rho}\frac{\partial u}{\partial \rho} + \frac{\partial^2 u}{\partial z^2} = 0.$$

★**27.** Show that Laplace's equation in three rectangular coordinates,
$$\frac{\partial^2 u}{\partial x^2} + \frac{\partial^2 u}{\partial y^2} + \frac{\partial^2 u}{\partial z^2} = 0,$$
can be written as follows in spherical coordinates:
$$\frac{1}{r^2 \sin^2 \phi}\left[\sin^2 \phi \frac{\partial}{\partial r}\left(r^2 \frac{\partial u}{\partial r}\right) + \sin \phi \frac{\partial}{\partial \phi}\left(\sin \phi \frac{\partial u}{\partial \phi}\right) + \frac{\partial^2 u}{\partial \theta^2}\right] = 0.$$

Hints: By solving the three equations resulting from differentiating $x = r \sin \phi \cos \theta$, $y = r \sin \phi \sin \theta$, $z = r \cos \phi$, obtain

(3) $\begin{cases} \dfrac{\partial u}{\partial x} = \dfrac{\partial u}{\partial r}\sin \phi \cos \theta + \dfrac{\partial u}{\partial \phi}\dfrac{\cos \phi \cos \theta}{r} - \dfrac{\partial u}{\partial \theta}\dfrac{\sin \theta}{r \sin \phi}, \\[6pt] \dfrac{\partial u}{\partial y} = \dfrac{\partial u}{\partial r}\sin \phi \sin \theta + \dfrac{\partial u}{\partial \phi}\dfrac{\cos \phi \sin \theta}{r} + \dfrac{\partial u}{\partial \theta}\dfrac{\cos \theta}{r \sin \phi}, \\[6pt] \dfrac{\partial u}{\partial z} = \dfrac{\partial u}{\partial r}\cos \phi \qquad - \dfrac{\partial u}{\partial \phi}\dfrac{\sin \phi}{r}. \end{cases}$

Then, following the model of Example 3, § 1223, put each equation of (3) to double service, and add. This is rather tedious, but it is finite—41 terms to combine.

★**28.** Prove the following theorem on "Lagrange multipliers": Assume that $f(x, y)$ is differentiable in a neighborhood $N_{(a, b)}$ of a point (a, b) and that it has an extreme value (maximum or minimum) there when the variables x and y are subjected to a constraint of the form $\phi(x, y) = 0$, where (i) ϕ is differentiable in $N_{(a, b)}$, (ii) the equation $\phi = 0$ defines y as a differentiable function $g(x)$ in a neighborhood of a such that $g(a) = b$, and (iii) $\phi_2(a, b) \neq 0$. Then there must exist a number λ (called a **Lagrange multiplier**) such that $f_1(a, b) + \lambda \phi_1(a, b) = f_2(a, b) + \lambda \phi_2(a, b) = 0$. (Cf. Exs. 29-37.) *Hint:* The function $h(x) \equiv f(x, g(x))$ has an extremum at $x = a$. Write out $h'(a) = 0$ and define $\lambda \equiv -f_2(a, b)/\phi_2(a, b)$.

★**29.** State and prove a theorem on Lagrange multipliers having to do with extrema of a function $f(x, y, \cdots, u)$ with a constraint $\phi(x, y, \cdots, u) = 0$, and leading to a set of equations

$$f_1(a, b, \cdots, u) + \lambda\phi_1(a, b, \cdots, u) = f_2(a, b, \cdots, u) + \lambda\phi_2(a, b, \cdots, u) = \cdots = 0.$$

(Cf. Ex. 28.)

★**30.** Prove the following theorem on Lagrange multipliers: Assume that $f(x, y, u, v)$ is differentiable in a neighborhood N of a point (a, b, c, d) and that it has an extreme value there when the variables x and y are subjected to the two constraints $\phi(x, y, u, v) = 0$, $\psi(x, y, u, v) = 0$, where (i) ϕ and ψ are differentiable in N, (ii) the equations $\phi = 0$, $\psi = 0$ define u and v as differentiable functions $u(x, y)$, $v(x, y)$ in a neighborhood of (a, b) such that $u(a, b) = c$, $v(a, b) = d$, and (iii) $\partial(\phi, \psi)/\partial(u, v)$ is nonzero at (a, b, c, d). Then there must exist two constants λ and μ (called **Lagrange multipliers**) such that at the point (a, b, c, d)

(4) $f_1 + \lambda\phi_1 + \mu\psi_1 = f_2 + \lambda\phi_2 + \mu\psi_2 = f_3 + \lambda\phi_3 + \mu\psi_3 = f_4 + \lambda\phi_4 + \mu\psi_4 = 0.$

(Cf. Exs. 28-29, 31.) *Hint*: Define λ and μ by the last two of equations (4). Then show that the results of substituting these values in the first two equations are equivalent to the two equations $f_1 + f_3 \frac{\partial u}{\partial x} + f_4 \frac{\partial v}{\partial x} = f_2 + f_3 \frac{\partial u}{\partial y} + f_4 \frac{\partial v}{\partial y} = 0,$
which result from the extremum character of $f(x, y, u(x, y), v(x, y))$.

★**31.** State a general form of the theorem on Lagrange multipliers (cf. Exs. 28-30), and prove it for the case of a function of five variables, with three constraints.

★**32.** Solve Example 3, § 1214, by the method of Lagrange multipliers (cf. Ex. 28).

★**33.** Use the method of Lagrange multipliers (Ex. 28) to show that the lengths of the semiaxes of the ellipse $ax^2 + 2bxy + cy^2 = 1$ are the square roots of the roots λ_1 and λ_2 of the equation

(5) $\begin{vmatrix} a\lambda - 1 & b\lambda \\ b\lambda & c\lambda - 1 \end{vmatrix} = 0,$†

and that the equation of the axis corresponding to $\lambda = \lambda_i$ is $(a\lambda_i - 1)x + b\lambda_i y = 0$ (or $b\lambda_i x + (c\lambda_i - 1)y = 0$), $i = 1, 2$. *Hint*: Maximize or minimize $f(x, y) = x^2 + y^2$ subject to the constraint $-(ax^2 + 2bxy + cy^2) + 1 = 0$, and observe that if x, y, λ satisfy Lagrange's equations, then $ax^2 + 2bxy + cy^2 = x(ax + by) + y(bx + cy) = (x^2 + y^2)/\lambda = 1.$

★**34.** Use the method of Exercise 33 to find the equations of the axes of the ellipse $8x^2 + 4xy + 5y^2 = 36$, and the simplified equation resulting from the elimination of the xy-term by rotation of coordinate axes.

★**35.** Show that the lengths of the semiaxes of the ellipsoid $ax^2 + by^2 + cz^2 + 2fyz + 2gxz + 2hxy = 1$ are the square roots of the roots of the equation

(6) $\begin{vmatrix} a\lambda - 1 & h\lambda & g\lambda \\ h\lambda & b\lambda - 1 & f\lambda \\ g\lambda & f\lambda & c\lambda - 1 \end{vmatrix} = 0.$

(Cf. Ex. 33.)

★**36.** Use the method of Exercise 35 to find the lengths of the semiaxes of the ellipsoid $5x^2 + 3y^2 + 3z^2 + 2xy - 2xz - 2yz - 12 = 0$.

† The roots of (5) are the reciprocals of the roots of the **characteristic equation** $\begin{vmatrix} a - \lambda & b \\ b & c - \lambda \end{vmatrix} = 0$ of the matrix $\begin{pmatrix} a & b \\ b & c \end{pmatrix}$ of the "quadratic form" $ax^2 + 2bxy + cy^2$

★★37. Prove that if $x \geq 0, y \geq 0, 0 < \alpha < 1, 0 < \beta < 1, \alpha + \beta = 1$, then $x^\alpha y^\beta \leq \alpha x + \beta y$. *Hints:* First, show that it may be assumed without loss of generality that $\alpha x + \beta y = 1$. Second, show that subject to this constraint the function $x^\alpha y^\beta$ is nonnegative, vanishing for $x = 0$ or $y = 0$, and therefore has a maximum value. Third, use a Lagrange multiplier to show that this maximum must be given by $x = y = 1$.

★★38. Prove the **Hölder inequality** for finite sums, where $p > 1, p > 1$, and $\frac{1}{p} + \frac{1}{q} = 1$:

(7) $$\left|\sum_{i=1}^{n} a_i b_i\right| \leq \left[\sum_{i=1}^{n} |a_i|^p\right]^{\frac{1}{p}} \left[\sum_{i=1}^{n} |b_i|^q\right]^{\frac{1}{q}}.$$

(This reduces to the Schwarz inequality (Ex. 43, § 107) when $p = q = 2$.) Extend to infinite series (cf. Ex. 14, § 717). *Hint:* First, show that it may be assumed without loss of generality that $a_i \geq 0$, $b_i \geq 0$, and $\sum a_i^p = \sum b_i^q = 1$. Second, use Ex. 37 with $x = a_i^p$, $y = b_i^q$, $\alpha = \frac{1}{p}$, $\beta = \frac{1}{q}$. Third, add the resulting inequalities.

★★39. Prove the **Minkowski inequality** for finite sums, where $p \geq 1$:

(8) $$\left[\sum_{i=1}^{n} |a_i + b_i|^p\right]^{\frac{1}{p}} \leq \left[\sum_{i=1}^{n} |a_i|^p\right]^{\frac{1}{p}} + \left[\sum_{i=1}^{n} |b_i|^p\right]^{\frac{1}{p}}.$$

Extend to infinite series. (Cf. Ex. 44, § 107, Ex. 14, § 717.) *Hint:* For $p > 1$ define $q = p/(p-1)$, assume $a_i \geq 0, b_i \geq 0$, and apply the Hölder inequality (Ex. 38):

$\sum(a_i + b_i)^p = \sum a_i(a_i + b_i)^{p-1} + \sum b_i(a_i + b_i)^{p-1}$

$\leq [\sum a_i^p]^{\frac{1}{p}}[\sum(a_i + b_i)^p]^{\frac{1}{q}} + [\sum b_i^p]^{\frac{1}{p}}[\sum(a_i + b_i)^p]^{\frac{1}{q}}.$

★★40. Prove the **Hölder inequality** and the **Minkowski inequality** for Riemann integrals:

(9) $$\left|\int_a^b f(x)g(x)\,dx\right| \leq \left[\int_a^b |f(x)|^p\,dx\right]^{\frac{1}{p}} \left[\int_a^b |g(x)|^q\,dx\right]^{\frac{1}{q}}$$
$$\left(p > 1, q > 1, \frac{1}{p} + \frac{1}{q} = 1\right),$$

(10) $$\left[\int_a^b |f(x) + g(x)|^p\,dx\right]^{\frac{1}{p}} \leq \left[\int_a^b |f(x)|^p\,dx\right]^{\frac{1}{p}} + \left[\int_a^b |g(x)|^p\,dx\right]^{\frac{1}{p}}$$
$$(p \geq 1).$$

State the hypotheses and conclusions precisely. (Cf. Exs. 29-30, 55, § 503.) *Hint:* First assume without loss of generality that $f(x) \geq 0, g(x) \geq 0$. Second, prove the inequalities for step-functions where $f(x)$ and $g(x)$ have the values α_i and β_i on a subinterval of length Δx_i, using for the Hölder inequality of Ex. 38, $a_i = \alpha_i \Delta x_i^{\frac{1}{p}}, b_i = \beta_i \Delta x_i^{\frac{1}{q}}$, and for the Minkowski inequality of Ex. 39, $a_i = \alpha_i \Delta x_i^{\frac{1}{p}}$, $b_i = \beta_i \Delta x_i^{\frac{1}{p}}$. Third, use approximations by step functions to infer (9) and (10). Alternatively, use the methods of Exs. 38-39 directly, with the aid of Ex. 57, § 503.

★1226. DIFFERENTIATION UNDER THE INTEGRAL SIGN. LEIBNITZ'S RULE

In this section we shall study functions of the form

$$(1) \qquad F(x) = \int_{\phi(x)}^{\psi(x)} f(x, y) \, dy,$$

and investigate their continuity and differentiability properties. We shall first assume that ϕ and ψ are constants, and then permit them to be more general functions of x. It should be appreciated that (1) is a function of x alone, and not of y.

Theorem I. *If $f(x, y)$ is continuous on the closed rectangle $A : a \leq x \leq b$, $c \leq y \leq d$, then $F(x) = \int_c^d f(x, y) \, dy$ exists and is continuous for $a \leq x \leq b$.*
(Cf. Ex. 7, § 1227.)

Proof. Existence follows from continuity of $f(x, y)$. Since A is compact $f(x, y)$ is uniformly continuous on A (§ 1017). Therefore, if ϵ is an arbitrary positive number, we can find a positive number δ such that whenever $|x_1 - x_2| < \delta$ and $|y_1 - y_2| < \delta$ (and *a fortiori* when $|x_1 - x_2| < \delta$ and $y_1 = y_2$), $|f(x_2, y_2) - f(x_1, y_1)| < \epsilon/(d - c)$. Accordingly, we let $|x_1 - x_2| < \delta$ and find (cf. Ex. 2, § 503):

$$|F(x_2) - F(x_1)| \leq \int_c^d |f(x_2, y) - f(x_1, y)| dy < \int_c^d \frac{\epsilon}{d - c} dy = \epsilon.$$

In discussing the differentiability of $F(x)$, defined in (1), we shall consider the domain of definition of the function $f(x, y)$ to be limited to the rectangle $A : a \leq x \leq b$, $c \leq y \leq d$. The partial derivative $f_1(x, y)$, therefore, when considered along the vertical sides of A, $x = a$, $x = b$, is given by a one-sided limit. For example:

$$f_1(a, c) = \lim_{h \to 0+} [f(a + h, c) - f(a, c)]/h.$$

Theorem II. Leibnitz's Rule. *If $f(x, y)$ and $f_1(x, y)$ exist and are continuous on the closed rectangle $A : a \leq x \leq b$, $c \leq y \leq d$, then the function $F(x) = \int_c^d f(x, y) \, dy$ is differentiable for $a \leq x \leq b$, and*

$$(2) \qquad F'(x) = \int_c^d f_1(x, y) \, dy.$$

(Cf. Ex. 7, § 1227.)

Proof. Since A is compact, $f_1(x, y)$ is uniformly continuous there. Let a positive number ϵ be given, and choose $\delta > 0$ so that $|x_1 - x_2| < \delta$ and $|y_1 - y_2| < \delta$ imply $|f_1(x_2, y_2) - f_1(x_1, y_1)| < \epsilon/(d - c)$ (and then let $y_1 = y_2$). Letting h be an arbitrary number such that $0 < |h| \leq \delta$, and

§ 1226] DIFFERENTIATION UNDER INTEGRAL SIGN   417

letting x_0 be an arbitrary point of the interval $[a, b]$, we form a difference which we wish to show is less than ϵ (once this is established, the proof is complete):

(3) $$\left| \frac{F(x_0 + h) - F(x_0)}{h} - \int_c^d f_1(x_0, y) \, dy \right|.$$

Using the definition of $F(x)$ and the law of the mean for a function of one variable (the θ depending on y), we have for (3):

$$\left| \frac{1}{h} \int_c^d [f(x_0 + h, y) - f(x_0, y)] \, dy - \int_c^d f_1(x_0, y) \, dy \right|$$

$$= \left| \int_c^d f_1(x_0 + \theta h, y) \, dy - \int_c^d f_1(x_0, y) \, dy \right|$$

$$\leq \int_c^d |f_1(x_0 + \theta h, y) - f_1(x_0, y)| \, dy < \int_c^d \frac{\epsilon}{d - c} \, dy = \epsilon.$$

Note. The function $f_1(x_0 + \theta(y)h, y)$, in spite of the unknown nature of $\theta(y)$, is an integrable function of y since it is equal to $\frac{1}{h}[f(x_0 + h, y) - f(x_0, y)]$, a continuous function of y. Consequently, $f_1(x_0 + \theta h, y) - f_1(x_0, y)$ is also integrable, and all integrals appearing in the preceding proof do exist.

Example 1. Let $F(x) = \int_1^2 \sin(xe^y) \, dy$. Then $F'(x) = \int_1^2 e^y \cos(xe^y) \, dy$. In this case the resulting integration can be performed explicitly:

$$F'(x) = \frac{1}{x} \int_1^2 \cos(xe^y) \, d(xe^y) = \frac{1}{x} [\sin(xe^y)]_1^2 = \frac{\sin(e^2 x) - \sin(ex)}{x}.$$

Theorems I and II can be generalized as follows (for hints on the proof, cf. Exs. 8, 9, § 1227; for generalizations, cf. Exs. 10, 11, § 1227):

Theorem III. *Let $\phi(x)$ and $\psi(x)$ be continuous for $a \leq x \leq b$ and for such x let $\phi(x) \leq \psi(x)$. Let A designate the set of all points such that $a \leq x \leq b$ and $\phi(x) \leq y \leq \psi(x)$. If $f(x, y)$ is continuous in a region R containing A, then $F(x) = \int_{\phi(x)}^{\psi(x)} f(x, y) \, dy$ is continuous for $a \leq x \leq b$.* (Cf. Figure 1204.)

Theorem IV. General form of Leibnitz's Rule. *Under the assumptions of Theorem III and the additional hypotheses that $\phi(x)$ and $\psi(x)$ are continuously differentiable for $a \leq x \leq b$, and that $f_1(x, y)$ exists and is continuous in R, the function $F(x) = \int_{\phi(x)}^{\psi(x)} f(x, y) \, dy$ is continuously differentiable, and*

(4) $$F'(x) = f(x, \psi(x))\psi'(x) - f(x, \phi(x))\phi'(x) + \int_{\phi(x)}^{\psi(x)} f_1(x, y) \, dy.$$

A special case of Theorem IV was given in Exercise 2, § 506.

The extension of Leibnitz's Rule (Theorem II) to improper integrals is presented in Chapter 14.

Example 2. If $F(x) = \int_{x^2}^{x^3} \tan(xy^2)\, dy$,

$$F'(x) = 3x^2 \tan(x^7) - 2x \tan(x^5) + \int_{x^2}^{x^3} y^2 \sec^2(xy^2)\, dy.$$

★1227. EXERCISES

In Exercises 1-2, evaluate the indicated derivative.

★1. $\dfrac{d}{dx}\displaystyle\int_0^x \dfrac{\sin xy}{y}\, dy.$ ★2. $\dfrac{d}{dx}\displaystyle\int_{-x}^x \dfrac{1 - e^{-xy}}{y}\, dy.$

In Exercises 3-4, verify Leibnitz's Rule.

★3. $\displaystyle\int_x^{x^2} (x^2 + y^2)\, dy.$ ★4. $\displaystyle\int_0^x \sqrt{x^2 - y^2}\, dy.$

★5. Define $I_n(x) = \displaystyle\int_a^x (x-t)^{n-1} f(t)\, dt$, $n = 1, 2, \cdots$. Prove that

$$\frac{d^n I_n}{dx^n} = (n-1)!\, f(x).$$

★6. Show that if $y(x)$ satisfies the "integral equation"

$$y(x) = 4\int_0^x (t-x)y(t)\, dt - \int_0^x (t-x)f(t)\, dt,$$

then $y(x)$ satisfies the differential equation and boundary conditions

$$\frac{d^2 y}{dx^2} + 4y = f(x),$$
$$y(0) = 0, \quad y'(0) = 0.$$

★7. Prove that Theorems I and II, § 1226, remain valid if the interval $[a, b]$ for x is replaced by an arbitrary interval, finite or infinite.

§ 1228]   THE IMPLICIT FUNCTION THEOREM   419

★8. Prove Theorem III, § 1226. *Hints:* For a given $x_0 \in [a, b]$ let $\eta > 0$ be such that the rectangle $B: |x - x_0| \leq \eta$, $\phi(x_0) - \eta \leq y \leq \psi(x_0) + \eta$ is contained in R, and let γ be a constant between $\phi(x_0)$ and $\psi(x_0)$. First prove that $G(x, u) \equiv \int_\gamma^u f(x, y)\, dy$ is continuous in B. Then write $\int_u^v f(x, y)\, dy = G(x, v) - G(x, u)$.

★9. Prove Theorem IV, § 1226. *Hints:* Proceed as in the hints of Ex. 8, and prove that $G(x, u)$ is continuously differentiable. Then use the chain rule.

★10. Same as Exercise 7 for Theorems III and IV, § 1226.

★★11. Extend the Theorems of § 1226 to functions of several variables (the integration applying to only one).

★1228. THE IMPLICIT FUNCTION THEOREM

In this section we study conditions which guarantee the existence of an implicitly defined function, when one equation is given, and an implicitly defined set of functions, when a system of equations is given. We start with the simplest case, that of a function of one variable, $y(x)$ defined by an equation in two variables, $f(x, y) = 0$. To illustrate some of the difficulties that may arise we consider first a few simple examples.

Example 1. If the given equation is $f(x, y) \equiv x - e^y = 0$, the function $f(x, y)$ has all of the continuity and differentiability conditions one could desire, for *all* real values of x and y. This equation defines a single-valued function $y = \phi(x) \equiv \ln x$, satisfying the given equation identically, but this function is defined only for *positive* x.

Example 2. The function $f(x, y) \equiv 2x - \sin x$ is well-behaved (in some senses at least), but the equation $f(x, y) = 0$ defines y as a function of x only in the sense of a completely unrestricted range!

Example 3. Let $f(x, y) \equiv x^4 - y^2$. Then the equation $f(x, y) = 0$ defines y as a *double-valued* function of $x(x \neq 0)$. It also defines y as a *single-valued* function of x, but it does so in many ways. Four of these are $y = x^2$, $y = -x^2$, $y = x|x|$, and $y = -x|x|$, and if discontinuities are permitted, many more solutions are possible (for example, $y = x^2$ if x is rational, and $y = -x^2$ if x is irrational). However, if we choose a point on the original curve *not the origin*, there will be determined *in a suitably restricted neighborhood* of this point the graph of a unique function $y = \phi(x)$ satisfying the identity $f(x, \phi(x)) = 0$. (Cf. the Note below.)

We shall state and prove two special cases of the Implicit Function Theorem. The general case, and its proof, should then be clear (cf. Ex. 2, § 1231). For an explicit statement and proof of the general theorem, see C. Carathéodory, *Variationsrechnung und Partielle Differentialgleichungen Erster Ordnung* (Leipzig, B. G. Teubner, 1935), pp. 9-12.

The details in the two proofs that follow are designed for generalization and are therefore not necessarily the most economical for either specific theorem. Notice how hypothesis *(iii)*, in the simplest case (Theorem 1), precludes the situation of Example 2, where the variable y does not appear, and the difficulty associated with the origin in Example 3, where uniqueness fails.

Theorem I. Implicit Function Theorem. *If*

(i) $f(x, y)$ *and* $f_2(x, y)$ *are continuous in a neighborhood* $N_{(a, b)}$ *of the point* (a, b) *(in* E_2*),*
(ii) $f(a, b) = 0$,
(iii) $f_2(a, b) \neq 0$,

then there exist neighborhoods N_a *of* a *and* N_b *of* b *(in* E_1*) and a function* $\phi(x)$ *defined in* N_a *such that*

(iv) *for all x in the neighborhood* N_a, $\phi(x) \in N_b$, $(x, \phi(x)) \in N_{(a,b)}$, *and* $f(x, \phi(x)) = 0$,
(v) $\phi(x)$ *is uniquely determined by* (iv); *that is, if* $\Phi(x)$ *is defined in* N_a *and has properties* (iv), *then* $\Phi(x) = \phi(x)$ *throughout* N_a,
(vi) $\phi(a) = b$,
(vii) $\phi(x)$ *is continuous in* N_a,
(viii) *if* $f_1(x, y)$ *exists and is continuous in* $N_{(a, b)}$, *then* $\phi'(x)$ *exists and is continuous in* N_a, *and*

$$\phi'(x) = -\frac{f_1(x, \phi(x))}{f_2(x, \phi(x))}.$$

Proof. By continuity and (iii), there exist closed neighborhoods of the points a and b:

$$A: |x - a| \leq \alpha, \quad B: |y - b| \leq \beta,$$

such that $x \in A$ and $y \in B$ imply $(x, y) \in N_{(a, b)}$ and $f_2(x, y) \neq 0$.

We first establish uniqueness or, more precisely, that for each $x \in A$ there is *at most one* $y \in B$ such that $f(x, y) = 0$. For, if y' and y'' were two such y, for some x, then by the Law of the Mean (§ 405):

$$0 = f(x, y'') - f(x, y') = f_2(x, y_1)(y'' - y'),$$

where y_1 is between y' and y'' and therefore belongs to B. This contradicts the basic property assumed for the sets A and B.

We now denote by C the compact two-point set of numbers η such that $|\eta - b| = \beta$ (that is, $\eta = b \pm \beta$), and define the nonnegative function

(1) $$F(x, y) \equiv (f(x, y))^2.$$

Then, since $C \subset B$, and by the uniqueness just established, $\eta \in C$ implies $f(a, \eta) \neq 0$. Therefore there exists a positive number ϵ such that for $\eta \in C$,

(2) $$\begin{cases} F(a, \eta) \geq \epsilon, \\ F(a, b) = 0. \end{cases}$$

We infer from (2) and the continuity of $F(x, y)$, as a function of x, at the three values of y: $y = b$, $y = \eta \in C$, the existence of a positive number $\delta \leq \alpha$ such that $|x - a| \leq \delta$ implies

(3) $$\begin{cases} F(x, \eta) > \tfrac{3}{4}\epsilon, \\ F(x, b) < \tfrac{1}{2}\epsilon. \end{cases}$$

If N_δ denotes the neighborhood of a: $|x - a| < \delta$, and D its closure, we can infer from (3) that for each fixed $x \in D$, $F(x, y)$ as a function of y on the compact set B has a minimum value attained at an *interior* point $\bar{y}$ of B. By Theorem I, § 409,

$$F_2(x, \bar{y}) = 2f(x, \bar{y}) f_2(x, \bar{y}) = 0,$$

and hence $f(x, \bar{y}) = 0$.

With the uniqueness already established as the first part of this proof, and with N_b defined as the interior of B, we now have proved the existence of a function $y = \bar{y} = \phi(x)$ satisfying conclusions (*iv*), (*v*), and (*vi*) of the theorem.

To prove continuity of $\phi(x)$, we write, with x and $x + \Delta x$ in N_a, and with $\phi = \phi(x)$ and $\Delta\phi = \phi(x + \Delta x) - \phi(x)$:

$0 = f(x + \Delta x, \phi + \Delta\phi) - f(x, \phi)$

$= [f(x + \Delta x, \phi + \Delta\phi) - f(x, \phi + \Delta\phi)] + [f(x, \phi + \Delta\phi) - f(x, \phi)].$

By the Law of the Mean, we can write this last bracketed expression as $f_2(x, y_1) \Delta\phi$, where y_1 is between ϕ and $\phi + \Delta\phi$, and solve for $\Delta\phi$:

(4) $$\Delta\phi = -\frac{f(x + \Delta x, \phi + \Delta\phi) - f(x, \phi + \Delta\phi)}{f_2(x, y_1)}.$$

By the uniform continuity of $f(x, y)$ on the closed rectangle: $x \in D, y \in B$, we can conclude that the numerator on the right of (4) tends to 0 with Δx, while the denominator is bounded from 0. Therefore $\Delta\phi \to 0$ and ϕ is continuous.

Finally, the formula of (*viii*) follows from (4) by one further application of the Law of the Mean (to the numerator of the fraction). The continuity of $\phi'(x)$ is implied by its representation as a quotient of continuous functions.

Theorem II. Implicit Function Theorem. *Let*

(5) $$\begin{cases} f(x, y, z, u, v) = 0, \\ g(x, y, z, u, v) = 0, \end{cases}$$

be a system of two equations, to be solved for the two unknowns u and v in terms of the remaining variables. If

(i) f, g, f_4, f_5, g_4, and g_5 are continuous in a neighborhood $N_{(a, b, c, d, e)}$ of the point (a, b, c, d, e) (in E_5),

(ii) $f(a, b, c, d, e) = g(a, b, c, d, e) = 0$.

(iii) the Jacobian $J(x, y, z, u, v) = \partial(f, g)/\partial(u, v) = f_4 g_5 - f_5 g_4$ is nonzero at (a, b, c, d, e),

then there exist neighborhoods $N_{(a,b,c)}$ of (a, b, c) (in E_3) and $N_{(d,e)}$ of (d, e) (in E_2) and functions $u = \phi(x, y, z)$, $v = \psi(x, y, z)$ defined in $N_{(a,b,c)}$ such that

(iv) for all (x, y, z) in the neighborhood $N_{(a,b,c)}$, $(\phi(x, y, z), \psi(x, y, z))$ $\in N_{(d,e)}$, $(x, y, z, \phi(x, y, z), \psi(x, y, z)) \in N_{(a,b,c,d,e)}$, and $f(x, y, z, \phi(x, y, z), \psi(x, y, z)) = g(x, y, z, \phi(x, y, z), \psi(x, y, z)) = 0$,

(v) ϕ and ψ are uniquely determined by (iv); that is, if Φ and Ψ are defined in $N_{(a,b,c)}$ and have properties (iv), then $\Phi = \phi$, $\Psi = \psi$ throughout $N_{(a,b,c)}$,

(vi) $\phi(a, b, c) = d$, $\psi(a, b, c) = e$,

(vii) ϕ and ψ are continuous in $N_{(a,b,c)}$,

(viii) if f_1, f_2, f_3, g_1, g_2, and g_3 exist and are continuous in $N_{(a,b,c,d,e)}$, then ϕ and ψ have continuous first derivatives, which are given by formulas of type (7) and (8), § 1222.

Proof. By continuity and (iii), there exist closed neighborhoods of the points (a, b, c) and (d, e):

$A: (x - a)^2 + (y - b)^2 + (z - c)^2 \leq \alpha^2$, $B: (u - d)^2 + (v - e)^2 \leq \beta^2$,

such that $(x, y, z) \in A$, $(u_i, v_i) \in B$, $i = 1, 2$, imply $(x, y, z, u_i, v_i) \in N_{(a,b,c,d,e)}$, $i = 1, 2$, and

(6) $\quad \begin{vmatrix} f_4(x, y, z, u_1, v_1) & f_5(x, y, z, u_1, v_1) \\ g_4(x, y, z, u_2, v_2) & g_5(x, y, z, u_2, v_2) \end{vmatrix} \neq 0.$

(The determinant $qt - rs$ is a continuous function of the four variables q, r, s, t.)

In order to prove that for any $(x, y, z) \in A$ there is *at most* one $(u, v) \in B$ such that

(7) $\quad F(x, y, z, u, v) \equiv [f(x, y, z, u, v)]^2 + [g(x, y, z, u, v)]^2$

vanishes, we assume there exist *two* such points (u', v') and (u'', v''), for some $(x, y, z) \in A$. The Law of the Mean (§ 1212), applied to the two equations

(8) $\quad f(x, y, z, u'', v'') - f(x, y, z, u', v')$
$\qquad\qquad = g(x, y, z, u'', v'') - g(x, y, z, u', v') = 0,$

gives, for suitable points (u_1, v_1) and (u_2, v_2) of the straight line segment joining (u', v') and (u'', v'') (and hence belonging to the convex closed neighborhood B—cf. Ex. 38, § 1025):

(9) $\quad \begin{cases} f_4(x, y, z, u_1, v_1)(u'' - u') + f_5(x, y, z, u_1, v_1)(v'' - v') = 0, \\ g_4(x, y, z, u_2, v_2)(u'' - u') + g_5(x, y, z, u_2, v_2)(v'' - v') = 0. \end{cases}$

Since $(u'' - u')^2 + (v'' - v')^2 \neq 0$, this contradicts (6).

We now denote by C the compact set of all points (η, ζ) in E_2 such that

(10) $$(\eta - d)^2 + (\zeta - e)^2 = \beta^2.$$

Then, since $C \subset B$, and by the uniqueness just established, $(\eta, \zeta) \in C$ implies $F(a, b, c, \eta, \zeta) > 0$. Therefore, by the compactness of C, there exists a positive number ϵ such that for all $(\eta, \zeta) \in C$:

(11) $$\begin{cases} F(a, b, c, \eta, \zeta) \geq \epsilon, \\ F(a, b, c, d, e) = 0. \end{cases}$$

We infer from (11) and the continuity of $F(x, y, z, u, v)$ (Ex. 1, § 1231) the existence of a positive number $\delta \leq \alpha$ such that if D is the closed neighborhood (with interior $N_{(a, b, c)}$) defined by

(12) $$D: (x - a)^2 + (y - b)^2 + (z - c)^2 \leq \delta^2,$$

then $(x, y, z) \in D$ and $(\eta, \zeta) \in C$ imply

(13) $$\begin{cases} F(x, y, z, \eta, \zeta) > \tfrac{1}{2}\epsilon, \\ F(x, y, z, d, e) < \tfrac{1}{2}\epsilon. \end{cases}$$

Therefore, for each fixed $(x, y, z) \in D$, $F(x, y, z, u, v)$ as a function of (u, v) on the compact set B has a minimum value attained at an *interior* point $(\bar{u}, \bar{v})$ of B. By § 1214, $F_4(x, y, z, \bar{u}, \bar{v}) = F_5(x, y, z, \bar{u}, \bar{v}) = 0$, and

(14) $$\begin{cases} f(x, y, z, \bar{u}, \bar{v}) f_4(x, y, z, \bar{u}, \bar{v}) + g(x, y, z, \bar{u}, \bar{v}) g_4(x, y, z, \bar{u}, \bar{v}) = 0, \\ f(x, y, z, \bar{u}, \bar{v}) f_5(x, y, z, \bar{u}, \bar{v}) + g(x, y, z, \bar{u}, \bar{v}) g_5(x, y, z, \bar{u}, \bar{v}) = 0. \end{cases}$$

Since $J(x, y, z, \bar{u}, \bar{v}) \neq 0$ for $(x, y, z) \in A$ and $(\bar{u}, \bar{v}) \in B$, it follows that

(15) $$F(x, y, z, \bar{u}, \bar{v}) = f(x, y, z, \bar{u}, \bar{v}) = g(x, y, z, \bar{u}, \bar{v}) = 0.$$

With the uniqueness already established as the first part of this proof, and with $N_{(d, e)}$ defined as the interior of B, we now have proved the existence of two functions

(16) $$u = \bar{u} = \phi(x, y, z), \quad v = \bar{v} = \psi(x, y, z)$$

satisfying conclusions (iv), (v), and (vi) of the theorem.

To prove continuity of ϕ and ψ, we write

$$\begin{aligned} 0 &= [f(x + \Delta x, y + \Delta y, z + \Delta z, \phi + \Delta\phi, \psi + \Delta\psi) \\ &\quad - f(x, y, z, \phi + \Delta\phi, \psi + \Delta\psi)] \\ &\quad + [f(x, y, z, \phi + \Delta\phi, \psi + \Delta\psi) - f(x, y, z, \phi, \psi + \Delta\psi)] \\ &\quad + [f(x, y, z, \phi, \psi + \Delta\psi) - f(x, y, z, \phi, \psi)] \\ &= [f(x + \Delta x, y + \Delta y, z + \Delta z, \phi + \Delta\phi, \psi + \Delta\psi) \\ &\quad - f(x, y, z, \phi + \Delta\phi, \psi + \Delta\psi)] \\ &\quad + f_4(x, y, z, u_1, \psi + \Delta\psi) \Delta\phi + f_5(x, y, z, \phi, v_2) \Delta\psi, \end{aligned}$$

where u_1 and v_2 lie between ϕ and $\phi + \Delta\phi$, and ψ and $\psi + \Delta\psi$, respectively, with a similar expression for g. By (6), Cramer's Rule gives $\Delta \phi$ and $\Delta \psi$ as quotients of determinants, and these quotients $\to 0$ as $\Delta x^2 + \Delta y^2 + \Delta z^2 \to 0$, by uniform continuity of f on the compact set $(x, y, z) \in D$, $(u, v) \in B$.

Finally, the formulas of $(viii)$ follow from the preceding application of Cramer's Rule if one sets in turn $\Delta y = \Delta z = 0$, $\Delta x = \Delta z = 0$, and $\Delta x = \Delta y = 0$, with suitable use of the Law of the Mean. The continuity of the partial derivatives is implied by their representation as quotients of continuous functions.

NOTE. An alternative formulation of the Implicit Function Theorem is obtained, as a Corollary to the Theorem as stated above, by expressing the uniqueness of the solution in terms of *continuity*. Specifically, in the statement of Theorem I (II) all reference to the neighborhood N_δ ($N_{(d,e)}$) may be omitted if the assumption of properties (vi) and (vii) for the function Φ (functions Φ and Ψ) of part (v) is made. The reason for this, for Theorem II to be precise, is that if $\phi, \Phi, \psi,$ and Ψ are *all* continuous in $N_{(a, b, c)}$ and if $\phi(a, b, c) = \Phi(a, b, c) = d$ and $\psi(a, b, c) = \Psi(a, b, c) = e$, then the membership $(\phi(x, y, z), \psi(x, y, z)) \in N_{(d, e)}$ implies the membership $(\Phi(x, y, z), \Psi(x, y, z)) \in N_{(d,e)}$, for all points (x, y, z) of $N_{(a, b, c)}$. Detailed hints are given in Exercise 3, § 1231. In Example 3, above, for any point on the original curve, not the origin, a unique *continuous* solution exists if the values of x alone, without regard to y, are sufficiently restricted.

★1229. EXISTENCE THEOREM FOR INVERSE TRANSFORMATIONS

We shall restrict the details of our discussion in this section to transformations from a two-dimensional space E_2 to a two-dimensional space E_2. The general existence theorem for n dimensions, and its proof, are completely analogous to those for the case $n = 2$, and need not be given explicitly. (Cf. Ex. 4, § 1231.)

Consider a transformation, then, of the form

$$(1) \quad \begin{cases} x = x(u, v), \\ y = y(u, v). \end{cases}$$

Certainly it is not *always* possible to solve such a system for u and v in terms of x and y. For example, the system $x = u + v$, $y = 2u + 2v$ has no solution at all unless $2x = y$, and then not a unique solution in any sense. We now state a theorem that guarantees, under certain conditions, the existence of an inverse transformation.

Theorem. *If*

(i) $x(u, v)$ and $y(u, v)$ are continuously differentiable in a neighborhood of the point (c, d),

(ii) the Jacobian $J = \partial(x, y)/\partial(u, v)$ is nonzero at (c, d),

§ 1229]   INVERSE TRANSFORMATIONS

then there exist a neighborhood $N_{(a,b)}$ of the point $(a, b) = (x(c, d), y(c, d))$ (in E_2) and functions $u = u(x, y)$, $v = v(x, y)$, defined in $N_{(a,b)}$ and such that
(iii) *the following two identities hold throughout $N_{(a,b)}$:*

$$x(u(x, y), v(x, y)) = x,$$
$$y(u(x, y), v(x, y)) = y,$$

(iv) $u(a, b) = c$ and $v(a, b) = d$,
(v) $u(x, y)$ and $v(x, y)$ are continuous in $N_{(a,b)}$,
(vi) *if $\phi(x, y)$ and $\psi(x, y)$ (in place of $u(x, y)$ and $v(x, y)$, respectively) have properties (iii), (iv), (v), then $\phi(x, y) = u(x, y)$ and $\psi(x, y) = v(x, y)$ in $N_{(a,b)}$,*
(vii) $u(x, y)$ and $v(x, y)$ *are continuously differentiable in $N_{(a,b)}$, with partial derivatives*

$$u_x = y_v/J, \quad u_y = -x_v/J, \quad v_x = -y_u/J, \quad v_y = x_u/J,$$

(viii) *the set B of all points in the uv-plane that are images of points of $N_{(a,b)}$ under the inverse transformation is an open set, and the correspondence between points of B and points of $N_{(a,b)}$ is one-to-one.*

Proof. We define the system

(2) $\quad\begin{cases} f(x, y, u, v) \equiv x - x(u, v) = 0, \\ g(x, y, u, v) \equiv y - y(u, v) = 0, \end{cases}$

and apply the Implicit Function Theorem for a system of two equations, which is applicable since the two Jacobians $\partial(f, g)/\partial(u, v)$ and $\partial(x, y)/\partial(u, v)$ are identical. That theorem, in the form of the Note, § 1228, gives (iii), (iv), (v), (vi), and part of (vii) immediately, while the formulas of (vii) were obtained in § 1223. Conclusion (viii) is somewhat more difficult, but not much. Notice that the part of the theorem already established states, in part, that the point (a, b) into which (c, d) is carried by the transformation is completely surrounded by a neighborhood $N_{(a,b)}$ of points that are images of points in the original neighborhood of (c, d), under the given transformation. This fact, applied to the inverse transformation at an arbitrary point of B establishes the desired openness. The fact that the transformation is one-to-one follows from the fact that all of the transformation functions considered are single-valued.

The argument used in proving conclusion (viii) also establishes the following (cf. Ex. 16, § 1028):

Corollary. *If the transformation functions (1) are continuously differentiable over an open set B in the uv-plane, and if the Jacobian $\partial(x, y)/\partial(u, v)$ of this transformation does not vanish in B, then the image A of B by this transformation is an open set in the xy-plane. In other words, if we define an **open mapping**, defined on a region, to be one that always maps open sets onto open sets, then any transformation of the type just considered is open.* (Cf. Ex. 5, § 1231.)

NOTE 1. A transformation satisfying the conditions of the Corollary need not be one-to-one. Example: $x = u \cos v, y = u \sin v$, B the rectangle $1 < u < 3$, $0 < v < 3\pi$.

NOTE 2. If the Jacobian $\partial(x, y)/\partial(u, v)$ vanishes, the set A of the Corollary need not be open. Example: $x = u \cos v, y = u \sin v$, B the rectangle $-1 < u < 1, 0 < v < \pi/2$.

★1230. SUFFICIENCY CONDITIONS FOR FUNCTIONAL DEPENDENCE

A set of necessary conditions for functional dependence was obtained in § 1224. In a sense which we shall discuss in this section these conditions, which involve the identical vanishing of certain Jacobians, are also sufficient. The simplest case, that of *one* function, was established in Theorem II, § 1212, and alluded to at the end of § 1224. In the present section we shall give full details for only one case, that of two functions of two variables. (Cf. Exs. 6, 7, § 1231.) We first look at a simple example:

Example. Let $u = f(x, y) = x + y, v = g(x, y) = 2x + 2y$. The Jacobian $\frac{\partial(u, v)}{\partial(x, y)} = \begin{vmatrix} 1 & 1 \\ 2 & 2 \end{vmatrix} = 0$. The two functions, u and v, are related by the equation $v = 2u$. The pair of equations $u = x + y, v = 2x + 2y$ defines a transformation that carries the entire xy-plane onto the line $v = 2u$ in the uv-plane. The transformation is, in this sense, *degenerate*.

Theorem. *If*

(i) *the functions* $u = f(x, y)$ *and* $v = g(x, y)$ *are continuously differentiable in a region R in the xy-plane,*

(ii) *the Jacobian $\partial(u, v)/\partial(x, y)$ vanishes identically in R,*

(iii) *(a, b) is a point of R where not all first partial derivatives u_x, u_y, v_x, v_y vanish,*

then there exist a neighborhood $N_{(c, d)}$ of the point $(c, d) \equiv (f(a, b), g(a, b))$ and a function $F(u, v)$ defined in $N_{(c, d)}$ such that

(iv) *F is continuously differentiable in $N_{(c, d)}$,*

(v) *at least one of the two partial derivatives F_u and F_v is nonzero throughout $N_{(c, d)}$,*

(vi) *there exists a neighborhood $N_{(a, b)}$ in which $f(x, y)$ and $g(x, y)$ are functionally dependent by means of F:*

$$F(u, v) = F(f(x, y), g(x, y)) = 0.$$

Proof. The idea of the proof is that if we can solve for one of the variables (say y) in one of the equations (say the first) and substitute the result in the other, then the other independent variable (x in this case) drops out, leaving a relationship between u and v. For the details, then, assume for definiteness that u_y is nonzero at (a, b): $f_2(a, b) \neq 0$. The Implicit Function Theorem for a single equation guarantees that we can solve the equation $u = f(x, y)$ for y in the sense that there exists a neigh-

§ 1230] FUNCTIONAL DEPENDENCE

borhood $N_{(a,c)}$ of the point (a, c), and a function $y = \phi(x, u)$, defined and continuously differentiable in $N_{(a,c)}$ such that $b = \phi(a, c)$ and, for all (x, u) in $N_{(a,c)}$, $(x, \phi(x, u)) \subset R$ and $f(x, \phi(x, u)) = u$. Furthermore,

$$\frac{\partial \phi}{\partial x} = -\frac{f_1(x, \phi(x, u))}{f_2(x, \phi(x, u))}, \quad \frac{\partial \phi}{\partial u} = \frac{1}{f_2(x, \phi(x, u))}.$$

We define a new function in $N_{(a,c)}$: $\psi(x, u) \equiv g(x, \phi(x, u))$. In $N_{(a,c)}$ this function is continuously differentiable, and

$$\frac{\partial \psi}{\partial x} = \frac{\partial g}{\partial x} + \frac{\partial g}{\partial y}\frac{\partial \phi}{\partial x} = \frac{f_2 g_1 - f_1 g_2}{f_2} = \frac{1}{f_2}\frac{\partial(u, v)}{\partial(x, y)} = 0,$$

because of the vanishing of the Jacobian, for all (x, u) in $N_{(a,c)}$. This means that for any x sufficiently near a, $\psi(x, u)$ is actually only a function of u (cf. Ex. 19, § 1216), defined and differentiable for u in a neighborhood N_c of c: $G(u) \equiv g(x, \phi(x, u))$. An interpolated comment is now in order: if (x, y) is sufficiently near (a, b), $f_2(x, y)$ is of constant sign, and hence, since *for any such fixed x* the two functions $u = f(x, y)$ and $y = \phi(x, u)$ can be considered as *inverse functions*, the relation $\phi(x, f(x, y)) = y$ (as well as the relation $f(x, \phi(x, u)) = u$) is satisfied. We now conclude that if (x, y) is in a sufficiently restricted neighborhood $N_{(a,b)}$ of (a, b), $u = f(x, y)$ will be in N_c, so that, when $\phi(x, u)$ is replaced by y and u by $f(x, y)$, the equation defining G becomes $G(f(x, y)) = g(x, \phi(x, f(x, y))) = g(x, y)$, which must hold in $N_{(a,b)}$. Finally, we let $F(u, v) \equiv G(u) - v$, and the proof is complete.

NOTE 1. Hypothesis (iii) does not impose much, for if the four partial derivatives all vanished throughout R (or any subregion of R), both functions u and v would be constant (Theorem II, § 1212) and therefore trivially dependent in R (or the subregion).

NOTE 2. The proof just given indicates what happens geometrically, and shows that the Example comes close to being representative. The equation $v = G(u)$ determines a curve C in the uv-plane having the property that an entire neighborhood of the point (a, b) is mapped by the transformation $u = f(x, y)$, $v = g(x, y)$ onto a part of C. In case u and v are constants (cf. Note 1), the transformation carries a two-dimensional set into a single point. In other words, a *degenerate* transformation (one with identically vanishing Jacobian) is associated with a *lowering* of dimension.

The case of functional dependence of two functions of more than two variables is entirely analogous. For example, if $u = f(x, y, z)$ and $v = g(x, y, z)$ have identically vanishing Jacobians but have partial derivatives that are not all zero at a particular point (say $f_3 \neq 0$), then the details of the preceding proof would be altered essentially only by solving for $z = \phi(x, y, u)$, such that $f(x, y, \phi(x, y, u)) = u$, where $\partial \phi / \partial x = -f_1/f_3$ and $\partial \phi / \partial y = -f_2/f_3$, and by defining $\psi(x, y, u) \equiv g(x, y, \phi(x, y, u))$. Then the vanishing of the Jacobians implies that ψ is a function of u alone.

1231. EXERCISES

★1. Prove the existence of a positive number δ as claimed for the implication involving (12) and (13), § 1228. *Hint:* If there were no such δ, then there would be a sequence $\{(x_n, y_n, z_n, \eta_n, \zeta_n)\}$ such that $(x_n - a)^2 + (y_n - b)^2 + (z_n - c)^2 \leq 1/n$, $(\eta_n, \zeta_n) \in C$, and $F(x_n, y_n, z_n, \eta_n, \zeta_n) \leq \frac{1}{2}\epsilon$. Choose a convergent subsequence which $\to (a, b, c, \bar{\eta}, \bar{\zeta})$.

★2. State and prove a form of the Implicit Function Theorem appropriate for solving the system of equations $f(x, y, u, v, w) = g(x, y, u, v, w) = h(x, y, u, v, w) = 0$ for u, v, and w in terms of x and y.

★3. Prove the statements of the Note, § 1228. *Hints:* For Theorem II, § 1228, consider a point (λ, μ, ν) nearest (a, b, c) which is a limit point of points (x, y, z) for which the points $(\Phi(x, y, z), \Psi(x, y, z))$ and $(\phi(x, y, z), \psi(x, y, z))$ are distinct. Then $(\Phi(\lambda, \mu, \nu), \Psi(\lambda, \mu, \nu)) = (\phi(\lambda, \mu, \nu), \psi(\lambda, \mu, \nu)) \in N_{(a,c)}$, and the two mappings must agree in a neighborhood of (λ, μ, ν). (Contradiction.)

★4. State and prove an existence theorem for the inverse of a transformation from E_3 into E_3.

★5. State and prove the analogue of the Corollary, § 1229, for three dimensions.

★6. State and prove the theorem on functional dependence of two functions of three variables, alluded to in the penultimate paragraph of § 1230.

★7. State and prove a theorem on functional dependence of three functions of three variables.

In Exercises 8–11, use the principle of the Theorem of § 1230 to establish functional dependence for the given functions. Then find an explicit relation between them.

★8. $u = \dfrac{x^2 - y^2}{x^2 + y^2}$, $v = \dfrac{2xy}{x^2 + y^2}$.

★9. $u = \dfrac{xe^z}{y^2} + \dfrac{y^2 e^{-z}}{x}$, $v = \dfrac{xe^z}{y^2} - \dfrac{y^2 e^{-z}}{x}$.

★10. $u = xe^y \sin z$, $v = xe^y \cos z$, $w = x^2 e^{2y}$.

★11. $u = t^2 + x^2 + y^2 + z^2$, $v = tx + ty + tz + xy + xz + yz$, $w = t + x + y + z$.

13

Multiple Integrals

1301. INTRODUCTION

The Riemann integral of a function of one variable is defined over an interval $[a, b]$. Since the set over which the function is integrated is always a closed interval, conditions for integrability concern only the function itself, and not its domain. For example, all continuous functions are integrable over $[a, b]$.

In higher dimensional spaces, however, one is interested in integrating functions over more general sets than rectangles (and other higher dimensional analogues of intervals). Integrability conditions, therefore, involve not only the function but the set over which it is integrated. A function may fail to be integrable either by having "too many" discontinuities or by being integrated over "too pathological" a set. Not even a constant function is integrable if the set over which it is integrated is too irregular.

We shall start by defining the double integral in terms of rectangular subdivisions or *nets*. This permits a simple definition of *area* of a set and, in terms of area, a second formulation of the double integral, based on more general subdivisions than rectangles.

Multiple integrals of functions of more than two variables are treated similarly, but we shall not attempt to present all of the details. Triple integrals are discussed specifically in later sections.

1302. DOUBLE INTEGRALS

We begin by defining the double integral of a function $f(x, y)$ over a closed rectangle R, $a \leq x \leq b$, $c \leq y \leq d$, assuming that f is defined there. A **net** $\mathfrak{N}$ on R is a finite system of lines parallel to the coordinate axes, at abscissas $a = a_0 < a_1 < a_2 < \cdots < a_p = b$ and at ordinates $c = c_0 < c_1 < c_2 < \cdots < c_q = d$ (cf. Figure 1301). The rectangle R is thus cut by a net into a finite number of smaller closed rectangles R_1, $R_2, \cdots, R_n$ $(n = pq)$, with areas $\Delta A_1, \Delta A_2, \cdots, \Delta A_n$ (where the area of a rectangle is *defined* to be the product of its width and its length). The

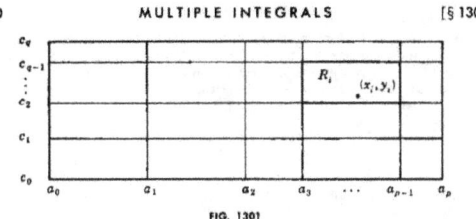

FIG. 1301

norm $|\mathfrak{N}|$ of a net $\mathfrak{N}$ is the greatest length of the diagonals of the rectangles $R_1, R_2, \cdots, R_n$. Let (x_i, y_i) be an arbitrary point of R_i, $i = 1, 2, \cdots, n$, and form the sum

$$(1) \qquad \sum_{i=1}^{n} f(x_i, y_i)\, \Delta A_i = f(x_1, y_1)\, \Delta A_1 + \cdots + f(x_n, y_n)\, \Delta A_n.$$

The integral of f is defined as the limit of the sums (1):

Definition I. *A function $f(x, y)$, defined on the closed rectangle R: $a \leq x \leq b, c \leq y \leq d$, is **integrable** there if and only if the limit*

$$(2) \qquad \lim_{|\mathfrak{N}| \to 0} \sum_{i=1}^{n} f(x_i, y_i)\, \Delta A_i,$$

*in the sense of Definition I, § 501, and with the preceding notation, exists and is finite. In case of integrability the **integral** or **double integral** of f over R is the limit (2):*

$$(3) \qquad \iint_R f(x, y)\, dA \equiv \lim_{|\mathfrak{N}| \to 0} \sum_{i=1}^{n} f(x_i, y_i)\, \Delta A_i.$$

Integration over an arbitrary (nonempty) bounded set is defined as follows:

Definition II. *If a function $f(x, y)$ is defined on a bounded set S, let the function $f_S(x, y)$ be defined to be equal to $f(x, y)$ if (x, y) is a point of S, and zero otherwise, and let R be any closed rectangle, with sides parallel to the coordinate axes, containing S. Then f is **integrable on** (or **over**) S if and only if f_S is integrable on R, and (in case of integrability) the **integral** or **double integral** of f over S is defined by the equation*

$$(4) \qquad \iint_S f(x, y)\, dA \equiv \iint_R f_S(x, y)\, dA.$$

NOTE. Definition II is independent of the choice of the rectangle R (cf. Ex. 17, § 1310).

As in the case of a function of one variable (§ 501), every integrable function is bounded and its integral is unique (cf. Ex. 18, § 1310, Theorem II, § 1306).

Under suitable restrictions on the function $f(x, y)$, the integral of Definition I exists, and under further restrictions on the set S the integral of Definition II exists (Theorem II, § 1303, Theorem XI, § 1306).

1303. AREA

Area has at this point been defined only for rectangles. We now extend its definition:

Definition I. *A bounded set S **has area** if and only if the function $f(x, y) \equiv 1$ is integrable over S. In case S has area, its **area** is*

$$A(S) \equiv \iint_S 1\, dA = \iint_S dA.$$

NOTE 1. The concepts of *having area* and of *area* are independent of the rectangular coordinate system. (Cf. Ex. 29, § 1310.)

NOTE 2. For rectangles the new definition of area is consistent with the original. That is, according to Definition I, the area of any rectangle is equal to the product of its width and its length. (Cf. Note 1.)

NOTE 3. If $f(x)$ is integrable on $[a, b]$ and if $f(x) \geqq 0$ there, then the integral $\int_a^b f(x)\, dx$ can be interpreted as the area under the curve $y = f(x)$, above the x-axis, between $x = a$ and $x = b$. (Cf. Theorem III, § 1308.)

★NOTE 4. The concept defined above as *area*, when extended to Euclidean spaces of arbitrary dimension, is known as **content** (or **Jordan content**, after the French mathematician C. Jordan (1838-1922)). In E_3 it is known as **volume**. Its most valuable generalization is *Lebesgue measure*. Any finite set in any Euclidean space clearly has content zero (and *a fortiori* measure zero). Any subset of a set of content zero is a set of content zero. The set of rational numbers of the unit interval $[0, 1]$, as a set in E_1, has no Jordan content, zero or otherwise, by virtue of the nonintegrability of the function of Example 6, § 201, discussed in § 501, although, being denumerable, it does have measure zero (cf. Ex. 52, § 503). This same set, when considered as a set in E_2, has content or area zero (cf. Ex. 19, § 1310).

It is convenient to have a simple method of establishing whether or not a given set has area. We shall show in Theorem IV, § 1305, that a necessary and sufficient condition for the existence of area for a bounded set is that *its frontier have zero area*. By means of this result we shall be able to prove the following theorem (cf. § 1305):

Theorem I. *Any bounded set whose frontier consists of a finite number of rectifiable simple arcs has area.*

In terms of area we can now state a convenient two-dimensional analogue of the integrability of a continuous function of one variable over a closed interval (Theorem VIII, § 501):

Theorem II. *A function $f(x, y)$ bounded and continuous on a set with area is integrable over that set.*

This theorem is a corollary of Theorem XI, § 1306.

Some of the most important properties of area are contained in the theorem:

★**Theorem III.** *Any subset of a set with zero area is a set with zero area. If S_1 and S_2 are bounded sets with area, then so are their union $S_1 \cup S_2$, intersection $S_1 \cap S_2$, and differences $S_1 - S_2$ and $S_2 - S_1$. Furthermore,*

(1) $\quad A(S_1 \cup S_2) = A(S_1 - S_2) + A(S_1 \cap S_2) + A(S_2 - S_1).$

Similar statements hold for any finite number of sets. In general, if S_1, $S_2, \cdots, S_n$ are bounded sets with area such that every pair has an intersection of zero area, then

(2) $\quad A(S_1 \cup S_2 \cup \cdots \cup S_n) = A(S_1) + A(S_2) + \cdots + A(S_n).$

(Cf. § 1022.)

The proof is left to the reader in Exercise 20, § 1310, where hints are provided.

We conclude this section with an example of a bounded set without area.

Example. Let S consist of the rational points in the unit square $0 \leq x \leq 1$, $0 \leq y \leq 1$ (that is, both coordinates are rational; cf. Example 10, § 1001). Let $f(x, y) \equiv 1$, and let $\mathfrak{N}$ be an arbitrary net over the unit square. Then in every rectangle R_i of $\mathfrak{N}$ there are points where $f_S(x, y) = 1$ and points where $f_S(x, y) = 0$. Therefore the sums $\sum_{i=1}^{n} f_S(x_i, y_i) \Delta A_i$ may have values ranging from 0 (in case *all* of the points (x_i, y_i) are chosen so that $f_S(x_i, y_i) = 0$) to 1 (in case *all* of the points are chosen so that $f_S(x_i, y_i) = 1$). This is true however small the norm of the net may be, so that a limit cannot exist. Therefore f_S is not integrable over the unit square, f is not integrable over S, and S does not have area. The frontier of S is the closed unit square, whose area is equal to 1.

1304. SECOND FORMULATION OF THE DOUBLE INTEGRAL

In § 1302 the double integral of a function over a closed rectangle was formulated by means of cutting this rectangle up into smaller closed rectangles, where any two of these rectangles have at most one side in common. A very natural extension of this process (sometimes informally called the **cracked china** definition of the double integral) is a **partition** Π of a bounded set S with area into a finite number of closed **pieces** S_1,

§ 1305] INNER AND OUTER AREA

$S_2, \cdots, S_n$, each with area, such that any two of these pieces have at most a set of zero area in common. (Cf. Figure 1302.) The **diameter** $\delta(A)$ of a compact set A is the maximum distance between any two of its points (cf. Fig. 1302, and Ex. 33, § 1025). The **norm** $|\Pi|$ of a partition Π

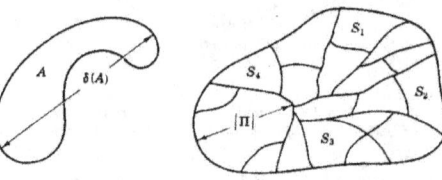

FIG. 1302

is the greatest of the diameters $\delta(S_1), \cdots, \delta(S_n)$ of its pieces. If (x_i, y_i) is an arbitrary point of S_i and if ΔA_i denotes its area, we can form sums similar to (1), § 1302:

(1) $$\sum_{i=1}^{n} f(x_i, y_i) \, \Delta A_i = f(x_1, y_1) \, \Delta A_1 + \cdots + f(x_n, y_n) \, \Delta A_n.$$

The matter of special importance is the existence of the limit of these sums:

Theorem. *With the preceding notation, if $f(x, y)$ is integrable over S, the limit of the sums (1), as the norm of the partition tends toward zero, exists and is equal to the integral of f over S:*

(2) $$\lim_{|\Pi| \to 0} \sum_{i=1}^{n} f(x_i, y_i) \, \Delta A_i = \iint_S f(x, y) \, dA.$$

A proof of this theorem is given in § 1307. For a proof of a simple but useful special case, see Exercise 25, § 1310.

★**1305. INNER AND OUTER AREA. CRITERION FOR AREA**

Let S be a bounded set contained in a closed rectangle R with sides parallel to the coordinate axes, let $f(x, y) \equiv 1$ in S, let a net $\mathfrak{N}$ be imposed on R, and consider the sum

(1) $$\sum_{i=1}^{n} f_S(x_i, y_i) \, \Delta A_i,$$

whose limit, as $|\mathfrak{N}| \to 0$, defines the area $A(S)$. Our immediate goal is to determine the extreme values of (1), for a fixed $\mathfrak{N}$, as the points (x_i, y_i) vary.

Accordingly, we look at an arbitrary term of (1) and observe that there are three possibilities: (i) if the rectangle R_i is completely contained in S, then $f_S(x_i, y_i) \Delta A_i = \Delta A_i$ for all choices of points (x_i, y_i); (ii) if R_i intersects both S and its complement, then $f_S(x_i, y_i) \Delta A_i = \Delta A_i$ or 0 according as (x_i, y_i) belongs to S or to the complement; (iii) if R_i has no points in common with S, then $f_S(x_i, y_i) \Delta A_i = 0$. Therefore (cf. Fig. 1303) the

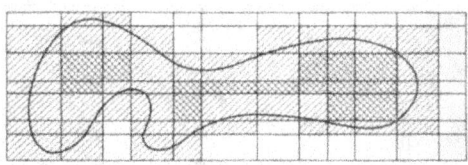

FIG. 1303

smallest possible value of (1), obtained by choosing the points (x_i, y_i) outside S whenever possible, is the total area of the rectangles of $\mathfrak{R}$ contained in S (doubly cross-hatched in Figure 1303), while the largest possible value, obtained by choosing (x_i, y_i) in S whenever possible, is the total area of the rectangles of $\mathfrak{R}$ that intersect S (singly or doubly cross-hatched in Figure 1303). That is, if we define:

(2) $\qquad a(\mathfrak{R}) \equiv$ total area of all $R_i \subset S$,

(3) $\qquad A(\mathfrak{R}) \equiv$ total area of all R_i intersecting S,

then

(4) $\qquad a(\mathfrak{R}) \leq \sum_{i=1}^{n} f_S(x_i, y_i) \Delta A_i \leq A(\mathfrak{R})$.

In terms of (2) and (3) we define two important concepts:

Definition. *The **inner area** and **outer area** of a bounded set S, denoted $\underline{A}(S)$ and $\bar{A}(S)$, respectively, are defined:*

$$\underline{A}(S) \equiv \sup_{\mathfrak{R}} a(\mathfrak{R}), \quad \bar{A}(S) \equiv \inf_{\mathfrak{R}} A(\mathfrak{R}),$$

formed with respect to all nets $\mathfrak{R}$ over a closed rectangle containing S.

We observe in passing that whether a bounded set has area or not, it always has both inner and outer area. A limit formulation for inner and outer area follows:

Theorem I. *If S is a bounded set, the limits of $a(\mathfrak{R})$ and $A(\mathfrak{R})$, as $|\mathfrak{R}| \to 0$, exist and equal the inner and outer area, respectively:*

$$\lim_{|\mathfrak{R}| \to 0} a(\mathfrak{R}) = \underline{A}(S), \quad \lim_{|\mathfrak{R}| \to 0} A(\mathfrak{R}) = \bar{A}(S).$$

§ 1305] INNER AND OUTER AREA

Outline of proof. We shall indicate only the proof that $a(\mathfrak{R}) \to \underline{A} \equiv \underline{A}(S)$. If $\epsilon > 0$, first choose a net $\mathfrak{R}_1$ such that $a(\mathfrak{R}_1) > \underline{A} - \frac{\epsilon}{2}$. Then let $\delta > 0$ be such that $|\mathfrak{R}| < \delta$ implies that the total area of those rectangles of $\mathfrak{R}$ that intersect the lines of $\mathfrak{R}_1$ is less than $\epsilon/2$. It follows that $a(\mathfrak{R}) > \underline{A} - \epsilon$. Complete details are requested in Exercise 21, § 1310.

Proofs of the following two theorems are left to the reader (Ex. 22, § 1310):

Theorem II. *If S is a bounded set,*

(5) $$\underline{A}(S) \leq \bar{A}(S),$$

equality holding if and only if S has area. In case of equality, $A(S) = \underline{A}(S) = \bar{A}(S)$.

Theorem III. *A bounded set has zero area if and only if its outer area is zero; that is, if and only if corresponding to $\epsilon > 0$ there exists a net $\mathfrak{R}$ such that $A(\mathfrak{R}) < \epsilon$.*

Example. The set of the Example, § 1303, has inner area 0 and outer area 1.

Since $A(\mathfrak{R}) - a(\mathfrak{R})$, for a given net $\mathfrak{R}$, is the total area of the rectangles of $\mathfrak{R}$ that intersect both S and its complement, and since the frontier $F(S)$ of S consists of points p such that every neighborhood of p intersects both S and its complement, the following theorem should seem plausible:

Theorem IV. *A bounded set has area if and only if its frontier has zero area.*

Proof. (i) Assume that S has area, and let the containing rectangle R be sufficiently large to contain the closure of S in its interior. If ϵ is a given positive number, choose a net $\mathfrak{R}$ such that $A(\mathfrak{R}) - a(\mathfrak{R}) < \epsilon$, and consider the rectangles $\{R_{i_k}\}$ that intersect both S and S'. We shall show that every point $F(S)$ belongs to at least one R_{i_k}, so that $\bar{A}(F(S)) < \epsilon$, and consequently $A(F(S)) = 0$. If $p \in F(S)$, then three possibilities present themselves. 1: p is in the interior of some R_j of $\mathfrak{R}$; then R_j is one of $\{R_{i_k}\}$. 2: p is a point of a common edge of exactly two rectangles R_j and R_l; then at least one of R_j and R_l must be one of $\{R_{i_k}\}$. 3: p is a common vertex of four rectangles of $\mathfrak{R}$; then at least one of these four must be one of $\{R_{i_k}\}$. (These last three parts follow readily by the indirect method of proof.)

(ii) Assume that $A(F(S)) = 0$, let $\epsilon > 0$, and choose a net $\mathfrak{R}$ such that the total area of all rectangles $\{R_{i_k}\}$ of $\mathfrak{R}$ that intersect $F(S)$ is less than ϵ. We shall show that every rectangle of $\mathfrak{R}$ that intersects both S and S' is one of $\{R_{i_k}\}$. Assume the contrary: let R_j be a rectangle containing a point $p:(x_0, y_0)$ of S and a point $q:(x_1, y_1)$ of S', but no point of $F(S)$. For the segment $I:(x_0 + (x_1 - x_0)t, y_0 + (y_1 - y_0)t), 0 \leq t \leq 1$, joining p and q, let t_0 be the least upper bound of the values of t for points of I that belong to S, and let r be the corresponding point of I. Then

$r \in F(S)$, since if $r \in S$, $r \in D(S')$, and if $r \in S'$, $r \in D(S)$. (Contradiction.)

Theorem I, § 1303, is an immediate corollary of the preceding theorem and the following:

Theorem V. *Any rectifiable simple arc has zero area.*

Proof. Let the simple arc C have length L, and let $\epsilon > 0$. Enclose C in a closed square and impose on this square a net $\mathfrak{N}$ that cuts the square into smaller squares, all of side δ, where $0 < \delta < \min\left(L, \frac{\epsilon}{8L}\right)$. Let $k = \left[\frac{L}{\delta}\right] + 1$ (cf. Example 5, § 201), so that $\frac{L}{\delta} < k \leq \frac{L}{\delta} + 1$, and cut C into k pieces, each of length less than δ (this is possible since $L/k < \delta$). Each of these pieces can intersect at most 4 squares of $\mathfrak{N}$ (this is true since each has length less than the side of each square). Since the area of four squares is $4\delta^2$, the total area of the squares containing points of C is not greater than

$$4k\delta^2 \leq 4\delta^2\left(\frac{L+\delta}{\delta}\right) < 4\delta(L+L) < 8L(\epsilon/8L) = \epsilon.$$

With the aid of Theorem III, the proof is complete.

NOTE 1. *A simple arc of infinite length may have zero area.* The graph of $y = x \sin(1/x)$ for $0 \leq x \leq 1$ is an example. (Cf. Example 1, § 1105.)

★★NOTE 2. *A simple arc may fail to have area.* Let us observe first that the remarks in Note 2, § 1104, show that the inner area of every simple arc is zero. Therefore, if a simple arc has positive outer area it has no area. An example of a simple arc with positive outer area can be constructed by modifying the process of § 1104 defining a Peano space-filling curve. The principal modification is the deletion of open "channels" or "paths" straddling the common edges of subsquares that correspond to nonadjacent subintervals. If this is done correctly, the resulting mapping is one-to-one, and the arc is simple. Furthermore, it is "almost" space-filling in the sense that its outer area (which, incidentally, is also its Lebesgue measure) can be made arbitrarily close to 1.

★★NOTE 3. The example of Note 2 shows that not all compact sets have area. If the simple arc of Note 2 is placed in the unit square $0 \leq x \leq 1$, $0 \leq y \leq 1$, with end points $(0, 0)$ and $(1, 0)$, and combined with a simple arc with the same end points and lying in the fourth quadrant, a simple closed curve results. By means of the Jordan Curve Theorem (cf. the book by M. H. A. Newman referred to on page 477), one can produce a bounded region and a compact region neither of which has area.

★1306. THEOREMS ON DOUBLE INTEGRALS

In this section we consider some of the most important properties of the double integral. A few are direct extensions of theorems of §§ 501 and 502 and need virtually no change in proof. Others require an entirely new formulation. Theorem XI states sufficient conditions for integrability of a function f over a set S in terms of the continuity of f and the nature of S. All of these theorems can be immediately translated into theorems for any Euclidean space E_n, $n = 1, 2, 3, \cdots$, the word *area* being replaced by the word *content* (or, for E_3, *volume*). Discussion in this section is limited to the plane E_2.

We start by introducing a new concept, which is helpful in simplifying many statements:

Definition I. *A property $P(p)$ which for each point p of a set S is either true or false is said to hold **very nearly everywhere** in S if and only if the subset of S for which P is false is of zero area.*

Examples. The function $f(x, y) = x - y$ is very nearly everywhere nonzero in the unit square $0 \le x \le 1$, $0 \le y \le 1$. The function $f(x, y) = 2x^2 + 3y^2$ is very nearly everywhere positive in the unit disk $x^2 + y^2 \le 1$. The function $f(x, y) = xy/(x^2 + y^2)$, $x^2 + y^2 \ne 0$, $f(0, 0) = 0$, is very nearly everywhere continuous in the unit disk.

★★Note 1. A well-known generalization of *very nearly everywhere* is *almost everywhere*, based on Lebesgue measure zero instead of zero area; it is presented for E_1 in Exercises 52-54, § 503. If a property holds very nearly everywhere, it holds almost everywhere, but not conversely. For instance, the function f_S of the Example, § 1303, is equal to zero almost everywhere but not very nearly everywhere.

Our first theorem makes use of Definition I.

Theorem I. *A bounded function that is equal to zero very nearly everywhere on a bounded set S is integrable on S, and its integral over S is zero.*

Proof. Assume $|f(x, y)| < K$ on S, let Z be a subset of S with zero area such that $f(x, y) = 0$ on $S - Z$, and let $\epsilon > 0$. Extend f to a closed rectangle $R \supset S$ by means of the function f_S, and let $\mathfrak{N}$ be a net over R such that relative to the set Z, $A(\mathfrak{N}) < \epsilon/K$. Then, since for every point (x, y) belonging to a rectangle not contributing to $A(\mathfrak{N})$, $f_S(x, y) = 0$, we have the inequalities

$$\left| \sum_{i=1}^{n} f_S(x_i, y_i) \, \Delta A_i \right| \le K \cdot A(\mathfrak{N}) < \epsilon,$$

and the proof is complete.

Theorems II-V are immediate extensions of Theorems I-IV, § 501, and have analogous proofs (cf. Ex. 23, § 1310). The set S is assumed to be bounded.

Theorem II. *If $\lim_{|\mathfrak{N}| \to 0} \sum_{i=1}^{n} f(x_i, y_i) \, \Delta A_i$ exists, the limit is unique.*

Theorem III. *If $f(x, y)$ and $g(x, y)$ are integrable over S, and if $f(x, y) \le$*

$$\iint_S f(x, y)\, dA \leq \iint_S g(x, y)\, dA.$$

Theorem IV. *If $f(x, y)$ is integrable over S and if k is a constant, then $k f(x, y)$ is integrable over S and*

$$\iint_S k f(x, y)\, dA = k \iint_S f(x, y)\, dA.$$

Theorem V. *If $f(x, y)$ and $g(x, y)$ are integrable over S, then so are their sum and difference, and*

$$\iint_S [f(x, y) \pm g(x, y)]\, dA = \iint_S f(x, y)\, dA \pm \iint_S g(x, y)\, dA.$$

A similar statement holds for an arbitrary finite number of functions.

The extension of Theorem VI, § 501, is:

Theorem VI. *If $f(x, y)$ and $g(x, y)$ are bounded on a bounded set S and equal very nearly everywhere there, then the integrability of either implies that of the other, and the equality of their integrals.*

Proof. Assume that $f(x, y)$ is integrable on S and let $h(x, y) = g(x, y) - f(x, y)$. Then by Theorem I, $h(x, y)$ is integrable over S, and since $g(x, y) = f(x, y) + h(x, y)$ the result follows from Theorem V.

Theorem V, § 501, takes the form:

Theorem VII. *If S_1 and S_2 are bounded sets having an intersection $Z = S_1 \cap S_2$ of zero area, and if $f(x, y)$ is integrable over S_1 and integrable over S_2, then $f(x, y)$ is integrable over their union $S_1 \cup S_2$, and*

$$\iint_{S_1 \cup S_2} f(x, y)\, dA = \iint_{S_1} f(x, y)\, dA + \iint_{S_2} f(x, y)\, dA.$$

A similar result holds for any finite number of bounded sets each pair of which has an intersection of zero area.

Proof. Let R be a closed rectangle containing $S_1 \cup S_2$, and define $f_i(x, y)$ $(i = 1, 2)$ over R to be equal to $f(x, y)$ if $(x, y) \in S_i$ and equal to 0 otherwise. Also define $g(x, y)$ to be equal to $f(x, y)$ if $(x, y) \in S_1 \cup S_2$ and equal to 0 otherwise. Since f_1 and f_2 are integrable over R, then (Theorem V) so is $f_1 + f_2$, and

$$\iint_R (f_1 + f_2)\, dA = \iint_R f_1\, dA + \iint_R f_2\, dA = \iint_{S_1} f\, dA + \iint_{S_2} f\, dA.$$

On the other hand, $g(x, y) = f_1(x, y) + f_2(x, y)$ except possibly on the set

Theorem VIII. *If $f(x, y)$ is constant, $f(x, y) = k$, on a bounded set S with area, then $f(x, y)$ is integrable there and*

$$\iint_S f(x, y)\, dA = kA(S).$$

Proof. When $k = 1$ the theorem is true by definition, and for other values of k it follows from Theorem IV.

Sufficiency conditions for integrability can be obtained with the aid of step-functions (cf. § 502):

Definition II. *A **step-function** is a bounded function that is constant in the interior of each rectangle of some net over a closed rectangle.*

Theorem IX. *Any step-function $\sigma(x, y)$ is integrable, and if $\sigma(x, y)$ has the values $\sigma_1, \cdots, \sigma_n$ in the interiors of the rectangles $R_1, \cdots, R_n$, respectively, then*

$$\iint_R \sigma(x, y)\, dA = \sum_{i=1}^{n} \sigma_i\, \Delta A_i.$$

Proof. This is an immediate consequence of Theorems VI, VII, and VIII.

Theorem X. *A function $f(x, y)$, defined on a bounded set S, is integrable there if and only if corresponding to $\epsilon > 0$ there exist step-functions $\sigma(x, y)$ and $\tau(x, y)$ defined on a closed rectangle $R \supset S$ such that*

(1) $$\sigma(x, y) \leq f(x, y) \leq \tau(x, y)$$

on S, $\sigma(x, y) \leq \tau(x, y)$ on R, and

(2) $$\iint_R [\tau(x, y) - \sigma(x, y)]\, dA < \epsilon.$$

Proof. The proof is identical (*mutatis mutandis*) with that of Theorem II, § 502. (Give the details in Ex. 24, § 1310.)

Theorem XI. *A function $f(x, y)$, defined and bounded on a bounded set S with area, and continuous very nearly everywhere there, is integrable over S.*

Proof. As a preliminary we shall show that it can be assumed without loss of generality that S is a closed rectangle with sides parallel to the coordinate axes. This is done by showing that under the assumptions of the theorem, the extension f_S of f to a closed rectangle R containing S is continuous very nearly everywhere on R. (The integrability of f_S over R, then, implies that of f over S.) This is true since if Z is a subset of S

discontinuity of f_3 are members either of Z or of $F(S)$, the frontier of S (prove this!). Since both Z and $F(S)$ have zero area, so does their union $Z \cup F(S)$ (prove this!).

In the remaining parts of the proof we shall assume that S is a closed rectangle with sides parallel to the coordinate axes.

Let Z be a subset of the rectangle S, of zero area, such that f is continuous on $S - Z$, and let $\epsilon > 0$. We seek step-functions $\sigma(x, y)$ and $\tau(x, y)$ satisfying (1) and (2) for S. The first step in this direction is to subdivide S by a net $\mathfrak{N}_0$, to be held fixed, such that the total area of the rectangles of $\mathfrak{N}_0$ that intersect Z is less than $\epsilon/4K$, where $|f(x, y)| < K$ on S. Define $\sigma(x, y)$ and $\tau(x, y)$ on *these* rectangles to be equal to $-K$ and K, respectively. Then $\sigma(x, y) < f(x, y) < \tau(x, y)$ there, and the integral $\iint [\tau(x, y) - \sigma(x, y)]\, dA$ over these rectangles is less than $2K(\epsilon/4K) = \epsilon/2$.

Finally, in each closed rectangle $R_i^{(2)}$ of $\mathfrak{N}_0$ that lies in $S - Z$, f is continuous, and therefore uniformly continuous, so that there exists a $\delta_i > 0$ such that $\sqrt{(x - x')^2 + (y - y')^2} < \delta_i$ implies $|f(x, y) - f(x', y')| < \epsilon/2A$, where $A = A(S)$. Let δ be the least of these δ_i, and adjoin further lines to those of $\mathfrak{N}_0$ to obtain a net $\mathfrak{N}$ of norm $< \delta$. For *this* net $\mathfrak{N}$, define σ and τ, where they are not already defined, as follows: on the lines of $\mathfrak{N}$, $\sigma \equiv \tau \equiv f$. In the interior of a rectangle R_i of $\mathfrak{N}$, define $\sigma(x, y) \equiv \sigma_i \equiv \inf_{(x, y) \in R_i} f(x, y)$ and $\tau(x, y) \equiv \tau_i \equiv \sup_{(x, y) \in R_i} f(x, y)$. Then (1) is clearly satisfied, and (2) follows from $\iint_S [\tau(x, y) - \sigma(x, y)]\, dA < \frac{2K\epsilon}{4K} + \frac{\epsilon}{2A} A = \epsilon$.

This completes the proof.

Theorem XII. *If f and g are continuous on a set S of positive area, and if $f \leq g$ there and f and g are not identically equal in $I(S)$, then*
$$\iint_S f\, dA < \iint_S g\, dA.$$

Proof. Let $h(x, y) \equiv g(x, y) - f(x, y)$, and assume $h(a, b) > 0$, where $(a, b) \in I(S)$. Let ϵ be a positive number and $N_{(a, b)} \subset S$ be a neighborhood of (a, b) such that $h(x, y) \geq \epsilon$ in $N_{(a, b)}$ (cf. Ex. 18, § 212), and define $\phi(x, y) \equiv \epsilon$ if $(x, y) \in N_{(a, b)}$ and $\phi(x, y) \equiv 0$ otherwise. Then (Theorem III) $h \geq \phi$, and
$$\iint_S g\, dA - \iint_S f\, dA = \iint_S h\, dA \geq \iint_S \phi\, dA = \epsilon A(N_{(a, b)}) > 0.$$

Theorem XIII. Mean Value Theorem for Double Integrals. *If $f(x, y)$ is continuous on a compact region R with area, there exists a point (ξ, η) in the interior of R such that*

§ 1307] PROOF OF THE SECOND FORMULATION

$$\iint_R f(x, y)\, dA = f(\xi, \eta) \cdot A(R).$$

Proof. Let $m = \min_{(x,y)\in R} f(x, y)$, $M = \max_{(x,y)\in R} f(x, y)$, and assume $m < M$ (otherwise f is constant). Then $m \leq f \leq M$, and f is not identically equal to either m or M in $I(R)$. Hence (Theorem XII),

$$mA(R) < \iint_R f\, dA < MA(R).$$

Therefore (Theorem IV, § 1010) there exists a point $(\xi, \eta) \in I(R)$ such that $f(\xi, \eta) = \left[\iint_R f\, dA\right]/A(R)$.

★★1307. PROOF OF THE SECOND FORMULATION

In this section we shall establish a sequence of lemmas leading to a proof of the Theorem of § 1304, regarding an alternative formulation of the double integral:

(1) $$\lim_{|\pi|\to 0} \sum_{i=1}^{n} f(x_i, y_i)\, \Delta A_i = \iint_S f(x, y)\, dA.$$

Some of the details are left as an exercise for the reader (Ex. 26, § 1310). (Also cf. Ex. 25, § 1310.)

First we introduce a definition and notation of only temporary interest and utility for this section:

Definition. *A function $f(x, y)$ defined and bounded over a bounded set S with area is **I-integrable** there if and only if the limit (1) exists and is finite. In case of I-integrability the **I-integral** of f over S, denoted $I_S f(x, y)\, dA$, is defined to be this limit:*

(2) $$I_S f(x, y)\, dA \equiv \lim_{|\pi|\to 0} \sum_{i=1}^{n} f(x_i, y_i)\, \Delta A_i.$$

Our ultimate objective is to prove that the existence of $\iint_S f(x, y)\, dA$ implies that of $I_S f(x, y)\, dA$, and their equality.† As a preliminary step, let us show that it can be assumed without loss of generality that S is a closed rectangle with sides parallel to the coordinate axes. That is, assume for the moment that the Theorem is established for all such rectangles. If a bounded set S with area is given, enclose it in a rectangle R. Then the existence of $\iint_S f(x, y)\, dA$ implies that of $\iint_R f_S(x, y)\, dA$ which, by assumption, implies that of $I_R f_S(x, y)\, dA$, equality holding throughout.

† The reverse implication is trivial. (Proof?)

The final implication $\lim \sum f(x_i, y_i) \Delta A_i = I_R f_S \, dA$ follows from the fact that *every* sum $\sum f(x_i, y_i) \Delta A_i$ for a partition of S is *one* of the sums for a partition of R having the same norm as the given partition of S (obtained by cutting up the difference $R - S$ by means of a net of small norm).

For the remainder of this section we shall assume that S is a closed rectangle with sides parallel to the coordinate axes. For convenient comparison with Theorems of § 1306 we start the lemma numbering with 0:

Lemma 0. *If S and R are closed rectangles with sides parallel to the coordinate axes, if $S \subset R$, and if f is defined and bounded over S, then the I-integrability of f over S is equivalent to the I-integrability of its extension f_S to R ($f_S(x, y) = 0$ on $R - S$), and the equality of their I-integrals.*

Proof. The existence of $I_R f_S \, dA$ implies that of $I_S f \, dA$ and their equality, by the same argument as that of the concluding sentence of the second preceding paragraph. Now assume the existence of $I_S f \, dA$. If $|f(x, y)| < K$ on S and if $\epsilon > 0$, choose $\delta > 0$ so that (i) the total area of strips of width 2δ, parallel to the axes, running across R and having the sides of S as center lines is less than $\epsilon/4K$, and (ii) every sum $\sum f \Delta A$ for a partition of S of norm less than δ approximates $I_S f \, dA$ within $\epsilon/2$. Let Π be any partition of R with norm $< \delta$, and consider, for this partition, an arbitrary sum $\sum f_S(x_i, y_i) \Delta A_i$. For every piece S_i of Π such that $S_i \cap S = \emptyset$, the term $f_S(x_i, y_i) \Delta A_i$ is equal to 0 and can be deleted from the sum. For every piece S_i such that $S_i \subset S$, define $(x_i', y_i') = (x_i, y_i)$, and $\Delta B_i = \Delta A_i$ (and renumber all S_i so that *these* ΔB_i have subscripts $i = 1, 2, \cdots, m$). For every piece S_i that intersects both S and S', let $(x_i', y_i') \in S_i \cap S$ and $\Delta B_i = A(S_i \cap S)$, $i = m+1, \cdots, n$. In this way the sum $\sum f_S(x_i, y_i) \Delta A_i$ (over R) is made to correspond to a sum $\sum_{i=1}^{n} f(x_i', y_i') \Delta B_i$ (over S) such that their difference contains only terms where S_i intersects both S and S'. Thus the absolute value of this difference, involving *only* such terms, reduces to

$$\left| \sum_{m+1}^{n} f_S(x_i, y_i) \Delta A_i - \sum_{m+1}^{n} f(x_i', y_i') \Delta B_i \right| < \sum_{m+1}^{n} K \Delta A_i + \sum_{m+1}^{n} K \Delta B_i$$

$$\leq 2K \sum_{m+1}^{n} \Delta A_i < 2K(\epsilon/4K) = \epsilon/2.$$

Finally,

$$\left| \sum_{i=1}^{n} f_S \Delta A_i - I_S f \, dA \right| \leq \left| \sum_{i=1}^{n} f_S \Delta A_i - \sum_{i=1}^{n} f \Delta B_i \right| + \left| \sum_{i=1}^{n} f \Delta B_i - I_S f \, dA \right|$$

$$= \left| \sum_{m+1}^{n} f_S \Delta A_i - \sum_{m+1}^{n} f \Delta B_i \right| + \left| \sum_{i=1}^{n} f \Delta B_i - I_S f \, dA \right| < \frac{\epsilon}{2} + \frac{\epsilon}{2} = \epsilon.$$

Lemma 1. *A bounded function f that is equal to zero on S except for the points of a net $\mathfrak{N}_n$ over S is I-integrable over S, and $I_S f \, dA = 0$.*

§ 1307] PROOF OF THE SECOND FORMULATION 443

Proof. If $|f| < K$ on S, if $\epsilon > 0$, and if $\delta > 0$ is such that strips of width 2δ containing the lines of $\mathfrak{N}_0$ have total area $< \epsilon/K$, then $|\Pi| < \delta$ implies $|\sum f(x_i, y_i) \Delta A_i| < K \cdot [\text{total area of strips}] < \epsilon$.

Lemmas 2–5 are identical with Theorems II–V, § 1306, *mutatis mutandis* (that is, change "integrable" to "I-integrable," $\iint_S$ to I_S, and "$|\mathfrak{N}| \to 0$" to "$|\Pi| \to 0$").

Lemma 6. *If f and g are bounded on S, and equal there except for the points of a net $\mathfrak{N}_0$ over S, then the I-integrability of either implies that of the other, and the equality of their I-integrals.*

Proof. Identical with that of Theorem VI, § 1306 (*mutatis mutandis*).

Lemma 7. *If S_1 and S_2 are rectangles with at most a line segment in common, and if f is I-integrable over each, then f is I-integrable over their union, and*
$$I_{S_1 \cup S_2} f\, dA = I_{S_1} f\, dA + I_{S_2} f\, dA.$$
A similar result holds for any finite number of rectangles.

Proof. Extend f to a closed rectangle R containing $S_1 \cup S_2$ ($f \equiv 0$ in $R - S_1 \cup S_2$). Then, by Lemma 0, the remaining details parallel those of Theorem VII, § 1306.

Lemma 8. *If f is constant, $f \equiv k$, on S, then f is I-integrable there and*
$$I_S f\, dA = kA(S).$$

Proof. This is a consequence of Theorem III, § 1303.

Lemma 9. *Any step-function is I-integrable, and $I_S f\, dA = \iint_S f\, dA$.*

Proof. This is a consequence of Lemmas 0, 6, 7, and 8.

Proof of the Theorem, § 1304. Define
$$J \equiv \iint_S f(x, y)\, dA = \sup_{\sigma \leq f} \iint_S \sigma(x, y)\, dA = \inf_{\tau \geq f} \iint_S \tau(x, y)\, dA,$$
formed for all step-functions $\sigma(x, y) \leq f(x, y)$ and $\tau(x, y) \geq f(x, y)$ on S, and let $\epsilon > 0$. Choose two step-functions σ and τ such that $\sigma(x, y) \leq f(x, y) \leq \tau(x, y)$ on S and
$$\iint_S \sigma(x, y)\, dA > J - \frac{\epsilon}{2}, \quad \iint_S \tau(x, y)\, dA < J + \frac{\epsilon}{2}.$$
Choose $\delta > 0$ such that $|\Pi| < \delta$ implies
$$\sum_{i=1}^n \sigma(x_i, y_i) \Delta A_i > \iint_S \sigma\, dA - \frac{\epsilon}{2}, \quad \sum_{i=1}^n \tau(x_i, y_i) \Delta A_i < \iint_S \tau\, dA + \frac{\epsilon}{2},$$

Then, for any Π such that $|\Pi| < \delta$,

$$J - \epsilon < \iint_S \sigma\, dA - \frac{\epsilon}{2} < \sum_{i=1}^n \sigma(x_i, y_i)\, \Delta A_i \leq \sum_{i=1}^n f(x_i, y_i)\, \Delta A_i$$

$$\leq \sum_{i=1}^n \tau(x_i, y_i)\, \Delta A_i < \iint_S \tau\, dA + \frac{\epsilon}{2} < J + \epsilon.$$

By the meaning of (2), § 1304, the proof is complete.

1308. ITERATED INTEGRALS, TWO VARIABLES

Under suitable conditions on the functions $f(x, y)$, $\phi(x)$, and $\psi(x)$, the integral

$$(1) \qquad \int_{\phi(x)}^{\psi(x)} f(x, y)\, dy,$$

which is a function of x (and not of y) (cf. § 1226), is integrable over an interval $[a, b]$. In this case the integral

$$(2) \qquad \int_a^b \left\{ \int_{\phi(x)}^{\psi(x)} f(x, y)\, dy \right\} dx = \int_a^b \int_{\phi(x)}^{\psi(x)} f(x, y)\, dy\, dx$$

is called an **iterated integral** or, more completely, a **two-fold iterated integral**.

Similarly, a two-fold iterated integral with the order of integration reversed may sometimes be formed:

$$(3) \qquad \int_c^d \left\{ \int_{\alpha(y)}^{\beta(y)} f(x, y)\, dx \right\} dy = \int_c^d \int_{\alpha(y)}^{\beta(y)} f(x, y)\, dx\, dy.$$

Example 1. $\int_0^1 \int_0^x (x^2 + 4xy)\, dy\, dx = \int_0^1 [x^2 y + 2xy^2]_0^x\, dx$

$$= \int_0^1 (x^3 + 2x^3)\, dx = \int_0^1 3x^3\, dx = \tfrac{3}{4}.$$

The importance of iterated integrals lies in their use for the evaluation of double integrals. This corresponds to the evaluation of definite integrals of functions of one variable by means of indefinite integrals (the Fundamental Theorem of Integral Calculus, § 504). We state a corresponding Fundamental Theorem now, and prove it (in a more general form) in the following section.

Theorem I. Fundamental Theorem for Double Integrals. *If $f(x, y)$ is defined over a bounded set S, bounded below and above by the graphs of functions $\phi(x)$ and $\psi(x)$, respectively, and at the extreme left and right by the lines $x = a$ and $x = b$ (that is, S consists of all (x, y) such that $a \leq x \leq b$ and $\phi(x) \leq y \leq \psi(x)$; cf. Fig. 1304), then the following relation holds whenever both integrals exist:*

§ 1308]   ITERATED INTEGRALS   445

$$(4) \qquad \iint_S f(x, y)\, dA = \int_a^b \int_{\phi(x)}^{\psi(x)} f(x, y)\, dy\, dx.$$

A similar statement holds with an interchange in the roles of x and y (cf. Fig. 1304):

$$(5) \qquad \iint_S f(x, y)\, dA = \int_c^d \int_{\alpha(y)}^{\beta(y)} f(x, y)\, dx\, dy.$$

If S is bounded above and below and to the left and right by graphs of bounded functions (cf. Fig. 1304), then the double integral and the iterated integrals of (4) and (5) are all equal.

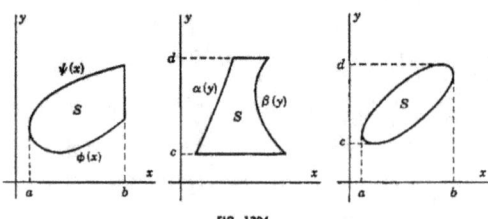

FIG. 1304

Example 2. Evaluate $I = \iint_S x^2 y\, dA$ in two ways by means of iterated integrals, where S is the region between the parabola $x = y^2$ and the line $x + y = 2$.

Solution. The simpler *formulation* uses x as the first variable of integration:

$$I = \int_{-2}^{1} \int_{y^2}^{2-y} x^2 y\, dx\, dy = \int_{-2}^{1} \frac{y}{3}\left[x^3\right]_{y^2}^{2-y} dy = -15\tfrac{3}{20}.$$

With y as the first variable of integration, the iterated integral must be expressed as the sum of two:

$$I = \int_0^1 \int_{-\sqrt{x}}^{\sqrt{x}} x^2 y\, dy\, dx + \int_1^4 \int_{-\sqrt{x}}^{2-x} x^2 y\, dy\, dx = 0 - 15\tfrac{3}{20} = -15\tfrac{3}{20}.$$

The *evaluation* details turn out to be easier with the second formulation.

The following relation between integrals and area between curves is a simple consequence of the Fundamental Theorem of this section (cf. (xii), § 1311, for a corresponding formula for volume between surfaces):

Theorem II. *If the set S of Theorem I has area, and if $\psi(x) - \phi(x)$ is*

(6) $$A(S) = \int_a^b [\psi(x) - \phi(x)] \, dx.$$

Proof. In formula (4), let $f(x, y) \equiv 1$.

NOTE. It is proved in § 1309 that the existence of the left-hand member of (6) implies that of the right-hand member, and that the existence of the right-hand member of (6) implies that of the left-hand member if and only if ϕ and ψ are both integrable.

As a consequence of the statements of the preceding Note we have:

Theorem III. *If $f(x)$ is a nonnegative function on (a, b), and if S is the set of points (x, y) such that $a \leq x \leq b$ and $0 \leq y \leq f(x)$, then S has area if and only if $f(x)$ is integrable and, in case of existence,*

$$A(S) = \int_a^b f(x) \, dx.$$

★1309. PROOF OF THE FUNDAMENTAL THEOREM

We establish first a basic theorem, whose title is borrowed from a similar but more elegant theorem in Lebesgue Theory, proved by the Italian mathematician G. Fubini in 1910. (Cf. Theorem III for a generalization.)

Theorem I. Fubini's Theorem. *If $f(x, y)$ is defined over the closed rectangle $R: a \leq x \leq b, c \leq y \leq d$, then*

(1) $$\iint_R f(x, y) \, dA = \int_a^b \int_c^d f(x, y) \, dy \, dx,$$

whenever both of these integrals exist.

Proof. The first step is to observe that this theorem is true for step-functions (Ex. 27, § 1310). Next assume that $f(x, y)$ is integrable over R, let $\epsilon > 0$, and form the related step-functions $\sigma(x, y)$ and $\tau(x, y)$ (§ 1306) such that $\sigma \leq f \leq \tau$, and $\int_a^b [\tau - \sigma] \, dA < \epsilon$. The equality (1) results from the two sets of inequalities:

(2) $$\iint_R \sigma(x, y) \, dA \leq \iint_R f(x, y) \, dA \leq \iint_R \tau(x, y) \, dA,$$

(3) $$\int_a^b \int_c^d \sigma(x, y) \, dy \, dx \leq \int_a^b \int_c^d f(x, y) \, dy \, dx \leq \int_a^b \int_c^d \tau(x, y) \, dy \, dx.$$

The Fundamental Theorem for Double Integrals (Theorem I, § 1308) is a direct consequence of Fubini's Theorem, obtained by extending the domain of definition of $f(x, y)$ from the given set S of Theorem I, § 1308, to a closed rectangle R by defining $f(x, y)$ to be identically zero outside S.

§ 1309] PROOF OF THE FUNDAMENTAL THEOREM   447

An immediate corollary to Fubini's Theorem is the interchangeability of the order of integration in an iterated integral:

Theorem II. *If $f(x, y)$ is integrable over the closed rectangle $R: a \leq x \leq b$, $c \leq y \leq d$, then the two iterated integrals $\int_a^b \int_c^d f(x, y)\, dy\, dx$ and $\int_c^d \int_a^b f(x, y)\, dx\, dy$ are equal whenever they both exist.*

The hypotheses of Fubini's Theorem can be weakened:

Theorem III. *If $f(x, y)$ is integrable over the closed rectangle $R: a \leq x \leq b$, $c \leq y \leq d$, and if $g(x)$ is any function satisfying the inequalities*

$$(4) \qquad \underline{\int_c^d} f(x, y)\, dy \leq g(x) \leq \overline{\int_c^d} f(x, y)\, dy$$

(the extreme members of (4) being the lower and upper integrals of $f(x, y)$ over $c \leq y \leq d$; cf. Ex. 44, § 503), for $a \leq x \leq b$, then $g(x)$ is integrable on $[a, b]$ and

$$(5) \qquad \iint_R f(x, y)\, dA = \int_a^b g(x)\, dx.$$

Consequently, the existence of $\iint_R f(x, y)\, dA$ and of $\int_c^d f(x, y)\, dy$ for $a \leq x \leq b$ implies that of $\int_a^b \int_c^d f(x, y)\, dy\, dx$, and equation (1).

Proof. Proceeding as in the proof of Theorem I, we obtain inequalities (2) and

$$(4) \qquad \int_a^b \int_c^d \sigma\, dy\, dx \leq \underline{\int_a^b} g\, dx \leq \overline{\int_a^b} g\, dx \leq \int_a^b \int_c^d \tau\, dy\, dx$$

and the desired conclusion.

Proof of the Note, § 1308. Assume that the set S has area, enclose it in a closed rectangle $R: a \leq x \leq b$, $c \leq x \leq d$, and define $f(x, y) \equiv 1$ on S and $f(x, y) \equiv 0$ on $R - S$. Then f is integrable over R and for each x of $[a, b]$ $\int_c^d f(x, y)\, dy = \psi(x) - \phi(x)$. The first statement of the Note follows from the last statement of Theorem III of this section. Now assume that $\psi(x)$ and $\phi(x)$ are both integrable. By introducing a translation parallel to the y-axis and subtracting areas and integrals we can reduce the problem of showing that the existence of the right-hand member of (6), § 1308, implies that of the left-hand member to proving the special case contained in Theorem III, § 1308. But this follows by approximating $f(x)$ of that theorem by means of step-functions. The details are left as an exercise for the reader (Ex. 28, § 1310). Finally, the Example, below, completes the proof.

Example. On the unit interval $[0, 1]$ define:

$$\phi(x) = \begin{cases} 0 \text{ for } x \text{ irrational,} \\ 1 \text{ for } x \text{ rational,} \end{cases} \quad \psi(x) = \begin{cases} 1 \text{ for } x \text{ irrational,} \\ 2 \text{ for } x \text{ rational,} \end{cases}$$

and let S be the set of all (x, y) such that $0 \leq x \leq 1$ and $\phi(x) \leq y \leq \psi(x)$. Then S is without area, since its inner area is 0 and its outer area is 2, although the function $\psi(x) - \phi(x)$ is identically equal to 1 and hence is integrable on $[0, 1]$.

1310. EXERCISES

In Exercises 1-4, write the double integral $\iint_S f(x, y) \, dA$ in two ways in terms of iterated integrals, where S is the given set. Compute the area of S by setting $f(x, y) \equiv 1$.

1. Bounded by the third degree curves $y^2 = x^3$ and $y = x^3$.
2. Inside both the circle $x^2 + y^2 = 2ax$ and the parabola $x^2 = ay$, $a > 0$.
3. Inside the circle $x^2 + y^2 = 25$ and above the line $3x = 4y$.
4. Inside both parabolas $y = x^2$ and $10x = y^2 - 15y + 24$.

In Exercises 5-8, determine the region over which the integration extends.

5. $\int_0^1 \int_0^{x+1} f(x, y) \, dy \, dx$.
6. $\int_0^2 \int_{y^2-3}^{4y} f(x, y) \, dx \, dy$.
7. $\int_0^1 \int_0^{(1-y^3)^{\frac{1}{3}}} f(x, y) \, dx \, dy$.
8. $\int_0^2 \int_{\sqrt{x}}^{\sqrt{20-x^2}} f(x, y) \, dy \, dx$.

In Exercises 9-12, transform the given iterated integral to one (or more) where the order of integration is reversed.

9. $\int_{-1}^2 \int_{-x}^{2-x^2} f(x, y) \, dy \, dx$.
10. $\int_0^2 \int_{y^2}^{4\sqrt{2y}} f(x, y) \, dx \, dy$.
11. $\int_1^4 \int_{y/2}^y f(x, y) \, dx \, dy$.
12. $\int_0^1 \int_0^{x+e^x} f(x, y) \, dy \, dx$.

In Exercises 13-16, evaluate the double integral $\iint_S f(x, y) \, dA$ by means of an iterated integral, for the given function $f(x, y)$ and set S.

13. $f(x, y) = x^2 + xy$; S is the square: $0 \leq x \leq 1, 0 \leq y \leq 1$.
14. $f(x, y) = x + y$; S is the triangular region bounded by $x = 2y, x = 6, y = 0$.
15. $f(x, y) = xy$; S is the first quadrant portion inside the circle $x^2 + y^2 = a^2$.
16. $f(x, y) = x^2 + y^2$; S is the region bounded by $x = 0, y = x$, and $y = e^{-x}$.
★17. Prove that Definition II, § 1302, is independent of the rectangle R. *Hints:* Show first that it is sufficient to establish this result for rectangles R_1 and R_2 such that $R_1 \subset R_2$. It is almost trivial to show that the existence of $\iint_{R_2} f_S(x, y) \, dA$

§ 1310] EXERCISES 449

implies that of $\iint_{R_1} f_x(x, y)\, dA$, and their equality. Proof of the reverse implication involves constructing thin strips enclosing the edges of R_1 as midlines, and of sufficiently small total area.

★18. Prove that every integrable function $f(x, y)$ is bounded.
★19. Prove the statement in the last sentence of Note 4, § 1303.
★20. Prove Theorem III, § 1303. *Hint:* Use Exs. 29 and 30, § 1025, and Theorem VII, § 1306.
★21. Complete the details of the proof of Theorem I, § 1305.
★22. Prove Theorems II and III, § 1305.
★23. Prove Theorems II-V, § 1306.
★24. Give the details of the proof of Theorem X, § 1306.
★25. Prove the Theorem of § 1310 for the special case where f is continuous and the partitions Π consist of pieces that are compact regions. *Hint:* By Theorem XIII, § 1306,

$$|\sum f(x_i, y_i)\, \Delta A_i - \iint f(x, y)\, dA|$$
$$= |\sum f(x_i, y_i)\, \Delta A_i - \sum f(\xi_i, \eta_i)\, \Delta A_i| \leq \sum |f(x_i, y_i) - f(\xi_i, \eta_i)|\, \Delta A_i.$$

★26. Give the details of the proofs of Lemmas 2-6, § 1307.
★27. Prove Fubini's Theorem, § 1309, for step-functions.
★28. Complete the details in the Proof of the Note of § 1308, given in § 1309.
★29. Prove Notes 1 and 2, § 1303. *Hints:* (i) First prove that since *any* rectangle has area (its frontier is rectifiable), the concept of *having area* is independent of the coordinate system. (ii) Then show that *area* is invariant under translations of axes. (iii) For a given rectangle R whose sides are not parallel to the coordinate axes, translate so that one vertex is at the origin and the adjacent sides have lengths a and b and direction cosines (λ_1, μ_1) and (λ_2, μ_2), respectively, where $\mu_1 > 0$ and $\mu_2 > 0$. Then (cf. Fig. 1305) by translation of the parts ψ_1 and ψ_2 of the upper curve of R to the functions χ_1 and χ_2 show that the total area of R is equal to $\left|\int_0^{a\lambda_1}(\chi_1 - \phi_1)\, dx\right| + \left|\int_0^{b\lambda_2}(\chi_2 - \phi_2)\, dx\right| = |a\lambda_1| \cdot b\mu_2 + |b\lambda_2| \cdot a\mu_1 =$

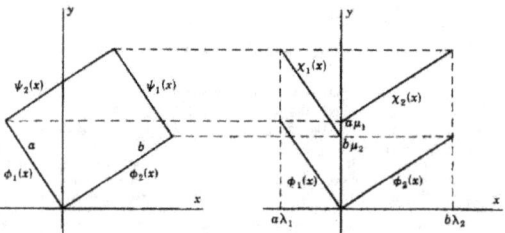

FIG. 1305

$ab(|\lambda_1|\mu_2 + |\lambda_2|\mu_1) = ab(\mu_1{}^2 + \mu_2{}^2) = ab$. (iv) Conclude that *area* is invariant under rotations.

★30. Prove that the area of a parallelogram three of whose vertices are (x_i, y_i), $i = 1, 2, 3$, is the absolute value of the determinant

$$(1) \qquad \begin{vmatrix} x_1 & y_1 & 1 \\ x_2 & y_2 & 1 \\ x_3 & y_3 & 1 \end{vmatrix}.$$

Hints: First show that (1) is invariant under translations, so that it can be assumed that $x_3 = y_3 = 0$ and that (1) has the form

$$(2) \qquad \begin{vmatrix} x_1 & y_1 \\ x_2 & y_2 \end{vmatrix}.$$

Then show that (2) is invariant under rotations, so that it can be assumed that (2) has the form

$$\begin{vmatrix} x_1 & 0 \\ x_2 & y_2 \end{vmatrix} = x_1 y_2.$$

Then compute the area by appropriate subdivision and translation of the parts of the upper curve, as in Exercise 29.

★31. Prove that if S is a bounded set in E_2 with area, then $I(S)$ and $\bar{S}$ have area, and $A(I(S)) = A(\bar{S}) = A(S)$. *Hint:* Show that $I(S) = S - F(S)$ and $\bar{S} = S \cup F(S)$. Cf. Theorem III, § 1303.

★32. Prove that a function integrable on a bounded set is integrable on any subset that has area. (Cf. Ex. 31, § 503.)

★33. Prove that $f(x, y)$ is integrable on a bounded set S with area if and only if corresponding to $\epsilon > 0$ there exists $\delta > 0$ such that for every net of norm $< \delta$ and every choice of points (x_i, y_i) and (x_i', y_i') of R_i, $i = 1, 2, \cdots, n$

$$\sum_{i=1}^{n} |f_S(x_i, y_i) - f_S(x_i', y_i')| \, \Delta A_i < \epsilon.$$

(Cf. Ex. 34, § 503.)

★34. Prove that the product of two functions integrable on a bounded set with area is integrable on that set. (Cf. Ex. 35, § 503, Ex. 33 above.)

★35. Prove that if $f(x, y)$ is integrable on a bounded set S with area, then so is $|f(x, y)|$. (Cf. Ex. 38, § 503.)

★36. State and prove the two-dimensional analogues of Exercises 39 and 40, § 503.

★37. Prove **Bliss's Theorem** for functions of two variables: *If* $f(x, y)$ *and* $g(x, y)$ *are integrable on a bounded set* S *with area, then the limit*

$$\lim_{|\mathfrak{N}| \to 0} \sum_{i=1}^{n} f_S(x_i, y_i) g_S(x_i', y_i') \, \Delta A_i$$

exists and is equal to $\iint_S f(x, y) g(x, y) \, dA$. (Cf. Ex. 41, § 503, Ex. 33 above.)

★38. Define the **lower** and **upper integrals** of a function $f(x, y)$ on a set S, and state and prove the analogues of the statements of Exercise 44, § 503.

★39. State and prove the analogue of Duhamel's Principle, § 1101, for functions of two variables, assuming continuity for all functions involved, and that the set of integration is compact with area.

★★40. Discuss Lebesgue measure zero for E_2. (Cf. Exs. 52 and 53, § 503.)

★★41. Prove that a function $f(x, y)$, defined on a bounded set with area, is integrable there if and only if it is bounded and almost everywhere continuous. (Cf. Ex. 54, § 503.)

1311. TRIPLE INTEGRALS. VOLUME

The **triple integral** of a function $f(x, y, z)$ over a closed rectangular parallelepiped R and, more generally, over a bounded set W, is defined (if it exists) in a manner similar to that used for a double integral. We shall omit all details in the text, but call on the student for a few in the exercises of § 1312. A few of the more important facts are listed below:

(*i*) The **integral** of $f(x, y, z)$ over W (if it exists) is the same as the integral of its extension f_W over $R \supset W$, defined in terms of a **net** $\mathfrak{N}$ of planes parallel to the coordinate planes, which subdivide R into smaller closed rectangular parallelepipeds R_i of volume ΔV_i, $i = 1, 2, \cdots, n$:

$$(1) \qquad \iiint_W f\, dV = \iiint_R f_W\, dV = \lim_{|\mathfrak{N}| \to 0} \sum_{i=1}^n f_W(x_i, y_i, z_i)\, \Delta V_i.$$

(*ii*) The **volume** of W (if it exists) is defined to be the integral of the function 1 over W:

$$(2) \qquad V(W) = \iiint_W 1\, dV.$$

(*iii*) The definition of (1) is independent of the containing R, and that of (2) is independent of the rectangular coordinate system. (Cf. Ex. 15, § 1312.)

(*iv*) A bounded set has volume if and only if its frontier has zero volume. The analogue of Theorem III, § 1303, holds in E_3. (Cf. Ex. 7, § 1312.)

(*v*) Any compact smooth surface has zero volume. Therefore any bounded set whose frontier consists of a finite number of compact smooth surfaces has volume. (Cf. Ex. 16, § 1312.)

(*vi*) A second formulation of the triple integral of f over W (bounded and with volume) is given by partitions Π of W into closed pieces W_i with volume V_i, $i = 1, \cdots, n$:

$$\iiint_W f\, dV = \lim_{|\Pi| \to 0} \sum_{i=1}^n f(x_i, y_i, z_i)\, \Delta V_i.$$

(Cf. Exs. 8, 31, § 1312.)

★(*vii*) The Theorems of § 1306 have analogues in E_3. In particular, step-functions in E_3 have definition and properties similar to those for step-functions in E_2. A function $f(x, y, z)$ is integrable over a bounded set W if and only if corresponding to $\epsilon > 0$ there exist step-functions

$\sigma(x, y, z)$ and $\tau(x, y, z)$ defined on a closed rectangular parallelepiped $R \supset W$ such that $\sigma \leq f \leq \tau$ on W, $\sigma \leq \tau$ on R, and $\iiint_R [\tau - \sigma] \, dV < \epsilon$. (Cf. Ex. 9, § 1312.)

(*viii*) A function bounded and very nearly everywhere continuous on a bounded set with volume is integrable there. (Cf. Ex. 10, § 1312.)

(*ix*) Under suitable conditions a triple integral can be evaluated by means of a (three-fold) **iterated integral** of the form

$$(3) \quad \int_{a_1}^{a_2} \int_{y_1(x)}^{y_2(x)} \int_{z_1(x, y)}^{z_2(x, y)} f(x, y, z) \, dz \, dy \, dx,$$

or one of the other 5 forms got by permuting the variables.

★(*x*) **Fubini's Theorem.** *If $f(x, y, z)$ is defined over R: $a_1 \leq x \leq a_2$, $b_1 \leq y \leq b_2$, $c_1 \leq z \leq c_2$, then*

$$(4) \quad \iiint_R f(x, y, z) \, dV = \int_{a_1}^{a_2} \int_{b_1}^{b_2} \int_{c_1}^{c_2} f(x, y, z) \, dz \, dy \, dx$$

whenever both integrals exist. (Cf. Ex. 11, § 1312.)

(*xi*) **Fundamental Theorem for Triple Integrals.** *If $f(x, y, z)$ is defined over a bounded set W consisting of all (x, y, z) such that $a_1 \leq x \leq a_2$, $y_1(x) \leq y \leq y_2(x)$, and $z_1(x, y) \leq z \leq z_2(x, y)$, then*

$$(5) \quad \iiint_W f(x, y, z) \, dV = \int_{a_1}^{a_2} \int_{y_1(x)}^{y_2(x)} \int_{z_1(x, y)}^{z_2(x, y)} f(x, y, z) \, dz \, dy \, dx$$

whenever both integrals exist. (Cf. Fig. 1306, Ex. 12, § 1312.)

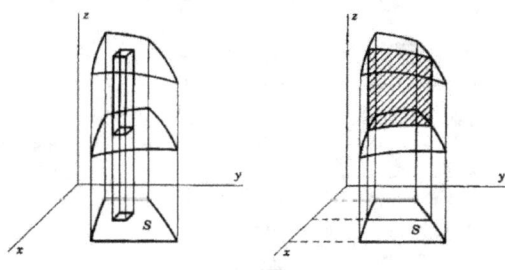

FIG. 1306

§ 1312] EXERCISES 453

(xii) If S is a bounded set with area in the xy-plane, and if W is a bounded set consisting of all (x, y, z) such that $(x, y) \in S$ and $z_1(x, y) \leq z \leq z_2(x, y)$, then

(6) $$V(W) = \iint_S [z_2(x, y) - z_1(x, y)] \, dA$$

whenever both members exist. (Cf. Fig. 1306, Ex. 13, § 1312.)

(xiii) If S is a bounded set with area in the xy-plane, if $f(x, y)$ is defined, bounded, and nonnegative over S, and if W is the set of all (x, y, z) such that $(x, y) \in S$ and $0 \leq z \leq f(x, y)$, then W has volume if and only if $f(x, y)$ is integrable over S and, in case of existence,

$$V(W) = \iint_S f(x, y) \, dA.$$

(Cf. Ex. 14, § 1312.)

Example. Write the triple integral $\iiint_W f(x, y, z) \, dV$ as an iterated integral, and compute the volume of W by setting $f(x, y, z) = 1$, where W is the region bounded by the paraboloids $z = x^2 + y^2$ and $2z = 12 - x^2 - y^2$.

Solution. The curve of intersection is the circle $x^2 + y^2 = 4$, $z = 4$. Therefore the integral can be written

$$\int_{-2}^{2} \int_{-\sqrt{4-x^2}}^{\sqrt{4-x^2}} \int_{x^2+y^2}^{\frac{12-x^2-y^2}{2}} f(x, y, z) \, dz \, dy \, dx.$$

If $f(x, y, z) = 1$, properties of symmetry (that is, of even functions) can be used to simplify this to

$$V = 6 \int_0^2 \int_0^{\sqrt{4-x^2}} (4 - x^2 - y^2) \, dy \, dx = 4 \int_0^2 (4 - x^2)^{\frac{3}{2}} \, dx = 12\pi.$$

1312. EXERCISES

In Exercises 1 and 2, write the triple integral $\iiint_W f(x, y, z) \, dV$ as an iterated integral. Compute the volume of W by setting $f(x, y, z) = 1$.

1. W is the region bounded by the plane $z = 7$ and the paraboloid $z = 23 - x^2 - y^2$.

2. W is the region inside the octahedron $|x| + |y| + |z| = 1$.

In Exercises 3 and 4, find the region of integration.

3. $\int_0^1 \int_{y^2}^{\sqrt{y}} \int_0^{9-x^2-y^2} f(x, y, z) \, dz \, dx \, dy.$

4. $\int_1^e \int_x^{x^2} \int_0^{\ln x} f(x, y, z) \, dz \, dy \, dx.$

5. If W is the region bounded below by the xy-plane and above by the sphere $x^2 + y^2 + z^2 = 2z$, express the triple integral $\iiint_W f(x, y, z)\, dV$ in six ways as iterated integrals.

6. If W is the region bounded by the planes $x = 1$, $y = 0$, $z = 0$, and $x - y - z = 0$, evaluate $\iiint_W x^2 y^2 z\, dV$ in at least 3 of the 6 possible ways in terms of iterated integrals.

In Exercises 7–14, discuss and prove the statement in the indicated paragraph of § 1311.

★**7.** (iv).   ★**8.** (vi), corresponding to Ex. 25, § 1310.
★**9.** (vii).   ★**10.** $(viii)$.
★**11.** (x).   ★**12.** (xi).
★**13.** (xii).   ★**14.** $(xiii)$.

★**15.** Discuss and prove the statements of (iii), § 1311. *Hint:* In order to prove that the volume of any rectangular parallelepiped with edges a, b, and c is abc, translate one vertex to the origin and let the neighboring vertices be $(a\lambda_1, a\mu_1, a\nu_1)$, $(b\lambda_2, b\mu_2, b\nu_2)$, and $(c\lambda_3, c\mu_3, c\nu_3)$, where $\nu_i \geq 0$, $i = 1, 2, 3$. As in Ex. 29, § 1310, translate the upper planes to obtain an integral of constant differences over parallelograms in the xy-plane. Use Ex. 30, § 1310, to obtain

$$V = a\nu_1 \begin{Vmatrix} b\lambda_2 & b\mu_2 \\ c\lambda_3 & c\mu_3 \end{Vmatrix} + b\nu_2 \begin{Vmatrix} c\lambda_3 & c\mu_3 \\ a\lambda_1 & a\mu_1 \end{Vmatrix} + c\nu_3 \begin{Vmatrix} a\lambda_1 & a\mu_1 \\ b\lambda_2 & b\mu_2 \end{Vmatrix}$$

$$= abc \left\{ \nu_1 \begin{Vmatrix} \lambda_2 & \mu_2 \\ \lambda_3 & \mu_3 \end{Vmatrix} + \cdots \right\} = abc(\nu_1^2 + \nu_2^2 + \nu_3^2) = abc.$$

★**16.** Prove that any compact smooth surface has zero volume (v), § 1311). *Hints:* By use of the Heine-Borel Theorem (Ex. 24, § 1021) and the Implicit Function Theorem (§ 1228) show that the problem can be reduced to showing that if $z = f(x, y)$ is continuous on a closed rectangle, its graph has zero volume.

★**17.** Prove that the volume of a parallelepiped four of whose noncoplanar vertices are (x_i, y_i, z_i), $i = 1, 2, 3, 4$, is the absolute value of the determinant

$$(1) \qquad \begin{vmatrix} x_1 & y_1 & z_1 & 1 \\ x_2 & y_2 & z_2 & 1 \\ x_3 & y_3 & z_3 & 1 \\ x_4 & y_4 & z_4 & 1 \end{vmatrix}.$$

(Cf. Ex. 30, § 1310.)

★**18.** Justify the method used in Integral Calculus for obtaining a volume by integrating a cross-section area. Discuss in particular volumes of revolution.

★**19.** Justify the method of "cylindrical shells" used in Integral Calculus for obtaining volumes of revolution.

In Exercises 20–30, state and prove the three-dimensional analogue of the statement of the indicated Exercise of § 1310.

★**20.** Ex. 31.   ★**21.** Ex. 32.   ★**22.** Ex. 33.
★**23.** Ex. 34.   ★**24.** Ex. 35.   ★**25.** Ex. 36.
★**26.** Ex. 37.   ★**27.** Ex. 38.   ★**28.** Ex. 39.

§ 1313] DOUBLE INTEGRALS IN POLAR COORDINATES 455

★★**29.** Ex. 40.   ★★**30.** Ex. 41.
★★**31.** Prove (vi), § 1311, in general (cf. § 1307).

1313. DOUBLE INTEGRALS IN POLAR COORDINATES

Frequently a problem whose solution depends upon the evaluation of a double integral is formulated most naturally in terms of polar coordinates. In many other cases the evaluation of a double integral in rectangular coordinates is greatly simplified by means of a change of coordinates, from rectangular to polar. Let S be a set with area contained in a **polar rectangle** R, $0 < \mu \leq \rho \leq \nu$, $\alpha \leq \theta \leq \beta$ (cf. Fig. 1307), over which is imposed a

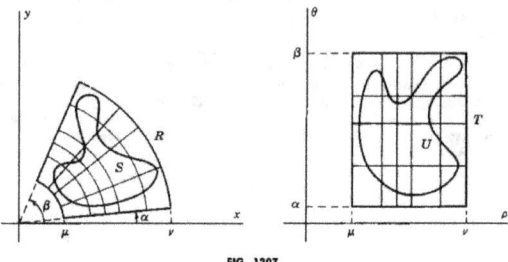

FIG. 1307

polar net $\mathfrak{N}$, consisting of circles $\rho =$ constant and rays $\theta =$ constant, which cut R into smaller polar rectangles $R_1, R_2, \cdots, R_n$. This corresponds to a rectangular net in the $\rho\theta$-plane (cf. Fig. 1307), which cuts the set U that corresponds to S and the set T that corresponds to R into smaller pieces. Let $f(x, y)$ be a function integrable on S, define f_S to be equal to f on S and equal to 0 outside S, and let $F(\rho, \theta) \equiv f(\rho \cos \theta, \rho \sin \theta)$, $F_1(\rho, \theta) \equiv f_S(\rho \cos \theta, \rho \sin \theta)$. Then the integral of $f(x, y)$ over S can be expressed, by the second formulation of the double integral, § 1304, as the limit

$$(1) \qquad \lim_{|\mathfrak{N}| \to 0} \sum_{i=1}^{n} f_S(x_i, y_i) \, \Delta A_i,$$

where (x_i, y_i) is an arbitrary point and ΔA_i is the area of the polar rectangle R_i.

It can be shown (Ex. 14, § 1315) that the area of a polar rectangle $\rho_1 \leq \rho \leq \rho_2$, $\theta_1 \leq \theta \leq \theta_2$ ($\rho_1 > 0$) is equal to $\bar{\rho} \, \Delta\rho \, \Delta\theta$, where $\bar{\rho}$ is the average radius $\frac{1}{2}(\rho_1 + \rho_2)$, $\Delta\rho \equiv \rho_2 - \rho_1$, and $\Delta\theta \equiv \theta_2 - \theta_1$. The limit (1) can therefore be written in the form

486   MULTIPLE INTEGRALS   [§ 1313

(2) $$\lim_{|\mathfrak{N}|\to 0} \sum_{i=1}^{n} f_S(x_i, y_i)\bar{\rho}_i \, \Delta\rho_i \, \Delta\theta_i,$$

where $\bar{\rho}_i$, $\Delta\rho_i$, and $\Delta\theta_i$ have the meanings just prescribed, for the polar rectangle R_i, and for expediency about to be realized $f_S(x, y)$ is evaluated at a point (x_i, y_i) of R_i whose ρ-coordinate is $\bar{\rho}_i$.

It now becomes apparent that the limit (2) can be interpreted as the double integral over the rectangle T (the variables being ρ and θ) of the *new* function $F_U(\rho, \theta) \cdot \rho$. If we write this new double integral as an iterated integral (assuming of course that the integrals exist), we obtain:

(3) $$\iint_S f(x, y) \, dA = \iint_R f_S(x, y) \, dA = \iint_T F_U(\rho, \theta)\rho \, dA$$
$$= \iint_U F(\rho, \theta)\rho \, dA = \int_\alpha^\beta \int_\mu^\nu F_U(\rho, \theta)\rho \, d\rho \, d\theta = \int_\mu^\nu \int_\alpha^\beta F_U(\rho, \theta)\rho \, d\theta \, d\rho.$$

Before formalizing our results we observe:

NOTE. The restriction that μ be *positive* may be replaced by the assumption $\mu \geqq 0$. To be sure, the correspondence between R and T in Figure 1308 is no longer one-to-one (the origin in the xy-plane corresponds to the entire θ-axis in the $\rho\theta$-plane), but if ρ is sufficiently small, the two corresponding shaded regions in Figure 1308 are both of area less than any preassigned positive number. In the

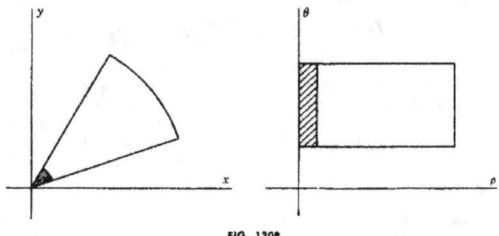

FIG. 1308

limit they disappear. Therefore, equality between two integrals which exists for the two corresponding unshaded regions of Figure 1308 is also valid for the entire sector (extending to the origin) and rectangle (extending to the θ-axis).

By virtue of (3), the preceding Note, and the Fundamental Theorem, § 1308, we have:

Theorem. *If $f(x, y)$ is defined over a bounded set S, bounded toward the origin and away from the origin by the graphs of functions $\rho = \rho_1(\theta)$ and*

§1314] VOLUMES WITH POLAR COORDINATES 457

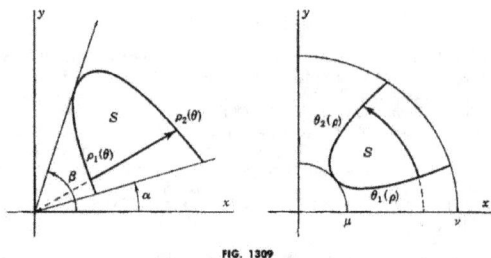

FIG. 1309

$\rho = \rho_1(\theta)$, respectively, and by the two rays $\theta = \alpha$ and $\theta = \beta$ (cf. Fig. 1309), and if $F(\rho, \theta) = f(\rho \cos \theta, \rho \sin \theta)$ then the following relation holds whenever the two integrals exist:

(4) $$\iint_S f(x, y) \, dA = \int_\alpha^\beta \int_{\rho_1(\theta)}^{\rho_2(\theta)} F(\rho, \theta) \rho \, d\rho \, d\theta.$$

A similar statement holds for the reverse order of integration (cf. Fig. 1309):

(5) $$\iint_S f(x, y) \, dA = \int_\mu^\nu \int_{\theta_1(\rho)}^{\theta_2(\rho)} F(\rho, \theta) \rho \, d\theta \, d\rho.$$

Example 1. Find the total area inside the lemniscate $\rho^2 = a^2 \cos 2\theta$.

Solution. Integrating the function 1, and using symmetry, we have

$$A = 4 \int_0^{\frac{\pi}{4}} \int_0^{a\sqrt{\cos 2\theta}} \rho \, d\rho \, d\theta = 2a^2 \int_0^{\frac{\pi}{4}} \cos 2\theta \, d\theta = a^2.$$

Example 2. Evaluate $\int_0^2 \int_0^{\sqrt{4-x^2}} (x^2 + y^2) \, dy \, dx$.

Solution. The region of integration is bounded in the first quadrant by the coordinate planes and the circle $x^2 + y^2 = 4$. Transforming the given integral I to polar coordinates, we have

$$I = \int_0^{\frac{\pi}{2}} \int_0^2 \rho^2 \cdot \rho \, d\rho \, d\theta = \int_0^{\frac{\pi}{2}} \left[\frac{\rho^4}{4} \right]_0^2 d\theta = 2\pi.$$

1314. VOLUMES WITH DOUBLE INTEGRALS IN POLAR COORDINATES

Combining statements of §§ 1311 and 1313 regarding volumes and the method of evaluating double integrals by means of polar coordinates, we

458   MULTIPLE INTEGRALS   [§ 1314

have for the volume under the surface $z = f(x, y)$ $(f \geqq 0)$, with $F(\rho, \theta) \equiv f(\rho \cos \theta, \rho \sin \theta)$:

$$V = \int_\alpha^\beta \int_{\rho_1(\theta)}^{\rho_2(\theta)} F(\rho, \theta)\rho \, d\rho \, d\theta \quad \text{or} \quad \int_\mu^\nu \int_{\theta_1(\rho)}^{\theta_2(\rho)} F(\rho, \theta)\rho \, d\theta \, d\rho,$$

and for the volume between the surfaces $z = f(x, y)$ and $z = g(x, y)$ $(f \leqq g)$, with $F(\rho, \theta) \equiv f(\rho \cos \theta, \rho \sin \theta)$, $G(\rho, \theta) \equiv g(\rho \cos \theta, \rho \sin \theta)$:

$$\int_\alpha^\beta \int_{\rho_1(\theta)}^{\rho_2(\theta)} [G(\rho, \theta) - F(\rho, \theta)]\rho \, d\rho \, d\theta \quad \text{or} \quad \int_\mu^\nu \int_{\theta_1(\rho)}^{\theta_2(\rho)} [G(\rho, \theta) - F(\rho, \theta)]\rho \, d\rho \, d\theta,$$

where the limits of integration are determined by the region of integration.

Example 1. Find the volume bounded by the half-cone $z = 2\sqrt{5(x^2 + y^2)}$ and the paraboloid $z = 20 - 2(x^2 + y^2)$.

Solution. The two surfaces meet in the circle $x^2 + y^2 = 5, z = 10$. Therefore the volume can be expressed

$$V = \int_{-\sqrt{5}}^{\sqrt{5}} \int_{-\sqrt{5-x^2}}^{\sqrt{5-x^2}} [20 - 2(x^2 + y^2) - 2\sqrt{5(x^2 + y^2)}] \, dy \, dx$$

$$= \int_0^{2\pi} \int_0^{\sqrt{5}} [20 - 2\rho^2 - 2\sqrt{5}\rho]\rho \, d\rho \, d\theta = \frac{125\pi}{3}.$$

Example 2. Find the volume common to a sphere of radius a, a right circular half-cone of generating angle η, and a dihedral wedge of angle α, if the vertex of the half-cone is at the center of the sphere and the two half-planes of the wedge contain the axis of the cone.

Solution. Let the positive half of the z-axis be the axis of the half-cone, with vertex at the origin and let the half-planes be $\theta = 0$ and $\theta = \alpha$ (cf. Fig. 1310).

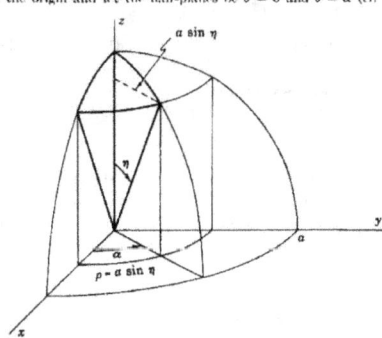

FIG. 1310

§ 1316] MASS OF A PLANE REGION

The equations of the cone and sphere, in cylindrical coordinates (§ 1111), are $z = \rho \cot \eta$ and $\rho^2 + z^2 = a^2$, respectively, so that the desired volume is $\int_0^\alpha \int_0^{a \sin \eta} [\sqrt{a^2 - \rho^2} - \rho \cot \eta] \rho \, d\rho \, d\theta = \frac{1}{3}\alpha a^3(1 - \cos \eta)$. The special case of a hemisphere is given by $\alpha = 2\pi$, $\eta = \pi/2$; $V = \frac{2}{3}\pi a^3$.

1315. EXERCISES

In Exercises 1-4, evaluate the given integral by changing to polar coordinates.

1. $\int_0^a \int_0^{\sqrt{a^2-x^2}} x^2 \, dy \, dx$.
2. $\int_0^a \int_0^{\sqrt{ax-x^2}} x^2 \, dy \, dx$.
3. $\int_0^1 \int_0^x (x^2 + y^2)^{\frac{1}{2}} \, dy \, dx$.
4. $\int_0^2 \int_0^{2\sqrt{4-x^2}} \sqrt{4x^2 + y^2} \, dy \, dx$.

Hint for Ex. 4: Replace the letter y by the letter z, and then substitute $z = 2y$ to obtain $4 \int_0^2 \int_0^{\sqrt{4-x^2}} \sqrt{x^2 + y^2} \, dy \, dx$.

In Exercises 5-8, find the area of the given region.

5. Inside the cardioid $\rho = 1 + \cos \theta$.
6. One loop of the four-leaved rose $\rho = a \sin 2\theta$.
7. Inside the circle $\rho = 2a \cos \theta$, outside the circle $\rho = a$.
8. Inside the small loop of the limaçon $\rho = 1 + 2 \cos \theta$.

In Exercises 9-12, find the volume of the given region.

9. Above the plane $z = 0$, below the cone $z = \rho$, inside the cylinder $\rho = 2a \cos \theta$.
10. Inside both the sphere $x^2 + y^2 + z^2 = a^2$ and the cylinder $\rho = a \cos \theta$.
11. Bounded by the sphere $x^2 + y^2 + z^2 = 20$ and the paraboloid $z = x^2 + y^2$.
12. Bounded by the plane $x + y + z = \frac{1}{2}$ and the paraboloid $z = x^2 + y^2$.

13. Express the integral $\int_0^{\pi/4} \int_0^{a \sec \theta} F(\rho, \theta) \rho \, d\rho \, d\theta$ by means of iterated integrals where the order of integration is reversed.

14. Show that the area of the polar rectangle $\mu \leq \rho \leq \nu$, $\alpha \leq \theta \leq \beta$ ($\mu \geq 0$) is equal to $\frac{1}{2}(\mu + \nu)(\nu - \mu)(\beta - \alpha)$, by means of rectangular coordinates. *Hint:* First show that the formula is correct when $\mu = 0$, $\beta = \pi/2$, and $\beta - \alpha$ is an acute angle.

1316. MASS OF A PLANE REGION OF VARIABLE DENSITY

A continuous positive function $\delta(x, y)$ over a bounded set S with positive area can be interpreted as a density function.† The **mass** of S is the integral of the density function:

$$(1) \qquad M(S) \equiv \iint_S \delta(x, y) \, dA.$$

† Such a set in E_2 with a density function is sometimes called a **lamina**. It is a mathematical idealization of a portion of a thin metal sheet of variable density or thick-

Justification for these definitions lies primarily in two facts: (i) If the density is constant over S, the mass is equal to the product of the area and the density: $M(S) = \delta \cdot A(S)$. In other words, *constant density is mass per unit area*. (ii) The ratio of the mass of a compact region R to its area is equal, by the Mean Value Theorem for double integrals (Theorem XIII, § 1306), to an *average density* $\delta(\xi, \eta)$ evaluated at a point (ξ, η) of R. Therefore, by permitting R to shrink to any fixed point p of R, we conclude that *density at a point can be regarded as a limit of mass per unit area*.

Example. Find the mass of a circle of radius a if its density is k times the distance from the center.

Solution. With the aid of polar coordinates we have

$$M(S) = k \iint_S \sqrt{x^2 + y^2}\, dA = k \int_0^{2\pi} \int_0^a \rho^2\, d\rho\, d\theta = \frac{2k\pi a^3}{3}.$$

1317. MOMENTS AND CENTROID OF A PLANE REGION

For a bounded set S with positive area, and a density function $\delta(x, y)$, we have the following *definitions*, whose justification (Ex. 13, § 1318) lies in a comparison of the approximating sums for the double integrals with the discrete case of a finite collection of point masses:

First moment with respect to the y-axis: $\quad M_y \equiv \iint_S \delta(x, y) x\, dA.$

First moment with respect to the x-axis: $\quad M_x \equiv \iint_S \delta(x, y) y\, dA.$

Second moment with respect to the y-axis: $\quad I_y \equiv \iint_S \delta(x, y) x^2\, dA.$

Second moment with respect to the x-axis: $\quad I_x \equiv \iint_S \delta(x, y) y^2\, dA.$

Polar second moment with respect to the origin:

$$I_0 \equiv \iint_S \delta(x, y)(x^2 + y^2)\, dA.$$

Centroid: $(\bar{x}, \bar{y}) \equiv \left(\dfrac{M_y}{M}, \dfrac{M_x}{M} \right).$

NOTE 1. In case of a density function δ identically equal to 1, the concepts defined above are purely geometric in character. However, they retain considerable interest and practicality in many fields, including Engineering and Statistics. The formulas for centroid become

$$(\bar{x}, \bar{y}) = \left(\frac{M_y}{A}, \frac{M_x}{A} \right).$$

NOTE 2. With a variable density function associated with a mass distribution, second moments are also called **moments of inertia** (this explains the notation I), and the centroid is also called the **center of mass**.

NOTE 3. The following relation always obtains:

$$I_0 = I_x + I_y.$$

More generally, first and second moments of a set S can be defined with respect to an arbitrary straight line L:

$$M_L = \iint_S \delta(x,y) D(x,y) \, dA,$$

$$I_L = \iint_S \delta(x,y)[D(x,y)]^2 \, dA,$$

where $D(x,y)$ is either of the two possible directed (perpendicular) distances from the line L to the point (x,y).

Similarly the polar second moment can be extended to an arbitrary fixed point $p_0 : (x_0, y_0)$:

$$I_{p_0} = \iint_S \delta(x,y)[(x - x_0)^2 + (y - y_0)^2] \, dA.$$

NOTE 4. The centroid has the important property that for any line L through the centroid the first moment M_L is zero. This means that the centroid can be uniquely defined as the point that has this property, and is therefore independent of the coordinate system. (Cf. Ex. 14, § 1318.)

NOTE 5. The centroid can also be defined as the point p that minimizes the polar moment I_p. (Cf. Ex. 15, § 1318.)

Example. Find M_y, M_x, I_y, I_x, I_0, and the centroid for the quarter-circle bounded in the first quadrant by the coordinate axes and the circle $x^2 + y^2 = a^2$. (Assume constant density $\delta = 1$.)

Solution. By symmetry and Note 3,

$$M_y = M_x = \iint_S x \, dA = \int_0^{\frac{\pi}{2}} \int_0^a \rho^2 \cos\theta \, d\rho \, d\theta = \frac{a^3}{3},$$

$$I_x = I_y = \frac{1}{2} I_0 = \frac{1}{2} \int_0^{\frac{\pi}{2}} \int_0^a \rho^3 \, d\rho \, d\theta = \frac{\pi a^4}{16}.$$

The centroid is $\left(\dfrac{4a}{3\pi}, \dfrac{4a}{3\pi}\right)$.

1318. EXERCISES

In Exercises 1–4, find the centroid of the given region. Assume $\delta = 1$.

1. Bounded in the first quadrant by the parabola $y^2 = 2px$ and the lines $x = b$ and $y = 0$.
2. First quadrant portion inside the hypocycloid $x^{\frac{2}{3}} + y^{\frac{2}{3}} = a^{\frac{2}{3}}$.
3. Inside the circle $x^2 + y^2 = 2ay$ and outside the circle $x^2 + y^2 = a^2$.
4. First-quadrant loop of the four-leaved rose $\rho = a \sin 2\theta$.

In Exercises 5-8, find the centroid of the given region with the given density function.

5. The square $0 \leq x \leq 1, 0 \leq y \leq 1; \delta(x, y) = kx$.
6. First quadrant portion inside the circle $x^2 + y^2 = a^2; \delta(x, y) = k(x^2 + y^2)$.
7. Inside the circle $x^2 + y^2 = 2ax; \delta(x, y) = k\sqrt{x^2 + y^2}$.
8. First-quadrant loop of the four-leaved rose $\rho = a \sin 2\theta; \delta(\rho, \theta) = k\rho$.

In Exercises 9-12, find the specified second moment for the given set and density function.

9. I_x; region bounded by the ellipse $b^2x^2 + a^2y^2 = a^2b^2; \delta = 1$.
10. I_y; inside the lemniscate $\rho^2 = a^2 \cos 2\theta; \delta = 1$.
11. I_y; triangle with vertices $(0, 0), (h, a), (h, a + b), h > 0, b > 0; \delta = kx$.
12. I_0; inside the cardioid $\rho = a(1 + \cos \theta), a > 0; \delta = k\rho$.
13. Give the details in the justification of the definitions of § 1317, called for in the first paragraph, § 1317.
14. Prove the statements of Note 4, § 1317, paying special attention to uniqueness.
15. Let $(\bar{x}, \bar{y})$ be the centroid of a bounded set S with area and a density function, and let $I_{x=\bar{x}}, I_{y=\bar{y}}$, and $I_{(\bar{x}, \bar{y})}$ denote the second moments of S with respect to the lines $x = \bar{x}$ and $y = \bar{y}$, and the point $(\bar{x}, \bar{y})$, respectively. Establish the formulas

$$I_y = I_{x=\bar{x}} + M(S)\bar{x}^2,$$
$$I_x = I_{y=\bar{y}} + M(S)\bar{y}^2,$$
$$I_0 = I_{(\bar{x}, \bar{y})} + M(S)(\bar{x}^2 + \bar{y}^2).$$

Hence prove Note 5, § 1317.

16. Show that the centroid of a triangle (of constant density) is the point of intersection of its medians.

1319. TRIPLE INTEGRALS, CYLINDRICAL COORDINATES

In order to evaluate a triple integral in terms of cylindrical coordinates (cf. § 1111), we proceed as with double integrals in polar coordinates, and use the second formulation for the triple integral (cf. (vi), § 1311), with a decomposition of a master "polar rectangular parallelepiped" into smaller ones. A typical element of volume is shown in Figure 1311. The actual volume of this element of volume is equal (by § 1313) to $\bar{\rho} \Delta z \Delta \rho \Delta \theta$, where $\bar{\rho}$ is the average of the two extreme values of ρ. Therefore (cf. § 1313), the triple integral of a function $F(\rho, \theta, z) = f(x, y, z)$ over a bounded set W, expressed as the triple integral of the extended function F_W over a polar rectangular parallelepiped R containing W (cf. § 1311), can be written

$$(1) \quad \iiint_W f \, dV = \lim_{|\Re| \to 0} \sum_{i=1}^{n} f_W(x_i, y_i, z_i) \Delta V_i$$

$$= \lim_{|\Re| \to 0} \sum_{i=1}^{n} F_{W^*}(\bar{\rho}_i, \theta_i, z_i) \bar{\rho}_i \Delta z_i \Delta \rho_i \Delta \theta_i$$

$$= \int_a^b \int_\mu^\nu \int_c^d F_{W^*}(\rho, \theta, z) \rho \, dz \, d\rho \, d\theta,$$

§ 1320] SPHERICAL COORDINATES 463

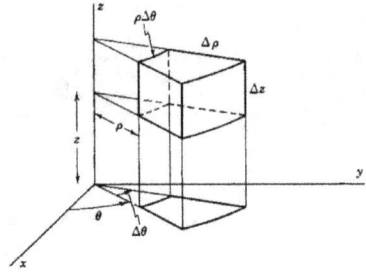

FIG. 1311

the iterated integral being extended over the rectangular parallelepiped in the $\rho\theta z$-space that corresponds to R, W^* being the set that corresponds to W.

By the Fundamental Theorem for triple integrals (cf. (xi), § 1311), the iterated integral of (1) can be written with six orders of integration. For a suitably defined set W, one of these integrals becomes

(2) $\quad \iiint_W f\, dV = \int_\alpha^\beta \int_{\rho_1(\theta)}^{\rho_2(\theta)} \int_{z_1(\rho,\theta)}^{z_2(\rho,\theta)} F(\rho, \theta, z)\rho\, dz\, d\rho\, d\theta.$

NOTE. The assumption $\rho > 0$ can be replaced by the weaker inequality $\rho \geqq 0$, for the same reasons in E_3 that apply in E_2. (Cf. the Note, § 1313.)

Example. Use cylindrical coordinates to compute the volume of a right circular cone of base radius a and altitude h.

Solution. Let the equation of the cone be $az = h\rho$. Then the desired volume is

$$\int_0^{2\pi} \int_0^a \int_{\frac{h\rho}{a}}^h \rho\, dz\, d\rho\, d\theta = \frac{h}{a}\int_0^{2\pi}\int_0^a \rho(a-\rho)\, d\rho\, d\theta = \frac{\pi a^2 h}{3}.$$

1320. TRIPLE INTEGRALS, SPHERICAL COORDINATES

Following the pattern used for cylindrical coordinates in § 1319, we seek suitable specifications for the second formulation of the triple integral in spherical coordinates. The type of *"polar rectangular parallelepiped"* used in this case is one that is bounded by two spheres with centers at the origin ($r =$ constant), two half-cones with axes on the z-axis ($\phi =$ constant), and two half-planes through the z-axis ($\theta =$ constant) (cf. Figure 1312). If the increments are small it appears that the volume of this *element of*

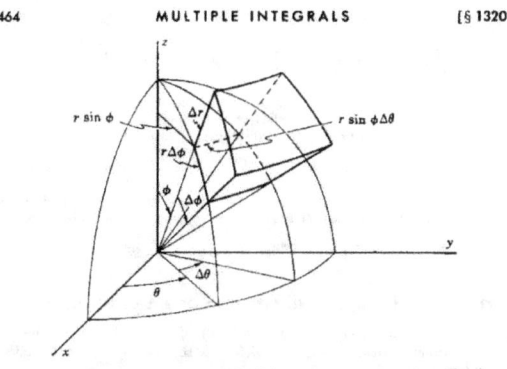

FIG. 1312

volume should be approximately equal to the product of the lengths of the three mutually orthogonal "sides," Δr, $r\,\Delta\phi$, and $r\sin\phi\,\Delta\theta$. That is, we should expect to have (approximately) $\Delta V = r^2 \sin\phi\,\Delta r\,\Delta\phi\,\Delta\theta$.

We shall now find the *precise* value of the element of volume. In Example 2, § 1314, the volume common to a sphere of radius a, a right-circular half-cone of generating angle η, and a dihedral wedge of angle α, if the vertex of the half-cone is at the center of the sphere and the two half-planes of the wedge contain the axis of the cone, was found to be $\frac{1}{3}\alpha a^3(1 - \cos\eta)$. Therefore (as we see by subtraction) the volume bounded by such a wedge, between two such concentric spheres of radius a and b, $a < b$, and two such half-cones of generating angles η and ζ, $\eta < \zeta$, is $\frac{1}{3}\alpha(b^3 - a^3)(\cos\eta - \cos\zeta)$. By the Law of the Mean (§ 405), $(b^3 - a^3) = (b - a)\cdot 3\bar{r}^2$ where $a < \bar{r} < b$, and $(\cos\eta - \cos\zeta) = (\zeta - \eta)\cdot\sin\bar{\phi}$ where $\eta < \bar{\phi} < \zeta$. Therefore, with the notation indicated in Figure 1312, the element of volume is exactly equal to

(1) $$\Delta V = \bar{r}^2 \sin\bar{\phi}\,\Delta r\,\Delta\phi\,\Delta\theta,$$

where $\bar{r}$ is between r and $r + \Delta r$, and $\bar{\phi}$ is between ϕ and $\phi + \Delta\phi$. Therefore the approximating sums for $\iiint_W f(x, y, z)\,dV$ can be written

$\sum_{i=1}^{n} f_W(x_i, y_i, z_i)\bar{r}_i^2 \sin\bar{\phi}_i\,\Delta r_i\,\Delta\phi_i\,\Delta\theta_i$, where $x_i = \bar{r}_i \sin\bar{\phi}_i \cos\theta_i$, $y_i = \bar{r}_i \sin\bar{\phi}_i \sin\theta_i$, $z_i = \bar{r}_i \cos\bar{\phi}_i$, and the iterated integral takes the form (for one of the six possible orders of integration):

§ 1321] MASS, MOMENTS, AND CENTROID 465

(2) $\iiint_W f\, dV = \int_\alpha^\beta \int_{\phi_1(\theta)}^{\phi_2(\theta)} \int_{r_1(\phi,\theta)}^{r_2(\phi,\theta)} F(r, \phi, \theta)\, r^2 \sin\phi\, dr\, d\phi\, d\theta.$

NOTE. As with the variable ρ in polar coordinates in E_2 (and also in E_3) so with the variable r in spherical coordinates, the assumption $r > 0$ can be replaced by the weaker inequality $r \geqq 0$. (Cf. the Notes, §§ 1313, 1319.)

Example. Evaluate $\int_{-a}^{a} \int_{-\sqrt{a^2-x^2}}^{\sqrt{a^2-x^2}} \int_{a-\sqrt{a^2-x^2-y^2}}^{a+\sqrt{a^2-x^2-y^2}} \sqrt{x^2+y^2+z^2}\, dz\, dy\, dx.$

Solution. The region of integration is the interior of the sphere $x^2 + y^2 + z^2 = 2az$. In spherical coordinates this is $r = 2a\cos\phi$, and the integral becomes

$$\int_0^{2\pi} \int_0^{\pi/2} \int_0^{2a\cos\phi} r^3 \sin\phi\, dr\, d\phi\, d\theta = \frac{8\pi a^4}{5}.$$

1321. MASS, MOMENTS, AND CENTROID OF A SPACE REGION

As in the two-dimensional case (§ 1316), if $\delta(x, y, z)$ is a continuous positive function over a compact region W with volume, the **mass** is given by integration:

(1) $\qquad M \equiv \iiint_W \delta(x, y, z)\, dV.$

The first moment of W is defined with respect to any plane, and the second moment (or moment of inertia) with respect to any plane, line or point. We give here only typical formulas for coordinate planes and axes and the origin (cf. § 1317):

First moment with respect to the yz-plane: $M_{yz} \equiv \iiint_W \delta(x, y, z) x\, dV.$

Second moment with respect to the yz-plane: $I_{yz} \equiv \iiint_W \delta(x, y, z) x^2\, dV.$

Second moment with respect to the x-axis: $I_x \equiv \iiint_W \delta(x, y, z)(y^2 + z^2)\, dV.$

Polar second moment with respect to the origin:

$$I_0 \equiv \iiint_W \delta(x, y, z)(x^2 + y^2 + z^2)\, dV.$$

Centroid: $(\bar{x}, \bar{y}, \bar{z}) \equiv \left(\dfrac{M_{yz}}{M}, \dfrac{M_{xz}}{M}, \dfrac{M_{xy}}{M} \right).$

NOTE 1. The second moments satisfy such relations as

$$I_{xy} + I_{xz} = I_x,$$
$$I_{xy} + I_{xz} + I_{yz} = I_0,$$

466   MULTIPLE INTEGRALS   [§ 1322

NOTE 2. Although the preceding formulas are expressed in terms of rectangular coordinates, they are immediately available for cylindrical and spherical coordinates.

Example 1. Find the centroid of the portion of the sphere $x^2 + y^2 + z^2 \leq a^2$ in the first octant, assuming constant density.

Solution. There is no loss of generality in assuming $\delta = 1$, and since $\bar{x} = \bar{y} = \bar{z}$, we need compute only

$$M_{xy} = \int_0^{\frac{\pi}{2}} \int_0^{\frac{\pi}{2}} \int_0^a (r \cos \phi) r^2 \sin \phi \, dr \, d\phi \, d\theta = \frac{\pi a^4}{16}.$$

Since $V = \frac{\pi a^3}{6}$, the centroid has coordinates $\bar{x} = \bar{y} = \bar{z} = \tfrac{3}{8}a$.

Example 2. Find the moment of inertia I_L of a right circular cylinder of base radius a, altitude h, and density proportional to the distance from the axis of the cylinder, with respect to a line L parallel to the axis of the cylinder and at a distance b from it.

Solution. Let the cylinder be described by the inequalities $0 \leq \rho \leq a$, $0 \leq z \leq h$, and the line L by $x = b, y = 0$. If the density is $\delta = k\rho$,

$$I_L = \int_0^{2\pi} \int_0^a \int_0^h k\rho[(x-b)^2 + y^2]\rho \, dz \, d\rho \, d\theta$$

$$= \int_0^{2\pi} \int_0^a \int_0^h k\rho^2(\rho^2 + b^2) \, dz \, d\rho \, d\theta + 0$$

$$= 2\pi k a^3 h \left(\frac{a^2}{5} + \frac{b^2}{3} \right).$$

If M is the mass of the cylinder and I_z its moment of inertia with respect to its axis, we have

$$M = \frac{2\pi k a^3 h}{3} \quad \text{and} \quad I_z = \frac{2\pi k a^5 h}{5} = \tfrac{3}{5} a^2 M,$$

so that

$$I_L = I_z + b^2 M = (\tfrac{3}{5}a^2 + b^2)M.$$

Example 3. Find the moment of inertia of a sphere of constant density with respect to a diameter.

Solution. Let the sphere be $x^2 + y^2 + z^2 \leq a^2$, with density k. Then by symmetry and Note 1,

$$I_x = I_y = I_z = \tfrac{2}{3} I_0 = \tfrac{2}{3} \int_0^{2\pi} \int_0^{\pi} \int_0^a kr^4 \sin \phi \, dr \, d\phi \, d\theta = \frac{8\pi k a^5}{15}.$$

Since the mass of the sphere is $\frac{4\pi k a^3}{3}$, the answer can be expressed $\tfrac{2}{5} a^2 M$.

1322. EXERCISES

1. Find the volume bounded by the sphere $x^2 + y^2 + z^2 = 8$ and the paraboloid $4z = x^2 + y^2 + 4$.

§ 1322] EXERCISES

2. Find the volume and centroid of the region bounded by the hyperboloid $x^2 + y^2 - z^2 + a^2 = 0$ and the cone $2x^2 + 2y^2 - z^2 = 0$, for $z \geq 0$.

3. Find the volume of the region bounded by the sphere $x^2 + y^2 + z^2 = 13$ and the cone $x^2 + y^2 = (z-1)^2$, for $z \geq 1$.

4. Find the moment of inertia of a homogeneous solid right circular cylinder (base radius a, altitude h, density k) with respect to its axis.

5. Find the centroid of the wedge-like region in the first octant bounded by the coordinate planes $x = 0$ and $z = 0$, the plane $z = by$ ($b > 0$), and the cylinder $x^2 + y^2 = a^2$.

6. Find the moment of inertia of a homogeneous right circular cone (base radius a, altitude h, density k) with respect to its axis.

7. Find the centroid and the moment of inertia with respect to its axis of symmetry of a homogeneous hemisphere ($\delta = k$).

8. Find M and $\bar{z}$ for the region bounded by the sphere $r = a$ and the half-cone $\phi = \alpha$ ($0 < \alpha \leq \frac{1}{2}\pi$), if the density is constant ($\delta = k$).

9. Find M, $\bar{z}$, and I_z for the hemisphere $0 \leq r \leq a$, $0 \leq \phi \leq \frac{1}{2}\pi$, if $\delta = k\rho$.

10. Find the mass and the moment of inertia of a sphere of radius a with respect to a diameter if the density is proportional to the distance from the center ($\delta = kr$).

11. Let W be the rectangular parallelepiped $0 \leq x \leq a$, $0 \leq y \leq b$, $0 \leq z \leq c$, with constant density k. Find M, I_{yz}, I_z, and I_0.

12. Find the centroid and I_z for the tetrahedron with vertices $(0, 0, 0)$, $(a, 0, 0)$, $(0, b, 0)$, $(0, 0, c)$, and constant density. *Hint:* For I_z let z be the final variable of integration.

13. Find the centroid of the homogeneous hemispherical shell $a \leq r \leq b$, $0 \leq \phi \leq \frac{1}{2}\pi$.

14. The density of a cube is proportional to the distance from one edge. Find its mass and its moment of inertia with respect to that edge. (Let the edges $= a$ and the constant of proportionality $= k$.)

15. Discuss the *Theorem of Pappus* (due to the Greek geometer Pappus, who lived during the second half of the third century A.D.): *Let S be a plane region in E_2 lying entirely to the right of the y-axis, with centroid $(\bar{x}, \bar{y})$. Let W be the solid region in E_3 obtained by revolving A about the y-axis. Then*
$$V(W) = 2\pi \bar{x} A(S).$$
Extend this to the formula $M(W) = 2\pi \bar{x} M(S)$, assuming a variable density for S.

16. Use the Theorem of Pappus (Ex. 15) to determine the centroid of a plane semicircle, and the volume of a solid torus obtained by revolving the circle $(x-b)^2 + y^2 \leq a^2$ ($a < b$) about the y-axis.

17. Let a solid torus (doughnut) of constant density be generated by revolving the circle $(x-b)^2 + y^2 \leq a^2$ ($a < b$) about the y-axis. Prove that the moment of inertia of this torus with respect to its axis of revolution is $\frac{1}{4}(3a^2 + 4b^2)M$, where M is its mass. (Cf. Ex. 16.)

18. Let I_c be the moment of inertia of a body of mass M with respect to a line L_c through its centroid, and let L be a line parallel to L_c and at a distance b from it. If I_L denotes the moment of inertia of the body with respect to the line L, show that $I_L = I_c + b^2 M$. Use this formula to check the result of Example 2, § 1321.

19. State and prove the analogue for E_2 of Note 4, § 1317.

1323. MASS, MOMENTS, AND CENTROID OF AN ARC

The concept of a mass determined by a density $\delta(s)$ as a function of arc length s on a rectifiable arc is similar to the two- and three-dimensional mass-distribution analogues. We state typical formulas for mass, moments, and centroid for a plane curve. Similar formulas hold for a space curve.

Mass $M = \int_{s_1}^{s_2} \delta(s)\, ds.$

First moment with respect to the y-axis: $M_y = \int_{s_1}^{s_2} \delta(s) x\, ds.$

Second moment with respect to the y-axis: $I_y = \int_{s_1}^{s_2} \delta(s) x^2\, ds.$

Polar second moment with respect to the origin: $I_0 = \int_{s_1}^{s_2} \delta(s)(x^2 + y^2)\, ds.$

Centroid: $(\bar{x}, \bar{y}) = \left(\dfrac{M_y}{M}, \dfrac{M_x}{M}\right).$

Example. Find the centroid of a homogeneous semicircular thin rod.

Solution. Let the rod be described in polar coordinates by $\rho = a$, $0 \leq \theta \leq \pi$, with $\delta = k$. Then $s = a\theta$, so that

$$M = \int_0^\pi k(a\,d\theta) = \pi a k,$$

$$M_x = \int_0^\pi k(a \sin \theta)(a\,d\theta) = 2a^2 k.$$

Therefore the centroid is $\left(0, \dfrac{2a}{\pi}\right).$

1324. ATTRACTION

According to the Newtonian law of gravitational attraction, the force between two point masses, m_1 and m_2, is equal to Gm_1m_2/d^2, where G is the universal constant of gravitation and d is the distance between the two masses. Since force is a *vector*, total forces between masses that are not point masses, but are distributed through a region, in one, two, or three dimensions, are found by determining their *components*. After the component of the force is found for an *element of mass*, the total force component results from integration over the region. This principle is illustrated in the Examples below.

NOTE. There is one theoretical difficulty that should be mentioned, although we shall not discuss it in detail. Suppose (for definiteness) we are interested in setting up the magnitude of the force between an element of mass, expressed in spherical coordinates and a unit mass at a distance d. In § 1320 we saw that the

§ 1324] ATTRACTION 469

element of volume ΔV is given *precisely* by an expression of the form $\bar{r}^2 \sin \bar{\phi} \, \Delta r \, \Delta \phi \, \Delta \theta$, where $\bar{r}$ and $\bar{\phi}$ are appropriately chosen intermediate values of the variables r and ϕ. If the mass of a region is determined by a continuous *variable* density, $\delta(r, \phi, \theta)$, then the element of mass, ΔM, is given by the product $\delta(r', \phi', \theta') \Delta V$, where (r', ϕ', θ') is a point in the element of volume whose existence is guaranteed by the Mean Value Theorem (Theorem XIII, § 1306) applied to the function $\delta(r, \phi, \theta)$. But r' and ϕ' are certainly not in general the same as $\bar{r}$ and $\bar{\phi}$! Furthermore, the distance $d(r, \phi, \theta)$ must also be evaluated at some intermediate point (r'', ϕ'', θ''), which cannot be expected to be the same as either of the first two points. After the appropriate component is formed, the total force is represented as the limit of a sum which is of the type involved in a multiple integral except that the functions whose product forms the integrand are evaluated at *different points* instead of the *same point*. This whole problem can be resolved without too much difficulty almost precisely as in the proof of Bliss's Theorem (Ex. 41, § 503), which is a special case of Duhamel's Principle for Integrals (§ 1101). (Cf. Exs. 37, 39, § 1310; Exs. 26, 28, § 1312.) In Exercise 10, § 1325, hints are given for the details of a proof assuming continuity of all functions involved, and simple conditions on the region of integration.

In the Examples below we avoid all of these considerations, and use just *one* representative point from each element of volume.

Example 1. Find the attractive force between a rod of uniform density and a unit mass (not a part of the rod).

Solution. Let the density and length of the rod be $\delta = k$ and L, respectively. Establish a coordinate system, as in Figure 1313, with the rod between $(0, b)$ and

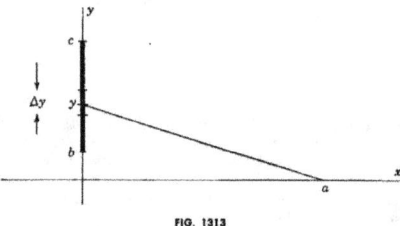

FIG. 1313

$(0, c)$, and the unit mass at $(a, 0)$, with $b < c$ and $a \geqq 0$. The distance between the unit mass at $(a, 0)$ and the element of mass of the rod, $k \Delta y$, at $(0, y)$ is $\sqrt{a^2 + y^2}$. Therefore the magnitude of the x-component of the element of force is $\dfrac{Ga}{(a^2 + y^2)^{\frac{3}{2}}} k \Delta y$, and that of the y-component is $\dfrac{Gy}{(a^2 + y^2)^{\frac{3}{2}}} k \Delta y$. Upon integration, we find

$$|F_x| = Gak \int_b^c \frac{dy}{(a^2 + y^2)^{\frac{3}{2}}} = \frac{Gk}{a} \left\{ \frac{c}{\sqrt{a^2 + c^2}} - \frac{b}{\sqrt{a^2 + b^2}} \right\},$$

$$|F_y| = Gk \int_b^c \frac{y\, dy}{(a^2 + y^2)^{\frac{3}{2}}} = Gk \left\{ \frac{1}{\sqrt{a^2 + b^2}} - \frac{1}{\sqrt{a^2 + c^2}} \right\}.$$

From these components the direction and magnitude of the force vector can be determined. Special cases are of interest. One is: $a = 0, 0 < b < c$. Then $|F_y| = \frac{Gk}{bc}(c - b) = \frac{GM}{bc}$. Another is obtained by letting $b = 0$ and $c \to +\infty$. Then $|F_x| = |F_y| = Gk/a$. Finally, let $c \to +\infty$ and $b \to -\infty$. Then $|F_x| = 2Gk/a$.

Example 2. Find the attractive force between a homogeneous plane semi-circular lamina and a unit mass situated on the axis of symmetry of the lamina and on the circumference (extended).

First Solution. Let the lamina be bounded above by the circle $\rho = 2a \sin \theta$ and below by the horizontal line $y = \rho \sin \theta = a$, the unit mass being located at the origin, and let the density be $\delta = k$. Since the force between an element of area at (ρ, θ) and the unit mass is $Gk\rho\, d\rho\, d\theta / \rho^2$, the total force is obtained by integrating the y-component of this:

$$F = Gk \int_{\frac{\pi}{4}}^{\frac{3\pi}{4}} \int_{a \csc \theta}^{2a \sin \theta} \frac{\sin \theta\, d\rho\, d\theta}{\rho} = 2Gk \int_{\frac{\pi}{4}}^{\frac{\pi}{2}} \sin \theta \ln (2 \sin^2 \theta)\, d\theta.$$

Integration by parts gives the result $2Gk[\ln (3 + 2\sqrt{2}) - \sqrt{2}]$. We notice (as we should expect) that this result is independent of the radius a. Computation shows that this force is equal (approximately) to $MG/(1.501a)^2$. In other words, the attraction is the same as it would be if the entire mass were concentrated at the point $\rho = 1.501a$, $\theta = \frac{1}{2}\pi$.

Second Solution. Let the lamina be the right-hand half of the circle $x^2 + y^2 = a^2$, and consider it to be cut into vertical strips of thickness Δx. Then, by the result of Example 1, the force on the unit mass at $(-a, 0)$ contributed by the strip at abscissa x is (approximately) equal to $2Gk\, \Delta x \sqrt{a^2 - x^2}/(x + a)\sqrt{2a^2 + 2ax} = \sqrt{2}Gk\, \Delta x \sqrt{a - x}/\sqrt{a(a + x)}$. Therefore the total force is $\frac{\sqrt{2}Gk}{\sqrt{a}} \int_0^a \frac{\sqrt{a - x}}{a + x}\, dx$, which is equal to the result given by the first solution.

Example 3. Show that a solid sphere whose density is a function of the distance from the center of the sphere only, attracts a particle outside the sphere as if all of its mass were concentrated at its center.

Solution. Center the sphere, of radius a, at the origin, and let a unit mass be located on the positive half of the z-axis, a distance $c > a$ from the origin (cf. Fig. 1314). Only the component of the force in the z-direction is of interest. This is represented by the integral

$$F_z = \int_0^{2\pi} \int_0^\pi \int_0^a \frac{G\delta(r)(c - r\cos\phi)}{(c^2 + r^2 - 2cr\cos\phi)^{\frac{3}{2}}} r^2 \sin\phi \, dr \, d\phi \, d\theta. \quad (1)$$

Since the limits of integration in (1) are constants, the order of integration is

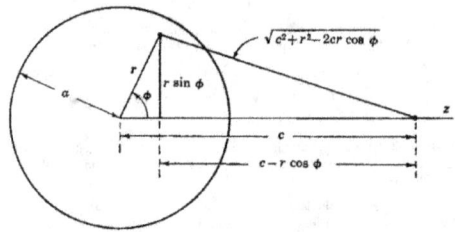

FIG. 1314

immaterial. Integrating first with respect to θ and then with respect to ϕ, we find

$$F_z = \frac{4\pi G}{c^2} \int_0^a r^2 \delta(r) \, dr = \frac{GM}{c^2}, \quad (2)$$

where M is the mass of the sphere:

$$M = \int_0^{2\pi} \int_0^\pi \int_0^a \delta(r) r^2 \sin\phi \, dr \, d\phi \, d\theta = 4\pi \int_0^a r^2 \delta(r) \, dr. \quad (3)$$

Formula (2) is that sought.

1325. EXERCISES

In Exercises 1-6, find the attractive force between the given set with constant density k and a unit mass at the given point.

1. The circular wire $x^2 + y^2 = a^2$, $z = 0$; $(0, 0, h)$.
2. The circular disk $x^2 + y^2 \leq a^2$, $z = 0$; $(0, 0, h)$.
3. The rectangular lamina $|x| \leq a$, $|y| \leq b$; $(c, 0)$ $(c > a)$.
4. The solid cylinder $x^2 + y^2 \leq a^2$, $0 \leq z \leq h$; $(0, 0, c)$ $(c > h)$.
5. The solid hemisphere $r \leq 2a \cos\phi$, $z \geq a$; $(0, 0, 0)$.
6. The solid cone $\phi \leq \alpha = \text{Arctan}\,(a/h)$, $0 < z \leq h$; $(0, 0, 0)$. Discuss the convergence of any improper integral concerned.
7. Show that the attractive force between two solid spheres, for each of which the density is a function of the distance from the center of the sphere only, is the same as if the total mass of each were concentrated at its center.
8. Show that the attractive force exerted by a spherical shell on a point mass inside the shell, if the density is a function of the distance from the center of the sphere only, is zero.

9. Discuss the attraction between the sphere of Example 3, § 1324, and a point mass inside the sphere. Show that if the sphere is homogeneous the attraction varies directly as the distance of the point mass from the center of the sphere. Discuss the behavior of any improper integrals concerned.

★10. Let R be a closed rectangular parallelepiped with faces parallel to the coordinate planes, and let $f(x, y, z)$, $g(x, y, z)$, and $h(x, y, z)$ be continuous on R. If $\mathfrak{N}$ is a net that subdivides R into pieces $S_1, \cdots, S_n$ of volume $\Delta V_1, \cdots, \Delta V_n$, and if (x_i, y_i, z_i), (x_i', y_i', z_i'), and (x_i'', y_i'', z_i'') are arbitrary points of S_i, $i = 1, \cdots, n$, show that

$$\lim_{|\mathfrak{N}| \to 0} \sum_{i=1}^{n} f(x_i, y_i, z_i) g(x_i', y_i', z_i') h(x_i'', y_i'', z_i'') \Delta V_i$$
$$= \iiint_R f(x, y, z) g(x, y, z) h(x, y, z) \, dV.$$

(Cf. Ex. 41, § 503; also §§ 1101, 1102, Exs. 37, 39, § 1310.) *Hint:* Write
$f(p)g(p')g(p'') - f(p)g(p)h(p) = f(p)g(p')h(p'') - f(p)g(p')h(p) + f(p)g(p')h(p) - f(p)g(p)h(p).$

1326. JACOBIANS AND TRANSFORMATIONS OF MULTIPLE INTEGRALS

The question to which we now address ourselves is the following: "What is the effect on a multiple integral

$$\iint_R \cdots \int f(x, y, \cdots) \, dx \, dy \cdots$$

of a change of variables, from $x, y, \cdots$ to $u, v, \cdots$?"

This question has been partially answered (§ 505) for the case of an integral of a function of a single variable: If $f(x)$ is continuous and if $x(u)$ is continuously differentiable, then

(1) $$\int_a^b f(x) \, dx = \int_c^d f(x(u)) \frac{dx}{du} \, du,$$

where $a = x(c)$ and $b = x(d)$.

It has also been answered to some extent for transformations to polar coordinates (§ 1313), cylindrical coordinates (§ 1319), and spherical coordinates (§ 1320), with the formulas

(2) $$\iint_S f(x, y) \, dA = \int_\alpha^\beta \int_\mu^\nu f_S(x(\rho, \theta), y(\rho, \theta)) \rho \, d\rho \, d\theta,$$

(3) $$\iiint_W f(x, y, z) \, dV = \int_\alpha^\beta \int_\mu^\nu \int_c^d f_{W*}(x(\rho, \theta, z), y(\rho, \theta, z), z) \rho \, dz \, d\rho \, d\theta,$$

(4) $$\iiint_W f(x, y, z) \, dV =$$
$$\int_\alpha^\beta \int_{\phi_1}^{\phi_2} \int_{r_1}^{r_2} f_{W*}(x(r, \phi, \theta), y(r, \phi, \theta), z(r, \phi, \theta)) r^2 \sin \phi \, dr \, d\phi \, d\theta.$$

§ 1326] JACOBIANS AND MULTIPLE INTEGRALS

In each of these cases the transformation of the multiple integral is achieved by (i) substituting for x, y, and z in the integrand their expressions in terms of the new variables, and (ii) introducing an extra factor in the new integrand $\left(\frac{dx}{du}, \rho, \rho, \text{ and } r^2 \sin \phi, \text{ respectively}\right)$. We notice that, in analogy with (1) where the extra factor is the *derivative* of the original variable with respect to the new, the extra factor in (2), (3), and (4) is the *Jacobian* of the original variables with respect to the new (cf. § 1223):

In (2): $\frac{\partial(x, y)}{\partial(\rho, \theta)} = \rho$,

in (3): $\frac{\partial(x, y, z)}{\partial(\rho, \theta, z)} = \rho$,

in (4): $\frac{\partial(x, y, z)}{\partial(r, \phi, \theta)} = r^2 \sin \phi$.

All of these transformation formulas, (1)-(4), are special instances of a general rule, involving Jacobians. At present we shall merely give the statement of the Transformation Theorem, for the case of two variables (cf. Ex. 10, § 1330). In the remaining sections of this chapter we shall present a general discussion (§ 1327), a proof (§ 1328), and a brief statement for improper multiple integrals (§ 1329).

Theorem. Transformation Theorem for Double Integrals. *If*

(i) *T is a one-to-one transformation between open sets Ω in the uv-plane and $\Psi = T(\Omega)$ in the xy-plane, defined by continuously differentiable functions $x = x(u, v)$ and $y = y(u, v)$;*

(ii) *the Jacobian $J(u, v) = \frac{\partial(x, y)}{\partial(u, v)}$ is nonzero throughout Ω;*

(iii) *R and S are bounded sets in the xy-plane and uv-plane, respectively, whose closures are subsets of Ψ and Ω, respectively ($\overline{R} \subset \Psi, \overline{S} \subset \Omega$), such that $R = T(S)$;*

(iv) *$f(x, y)$ is defined over R and $g(u, v) = f(x(u, v), y(u, v))$;*

then

(v) *R has area if and only if S has area;*

(vi) *assuming R and S have area, f is integrable over R if and only if $g|J|$ is integrable over S, and in case of integrability*

(5) $$\iint_R f(x, y) \, dA = \iint_S g(u, v) \, |J(u, v)| \, dA.$$

(Cf. Fig. 1315.)

A similar theorem holds for multiple integrals of any number of variables.

474   MULTIPLE INTEGRALS   [§ 1327

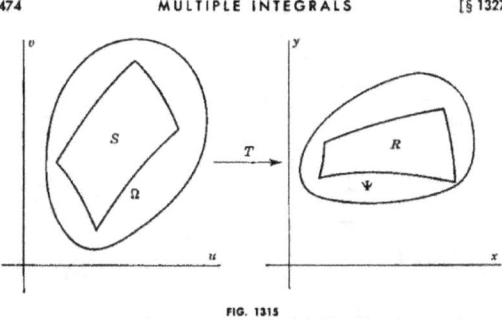

FIG. 1315

1327. GENERAL DISCUSSION

In this section we hope to appeal to the reader's intuition, the goal being to explain why the appearance of a Jacobian in the transformation of a multiple integral should be expected, and to develop a feeling for the geometric significance of the Jacobian in its present role. No attempt at mathematical rigor will be made until the following section, but it is hoped that the present discussion may suggest ideas that point toward a proof.

Let us examine the transformation T prescribed (in the Theorem, § 1326) by the functions $x(u, v)$ and $y(u, v)$, from the point of view of the *element of area* dA in the xy-plane that corresponds to positive differential increments du and dv in the variables u and v. (In the case of polar coordinates ρ and θ, for instance, $dA = \rho \, d\rho \, d\theta$.)

Accordingly, let S be the closed rectangle in the uv-plane with sides parallel to the coordinate axes and opposite vertices (u_0, v_0) and $(u_0 + du, v_0 + dv)$, and let R be the corresponding set $(R = T(S))$ in the xy-plane (cf. Fig. 1316). The secant vector joining the points in the xy-plane that correspond to (u_0, v_0) and $(u_0 + du, v_0)$ (shown by a broken line arrow from 1 to 2 in Figure 1316) has components

$$x(u_0 + du, v_0) - x(u_0, v_0) \quad \text{and} \quad y(u_0 + du, v_0) - y(u_0, v_0)$$

or, by the Law of the Mean (§ 1212),

$$x_1(u_0 + \theta_1 \, du, v_0) \, du \quad \text{and} \quad y_1(u_0 + \theta_2 \, du, v_0) \, du,$$

where $0 < \theta_1 < 1$ and $0 < \theta_2 < 1$. This vector approximates the corresponding *tangent* differential vector $\vec{t_u}$ with components

§ 1327]   GENERAL DISCUSSION   475

(1) $$\frac{\partial x}{\partial u}du \quad \text{and} \quad \frac{\partial y}{\partial u}du,$$

evaluated at the point (u_0, v_0).

Similarly, the segment in the xy-plane that corresponds to the one joining (u_0, v_0) to $(u_0, v_0 + dv)$ (shown by a broken line arrow from 1 to 4

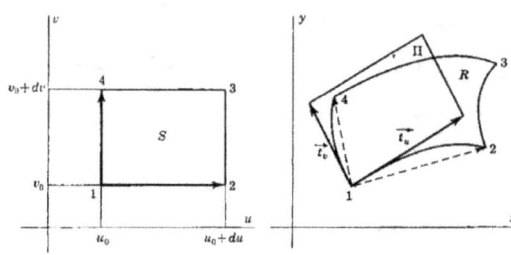

FIG. 1316

in Figure 1316) is approximately equal to the tangent differential vector $\vec{t}_v$ with components

(2) $$\frac{\partial x}{\partial v}dv \quad \text{and} \quad \frac{\partial y}{\partial v}dv,$$

evaluated at the point (u_0, v_0).

The set $R = T(S)$, therefore, is approximated by the parallelogram Π determined by the vectors (1) and (2). Using a formula from Plane Analytic Geometry (cf. Ex. 30, § 1310), we have, for the area of Π:

(3) $$A(\Pi) = \text{absolute value of} \begin{vmatrix} \frac{\partial x}{\partial u}du & \frac{\partial y}{\partial u}du \\ \frac{\partial x}{\partial v}dv & \frac{\partial y}{\partial v}dv \end{vmatrix} = \left|\frac{\partial(x, y)}{\partial(u, v)}\right| du\, dv.$$

Before taking the next step, let us recall that all of the differential expressions that we have encountered in earlier sections (§§ 411, 1108, 1112, 1206) have been associated with a *tangent* line to a curve (or tangent plane to a surface) rather than with the curve (or surface) itself. With this background in mind it seems appropriate now to *define* **the differential element of area,** corresponding to a point (u_0, v_0) and differentials du and dv, by the equation

(4) $$dA = |J(u, v)|\, du\, dv,$$

where $J(u, v)$ is the Jacobian $\partial(x, y)/\partial(u, v)$ of the transformation

The relation (4) leads directly to the formula for change of variables in the Transformation Theorem, § 1326.

$$(5) \quad \iint_R f(x, y)\, dA = \iint_S f(x(u, v), y(u, v))|J(u, v)|\, du\, dv.$$

The mapping T transforms a rectangle of area $du\, dv$ into a set that is approximated by one of area $|J|\, du\, dv$. Expressed slightly differently, this says that the transformation produces a change of area near any point by multiplying that area by a scale factor of proportionality, this factor being the magnitude of the Jacobian of the transformation evaluated at the point. Thus, whenever $|J| > 1$ a magnification or stretching takes place, and whenever $|J| < 1$ a diminution or shrinking is produced—the magnitude of this stretching or shrinking being equal to $|J|$. It should be appreciated, however, that this change of scale near a point is not uniform, but depends on the direction. For example, the transformation $x = 3u$, $y = \frac{1}{2}v$ stretches by a factor of 3 in the x-direction and shrinks by a factor of $\frac{1}{2}$ in the y-direction, the net effect on the area being an increase by a factor of $\frac{3}{2}$.

The next question is the significance of the *sign* of the Jacobian. This rests on the meaning of the sign of the determinant that gives the area of a parallelogram. It is thus a matter of Plane Analytic Geometry to show that a *positive* Jacobian corresponds to a *preservation* of orientation (that is, the smallest-angle rotation of $\vec{t_u}$ into the direction of $\vec{t_v}$, in Figure 1316, is counter-clockwise), and a *negative* Jacobian corresponds to a *reversal* of orientation.

In three (or more) dimensions, an analysis similar to the preceding is possible. The student is asked in Exercise 11, § 1330, to carry out the details for E_3. In this case *three* tangent vectors determine an approximating *parallelepiped*, whose volume is $|J(u, v, w)|\, du\, dv\, dw$ (cf. Ex. 17, § 1312). The sign of the Jacobian again has the significance that it is positive or negative according as the orientation is preserved or reversed.

NOTE 1. The assumptions (i) and (ii) of the Transformation Theorem, § 1326, that the transformation T is one-to-one and that the Jacobian never vanishes can be relaxed to the extent of permitting exceptional behavior in either regard on a set of zero area. A precise statement, with hints, is given in Exercise 12, § 1330. This principle was discussed in the Note, § 1313, for polar coordinates, and is illustrated again in Example 3, below.

NOTE 2. In practice, in order to determine the image of a closed region under a given transformation it is often sufficient to find the image of the frontier only. Complete statements follow in Theorems I and II (proofs are requested in Exercises 14 and 15, § 1330, with hints), with the aid of the Definition:

Definition. *A **Jordan curve** in E_2 is a simple closed curve C such that its complement C' consists of two disjoint regions, one bounded and one un-*

bounded, the frontier of each of which is C. The bounded region is called the **inside** of C, and the unbounded region is called the **outside** of C†. Similar formulations hold in higher dimensions.

Theorem I. *A region R whose frontier is a Jordan curve C is either the inside of C or the outside of C. Any such region is the interior of its closure.*

Theorem II. *Let R be a bounded region with a Jordan curve C as its frontier, and let M be a mapping that is continuous on its closure $\bar{R}$. If the image of R, under M, is an open set S, and that of C is a Jordan curve D, then S is the inside of D.*

A similar statement holds in higher dimensions.

We shall now illustrate by means of a few examples some typical simplifications of multiple integrals that can be effected through a change of variables.

1. *Simplification of the domain of integration.*

Example 1. Simplify $\iint_R f(x, y)\, dA$, where R is the interior of the parallelogram with vertices $(0, 0)$, $(1, 2)$, $(-4, 3)$, and $(-3, 5)$.

Solution. The linear transformation that maps $(0, 0) \to (0, 0)$, $(1, 2) \to (1, 0)$, and $(-4, 3) \to (0, 1)$ is
$$\begin{cases} x = u - 4v, \\ y = 2u + 3v. \end{cases}$$
Since $\partial(x, y)/\partial(u, v) = 11$,
$$\iint_R f(x, y)\, dA = 11 \int_0^1 \int_0^1 f(u - 4v, 2u + 3v)\, dv\, du.$$

Example 2. Let R be the region bounded in the first quadrant by the hyperbolas $x^2 - y^2 = a$, $x^2 - y^2 = b$, $2xy = c$, $2xy = d$, where $0 < a < b$, $0 < c < d$. By means of the transformation
$$\begin{cases} u = x^2 - y^2, \\ v = 2xy, \end{cases}$$
since $\partial(u, v)/\partial(x, y) = 4(x^2 + y^2) = 4\sqrt{u^2 + v^2}$,
$$\iint_R f(x, y)\, dA = \frac{1}{4} \int_a^b \int_c^d \frac{g(u, v)}{\sqrt{u^2 + v^2}}\, dv\, du.$$
If $f(x, y) = x^2 + y^2$, the integral has the value $\frac{1}{4}(b - a)(d - c)$.

2. *Simplification of the integrand. Separation of variables.*

---

† The **Jordan curve theorem** states that every simple closed curve in E_2 is a Jordan curve. For a proof see M. H. A. Newman, *Elements of the Topology of Plane Sets of Points* (Cambridge University Press, 1939).

478   MULTIPLE INTEGRALS   [§ 1328

Example 3. Evaluate $\int_0^1 \int_0^{1-x} e^{\frac{y-x}{y+x}} \, dy \, dx$.

Solution. Let R be the interior of the triangle in the xy-plane with vertices $(0,0)$, $(1,0)$, $(0,1)$, and let S be the interior of the square in the uv-plane with vertices $(0,0)$, $(1,0)$, $(0,1)$, and $(1,1)$, and map R onto S by the transformation

$$\begin{cases} u = x + y, \\ v = y/(x+y), \end{cases} \quad \begin{cases} x = u - uv, \\ y = uv. \end{cases}$$

Then $\partial(x, y)/\partial(u, v) = u$, and

$$\iint_R e^{\frac{y-x}{y+x}} \, dy \, dx = \iint_S u e^{2v-1} \, dv \, du = \int_0^1 u \, du \int_0^1 e^{2v-1} \, dv$$

$$= \tfrac{1}{4}(e - e^{-1}) = \tfrac{1}{2} \sinh 1.$$

The mapping is not one-to-one on the frontiers of the sets, and the Jacobian vanishes on the v-axis, which corresponds to the point $x = 0$, $y = 0$. However, the result is valid by Note 1.

★★1328. PROOF OF THE TRANSFORMATION THEOREM

In this section we give a proof of the Theorem of § 1326 for the case of double integrals. The details for three or more variables are quite similar and should present no difficulty (cf. Ex. 18, § 1330).

The method of proof that we have chosen is to make effective use of three kinds of approximations: (*i*) a set with area by a union of closed squares, (*ii*) a transformation by a linear transformation, and (*iii*) an integrable function by a step-function. Curiously enough, by far the major portion of our efforts will be spent in showing that the transformation formula (5), § 1326, is valid for the special case when the function $f(x, y)$ is identically 1 and the set S is a closed square with sides parallel to the coordinate axes (Lemma 10).

Before proceeding with the details, we wish to mention two other methods for arriving at the Transformation Theorem. One is to represent the original transformation as a sequence of simpler transformations obtained by changing one variable at a time. We have avoided this method because the individual transformations of this iterative process may not enjoy the pleasant properties assumed for the original transformation T, such as one-to-oneness and finite existence of the Jacobian. (This pathology exists even with a transformation from rectangular to polar coordinates in two dimensions.) The details associated with the consequent necessity for decomposing a region into pieces in each of which the procedure can be validated are awesome. The other alternative method for double integrals is to use Green's Theorem in the plane. However, the higher dimensional cases, involving highly sophisticated extensions of Green's Theorem, are

PROOF OF TRANSFORMATION THEOREM

difficult and far from elementary. For further discussion the reader is referred to the article by J. Schwartz, "The Formula for Change in Variables in a Multiple Integral," in the *American Mathematical Monthly*, Vol. 61, No. 2 (Feb., 1954), pp. 81-85, and the book by R. C. Buck, *Advanced Calculus* (New York, McGraw-Hill Book Company, Inc., 1956).

We are ready to start the sequence of lemmas that lead to a proof of the Transformation Theorem. All numbered hypotheses refer to those of the Theorem, § 1326. All squares considered will be assumed to have sides parallel to the coordinate axes.

The first lemma is a simple consequence of the continuity of the transformation T and its inverse T^{-1}, and the proof is left to the student in Exercise 16, § 1330.

Lemma 1. *Under hypotheses* (i)-(iii), *interior, frontier, and closure are preserved; that is,*

$$I(R) = T(I(S)), \quad F(R) = T(F(S)), \quad \bar{R} = T(\bar{S}),$$

and similarly for T^{-1}.

The next lemma involves only sets (in the uv-plane), and not the transformation T:

Lemma 2. *If S is a bounded set whose closure is contained in an open set Ω: $\bar{S} \subset \Omega$, there exists an open set U such that*

(1) $$\bar{S} \subset U \subset \bar{U} \subset \Omega.$$

Proof. Since $\bar{S}$ is compact and disjoint from the closed set Ω', the two sets $\bar{S}$ and Ω' are separated by a positive distance ρ (Ex. 26, § 1021). For each point $q \in \bar{S}$ let N_q be the circular neighborhood of radius $\frac{1}{2}\rho$, and let $U = \bigcup_q N_q$, for all $q \in \bar{S}$. (Complete the details.)

For later use we record a fact concerning distances and a linear transformation

(2) $$\begin{cases} x = au + bv + e, \\ y = cu + dv + f. \end{cases}$$

Lemma 3. *If L is the linear transformation* (2), *if $M \equiv \max(|a|, |b|, |c|, |d|)$, if $q_i = (u_i, v_i)$, $i = 1, 2$, are any two points in the uv-plane, and if $p_i \equiv T(q_i)$, $i = 1, 2$, then*

(3) $$d(p_1, p_2) \leq 2Md(q_1, q_2).$$

Proof. The left-hand member of (3) is

$$\sqrt{[a(u_2 - u_1) + b(v_2 - v_1)]^2 + [c(u_2 - u_1) + d(v_2 - v_1)]^2}$$
$$\leq M\sqrt{(|u_2 - u_1| + |v_2 - v_1|)^2 + (|u_2 - u_1| + |v_2 - v_1|)^2}$$
$$= \sqrt{2}M(|u_2 - u_1| + |v_2 - v_1|) \leq 2M\sqrt{(u_2 - u_1)^2 + (v_2 - v_1)^2}.$$

In order to maintain a control of the behavior of transformed sets, we shall need the inequality of the following lemma:

Lemma 4. *Under hypotheses (i)-(iii), corresponding to any open set U satisfying (1) there exists a constant K having the property that if S_0 is any set such that $\bar{S}_0 \subset U$ and if $R_0 = T(S_0)$, then the outer areas of these sets satisfy the inequality*

$$(4) \qquad \bar{A}(R_0) \leq K\bar{A}(S_0).$$

Proof. Let M be an upper bound for the four quantities $|x_i(u, v)|$, $|y_i(u, v)|$, $i = 1, 2$, for $(u, v) \in \bar{U}'$ (M exists since the four partial derivatives are continuous on the compact set $\bar{U}'$). We shall prove (4) for $K = 8M^2$. Accordingly, let S_0 be any set such that $\bar{S}_0 \subset U$, let $\delta > 0$ be one-half the distance between $\bar{S}_0$ and U', and let $\epsilon > 0$ be given. Then, by the Law of the Mean (§ 1212), if q_1 and $q_2 \in S_0$ and if $p_i = T(q_i)$, $i = 1, 2$, the inequality $d(q_1, q_2) < 2\delta$ implies $d(p_1, p_2) < 2Md(q_1, q_2)$. (This follows, since the conditions for the Law of the Mean are satisfied, by the same argument as that used in the proof of Lemma 3.) We now lay over S_0 a square net $\mathfrak{N}$ of equally spaced lines such that $|\mathfrak{N}| < \delta$ and the total area of the closed squares $\mathcal{S}_1, \cdots, \mathcal{S}_n$ of $\mathfrak{N}$ that intersect S_0 is less than $\bar{A}(S_0) + (\epsilon/8M^2)$. If we define $S_i = \bar{S}_0 \cap \mathcal{S}_i$, $i = 1, \cdots, n$, the diameter of S_i (cf. Ex. 33, § 1025) must then be less than $\left\{\frac{2}{n}[\bar{A}(S_0) + (\epsilon/8M^2)]\right\}^{\frac{1}{2}}$, and consequently the diameter of $T(S_i)$ must be less than $[\{8M^2\bar{A}(S_0)/n\} + \{\epsilon/n\}]^{\frac{1}{2}}$. Each $T(S_i)$ can thus be enclosed in a closed square with area less than the square of this last quantity. The area of the union of such squares (n in number) is therefore less than $8M^2\bar{A}(S_0) + \epsilon$. Since this union contains R_0, and ϵ is arbitrary, we have the desired result (4).

We are now ready for one of the conclusions of the Transformation Theorem:

Lemma 5. *Under hypotheses (i)-(iii), the transformations T and T^{-1} preserve the properties of having area and of having zero area.*

Proof. By Lemma 1, we need prove this only for zero area. For the transformation T this follows immediately from Lemma 4. Since T and T^{-1} play symmetric roles, the result also applies to T^{-1}.

We shall use the approximation of the given transformation by a linear transformation by means of the lemma:

Lemma 6. *Assume hypotheses (i)-(iii), let $q_0 = (u_0, v_0)$, and define the linear transformation L:*

$$(5) \qquad \begin{cases} x = x(u_0, v_0) + x_1(u_0, v_0)(u - u_0) + x_2(u_0, v_0)(v - v_0), \\ y = y(u_0, v_0) + y_1(u_0, v_0)(u - u_0) + y_2(u_0, v_0)(v - v_0), \end{cases}$$

§ 1328] PROOF OF TRANSFORMATION THEOREM

with Jacobian $J_0 = J(u_0, v_0)$. Define the composite transformation of Ω into the uv-plane:

(6) $\qquad G \equiv L^{-1}T.$

(That is, $G(q) = L^{-1}(T(q))$, where L^{-1} is the inverse of L.) Then there exist a positive number α and an infinitesimal function $\phi(\eta)$ ($\lim_{\eta \to 0+} \phi(\eta) = 0$) defined for $0 < \eta < \alpha$, such that

(7) $\qquad d(q, q_0) \leqq \eta < \alpha$ implies $d(G(q), q) < \eta\phi(\eta).$

Proof. Let α be the radius of any circular neighborhood N of q_0 that lies in Ω, and for any $q = (u, v) \in N$ consider the two points $(x, y) \equiv T(q)$ and $(X, Y) \equiv L(q)$. By the Law of the Mean (§ 1212) we can write

(8) $\begin{cases} x - X = [x_1(u_0 + \theta_1(u - u_0), v_0 + \theta_1(v - v_0)) - x_1(u_0, v_0)](u - u_0) \\ \qquad + [x_2(u_0 + \theta_1(u - u_0), v_0 + \theta_1(v - v_0)) - x_2(u_0, v_0)](v - v_0), \\ y - Y = [y_1(u_0 + \theta_2(u - u_0), v_0 + \theta_2(v - v_0)) - y_1(u_0, v_0)](u - u_0) \\ \qquad + [y_2(u_0 + \theta_2(u - u_0), v_0 + \theta_2(v - v_0)) - y_2(u_0, v_0)](v - v_0), \end{cases}$

where $0 < \theta_i < 1$, $i = 1, 2$. By the continuity of the partial derivatives at (u_0, v_0), each bracketed expression in (8) is an infinitesimal, having the limit 0 as $q \to q_0$. For a given η ($0 < \eta < \alpha$) define $\psi(\eta)$ to be three times the supremum, for all points q such that $d(q, q_0) \leqq \eta$, of the absolute values of the four bracketed quantities in (8). Then $\lim_{\eta \to 0+} \psi(\eta) = 0$, and the inequality $d(q, q_0) \leqq \eta$ implies

$d(T(q), L(q)) \leqq |x - X| + |y - Y|$

$\qquad \leqq \frac{2\psi(\eta)}{3} [|u - u_0| + |v - v_0|] < \psi(\eta) d(q, q_0) \leqq \eta\psi(\eta).$

Finally, since $G(q) = L^{-1}(T(q))$ and $q = L^{-1}(L(q))$, we can obtain the desired conclusion by using Lemma 3, as applied to the linear transformation L^{-1}, defining $\phi(\eta) \equiv 2M\psi(\eta)$.

The next step is to see what happens to a closed square if each of its points is moved a short distance.

Lemma 7. *Let G be a transformation which maps an open set Ω of the uv-plane into the uv-plane, and assume that G is continuous and open (§ 1229) on Ω. Let S be a closed square contained in Ω, with sides of length s, and assume that for every $q \in S$, $d(G(q), q) < \sigma < \frac{1}{2}s$. If S_0 is the closed square concentric with S and having sides parallel to those of S and of length $s_0 = s - 2\sigma$, then $S_0 \subset G(S)$.* (Cf. Fig. 1317.)

Proof. Split the set S_0 into two parts:

$A \equiv S_0 \cap G(S), \quad B \equiv S_0 - A.$

Since S is compact and G is continuous, $G(S)$ is also compact, and hence closed. Therefore A is closed. On the other hand, B is also closed for

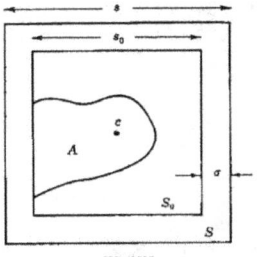

FIG. 1317

the following reason: if q is a limit point of B but not a point of B, q must belong to S_0 and hence to A, and therefore, since G is an open mapping, q must have a neighborhood consisting entirely of points of $G(S)$ and thus excluding all points of B. This contradiction proves that B is closed. We can now infer that since S_0 is connected, either A or B must be empty. Since A contains (at least) the image point of the common center of S and S_0 ($\sigma < \tfrac{1}{2}s_0$), B must be empty, and $A = S_0 \subset G(S)$.

At this point it is expedient to interpolate an implication out of natural order.

Lemma 8. *Assuming hypotheses (i)-(iii), the further assumption that the equation*

(9) $$A(R) = \iint_S |J(u, v)| \, dA$$

holds for every closed square $S \subset \Omega$, implies that the same equation holds for an arbitrary set S with area, subject to hypothesis (iii).

Proof. Assume $A(R) \neq \iint_S |J(u, v)| \, dA$ for a particular pair of sets R and S with area, subject to hypothesis (iii), and let

$$\epsilon \equiv \left| A(R) - \iint_S |J(u, v)| \, dA \right|.$$

Let $|J(u, v)| < P$ on $\bar{S}$, and for some open set U prescribed by Lemma 2, let K be a constant satisfying the inequality (4) of Lemma 4. Choose a net $\mathfrak{N}$ of such small norm that the sum of the areas of those closed squares of $\mathfrak{N}$ containing points of both S and S' is less than $\epsilon/(K + P)$. Denoting by S_1 the union of the closed squares of $\mathfrak{N}$ contained in S, we have

§1328] PROOF OF TRANSFORMATION THEOREM   483

$A(S - S_1) < \epsilon/(K + P)$. If $R_1 \equiv T(S_1)$, we have by finite addition of (9), for the squares of S_1, $A(R_1) = \iint_{S_1} |J(u, v)|\, dA$, and thus, by subtraction:

$$\epsilon = \left| A(R - R_1) - \iint_{S - S_1} |J(u, v)|\, dA \right| \leq A(R - R_1) + \iint_{S - S_1} |J|\, dA$$

$$\leq K \cdot A(S - S_1) + P \cdot A(S - S_1) = (K + P)A(S - S_1) < \epsilon.$$

This contradiction establishes the Lemma.

Some of the preceding results can be combined to give a useful limit statement:

Lemma 9. *Assuming hypotheses (i)-(iii), let $q_0 = (u_0, v_0) \in \Omega$, let L and G be the transformations (5) and (6) of Lemma 6, and let $\{S_n\}$ be a sequence of closed squares all containing q_0, and let $R_n \equiv T(S_n)$. Assume furthermore, if s_n is the length of a side of S_n, that $s_n \to 0$. Then the following two equivalent limit statements concerning areas hold:*

(10) $$\frac{A(G(S_n))}{A(S_n)} = \frac{A(L^{-1}(R_n))}{A(S_n)} \to 1,$$

(11) $$\frac{A(T(S_n))}{A(S_n)} = \frac{A(R_n)}{A(S_n)} \to |J_0|,$$

where $J_0 = J(u_0, v_0)$.

Proof. To begin, we show that (10) and (11) are equivalent by establishing the equation

(12) $$A(L^{-1}(R_n)) = A(R_n)/|J_0|.$$

The following three steps prove (12): First, any linear transformation with Jacobian J maps a closed square of area A onto a parallelogram of

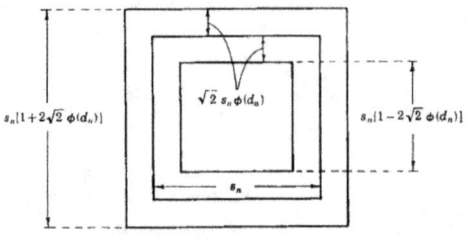

FIG. 1318

area $|J|\cdot A$ (cf. Ex. 17, § 1330). Second, the equation (9) is thus established for any linear transformation and any closed square S and therefore, by Lemma 8, for any set S with area. Third, this final result, when applied to the linear transformation L^{-1}, with Jacobian J_0^{-1}, yields (12).

We now proceed to establish (10). Let the diagonal of S_n be d_n. Since $d_n \to 0$, it follows that for n sufficiently large the conditions of Lemma 6 apply and for any point $q_n \in S_n$,

(13) $$d(G(q_n), q_n) < \sqrt{2} s_n \phi(d_n) < \tfrac{1}{2} s_n.$$

By Lemma 7, $G(S_n)$ contains a (smaller) square concentric with S_n of side $s_n[1 - 2\sqrt{2}\,\phi(d_n)]$. (Cf. Fig. 1318.) By (13) we immediately infer that $G(S_n)$ is a subset of a (larger) square concentric with S_n of side $s_n[1 + 2\sqrt{2}\,\phi(d_n)]$. Thus we have the inequalities

$$\{s_n[1 - 2\sqrt{2}\,\phi(d_n)]\}^2 \leq A(G(S_n)) \leq \{s_n[1 + 2\sqrt{2}\,\phi(d_n)]\}^2,$$

or

(14) $$[1 - 2\sqrt{2}\,\phi(d_n)]^2 \leq \frac{A(G(S_n))}{A(S_n)} \leq [1 + 2\sqrt{2}\,\phi(d_n)]^2.$$

Since $\phi(d_n) \to 0$, the extreme members of (14) both $\to 1$, and (10) is therefore established.

The time has come to turn our attention to the principal objective.

Lemma 10. *The Transformation Theorem holds for the function identically* $1 : f(x, y) \equiv 1$.

Proof. Thanks to Lemma 8, we need establish (9) only for an arbitrary closed square in Ω. We shall do this by obtaining a contradiction to the assumption that there exists a closed square S in Ω such that $A(R) \neq \iint_S |J(u,v)|\,dA$. Let the positive number ϵ be defined by the equation:

(15) $$\left| A(R) - \iint_S |J(u,v)|\,dA \right| = \epsilon A(S).$$

Subdivide S into four congruent closed subsquares. For at least one of these, call it S_1 and let $R_1 = T(S_1)$:

$$\left| A(R_1) - \iint_{S_1} |J(u,v)|\,dA \right| \geq \epsilon A(S_1).$$

Subdivide repeatedly, in this manner, to obtain for $n = 1, 2, \cdots$:

$$\left| A(R_n) - \iint_{S_n} |J(u,v)|\,dA \right| \geq \epsilon A(S_n).$$

By the Mean Value Theorem (Theorem XIII, § 1306) this can be written

$$\left| \frac{A(R_n)}{A(S_n)} - |J(u_n, v_n)| \right| \geq \epsilon,$$

where $(u_n, v_n) \subset S_n$. A contradiction now results from the two facts that $A(R_n)/A(S_n) \to |J_0|$ (by (11) of Lemma 9) and $|J(u_n, v_n)| \to |J_0|$ (by continuity of $J(u, v)$).

Lemma 11. *The Transformation Theorem holds whenever $f(x, y)$ is a step-function.*

Proof. Let $f(x, y)$ have values $\sigma_1, \cdots, \sigma_n$ in the interiors of rectangles $R_1, \cdots, R_n$ of a net, and vanish outside their union. Define $S_i = T^{-1}(R_i)$, $i = 1, \cdots, n$. Then $g(u, v)$ has the values $\sigma_1, \cdots, \sigma_n$ in the interiors of the sets $S_1, \cdots, S_n$, and vanishes outside their union. By Lemma 10,

$$\iint_R f\, dA = \sum_{i=1}^{n} \iint_{R \cap R_i} f\, dA = \sum_{i=1}^{n} \sigma_i A(R \cap R_i)$$

$$= \sum_{i=1}^{n} \sigma_i \iint_{S \cap S_i} |J|\, dA = \sum_{i=1}^{n} \iint_{S \cap S_i} g \cdot |J|\, dA = \iint_S g \cdot |J|\, dA.$$

Finally, we are within grasp of our goal:

Proof of the Transformation Theorem. Assume hypotheses (i)-(iv), that R and S have area, and that f is integrable over R. If $\epsilon > 0$ is given let $\sigma(x, y)$ and $\tau(x, y)$ be step-functions defined over R such that $\sigma(x, y) \leq f(x, y) \leq \tau(x, y)$ on R, and

(16) $$\iint_R [\tau(x, y) - \sigma(x, y)]\, dA < \epsilon.$$

Defining the functions over S:

$$s(u, v) \equiv \sigma((x(u, v), y(u, v)), \quad t(u, v) \equiv \tau((x(u, v), y(u, v)),$$

we have (by Lemma 11) the inequalities

$$\iint_R \sigma\, dA = \iint_S s|J|\, dA, \quad \iint_R \tau\, dA = \iint_S t|J|\, dA,$$

and the inequalities

$$\iint_R \sigma\, dA \leq \iint_R f\, dA \leq \iint_R \tau\, dA, \quad s|J| \leq g|J| \leq t|J| \text{ on } S,$$

and $$\iint_S [t(u, v) - s(u, v)]|J(u, v)|\, dA < \epsilon.$$

These inequalities establish the integrability over S of $g|J|$ and, because of the following inequality, obtained by combining the preceding expressions:

$$\left| \iint_R f(x,y)\, dA - \iint_S g(u,v)|J(u,v)|\, dA \right| < \epsilon,$$

the transformation equality desired.

Finally, the integrability of $g|J|$ over S implies that of f over R, by direct application of the result just obtained to the inverse transformation T^{-1}, with Jacobian J^{-1}.

★1329. IMPROPER MULTIPLE INTEGRALS

The definition of multiple integral, given earlier in this chapter, can be extended to apply to unbounded functions or domains of integration. As might be expected, by comparison with improper integrals of functions of a single variable, this involves a process equivalent to taking a limit of a (proper) multiple integral. One important distinction between improper multiple integrals and improper single integrals is that the variety of ways in which a limit can be formed makes it impractical to formulate improper multiple integrals in such a way as to include conditional convergence (cf. Ex. 45, § 516). For this reason we first assume the integrand to be nonnegative (Definition I), and then, for a more general function, consider its positive and negative parts (Definition II).

Definition I. *Let $f(x,y)$ be defined and nonnegative on a set R, and integrable over every compact subset R_1 of R that has area. Then the **double integral** of f over R is defined:*

$$(1) \qquad \iint_R f(x,y)\, dA \equiv \sup_{R_1 \subset R} \iint_{R_1} f(x,y)\, dA,$$

*taken over all compact subsets R_1 of R that have area. If f or R is unbounded, the integral (1) is **improper**. If its value is finite it is **convergent**.*

Similar statements hold for higher dimensions.

Definition II. *Let $f(x,y)$ be defined on a set R, and let f^+ and f^- be its positive and negative parts, respectively:*

$$(2) \qquad f^+(x,y) \equiv \max(f(x,y), 0), \quad f^-(x,y) \equiv \max(-f(x,y), 0)$$

*(cf. Exs. 39, 40, § 503, Ex. 36, § 1310). Then f is **integrable** over R if and only if both f^+ and f^- have finite integrals (1), and*

$$(3) \qquad \iint_R f(x,y)\, dA \equiv \iint_R f^+(x,y)\, dA - \iint_R f^-(x,y)\, dA.$$

*If f or R is unbounded, the integral (3) is **improper**. Under the finiteness conditions prescribed, the integral is **convergent**.*

Similar statements hold for higher dimensions.

§ 1329]   IMPROPER MULTIPLE INTEGRALS   487

Note 1. Formulas (1) and (3) are equivalent to definitions given before in case the appropriate conditions of integrability and having area are satisfied. (Cf. Ex. 27, § 1330.)

Note 2. Many of the familiar properties of multiple integrals, established for previous definitions, hold for Definitions I and II. For example, if f and g are integrable over R, according to Definition II, then so is $f + g$, and
$$\iint_R (f + g)\, dA = \iint_R f\, dA + \iint_R g\, dA. \quad \text{(Cf. Ex. 28, § 1330.)}$$

A method for the evaluation of a multiple integral that is often convenient is given in the theorem (for hints on a proof, cf. Ex. 29, § 1330):

Theorem. *Assume that $\iint_R f(x, y)$ exists (finite or infinite) according to Definition I or II, and let $\{R_n\}$ be an increasing sequence of compact sets with area the union of whose interiors is the interior of R. Then*
$$\lim_{n \to +\infty} \iint_{R_n} f\, dA = \iint_R f\, dA.$$

For improper multiple integrals, the Transformation Theorem of § 1326 can be adapted by making the following changes:

In (*iii*), let R and S be any subsets of Ψ and Ω, respectively.

Omit (*v*).

In (*vi*), omit the first phrase, and let integrability be interpreted in the sense of Definition I or II, above.

The proof follows immediately from the Definition, properties of compact sets and homeomorphisms, and the Transformation Theorem, § 1326 (cf. Ex. 30, § 1330).

Example 1. Use the ideas of this section to evaluate
$$\int_0^{+\infty} e^{-x^2}\, dx \quad \text{and} \quad \int_{-\infty}^{+\infty} e^{-x^2}\, dx.$$

Solution. Let S_n be the closed square $0 \leq x \leq n, 0 \leq y \leq n$, and C_n the closed quarter-circle $x \geq 0, y \geq 0, x^2 + y^2 \leq n^2$. Integrating the positive function $e^{-(x^2+y^2)}$ over S_n and C_n, we have

$$(4) \quad J_n \equiv \iint_{S_n} e^{-(x^2+y^2)}\, dA = \int_0^n e^{-x^2}\, dx \int_0^n e^{-y^2}\, dy = \left[\int_0^n e^{-x^2}\, dx\right]^2.$$

$$(5) \quad K_n \equiv \iint_{C_n} e^{-(x^2+y^2)}\, dA = \int_0^{\pi/2}\!\!\int_0^n e^{-\rho^2} \rho\, d\rho\, d\theta = \frac{\pi}{4}(1 - e^{-n^2}).$$

Since the integrand in (4) and (5) is positive, $J_n \uparrow$ and $K_n \uparrow$ as $n \uparrow$, and hence $J \equiv \lim J_n$ and $K \equiv \lim K_n$ exist. From the inclusions

$$(6) \quad C_n \subset S_n \subset C_{2n}$$

NOTE 1. Formulas (1) and (3) are equivalent to definitions given before in case the appropriate conditions of integrability and having area are satisfied. (Cf. Ex. 27, § 1330.)

NOTE 2. Many of the familiar properties of multiple integrals, established for previous definitions, hold for Definitions I and II. For example, if f and g are integrable over R, according to Definition II, then so is $f + g$, and
$$\iint_R (f+g)\,dA = \iint_R f\,dA + \iint_R g\,dA. \quad \text{(Cf. Ex. 28, § 1330.)}$$

A method for the evaluation of a multiple integral that is often convenient is given in the theorem (for hints on a proof, cf. Ex. 29, § 1330):

Theorem. *Assume that* $\iint_R f(x, y)$ *exists (finite or infinite) according to Definition I or II, and let* $\{R_n\}$ *be an increasing sequence of compact sets with area the union of whose interiors is the interior of* R. *Then*
$$\lim_{n \to +\infty} \iint_{R_n} f\,dA = \iint_R f\,dA.$$

For improper multiple integrals, the Transformation Theorem of § 1326 can be adapted by making the following changes:

In (iii), let R and S be any subsets of Ψ and Ω, respectively.

Omit (v).

In (vi), omit the first phrase, and let integrability be interpreted in the sense of Definition I or II, above.

The proof follows immediately from the Definition, properties of compact sets and homeomorphisms, and the Transformation Theorem, § 1326 (cf. Ex. 30, § 1330).

Example 1. Use the ideas of this section to evaluate
$$\int_0^{+\infty} e^{-x^2}\,dx \text{ and } \int_{-\infty}^{+\infty} e^{-x^2}\,dx.$$

Solution. Let S_n be the closed square $0 \le x \le n, 0 \le y \le n$, and C_n the closed quarter-circle $x \ge 0, y \ge 0, x^2 + y^2 \le n^2$. Integrating the positive function $e^{-(x^2+y^2)}$ over S_n and C_n, we have

$$(4) \quad J_n \equiv \iint_{S_n} e^{-(x^2+y^2)}\,dA = \int_0^n e^{-x^2}\,dx \int_0^n e^{-y^2}\,dy = \left[\int_0^n e^{-x^2}\,dx\right]^2,$$

$$(5) \quad K_n \equiv \iint_{C_n} e^{-(x^2+y^2)}\,dA = \int_0^{\frac{\pi}{2}}\int_0^n e^{-\rho^2}\rho\,d\rho\,d\theta = \frac{\pi}{4}(1 - e^{-n^2}).$$

Since the integrand in (4) and (5) is positive, $J_n \uparrow$ and $K_n \uparrow$ as $n \uparrow$, and hence $J = \lim J_n$ and $K = \lim K_n$ exist. From the inclusions

$$(6) \quad C_n \subset S_n \subset C_{2n}$$

we infer the inequalities

(7) $$K_n \leq J_n \leq K_{2n},$$

and the equality $J = K$. From (5) we have $K_n \to \pi/4$. Therefore $J_n \to \pi/4$. Taking square roots and using elementary properties of symmetry, we have:

(8) $$\int_0^{+\infty} e^{-x^2}\,dx = \tfrac{1}{2}\sqrt{\pi}, \quad \int_{-\infty}^{+\infty} e^{-x^2}\,dx = \sqrt{\pi}.$$

By the Theorem of this section, the number $J = K = \pi/4$ is the value of the improper double integral of $e^{-(x^2+y^2)}$ extended over the closed first quadrant.

Example 2. Determine the values of k such that

(9) $$\iint_{x^2+y^2 \leq 1} \frac{1}{(x^2+y^2)^k}\,dA$$

converges.

Solution. The domain of integration R is the set of (x,y) such that $0 < x^2 + y^2 \leq 1$. Letting R_n be the set such that $\frac{1}{n^2} \leq x^2 + y^2 \leq 1$, we evaluate

$$\iint_{R_n} (x^2+y^2)^{-k}\,dA = \int_0^{2\pi}\int_{\frac{1}{n}}^{1} \rho^{-2k+1}\,d\rho\,d\theta = \frac{\pi}{1-k}[1 - n^{2(k-1)}],$$

if $k \neq 1$, a formula involving $\ln n$ applying in case $k = 1$. From this it follows (from letting $n \to +\infty$) that (9) converges if and only if $k < 1$, and in case of convergence the value of (9) is $\pi/(1-k)$.

Example 3. Show that

(10) $$\iiint_{x^2+y^2+z^2 \geq 1} \frac{(x^2+y^2)\ln(x^2+y^2+z^2)}{(x^2+y^2+z^2)^k}\,dA$$

is convergent if $k > \tfrac{5}{2}$.

Solution. Using spherical coordinates and letting R_n be the set of points such that $1 \leq r \leq n$, we have

$$\iiint_{R_n} \frac{r^2 \sin^2\phi \ln(r^2)}{r^{2k}} r^2 \sin\phi\,dr\,d\phi\,d\theta = \frac{16\pi}{3}\int_1^n r^{4-2k}\ln r\,dr.$$

Integration by parts establishes convergence, and the value of the integral. Thus (10) is equal to $16\pi/3(2k-5)^2$.

1330. EXERCISES

1. Evaluate $\iint_R \sqrt{x+y}\,dA$, where R is the parallelogram bounded by the lines $x+y=0, x+y=1, 2x-3y=0, 2x-3y=4$.

2. Evaluate $\int_0^1 \int_0^{1-x} \frac{1}{(1+x^2+y^2)^2}\,dy\,dx$ by means of a suitable rotation of

3. Evaluate $\iint_R xy\, dA$, where R is the region in the first quadrant bounded by the hyperbolas $x^2 - y^2 = a$ and $x^2 - y^2 = b$, where $0 < a < b$, and the circles $x^2 + y^2 = c$ and $x^2 + y^2 = d$, where $0 < c < d$.

4. Evaluate $\iint_R (x-y)^4 e^{x+y}\, dA$, where R is the square with vertices $(1, 0)$, $(2, 1)$, $(1, 2)$, and $(0, 1)$.

5. Evaluate $\int_0^1 \int_0^x \sqrt{x^2 + y^2}\, dy\, dx$ by means of the transformation $x = u$, $y = uv$.

★6. Compute $\iint_R (ax + by)^n\, dA$, where R is the unit disk $x^2 + y^2 \leq 1$, by using a suitable rotation ($a^2 + b^2 > 0$, n a positive integer). (Cf. Ex. 36, § 515.)

★7. Evaluate $\int_0^1 \int_0^{1-x} y \ln(1 - x - y)\, dy\, dx$, using the transformation $x = u - uv$, $y = uv$.

★8. Show that the mapping
$$x = u - uv, \quad y = uv - uvw, \quad z = uvw$$
transforms the interior of the tetrahedron bounded by the planes $x = 0$, $y = 0$, $z = 0$, $x + y + z = 1$ into the interior of the cube $0 < u < 1$, $0 < v < 1$, $0 < w < 1$. Use this transformation to compute the coordinates of the centroid of the tetrahedron.

9. Explain why the transformation equation (1), § 1326, for a function of one variable has no absolute value signs.

10. State the analogue of the Transformation Theorem, § 1326, for functions of n variables.

11. Carry out the detailed discussion requested in the paragraph preceding Note 1, § 1327.

★★12. Let R and S be compact sets with area, in the xy-plane and uv-plane, respectively, let A and B be compact sets with zero area, such that $F(R) \subset A \subset R$ and $F(S) \subset B \subset S$. Let T be a one-to-one transformation between $S - B$ and $R - A$ given by continuously differentiable functions $x(u, v)$, $y(u, v)$ with nonzero Jacobian $J(u, v)$ throughout $S - B$. Let $f(x, y)$ be continuous and bounded in $R - A$, and let $g(u, v)|J(u, v)| \equiv f(x(u, v), y(u, v))|J(u, v)|$ be (continuous and) bounded in $S - B$. Finally, let $f(x, y)$ and $J(u, v)$ be defined (or redefined) on A and B, respectively, in any manner that preserves the boundedness of f on R and $g|J|$ on S. Prove that
$$\iint_R f(x, y)\, dA = \iint_S g(u, v)|J(u, v)|\, dA.$$

Hint: If $|f| < K$, $|gJ| < K$, $\epsilon > 0$, let R_1 and S_1 be compact sets with area such that $R_1 \subset R$, $S_1 \subset S$, $R_1 = T(S_1)$, $A(R - R_1) < \epsilon/2K$, $A(S - S_1) < \epsilon/2K$.

★★13. Let C be a Jordan curve in E_2, and assume that A is a region such that $A \cap C = \emptyset$ and $F(A) \subset C$. Prove that A is either the inside of C or the outside of C. *Hint:* Since A is connected, A must be contained entirely in the inside of

490   MULTIPLE INTEGRALS   [§ 1331

C or entirely in the outside of C. (Cf. Lemma 1, § 1020.) If $A \subset \{\text{inside of } C\}$, then $\{\text{inside of } C\} - A$ is open and hence empty.

★★**14.** Prove Theorem I, § 1327. *Hint:* Since $R \subset I(\overline{R}) \subset \overline{R}$, $I(\overline{R})$ is a region (cf. Ex. 11, § 1025). Use Ex. 13.

★★**15.** Prove Theorem II, § 1327. *Hint:* Since $\overline{R} = R \cup C$ is compact, $M(R \cup C) = S \cup D$ is a compact set containing S, and hence its closure $S \cup F(S)$. Use the ideas of the hint of Ex. 13.

★★**16.** Prove Lemma 1, § 1328.

★★**17.** Prove that any linear transformation with Jacobian J maps a closed square of area A onto a parallelogram of area $|J|A$. (Cf. Ex. 30, § 1310.)

★★**18.** Prove the Transformation Theorem for multiple integrals for the case $n = 3$.

In Exercises 19-26 determine whether the intergral is convergent or divergent, and if it is convergent evaluate it.

★**19.** $\iint\limits_{x^2+y^2 \leq 1} \frac{x^2 \, dA}{(x^2+y^2)^{\frac{3}{2}}}$.   ★**20.** $\iint\limits_{x^2+y^2 \leq 1} \frac{\ln(x^2+y^2) \, dA}{(x^2+y^2)^{\frac{1}{2}}}$.

★**21.** $\iint\limits_{x^2+y^2 \geq 1} \frac{\ln(x^2+y^2) \, dA}{(x^2+y^2)}$.   ★**22.** $\iint\limits_{\text{Entire plane}} \frac{dA}{(1+x^2+y^2)^{\frac{3}{2}}}$.

★**23.** $\iiint\limits_{x^2+y^2+z^2 \leq 1} \frac{x^2 \, dV}{(x^2+y^2+z^2)^2}$.   ★**24.** $\iiint\limits_{x^2+y^2+z^2 \leq 1} \frac{\ln(x^2+y^2+z^2) \, dV}{x^2+y^2+z^2}$.

★**25.** $\iiint\limits_{x^2+y^2+z^2 \geq 1} \frac{\ln(x^2+y^2+z^2) \, dV}{(x^2+y^2+z^2)^2}$.   ★**26.** $\iiint\limits_{\text{Entire space}} \frac{dV}{(1+x^2+y^2+z^2)^{\frac{3}{2}}}$.

★★**27.** Prove Note 1, § 1329.

★★**28.** Prove the statement of Note 2, § 1329, regarding $\iint (f+g) \, dA$. State and prove two more theorems of similar nature. *Hints:* First establish the desired relation for nonnegative functions. Then establish and use the inequality $(f+g)^+ \leq f^+ + g^+$ and the equation $(f+g)^+ + f^- + g^- = (f+g)^- + f^+ + g^+$.

★★**29.** Prove the Theorem, § 1329. *Hints:* First establish the result assuming $f \geq 0$ by letting R_0 be an arbitrary compact set with area lying in the *interior* of R. Show that the interiors of R_n cover R_0, and use the Heine-Borel theorem (Ex. 24, § 1021).

★★**30.** Prove the Transformation Theorem for improper integrals, as stated in § 1329.

★★**1331. LINE AND SURFACE INTEGRALS. EXERCISES**

★★**1.** If C is a rectifiable arc in E_2 with parametrization $x = f(t)$, $y = g(t)$, $a \leq t \leq b$, and if $F(x, y)$ is continuous on C, the **line integrals** $\int_C F \, dx$ and $\int_C F \, dy$ are defined as the Riemann-Stieltjes integrals:

(1) $\int_C F \, dx = \int_a^b F(f(t), g(t)) \, df(t), \quad \int_C F \, dy = \int_a^b F(f(t), g(t)) \, dg(t).$

Prove that if C is a *simple* arc the values of (1) are independent of the parametrization provided the orientation is preserved, and that the effect of reversing the orientation is to change their signs. Define and discuss, in a similar way, the line integral $\int_C F\,ds$, where s represents arc length as a monotonically increasing function of the parameter t. Illustrate with examples. Extend to E_3.

★★2. Assume that P and Q are continuous functions of x and y in a region R of E_2 and that C is a rectifiable arc lying in R. Then the line integral $\int_C P\,dx + Q\,dy$ is *defined* as $\int_C P\,dx + \int_C Q\,dy$ (cf. Ex. 1). Prove that $\int_C P\,dx + Q\,dy$ is independent of the path C joining an arbitrary pair of points in R if and only if $\int_C P\,dx + Q\,dy$ vanishes for all closed curves C.

★★3. The expression $P\,dx + Q\,dy$ is **exact** in a region R of E_2 if and only if there exists a differentiable function $\phi = \phi(x, y)$ such that $\partial\phi/\partial x = P$ and $\partial\phi/\partial y = Q$ throughout R. If P and Q are continuous, prove that $P\,dx + Q\,dy$ is exact in R if and only if $\int_C P\,dx + Q\,dy$ is independent of the path in R (cf. Ex. 2). *Hints:* (i) Assuming exactness, write
$$\sum [\phi(x(t_i), y(t_i)) - \phi(x(t_{i-1}), y(t_i)) + \phi(x(t_{i-1}), t_i)) - \phi(x(t_{i-1}), y(t_{i-1}))]$$
$$= \sum [P(x(t_i'), y(t_i))(x(t_i) - x(t_{i-1})) + Q(x(t_{i-1}), y(t_i''))(y(t_i) - y(t_{i-1}))],$$
and use the ideas of Duhamel's Principle (§ 1101) applied to two Riemann-Stieltjes integrals. (ii) Assuming independence of path when C is made up of segments parallel to the coordinate axes, *define* $\phi(x, y)$ as the integral from a fixed point (x_0, y_0) to a variable point (x, y).

★★★4. Let a **simple** compact region R in E_2 be defined as one that is bounded above and below and to the right and left by rectifiable arcs. Explain what is meant by orienting the simple closed curve C that bounds R in such a way that "R is on the left." Let a **finitely decomposable** compact region R in E_2 be defined as one that is a finite union of simple compact regions $R_1, R_2, \cdots, R_n$, any two of which have at most frontier points in common. Discuss the "boundary" or "oriented frontier" C of R, and the manner in which it is obtained from those of $R_1, R_2, \cdots, R_n$ by cancellation of portions that lie in the interior of R. By examples show that C may consist of one or more simple closed curves.

★★5. Let R be a finitely decomposable compact region with boundary C (cf. Ex. 4). Assuming continuity of the functions P and Q, and their partial derivatives $\partial P/\partial y$ and $\partial Q/\partial x$, prove **Green's Theorem in the Plane**:

(2) $$\int_C P\,dx + Q\,dy = \iint_R \left(\frac{\partial Q}{\partial x} - \frac{\partial P}{\partial y}\right) dA.$$

Hint: Prove (2) for simple compact regions by use of iterated integrals.

★★6. Discuss the following formulas for the area of a finitely decomposable compact region R (cf. Ex. 5):

(3) $$A(R) = -\int_C y\,dx = \int_C x\,dy = \frac{1}{2}\int_C (-y\,dx + x\,dy).$$

★★7. The expression $P\,dx + Q\,dy$ is **locally exact** in a region R of E_2 if and only if it is exact (Ex. 3) in some neighborhood of every point of R. Assuming continuity of the functions P, Q, $\partial P/\partial y$, and $\partial Q/\partial x$, prove that $P\,dx + Q\,dy$ is locally exact in R if and only if $\partial P/\partial y = \partial Q/\partial x$ throughout R. *Hint for "if":* Use Green's Theorem (Ex. 5) to show that for any rectangle lying in R the line integral
$$\int_C P\,dx + Q\,dy$$
from any vertex to the opposite vertex is independent of the choice of the two sides of the rectangle that constitute the path. Then use the second hint of Ex. 3.

★★8. A region R of E_2 is **simply-connected** if and only if corresponding to every closed polygon C (simple or not) lying in R there exists a point p in R such that for every vertex p_i of C, $i = 1, 2, \cdots, n$, there exists a simple arc C_i from p to p_i lying in R and having the property that for every $i = 1, 2, \cdots, n$ the three arcs C_i, $\overline{p_i p_{i+1}}$, and $-C_{i+1}$ (where $C_{n+1} = C_1$, $p_{n+1} = p_1$, and the negative sign indicates a reversal of orientation) form the boundary (with "positive" or "negative" orientation) of a finitely decomposable region R_i lying in R. (Cf. Fig. 1319.)

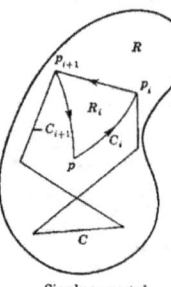 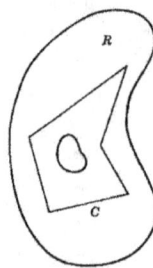

Simply-connected — Not simply-connected

FIG. 1319

Assuming continuity of P, Q, $\partial P/\partial y$, and $\partial Q/\partial x$, prove that $P\,dx + Q\,dy$ is exact in a simply-connected region if and only if it is locally exact there. Hence show the equivalence in a simply connected region of the following five statements: (i) $P\,dx + Q\,dy$ is locally exact; (ii) $P\,dx + Q\,dy$ is exact; (iii) $\partial P/\partial y = \partial Q/\partial x$; (iv) $\int_C P\,dx + Q\,dy$ is independent of the path; (v) $\int_C P\,dx + Q\,dy = 0$ for rectifiable closed curves. *Hint:* Use Exs. 7, 5, and the second hint of Ex. 3.

★★9. Let S be a portion of a surface prescribed parametrically: $x(u, v)$, $y(u, v)$, $z(u, v)$; $(u, v) \in R$, where R is a closed rectangle and the parametrization functions are continuously differentiable. *Define* the area of S by the formula:

(4) $$A(S) \equiv \iint_R \left[\left(\frac{\partial(y, z)}{\partial(u, v)}\right)^2 + \left(\frac{\partial(z, x)}{\partial(u, v)}\right)^2 + \left(\frac{\partial(x, y)}{\partial(u, v)}\right)^2\right]^{\frac{1}{2}} dA.$$

[§ 1331]   LINE AND SURFACE INTEGRALS   493

Justify this definition on the following three grounds: (i) *Invariance:* A change from the parameters u, v to new parameters $\bar{u}, \bar{v}$ does not affect the value of (4). (ii) *Intuitive appeal:* Corresponding to a "differential rectangle" dR, with sides du and dv, there is in space a "differential parallelogram" whose area is given by the expression that follows the integral signs in (4). (iii) *Parameters x and y:* If $u = x$ and $v = y$, formula (4) reduces to $\iint \sec \gamma \, dA$. (Cf. Ex. 18, § 1225; § 1327.)

★★**10.** If $F = F(x, y, z)$ is continuous on a portion of a surface S, then, with the notation of Exercise 9, **the surface integral** $\iint_S F \, dS$ is defined by the formula:

(5) $\iint_S F \, dS = \iint_R \phi(u, v) \left[\left(\frac{\partial(y, z)}{\partial(u, v)} \right)^2 + \left(\frac{\partial(z, x)}{\partial(u, v)} \right)^2 + \left(\frac{\partial(x, y)}{\partial(u, v)} \right)^2 \right]^{\frac{1}{2}} dA,$

where $\phi(u, v) = F(x(u, v), y(u, v), z(u, v))$. Discuss the surface integral (5) in the following three regards: (i) analogy with line integrals (Ex. 1); (ii) independence of parametrization; (iii) surfaces made up of more than one portion of the type prescribed.

★★**11.** Formulate for E_3 appropriate definitions of the two concepts of **simple compact region** and **finitely decomposable** compact region. Let the "boundary" or "oriented frontier" be defined in terms of the *outward normal*. (Explain.) (Cf. Exs. 4, 9; also Ex. 18, § 1225.)

★★**12.** Let P, Q, and R, and their partial derivatives $\partial P/\partial x$, $\partial Q/\partial y$, and $\partial R/\partial z$, be continuous on a finitely decomposable compact region Ω, and denote by $(\cos \alpha, \cos \beta, \cos \gamma)$ the outward normal unit vector, at each point of the "boundary" surface S (cf. Ex. 11). (Then the expression $P \cos \alpha + Q \cos \beta + R \cos \gamma$ represents the "outward normal component" of the vector (P, Q, R).) Prove the **divergence theorem** (so called because of the "divergence operator" of vector analysis occurring in the integrand on the left):

(6) $\iiint_\Omega \left(\frac{\partial P}{\partial x} + \frac{\partial Q}{\partial y} + \frac{\partial R}{\partial z} \right) dV = \iint_S (P \cos \alpha + Q \cos \beta + R \cos \gamma) \, dS.$

Hint: Establish the equality $\iiint (\partial R/\partial z) \, dV = \iint R \cos \gamma \, dS$ for a simple compact region by evaluating the left-hand member as an iterated integral, first integrating with respect to the variable z.

★★**13.** State and prove the analogue of Exercise 2 for line integrals $\int_C P \, dx + Q \, dy + R \, dz$ and a region in E_3.

★★**14.** State and prove the analogue of Exercise 3 for three dimensions.

★★**15.** For a portion S of a surface, as prescribed in Exercise 9, in terms of a rectangle $R: a \leq u \leq b, c \leq v \leq d$, define the **boundary** of S to be the closed curve C that is the image in space of the frontier of R in the uv-plane. The **orientations** of S and C are related as follows: If the orientation of C corresponds to the order $(a, c), (b, c), (b, d), (a, d)$ of the vertices of R, then the "orientation" of S (that is, of the *normal* to S) is that of the vector $(\partial(y, z)/\partial(u, v), \partial(z, x)/\partial(u, v), \partial(x, y)/\partial(u, v))$; if the orientation of C is reversed so is that of S. Discuss this

★★16. An **orientable surface** is a surface that is composed of a finite number of oriented portions (Ex. 15) in such a fashion that whenever two such portions have an arc in common, this arc belongs to the two boundaries of these portions and as such has opposite orientations. The boundary C of S may consist of one or more simple closed curves. Discuss this concept, with illustrations.

★★17. Let C be the boundary of an orientable surface S (Ex. 16) on which the functions P, Q, and R are continuously differentiable, and denote by $(\cos \alpha, \cos \beta, \cos \gamma)$ the unit normal vector to S. Prove **Stokes's Theorem**:

$$(7) \quad \iint_S \left[\left(\frac{\partial R}{\partial y} - \frac{\partial Q}{\partial z}\right) \cos \alpha + \left(\frac{\partial P}{\partial z} - \frac{\partial R}{\partial x}\right) \cos \beta + \left(\frac{\partial Q}{\partial x} - \frac{\partial P}{\partial y}\right) \cos \gamma \right] dS$$

$$= \int_C P \, dx + Q \, dy + R \, dz.$$

Show how Green's Theorem in the Plane (Ex. 5) can be considered as a special case of Stokes's Theorem. *Hints:* For a portion of a surface, as given in Ex. 9, the left-hand member of (7) becomes

$$(8) \quad \iint_R \left[\left(\frac{\partial R}{\partial y} - \frac{\partial Q}{\partial z}\right) \frac{\partial(y,z)}{\partial(u,v)} + \left(\frac{\partial P}{\partial z} - \frac{\partial R}{\partial x}\right) \frac{\partial(z,x)}{\partial(u,v)} + \left(\frac{\partial Q}{\partial x} - \frac{\partial P}{\partial y}\right) \frac{\partial(x,y)}{\partial(u,v)} \right] dA.$$

If Γ is the boundary of R, the right-hand member of (7) can be written

$$(9) \quad \int_\Gamma \left(P \frac{\partial x}{\partial u} + Q \frac{\partial y}{\partial u} + R \frac{\partial z}{\partial u} \right) du + \left(P \frac{\partial x}{\partial v} + Q \frac{\partial y}{\partial v} + R \frac{\partial z}{\partial v} \right) dv.$$

Apply Green's Theorem in the Plane to (9), expand the result, and compare with (8).

★★18. Prove the analogue of Exercise 7 for three dimensions: $P \, dx + Q \, dy + R \, dz$ is locally exact in a region Ω if and only if the following three identities hold throughout Ω:

$$(10) \quad \frac{\partial R}{\partial y} = \frac{\partial Q}{\partial z}, \quad \frac{\partial P}{\partial z} = \frac{\partial R}{\partial x}, \quad \frac{\partial Q}{\partial x} = \frac{\partial P}{\partial y}.$$

Hint: Adapt the hint of Ex. 7 to Stokes's Theorem (Ex. 17) and a rectangular parallelepiped.

★★★19. A region Ω of E_3 is **simply-connected** if and only if corresponding to every closed polygon C (simple or not) lying in Ω there exists a point p in Ω such that for every vertex p_i of C, $i = 1, 2, \cdots, n$, there exists a simple arc C_i from p to p_i lying in Ω and having the property that for every $i = 1, 2, \cdots, n$ the three arcs C_i, $\overline{p_i p_{i+1}}$, and $-C_{i+1}$ (where $C_{n+1} = C_1$, $p_{n+1} = p_1$ and the negative sign indicates a reversal of orientation) form the boundary of an orientable surface (Ex. 16) lying in Ω. Assuming continuity of all functions concerned, prove that $P \, dx + Q \, dy + R \, dz$ is exact in a simply-connected region if and only if it is locally exact there. Hence show the equivalence in a simply connected region of the following five statements: (i) $P \, dx + Q \, dy + R \, dz$ is locally exact; (ii) $P \, dx + Q \, dy + R \, dz$ is exact; (iii) the identities (10), Exercise 18, hold; (iv) $\int_C P \, dx + Q \, dy + R \, dz$ is independent of the path; (v) $\int_C P \, dx + Q \, dy + R \, dz = 0$ for rectifiable closed curves.

14

*Improper Integrals

★1401. INTRODUCTION. REVIEW

In Chapter 5 (§§ 511–515) improper integrals, for both finite and infinite intervals, were defined, and their elementary properties treated. The language of dominance and the "big O" and "little o" notation were introduced.

For convenience we repeat some statements and definitions given in the Exercises, § 515, formulated for a function $f(x)$ defined on a half-open interval $[a, b)$ or $[a, +\infty)$. Parentheses are used to achieve a compression of two alternative statements into one. Similar formulations hold for other intervals. Proofs of all statements of this section are left to the reader (e.g., cf. § 304, Ex. 31, § 503, and Note 2, § 501, for Theorem I, below).

We start by writing down explicitly an immediate consequence of convergence. If the function $f(x)$ is improperly integrable on the interval $[a, b)$ $([a, +\infty))$, and $\epsilon > 0$, then there exists a number $\gamma = \gamma(\epsilon)$ of the interval (a, b) $((a, +\infty))$ such that for any u of the interval (γ, b) $((\gamma, +\infty))$:

$$(1) \qquad \left| \int_u^b f(x)\, dx \right| < \epsilon \quad \left(\left| \int_u^{+\infty} f(x)\, dx \right| < \epsilon \right).$$

Theorem I. Cauchy Criterion. *If $f(x)$ is integrable on (a, β) for every β such that $a < \beta < b$ $(a < \beta)$, then the improper integral*

$$(2) \qquad \int_a^b f(x)\, dx \quad \left(\int_a^{+\infty} f(x)\, dx \right)$$

converges if and only if corresponding to $\epsilon > 0$ there exists a number $\gamma = \gamma(\epsilon)$ such that $a < \gamma < b$ $(a < \gamma)$ and such that whenever β_1 and β_2 belong to the interval (γ, b) $((\gamma, +\infty))$,

$$(3) \qquad \left| \int_{\beta_1}^{\beta_2} f(x)\, dx \right| < \epsilon.$$

Theorem II. Comparison Test. *If the nonnegative function $g(x)$ dominates the function $f(x)$ (whose values are not assumed to be of one sign)*

495

on the interval $I = [a, b)$ ($[a, +\infty)$), if $f(x)$ and $g(x)$ are integrable on $[a, \beta]$ for every $\beta \in I$, and if the integral $\int_a^b g(x)\, dx$ $\left(\int_a^{+\infty} g(x)\, dx \right)$ converges, so does the integral (2).

Definition. *If $f(x)$ is integrable on $[a, \beta]$ for every β such that $a < \beta < b$ ($a < \beta$), then the integral (2) is said to **converge absolutely** if and only if the improper integral*

(4) $$\int_a^b |f(x)|\, dx \quad \left(\int_a^{+\infty} |f(x)|\, dx \right)$$

*converges. The integral (1) **converges conditionally** if and only if it converges but does not converge absolutely.*

NOTE. From the fact that integrability of $f(x)$ on $[a, \beta]$ implies that of $|f(x)|$ there (Ex. 38, § 503) we infer that the integral (4) always exists in a finite or infinite sense. It is a simple consequence of the Comparison Test (Theorem II) that *absolute convergence implies convergence*.

Example 1. Show that $\int_1^{+\infty} \frac{\sin x}{x^p}\, dx$ converges absolutely for $p > 1$. Show that $\int_0^{+\infty} \frac{\sin x}{x^p}\, dx$ converges absolutely for $1 < p < 2$, and diverges for $p \geqq 2$. (Cf. Example 1, § 1402.)

Solution. Since for $x \geqq 1$, $\sin x/x^p$ is continuous and dominated by $1/x^p$, the first statement follows from Example 3, § 513. The remaining parts are consequences of the fact that, as $x \to 0+$, $\sin x/x^p$ is of the same order of magnitude as $1/x^{p-1}$. Thus, by Example 3, § 512, $\int_0^1 \frac{\sin x}{x^p}\, dx$ converges absolutely for $p < 2$.

Example 2. Show that if $P(x)$ is a polynomial and $p > 0$, then
$$\int_0^{+\infty} P(x) e^{-px}\, dx$$
converges absolutely.

Solution. $P(x)e^{-px} = O\left(\frac{1}{x^2}\right)$, as $x \to +\infty$.

★1402. ALTERNATING INTEGRALS. ABEL'S TEST

In case it has been determined that a given improper integral fails to converge absolutely, it is still possible to have (conditional) convergence through a fortuitous distribution of positiveness and negativeness. For example, the improper integral $\int_0^{+\infty} \frac{\sin x}{x^p}\, dx$, as we shall see in Example 1, converges conditionally for $0 < p \leqq 1$, thanks to the regular alternation of signs of $\sin x$. The situation is similar to that of alternating series (§ 715) where, for example, the alternating p-series $1 - \frac{1}{2^p} + \frac{1}{3^p} - \cdots$ converges

§ 1402] ALTERNATING INTEGRALS. ABEL'S TEST 497

conditionally for $0 < p \leq 1$ (Example 1, § 716). For integrals like the one just mentioned, whose integrands alternate signs over consecutive intervals, it is possible to relate convergence to that of an associated infinite series. Such a test is given in Exercise 13, § 1403.

In the present section, however, we shall establish convergence of such improper integrals (taken for definiteness for the interval $[a, +\infty)$) by means of a powerful test whose essential ideas are due to the Norwegian mathematician N. H. Abel (1802–1829) (cf. Ex. 22, § 717). The proof given here rests on a form of the Second Mean Value Theorem whose proof, as outlined in Exercises 28–29, § 518, is based on the concept of the Riemann-Stieltjes integral (§ 517). For a proof of Abel's Test which is independent of the Riemann-Stieltjes integral (but which requires more restrictive hypotheses), see the hint accompanying Exercise 9, § 1403.

Theorem I. Abel's Test. *Assume that $f(x)$ is continuous and $\phi(x)$ monotonically decreasing for $x \geq a$, that $\lim_{x \to +\infty} \phi(x) = 0$, and that $F(x) \equiv \int_a^x f(t)\, dt$ is bounded for all $x \geq a$. Then $\int_a^{+\infty} f(x)\phi(x)\, dx$ converges.*

Proof. We shall establish convergence by means of the Cauchy Criterion, § 1401. Let us observe first that for $a \leq u < v$ the integral $\int_u^v f(x)\phi(x)\, dx$ exists (cf. Theorems VIII, X, § 501, Ex. 35, § 503). Let $\epsilon > 0$, assume $|F(x)| < M$ for all $x \geq a$, and let u and v be arbitrary numbers such that $a \leq u < v$. By the Bonnet form of the Second Mean Value Theorem (Ex. 29, § 518), for a suitable number ξ such that $u \leq \xi \leq v$,
$$\left| \int_u^v f(x)\phi(x)\, dx \right| = \phi(u) \left| \int_u^\xi f(x)\, dx \right| = \phi(u)|F(\xi) - F(u)| \leq 2M\phi(u).$$
Accordingly, if N is chosen so large that $\phi(N) < \epsilon/2M$, and if $v > u > N$, then the conditions of the Cauchy Criterion are satisfied and the desired convergence is proved.

Theorem II. *If $\phi(x)$ is monotonically decreasing for $x \geq a$ and if $\lim_{x \to +\infty} \phi(x) = 0$, then*

(1) $$\int_a^{+\infty} \phi(x) \sin x\, dx \quad \text{and} \quad \int_a^{+\infty} \phi(x) \cos x\, dx$$

converge. The convergence is absolute or conditional according as $\int_a^{+\infty} \phi(x)\, dx$ converges or diverges.

Proof. The convergence is a corollary of Theorem I. The only non-trivial part remaining is to show that divergence of $\int^{+\infty} \phi(x)\, dx$ implies that of $\int_a^{+\infty} \phi(x) |\sin x| \, dx$ and $\int_a^{+\infty} \phi(x) |\cos x| \, dx$. We shall show this for the former only (proof for the other being essentially the same). Accordingly, let m and n be positive integers such that $n\pi > m\pi \geq a$. Then

$$\int_a^{n\pi} \phi(x) |\sin x| \, dx \geq \int_{m\pi}^{(m+1)\pi} \phi(x) |\sin x| \, dx + \cdots + \int_{(n-1)\pi}^{n\pi} \phi(x) |\sin x| \, dx$$

$$\geq \phi((m+1)\pi) \int_{m\pi}^{(m+1)\pi} |\sin x| \, dx + \cdots + \phi(n\pi) \int_{(n-1)\pi}^{n\pi} |\sin x| \, dx$$

$$= 2[\phi((m+1)\pi) + \cdots + \phi(n\pi)] \geq \frac{2}{\pi} \int_{(m+1)\pi}^{(n+1)\pi} \phi(x) \, dx.$$

These inequalities provide the desired implication.

Example 1. Show that $\int_0^{+\infty} \frac{\sin x}{x^p} \, dx$ converges conditionally for $0 < p \leq 1$. (Cf. Example 1, § 1401.)

Solution. By the solution of Example 1, § 1401, we need prove the statement only for the integral $\int_1^{+\infty} \frac{\sin x}{x^p} \, dx$. But for this integral the result follows from Theorem II (cf. Example 3, § 513).

Example 2. Show that $\int_0^{+\infty} \sin x^2 \, dx$ converges conditionally.

Solution. Making the substitution $x = \sqrt{y}, y = x^2$, we have

$$\int_0^b \sin x^2 \, dx = \frac{1}{2} \int_0^{b^2} \frac{\sin y}{\sqrt{y}} \, dy \quad (b > 0).$$

Therefore the result follows from Example 1 (why?). It is a curious fact that the given improper integral converges even though the integrand oscillates persistently between 1 and −1, and does not approach 0 (cf. Exs. 42, 46, § 515). The behavior that makes this phenomenon possible in this case is that the lengths of the intervals of oscillation shrink to zero. (Show this.)

Example 3. Show that $\int_0^{+\infty} \frac{\cos x}{\sqrt{x}} \, dx = \frac{1}{2} \int_0^{+\infty} \frac{\sin x}{x^{\frac{3}{2}}} \, dx$, and account for the remarkable fact that one integral converges conditionally, while the other converges absolutely.

Solution. Integration by parts gives

$$\int_a^b \frac{\cos x \, dx}{\sqrt{x}} = \int_a^b \frac{1}{\sqrt{x}} d(\sin x) = \frac{\sin x}{\sqrt{x}} \Big]_a^b + \frac{1}{2} \int_a^b \frac{\sin x}{x^{\frac{3}{2}}} \, dx.$$

If absolute value signs are introduced in the original integrand the process of integration by parts breaks down (for example, $\frac{d}{dx} |\sin x| \neq |\cos x|$). Thus no inference regarding absolute convergence can be drawn. The "remarkable fact" is no more remarkable than the fact that the series $\frac{1}{1 \cdot 2} + \frac{1}{3 \cdot 4} + \cdots$, obtained by pairing terms of the conditionally convergent alternating harmonic series, converges absolutely.

★1403. EXERCISES

In Exercises 1-4, determine whether the integral converges or diverges and, in case of convergence, whether it converges absolutely or conditionally.

★1. $\int_0^{+\infty} \frac{\cos 2x}{\sqrt{1+x^3}} dx.$ ★2. $\int_0^{+\infty} \frac{e^{-\frac{1}{2}x}\sin x}{x} dx.$

★3. $\int_0^{+\infty} \frac{x\sin x}{x^2-x-1} dx.$ ★4. $\int_2^{+\infty} \frac{\cos x}{\ln x} dx.$

In Exercises 5-8, give the values of p for which the integral converges absolutely, and those for which it converges conditionally.

★5. $\int_0^{+\infty} \frac{\cos x}{x^p} dx.$ ★6. $\int_0^{+\infty} \frac{\sin x}{(x+1)x^p} dx.$

★7. $\int_0^{+\infty} \frac{x^p \cos x}{1+x^2} dx.$ ★8. $\int_0^{+\infty} \frac{\sin^2 x}{x^p} dx.$

★9. Prove Abel's Test (Theorem I, § 1402), under the additional assumption that $\phi(x)$ is continuously differentiable, without benefit of the Bonnet form of the Second Mean Value Theorem. *Hint:* Assume that $|F(x)| < M$, integrate by parts, and show, for $a \leq u < v$: $\left|\int_u^v f(x)\phi(x)\,dx\right| \leq |F(v)\phi(v)| + |F(u)\phi(u)| + \int_u^v |F(x)|\{-\phi'(x)\}\,dx \leq 3M\phi(u).$

★10. Find two examples to prove that in Abel's Test (Theorem I, § 1402) neither assumption $\phi(x) \to 0$ nor $\phi(x) \downarrow$ can be omitted.

★11. Prove that if $f(x)$ is continuous and $\phi(x)$ bounded and monotonic for $x \geq a$, and if $\int_a^{+\infty} f(x)\,dx$ converges, then so does $\int_a^{+\infty} f(x)\phi(x)\,dx.$ *Hint:* If $\phi(x) \uparrow$ and $\lim_{x \to +\infty} \phi(x) = L$, write $f(x)\phi(x) = [-f(x)][L - \phi(x)] + Lf(x)$, and apply Abel's Test. This is an alternative form of Abel's Test.

★12. Assume that $\phi(x)$ is a positive monotonically decreasing function for $x \geq a$ and that $\int_a^{+\infty} \phi(x)\,dx$ converges. Prove that $\phi(x) = o(x^{-1})$ as $x \to +\infty$. *Hint:* $\int_x^{+\infty} \phi(t)\,dt \geq \int_x^{2x} \phi(t)\,dt \geq x\phi(x).$

★13. Assume that $f(x)$ is defined for $x \geq a$ and integrable on $[a, b]$ for every $b > a$, let $a = b_0 < b_1 < b_2 < \cdots, b_n \to +\infty$, and define $I_n = \int_{b_{n-1}}^{b_n} f(x)\,dx.$ Prove that the convergence of $\int_a^{+\infty} f(x)\,dx$ implies that of $\sum_{n=1}^{+\infty} I_n$, and the equality of their values. Show by an example that the converse implication is false. Prove that with the further assumption that $f(x)$ does not change sign on any one of the intervals (b_{n-1}, b_n), $n = 1, 2, \cdots$, the convergence of $\sum_{n=1}^{+\infty} I_n$ implies that of

$$\int_A^{+\infty} f(x)\,dx.$$ Use this result to establish convergence of $\int_0^{+\infty} \frac{\sin x}{x}\,dx$ and $\int_0^{+\infty} \sin x^2\,dx.$

★14. Show that $0 < \int_0^x \frac{\sin t}{t}\,dt < \pi$ for all positive x.

★1404. UNIFORM CONVERGENCE

Uniform convergence for an improper integral is similar to uniform convergence for an infinite series, and many of the consequences of uniform convergence that were established for infinite series in Chapter 9 have parallels for improper integrals, which we shall study in this chapter. Many of these are special cases of general theorems on uniform limits presented in §§ 1012–1015.

We start with the basic definition, formulated for definiteness for imimproper integrals on a half-open interval $[c, d)$ or an infinite interval $[c, +\infty)$, parentheses being used to express an alternative statement.

Definition. *Let $f(x, y)$ be defined for every point x of a set A and every y of the interval $I = [c, d)$ $([c, +\infty))$, and assume that for every $x \in A$ and every $\beta \in I$, $f(x, y)$ as a function of y is (Riemann) integrable on $[c, \beta]$. Then the integral*

$$(1) \qquad \int_c^d f(x, y)\,dy \; \left(\int_c^{+\infty} f(x, y)\,dy \right)$$

converges uniformly *to a function $F(x)$ for $x \in A$, written*

$$(2) \qquad \int_c^\beta f(x, y)\,dy \rightrightarrows F(x), \text{ as } \beta \to d- \; (\beta \to +\infty),$$

if and only if corresponding to $\epsilon > 0$ there exists a number $\gamma \equiv \gamma(\epsilon)$ belonging to the interval I such that the inequality $\gamma < \beta < d$ ($\gamma < \beta$) implies

$$(3) \qquad \left| \int_c^\beta f(x, y)\,dy - F(x) \right| < \epsilon$$

*for every $x \in A$. The variable x is called a **parameter**.*

NOTE. As with infinite series (§ 902), uniform convergence implies convergence, but not conversely. (Cf. the Example, below.)

The question before us is usually whether convergence, known to obtain, is uniform. It is therefore frequently convenient to formulate uniform convergence, and its negation, as follows (the reader may supply the proofs, which are immediate):

Theorem I. *Under the assumptions of the preceding Definition, and the further assumption that (1) converges for each $x \in A$, the convergence of (1) is uniform if and only if*

$$\int_\beta^d f(x, y)\, dy \rightrightarrows 0 \quad \left(\int_\beta^{+\infty} f(x, y)\, dy \rightrightarrows 0\right), \tag{4}$$

as $\beta \to d-$ $(\beta \to +\infty)$; in other words, if and only if corresponding to $\epsilon > 0$ there exists a number $\gamma = \gamma(\epsilon)$ belonging to the interval (c, d) $((c, +\infty))$ and such that the inequality $\gamma < \beta < d$ $(\gamma < \beta)$ implies

$$\left|\int_\beta^d f(x, y)\, dy\right| < \epsilon \quad \left(\left|\int_\beta^{+\infty} f(x, y)\, dy\right| < \epsilon\right), \tag{5}$$

for every $x \in A$.

Corollary. *Under the assumptions of the preceding Definition, if c' is an arbitrary point of I, the uniform convergence of (1) is equivalent to that of the integral obtained from (1) by replacing c by c'.*

Theorem II. Negation of uniform convergence. *Under the assumptions of Theorem I, the convergence of (1) fails to be uniform if and only if there exists a positive number ϵ such that corresponding to an arbitrary number γ belonging to the interval (c, d) $((c, +\infty))$ there exists a number β of the interval (γ, d) $((\gamma, +\infty))$ and a point $x \in A$ such that*

$$\left|\int_\beta^d f(x, y)\, dy\right| \geq \epsilon \quad \left(\left|\int_\beta^{+\infty} f(x, y)\, dy\right| \geq \epsilon\right). \tag{6}$$

Example. Show that the integral

$$\int_0^{+\infty} \frac{\sin xy}{y}\, dy \tag{7}$$

converges uniformly for $x \geq \delta > 0$, but not uniformly for $x > 0$.

Solution. The substitution $u = xy$ gives

$$\int_\beta^{+\infty} \frac{\sin xy}{y}\, dy = \int_{x\beta}^{+\infty} \frac{\sin u}{u}\, du. \tag{8}$$

The convergence of $\int_0^{+\infty} \frac{\sin u}{u}\, du$ means that corresponding to $\epsilon > 0$ there exists a number $\gamma > 0$ such that $\alpha > \gamma$ implies

$$\left|\int_\alpha^{+\infty} \frac{\sin u}{u}\, du\right| < \epsilon.$$

Therefore, if $x \geq \delta$ and $\beta > \gamma/\delta$, we conclude that the quantity in (8) is numerically less than ϵ, and (by (5)) uniform convergence of (7) is established.

We now let $\epsilon = \frac{1}{2}\int_0^{+\infty} \frac{\sin u}{u}\, du$. (This is easily shown to be positive; see Example 2, § 1408, for a specific evaluation.) Then however large β may be, we can find a value of x sufficiently near 0 to ensure that the value of (8) is close enough to $\int_0^{+\infty} \frac{\sin u}{u}\, du$ to be greater than ϵ. By Theorem II, this shows that the convergence is not uniform for $x > 0$.

★1405. DOMINANCE AND THE WEIERSTRASS M-TEST

Dominance in uniform convergence of improper integrals is similar to dominance in uniform convergence of series (cf. § 903). The formulations of this section are once more framed in terms of the intervals $[c, d)$ and $[c, +\infty)$.

Definition. *The statement that a function $g(x, y)$ **dominates** a function $f(x, y)$ for $x \in A$ and $y \in B$ means that for every $x \in A$ and $y \in B$,*
$$|f(x, y)| \leq g(x, y).$$

Theorem I. Comparison Test. *Let $f(x, y)$ and $g(x, y)$ be defined for every x of a set A and every y of the interval $I = [c, d)$ $([c, +\infty))$ and assume that for every $x \in A$ and every $\beta \in I$, $f(x, y)$ and $g(x, y)$ as functions of y are integrable on $[c, \beta]$. Furthermore, assume that $g(x, y)$ dominates $f(x, y)$ for $x \in A$ and $y \in I$, and that*

$$(1) \qquad \int_c^d g(x, y)\, dy \quad \left(\int_c^{+\infty} g(x, y)\, dy \right)$$

converges uniformly for $x \in A$. Then

$$(2) \qquad \int_c^d f(x, y)\, dy \quad \left(\int_c^{+\infty} f(x, y)\, dy \right)$$

converges uniformly for $x \in A$.

Proof. By the comparison test for improper integrals (§ 514) the integral (2) is absolutely convergent, and therefore (§ 1401) convergent, for each $x \in A$. If $\epsilon > 0$ is given, and if γ is a point of I such that the inequality $\gamma < \beta < d$ ($\gamma < \beta$) implies

$$(3) \qquad \int_\beta^d g(x, y)\, dy < \epsilon \quad \left(\int_\beta^{+\infty} g(x, y)\, dy < \epsilon \right),$$

for all $x \in A$, the appropriate corresponding inequalities for $f(x, y)$ follow from
$$\left| \int_\beta^{\beta'} f(x, y)\, dy \right| \leq \int_\beta^{\beta'} |f(x, y)|\, dy \leq \int_\beta^{\beta'} g(x, y)\, dy,$$
where $\beta' > \beta$ (let $\beta' \to d-$ or $+\infty$).

Since an integral of the form $\int_c^d M(y)\, dy$ $\left(\int_c^{+\infty} M(y)\, dy \right)$, where the integrand is independent of x, converges uniformly for x in any set A, whenever it converges at all, we have as a special case of Theorem I the analogue of the Weierstrass M-test for infinite series.

Theorem II. Weierstrass M-Test. *Let $f(x, y)$ and $M(y)$ be defined for every $x \in A$ and $y \in I = [c, d)$ $([c, +\infty))$, and assume that for every*

§ 1406] THE CAUCHY CRITERION AND ABEL'S TEST

$x \in A$ and every $\beta \in I$, $f(x, y)$ and $M(y)$ as functions of y are integrable on $[c, \beta]$. Furthermore, assume that for every $x \in A$,

(4) $$|f(x, y)| \leq M(y),$$

and that

(5) $$\int_c^d M(y)\, dy \; \left(\int_c^{+\infty} M(y)\, dy \right)$$

is convergent. Then

(6) $$\int_c^d f(x, y)\, dy \; \left(\int_c^{+\infty} f(x, y)\, dy \right)$$

converges uniformly for $x \in A$.

Example. Show that $\int_0^{+\infty} e^{-y} \cos xy \, dy$ converges uniformly for all real x.

Solution. $|e^{-y} \cos xy| \leq e^{-y}$, and $\int_0^{+\infty} e^{-y}\, dy$ converges.

★1406. THE CAUCHY CRITERION AND ABEL'S TEST FOR UNIFORM CONVERGENCE

The Cauchy Criterion for uniform convergence of a sequence of functions is given, with hints on a proof, in Exercise 45, § 904. It is restated for more general functions in Exercise 19, § 1015. We now state it, and give the proof, for uniform convergence of the improper integral

(1) $$\int_c^d f(x, y)\, dy \; \left(\int_c^{+\infty} f(x, y)\, dy \right).$$

Theorem I. Cauchy Criterion for Uniform Convergence. *Let $f(x, y)$ be defined for every $x \in A$ and $y \in I = [c, d)$ $([c, +\infty))$, and assume that for every $x \in A$ and $\beta \in I$, $f(x, y)$ as a function of y is integrable on $[c, \beta]$. Then the integral (1) converges uniformly for $x \in A$ if and only if corresponding to $\epsilon > 0$ there exists a number $\gamma = \gamma(\epsilon) \in I$ such that whenever β_1 and β_2 belong to the interval (γ, d) $((\gamma, +\infty))$,*

(2) $$\left| \int_{\beta_1}^{\beta_2} f(x, y)\, dy \right| < \epsilon,$$

for every $x \in A$.

Equivalently, the integral (1) fails to converge uniformly for $x \in A$ if and only if there exists a positive number ϵ such that corresponding to an arbitrary point $\gamma \in I$ there exist β_1 and β_2 of the interval (γ, d) $((\gamma, +\infty))$ and a point $x \in A$ such that

(3) $$\left| \int_{\beta_1}^{\beta_2} f(x, y)\, dy \right| \geq \epsilon.$$

Proof for $[c, +\infty)$: "Only if": Assuming uniform convergence, with $\epsilon > 0$ given, we know there exists a number $\gamma \in (c, +\infty)$ such that $\beta > \gamma$ implies (for every $x \in A$)

$$\left| \int_\beta^{+\infty} f(x, y) dy \right| < \frac{\epsilon}{2}.$$

Therefore $\beta_1 > \gamma$ and $\beta_2 > \gamma$ imply (for every $x \in A$):

$$\left| \int_{\beta_1}^{\beta_2} f(x, y) \, dy \right| = \left| \int_{\beta_1}^{+\infty} f(x, y) \, dy - \int_{\beta_2}^{+\infty} f(x, y) \, dy \right|$$
$$\leq \left| \int_{\beta_1}^{+\infty} f(x, y) \, dy \right| + \left| \int_{\beta_2}^{+\infty} f(x, y) \, dy \right| < \frac{\epsilon}{2} + \frac{\epsilon}{2} = \epsilon.$$

"If": Assuming the statement following the words "if and only if," we are assured that the integral (1) converges for every $x \in A$, by the Cauchy Criterion of § 1401. Let us assume, furthermore, that this convergence is *not* uniform on A, and obtain a contradiction. By the negation of uniform convergence (Theorem II, § 1404), there exists a positive number 2ϵ (a shift in notation!) such that for any number $\gamma > c$ there exist $x \in A$ and $\beta_1 > \gamma$ such that $\left| \int_{\beta_1}^{+\infty} f(x, y) \, dy \right| \geq 2\epsilon$. Now, corresponding to *this* number ϵ let γ be the number declared to exist in the statement of our theorem. Then for *this* γ let x and β_1 be the numbers whose existence we just discussed. Finally, since $2\epsilon > \epsilon$ and since

$$\lim_{\beta \to +\infty} \left| \int_{\beta_1}^{\beta} f(x, y) \, dy \right| = \left| \int_{\beta_1}^{+\infty} f(x, y) \, dy \right|,$$ there must exist a number $\beta_2 > \beta_1$

such that $\left| \int_{\beta_1}^{\beta_2} f(x, y) \, dy \right| > \epsilon$. This contradiction to (2) completes the proof.

A useful test for uniform convergence in cases when the convergence is *conditional*, is similar to the Abel test of § 1402, and to an analogous test for infinite series (Ex. 49, § 904). We state the theorem only for the domain of integration $[c, +\infty)$, other formulations being similar.

Theorem II. Abel's Test. *For every $x \in A$, let $f(x, y)$ and $\phi(x, y)$, as functions of y, be such that $f(x, y)$ is continuous and $\phi(x, y)$ is monotonically decreasing for $y \geq c$. Assume, furthermore that there exists a constant M such that*

(4) $$\left| \int_c^\beta f(x, y) \, dy \right| < M$$

for every $x \in A$ and every $\beta \geq c$, and that

(5) $$\phi(x, y) \rightrightarrows 0, \quad \text{as} \quad y \to +\infty$$

uniformly for $x \in A$. Then the integral

§ 1407] THREE THEOREMS ON CONVERGENCE

$$(6) \qquad \int_c^{+\infty} f(x, y)\phi(x, y)\, dy$$

converges uniformly for $x \in A$.

Proof. We use the Cauchy Criterion of the preceding Theorem, and the ideas of the proof of Theorem I, § 1402 (including the Bonnet form of the Second Mean Value Theorem), and write ($c < \beta_1 < \beta_2$):

$$\left| \int_{\beta_1}^{\beta_2} f(x, y)\phi(x, y)\, dy \right| = \left| \phi(x, \beta_1) \int_{\beta_1}^{\xi} f(x, y)\, dy \right| \leq 2M\phi(x, \beta_1).$$

We therefore chose $\gamma = \gamma(\epsilon)$ so that $\phi(x, y) < \epsilon/2M$ for $y > \gamma$.

Example. Show that

$$(7) \qquad \int_0^{+\infty} \frac{e^{-xy} \sin ay}{y}\, dy, \ a \neq 0,$$

converges uniformly for $x \geq 0$.

Solution. By the Corollary to Theorem I, § 1404, we consider in place of (7) the corresponding integral on the interval $1 \leq y < +\infty$. Uniform convergence for $x \geq \delta > 0$ is easily established by the principle of dominance (cf. the Example, § 1405). However, for $x = 0$ the integral converges *conditionally*, and we call upon Abel's test (Theorem II) to establish uniformity for x near 0. Accordingly, let $f(x, y) = \sin ay$, $\phi(x, y) = e^{-xy}/y$, $A = [0, +\infty)$, and $c = 1$. (Cf. the Corollary, Theorem I, § 1404.) The conditions of Theorem II are readily verified, so that the integral $\int_1^{+\infty} \frac{e^{-xy} \sin ay}{y}\, dy$ converges uniformly for $x \geq 0$.

★1407. THREE THEOREMS ON UNIFORM CONVERGENCE

The three theorems on uniform limits, § 1013, specialize to improper integrals, with few adjustments. We give the statements, together with essential features of the proofs that are peculiar to the present material. The interval of integration will be assumed for definiteness to be either $[c, d)$ or $[c, +\infty)$.

Theorem I. *Let I be an arbitrary interval (finite or infinite) and let J be the interval $[c, d)$ ($[c, +\infty)$). Assume that $f(x, y)$ is continuous for all x and y such that $x \in I$, $y \in J$, and that the integral*

$$(1) \qquad \int_c^d f(x, y)\, dy \ \left(\int_c^{+\infty} f(x, y)\, dy \right)$$

converges uniformly for $x \in I$. Then the limit function

$$(2) \qquad F(x) \equiv \int_c^d f(x, y)\, dy \ \left(F(x) \equiv \int_c^{+\infty} f(x, y)\, dy \right)$$

is continuous on I.

Proof. For each $\beta \in J$ the function
$$F_\beta(x) \equiv \int_c^\beta f(x, y)\, dy$$
is a continuous function of x, for $x \in I$ (Theorem I, § 1226, Ex. 7, § 1227), and as $\beta \to d- \ (\beta \to +\infty)$,
$$F_\beta(x) \rightrightarrows F(x),$$
for $x \in I$.

Theorem II. *Let I be a closed interval $[a, b]$ and let J be the interval $[c, d)$ ($[c, +\infty)$). For each $\beta \in J$ assume that $f(x, y)$ is integrable on the closed rectangle $x \in I$, $c \leq y \leq \beta$, and that the integrals $\int_c^\beta f(x, y)\, dy$, for $x \in I$, and $\int_a^b f(x, y)\, dx$, for $y \in J$, always exist. Furthermore, assume that the integral*

(1) $\qquad\qquad \int_c^d f(x, y)\, dy \ \left(\int_c^{+\infty} f(x, y)\, dy \right)$

converges uniformly for $x \in I$. Then the limit function is integrable on I, and

(3) $\qquad \int_a^b \int_c^d f(x, y)\, dy\, dx = \int_c^d \int_a^b f(x, y)\, dx\, dy$
$$\left(\int_a^b \int_c^{+\infty} f(x, y)\, dy\, dx = \int_c^{+\infty} \int_a^b f(x, y)\, dx\, dy \right).$$

Proof. By the Fubini Theorem (§ 1309), $\int_a^b \int_c^\beta f(x, y)\, dy\, dx = \int_c^\beta \int_a^b f(x, y)\, dx\, dy$, and (3) follows from Theorem II, § 1013, by uniform convergence.

Theorem III. *Let I be the interval $[a, b]$ and let J be the interval $[c, d)$ ($[c, +\infty)$). Assume that $f(x, y)$ and $f_1(x, y)$ exist and are continuous for all (x, y), where $x \in I$, $y \in J$. Assume, furthermore, that*

(i) $\int_c^d f(x, y)\, dy \ \left(\int_c^{+\infty} f(x, y)\, dy \right)$ *converges for some value x_0 of $x \in I$;*

(ii) $\int_c^d f_1(x, y)\, dy \ \left(\int_c^{+\infty} f_1(x, y)\, dy \right)$ *converges uniformly for $x \in I$.*

Then

(iii) *the integral in (i) converges uniformly for $x \in I$;*
(iv) *the limit function defined by (iii) is differentiable for $x \in I$;*

(v) $\qquad\qquad \dfrac{d}{dx} \int_c^d f(x, y)\, dy = \int_c^d f_1(x, y)\, dy$
$$\left(\dfrac{d}{dx} \int_c^{+\infty} f(x, y)\, dy = \int_c^{+\infty} f_1(x, y)\, dy \right).$$

§ 1407] THREE THEOREMS ON CONVERGENCE

Proof. For any $\beta \in J$ (by Theorem II, § 1226, Ex. 7, § 1227):

(4) $$\frac{d}{dx}\int_c^\beta f(x, y)\, dy = \int_c^\beta f_1(x, y)\, dy.$$

The remainder of the proof follows from Theorem III, § 1013.

The hypothesis of uniform convergence (or some substitute) is essential to each of the preceding theorems. The following Examples illustrate this fact.

Example 1. *The limit of the integral need not equal the integral of the limit.* Direct evaluation shows that

$$\int_0^{+\infty} 2xy e^{-xy^2}\, dy = \begin{cases} 1 & \text{if } x > 0, \\ 0 & \text{if } x = 0. \end{cases}$$

If $x \to 0+$, the limit of the integral is 1, but since the limit of the integrand is identically 0, the integral of the limit is 0. Similarly, if $x \to +\infty$, the limit of the integral is 1 and the integral of the limit is 0.

Example 2. *The value of an iterated integral may depend on the order of integration:* If $f(x, y) \equiv (2y - 2xy^2)e^{-xy^2}$,

$$\int_0^1 \int_0^{+\infty} f(x, y)\, dx\, dy = \int_0^1 \left[2xy e^{-xy^2}\right]_{x=0}^{x=+\infty} dy = \int_0^1 0\, dy = 0,$$

$$\int_0^{+\infty} \int_0^1 f(x, y)\, dy\, dx = \int_0^{+\infty} \left[y^2 e^{-xy^2}\right]_{y=0}^{y=1} dx = \int_0^{+\infty} e^{-x}\, dx = 1.$$

Example 3. *Differentiation under an integral sign may be meaningless.* Although

$$\int_0^{+\infty} \frac{\sin xy}{y}\, dy$$

converges uniformly for $x \geq \delta > 0$ (Example, § 1404), the integral of the derivative (with respect to x),

$$\int_0^{+\infty} \cos xy\, dy,$$

diverges for all x.

Example 4. *Differentiation under an integral sign may be incorrect.* The function

$$F(x) \equiv \int_0^{+\infty} x^2 e^{-x^2 y}\, dy$$

is equal to x for all x, including $x = 0$. Therefore $F'(x) = 1$ for all x, including $x = 0$. On the other hand, the integral of the derivative,

$$\int_0^{+\infty} (3x^2 - 2x^4 y)e^{-x^2 y}\, dy,$$

is equal to 1 if $x \neq 0$, and equal to 0 if $x = 0$. Thus formal differentiation gives an incorrect result for $x = 0$, although every integral considered converges (absolutely).

★1408. EVALUATION OF IMPROPER INTEGRALS

The Theorems of the preceding section can often be used to provide evaluations of interesting and important special improper integrals. We give several examples, and more are requested in § 1409.

Example 1. Establish the formulas

(1) $$\int_0^{+\infty} e^{-px} \sin qx\, dx = \frac{q}{p^2 + q^2},\ p > 0,$$

(2) $$\int_0^{+\infty} e^{-px} \cos qx\, dx = \frac{p}{p^2 + q^2},\ p > 0.$$

Then, by integrating with respect to q, between a and b, prove:

(3) $$\int_0^{+\infty} e^{-px} \frac{\cos ax - \cos bx}{x}\, dx = \tfrac{1}{2} \ln \frac{p^2 + b^2}{p^2 + a^2},\ p > 0,$$

(4) $$\int_0^{+\infty} e^{-px} \frac{\sin bx - \sin ax}{x}\, dx = \operatorname{Arctan} \frac{b}{p} - \operatorname{Arctan} \frac{a}{p},\ p > 0,$$

and thus ($b = 0$)

(5) $$\int_0^{+\infty} e^{-px} \frac{\sin ax}{x}\, dx = \operatorname{Arctan} \frac{a}{p},\ p > 0.$$

Solution. Equations (1) and (2) follow immediately from (4) and (5), § 506. The integration is justified since, for a fixed $p > 0$, the integrands in (1) and (2) are dominated by e^{-px}. Therefore the integrals (1) and (2) converge uniformly for all q, and Theorem II, § 1407, applies.

Example 2. By letting $p \to 0+$ in (3) and (5), prove:

(6) $$\int_0^{+\infty} \frac{\cos bx - \cos ax}{x}\, dx = \ln \left|\frac{a}{b}\right|,\ ab \neq 0.$$

(7) $$\int_0^{+\infty} \frac{\sin ax}{x}\, dx = \begin{cases} \tfrac{1}{2}\pi, & a > 0, \\ 0, & a = 0, \\ -\tfrac{1}{2}\pi, & a < 0. \end{cases}$$

Solution. In deriving (6) we may without loss of generality assume that both a and b are positive. In applying to (3) Abel's test for uniform convergence for $p \geq 0$, we let $f(p, x) = \cos bx - \cos ax$ and $\phi(p, x) = e^{-px}/x,\ A = [0, +\infty),\ c = 1$ (cf. the Example, § 1406). The conditions of Theorem II, § 1406, are satisfied, so that the integral (3) converges uniformly for $p \geq 0$. Therefore, by Theorem I, § 1407, the function $F(p)$ defined by the left-hand member of (3) is continuous for $p \geq 0$. Since the right-hand member of (3) is also continuous for $p \geq 0$, and equal to the left-hand member for $p > 0$, equation (3) must also be satisfied when $p = 0$. This gives (6).

Equation (7), for $a \neq 0$, follows from (5) in similar fashion, with the help of the uniform convergence established in the Example, § 1406.

Example 3. By integrating (2) with respect to p, between positive numbers a and b, prove:

(8) $$\int_0^{+\infty} \frac{e^{-ax} - e^{-bx}}{x} \cos qx\, dx = \tfrac{1}{2} \ln \frac{q^2 + b^2}{q^2 + a^2},\ a > 0, b > 0,$$

and thus:

(9) $$\int_0^{+\infty} \frac{e^{-ax} - e^{-bx}}{x}\, dx = \ln \frac{b}{a}, a > 0, b > 0.$$

Solution. The integration is justified by the uniform convergence of (2) for $p \geq \delta$, where $\delta = \min(a, b)$, which is true since the integrand is dominated by $e^{-\delta x}$, $x \geq 0$.

Example 4. Use differentiation and integration by parts to establish the formula

(10) $$\int_0^{+\infty} e^{-x^2} \cos rx\, dx = \tfrac{1}{2}\sqrt{\pi} e^{-\frac{1}{4}r^2}.$$

Solution. Let the function $y = \phi(r)$ be defined by the integral in (10). Then, by Theorem III, § 1407, and the uniform convergence of the resulting integral,

$$\phi'(r) = -\int_0^{+\infty} x e^{-x^2} \sin rx\, dx,$$

which, by integration by parts, gives

$$\tfrac{1}{2}\int_0^{+\infty} \sin rx\, d(e^{-x^2}) = \tfrac{1}{2} e^{-x^2} \sin rx \Big]_0^{+\infty} - \tfrac{1}{2}\int_0^{+\infty} e^{-x^2} d(\sin rx),$$

or

(11) $$\frac{dy}{dr} = -\tfrac{1}{2} ry.$$

This linear differential equation can be solved as follows: Since (11) is equivalent to

$$\frac{d}{dr}[y e^{\frac{1}{4}r^2}] = \left(\frac{dy}{dr} + \tfrac{1}{2}ry\right) e^{\frac{1}{4}r^2} = 0,$$

the solutions of (11) are all functions of the form $y = k e^{-\frac{1}{4}r^2}$. We can evaluate the constant k by setting $r = 0$ in the integral in (10), and using (8), § 1329.

Example 5. Use differentiation and a change of variable to establish the formula

(12) $$\int_0^{+\infty} e^{-x^2 - \frac{r^2}{x^2}}\, dx = \tfrac{1}{2}\sqrt{\pi} e^{-2|r|}.$$

Solution. Without loss of generality we shall assume that $r > 0$. Let the function $y = \phi(x)$, $r > 0$, be defined by the integral in (12) (which converges for all r since the integrand is dominated by e^{-x^2}). We shall now justify taking a derivative:

(13) $$\frac{dy}{dr} = \phi'(r) = \int_0^{+\infty} -\frac{2r}{x^2} e^{-x^2 - \frac{r^2}{x^2}}\, dx.$$

This is validated as follows: if B is an upper bound for pe^{-p}, for all $p > 0$, the integrand in (13) is dominated by $\frac{2B}{r} e^{-x^2}$. Therefore the integral of (13) converges uniformly for $0 < \alpha \leq r \leq \beta$, and thus (13) is valid for any positive r. The next step is to rewrite the integral in (13) by means of the substitution $x = r/u$,

(14) $$\frac{dy}{dr} = \phi'(r) = -2\int_0^{+\infty} e^{-\frac{r^2}{u^2}-u^2}\,du = -2\phi(r) = -2y.$$

Solving the linear differential equation (14) as in Example 4, we have $\frac{d}{dr}[ye^{2r}] = \left(\frac{dy}{dr} + 2y\right)e^{2r} = 0$, and therefore $y = ke^{-2r}$. Finally, this equation can be extended, by the uniform convergence in (12), to the range $r \geq 0$ (cf. Example 2), and the substitution of $r = 0$ gives an evaluation of k, and the formula (12).

Example 6. Using trigonometric identities and symmetry properties, show that

(15) $$\int_0^{\frac{1}{2}\pi} \ln \sin x\, dx = \int_0^{\frac{1}{2}\pi} \ln \cos x\, dx = -\tfrac{1}{2}\pi \ln 2.$$

Solution. The substitution $y = \tfrac{1}{2}\pi - x$ shows the equality of the two integrals; and the fact that $\ln \sin x = o(x^{-\frac{1}{2}})$ as $x \to 0+$ establishes the finiteness of their common value, which we shall denote by I. Adding the two integrals of (15), and using the substitution $y = 2x$, we have

$$2I = \int_0^{\frac{1}{2}\pi} \ln(\sin x \cos x)\, dx = \frac{1}{2}\int_0^{\pi} \ln \frac{\sin y}{2}\, dy.$$

The substitution $u = \pi - y$ shows that

$$\int_0^{\pi} \ln \sin y\, dy = \int_0^{\frac{1}{2}\pi} \ln \sin y\, dy + \int_{\frac{1}{2}\pi}^{\pi} \ln \sin y\, dy = 2I.$$

Therefore, from the preceding expression for $2I$, we have

$$2I = \frac{1}{2}\int_0^{\pi} \ln \sin y\, dy - \frac{1}{2}\int_0^{\pi} \ln 2\, dy = I - \frac{1}{2}\pi \ln 2,$$

and the desired evaluation.

★1409. EXERCISES

★1. Show that $\int_0^{+\infty} \frac{\cos xy}{1+y^2}\, dy$ converges uniformly for all x.

★2. Show that $\int_0^{+\infty} \frac{e^{-xy}\sin ay}{y}\, dy$, for a given $x > 0$, converges uniformly for all a (cf. Example, § 1406).

In Exercises 3–8, show that the given integral converges uniformly for the first of the two given intervals, and that it converges but not uniformly for the second of the two given intervals. The letter η denotes an arbitrarily small positive number, and the letter that is not the variable of integration denotes the parameter.

★3. $\int_0^1 \frac{dx}{x^p}$; $(-\infty, 1-\eta]$; $(-\infty, 1)$.

★4. $\int_1^{+\infty} \frac{dx}{x^p}$; $[1+\eta, +\infty)$; $(1, +\infty)$.

★5. $\int_0^{+\infty} e^{-ax}\, dx$; $[\eta, +\infty)$; $(0, +\infty)$.

§ 1409] EXERCISES 511

★6. $\int_0^{+\infty} \frac{\sin^2 ax}{ax^2} dx$; $[\eta, +\infty)$; $(0, +\infty)$.

★7. $\int_0^{+\infty} x^2 y e^{-xy} dy$; $[\eta, +\infty)$; $(0, +\infty)$.

★8. $\int_0^{+\infty} y^{5/2} e^{-xy} dy$; $[\eta, +\infty)$; $(0, +\infty)$.

★9. Show that $\int_0^{+\infty} x^2 e^{-xy} dy$ converges uniformly for $x > 0$ and that $\int_0^{+\infty} x e^{-xy} dy$ does not. Hence explain the phenomenon of Example 1, § 1407.

★10. Show that the integral for the gamma function, $\int_0^{+\infty} x^{\alpha-1} e^{-x} dx$ (Example 9, § 514), converges uniformly for $[\eta, b]$, but not uniformly for $[\eta, +\infty)$ or for $(0, b]$, where $0 < b < +\infty$.

In Exercises 11–20, establish the given evaluation.

★11. $\int_0^{+\infty} \frac{e^{-a^2x^2} - e^{-b^2x^2}}{x^2} dx = (b-a)\sqrt{\pi}, 0 \leqq a < b$.
Hint: Integrate $\int_0^{+\infty} \alpha e^{-\alpha^2 x^2} dx = \frac{1}{2}\sqrt{\pi}, \alpha > 0$.

★12. $\int_0^{+\infty} x^2 e^{-x^2} dx = \frac{\sqrt{\pi}}{4}$. *Hint:* Integrate by parts.

★13. $\int_0^{\frac{\pi}{2}} x \cot x \, dx = \frac{\pi \ln 2}{2}$. *Hint:* Integrate by parts.

★14. $\int_0^{+\infty} \frac{\sin^2 \alpha x}{x^2} dx = \frac{\pi}{2} |\alpha|$. *Hint:* Integrate by parts.

★15. $\int_0^{+\infty} \frac{1 - \cos \alpha x}{x^2} dx = \frac{\pi}{2} |\alpha|$. *Hint:* Use an identity.

★16. $\int_0^{+\infty} e^{-\left(x - \frac{a}{x}\right)^2} dx = \begin{cases} \frac{1}{2}\sqrt{\pi}, a \geqq 0, \\ \frac{1}{2}\sqrt{\pi} e^{4a}, a \leqq 0. \end{cases}$

★17. $\int_0^1 \frac{\ln(1+x)}{1+x^2} dx = \frac{\pi \ln 2}{8}$. *Hint:* Let $x = \tan \theta$, and obtain three integrals by use of the identity $\cos(\frac{1}{4}\pi - \theta) = \frac{1}{2}\sqrt{2}(\cos \theta + \sin \theta)$.

★18. $\int_0^{+\infty} \frac{\sin \alpha x \sin x}{x} dx = \frac{1}{2} \ln \left|\frac{\alpha+1}{\alpha-1}\right|, |\alpha| \neq 1$.

★19. $\int_0^{+\infty} \frac{\sin \alpha x \cos x}{x} dx = \frac{\pi}{2}$ if $\alpha > 1$, $\frac{\pi}{4}$ if $\alpha = 1$, 0 if $-1 < \alpha < 1$, $-\frac{\pi}{4}$ if $\alpha = -1$, and $-\frac{\pi}{2}$ if $\alpha < -1$.

★20. $\int_0^{+\infty} \frac{\cos rx}{1+x^2} dx = \frac{1}{2}\pi e^{-|r|}$. *Hints:* Let the integral define y as a function of $r > 0$, and show: $\frac{dy}{dr} = -\int_0^{+\infty} \frac{[(1+x^2) - 1] \sin rx}{x(1+x^2)} dx = -\frac{\pi}{2} +$

512   IMPROPER INTEGRALS   [§ 1409

$\int_0^{+\infty} \frac{\sin rx}{x(1+x^2)}\,dx$; $\frac{d^2y}{dr^2} = y$. Therefore $y = c_1 e^r + c_2 e^{-r}$ and $dy/dr = c_1 e^r - c_2 e^{-r}$.
Show that as $r \to 0+$: $y \to \frac{1}{2}\pi$ and $dy/dr \to -\frac{1}{2}\pi$.

★21. Show that the integral $\int_0^{+\infty} \frac{x\,dy}{x^2+y^2}$ converges to $\frac{\pi}{2}\operatorname{sgn} x$ (Example 1, § 206) for all x, uniformly on any bounded set. Since the signum function is discontinuous at $x = 0$, this seems to contradict Theorem I, § 1407. Explain.

★22. Using formula (2), § 1408, show that the nth derivative of $(1+x^2)^{-1}$ is equal to

$$(-1)^{\frac{n}{2}} \int_0^{+\infty} y^n e^{-y} \cos xy \, dy, \quad n = 0, 2, 4, \cdots,$$

$$(-1)^{\frac{n+1}{2}} \int_0^{+\infty} y^n e^{-y} \sin xy \, dy, \quad n = 1, 3, 5, \cdots.$$

Hence show, with the aid of Exercise 37, § 515:

$$\left|\frac{d^n}{dx^n}\frac{1}{1+x^2}\right| \leq n!, \quad n = 0, 1, 2, \cdots.$$

(Cf. the second solution of Example 5, § 813.)

★23. Prove that $\lim\limits_{x \to +\infty} \int_0^{+\infty} \frac{dy}{1 + xy^2} = 0$.

★24. Prove that $\lim\limits_{x \to +\infty} \int_0^{+\infty} e^{-xy^2}\,dy = 0$.

★25. Define
$$f(x, y) = \begin{cases} y^{-2} & \text{if } 0 < x < y < 1, \\ -x^{-2} & \text{if } 0 < y < x < 1, \\ 0 & \text{otherwise.} \end{cases}$$

Show that $\int_0^1 \int_0^1 f(x, y)\,dx\,dy = 1$, $\int_0^1 \int_0^1 f(x, y)\,dy\,dx = -1$. Explain how this is possible when every single integral involved is proper?

★26. Show that $\int_0^{+\infty} \int_0^{+\infty} (x-y)e^{-(x-y)^2}\,dx\,dy$ is equal to $\frac{1}{4}\sqrt{\pi}$, while the iterated integral in the reverse order is equal to $-\frac{1}{4}\sqrt{\pi}$. Explain this phenomenon.

★27. Assume that $f(x, y)$ is continuous for $x \geq a$, $y \geq c$, and that $\int_c^{+\infty} f(x, y)\,dy$ converges uniformly for $x > a$. Prove that $\lim\limits_{x \to a+} \int_c^{+\infty} f(x, y)\,dy = \int_c^{+\infty} f(a, y)\,dy$ whenever the right-hand member exists (cf. Ex. 28). *Hint*: If the integral converges at $x = a$ it converges uniformly for $x \geq a$. Use Theorem I, § 1407.

★★28. Use the Moore-Osgood theorem to prove that the right-hand member of the equation of Exercise 27 exists and is finite.

★★29. Assume that $\int_c^{+\infty} f(x, y)\,dy$ converges uniformly for $x \geq a$, and that the

§ 1410]   ITERATED IMPROPER INTEGRALS

limit $\lim_{x \to +\infty} f(x, y)$ exists and is uniform for any (finite) interval $c \leq y \leq \beta$. Prove that

$$\lim_{x \to +\infty} \int_c^{+\infty} f(x, y)\, dy = \int_c^{+\infty} \lim_{x \to +\infty} f(x, y)\, dy$$

(proving that both members exist and are finite). State and prove two other special cases of the general theorem of which this is an illustration. *Hint*: Cf. § 1014.

★★**30.** Adapt and prove the Abel test of Exercise 11, § 1403, for uniform convergence. (Cf. Ex. 23, § 717, Ex. 50, § 904.)

★**1410. ITERATED IMPROPER INTEGRALS**

Reversing the order of integration in an iterated integral where *both* integrals are separately improper can be a very delicate matter to justify. In this section we shall limit consideration to the interval $[0, +\infty]$, for simplicity, proving a theorem of a rather general character. It is possible, by means of this theorem, to evaluate the **Fresnel integrals**:

(1) $$\int_0^{+\infty} \sin x^2\, dx = \int_0^{+\infty} \cos x^2\, dx = \frac{\sqrt{\pi}}{2\sqrt{2}}$$

We shall give the details for the first integral of (1) in the Example, below, asking the student for the proof of the other in Exercise 2, § 1412.

NOTE. Other techniques that are frequently useful in connection with questions of interchange of limiting processes are (*i*) the Lebesgue bounded and dominated convergence theorems (Ex. 49, § 503, Ex. 47, § 515), (*ii*) the Moore-Osgood theorem (§ 1014), and (*iii*) the principle of iterated suprema (Theorem II, § 1411). An extremely potent method for evaluating many definite integrals, proper and improper, is that of *contour integration* in the theory of Analytic Functions of a Complex Variable.

Theorem. *Assume $f(x, y)$ is continuous for $x \geq 0$, $y \geq 0$, and that the following three integrals exist and are finite:*

(2) $$\int_0^{+\infty} \int_0^n f(x, y)\, dy\, dx, \quad n > 0,$$

(3) $$\int_0^{+\infty} \int_0^{+\infty} f(x, y)\, dy\, dx,$$

(4) $$\int_0^n \int_0^{+\infty} f(x, y)\, dx\, dy, \quad n > 0.$$

Assume, furthermore, that

(5) $$\lim_{n \to +\infty} \int_0^{+\infty} \int_n^{+\infty} f(x, y)\, dy\, dx = 0,$$

(6) $$\lim_{m \to +\infty} \int_0^n \int_m^{+\infty} f(x, y)\, dx\, dy = 0, \quad n > 0.$$

514   IMPROPER INTEGRALS   [§ 1410

Then

(7) $$\int_0^{+\infty}\int_0^{+\infty} f(x, y)\, dx\, dy$$

exists and is equal to (3).

Proof. The integral (7) is equal to

$$\lim_{n\to+\infty}\int_0^n\int_0^{+\infty} f(x,y)\,dx\,dy = \lim_{n\to+\infty}\left\{\lim_{m\to+\infty}\int_0^n\int_0^m f(x,y)\,dx\,dy\right\},$$

by (6). By the Fubini Theorem (§ 1309), this can be written

$$\lim_{n\to+\infty}\left\{\lim_{m\to+\infty}\int_0^m\int_0^n f(x,y)\,dy\,dx\right\} = \lim_{n\to+\infty}\int_0^{+\infty}\int_0^n f(x,y)\,dy\,dx.$$

Finally, by (5), this is equal to (3).

Example. Justify the equation

(8) $$\int_0^{+\infty}\int_0^{+\infty} e^{-x^2y}\sin y\,dx\,dy = \int_0^{+\infty}\int_0^{+\infty} e^{-x^2y}\sin y\,dy\,dx,$$

and thus obtain the formula

(9) $$\int_0^{+\infty}\frac{\sin y}{\sqrt{y}}\,dy = \sqrt{\frac{\pi}{2}},$$

as well as the first of (1).

Solution. The integrals (2), (3), and (4) are equal to

$$\int_0^{+\infty}\left[\frac{1}{1+x^4} - \frac{e^{-nx^2}}{1+x^4}(x^2\sin n + \cos n)\right]dx,$$

$$\int_0^{+\infty}\frac{dx}{1+x^4}, \quad \text{and} \quad \int_0^n \frac{\sqrt{\pi}}{2}\frac{\sin y}{\sqrt{y}}\,dy,$$

respectively (cf. (4), § 506, and (8), § 1329, with the substitution $u = x\sqrt{y}$). We shall now prove (5):

$$\left|\int_0^{+\infty}\int_n^{+\infty} e^{-x^2y}\sin y\,dy\,dx\right| = \left|\int_0^{+\infty}\frac{e^{-nx^2}}{1+x^4}(x^2\sin n + \cos n)\,dx\right|$$

$$\leq \int_0^{+\infty}\frac{1+x^2}{1+x^4}e^{-nx^2}\,dx \leq 2\int_0^{+\infty} e^{-nx^2}\,dx = \frac{\sqrt{\pi}}{\sqrt{n}} \to 0.$$

To prove (6) we use the inequality $e^{-u} < 1/u^2$:

$$\left|\int_0^n\int_m^{+\infty} e^{-x^2y}\sin y\,dx\,dy\right| \leq \int_0^n \frac{|\sin y|}{\sqrt{y}}\int_{m\sqrt{y}}^{+\infty} e^{-u^2}\,du\,dy$$

$$\leq \int_0^n \frac{|\sin y|}{\sqrt{y}}\int_{m\sqrt{y}}^{+\infty}\frac{du}{u^2}\,dy = \frac{1}{m}\int_0^n \frac{|\sin y|}{y}\,dy < \frac{n}{m} \to 0.$$

Having justified (8) we proceed to evaluate the two members. In the

(10) $$\int_0^{+\infty} \int_0^{+\infty} e^{-u^2} \frac{\sin y}{\sqrt{y}} \, du \, dy = \frac{\sqrt{\pi}}{2} \int_0^{+\infty} \frac{\sin y}{\sqrt{y}} \, dy,$$

which is known to converge by Theorem II, §1402. The right-hand member of (8) is equal to $\int_0^{+\infty} \frac{dx}{1+x^4}$, which can be integrated by the substitutions $x = \tan \theta$, $\alpha = 2\theta$, and $t = \tan \alpha$, as follows:

$$\int_0^{\frac{\pi}{2}} \frac{\sec^2 \theta \, d\theta}{1 + \tan^4 \theta} = \int_0^{\frac{\pi}{2}} \frac{\cos^2 \theta \, d\theta}{(\sin^2 \theta + \cos^2 \theta)^2 - 2\sin^2 \theta \cos^2 \theta}$$

$$= \int_0^{\frac{\pi}{2}} \frac{1 + \cos 2\theta}{2 - \sin^2 2\theta} \, d\theta = \frac{1}{2} \int_0^{\pi} \frac{(1 + \cos \alpha) \, d\alpha}{2 - \sin^2 \alpha} = \frac{1}{2} \int_0^{\pi} \frac{d\alpha}{2 - \sin^2 \alpha}$$

$$= \int_0^{\frac{\pi}{2}} \frac{d\alpha}{1 + \cos^2 \alpha} = \int_0^{\frac{\pi}{2}} \frac{\sec^2 \alpha \, d\alpha}{1 + \sec^2 \alpha} = \int_0^{+\infty} \frac{dt}{t^2 + 2} = \frac{\pi}{2\sqrt{2}}.$$

Equating this result and (10) gives (9).

Finally, substitution of $x = \sqrt{y}$ transforms the first integral of (1) to one-half the left-hand member of (9).

★1411. IMPROPER INTEGRALS OF INFINITE SERIES

The Theorem of §1410 can be adapted to forming improper integrals of infinite series, term by term. A more practical tool, for most purposes, is given in Theorem II, below, based on the **Principle of Iterated Suprema**:

Theorem I. *If $f(x, y)$ is a real-valued function, defined for every x in a set A and every y in a set B, then*

(1) $$\sup_{x \in A} \left\{ \sup_{y \in B} f(x, y) \right\} = \sup_{y \in B} \left\{ \sup_{x \in A} f(x, y) \right\},$$

in the finite or infinite sense.

Proof. By symmetry, we need only obtain a contradiction to the assumption that the left-hand member of (1) is finite and less than the right-hand member, which is finite or equal to $+\infty$. Accordingly, we assume there exists a number c such that

(2) $$\sup_{x \in A} \left\{ \sup_{y \in B} f(x, y) \right\} < c < \sup_{y \in B} \left\{ \sup_{x \in A} f(x, y) \right\}.$$

The first inequality of (2) implies that c is greater than $\sup_{y \in B} f(x, y)$, for every $x \in A$, and hence greater than $f(x, y)$, for every $x \in A$ and every $y \in B$. On the other hand, the right-hand inequality of (2) states that there exists a $y \in B$ such that $c < \sup_{x \in A} f(x, y)$ and hence, corresponding to this y there exists an $x \in A$ such that $c < f(x, y)$. This contradiction proves the theorem.

We now state a theorem on integrating a series term by term, assuming

Theorem II. *Assume that every term of the series $\sum_{n=1}^{+\infty} u_n(x)$ is defined and integrable on $[a, \beta]$ for every $\beta \in I$, and that for each such β the series $\sum_{n=1}^{+\infty} |u_n(x)|$ converges uniformly on $[a, \beta]$. Assume furthermore that*

$$(3) \quad \sum_{n=1}^{+\infty} \int_a^b |u_n(x)|\, dx < +\infty \quad \left(\sum_{n=1}^{+\infty} \int_a^{+\infty} |u_n(x)|\, dx < +\infty \right).$$

Then the following equation holds, both members existing and being finite:

$$(4) \quad \int_a^b \sum_{n=1}^{+\infty} u_n(x)\, dx = \sum_{n=1}^{+\infty} \int_a^b u_n(x)\, dx$$

$$\left(\int_a^{+\infty} \sum_{n=1}^{+\infty} u_n(x)\, dx = \sum_{n=1}^{+\infty} \int_a^{+\infty} u_n(x)\, dx \right).$$

Proof for $[a, +\infty)$: We first prove the Theorem for the case where every $u_n(x) \geq 0$ on $[a, b)$:

$$(5) \quad \int_a^{+\infty} \sum_{n=1}^{+\infty} u_n(x)\, dx = \sup_{\beta \in I} \int_a^{\beta} \sum_{n=1}^{+\infty} u_n(x)\, dx.$$

By uniform convergence, the last integral is finite, and

$$(6) \quad \int_a^{\beta} \sum_{n=1}^{+\infty} u_n(x)\, dx = \sum_{n=1}^{+\infty} \int_a^{\beta} u_n(x)\, dx = \sup_N \sum_{n=1}^{N} \int_a^{\beta} u_n(x)\, dx.$$

By (6), and Theorem I, the quantity (5) can be written

$$(7) \quad \sup_N \sup_{\beta \in I} \sum_{n=1}^{N} \int_a^{\beta} u_n(x)\, dx = \sup_N \lim_{\beta \to +\infty} \sum_{n=1}^{N} \int_a^{\beta} u_n(x)\, dx$$

$$= \sup_N \sum_{n=1}^{N} \lim_{\beta \to +\infty} \int_a^{\beta} u_n(x)\, dx = \sum_{n=1}^{+\infty} \int_a^{+\infty} u_n(x)\, dx.$$

We now remove the restriction that $u_n(x) \geq 0$, and define $u_n^+(x) \equiv \max(u_n(x), 0)$ and $u_n^-(x) \equiv \max(-u_n(x), 0)$ so that $u_n(x) = u_n^+(x) - u_n^-(x)$ and $|u_n(x)| = u_n^+(x) + u_n^-(x)$ (cf. Ex. 40, § 503). Then, by dominance of u_n^+ and u_n^- by $|u_n|$, and by the preceding argument, equation (4) holds for both series $\sum u_n^+$ and $\sum u_n^-$. Therefore

$$\int_a^{+\infty} \sum_{n=1}^{+\infty} u_n(x)\, dx = \int_a^{+\infty} \left\{ \sum_{n=1}^{+\infty} [u_n^+(x) - u_n^-(x)] \right\} dx$$

$$= \int_a^{+\infty} \left\{ \sum_{n=1}^{+\infty} u_n^+(x) - \sum_{n=1}^{+\infty} u_n^-(x) \right\} dx$$

$$= \int_a^{+\infty} \sum_{n=1}^{+\infty} u_n^+(x)\, dx - \int_a^{+\infty} \sum_{n=1}^{+\infty} u_n^-(x)\, dx$$

$$= \sum_{n=1}^{+\infty} \int_a^{+\infty} u_n^+(x)\, dx - \sum_{n=1}^{+\infty} \int_a^{+\infty} u_n^-(x)\, dx$$

$$= \sum_{n=1}^{+\infty} \int_a^{+\infty} [u_n^+(x) - u_n^-(x)]\, dx,$$

and the proof is complete.

§ 1412]   EXERCISES   517

Example 1. Justify the evaluation

$$\int_0^1 \frac{\ln x}{1-x}\,dx = \int_0^1 \ln x [1 + x + x^2 + \cdots]\,dx$$
$$= \left[x\ln x - x\right]_0^1 + \left[\frac{x^2}{2}\ln x - \frac{x^2}{2^2}\right]_0^1 + \left[\frac{x^3}{3}\ln x - \frac{x^3}{3^2}\right]_0^1 + \cdots$$
$$= -\left[1 + \frac{1}{2^2} + \frac{1}{3^2} + \frac{1}{4^2} + \cdots\right] = -\frac{\pi^2}{6}.$$

Solution. Since there are difficulties at both end-points (at 0 because of $\ln x$ and at 1 because the series $1 + x + x^2 + \cdots$ diverges there), we can split the interval $[0, 1]$ into two parts or, more simply, consider

(8) $$\int_\alpha^\beta \frac{\ln x}{1-x}\,dx = \left[x\ln x - x\right]_\alpha^\beta + \left[\frac{x^2}{2}\ln x - \frac{x^2}{2^2}\right]_\alpha^\beta + \cdots,$$

where $0 < \alpha < \beta < 1$. The equation (8) is justified by uniform convergence of $\sum |u_n(x)|$, where $u_n(x) = x^{n-1}\ln x$. In fact, inequality (3) becomes

$$\sum_{n=1}^{+\infty} \int_0^1 |x^{n-1}\ln x|\,dx = -\sum_{n=1}^{+\infty}\int_0^1 x^{n-1}\ln x\,dx = \sum_{n=1}^{+\infty}\frac{1}{n^2} < +\infty.$$

By Theorem II, the term-by-term integration is justified, and the series $-\sum 1/n^2$ results. For the value $-\pi^2/6$, see the Note, § 720.

Example 2. Justify the expansion, for $p > 0$:

$$\int_0^1 \frac{x^{p-1}\,dx}{1+x} = \int_0^1 [x^{p-1} - x^p + x^{p+1} - \cdots]\,dx$$
$$= \frac{1}{p} - \frac{1}{p+1} + \frac{1}{p+2} - \frac{1}{p+3} + \cdots.$$

Solution. By Theorem II, the following equation holds for $0 \leq y < 1$:

(9) $$\int_0^y \frac{x^{p-1}\,dx}{1+x} = \frac{y^p}{p} - \frac{y^{p+1}}{p+1} + \frac{y^{p+2}}{p+2} - \cdots.$$

The left-hand member of (9) defines a continuous function of y for $y \geq 0$. The desired equality for $y = 1$ will be proved as soon as we can establish the *uniform* convergence of the right-hand side of (9) for $0 \leq y \leq 1$. But this follows from the Abel test for uniform convergence (Ex. 49, § 904), with $u_n(y) = \pm 1$, $v_n(y) = \frac{y^{n+p}}{n+p}$.

★1412. EXERCISES

★1. Justify the interchange:

$$\int_0^{+\infty}\int_0^{+\infty} e^{-xy}\sin y\,dx\,dy = \int_0^{+\infty}\int_0^{+\infty} e^{-xy}\sin y\,dy\,dx,$$

and thus obtain a second method of evaluating $\int_0^{+\infty}\frac{\sin y}{y}\,dy$ (Example 2, § 1408).

★2. Establish the value given for the second Fresnel integral, (1), § 1410.

★3. Justify the interchange:

$$\int_0^{+\infty}\int_0^{+\infty} 2x\cos rye^{-x^2(1+y^2)}\,dx\,dy = \int_0^{+\infty}\int_0^{+\infty} 2x\cos rye^{-x^2(1+y^2)}\,dy\,dx,$$

and thus give a second derivation of the formula of Exercise 20, § 1409.

In Exercises 4-9, verify the given evaluations. (Cf. the Note, § 720, Exs. 11-12, § 721, and the footnote, p. 214.)

★4. $\int_0^{+\infty} \dfrac{x\,dx}{e^x-1} = -\int_0^1 \dfrac{\ln x}{1-x}\,dx = \dfrac{\pi^2}{6}.$

★5. $\int_0^{+\infty} \dfrac{x\,dx}{e^x+1} = -\int_0^1 \dfrac{\ln x}{1+x}\,dx = \dfrac{\pi^2}{12}.$

★6. $\int_0^{+\infty} \dfrac{x\,dx}{e^x-e^{-x}} = -\int_0^1 \dfrac{\ln x}{1-x^2}\,dx = \dfrac{\pi^2}{8}.$

★7. $\int_0^{+\infty} \ln\dfrac{e^x+1}{e^x-1}\,dx = \int_0^1 \dfrac{1}{x}\ln\dfrac{1+x}{1-x}\,dx = \dfrac{\pi^2}{4}.$

★8. $\int_0^1 \ln x \ln(1+x)\,dx = 2 - 2\ln 2 - \dfrac{\pi^2}{12}.$

Hint: $\dfrac{1}{n(n+1)^2} = \dfrac{1}{n} - \dfrac{1}{n+1} - \dfrac{1}{(n+1)^2}$, by partial fractions.

★9. $\int_0^{+\infty} \dfrac{x^2\,dx}{e^x-1} = \int_0^1 \dfrac{(\ln x)^2}{1-x}\,dx = 2\zeta(3).$

★10. Show that $\int_0^{+\infty}\left[\dfrac{1}{(x+1)^3}+\dfrac{1}{(x+2)^3}+\dfrac{1}{(x+3)^3}+\cdots\right]dx = \dfrac{\pi^2}{12}.$

★11. Show that the improper integral
$$\int_0^{+\infty}\left[\dfrac{1}{(x+1)^p}+\dfrac{1}{(x+2)^p}+\dfrac{1}{(x+3)^p}+\cdots\right]dx$$
converges if and only if $p > 2$. (Cf. Ex. 10.)

★1413. THE GAMMA FUNCTION

It was shown in Example 9, § 514, that the integral defining the **gamma function**:

(1) $$\Gamma(\alpha) = \int_0^{+\infty} x^{\alpha-1}e^{-x}\,dx$$

converges for $\alpha > 0$. In this section we shall investigate a few of the properties of the gamma function.

I. *Continuity and derivatives.* For $\alpha > 0$ the gamma function is continuous and, in fact, differentiable to any prescribed order. These derivatives are given by formal differentiation under the integral sign, the first two being:

(2) $$\Gamma'(\alpha) = \int_0^{+\infty} x^{\alpha-1}e^{-x}\ln x\,dx,$$

§ 1413] THE GAMMA FUNCTION

(3) $$\Gamma''(\alpha) = \int_0^{+\infty} x^{\alpha-1} e^{-x} (\ln x)^2 \, dx.$$

To prove these results, we establish the uniform convergence of the integrals

(4) $$\int_0^1 x^{\alpha-1} e^{-x} (\ln x)^n \, dx \quad \text{and} \quad \int_1^{+\infty} x^{\alpha-1} e^{-x} (\ln x)^n \, dx$$

for $0 < \delta \leq \alpha \leq \gamma$, for any fixed nonnegative integer n. For the first integral of (4) this is true since the integrand is dominated by

$$x^{\delta-1} e^{-x} |\ln x|^n = o(x^{\frac{1}{2}\delta-1}) \text{ as } x \to 0+,$$

and for the second integral of (4) this is true since the integrand is dominated by

$$x^{\gamma-1} e^{-x} (\ln x)^n = o(x^{-2}) \text{ as } x \to +\infty.$$

II. *The functional equation.* For $\alpha > 0$:

(5) $$\Gamma(\alpha + 1) = \alpha \Gamma(\alpha).$$

We show this by integrating by parts:

$$\int_\epsilon^b x^\alpha e^{-x} \, dx = \int_\epsilon^b -x^\alpha \, d(e^{-x}) = -\left[x^\alpha e^{-x}\right]_\epsilon^b + \int_\epsilon^b \alpha x^{\alpha-1} e^{-x} \, dx.$$

The formula results from letting $\epsilon \to 0+$ and $b \to +\infty$.

III. *Relation to the factorial function.* For every nonnegative integer n:

(6) $$\Gamma(n+1) = n!.$$

This is true for $n = 0$ (recall that $0!$ is *defined* to be equal to 1), since $\Gamma(1) = \int_0^{+\infty} e^{-x} \, dx = 1$. By the functional equation (5), $\Gamma(2) = \Gamma(1) = 1$, $\Gamma(3) = 2\Gamma(2) = 2!$, $\Gamma(4) = 3\Gamma(3) = 3!$, etc. The general statement (6) is obtained by mathematical induction.

IV. *Monotonic behavior.* Since the integrands in (1) and (3) are positive, $\Gamma(\alpha) > 0$ and $\Gamma''(\alpha) > 0$ for all $\alpha > 0$, and the graph of $\Gamma(\alpha)$ therefore lies above the α-axis and is concave upward for all $\alpha > 0$. Furthermore, $\Gamma(\alpha)$ has exactly one relative minimum value which, since $\Gamma(1) = \Gamma(2) = 1$, is attained for some $\alpha = \alpha_0$, $1 < \alpha_0 < 2$.† Furthermore, $\Gamma(\alpha)$ must be strictly decreasing for $0 < \alpha < \alpha_0$ and strictly increasing for $\alpha > \alpha_0$. (See Fig. 1401, below, for the graph of the gamma function.)

V. *Two limits.*

(7) $$\Gamma(0+) = +\infty, \quad \Gamma(+\infty) = +\infty.$$

The first of these follows from the continuity of the gamma function at $\alpha = 1$, the fact that $\Gamma(1) = 1$, and the functional relation (5), α being permitted to $\to 0+$. The second is a consequence of (6) and the monotonic behavior of $\Gamma(\alpha)$.

† The value of α_0 has been computed to be $1.461632 \cdots$, and $\Gamma(\alpha_0) = 0.885603 \cdots$.

520 IMPROPER INTEGRALS [§ 1413

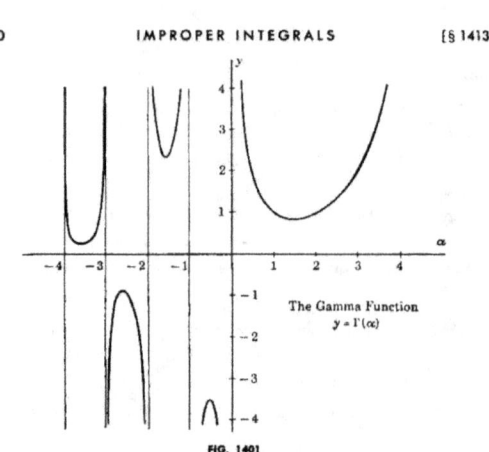

FIG. 1401

VI. *Evaluation of* $\Gamma(\frac{1}{2})$. By means of the substitution $y = \sqrt{x}$, $x = y^2$, we find:

$$\Gamma(\tfrac{1}{2}) = \int_0^{+\infty} x^{-\frac{1}{2}} e^{-x} dx = 2 \int_0^{+\infty} e^{-y^2} dy.$$

From Example 1, § 1329, we have

(8) $\Gamma(\tfrac{1}{2}) = \sqrt{\pi}.$

From (5) we deduce, furthermore:

(9) $\Gamma\left(\dfrac{3}{2}\right) = \dfrac{1}{2}\sqrt{\pi}, \quad \Gamma\left(\dfrac{5}{2}\right) = \dfrac{1\cdot 3}{2^2}\sqrt{\pi}, \quad \Gamma\left(\dfrac{7}{2}\right) = \dfrac{1\cdot 3\cdot 5}{2^3}\sqrt{\pi}, \cdots.$

VII. $\Gamma(\alpha)$ *for* $\alpha < 0$. Although the gamma function is defined by (1) only for $\alpha > 0$, the functional equation (5) permits extending the domain of definition to all nonintegral negative numbers, by means of the form

(10) $\Gamma(\alpha) = \dfrac{\Gamma(\alpha + 1)}{\alpha}.$

For example, $\Gamma(-\tfrac{1}{2}) = \dfrac{\Gamma(\tfrac{1}{2})}{-\tfrac{1}{2}} = -2\sqrt{\pi}.$ The general character of the extended gamma function is indicated by its graph, Figure 1401. This graph has vertical asymptotes for nonpositive integral values of α. Between successive asymptotes the graph of $\Gamma(\alpha)$ alternates between the upper and lower half planes, and the successive relative maxima and minima of $\Gamma(\alpha)$ approach 0 as $\alpha \to -\infty$ (Ex. 15, § 1415).

VIII. *Other expressions for* $\Gamma(\alpha)$. By means of the indicated changes of variable in formula (1), the following two forms for $\Gamma(\alpha)$ result:

(11) $$\Gamma(\alpha) = 2 \int_0^{+\infty} x^{2\alpha-1} e^{-x^2} dx, \quad \alpha > 0,$$

(set $x = y^2$),

(12) $$\Gamma(\alpha) = p^\alpha \int_0^{+\infty} x^{\alpha-1} e^{-px} dx, \quad \alpha > 0, \quad p > 0,$$

(set $x = py$).

Two other forms of the gamma function are discussed in § 1420.

IX. *Tables of the gamma function*. By virtue of the functional equation (5), any value of $\Gamma(\alpha)$ can be computed if its values are known for $1 \leq \alpha \leq 2$. Some tabulations of the gamma function are given in B. O. Peirce, *A Short Table of Integrals* (New York, Ginn and Company, 1956) and Eugen Jahnke and Fritz Emde, *Tables of Functions* (New York, Dover Publications, 1943).

★1414. THE BETA FUNCTION

It was shown in Example 8, § 514, that the integral defining the **beta function**:

(1) $$B(p, q) \equiv \int_0^1 x^{p-1} (1-x)^{q-1} dx$$

converges if and only if both p and q are positive. The substitution $y = 1 - x$ shows that the beta function is symmetric in p and q:

(2) $$B(p, q) = B(q, p), \quad p > 0, \quad q > 0.$$

The two substitutions $x = \sin^2 \theta$ and $x = y/(1 + y)$ give the following two alternative expressions ($p > 0, q > 0$):

(3) $$B(p, q) = 2 \int_0^{\frac{\pi}{2}} \sin^{2p-1} \theta \cos^{2q-1} \theta \, d\theta,$$

(4) $$B(p, q) = \int_0^{+\infty} \frac{x^{p-1}}{(1+x)^{p+q}} dx.$$

An important relation expressing the beta function in terms of the gamma function is:

(5) $$B(p, q) = \frac{\Gamma(p)\Gamma(q)}{\Gamma(p+q)}, \quad p > 0, \quad q > 0.$$

We shall now prove this.

Estimates will be obtained by integration of the function

(6) $$f(x, y) \equiv 4x^{2p-1} y^{2q-1} e^{-(x^2+y^2)}$$

over the two sets in the first quadrant (ρ and θ representing polar coordinates; cf. Fig. 1402):

$$R_n: \begin{cases} \frac{1}{n} \leq x \leq n, \\ \frac{1}{n} \leq y \leq n, \end{cases} \qquad S_n: \begin{cases} \frac{1}{n} \leq \rho \leq n, \\ \frac{1}{n} \leq \operatorname{Arctan} \theta \leq n. \end{cases}$$

It is a matter of elementary geometry to show the inclusions (Ex. 16, § 1415):

(7) $$R_n \subset S_{n^2}, \quad S_n \subset R_{n^2},$$

and hence, if

$$I_n \equiv \iint_{R_n} f(x, y) \, dA, \quad J_n \equiv \iint_{S_n} f(x, y) \, dA,$$

(8) $$I_n \leq J_{n^2}, \quad J_n \leq I_{n^2}.$$

Since $f(x, y) \geq 0$, $I_n \uparrow$ and $J_n \uparrow$, as n increases, and by the inequalities (8),

(9) $$I \equiv \lim_{n \to +\infty} I_n = \lim_{n \to +\infty} J_n.$$

We now evaluate I_n and J_n:

$$I_n = \left\{ 2 \int_{\frac{1}{n}}^n x^{2p-1} e^{-x^2} \, dx \right\} \left\{ 2 \int_{\frac{1}{n}}^n y^{2q-1} e^{-y^2} \, dy \right\},$$

$$J_n = 4 \int_{\operatorname{Arctan} \frac{1}{n}}^{\operatorname{Arctan} n} \int_{\frac{1}{n}}^n (\rho \cos \theta)^{2p-1} (\rho \sin \theta)^{2q-1} e^{-\rho^2} \rho \, d\rho \, d\theta$$

$$= \left\{ 2 \int_{\frac{1}{n}}^n \rho^{2(p+q)-1} e^{-\rho^2} \, d\rho \right\} \left\{ 2 \int_{\operatorname{Arctan} \frac{1}{n}}^{\operatorname{Arctan} n} \sin^{2q-1} \theta \cos^{2p-1} \theta \, d\theta \right\}.$$

Therefore, by (9) above, formula (11), § 1413, for the gamma function, and formulas (2) and (3), above, for the beta function, we have in the limit, as $n \to +\infty$:
$$I = \Gamma(p)\Gamma(q) = \Gamma(p+q)\,\mathrm{B}(p,q),$$
and formula (5), as desired.

★1415. EXERCISES

In Exercises 1-8, establish the given formula for the specified values of the variables.

★1. $\Gamma(\alpha) = \int_0^1 \left(\ln \frac{1}{x}\right)^{\alpha-1} dx,\ \alpha > 0.$

★2. $\Gamma(\alpha) = p^\alpha \int_0^1 x^{p-1} \left(\ln \frac{1}{x}\right)^{\alpha-1} dx,\ \alpha > 0,\ p > 0.$

★3. $\Gamma\left(n + \frac{1}{2}\right) = \frac{(2n)!\sqrt{\pi}}{4^n n!},\ n = 0, 1, 2, \cdots.$

★4. $\mathrm{B}(p+1, q) = \frac{p}{p+q}\,\mathrm{B}(p, q),\ p > 0,\ q > 0.$

★5. $\int_0^{+\infty} \frac{x^{\alpha-1}}{1+x}\,dx = \Gamma(\alpha)\Gamma(1-\alpha),\ 0 < \alpha < 1.$

★6. $\int_0^1 x^{p-1}(1-x^r)^{q-1}\,dx = \frac{1}{r}\,\mathrm{B}\!\left(\frac{p}{r}, q\right),\ p > 0,\ q > 0,\ r > 0.$

★7. $\int_0^1 \frac{x^n\,dx}{\sqrt{1-x^2}} = \frac{\sqrt{\pi}}{2} \frac{\Gamma\!\left(\frac{n+1}{2}\right)}{\Gamma\!\left(\frac{n+2}{2}\right)},\ n > 0.$ (Cf. Ex. 6.)

★8. $\int_0^1 \frac{dx}{\sqrt{1-x^n}} = \frac{\sqrt{\pi}}{n} \frac{\Gamma\!\left(\frac{1}{n}\right)}{\Gamma\!\left(\frac{1}{n}+\frac{1}{2}\right)},\ n > 0.$ (Cf. Ex. 6.)

In Exercises 9-12, evaluate by means of formulas already established.

★9. $\int_0^{+\infty} x^4 e^{-x^2}\,dx.$ ★10. $\int_0^{+\infty} \frac{e^{-5x}}{\sqrt{x}}\,dx.$

★11. $\int_0^1 \left[\frac{1}{x}\ln\frac{1}{x}\right]^{\frac{1}{2}} dx.$ ★12. $\int_0^1 \frac{x^3\,dx}{\sqrt{1-x^4}}.$

★13. Verify, by means of the substitution $u = 2\theta\ (\alpha > 0)$:
$$\mathrm{B}(\alpha, \alpha) = 2\int_0^{\frac{\pi}{2}} \sin^{2\alpha-1}\theta \cos^{2\alpha-1}\theta\,d\theta = 2^{-2\alpha+2}\int_0^{\frac{\pi}{2}} \sin^{2\alpha-1} 2\theta\,d\theta$$
$$= 2^{-2\alpha+2}\int_0^{\frac{\pi}{2}} \sin^{2\alpha-1} u\,du = 2^{-2\alpha+1}\mathrm{B}(\alpha, \tfrac{1}{2}),$$
and hence derive the duplication formula of A. M. Legendre (1752-1833):
$$\Gamma(2\alpha) = \frac{1}{\sqrt{\pi}}\,2^{2\alpha-1}\Gamma(\alpha)\Gamma\!\left(\alpha + \frac{1}{2}\right).$$

★14. Show that the substitution $x = \sqrt{\tan\theta}$ leads to the following evaluation of an integral occurring in the Example, § 1410:

$$\int_0^{+\infty} \frac{dx}{1+x^4} = \frac{1}{2}\int_0^{\frac{\pi}{2}} \cos^{-\frac{1}{2}}\theta \sin^{-\frac{1}{2}}\theta\, d\theta = \frac{1}{4} B\left(\frac{3}{4}, \frac{1}{4}\right)$$

$$= \frac{1}{4}\Gamma\left(\frac{3}{4}\right)\Gamma\left(\frac{1}{4}\right) = \frac{\sqrt{\pi}}{4}\sqrt{2}\Gamma\left(\frac{1}{2}\right) = \frac{\pi}{2\sqrt{2}}.$$

(Cf. Ex. 13.)

★15. Prove that the statements of the last sentence of VII, § 1413, regarding the graph of $\Gamma(\alpha)$ for $\alpha < 0$.

★16. Establish the inclusions (7), § 1414.

★1416. INFINITE PRODUCTS

An expression of the form

(1) $$\prod_{n=1}^{+\infty} p_n = p_1 p_2 p_3 \cdots,$$

or

(2) $$\prod_{n=1}^{+\infty} (1+a_n) = (1+a_1)(1+a_2)(1+a_3)\cdots,$$

where $1 + a_n = p_n$, $n = 1, 2, \cdots$, is called an **infinite product**. The **partial products** are

(3) $\quad P_n \equiv p_1 p_2 \cdots p_n = (1+a_1)(1+a_2)\cdots(1+a_n), n = 1, 2, \cdots.$

For simplicity, we shall assume in this chapter that *all factors p_n are nonzero*—in other words, that a_n *is never equal to* -1. Under this assumption, the infinite product (1) or (2) is said to **converge** if and only if the limit of the partial products exists and is finite and nonzero:

(4) $$\prod_{n=1}^{+\infty} p_n \equiv \lim_{n \to +\infty} P_n = P \neq 0.$$

In all other cases the infinite product **diverges**. For example, if

(5) $$\lim_{n \to +\infty} P_n = 0,$$

the infinite product is said to **diverge to zero**. In case of convergence the number P in (4) is called the **value** of the infinite product.

Many of the theorems on infinite series have their analogues in the theory of infinite products. We shall content ourselves here with establishing three basic theorems. For a more thorough treatment of infinite products, see E. T. Whittaker and G. N. Watson, *Modern Analysis* (Cambridge University Press, 1935), or E. W. Hobson, *The Theory of Functions of a Real Variable* (Washington, D. C., The Harren Press, 1950).

Theorem I. *If the infinite product* $\prod_{n=1}^{+\infty} p_n$ *converges, the* **general factor** *p_n tends toward 1 as a limit:*

$$\lim_{n \to +\infty} p_n = 1.$$

Proof. $\lim_{n \to +\infty} p_n = \dfrac{\lim P_n}{\lim P_{n-1}} = \dfrac{P}{P} = 1.$

Theorem II. *The infinite product* $\prod_{n=1}^{+\infty} p_n$ *converges if and only if the infinite series*

(6) $$\sum_{n=1}^{+\infty} \ln |p_n|.$$

converges.

Proof. By Theorem I, the general factor p_n is positive for sufficiently large n, and there is no loss of generality in assuming that $p_n > 0$ for *all* n. In this case the result follows from the relations

$$S_n \equiv \sum_{k=1}^{n} \ln p_k = \ln P_n, \quad P_n = e^{S_n},$$

and the continuity of $\ln x$ and e^x.

Theorem III. *Consider the infinite product*

(7) $$(1 + a_1)(1 + a_2)(1 + a_3) \cdots$$

and the two infinite series

(8) $$a_1 + a_2 + a_3 + \cdots,$$

(9) $$a_1^2 + a_2^2 + a_3^2 + \cdots.$$

(i) *The convergence of any two of (7)-(9) implies that of the third.*
(ii) *The absolute convergence of (8) implies the convergence of both (7) and (9).*
(iii) *In case (8) converges conditionally, (7) converges or diverges to zero according as (9) converges or diverges.*

Proof. Without loss of generality we shall assume that $|a_n| < \tfrac{1}{2}$ for all n, since the convergence of any one of (7)-(9) implies this inequality for n sufficiently large. Taylor's Formula with a Remainder in the Lagrange form (§ 804), applied to the function $f(x) = \ln(1 + x)$ for $n = 2$ is

$$\ln(1 + x) = x - \frac{x^2}{2(1 + \xi)^2},$$

where $0 \leq |\xi| \leq |x|$. Therefore, with the assumption $|a_n| < \tfrac{1}{2}$,

(10) $$c_n \equiv \ln(1 + a_n) = a_n - b_n,$$

where

(11) $$\tfrac{2}{9} a_n^2 \leq b_n \leq 2 a_n^2.$$

Since $\sum b_n$ is a positive series which, by (11), converges if and only if (9) converges, the statement (i) follows from the relation (10), thanks to

Theorem II. Conclusion (ii) follows from (i) and the inequality $a_n^2 \leq \frac{1}{2}|a_n|$. Conclusion (iii) follows in part from (i) and in part from the fact that if $\sum a_n$ converges and $\sum a_n^2 = +\infty$, $\sum \ln(1+a_n) = -\infty$ (from (10) and (11)).

Example 1. The infinite product

$$\left(1 - \frac{1}{4 \cdot 1^2}\right)\left(1 - \frac{1}{4 \cdot 2^2}\right)\left(1 - \frac{1}{4 \cdot 3^2}\right) \cdots \left(1 - \frac{1}{4n^2}\right) \cdots$$

converges, since the series $\sum 1/4n^2$ converges absolutely. Its value is $2/\pi$ (as can be seen by factoring each quantity in parentheses and comparing the result with (1), § 1417).

Example 2. The infinite product

$$\left(1 + \frac{1}{1}\right)\left(1 - \frac{1}{3}\right)\left(1 + \frac{1}{3}\right)\left(1 - \frac{1}{5}\right)\left(1 + \frac{1}{5}\right) \cdots$$

converges, since the series $1 - \frac{1}{3} + \frac{1}{3} - \frac{1}{5} + \frac{1}{5} - \cdots$ and $1 + \frac{1}{3^2} + \frac{1}{3^2} + \frac{1}{5^2} + \frac{1}{5^2} + \cdots$ both converge. Its value is $\frac{\pi}{2}$, as shown in § 1417.

Example 3. The infinite product

$$\left(1 + \frac{1}{1}\right)\left(1 - \frac{1}{\sqrt{2}}\right)\left(1 + \frac{1}{\sqrt{3}}\right)\left(1 - \frac{1}{\sqrt{4}}\right) \cdots$$

diverges to zero, by (iii), Theorem III.

★1417. WALLIS'S INFINITE PRODUCT FOR π

We wish to establish the following limit formula, due to the English mathematician J. Wallis (1616-1703) (cf. Ex. 19, § 1515):

(1) $$\frac{\pi}{2} = \frac{2}{1} \cdot \frac{2}{3} \cdot \frac{4}{3} \cdot \frac{4}{5} \cdot \frac{6}{5} \cdot \frac{6}{7} \cdots$$
$$= \lim_{n \to +\infty} \frac{2^{4n}(n!)^4}{[(2n)!]^2(2n+1)}.$$

To this end we recall the two Wallis formulas, obtained by integration by parts and mathematical induction (cf. Ex. 36, § 515):

(2) $$\begin{cases} I_{2n} \equiv \int_0^{\frac{\pi}{2}} \sin^{2n} x \, dx = \frac{\pi}{2} \cdot \frac{1 \cdot 3 \cdots (2n-1)}{2 \cdot 4 \cdots 2n}, \\ I_{2n+1} \equiv \int_0^{\frac{\pi}{2}} \sin^{2n+1} x \, dx = \frac{2 \cdot 4 \cdots (2n)}{3 \cdot 5 \cdots (2n+1)}, \end{cases}$$

for $n = 1, 2, \cdots$. Since, on the interval $[0, \frac{1}{2}\pi]$, $0 \leq \sin x \leq 1$, $I_{2n} \geq I_{2n+1} \geq I_{2n+2} > 0$, and therefore

(3) $$1 \geq \frac{I_{2n+1}}{I_{2n}} \geq \frac{I_{2n+2}}{I_{2n}} = \frac{2n+1}{2n+2} \to 1.$$

§ 1418] EULER'S CONSTANT

It follows that $I_{2n+1}/I_{2n} \to 1$, and therefore

(4) $\quad \dfrac{\pi}{2} \cdot \dfrac{I_{2n+1}}{I_{2n}} = \dfrac{2 \cdot 4 \cdots 2n}{1 \cdot 3 \cdots (2n-1)} \cdot \dfrac{2 \cdot 4 \cdots 2n}{3 \cdot 5 \cdots (2n+1)} \to \dfrac{\pi}{2}.$

But this states that the partial products of an even number of factors of the infinite product (1) have $\tfrac{1}{2}\pi$ as a limit. Since the general factor of (1) tends toward 1, the *general* partial product also has $\tfrac{1}{2}\pi$ as a limit. The second form of (1) is obtained by multiplying numerator and denominator of (4) by $(2 \cdot 4 \cdots 2n)^2$.

★1418. EULER'S CONSTANT

A constant important in number theory and parts of the theory of analytic functions of a complex variable, known as **Euler's constant**, is defined (cf. Ex. 19, § 721):

(1) $\qquad\qquad\qquad C \equiv \lim\limits_{n \to +\infty} C_n,$

where

(2) $\qquad\qquad C_n \equiv \left(1 + \dfrac{1}{2} + \cdots + \dfrac{1}{n}\right) - \ln n.$

We shall show that C exists and is positive by establishing:

(3) $\qquad\qquad\qquad C_n \downarrow,$

(4) $\qquad\qquad\qquad C_n > \tfrac{1}{2}.$

Proof of (3): The problem is to show that

$$C_n - C_{n-1} = \dfrac{1}{n} - \ln \dfrac{n}{n-1} < 0, \quad n > 1.$$

But this is equivalent to proving

$$\ln\left(1 - \dfrac{1}{n}\right) < -\dfrac{1}{n}, \quad n > 1.$$

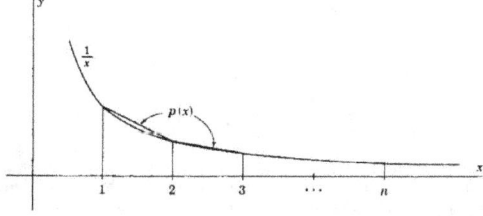

FIG. 1403

This was established by the Law of the Mean in Example 2, § 405.

Proof of (4): On the interval $[1, n]$ define the function $p(x)$ by linear interpolation between successive pairs of points $(1, 1)$, $(2, \frac{1}{2})$, $(3, \frac{1}{3})$, $\cdots$, $\left(n, \frac{1}{n}\right)$ (cf. Fig. 1403). Then since the graph of $y = \frac{1}{x}$ is everywhere concave upward, the inequality $p(x) - \frac{1}{x} \geq 0$ holds throughout $[1, n]$ (why?). The resulting inequality

$$\int_1^n \left[p(x) - \frac{1}{x} \right] dx > 0$$

becomes

$$\left[\frac{1}{2} + \frac{1}{2} + \frac{1}{3} + \cdots + \frac{1}{n-1} + \frac{1}{2n} \right] - \ln n = C_n - \frac{1}{2} - \frac{1}{2n} > 0,$$

and (4) follows.

To ten decimal places, $C = 0.5772156649$.

★1419. STIRLING'S FORMULA

An important "asymptotic formula" for $n!$, due to the eighteenth century English mathematician James Stirling, gives the expression

(1) $$S_n \equiv \sqrt{2\pi n} \left(\frac{n}{e} \right)^n$$

as an "approximation" to $n!$, for large values of n, in the sense that

(2) $$\lim_{n \to +\infty} \frac{n!}{S_n} = 1.$$

It is our purpose in this section to prove the inequality for $n = 3, 4, \cdots$:

(3) $$S_n \left[1 + \frac{1}{12(n+1)} \right] < n! < S_n \left[1 + \frac{1}{12(n-2)} \right].$$

This inequality implies (2) and, in addition, gives a measure of the accuracy of S_n as an approximation for $n!$. In particular, we can infer from (3) that

(4) $$\frac{n!}{S_n} = 1 + \frac{1}{12n} + O\left(\frac{1}{n^2}\right),$$

and that for $n > 13$

(5) $$n! - S_n > \frac{(n-1)!}{13}.$$

In other words, Stirling's formula can be interpreted as a good approximation of $n!$ only in the *relative* sense of (4), whereas the actual *difference* between $n!$ and S_n grows at a rapid rate—at least as rapidly as $(n-1)!/13$.

Our derivation of (3) is based on the following trapezoidal formula for the integral of a function $f(x)$ possessing a second derivative over an interval $[a, b]$ (cf. Ex. 22, § 503):

(6) $$\int_a^b f(x)\, dx = \tfrac{1}{2}[f(a) + f(b)](b - a) - \tfrac{1}{12} f''(\xi)(b - a)^3,$$

where ξ is a suitable number between a and b.

We now consider the function $f(x) = \ln x$ on the interval $[1, n]$, integrate, and use (6) on the successive intervals $[1, 2], \cdots, [n - 1, n]$:

$$\int_1^n \ln x\, dx = \int_1^2 \ln x\, dx + \int_2^3 \ln x\, dx + \cdots + \int_{n-1}^n \ln x\, dx$$
$$= \tfrac{1}{2}(\ln 1 + \ln 2) + \tfrac{1}{2}(\ln 2 + \ln 3) + \cdots \tfrac{1}{2}(\ln (n - 1) + \ln n)$$
$$+ \frac{1}{12}\left[\frac{1}{\xi_2^2} + \frac{1}{\xi_3^2} + \cdots + \frac{1}{\xi_n^2}\right],$$

where $1 < \xi_2 < 2, \cdots, n - 1 < \xi_n < n$. This relation can be written

(7) $$\left(n + \tfrac{1}{2}\right) \ln n - n + 1 - \ln n! = \frac{1}{12}\left[\frac{1}{\xi_2^2} + \cdots + \frac{1}{\xi_n^2}\right].$$

Since $\xi_n > n - 1$, the series $\sum \xi_n^{-2}$ converges. Define

$$s \equiv \sum_{k=2}^{+\infty} \frac{1}{\xi_k^2}, \quad r_n \equiv \sum_{k=n+1}^{+\infty} \frac{1}{\xi_k^2} = s - \sum_{k=2}^n \frac{1}{\xi_k^2}.$$

Equation (7) now becomes

(8) $$\ln n! = (n + \tfrac{1}{2}) \ln n - n + 1 - s + r_n,$$

or, with $\alpha \equiv e^{1-s}$:

(9) $$n! = \alpha e^{r_n} \sqrt{n} \left(\frac{n}{e}\right)^n.$$

Substitution of (9) in Wallis's product formula (1), § 1417, gives an evaluation of the constant α, since $r_n \to 0$:

$$\frac{\pi}{2} = \lim_{n \to +\infty} \frac{2^{4n} \alpha^4 e^{4r_n} n^2 (n/e)^{4n}}{\alpha^2 e^{2r_{2n}} \cdot 2n \cdot (2n/e)^{4n} (2n + 1)}$$
$$= \lim_{n \to +\infty} \alpha^2 \frac{e^{4r_n}}{e^{2r_{2n}}} \cdot \lim_{n \to +\infty} \frac{n}{2(2n + 1)} = \frac{\alpha^2}{4},$$

and $\alpha = \sqrt{2\pi}$. Therefore (9) can be written

(10) $$n! = S_n e^{r_n}.$$

Using the inequalities $n - 1 < \xi_n < n$, and the integral test estimate for infinite series (Theorem II, § 720), we have

(11) $$\frac{1}{12(n + 1)} < \frac{1}{12}\left[\frac{1}{(n + 1)^2} + \frac{1}{(n + 2)^2} + \cdots\right] < r_n$$
$$< \frac{1}{12}\left[\frac{1}{n^2} + \frac{1}{(n + 1)^2} + \cdots\right] < \frac{1}{12(n - 1)}.$$

By Taylor's Formula with a Remainder in the Lagrange form (§ 804), if $0 < x \leq \frac{1}{2}$:

(12) $$1 + x < e^x = 1 + x + \frac{e^\xi}{2} x^2 < 1 + x + x^2,$$

where $0 < \xi < x$. Therefore, with x equal to $1/12(n+1)$ and $1/12(n-1)$ in turn, we have from (11), if $n \geq 3$:

$$1 + \frac{1}{12(n+1)} < e^{c_n} < 1 + \frac{1}{12(n-1)} + \frac{1}{144(n-1)^2}$$

$$< 1 + \frac{1}{12(n-1)} + \frac{1}{12(n-1)(n-2)} = 1 + \frac{1}{12(n-2)},$$

and the derivation of (3) is complete.

★1420. WEIERSTRASS'S INFINITE PRODUCT FOR $1/\Gamma(\alpha)$

In this section we shall derive an important formula used by Weierstrass in his research on the gamma function:

(1) $$\frac{1}{\Gamma(\alpha)} = e^{C\alpha} \alpha \sum_{n=1}^{+\infty} \left(1 + \frac{\alpha}{n}\right) e^{-\frac{\alpha}{n}}, \quad \alpha > 0,$$

where C is Euler's constant (§ 1418). This product formula is important, in part, because it represents the (reciprocal of the) gamma function not only for $\alpha > 0$, but for all values of α (including complex numbers) except $\alpha = 0, -1, -2, \cdots$. We shall prove (1) only for real $\alpha > 0$. For a more complete discussion see E. T. Whittaker and G. N. Watson, *Modern Analysis* (Cambridge University Press, 1935), or L. V. Ahlfors, *Complex Analysis* (New York, McGraw-Hill Book Company, 1953).

We shall obtain (1) by first proving

(2) $$\Gamma(\alpha) = \lim_{n \to +\infty} \int_0^n \left(1 - \frac{x}{n}\right)^n x^{\alpha-1} dx, \quad \alpha > 0,$$

and then the formula due to L. Euler (1707-1783):

(3) $$\Gamma(\alpha) = \lim_{n \to +\infty} \frac{n! \, n^\alpha}{\alpha(\alpha+1) \cdots (\alpha+n)}, \quad \alpha > 0.$$

In order to prove (2), since $\Gamma(\alpha) = \lim_{n \to +\infty} \int_0^n e^{-x} x^{\alpha-1} dx$, we must show:

$$\lim_{n \to +\infty} \int_0^n \left[1 - e^x \left(1 - \frac{x}{n}\right)^n\right] e^{-x} x^{\alpha-1} dx = 0.$$

This follows from the inequalities

$$0 \leq 1 - e^x \left(1 - \frac{x}{n}\right)^n \leq \frac{x^2}{n}, \quad 0 \leq x \leq n,$$

which are consequences, in turn, of the three inequalities, established by the Law of the Mean (cf. Exs. 29, 39, § 408):

$$1 + h \leq e^h, \qquad h \geq 0$$
$$1 - h \leq e^{-h}, \qquad h \geq 0$$
$$1 - nh \leq (1-h)^n, \quad 0 \leq h \leq 1, \ n \geq 1,$$

as follows:

$$1 - e^x \left(1 - \frac{x}{n}\right)^n \geq 1 - e^x \left(e^{-\frac{x}{n}}\right)^n = 1 - e^x e^{-x} = 0,$$

$$1 - e^x \left(1 - \frac{x}{n}\right)^n = 1 - \left(e^{\frac{x}{n}}\right)^n \left(1 - \frac{x}{n}\right)^n \leq 1 - \left(1 + \frac{x}{n}\right)^n \left(1 - \frac{x}{n}\right)^n$$

$$= 1 - \left(1 - \frac{x^2}{n^2}\right)^n \leq 1 - \left(1 - n\frac{x^2}{n^2}\right) = \frac{x^2}{n}.$$

The substitution of $x = nu$ in the integral appearing in (2) gives

$$\int_0^n \left(1 - \frac{x}{n}\right)^n x^{\alpha-1} dx = n^\alpha \int_0^1 u^{\alpha-1} (1-u)^n du,$$

which reduces, by means of repeated integration by parts that reduce the exponent on $(1-u)$, to the quantity whose limit is formed in (3).

Finally, the reciprocal of (3) is equal to

$$\lim_{n \to +\infty} \alpha \left(1 + \frac{\alpha}{1}\right)\left(1 + \frac{\alpha}{2}\right) \cdots \left(1 + \frac{\alpha}{n}\right) e^{-\alpha \ln n}$$

$$= \lim_{n \to +\infty} \alpha \left(1 + \frac{\alpha}{1}\right) e^{-\alpha} \left(1 + \frac{\alpha}{2}\right) e^{-\frac{\alpha}{2}} \cdots \left(1 + \frac{\alpha}{n}\right) e^{-\frac{\alpha}{n}} e^{C_n \alpha},$$

where C_n is the quantity (2), § 1418, whose limit is equal, by definition, to Euler's constant. This fact leads immediately to the desired formula (1).

★1421. EXERCISES

★1. The infinite product $\prod_{n=1}^{+\infty} (1 + a_n)$ is said to be **absolutely convergent** if and only if the series $\sum_{n=1}^{+\infty} \ln |1 + a_n|$ is absolutely convergent. Prove that an absolutely convergent infinite product is convergent. Prove that $\prod_{n=1}^{+\infty} (1 + a_n)$ is absolutely convergent if and only if $\sum_{n=1}^{+\infty} a_n$ is absolutely convergent. *Hint:* Cf. (10) and (11), § 1416, assume $|a_n| < \frac{1}{2}$, and show that $|c_n| \leq |a_n| + |b_n| \leq 2|a_n|$, and $|a_n| \leq |c_n| + \frac{1}{2}|a_n|$.

★2. Prove that the factors of an absolutely convergent infinite product can be rearranged arbitrarily without affecting absolute convergence or the value of the product. (Cf. Ex. 1.)

★3. Use Stirling's formula to show that $\lim_{n \to +\infty} \sqrt[n]{\frac{n^n}{n!}} = e$. (Cf. Ex. 48, § 515, Ex. 36, § 711.)

★4. The infinite product $\prod_{n=1}^{+\infty} (1 + u_n(x))$ is said to **converge uniformly** for x in a set A if and only if it converges there and for $\epsilon > 0$ there exists $N(\epsilon)$ such that $n > N$ and $x \in A$ imply
$$\left| \prod_{k=1}^{n} (1 + u_k(x)) - \prod_{k=1}^{+\infty} (1 + u_k(x)) \right| < \epsilon.$$
Prove the **Cauchy Criterion** for uniform convergence: The infinite product $\prod_{n=1}^{+\infty} (1 + u_n(x))$ converges uniformly for $x \in A$ if and only if corresponding to $\epsilon > 0$ there exists $N(\epsilon)$ such that $n > m > N$ and $x \in A$ imply
$$\left| \prod_{k=1}^{n} (1 + u_k(x)) - \prod_{k=1}^{m} (1 + u_k(x)) \right| < \epsilon.$$

★5. Prove the **Weierstrass M-test** for uniform convergence of an infinite product: If $\{M_n\}$ is a sequence of positive constants such that $\sum_{1}^{+\infty} M_n$ converges and $|u_n(x)| < M_n$ for $x \in A$, $n = 1, 2, \cdots$, then the infinite product $\prod_{n=1}^{+\infty} (1 + u_n(x))$ converges uniformly on A. (Cf. Ex. 4.) *Hint:* Show that for large m and n:
$$\prod_{k=1}^{m} \left| 1 + u_k(x) \right| \cdot \left| \prod_{k=m+1}^{n} (1 + u_k(x)) - 1 \right| \leq \prod_{k=1}^{m} (1 + M_k) \left\{ \prod_{k=m+1}^{n} (1 + M_k) - 1 \right\}.$$

★6. Prove that Weierstrass's infinite product for $1/\Gamma(\alpha)$ ((1), § 1420) converges for every real number $\alpha \neq 0, -1, -2, \cdots$, and uniformly on any bounded set that excludes these nonpositive integers. (Cf. Exs. 4, 5.)

★★1422. IMPROPER RIEMANN-STIELTJES INTEGRALS

Let the functions $f(x)$ and $g(x)$ be defined for all real x, and assume that f is Riemann-Stieltjes integrable with respect to g on every finite closed interval $[a, b]$. Then the *improper Riemann-Stieltjes integral* of f with respect to g on the interval $(-\infty, +\infty)$ is defined to be the following limit, if it exists and is finite:

$$(1) \qquad \int_{-\infty}^{+\infty} f(x)\, dg(x) = \lim_{\substack{m \to +\infty \\ n \to +\infty}} \int_{-m}^{n} f(x)\, dg(x),$$

the limit being in the sense of Definition II, § 1006. Other types of improper integrals are defined similarly.

We shall consider in this section only integrals (1) where $f(x)$ is everywhere continuous, $g(x)$ is monotonically increasing on $(-\infty, +\infty)$, and both $f(x)$ and $g(x)$ are bounded: $|f(x)| < K$, $|g(x)| < K$. In this case both $g(+\infty) \equiv \lim_{x \to +\infty} g(x)$ and $g(-\infty) \equiv \lim_{x \to -\infty} g(x)$ exist and are finite, and the integral (1) exists and is finite.

To prove this last statement we invoke the Cauchy Criterion for the

§ 1422] RIEMANN-STIELTJES INTEGRALS 533

existence of a finite limit, and establish two inequalities, one of which has the form

$$\int_{m_1}^{m_2} |f(x)|\, dg(x) < \epsilon, \quad n_2 > n_1 > N(\epsilon).$$

(Cf. Ex. 22, § 518, Ex. 1, § 1423.)

Our principal goal in this section is to establish the "Helly-Bray convergence theorem," important in statistical theory. We first prove it for a finite interval (Theorem I), and then for the interval $(-\infty, +\infty)$ (Theorem II). Both proofs involve the Moore-Osgood theorem (§ 1014). For further discussion of convergence theorems for Riemann-Stieltjes integrals, see L. M. Graves, *The Theory of Functions of Real Variables* (New York, McGraw-Hill Book Company, 1946).

Theorem I. *Assume that the functions $f(x)$, $g(x)$, and $g_n(x)$, $n = 1, 2, \cdots$ are defined on $[a, b]$ and that*

(i) *$f(x)$ is continuous on $[a, b]$,*

(ii) *each of the functions $g(x)$ and $g_n(x)$, $n = 1, 2, \cdots$, is monotonically increasing on $[a, b]$,*

(iii) *there exists a set A dense in $[a, b]$ (that is, $\bar{A} = [a, b]$) and including the two points a and b such that for each point $x \in A$,*
$$\lim_{n \to +\infty} g_n(x) = g(x).$$

Then

(iv) $\displaystyle\lim_{n \to +\infty} \int_a^b f(x)\, dg_n(x) = \int_a^b f(x)\, dg(x).$

Proof. For $m = 1, 2, \cdots$ let $\{\mathfrak{N}_m\}$ be a sequence of nets on $[a, b]$ made up of points of A and such that $\lim_{m \to +\infty} |\mathfrak{N}_m| = 0$, and for each net $\mathfrak{N}_m$ and each $i = 1, 2, \cdots, m$, let x_i' and x_i'' be points such that $a_{i-1} \leq x_i \leq a_i$ and

$$f(x_i') = \max_{a_{i-1} \leq x \leq a_i} f(x), \quad f(x_i'') = \min_{a_{i-1} \leq x \leq a_i} f(x).$$

For every pair of positive integers, m and n, define

$$F(m, n) \equiv \sum_{i=1}^m f(x_i')(\Delta g_n)_i, \quad G(m, n) \equiv \sum_{i=1}^m f(x_i'')(\Delta g_n)_i,$$

$$F(m) \equiv \sum_{i=1}^m f(x_i')(\Delta g)_i, \quad G(m) \equiv \sum_{i=1}^m f(x_i'')(\Delta g)_i.$$

Then for every n the following inequalities hold:

(2) $\begin{cases} F(m, n) \geq \int_a^b f(x)\, dg_n(x) \geq G(m, n), \\ F(m) \geq \int_a^b f(x)\, dg(x) \geq G(m). \end{cases}$

Since $g_n(a) \to g(a)$ and $g_n(b) \to g(b)$, there exists a number K such that

$g(b) - g(a) < K$ and $g_n(b) - g_n(a) < K$ for $n = 1, 2, \cdots$, and since f is uniformly continuous on $[a, b]$, corresponding to $\epsilon > 0$ there exists a number M such that $m > M$ implies

$$f(x_i') - f(x_i'') < \epsilon/K, \quad i = 1, 2, \cdots, m.$$

Combining these facts with the inequalities (2), we infer that $m > M$ implies

$$\left| F(m, n) - \int_a^b f(x) \, dg_n(x) \right| \leq F(m, n) - G(m, n)$$
$$< \sum_{i=1}^m \frac{\epsilon}{K} (\triangle g_n)_i = \frac{\epsilon}{K} [g_n(b) - g_n(a)] < \epsilon,$$

and similarly,

$$\left| F(m) - \int_a^b f(x) \, dg(x) \right| < \epsilon.$$

In other words,

$$\lim_{m \to +\infty} F(m, n) = \int_a^b f(x) \, dg_n(x),$$

uniformly in $n = 1, 2, \cdots$, and

$$\lim_{m \to +\infty} F(m) = \int_a^b f(x) \, dg(x).$$

On the other hand, for each $m = 1, 2, \cdots$:

$$\lim_{n \to +\infty} F(m, n) = F(m).$$

Therefore, by the Moore-Osgood theorem, § 1014,

$$\lim_{n \to +\infty} \lim_{m \to +\infty} F(m, n) = \lim_{m \to +\infty} \lim_{n \to +\infty} F(m, n),$$

and (iv) results.

Theorem II. *Assume that the functions $f(x), g(x)$, and $g_n(x), n = 1, 2, \cdots$ are defined on $(-\infty, +\infty)$, and that*

(i) *$f(x)$ is bounded and continuous on $(-\infty, +\infty)$,*

(ii) *each of the functions $g(x)$ and $g_n(x), n = 1, 2, \cdots$, is bounded and monotonically increasing on $(-\infty, +\infty)$,*

(iii) *there exists a dense set A including both points at infinity, $+\infty$ and $-\infty$, such that for each point $x \in A$, $\lim_{n \to +\infty} g_n(x) = g(x)$ (in particular, $\lim_{n \to +\infty} g_n(\pm\infty) = g(\pm\infty)$).*

Then

(iv) $\lim_{n \to +\infty} \int_{-\infty}^{+\infty} f(x) \, dg_n(x) = \int_{-\infty}^{+\infty} f(x) \, dg(x).$

Proof. Let $a_1, a_2, \cdots, a_m, \cdots$ be an increasing sequence of positive

§ 1423] EXERCISES

members of A whose limit is $+\infty$, and for each pair of positive integers m and n, define

$$F(m, n) \equiv \int_{-a_m}^{+a_m} f(x)\, dg_n(x),$$

$$F(m) \equiv \int_{-a_m}^{a_m} f(x)\, dg(x).$$

By definition, $\lim_{m \to +\infty} F(m, n) = \int_{-\infty}^{+\infty} f(x)\, dg_n(x)$ and $\lim_{m \to +\infty} F(m) = \int_{-\infty}^{+\infty} f(x)\, dg(x)$, and by Theorem I, $\lim_{n \to +\infty} F(m, n) = F(m)$. The proof will be complete, then, as soon as it is established that the limit $\lim_{m \to +\infty} F(m, n)$ is uniform in $n = 1, 2, \cdots$. The problem, then, is to show that corresponding to $\epsilon > 0$ there exists a number $M = M(\epsilon)$ such that each of the two integrals

$$\int_{a_m}^{+\infty} f(x)\, dg_n(x), \quad \int_{-\infty}^{-a_m} f(x)\, dg_n(x)$$

is numerically less than $\tfrac{1}{2}\epsilon$ whenever $m > M$. We shall show the details only for the first of these two, observing first that if $|f(x)| < K$, then

(3) $\left| \int_{a_m}^{+\infty} f(x)\, dg_n(x) \right| < \int_{a_m}^{+\infty} K\, dg_n(x) = K[g_n(+\infty) - g_n(a_m)]$.

Let M_1 be chosen so that $g(a_{M_1}) > g(+\infty) - \epsilon/6K$, then N_1 such that $n > N_1$ implies $g_n(a_{M_1}) > g(a_{M_1}) - \epsilon/6K$, and finally $N_2 \geq N_1$ such that $n > N_2$ implies $g_n(+\infty) < g(+\infty) + \epsilon/6K$. Then $m > M_1$ and $n > N_2$ imply

$g_n(+\infty) - g_n(a_m) \leq g_n(+\infty) - g_n(a_{M_1})$
$\qquad < [g(+\infty) - g(a_{M_1})] + \epsilon/3K < \epsilon/2K.$

Next, we choose $M \geq M_1$ so that the inequality

(4) $\qquad g_n(+\infty) - g_n(a_m) < \epsilon/2K$

is satisfied for $n = 1, 2, \cdots, N_2$, and therefore, by the preceding inequality, for all $n = 1, 2, \cdots$. Finally, the inequality (4) implies that the left-hand member of (3) is less than $\tfrac{1}{2}\epsilon$, as we wished to show.

★★1423. EXERCISES

★★1. Complete the details of the proof of the existence of (1), § 1422.

★★2. Construct an example to show that if the points a and b are not included in the set A of Theorem I, § 1422, the conclusion (iv) is not necessarily true.

★★3. Construct an example to show that if the points ($\pm\infty$) are not included in the set A of Theorem II, § 1422, the conclusion (iv) is not necessarily true.

★★★4. Prove that if the functions $f(x), f_m(x), g(x),$ and $g_n(x), m, n = 1, 2, \cdots$, are subject to the assumptions of $f(x)$, $f_m(x)$, $g(x)$, and $g_n(x)$, respectively, of Theorem I, § 1422, and if $\lim_{m \to +\infty} f_m(x) = f(x)$ uniformly on $[a, b]$, then, in the sense of Definition II, § 1006:

$$\lim_{\substack{m \to +\infty \\ n \to +\infty}} \int_a^b f_m(x)\, dg_n(x) = \int_a^b f(x)\, dg(x).$$

Hint: Use the Moore-Osgood theorem, § 1014.

★★5. State and prove a theorem like that of Exercise 4 for the interval $(-\infty, +\infty)$.

★★6. Let I be the interval $[a, b]$ or $(-\infty, +\infty)$, and assume that $f(x)$ and $g_n(x)$, $n = 1, 2, \cdots$ are uniformly bounded on I: $|f(x)| < K$, $|g_n(x)| < K$, $n = 1, 2, \cdots$. Furthermore, assume that $f(x)$ is continuous on I and that for each $n = 1, 2, \cdots$, $g_n(x)$ is monotonically increasing on I. Prove that there exists a subsequence $\{g_{n_k}\}$ of $\{g_n\}$ and a monotonically increasing function $g(x)$ on I such that

$$\lim_{k \to +\infty} \int_I f(x)\, dg_{n_k}(x) = \int_I f(x)\, dg(x).$$

Hints: Let A be the set consisting of all rational numbers on I, together with a and b, or $\pm\infty$, arranged as a sequence $r_1, r_2, \cdots$. First, choose a subsequence of $\{g_n\}$, written $g_{11}, g_{12}, g_{13}, \cdots$ such that $\{g_{1n}(r_1)\}$ converges. Then pick a subsequence of this subsequence, written $g_{21}, g_{22}, g_{23}, \cdots$ such that $\{g_{2n}(r_2)\}$ converges. In general, let $\{g_{mn}(r_m)\}$ converge for each m. Then the diagonal sequence $g_{11}, g_{22}, \cdots$ converges for each $r_1, r_2, \cdots$, defining a function $g(x) \equiv \lim_{n \to +\infty} g_{nn}(x)$ for every $x \in A$. Show that $g(x)$ is monotonically increasing on A, and extend its definition to all real numbers x: $g(x) \equiv \inf_{r_n \geq x} g(r_n)$. Show that $g(x) \uparrow$, and use the Helly-Bray convergence theorem.

15

★ Fourier Series and Orthogonal Functions

★1501. INTRODUCTION

The concept of *infinite series* dates as far back as the ancient Greeks. (For example, Archimedes (287-212 B.C.) summed a geometric series to compute the area of a parabolic segment.) As early as the latter half of the seventeenth century A.D., power series expansions of functions were being investigated. (For example, Newton in 1676 wrote down certain binomial series for fractional exponents, and Leibnitz, late in the century, quoted the familiar although previously published Maclaurin series for e^x, $\ln(1+x)$, $\sin x$, and $\operatorname{Arctan} x$. In 1715 Brook Taylor (1685-1731, English) published his famous formula for expanding a given (analytic) function in a power series (cf. Chapter 8).

By the middle of the eighteenth century it became important to study the possibility of representing a function by series other than power series. This question arose in connection with the problem of the vibrating string (cf. § 1529). Following the initial attacks on this problem by d'Alembert (1717-1783, French), in 1747, and Euler (1707-1783, Swiss), in 1748, Daniel Bernoulli (1700-1782, Swiss) showed, in 1753, that the mathematical conditions imposed by physical considerations were at least formally satisfied by functions defined by infinite series whose terms involved sines and cosines of integral multiples of prescribed variables. Lagrange (1736-1813, French) continued to pursue the implied problem of representing a given function in terms of sines and cosines. Fourier (1758-1830, French), in his book, *Théorie analytique de la Chaleur* (1822), which contained results of his studies on the conduction of heat (cf. § 1530), announced the amazing result that an "arbitrary" function could be expanded in a series with general term $a_n \sin nx$. Some of the "proofs" given by Fourier were lacking in rigor. Indeed, as we shall see, he claimed far too much, since only certain functions can be expanded in the form which he prescribed. However, the extent of "arbitrariness" that can be permitted is nevertheless remarkable.

Some of the difficulties that beset the early investigators of what became known as *Fourier series* lay in the lack of precision applied at that time to such concepts as *function* and *convergence*. For some of these mathematicians, for example, the word *function* was restricted to a single analytic expression, and was not used even for a function with a broken line graph. The first rigorous proof for a fairly extensive class of functions that a Fourier series actually represents a given function was published in 1829 by Dirichlet (1805-1859, German). Such researches called for more carefully formulated statements than had traditionally been demanded. Aside from the enormous importance of Fourier series as a technique for solving "boundary value" problems (in such applied areas as vibration and heat conduction), the purely theoretical studies in this subject have had a total effect on the general theory of functions of a real variable, and on the theory of sets, that is incalculable.

In this chapter we shall present some of the most important basic facts regarding Fourier series, and generalizations to expansions in terms of "orthonormal systems." Many of the most beautiful results of the theory, however, cannot be included since they are expressed in terms of the Lebesgue, rather than the Riemann, integral.

★1502. LINEAR FUNCTION SPACES

For conceptual purposes it is often helpful, when considering a family of functions having a common domain of definition, to regard this family as a "space," and the individual functions as "points" of this space. Under these circumstances, such a space of functions is called a **function space**. A simple example is the set of all functions defined and continuous on the interval $[0, 1]$.

One of the most important methods of combining functions in a function space is by means of a *linear combination*:

Definition I. *A **linear combination** of a set $f_1, f_2, \cdots, f_n$ of functions defined on a set D is a function of the form*

$$(1) \qquad \alpha_1 f_1 + \alpha_2 f_2 + \cdots + \alpha_n f_n,$$

where $\alpha_1, \alpha_2, \cdots, \alpha_n$ are real numbers.

For many purposes it is important for a function space to be "closed under finite linear combinations," according to the following definition:

Definition II. *A function space S is a **linear space** if and only if whenever $f_1, f_2, \cdots, f_n$ are members of S and $\alpha_1, \alpha_2, \cdots, \alpha_n$ are arbitrary real numbers, the linear combination $\alpha_1 f_1 + \alpha_2 f_2 + \cdots + \alpha_n f_n$ is also a member of S.*

Theorem. *A function space S is a linear space if and only if it is closed under (i) addition, and (ii) multiplication by a constant; that is, if and only if (i) whenever f and g are members of S, so is $f + g$, and (ii) whenever f is a member of S, so is αf for any real number α.*

§ 1503] DISTANCE IN A FUNCTION SPACE

Proof. The "only if" part is trivial (let $n = 2$, $\alpha_1 = \alpha_2 = 1$, and $\alpha_1 = \alpha$, $\alpha_2 = 0$, in turn). The "if" part follows by induction from the fact that if $\alpha_1 f_1 \in S$ and $\alpha_2 f_2 \in S$, then $\alpha_1 f_1 + \alpha_2 f_2 \in S$.

Examples. Each of the following sets of functions, assumed to be single-valued, real-valued, and defined over a common closed interval $I = [a, b]$, is a linear space. Verification of this statement for each example is easy and is left to the reader (Ex. 7, § 1511). Other examples are given in §§ 1508, 1510, and 1519.

1. RV: all real-valued functions.
2. B: all bounded functions.
3. RI: all Riemann integrable functions.
4. SC: all sectionally continuous functions (cf. § 507).
5. C: all continuous functions.
6. C^n: all functions with a continuous nth derivative, where n is a fixed positive integer.
7. C^∞: all "infinitely differentiable" functions; that is, each function has derivatives of all orders.
8. Π: all polynomials.
9. Γ: all constant-valued functions.

NOTE 1. Each space of the preceding list contains all spaces that follow it in the list, and therefore is contained in all that precede it (Ex. 7, § 1511).

NOTE 2. An example of a function space that is not a linear space is the set of all monotonic functions on a given closed interval. For example, the two functions $\sin x - 2x$ and $2x$ are separately monotonic on the interval $[0, 2\pi]$, but their sum is not.

NOTE 3. The function space of all real-valued functions whose common domain of definition consists of the positive integers $1, 2, \cdots, n$ is identical with the set of all ordered n-tuples of real numbers, that is, the set of all points in the Euclidean space of n dimensions. In other words, *Euclidean spaces are special cases of function spaces.* Indeed, they are *linear* function spaces.

NOTE 4. The *functions* of a linear function space can be regarded as generalizations of *vectors* in a Euclidean space, the combination (1) corresponding to a linear combination of vectors. For example, if the domain of definition of the members of S consists of the integers 1 and 2, the sum $f + g$ of the two plane vectors f and g reduces to their resultant, obtained by the familiar completing of a parallelogram. For these reasons, linear spaces are alternatively called **vector spaces** and their members are called **vectors**. The coefficients $\alpha_1, \alpha_2, \cdots, \alpha_n$ of Definition I are called **scalars** to distinguish them from vectors. The concept of linear or vector space can be abstracted, and defined without regard to a function space. (Cf. the first Halmos reference at the end of the chapter.)

★1503. DISTANCE IN A FUNCTION SPACE

In a function space S it is often convenient to have some way of determining when one function f of the space is "near" another function g of the space, and to be able to measure the degree of "nearness" of the two functions. One of the most common methods of doing this is to define a "distance" between f and g, denoted $d(f, g)$, subject to the following four properties:

(i) *Distance is defined and real-valued for every pair, f and g, of members of S, and is **nonnegative**:*

(1) $$d(f, g) \geq 0.$$

(ii) *Every member f of S is at a zero distance from itself:*

(2) $$d(f, f) = 0.$$

(iii) *Distance is **symmetric**, for all f and g of S:*

(3) $$d(f, g) = d(g, f).$$

(iv) *Distance satisfies the **triangle inequality**, for all f, g, and h of S:*

(4) $$d(f, h) \leq d(f, g) + d(g, h).$$

A space with a distance function satisfying properties (i)-(iv) is called a **distance space** or a **pseudo-metric space**. If a distance space satisfies the additional property

(v) *Distance is **strictly positive**, for unequal members, f and g, of S:*

(5) $$d(f, g) > 0 \quad \text{if} \quad f \neq g,$$

*it is called a **metric space*** (cf. § 1026).

We shall give now a few examples of distance and metric spaces. The distance function described in the following section, § 1504, turns out to be the most important in the theory of Fourier series. For some results of this chapter (cf. §§ 1524, 1532) it will be sufficient to have a *distance* space of the functions, while for others (cf. § 1534) it will be important that the distance function satisfy property (v), in addition to (i)-(iv).

Example 1. In the space C of all continuous functions on a closed interval $I = [a, b]$ (Example 5, § 1502), define the distance between two functions f and g:

(6) $$d(f, g) = \max_{x \in I} |f(x) - g(x)|.$$

Then all five properties (i)-(v) are satisfied, so that the resulting space is a *metric space* (Ex. 9, § 1511). The reader may be interested in verifying (Ex. 10, § 1511) that convergence of a sequence according to this distance,

(7) $$d(f_n, f) \to 0,$$

is equivalent to uniform convergence on I (§ 901):

(8) $$f_n(x) \rightrightarrows f(x), \quad x \in I.$$

Example 2. In the space B of all bounded functions on I (Example 2, § 1502), a distance, which for continuous functions reduces to the one given above, can be defined:

(9) $$d(f, g) = \sup_{x \in I} |f(x) - g(x)|.$$

This space is again a metric space. The reader may observe (Ex. 12, § 1511) that

§ 1504] INNER PRODUCT. ORTHOGONALITY 541

the preceding space C of continuous functions is a closed subset of the space B in the sense that whenever a sequence $\{f_n\}$ of continuous functions converges to a function f in the sense of (7), this limit function must belong to the subspace C. (Cf. Theorem III, § 310.) Another closed subspace is the space RI of all Riemann integrable functions on I (Example 3, § 1502). (Cf. the Theorems, §§ 905, 906.)

Example 3. In the space RI of all Riemann integrable functions on a closed interval $I = [a, b]$ (Example 3, § 1502), define the distance between two functions f and g:

$$(10) \qquad d(f, g) = \int_a^b |f(x) - g(x)|\, dx.$$

Then the four properties (i)-(iv) are satisfied (Ex. 13, § 1511). However, this space is *not* a metric space since property (v) fails. For example, any two functions that differ only on a finite set of points are at a zero distance apart (cf. Theorem VI, § 501). This *pseudo-metric space* contains as a subspace the *metric space* C of continuous functions, with the same distance function (10) (cf. Ex. 3, § 503). In distinction to the situation of Example 1, however, the subspace C in this case is not closed, since in case of nonuniform convergence it is possible for a sequence of continuous functions to have a limit in the sense of (10) that is discontinuous. (Cf. the example of Fig. 903, § 905.) On the contrary, the subspace C is *dense* in RI in the sense that there always exists a function of the subspace arbitrarily "close" to any given function of the space; more precisely, if $f \in RI$ and $\epsilon > 0$, there exists $g \in C$ such that $d(f, g) < \epsilon$. (Give the proof in Ex. 13, § 1511.) The space RI with the distance (10) is often denoted R^1 (cf. Ex. 21, § 1511).

★1504. INNER PRODUCT. ORTHOGONALITY. THE SPACE R^2

For the space RI of all Riemann integrable functions on an interval $[a, b]$, a distance function can be defined in many ways. Two of these are described in Examples 2 and 3, § 1503, and an infinite family is defined in Exercise 21, § 1511, by the formula

$$(1) \qquad d(f, g) = \left\{ \int_a^b |f(x) - g(x)|^p\, dx \right\}^{\frac{1}{p}},$$

where $p \geq 1$. Of all these the most important for the theory of Fourier series is that provided by the number $p = 2$:

$$(2) \qquad d(f, g) = \left\{ \int_a^b [f(x) - g(x)]^2\, dx \right\}^{\frac{1}{2}}.$$

Basically, the reasons for this are (i) that the distance (2) is structurally similar to the distance in Euclidean spaces (square root of the sum of the squares of the differences of the coordinates) and (ii) that this particular distance can be obtained from an "inner product," which is the analogue in function space of the "dot product" or "scalar product" of vectors. The definition of this vital concept follows, all functions having a given closed interval $[a, b]$ as common domain:

542 FOURIER SERIES; ORTHOGONAL FUNCTIONS [§ 1504

Definition I. *If f and $g \in RI$ for the interval (a, b), their **inner product** (f, g) is defined:*

$$(3) \qquad (f, g) \equiv \int_a^b f(x)g(x)\, dx.$$

NOTE 1. Since the product of two integrable functions is integrable (Ex. 35, § 503) the inner product (3) always exists.

NOTE 2. The inner product (3) reduces to the scalar product of two vectors in case f and g have a finite domain and the integral in (3) is replaced by a finite sum.

The following properties of inner product are simple consequences of the definition and elementary properties of the Riemann integral (Ex. 14, § 1511):

Theorem I. (i) *The inner product is **symmetric**:*

$$(4) \qquad (f, g) = (g, f).$$

(ii) *The inner product is **linear** in each member:*

$$(5) \quad \begin{cases} (\alpha_1 f_1 + \cdots + \alpha_n f_n, g) = \alpha_1(f_1, g) + \cdots + \alpha_n(f_n, g), \\ (f, \beta_1 g_1 + \cdots + \beta_n g_n) = \beta_1(f, g_1) + \cdots + \beta_n(f, g_n). \end{cases}$$

Taking as point of departure the criterion for orthogonality of two vectors in terms of the vanishing of their scalar product (equivalently, two lines in E_3 having direction numbers l_1, m_1, n_1 and l_2, m_2, n_2 are perpendicular if and only if $l_1 l_2 + m_1 m_2 + n_1 n_2 = 0$), we introduce a definition:

Definition II. *Two integrable functions, f and g, are **orthogonal** on $[a, b]$ if and only if their inner product vanishes:*

$$(6) \qquad f \perp g \quad \text{if and only if} \quad (f, g) = 0.$$

*An **orthogonal family** of functions is a collection of functions every two distinct members of which are orthogonal.*

Continuing the analogy further, we define a new concept, which corresponds to the *magnitude* of a vector:

Definition III. *The **norm** of a function f integrable on (a, b) is defined:*

$$(7) \qquad \|f\| \equiv (f, f)^{\frac{1}{2}} = \left\{ \int_a^b [f(x)]^2\, dx \right\}^{\frac{1}{2}}.$$

The following properties of norm, for a function, are left for the reader to establish (Ex. 15, § 1511):

Theorem II. (i) *The norm is defined for every member of RI, and is **nonnegative**:*

$$(8) \qquad \|f\| \geqq 0.$$

§ 1504]   INNER PRODUCT. ORTHOGONALITY   543

(ii) *For the function* $0 = O(x)$ *that is identically zero the norm is zero:*

(9) $$\|0\| = 0.$$

(iii) *The **triangle inequality** holds, for all f and g of RI:*

(10) $$\|f + g\| \leq \|f\| + \|g\|.$$

(iv) *The norm is **positive-homogeneous**; that is, if f is integrable and α is a real number,*

(11) $$\|\alpha f\| = |\alpha| \cdot \|f\|.$$

If S is a linear subspace of RI such that all members of S except for the one that is identically zero have positive norm:

(12) $$\|f\| > 0 \quad \text{if} \quad f \neq 0,$$

then S is called a **normed linear space**, and the norm is called **strictly positive**.

NOTE 3. Throughout the remainder of this chapter, unless a specific statement to the contrary is made, the word *norm* will be used only in the sense of equation (7). However, it should be remarked that the terms *norm* and *normed linear space* are susceptible to considerable generalization. Indeed, *any* linear space for all points of which a "norm" having the five preceding properties, (8)-(12), exists is called a **normed linear space**. Euclidean spaces are examples (cf. Note 4, below).

NOTE 4. The **Schwarz inequality** (Ex. 29, § 503) can be written:

$$|(f, g)| \leq \|f\| \cdot \|g\|,$$

which takes the form for vectors (u_1, u_2, u_3) and (v_1, v_2, v_3) in E_3:

$$|u_1 v_1 + u_2 v_2 + u_3 v_3| \leq \sqrt{u_1^2 + u_2^2 + u_3^2} \sqrt{v_1^2 + v_2^2 + v_3^2}.$$

If (u_1, u_2, u_3) and (v_1, v_2, v_3) are nonzero vectors with an angle θ between them, this last inequality can be written

$$\frac{|u_1 v_1 + u_2 v_2 + u_3 v_3|}{\sqrt{u_1^2 + u_2^2 + u_3^2} \sqrt{v_1^2 + v_2^2 + v_3^2}} \leq 1 \quad \text{or} \quad |\cos \theta| \leq 1.$$

Thus the Schwarz inequality in function space can be considered as a generalization of the statement that the cosine of an angle never exceeds 1 numerically. For the significance of equality in the Schwarz inequality, see Exercise 20, § 1511.

Formula (10) is the Minkowski inequality of Exercise 30, § 503.

In terms of the *norm* of a function, we now define a new distance (distinct from those of Examples 2 and 3, § 1503) between functions belonging to the space RI, for a given closed interval $[a, b]$:

Definition IV. *The **distance** between two integrable functions f and g is the norm of their difference:*

(13) $$d(f, g) \equiv \|f - g\| = \left\{ \int_a^b [f(x) - g(x)]^2 \, dx \right\}^{\frac{1}{2}}.$$

From the properties of norm we have almost immediately (give the proof in Ex. 16, § 1511):

544 FOURIER SERIES; ORTHOGONAL FUNCTIONS [§ 1504

Theorem III. *The quantity* (13) *is a distance according to properties* (i)-(iv) *of* § 1503. *This distance is strictly positive* ((v), § 1503) *if and only if the norm (Definition III) is strictly positive* ((12), *above*).

NOTE 5. The space RI with the distance (13) will henceforth be denoted R^a whenever that particular distance is important.

Definition V. *If a function with nonzero norm is divided by plus or minus its norm the resulting function has unit norm (and therefore corresponds to a vector of unit length):*

$$(14) \qquad \left\| \frac{f}{\pm \|f\|} \right\| = 1.$$

*In this case the function f is said to be **normalized**, and either function $\pm f/\|f\|$ is called the **result of normalizing** f. An **orthonormal family** of functions is an orthogonal family each member of which has norm equal to unity.*

NOTE 6. It can be proved (Ex. 23, § 1511) that every orthonormal family of functions integrable on $[a, b]$ is either finite or denumerable, and therefore can be arranged in a sequence: $\phi_1, \phi_2, \cdots$.

Example 1. Prove the **Pythagorean theorem**: *If f and g are orthogonal members of R^a, then*

$$(15) \qquad \|f - g\|^2 = \|f\|^2 + \|g\|^2.$$

(Cf. Fig. 1501, and Ex. 24, § 1511.)

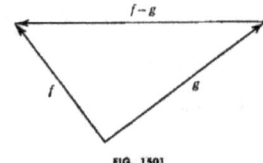

FIG. 1501

Solution. By Theorem I,
$$\|f - g\|^2 = (f - g, f - g) = (f, f) - 2(f, g) + (g, g)$$
$$= (f, f) + (g, g) = \|f\|^2 + \|g\|^2.$$

Example 2. For the interval $[-\pi, \pi]$, show that the functions

$$(16) \qquad \begin{cases} 1, \cos x, \cos 2x, \cos 3x, \cdots, \\ \sin x, \sin 2x, \sin 3x, \cdots \end{cases}$$

form an orthogonal family of functions of nonzero norm. Normalize these functions and thus obtain an orthonormal family.

Solution. Standard integration procedures (check the details) give

$$\int_{-\pi}^{\pi} 1\, dx = 2\pi,$$

$$\int_{-\pi}^{\pi} \cos^2 nx\, dx = \int_{-\pi}^{\pi} \sin^2 nx\, dx = \pi, \quad n = 1, 2, \cdots,$$

$$\int_{-\pi}^{\pi} \cos nx\, dx = \int_{-\pi}^{\pi} \sin nx\, dx = 0, \quad n = 1, 2, \cdots,$$

$$\int_{-\pi}^{\pi} \cos mx \cos nx\, dx = \int_{-\pi}^{\pi} \sin mx \sin nx\, dx = 0,$$
$$m = 1, 2, \cdots, \quad n = 1, 2, \cdots, \quad m \neq n,$$

$$\int_{-\pi}^{\pi} \cos mx \sin nx\, dx = 0, \quad m = 1, 2, \cdots, \quad n = 1, 2, \cdots.$$

Therefore the family (16) is orthogonal, the norm of the first number is $\sqrt{2\pi}$, and the norm of every other member is $\sqrt{\pi}$. An orthonormal family corresponding to (16) is

(17)
$$\begin{cases} \dfrac{1}{\sqrt{2\pi}},\ \dfrac{\cos x}{\sqrt{\pi}},\ \dfrac{\cos 2x}{\sqrt{\pi}},\ \dfrac{\cos 3x}{\sqrt{\pi}},\ \cdots, \\[4pt] \dfrac{\sin x}{\sqrt{\pi}},\ \dfrac{\sin 2x}{\sqrt{\pi}},\ \dfrac{\sin 3x}{\sqrt{\pi}},\ \cdots. \end{cases}$$

★1505. LEAST SQUARES. FOURIER COEFFICIENTS

Let $\Psi = (\psi_1, \psi_2, \cdots, \psi_n)$ be a finite orthonormal family of functions belonging to the space R^2 for an interval $[a, b]$, and let S be the linear function space of all linear combinations $\psi = \beta_1\psi_1 + \cdots + \beta_n\psi_n$ of the members of Ψ. If f is an arbitrary member of R^2, does there exist a minimum distance between f and a point ψ of S (where "distance" is that of (13), § 1504)? If so, is ψ uniquely determined and in what way? It is to these questions that we now address ourselves.

Our first objective, then, is to minimize the quantity

(1) $$[d(f, \psi)]^2 = \|f - (\beta_1\psi_1 + \cdots + \beta_n\psi_n)\|^2$$

as a function of the unknowns $\beta_1, \cdots, \beta_n$. We expand the right-hand member of (1) as the inner product $(f - \sum \beta_i\psi_i,\ f - \sum \beta_i\psi_i)$, and make use of the linearity properties of inner products stated in Theorem I, § 1504, together with the facts that $(\psi_i, \psi_j) = 0$, $i \neq j$, and $(\psi_i, \psi_i) = 1$, $i, j = 1, 2, \cdots, n$:

(2) $$[d(f, \psi)]^2 = \|f\|^2 + (\beta_1^2 + \beta_2^2 + \cdots + \beta_n^2)$$
$$- 2[\beta_1(f, \psi_1) + \beta_2(f, \psi_2) + \cdots + \beta_n(f, \psi_n)]$$
$$= \left\{ \sum_{i=1}^{n} [\beta_i - (f, \psi_i)]^2 \right\} + \left\{ \|f\|^2 - \sum_{i=1}^{n} (f, \psi_i)^2 \right\}.$$

The answers to the questions raised in the first paragraph are now accessible. Since the unknown β's occur only in the sum of squares $\sum [\beta_i - (f, \psi_i)]^2$, we infer that $d(f, \psi)$ is least when every term of that sum of squares vanishes. In other words, the minimum distance does exist and is given by a unique linear combination $\psi = \beta_1 \psi_1 + \cdots + \beta_n \psi_n$, where

(3) $$\beta_i = (f, \psi_i), i = 1, 2, \cdots, n.$$

The problem just solved can be thought of as that of finding the "best" approximation to a given function by means of a linear combination of the functions of a given finite orthonormal set—best in the sense of distance in the space R^2. This criterion of "best approximation" is also called that of **least squares**, since the distance formula is based on squaring.

We are ready for a basic definition and theorem (summarizing the preceding remarks):

Definition. *Let $\Phi = (\phi_1, \phi_2, \cdots)$ be an orthonormal family (finite or infinite) of functions $\phi_1, \phi_2, \cdots$ belonging to R^2 for the interval $[a, b]$ (cf. Note 6, § 1504) and let f be a given member of R^2. Then the **Fourier coefficients** of f with respect to the orthonormal family Φ are the numbers $\alpha_1, \alpha_2, \cdots$, defined:*

(4) $$\alpha_n = (f, \phi_n) = \int_a^b f(x) \phi_n(x)\, dx.†$$

Theorem. *Let $\Psi = (\psi_1, \psi_2, \cdots, \psi_n)$ be a finite subset of an orthonormal family $\Phi = (\phi_1, \phi_2, \cdots)$ of R^2, and let f be a given function of R^2. Then the linear combination $\beta_1 \psi_1 + \cdots + \beta_n \psi_n$ of the members of Ψ that best approximates f, in the sense of the distance in R^2, is uniquely given in case the coefficients $\beta_1, \cdots, \beta_n$ are the Fourier coefficients of f with respect to the corresponding members of Φ. The coefficient of any particular member of Ψ is independent of the remaining members of Ψ (or whether any other members of Φ are present at all).*

NOTE 1. In E_3, if Φ is the family of three unit vectors $\vec{i} = (1, 0, 0), \vec{j} = (0, 1, 0)$, and $\vec{k} = (0, 0, 1)$, and if $\vec{v}$ is a given vector with components $v_1, v_2,$ and v_3:

$$\vec{v} = (v_1, v_2, v_3),$$

then the Fourier coefficients of $\vec{v}$ are its three projections on the coordinate axes:

$$v_1 \cdot 1 + v_2 \cdot 0 + v_3 \cdot 0, \quad v_1 \cdot 0 + v_2 \cdot 1 + v_3 \cdot 0, \quad v_1 \cdot 0 + v_2 \cdot 0 + v_3 \cdot 1.$$

If Ψ consists of all *three* vectors of the orthonormal family Φ, the linear combination of $\vec{i}, \vec{j},$ and $\vec{k}$ that best approximates $\vec{v}$ has the three Fourier coefficients $v_1, v_2,$ and v_3:

† These numbers are also called *Fourier constants*. For the use of the proper name *Fourier* in this connection, see the first footnote on page 547.

$$\vec{v_1 i} + \vec{v_2 j} + \vec{v_3 k}.$$

In this case, the linear combination approximates $\vec{v}$ so well that it is equal to it. If Ψ consists of the *two* unit vectors $\vec{i}$ and $\vec{j}$, then the linear combination of these two vectors that best approximates $\vec{v}$ has the same Fourier coefficients (as far as they go):

$$\vec{v_1 i} + \vec{v_2 j}.$$

Finally, if Ψ consists of the *single* unit vector $\vec{i}$, the best approximation is

$$\vec{v_1 i}.$$

Note 2. If S is the linear space of all linear combinations of the finite orthonormal set $\Psi = (\psi_1, \psi_2, \cdots, \psi_n)$, and if ψ is the member of S that minimizes the distance $d(f, \psi) = \|f - \psi\|$, then ψ can be thought of as the foot of a perpendicular dropped from f to S or, equivalently, as the *orthogonal projection* of f on S. This is true since the "vector" $f - \psi$ is orthogonal to every ψ_i, $i = 1, 2, \cdots, n$, and hence to every member of S. (Discuss the details in Ex. 25, § 1511.)

★1506. GENERALIZED FOURIER SERIES. BESSEL'S INEQUALITY

Having in mind the "expansion" of a vector $\vec{v}$ in E_3 in terms of its components in an $\vec{i}, \vec{j}, \vec{k}$ system (Note 1, § 1505),

$$\vec{v} = v_1 \vec{i} + v_2 \vec{j} + v_3 \vec{k},$$

we give a definition:

Definition I. *Let $\Phi = (\phi_1, \phi_2, \cdots)$ be an orthonormal system in R^2, and let $f \in R^2$. Then the **generalized Fourier series** of f with respect to Φ† is the series whose general term is $\alpha_n \phi_n$, where α_n is the Fourier coefficient $\alpha_n = (f, \phi_n)$. We say that f is **expanded** in a generalized Fourier series with respect to Φ, and write*

$$(1) \qquad f \sim \alpha_1 \phi_1 + \alpha_2 \phi_2 + \cdots$$

without regard to any question of convergence.‡

Note 1. If a series of the form

$$\alpha_1 \phi_1 + \alpha_2 \phi_2 + \alpha_3 \phi_3 + \cdots$$

---

† There is no universal agreement on terminology. Some writers omit the word *generalized* and call these series "Fourier series with respect to an orthonormal system," while others use the proper name *Fourier* only for the trigonometric series originally studied by Fourier, and refer to the more general series of the present definition as "orthogonal expansions" or "orthogonal developments" of the given function. In this book we shall reserve the two-word expression *Fourier series* for the trigonometric Fourier series introduced in § 1509, which are a special case of the *generalized Fourier series* presented here.

‡ See §§ 1510, 1520, 1521, 1532 for statements regarding the manner in which the right-hand member of (1) can "represent" the function f.

converges *uniformly* on the interval under consideration, and if a function f is *defined* to be the sum of that series (so that (1) becomes an equality by definition), then the coefficients $\alpha_1, \alpha_2, \alpha_3, \cdots$ are the Fourier coefficients of f (so that the relation (1) remains valid *as printed*). This is true since the formal procedure of multiplying on both sides of the *equality* (1) by the quantity ϕ_n and then integrating term by term is valid, and produces the desired relation $\alpha_n = (f, \phi_n)$. (Check the details.)

An important relation between the Fourier coefficients of a function and its norm follows:

Theorem I. Bessel's Inequality. *If* $\Phi = (\phi_1, \phi_2, \cdots)$ *is an orthonormal family in* R^2, *for an interval* $[a, b]$, *and if* $\alpha_1, \alpha_2, \cdots$ *are the Fourier coefficients of a given function f with respect to* Φ *(that is, if* $\alpha_n = (f, \phi_n)$, $n = 1, 2, \cdots$*), then the series* $\sum \alpha_n^2$ *converges, and*

(2) $$\sum_{n=1}^{+\infty} \alpha_n^2 \leq \|f\|^2.\dagger$$

Proof. For any finite subfamily $\Psi = (\psi_1, \cdots, \psi_n)$ of Φ, the relation (2), § 1505, together with the inequality $[d(f, \psi)]^2 \geq 0$ for the particular Fourier coefficients $\beta_i = (f, \psi_i)$, $i = 1, \cdots, n$, gives directly an inequality of the form (2) for an arbitrary subset of the coefficients α_n. In particular, since all partial sums of the series in (2) satisfy the inequality $\sum \alpha_n^2 \leq \|f\|^2$, so does the complete sum.

NOTE 2. In the Euclidean space E_3 (cf. Note 1, § 1505), Bessel's inequality takes such forms as

$$v_1^2 + v_2^2 + v_3^2 \leq |\vec{v}|^2,$$
$$v_1^2 + v_2^2 \leq |\vec{v}|^2,$$
$$v_1^2 \leq |\vec{v}|^2,$$

depending on whether the orthonormal family Φ consists of the three vectors $\vec{i}, \vec{j},$ and $\vec{k}$, the two vectors $\vec{i}$ and $\vec{j}$, or the single vector $\vec{i}$. In the first case Bessel's inequality becomes an equality:

$$v_1^2 + v_2^2 + v_3^2 = |v|^2.$$

(Cf. Ex. 26, § 1511.)

As a corollary to Bessel's inequality we have:

Theorem II. Riemann's Lemma. *For any infinite orthonormal family* $\Phi = (\phi_1, \phi_2, \cdots)$ *in* R^2, *and any function f of R^2, the limit of the nth Fourier coefficient is zero:*

(3) $$\lim_{n \to +\infty} \alpha_n = \lim_{n \to +\infty} (f, \phi_n) = 0.$$

† For the significance of equality in (2), see §§ 1512 and 1532. Also cf. Ex. 26, § 1511.

★1507. PERIODIC FUNCTIONS

In this section we shall consider only real-valued functions defined for all real numbers. For such functions we have the definition:

Definition I. *A function $f(x)$ is **periodic with period p**, where p is a real number, if and only if the equation*

(1) $$f(x + p) = f(x)$$

*holds identically for all real numbers x. A function is **periodic** if and only if it is periodic with period p for some nonzero p.*

Example 1. The function $\sin x$ is periodic with period 2π. It is also periodic with period -16π and, more generally, with period $2n\pi$ for any integer $n = 0, \pm 1, \cdots$.

Example 2. Let $f(x)$ be the function that is equal to 1 if x is rational and equal to 0 if x is irrational. Then $f(x)$ is periodic with period p if and only if p is rational. (Why? Cf. Ex. 27, § 1511.)

The preceding two examples typify the general situation, as shown by the following two theorems:

Theorem I. *The set of all periods of a periodic function $f(x)$ either is dense (cf. § 112) or consists of all integral multiples of a positive number p. In the latter case the number p is the smallest positive period (often called **the period** of the given function).*

Proof. The theorem is a consequence of the following three lemmas (give proofs in Ex. 28, § 1511):

Lemma 1. *If A is the set of all periods of $f(x)$, then A is closed under addition, subtraction, and multiplication by an arbitrary integer; that is, if a_1 and $a_2 \in A$, then $a_1 \pm a_2 \in A$; and if $a \in A$ and n is an integer, then $na \in A$.*

Lemma 2. *If p is a positive period of $f(x)$, then the set B of periods of the form $\pm np$, $n = 0, 1, 2, \cdots$, has the property that for any real number x there exists a member b of B such that $|x - b| < p$.*

Lemma 3. *If x and y are distinct real numbers such that no period of $f(x)$ lies between them, then no finite interval can contain more than a finite set of periods of $f(x)$.*

Theorem II. *Any periodic nonconstant function either is everywhere discontinuous or has a smallest positive period.*

Proof. Assume $f(x)$ is continuous at $x = a$, and has a dense set of periods. If there exists a point b such that $f(b) \neq f(a)$, then there must exist points arbitrarily near a at which $f(x)$ has the value $f(b)$. By continuity of f at a, $f(a) = f(b)$. (Contradiction.)

NOTE. Any function $f(x)$ defined on a half-open interval of length p can be defined outside this interval in such a way that it becomes periodic with period p and with domain of definition $(-\infty, +\infty)$, as follows: If x is an arbitrary real number, let x be represented in the form $x = x_0 + np$, where x_0 belongs to the original interval, and n is an integer. (This representation exists and is unique. Why?) Finally, define $f(x) \equiv f(x_0)$. When this has been done, the original function is said to have been **extended periodically**. It should be observed that any periodic extension of a function having the properties specified above is unique. (Why?)

★1508. LINEAR SPACES OF PERIODIC FUNCTIONS

Since the values of any periodic function $f(x)$ of specified period $p > 0$ are completely determined for all real numbers x by its values on any half-open interval of length p, its properties as a whole can be studied by restricting its domain to such a finite interval. In particular, a function $f(x)$ of period 2π can be thought of as defined for $-\pi \leq x \leq \pi$, with $f(-\pi) = f(\pi)$, and then extended periodically, with period 2π, outside this basic interval. Equivalently, the basic interval could have been chosen to be $[0, 2\pi]$, or any other of length 2π.

Since the most important and familiar Fourier series (§ 1509) make use of sines and cosines all of which are periodic with period 2π, we shall limit our considerations of linear spaces of periodic functions principally to those of period 2π. For simplicity we shall adopt $[-\pi, \pi]$ as the basic interval, although the interval $[0, 2\pi]$ is often used.

Each of the linear spaces in the Examples of § 1502, when restricted by the further condition that all of the functions considered be periodic with period 2π (or any *fixed* positive number), becomes a new linear space, which is easily denoted by the subscript 2π. For instance, $RV_{2\pi}$ is the set of all real-valued periodic functions with period 2π, and $RI_{2\pi}$ is the subspace consisting of functions Riemann integrable on the interval $[-\pi, \pi]$ or, equivalently, on any interval of length 2π. Finally, $R_{2\pi}{}^2$ is the space $RI_{2\pi}$, with the distance of Definition IV, § 1504, for the interval $[a, b] = [-\pi, \pi]$:

$$d(f, g) = \|f - g\| = \left\{\int_{-\pi}^{\pi} [f(x) - g(x)]^2 \, dx\right\}^{\frac{1}{2}}.$$

The two notations $RI_{2\pi}$ and $R_{2\pi}{}^2$ will be used interchangeably, but the latter principally when the notion of *distance* in the space is important.

NOTE. It is a curious fact that, although for a specified period $p > 0$ all periodic functions with period p form a linear space, it is not true that all periodic functions form a linear space. For instance, the functions $\sin x$ and $\sin \alpha x$, where α is irrational, are both periodic, but their sum is not (cf. Ex. 29, § 1511). The desire to formulate a reasonably restricted linear space containing all suitably well-behaved periodic functions has led to the study of *almost periodic functions*.

§ 1509]   TRIGONOMETRIC FOURIER SERIES

For a treatment of this fruitful subject, see the book by H. Bohr in the references at the end of the chapter.

★1509. TRIGONOMETRIC FOURIER SERIES

An orthonormal sequence of special importance is that of Example 2, § 1504:

(1) $$\frac{1}{\sqrt{2\pi}}, \frac{\cos x}{\sqrt{\pi}}, \frac{\sin x}{\sqrt{\pi}}, \frac{\cos 2x}{\sqrt{\pi}}, \frac{\sin 2x}{\sqrt{\pi}}, \cdots$$

for the interval $[-\pi, \pi]$ (or any interval of length 2π). The generalized Fourier series for a given function $f \in R_{2\pi}{}^2$ with respect to the sequence (1) will be called its **trigonometric Fourier series** or (in case misinterpretation is unlikely), more simply, its **Fourier series,** with the notation:

(2) $$f(x) \sim \alpha_0 \cdot \frac{1}{\sqrt{2\pi}} + \alpha_1 \frac{\cos x}{\sqrt{\pi}} + \beta_1 \frac{\sin x}{\sqrt{\pi}} + \alpha_2 \frac{\cos 2x}{\sqrt{\pi}} + \beta_2 \frac{\sin 2x}{\sqrt{\pi}} + \cdots,$$

where

$$\alpha_0 = \left(f, \frac{1}{\sqrt{2\pi}}\right) = \frac{1}{\sqrt{2\pi}} \int_{-\pi}^{\pi} f(x)\, dx,$$

$$\alpha_n = \left(f, \frac{\cos nx}{\sqrt{\pi}}\right) = \frac{1}{\sqrt{\pi}} \int_{-\pi}^{\pi} f(x) \cos nx\, dx, \quad n = 1, 2, \cdots,$$

$$\beta_n = \left(f, \frac{\sin nx}{\sqrt{\pi}}\right) = \frac{1}{\sqrt{\pi}} \int_{-\pi}^{\pi} f(x) \sin nx\, dx, \quad n = 1, 2, \cdots.$$

Substitution of these coefficients in (2) leads to the following form for writing the Fourier series of $f(x)$:

(3) $$f(x) \sim \tfrac{1}{2}a_0 + a_1 \cos x + b_1 \sin x + a_2 \cos 2x + b_2 \sin 2x + \cdots,$$

where

(4) $$\begin{cases} a_n = \dfrac{1}{\pi} \int_{-\pi}^{\pi} f(x) \cos nx\, dx, & n = 0, 1, 2, \cdots, \\ b_n = \dfrac{1}{\pi} \int_{-\pi}^{\pi} f(x) \sin nx\, dx, & n = 1, 2, \cdots. \end{cases}$$

(Denoting the constant term by $\tfrac{1}{2}a_0$ rather than a_0 permits an economy in formulas (4), by making the formula for a_n valid when $n = 0$.)

Any series of the form

(5) $$\frac{A_0}{2} + A_1 \cos x + B_1 \sin x + A_2 \cos 2x + B_2 \sin 2x + \cdots$$

is called a **trigonometric series,** but only when the coefficients are obtained in terms of a given function by formulas of the form (4) is the series (5) a Fourier series. (Cf. Example 3.)

552 FOURIER SERIES; ORTHOGONAL FUNCTIONS [§ 1509

Example 1. Find the Fourier series of the function $f \in R_{2\pi}{}^2$, defined for the interval $[-\pi, \pi]$:

$$f(x) = \begin{cases} x, & -\pi < x < \pi, \\ 0, & x = -\pi, \pi. \end{cases}$$

Solution. Substitution in formulas (4) gives $a_n = 0$, $n = 0, 1, 2, \cdots$ (since $f(x)$ is an odd function), and

$$b_n = \frac{1}{\pi}\int_{-\pi}^{\pi} x \sin nx \, dx = \frac{2}{\pi}\int_0^{\pi} x \, d\left(-\frac{\cos nx}{n}\right)$$

$$= -\frac{2x \cos nx}{n\pi}\Big]_0^{\pi} = \begin{cases} 2/n, & n = 1, 3, 5, \cdots, \\ -2/n, & n = 2, 4, 6, \cdots. \end{cases}$$

Therefore,

$$f(x) \sim 2\left[\frac{\sin x}{1} - \frac{\sin 2x}{2} + \frac{\sin 3x}{3} - \cdots\right].$$

The graphs of the partial sums $S_1 = b_1 \sin x$ and $S_3 = \sum_{n=1}^{3} b_n \sin nx$ are shown in Figure 1502.

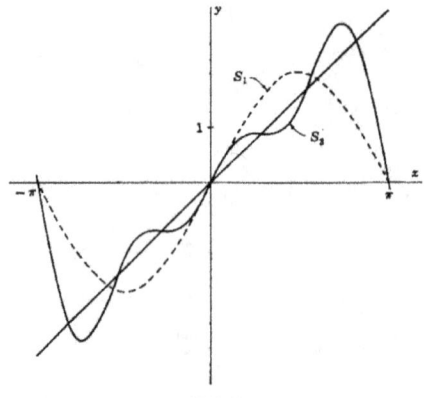

FIG. 1502

Example 2. Find the Fourier series of the function $f \in R_{2\pi}{}^2$, defined for the interval $[-\pi, \pi]$:

$$f(x) = |x|, \quad -\pi \leq x \leq \pi.$$

§ 1509]   TRIGONOMETRIC FOURIER SERIES

Solution. Since $f(x)$ is even, $b_n = 0$, $n = 1, 2, \cdots$,

$$a_0 = \frac{1}{\pi}\int_{-\pi}^{\pi} |x|\, dx = \frac{2}{\pi}\int_0^{\pi} x\, dx = \pi,$$

$$a_n = \frac{2}{\pi}\int_0^{\pi} x\, d\left(\frac{\sin nx}{n}\right) = -\frac{2\cos nx}{n^2\pi}\Big]_0^{\pi} = \begin{cases} -4/n^2\pi, & n = 1, 3, \cdots, \\ 0, & n = 2, 4, 6, \cdots. \end{cases}$$

Therefore

$$f(x) \sim \frac{\pi}{2} - \frac{4}{\pi}\left[\frac{\cos x}{1^2} + \frac{\cos 3x}{3^2} + \frac{\cos 5x}{5^2} + \cdots\right].$$

The graphs of the partial sums $S_1 = \frac{1}{2}a_0 + a_1\cos x$ and $S_2 = \frac{1}{2}a_0 + a_1\cos x + a_3\cos 3x$ are shown in Figure 1503.

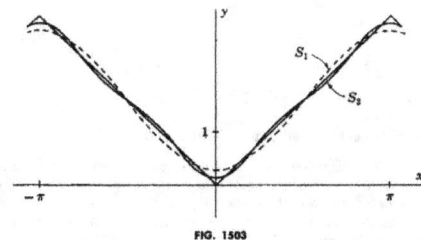

FIG. 1503

NOTE 1. In terms of the notation of (3) and (4), Bessel's inequality ((2), § 1506) becomes

(6) $$\frac{1}{2}a_0^2 + a_1^2 + b_1^2 + a_2^2 + b_2^2 + \cdots \leq \frac{1}{\pi}\int_{-\pi}^{\pi}[f(x)]^2\, dx,$$

and Riemann's Lemma (Theorem II, § 1506) states

(7) $$\lim_{n\to+\infty} a_n = \lim_{n\to+\infty} b_n = 0.$$

As will be discussed in § 1512, (6) is actually an equality, known as **Parseval's equation**.

Example 3. The trigonometric series

$$\sin x + \sin 2x + \sin 3x + \cdots$$

is not a Fourier series because (7) fails. The series

(8) $$\sin x + \frac{\sin 2x}{\sqrt{2}} + \frac{\sin 3x}{\sqrt{3}} + \cdots,$$

although (7) holds, is not a Fourier series (of a Riemann integrable function) since the series $\sum b_n^2$ is divergent (cf. (6)). (Cf. Note 3, below.)

NOTE 2. If Lebesgue (instead of Riemann) integration is used, it can be shown that a trigonometric series is a Fourier series of a suitably restricted function *if and only if* the infinite series on the left of (6) converges. This fact is the principal

554 FOURIER SERIES; ORTHOGONAL FUNCTIONS [§ 1510

substance of the celebrated Riesz-Fischer theorem (cf. the book by Hardy and Rogosinski, p. 16, listed in the references at the end of the chapter).

✶✶ NOTE 3. It is of interest that although the trigonometric series (8) is not a Fourier series, it does converge for every value of x. This can be seen (assuming that x is not a multiple of π, in which case the result is trivial) by invoking the Abel test of Exercise 22, § 717, with $a_n = \sin nx$ and $b_n = 1/\sqrt{n}$. In this case

$$\sum_{i=1}^{n} a_i = \frac{2\sin x \sin \tfrac{1}{2}x + 2\sin 2x \sin \tfrac{1}{2}x + \cdots + 2\sin nx \sin \tfrac{1}{2}x}{2\sin \tfrac{1}{2}x}.$$

With the aid of the identity $2 \sin \alpha \sin \beta = \cos(\alpha - \beta) - \cos(\alpha + \beta)$, this can be written

$$\frac{1}{2\sin \tfrac{1}{2}x} \{\cos \tfrac{1}{2}x - \cos \tfrac{3}{2}x + \cos \tfrac{3}{2}x - \cos \tfrac{5}{2}x + \cdots$$
$$+ \cos(n - \tfrac{1}{2})x - \cos(n + \tfrac{1}{2})x\},$$

so that

$$\left| \sum_{i=1}^{n} a_i \right| = \tfrac{1}{2}|\csc \tfrac{1}{2}x| \cdot |\cos \tfrac{1}{2}x - \cos(n + \tfrac{1}{2})x| \leq |\csc \tfrac{1}{2}x|,$$

and the partial sums of the series $\sum \sin nx$ are bounded.

★1510. A SPECIAL CONVERGENCE THEOREM

The question of *convergence* of the (trigonometric) Fourier series of a given function f of $R_{2\pi}^2$ has not yet been raised. (Cf. § 1532.) One reason is that our approach to the topic of Fourier series has been based on approximation in the sense of least squares (§ 1505) rather than in the sense of the convergence of a series of functions. That is, we have considered, so far, approximation based only on integration over an entire interval, and not on functional values at particular values of the independent variable. Another reason for deferring consideration of this question of convergence is that it is an extremely difficult one to answer. Even if the given function is everywhere continuous the full story is not understood, although this much is known: *it is possible for the Fourier series of a continuous function to be divergent at some point.*[†] The twentieth century Russian mathematician Kolmogoroff has given an example of a function (integrable in the Lebesgue sense) whose Fourier series is *everywhere divergent.*[‡]

There are really two parts to the question of convergence: (i) Does the series converge at all? (ii) If the series converges does it represent the given function; that is, does it converge to $f(x)$? A trivial example shows that the Fourier series of a given function may converge everywhere, but fail to represent the function everywhere. Such an example is provided by any function that vanishes except on a finite set of points.

[†] For a discussion of such an example, due to L. Fejér, see the book by Titchmarsh, p. 416, or that by Hobson, vol. 2, p. 540, in the references at the end of the chapter.
[‡] This example is given in the book by Zygmund, p. 175, listed in the references at the end of the chapter.

§ 1510] A SPECIAL CONVERGENCE THEOREM   555

A fairly general convergence theorem is established in § 1520. In the present section we shall content ourselves with stating a special case of that theorem:

Theorem. *Let $f(x)$ be periodic with period 2π, and sectionally smooth (§ 507) on the interval $[-\pi, \pi]$. Then the Fourier series of $f(x)$ converges for every real number x to the limit*

(1) $$\frac{f(x+) + f(x-)}{2}.$$

In particular, the series converges to the value of the given function at every point of continuity of the function, and if at every point of discontinuity the value of the function is defined as the average of its two one-sided limits there:

(2) $$f(x) = \frac{f(x+) + f(x-)}{2},$$

then its Fourier series represents the function everywhere. (Cf. Fig. 1504.)

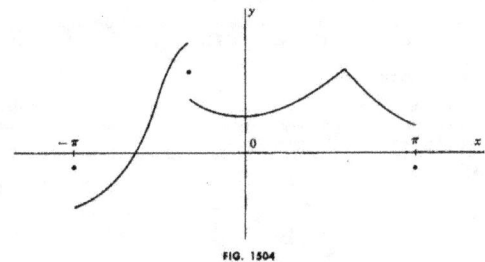

FIG. 1504

Note 1. Henceforth in this chapter, whenever a sectionally smooth function is under discussion, the "averaging" condition (2) will be assumed without specific detailing in every instance.

Note 2. A special case of the averaging condition (2) follows from (2) and periodicity: $f(-\pi) = f(\pi) = \frac{1}{2}[f(-\pi+) + f(\pi-)]$.

Note 3. The set SS of all sectionally smooth functions, and the set $SS_{2\pi}$ of those of period 2π are linear spaces. (Why? Cf. Ex. 6, § 508; § 1519.)

Example 1. Since the function $f(x)$ of Example 1, § 1509, satisfies the conditions of the preceding Theorem, it is represented everywhere by its Fourier series. In particular, for the interval $(-\pi, \pi)$,

(3) $$x = 2\left[\frac{\sin x}{1} - \frac{\sin 2x}{2} + \frac{\sin 3x}{3} - \cdots\right].$$

556 FOURIER SERIES; ORTHOGONAL FUNCTIONS [§ 1511

Substitution of $x = \frac{1}{2}\pi$ gives a familiar evaluation (cf. Ex. 31, § 811):

$$(4) \qquad 1 - \frac{1}{3} + \frac{1}{5} - \frac{1}{7} + \cdots = \frac{\pi}{4}.$$

Example 2. As in the preceding Example, we have from Example 2, § 1509, for the interval $[-\pi, \pi]$:

$$(5) \qquad |x| = \frac{\pi}{2} - \frac{4}{\pi}\left[\frac{\cos x}{1^2} + \frac{\cos 3x}{3^2} + \frac{\cos 5x}{5^2} + \cdots\right].$$

Substitution of $x = 0$ gives the evaluation (cf. Ex. 11, § 721):

$$(6) \qquad 1 + \frac{1}{3^2} + \frac{1}{5^2} + \frac{1}{7^2} + \cdots = \frac{\pi^2}{8}.$$

From this we can derive an important evaluation of a value of the Zeta-function (cf. the footnote, p. 214, and the Note, § 720),

$$(7) \qquad \zeta(2) = 1 + \frac{1}{2^2} + \frac{1}{3^2} + \frac{1}{4^2} + \cdots = \frac{\pi^2}{6}.$$

Since the sum of the even-numbered terms of (7) is

$$\frac{1}{2^2} + \frac{1}{4^2} + \cdots = \frac{1}{4}\left[\frac{1}{1^2} + \frac{1}{2^2} + \cdots\right] = \frac{1}{4}\zeta(2),$$

we have, by subtraction, the sum of the odd-numbered terms as given in (6): $\frac{3}{4}\zeta(2) = \frac{1}{8}\pi^2$; (7) follows. (Cf. Example 1, § 1512.)

★1511. EXERCISES

In Exercises 1-6, find the trigonometric Fourier series of the function $f(x)$ of the space $RI_{2\pi}$, with the averaging condition (2), § 1510. Obtain an evaluation of an infinite series from the indicated substitution.

★1. $f(x) = \begin{cases} -1, & -\pi < x < 0, \\ 1, & 0 < x < \pi; \end{cases}$ $x = \frac{\pi}{2}.$

★2. $f(x) = \begin{cases} 0, & -\pi < x < 0, \\ 1, & 0 < x < \pi; \end{cases}$ $x = \frac{\pi}{2}.$

★3. $f(x) = \begin{cases} 0, & -\pi < x < 0, \\ 1, & 0 < x < \frac{1}{2}\pi, \\ 0, & \frac{1}{2}\pi < x < \pi; \end{cases}$ $x = \frac{\pi}{2}.$

★4. $f(x) = \begin{cases} 0, & -\pi < x < \frac{1}{2}\pi, \\ 1, & \frac{1}{2}\pi < x < \pi; \end{cases}$ $x = \frac{\pi}{2}.$

★5. $f(x) = \begin{cases} 0, & -\pi < x < 0, \\ x, & 0 < x < \pi; \end{cases}$ $x = \pi.$

★6. $f(x) = \begin{cases} 0, & -\pi < x < 0, \\ x, & 0 < x < \frac{1}{2}\pi, \\ 0, & \frac{1}{2}\pi < x < \pi; \end{cases}$ $x = \frac{\pi}{2}.$

★7. Verify that each of the Examples, § 1502, is a linear space, and prove the statement of Note 1, § 1502.

§ 1511] EXERCISES 557

*8. Show that the set of all real-valued functions defined on a fixed interval and possessing there both maximum and minimum values is not a linear space.

*9. Prove that the space C of Example 1, § 1503, is a metric space.

*10. Prove that convergence in the space C of Example 1, § 1503, is equivalent to uniform convergence on the interval I.

*11. Prove that the space B of Example 2, § 1503, is a metric space, and that convergence in that space is equivalent to uniform convergence on the interval I.

*12. Prove the statements given in the last two sentences of the discussion of Example 2, § 1503.

*13. Prove that the space R^1 of Example 3, § 1503, is a distance space, and that its subspace C is a metric space dense in R^1. *Hints:* For (iv), § 1503,
$$d(f, h) = \int |(f - g) + (g - h)| \, dx. \text{ Cf. Ex. 45, § 503.}$$

*14. Prove Theorem I, § 1504.

*15. Prove Theorem II, § 1504. (Cf. Ex. 30, § 503, Ex. 22, below.)

*16. Prove Theorem III, § 1504. *Hint:* For (iv), § 1503, write $d(f, h) = \|(f - g) + (g - h)\|$.

*17. Prove that any set dense in the space R^2 is dense in the space R^1 (Example 3, § 1503). *Hint:* Use the Schwarz inequality: $\int_0^1 |f_n - f| \cdot 1 \, dx \leq \|f_n - f\| \cdot \|1\|$ (Note 4, § 1504).

*18. The **unit sphere** of R^2 is defined to be the set of all $f \in R^2$ such that $\|f\| \leq 1$. Prove that its complement S (the set of all f such that $\|f\| > 1$) is dense in R^1—although it is clearly not dense in R^2 (cf. Ex. 17). *Hint:* Assume without loss of generality that the interval is $[0, 1]$, define $g_n(x)$ to be $n^3 - n^5x$ for $0 \leq x \leq n^{-2}$, and 0 for $n^{-2} \leq x \leq 1$. Show that in the space R^1, $d(f + g_n, f) \to 0$, while in the space R^2, $\|f + g_n\| \to +\infty$.

**19. Prove that the space C as a subspace of the space R^2 is a metric space and that it is dense in R^2. *Hint:* Prove that the step-functions are dense in R^2 and that for any step-function σ and $\epsilon > 0$ there exists a continuous function g such that $d(g, \sigma) < \epsilon$. For these results, use the fact that if $f \in R^2$, if $|f(x)| \leq K$, if $|\sigma(x)| \leq K$, and if $|g(x)| \leq K$, then $\int |f - \sigma|^2 \, dx \leq 2K \int |f - \sigma| \, dx$ and $\int |g - \sigma|^2 \, dx \leq 2K \int |g - \sigma| \, dx$. (Cf. Ex. 13.)

**20. Prove that equality, in the Schwarz inequality of Note 4, § 1504, holds if and only if one of the two functions is at a zero distance from some scalar multiple of the other. Hence prove that for the space R^2 the triangle inequality becomes an equality if and only if one of the two functions is at a zero distance from some nonnegative scalar multiple of the other. *Hint for "only if":* Assume without loss of generality that $\|f\| = \|g\| = (f, g) = 1$. Then $\|f - g\| = 0$.

**21. Prove that the norm of a function f integrable on $[a, b]$, defined:
$$\|f\| = \left\{ \int_a^b |f(x)|^p \, dx \right\}^{\frac{1}{p}}, \quad p \geq 1,$$
satisfies all four properties of norm of Theorem II, § 1504. Hence show that (1), § 1504, is a distance function. (The space RI with this distance function is often denoted R^p.) *Hint:* Cf. Ex. 55, § 503, Ex. 40, § 1225.

558 FOURIER SERIES; ORTHOGONAL FUNCTIONS [§ 1512

★★**22.** Prove that whenever a linear space has an operation (f, g) on pairs of its members subject to the two properties of Theorem I, § 1504, and the third property that $(f, f) \geq 0$ for all members f of the space, then a "norm" defined $\|f\| = (f, f)^{\frac{1}{2}}$ has the four properties listed in Theorem II, § 1504. Furthermore, prove that the Schwarz inequality $|(f, g)| \leq \|f\| \cdot \|g\|$ is a further consequence. *Hint:* Prove the Schwarz inequality first. Cf. Ex. 29, § 503.

★★**23.** Prove Note 6, § 1504. *Hint:* Show that the set of all step-functions σ on I whose values are rational numbers and whose points of discontinuity are rational numbers is both dense in R^2 and denumerable. Assuming that a nondenumerable orthonormal set $\{\psi\}$ of functions exists, for each ψ find a σ such that $\|\psi - \sigma\| < \frac{1}{2}$, and obtain a contradiction from the fact that for any two ψ_1 and ψ_2, $\|\psi_1 - \psi_2\| = \sqrt{2}$. (Cf. Ex. 19.)

★**24.** Establish the following *law of cosines* (cf. Example 1, § 1504):

$$\|f - g\|^2 = \|f\|^2 + \|g\|^2 - 2(\|f\| \cdot \|g\|)(f \|f\|, g \|g\|)$$

for any functions f and g of R^2 for which $\|f\| \cdot \|g\| \neq 0$. Also prove the *parallelogram law* for arbitrary $f, g \in R^2$:

$$\|f + g\|^2 + \|f - g\|^2 = 2\|f\|^2 + 2\|g\|^2,$$

and that f and g are orthogonal if and only if

$$\|f + g\|^2 = \|f\|^2 + \|g\|^2.$$

★**25.** Justify the statements of Note 2, § 1505.

★**26.** Let $\Phi = (\phi_1, \phi_2, \cdots, \phi_n)$ be a finite orthonormal family in R^2, and let $f \in R^2$. Prove that Bessel's inequality $\sum_{i=1}^{n} \alpha_i^2 \leq \|f\|^2$ is an equality if and only if f is at a zero distance from some linear combination of $\phi_1, \cdots, \phi_n$. (Cf. Note 2, § 1505.)

★**27.** Verify the statements of Example 2, § 1507.

★**28.** Complete the proof of Theorem I, § 1507.

★**29.** Prove that $\sin x + \sin \alpha x$ is periodic if and only if α is rational. *Hint for "only if":* The identity $\sin (x + p) + \sin (\alpha x + \alpha p) = \sin x + \sin \alpha x$ is equivalent to $\cos (x + \frac{1}{2}p) \sin (\frac{1}{2}p) = -\cos (\alpha x + \frac{1}{2}\alpha p) \sin (\frac{1}{2}\alpha p)$. Set $x = \frac{1}{2}(n\pi - p)$.

★★**30.** Let $f(x)$ be periodic with period 2π. In each of the following parts prove that the stated properties, assumed to hold for *arbitrary finite intervals*, imply the specified order of magnitude behavior for the coefficients a_n and b_n of the trigonometric Fourier series of f:

(a) f sectionally smooth; $O(1/n)$.
(b) f continuous and sectionally smooth; $o(1/n)$.
(c) f continuous, f' sectionally smooth; $O(1/n^2)$.
(d) f' continuous and sectionally smooth; $o(1/n^2)$.

Generalize to higher order derivatives. *Hint:* Cf. Ex. 7, § 508.

★**1512. PARSEVAL'S EQUATION**

It will be proved in § 1532 that for the *trigonometric* Fourier series (3), § 1509, of an integrable function $f(x)$, Bessel's inequality (cf. Note 1, § 1509) becomes an equality:

§1513] COSINE SERIES. SINE SERIES

Theorem. Parseval's Equation. *If $f \in RI_{2\pi}$, and if $\frac{1}{2}a_0 + a_1 \cos x + b_1 \sin x + \cdots$ is the trigonometric Fourier series of f, for the interval $[-\pi, \pi]$, then*

(1) $$\frac{1}{2}a_0^2 + a_1^2 + b_1^2 + a_2^2 + b_2^2 + \cdots = \frac{1}{\pi}\int_{-\pi}^{\pi}[f(x)]^2\,dx.$$

Example 1. From Example 1, §1509,
$$4\left[\frac{1}{1^2} + \frac{1}{2^2} + \frac{1}{3^2} + \cdots\right] = \frac{1}{\pi}\int_{-\pi}^{\pi} x^2\,dx = \frac{2\pi^2}{3}.$$

This is equivalent (cf. Example 2, §1510) to:

(2) $$\zeta(2) = \frac{1}{1^2} + \frac{1}{2^2} + \frac{1}{3^2} + \cdots + \frac{1}{n^2} + \cdots = \frac{\pi^2}{6}.$$

Example 2. Use Parseval's equation and Example 2, §1509, to obtain the evaluation (cf. the footnote, p. 214):

(3) $$\zeta(4) = \frac{1}{1^4} + \frac{1}{2^4} + \frac{1}{3^4} + \cdots + \frac{1}{n^4} + \cdots = \frac{\pi^4}{90}.$$

Solution. Parseval's equation, for this example, is

(4) $$\frac{\pi^2}{2} + \frac{16}{\pi^2}\left[\frac{1}{1^4} + \frac{1}{3^4} + \frac{1}{5^4} + \cdots\right] = \frac{2\pi^2}{3},$$

whence

(5) $$\frac{1}{1^4} + \frac{1}{3^4} + \frac{1}{5^4} + \cdots = \frac{\pi^4}{96}.$$

By the method used in Example 2, §1510, to evaluate $\zeta(2)$, we infer (3) from (5).

★1513. COSINE SERIES. SINE SERIES

If $f(x) \in RI_{2\pi}$, and if $f(x)$ is *even* (that is, if $f(-x) = f(x)$ for all x), then every coefficient b_n of its Fourier series, (3), §1509, vanishes (why?), and the series becomes a pure **cosine series**:

(1) $$f(x) \sim \tfrac{1}{2}a_0 + a_1 \cos x + a_2 \cos 2x + \cdots,$$

where the coefficients can be written (Ex. 13, §1515):

(2) $$a_n = \frac{2}{\pi}\int_0^{\pi} f(x)\cos nx\,dx, \quad n = 0, 1, 2, \cdots.$$

If $f(x)$ is a given function defined only on $[0, \pi]$, and integrable there, its domain of definition can be extended immediately to the full interval $[-\pi, \pi]$ by requiring that it be even there, and then to $(-\infty, +\infty)$ by periodicity (with period 2π). The series (1), subject to the formula (2), is called the cosine series of the given function defined on $[0, \pi]$.

In a similar fashion, if $f(x) \in RI_{2\pi}$, and if $f(x)$ is *odd* (that is, if $f(-x) = -f(x)$ for all x), then the Fourier series of f becomes a pure **sine series**:

(3) $$f(x) \sim b_1 \sin x + b_2 \sin 2x + \cdots,$$

where the coefficients can be written (Ex. 13, § 1515):

$$(4) \quad b_n = \frac{2}{\pi} \int_0^\pi f(x) \sin nx \, dx, \quad n = 1, 2, \cdots.$$

If $f(x)$ is a given function defined only on $[0, \pi]$, and integrable there, and if it is redefined (if necessary) to have the value 0 at $x = 0$ and $x = \pi$ ($f(0) = f(\pi) = 0$), then its domain of definition can be extended immediately to the interval $[-\pi, \pi]$ by requiring that it be odd there, and then to $(-\infty, +\infty)$ by periodicity. The series (3), subject to (4), is called the *sine series* of the given function defined on $[0, \pi]$.

Example 1. Examples 1 and 2, § 1509, give the sine and cosine series respectively, for the function x, defined on $[0, \pi]$.

Example 2. Find the cosine and sine series for the function x^2 on $[0, \pi]$. Then use Parseval's equation to verify the evaluation of $\zeta(4)$ (Example 2, § 1512) and to obtain an evaluation of $\zeta(6)$.

Solution. Integration by parts gives, for (2) and (4):

$$a_0 = \frac{2\pi^2}{3}, \quad a_n = (-1)^n \frac{4}{n^2}, \quad n = 1, 2, \cdots,$$

$$b_n = \frac{(-1)^{n+1} 2\pi}{n} - \frac{4[1 + (-1)^{n+1}]}{n^3 \pi}, \quad n = 1, 2, \cdots.$$

Parseval's equation for the resulting cosine series for x^2 on the interval $[-\pi, \pi]$ is

$$\frac{2\pi^4}{9} + 16\left(\frac{1}{1^4} + \frac{1}{2^4} + \cdots\right) = \frac{2}{\pi} \int_0^\pi x^4 \, dx = \frac{2\pi^4}{5},$$

and the desired evaluation $\zeta(4) = \pi^4/90$ follows immediately.

For the sine series, Parseval's equation gives

$$4\pi^2 \zeta(2) - 32 \cdot \frac{15}{16} \zeta(4) + \frac{64}{\pi^2} \cdot \frac{63}{64} \zeta(6) = \frac{2\pi^4}{5},$$

whence

$$(5) \quad \zeta(6) = \frac{1}{1^6} + \frac{1}{2^6} + \frac{1}{3^6} + \frac{1}{4^6} + \cdots = \frac{\pi^6}{945}.$$

Note. Since any function $f(x)$ can be written as the sum of an even function and an odd function:

$$(6) \quad f(x) = \frac{f(x) + f(-x)}{2} + \frac{f(x) - f(-x)}{2},$$

any trigonometric Fourier series can be obtained by "adding" a cosine series and a sine series.

Example 3. Find the Fourier series of $f(x)$, defined:

$$f(x) = \begin{cases} 0, & -\pi < x < 0, \\ x, & 0 < x < \pi. \end{cases}$$

Solution. Using (6), we have
$$f(x) = \frac{|x|}{2} + \frac{x}{2},$$
and therefore, from Examples 1 and 2, § 1509, the Fourier series is
$$\frac{\pi}{4} + \sum_{n=1}^{+\infty} \left\{ -\frac{1}{n^2\pi}[1 + (-1)^{n+1}] \cos nx + \frac{(-1)^{n+1}}{n} \sin nx \right\}.$$

★1514. OTHER INTERVALS

Let $f(x)$ be a periodic function of period $2p$ (notationally, $2p$ is more convenient here than p), integrable on the interval $[-p, p]$ (or any interval of length $2p$). It is merely a matter of a change of scale associated with the positive factor π/p to adapt the expansion formulas of § 1509, and their derivation, to the Fourier series of $f(x)$ for the interval $[-p, p]$ (Ex. 14, § 1515):

(1) $\quad f(x) \sim \frac{1}{2}a_0 + a_1 \cos \frac{\pi x}{p} + b_1 \sin \frac{\pi x}{p} + a_2 \cos \frac{2\pi x}{p} + b_2 \sin \frac{2\pi x}{p} + \cdots,$

where

(2) $\quad \begin{cases} a_n = \dfrac{1}{p} \displaystyle\int_{-p}^{p} f(x) \cos \dfrac{n\pi x}{p}\, dx, & n = 0, 1, 2, \cdots, \\[2mm] b_n = \dfrac{1}{p} \displaystyle\int_{-p}^{p} f(x) \sin \dfrac{n\pi x}{p}\, dx, & n = 1, 2, \cdots. \end{cases}$

NOTE 1. A function $f(x)$ defined and integrable on $[0, p]$ can be expanded in either a cosine series or a sine series. (Cf. Ex. 15, § 1515.)

Example. Find the Fourier series for the function $f(x) = |x|$ on the interval $[-10, 10]$, as a function of RI_{20}.

First Solution. Substitution in (2), with $p = 10$, gives
$$a_0 = 10, \quad a_n = -\frac{20}{n^2\pi^2}[1 + (-1)^{n+1}], \quad b_n = 0, \quad n = 1, 2, 3, \cdots,$$
and therefore, for $|x| \leq 10$,

(3) $\quad |x| = 5 - \dfrac{40}{\pi^2}\left[\dfrac{1}{1^2}\cos\dfrac{\pi x}{10} + \dfrac{1}{3^2}\cos\dfrac{3\pi x}{10} + \dfrac{1}{5^2}\cos\dfrac{5\pi x}{10} + \cdots\right].$

Second Solution. Substitution of $x = \pi t/10$ in the Fourier series obtained in Example 2, § 1509, gives
$$\frac{\pi}{10}|t| = \frac{\pi}{2} - \frac{4}{\pi}\left[\frac{1}{1^2}\cos\frac{\pi t}{10} + \frac{1}{3^2}\cos\frac{3\pi t}{10} + \frac{1}{5^2}\cos\frac{5\pi t}{10} + \cdots\right],$$
which reduces immediately to (3) except for notation.

NOTE 2. For simplicity, most of the results in the remaining portions of this chapter are restricted to periodic functions with period 2π (instead of period $2p$). There is no essential loss of generality in this restriction, which is made for the sake

562   FOURIER SERIES; ORTHOGONAL FUNCTIONS   [§ 1515

of manipulative simplification. A simple scale factor converts statements in terms of period 2π to apply to the appropriate period.

★1515. EXERCISES

★1. Show that Parseval's equation for each of Exercises 1 and 2, § 1511, reduces to (6), § 1510.

In Exercises 2-4, find the Fourier series for the given function of the space $RI_{2\pi}$, with the averaging condition (2), § 1510. Then use Parseval's equation to obtain the indicated evaluation of the Zeta-function. (Cf. the footnote, p. 214; also Exs. 20, 22 below; Ex. 18, § 1528.)

★2. $f(x) = x^3$; $\zeta(6) = \pi^6/945$.
★3. $f(x) = x^4$; $\zeta(8) = \pi^8/9450$.
★4. $f(x) = x^5$; $\zeta(10) = \pi^{10}/93555$.

In Exercises 5-8, find the cosine series and the sine series for the given function of the space $RI_{2\pi}$, with the averaging condition (2), § 1510.

★5. $f(x) = \begin{cases} 0, & 0 < x < \tfrac{1}{2}\pi, \\ 1, & \tfrac{1}{2}\pi < x < \pi. \end{cases}$ ★6. $f(x) = \begin{cases} x, & 0 < x < \tfrac{1}{2}\pi, \\ \pi - x, & \tfrac{1}{2}\pi < x < \pi. \end{cases}$

★7. $f(x) = \cos x$.   ★8. $f(x) = \sin x$.

In Exercises 9-12, find the Fourier series of the function for the given interval $[-p, p]$, considered as a member of RI_{2p}, with the averaging condition (2), § 1510.

★9. $f(x) = \begin{cases} -1, & -1 < x < 0, \\ 1, & 0 < x < 1, \end{cases}$ where $p = 1$.

★10. $f(x) = x$, $-2 < x < 2$, where $p = 2$.

★11. $f(x) = \begin{cases} 0, & -4 < x < 0, \\ x, & 0 < x < 4, \end{cases}$ where $p = 4$.

★12. $\begin{cases} 0, & -6 < x < 0, \\ x, & 0 < x < 3, \\ 0, & 3 < x < 6, \end{cases}$ where $p = 6$.

★13. Establish formulas (2) and (4), § 1513.
★14. Establish formulas (1) and (2), § 1514.
★15. Establish formulas for cosine series and sine series coefficients, for functions of period $2p$.
★★16. Find the Fourier series for the function $\cos ax$, where a is not an integer. Then, by substituting $x = 0$ and $x = \pi$ and finally relabeling the variable, obtain the following two *partial fractions expansions* well known in complex variable theory:

$$\csc \pi x = \frac{1}{\pi x} + \frac{2x}{\pi} \sum_{n=1}^{+\infty} \frac{(-1)^n}{x^2 - n^2},$$

$$\cot \pi x = \frac{1}{\pi x} + \frac{2x}{\pi} \sum_{n=1}^{+\infty} \frac{1}{x^2 - n^2},$$

where x is nonintegral.

★★17. Same as Exercise 16 for $\cosh ax$,
$$\operatorname{csch} \pi x = \frac{1}{\pi x} + \frac{2x}{\pi} \sum_{n=1}^{+\infty} \frac{(-1)^n}{x^2 + n^2},$$
$$\coth \pi x = \frac{1}{\pi x} + \frac{2x}{\pi} \sum_{n=1}^{+\infty} \frac{1}{x^2 + n^2},$$
where x is nonzero.

★★18. Justify term by term differentiation of the second series of Exercise 16. Then interpret and establish the following for nonintegral x:
$$\frac{\pi^2}{\sin^2 \pi x} = \sum_{-\infty}^{+\infty} \frac{1}{(x-n)^2}.$$

★★19. Justify term by term integration of the second series of Exercise 16 over an interval $[0, a]$, where $0 < a < 1$. Thus establish the infinite product expansion, for $0 < x < 1$ (which can be shown to be valid for all complex x):
$$\sin \pi x = \pi x \prod_{n=1}^{+\infty} \left(1 - \frac{x^2}{n^2}\right).$$
(Cf. § 1416.) Thus obtain a second derivation of Wallis's infinite product for π (§ 1417).

★★20. Write the Maclaurin series for each term of the partial fractions expansion of $\pi x \cot \pi x$, obtained from Exercise 16. By collecting terms and comparing coefficients with those of the Maclaurin series of $\pi x \cot \pi x$ (cf. §§ 809, 810), derive a new method for evaluating $\zeta(2)$, $\zeta(4)$, $\zeta(6)$, $\cdots$. (Cf. Example 2, § 1510, Example 2, § 1512, Exs. 2-4, above, Ex. 22, below.)

★★21. The **Bernoulli polynomials**, $B_0(x), B_1(x), B_2(x), \cdots$, are defined by the following expansion, where the numbers $B_0, B_1, B_2, \cdots$ are the Bernoulli numbers (cf. Ex. 35, § 811):

(1) $$e^{xt} \cdot \frac{t}{e^t - 1} = \sum_{n=0}^{+\infty} \frac{(xt)^n}{n!} \cdot \sum_{n=0}^{+\infty} \frac{B_n}{n!} t^n = \sum_{n=0}^{+\infty} \frac{B_n(x)}{n!} t^n.$$

Prove that $B_n(x)$ is a polynomial of degree n given by the formula

(2) $$B_n(x) = \binom{n}{0} B_0 x^n + \binom{n}{1} B_1 x^{n-1} + \cdots + \binom{n}{n-1} B_{n-1} x + \binom{n}{n} B_n.$$

Show that $B_n(0) = B_n$, $n \geq 0$, and by (2), Ex. 35, § 811, and (2) above, that $B_n(1) = B_n$, $n \geq 2$. Furthermore, show that for $n \geq 1$:

(3) $$\begin{cases} B_{n+1}(x) = B_{n+1} + (n+1) \int_0^x B_n(t)\, dt, \\ B_{n+1}'(x) = (n+1) B_n(x), \\ \int_0^1 B_n(x)\, dx = 0. \end{cases}$$

Obtain the first five Bernoulli polynomials.

★★22. Obtain the following Fourier cosine and sine series for Bernoulli polynomials on the closed interval $0 \leq x \leq 1$ (cf. Ex. 21):

(4) $$B_{2n}(x) = (-1)^{n+1} \frac{2(2n)!}{(2\pi)^{2n}} \sum_{k=1}^{+\infty} \frac{\cos 2k\pi x}{k^{2n}}, \quad n \geq 1,$$

(5) $$B_{2n+1}(x) = (-1)^{n+1} \frac{2(2n+1)!}{(2\pi)^{2n+1}} \sum_{k=1}^{+\infty} \frac{\sin 2k\pi x}{k^{2n+1}}, \quad n \geq 1.$$

564 FOURIER SERIES; ORTHOGONAL FUNCTIONS [§ 1516

From (4) derive the formula for values of the Riemann Zeta-function (cf. § 708) for positive even integers:

(6) $$\zeta(2n) = \sum_{k=1}^{+\infty} \frac{1}{k^{2n}} = (-1)^{n+1}\frac{(2\pi)^{2n}B_{2n}}{2(2n)!}.$$

Infer that the Bernoulli numbers alternate in sign. *Hint:* Show that (5) holds for $n = 0$ in the open interval $0 < x < 1$. Then use mathematical induction, with the aid of (3) and integration by parts.

★★1516. PARTIAL SUMS OF FOURIER SERIES

Since, by Riemann's Lemma (§ 1506), the general term of the Fourier series of any integrable function f tends toward 0, it is sufficient in any considerations of convergence to limit our investigations to partial sums of an odd number of terms. We introduce the notation:

(1) $$S_n(x) \equiv \frac{a_0}{2} + \sum_{k=1}^{n}(a_k \cos kx + b_k \sin kx).$$

In terms of the given function f of $RI_{2\pi}$, $S_n(x)$ can be written:

(2) $$S_n(x) = \frac{1}{2\pi}\int_{-\pi}^{\pi} f(t)\,dt + \frac{1}{\pi}\sum_{k=1}^{n}\left\{\cos kx \int_{-\pi}^{\pi} f(t)\cos kt\,dt + \sin kx \int_{-\pi}^{\pi} f(t)\sin kt\,dt\right\}$$

$$= \frac{1}{\pi}\int_{-\pi}^{\pi}\left\{\frac{1}{2} + \sum_{k=1}^{n}\cos k(t-x)\right\}f(t)\,dt.$$

We now use the identity (Ex. 1, § 1528)

(3) $$\frac{1}{2} + \sum_{k=1}^{n}\cos 2k\alpha = \frac{\sin(2n+1)\alpha}{2\sin \alpha}.$$

After substituting first $\alpha = \frac{1}{2}(t-x)$ and then $t - x = u$, we can write the right-hand member of (2) as follows:

(4) $$\frac{1}{2\pi}\int_{-\pi}^{\pi}\frac{\sin(n+\frac{1}{2})(t-x)}{\sin \frac{1}{2}(t-x)}f(t)\,dt = \frac{1}{2\pi}\int_{-\pi-x}^{\pi-x}\frac{\sin(n+\frac{1}{2})u}{\sin \frac{1}{2}u}f(x+u)\,du.$$

By periodicity, we have **Dirichlet's Integral**

(5) $$S_n(x) = \frac{1}{2\pi}\int_{-\pi}^{\pi}\frac{\sin(n+\frac{1}{2})u}{\sin \frac{1}{2}u}f(x+u)\,du.$$

By writing this integral as the sum of two, and finally using the substitution $v = -u$ to convert the second integral, below, we obtain for $S_n(x)$:

$$\frac{1}{2\pi}\int_{0}^{\pi}\frac{\sin(n+\frac{1}{2})u}{\sin \frac{1}{2}u}f(x+u)\,du + \frac{1}{2\pi}\int_{-\pi}^{0}\frac{\sin(n+\frac{1}{2})v}{\sin \frac{1}{2}v}f(x+v)\,dv,$$

or

(6) $$S_n(x) = \frac{1}{2\pi}\int_{0}^{\pi}\frac{\sin(n+\frac{1}{2})u}{\sin \frac{1}{2}u}[f(x+u) + f(x-u)]\,du.$$

§ 1518] THE RIEMANN-LEBESGUE THEOREM

Any question of convergence of a Fourier series can thus be formulated in terms of the limit of the partial sums $S_n(x)$ in equation (6).

★★1517. FUNCTIONS WITH ONE-SIDE LIMITS

Under the assumption that the function $f(x)$, whose Fourier series is under consideration, has one-sided limits at a particular point x, we are now in a position to establish a criterion for the partial sums of the Fourier series to converge at x to the average of these two limits:

(1) $$S_n(x) \to \frac{f(x+) + f(x-)}{2}.$$

We start by writing equation (6), § 1516, for the function $f(x)$ that is identically 1 (cf. Ex. 2, § 1528):

(2) $$1 = \frac{1}{\pi} \int_0^\pi \frac{\sin(n+\tfrac{1}{2})u}{\sin \tfrac{1}{2}u}\, du.$$

Multiplying both members of (2) by the constant $\tfrac{1}{2}[f(x+) + f(x-)]$, and subtracting the resulting expressions from the two members of (6), § 1516, we have:

(3) $$S_n(x) - \frac{f(x+) + f(x-)}{2} = \frac{1}{2\pi} \int_0^\pi \frac{\sin(n+\tfrac{1}{2})u}{\sin \tfrac{1}{2}u}[f(x+u) - f(x+)]\, du$$
$$+ \frac{1}{2\pi} \int_0^\pi \frac{\sin(n+\tfrac{1}{2})u}{\sin \tfrac{1}{2}u}[f(x-u) - f(x-)]\, du.$$

Therefore, a necessary and sufficient condition for the limit relation (1) to hold is that the right-hand member of (3) converge to 0 as $n \to +\infty$. We are thus led to a study of the two integrals on the right of (3).

★★1518. THE RIEMANN-LEBESGUE THEOREM

The following generalization of Riemann's Lemma (Theorem II, § 1506) will be needed:

Theorem. Riemann-Lebesgue Theorem. *If $g(x)$ is integrable on $[a, b]$, then*

$$\lim_{\lambda \to +\infty} \int_a^b \sin \lambda x\, g(x)\, dx = 0.$$

Proof. We observe first that the theorem is true whenever $g(x)$ is identically equal to 1, since the integral then has the value $[\cos \lambda a - \cos \lambda b]/\lambda$. It follows that the theorem is true whenever $g(x)$ is a step-function, since the integral under consideration can then be written as a finite sum of quantities of the form $\sigma_k[\cos \lambda a_{k-1} - \cos \lambda a_k]/\lambda$, each of which tends toward

For a given integrable function $g(x)$, and $\epsilon > 0$, let $\sigma(x)$ be a step-function such that $\sigma(x) \leq g(x)$, and

$$\int_a^b [g(x) - \sigma(x)] \, dx < \frac{\epsilon}{2}.$$

Then let Λ be chosen such that $\lambda > \Lambda$ implies

$$\left| \int_a^b \sin \lambda x \, \sigma(x) \, dx \right| < \frac{\epsilon}{2}.$$

Then for any such λ,

$$\left| \int_a^b \sin \lambda x \, g(x) \, dx \right| \leq \int_a^b |\sin \lambda x| \cdot [g(x) - \sigma(x)] \, dx + \left| \int_a^b \sin \lambda x \, \sigma(x) \, dx \right| < \frac{\epsilon}{2} + \frac{\epsilon}{2} = \epsilon.$$

★★1519. FUNCTIONS OF BOUNDED VARIATION. THE SPACE $BV_{2\pi}$

In order to take optimal advantage of the convergence criterion obtained in § 1517, for functions with one-sided limits, we turn our attention to a class of functions introduced in § 516, called *functions of bounded variation*. For present purposes the most important property of these functions is their representation in terms of monotonic functions (Theorem VIII, § 516): *A function $f(x)$ defined on an interval $[a, b]$ is of bounded variation there if and only if it can be represented there as the difference between two monotonically increasing functions*. As a consequence of this representation, two important properties of monotonic functions (§ 215; Ex. 27, § 216; Theorem X, § 501) are immediately inherited by all functions of bounded variation:

Theorem. *Every function of bounded variation on an interval $[a, b]$ has one-sided limits at each point of $[a, b]$ and is Riemann integrable there.*

It is not difficult to show that the space **BV** of functions defined on an interval $[a, b]$ and of bounded variation there form a linear space (Ex. 3, § 1528). The same is true of the space $BV_{2\pi}$ of all periodic functions with period 2π that are of bounded variation on every finite interval.

Probably the most important subsets of the space BV, for a given interval $[a, b]$, are:

(*i*) the set of all monotonic functions on $[a, b]$ (these do not form a linear space—cf. Note 3, § 1502);

(*ii*) the set of all sectionally smooth functions on $[a, b]$ (these form a linear space—cf. Ex. 6, § 508, Ex. 43, § 518);

(*iii*) the set of all sectionally continuous functions on $[a, b]$ having relative maxima and minima at only a finite number of points there (these do not form a linear space—cf. Ex. 4, § 1528).

★★1520. JORDAN CONVERGENCE THEOREM

We shall establish in this section a generalization of the special convergence theorem of § 1510, obtained by replacing in the statement of that theorem the words *sectionally smooth* by the words *of bounded variation*. This theorem is known as the *Jordan Convergence Theorem* (C. Jordan, 1838-1922, French).

The first step in the proof is to observe that, because of the form of each member on the right-hand side of (3), § 1517, we need only prove that if $\phi(u)$ is a function of bounded variation on the interval $[0, \pi]$, such that $\phi(0+) = 0$, then

$$(1) \qquad \lim_{\lambda \to +\infty} \int_0^\pi \frac{\sin \lambda u}{\sin \frac{1}{2}u} \phi(u) \, du = 0.$$

(We let $\phi(u) = f(x + u) - f(x+)$ and $\phi(u) = f(x - u) - f(x-)$, in turn, with $\lambda = n + \frac{1}{2}$.)

We now note that since $\phi(u)$ can be represented as the difference between two monotonically increasing functions, we can assume without loss of generality that $\phi(u)$ is itself monotonically increasing on $[0, \pi]$. Furthermore, we can assume that $\phi(u)$ is nonnegative there, is continuous from the right at $u = 0$, and vanishes at $u = 0$ (give the details in Ex. 5, § 1528):

$$(2) \qquad \phi(u) \uparrow, \quad \phi(u) \geqq 0, \quad \phi(0) = \phi(0+) = 0.$$

The next step is to show that a relation equivalent to (1) is

$$(3) \qquad \lim_{\lambda \to +\infty} \int_0^\pi \frac{\sin \lambda u}{u} \phi(u) \, du.$$

The reason for this is that

$$\int_0^\pi \frac{\sin \lambda u}{\frac{1}{2}u} \phi(u) \, du - \int_0^\pi \frac{\sin \lambda u}{\sin \frac{1}{2}u} \phi(u) \, du = \int_0^\pi \sin \lambda u \, \psi(u) \, du.$$

where

$$\psi(u) = \begin{cases} \dfrac{\sin \frac{1}{2}u - \frac{1}{2}u}{\frac{1}{2}u \sin \frac{1}{2}u} \phi(u), & 0 < u \leqq \pi, \\ 0, & u = 0. \end{cases}$$

Since $\psi(u)$ is integrable on $[0, \pi]$, the result follows from the Riemann-Lebesgue Theorem (§ 1518).

To prove (3), let $\epsilon > 0$ be given, and let $\delta > 0$ be such that $\phi(\delta) < \epsilon/4\pi$. Then, by the Bonnet form for the Second Mean Value Theorem for Integrals (Ex. 29, § 518), there exists a number ξ between 0 and δ such that

$$\int_0^\delta \frac{\sin \lambda u}{u} \phi(u) \, du = \phi(\delta) \int_\xi^\delta \frac{\sin \lambda u}{u} \, du.$$

568 FOURIER SERIES; ORTHOGONAL FUNCTIONS [§ 1521

Therefore, with $v = \lambda u$,

$$\left| \int_0^\delta \frac{\sin \lambda u}{u} \phi(u)\, du \right| < \frac{\epsilon}{4\pi} \left| \int_{\lambda\delta}^{\lambda\delta} \frac{\sin v}{v}\, dv \right|.$$

Since $\left| \int_\alpha^\beta \frac{\sin x}{x}\, dx \right| < 2\pi$ for all α and β (cf. Ex. 14, § 1403), we conclude:

(4) $$\left| \int_0^\delta \frac{\sin \lambda u}{u} \phi(u)\, du \right| < \frac{\epsilon}{2}.$$

Our final objective is to show the existence (corresponding to the given ϵ and the determined δ) of a number Λ such that $\lambda > \Lambda$ implies

(5) $$\left| \int_\delta^\pi \sin \lambda u \, \frac{\phi(u)}{u}\, du \right| < \frac{\epsilon}{2}.$$

But this follows from the Riemann-Lebesgue theorems (§ 1518), applied to the interval $[\delta, \pi]$, the function $\phi(u)/u$ playing the role of $g(x)$.

★★1521. FEJÉR'S SUMMABILITY THEOREM

If the condition of bounded variation of the function $f(x)$ is dropped from the hypotheses of Jordan's Convergence Theorem, convergence of the Fourier series cannot be inferred (cf. § 1510). However, as long as the existence of the one-sided limits at the point in question is retained, a useful and striking result can be obtained if, instead of considering *convergence*, we turn our attention to *summability* $(C, 1)$ (cf. Ex. 15, § 717).

Theorem. Fejér's Summability Theorem. *If* $f \in RI_{2\pi}$, *if* $\frac{1}{2}a_0 + a_1 \cos x + b_1 \sin x + \cdots$ *is its Fourier series, if*

(1) $S_n(x) \equiv \frac{1}{2}a_0 + a_1 \cos x + b_1 \sin x + \cdots + a_n \cos nx + b_n \sin nx$, *and if*

(2) $$\sigma_n(x) \equiv \frac{S_0(x) + S_1(x) + \cdots + S_{n-1}(x)}{n},$$

then

(3) $$\sigma_n(x) \to \frac{f(x+) + f(x-)}{2}$$

at any point x at which the function f has two one-sided limits. In particular, the Fourier series is summable to the value $f(x)$ at every point x at which $f(x) = \frac{1}{2}[f(x+) + f(x-)]$, and therefore at every point of continuity of the function f.

Proof. Defining

(4) $\phi(x, u) \equiv [f(x + u) - f(x+)] + [f(x - u) - f(x-)],$

we can write (3), § 1517, in the form

(5) $$S_n(x) - \frac{f(x+)+f(x-)}{2} = \frac{1}{2\pi}\int_0^\pi \frac{\sin(n+\tfrac{1}{2})u}{\sin\tfrac{1}{2}u}\phi(x,u)\,du.$$

Consequently,

(6) $$\sigma_n(x) - \frac{f(x+)+f(x-)}{2} = \sum_{k=0}^{n-1}\frac{1}{n}\left[S_k(x)-\frac{f(x+)+f(x-)}{2}\right]$$
$$= \frac{1}{2n\pi}\int_0^\pi \frac{1}{\sin\tfrac{1}{2}u}\left\{\sum_{k=0}^{n-1}\sin(k+\tfrac{1}{2})u\right\}\phi(x,u)\,du.$$

By means of the identity $\sum_{k=0}^{n-1}\sin(k+\tfrac{1}{2})u = \sin^2\tfrac{1}{2}nu/\sin\tfrac{1}{2}u$ (Ex. 6, § 1528), equation (6) can be written

(7) $$\sigma_n(x) - \frac{f(x+)+f(x-)}{2} = \frac{1}{2n\pi}\int_0^\pi \frac{\sin^2\tfrac{1}{2}nu}{\sin^2\tfrac{1}{2}u}\phi(x,u)\,du,$$

and the proof of the theorem reduces to showing

(8) $$\frac{1}{n}\int_0^\pi \frac{\sin^2\tfrac{1}{2}nu}{\sin^2\tfrac{1}{2}u}\phi(x,u)\,du \to 0.$$

The next step is to show that (8) can be replaced by

(9) $$\frac{1}{n}\int_0^\pi \frac{\sin^2\tfrac{1}{2}nu}{u^2}\phi(x,u)\,du \to 0.$$

The equivalence of (8) and (9) follows from the following inequality (where it is assumed that for all x, $|f(x)| < K$):

$$\left|\int_0^\pi \sin^2\tfrac{1}{2}nu\left\{\frac{1}{\sin^2\tfrac{1}{2}u}-\frac{1}{(\tfrac{1}{2}u)^2}\right\}\phi(x,u)\,du\right|$$
$$\leq \int_0^\pi \left|\csc^2\tfrac{1}{2}u - \frac{4}{u^2}\right|\cdot 4K\,du < +\infty.$$

(Give the details.)

The proof of (9) is simpler than the proof of (3), § 1520, since the first factor of the present integrand is nonnegative. Let x be a point at which $f(x+)$ and $f(x-)$ exist. Then, since

(10) $$\lim_{u\to 0+}\phi(x,u) = 0,$$

corresponding to $\epsilon > 0$ there exists a $\delta > 0$ such that $|\phi(x,\delta)| < 2\epsilon/\pi$. Then (with $v = \tfrac{1}{2}nu$)

(11) $$\left|\frac{1}{n}\int_0^\delta \frac{\sin^2\tfrac{1}{2}nu}{u^2}\phi(x,u)\,du\right| < \frac{2\epsilon}{n\pi}\int_0^\delta \frac{\sin^2\tfrac{1}{2}nu}{u^2}\,du$$
$$= \frac{\epsilon}{\pi}\int_0^{\tfrac{1}{2}n\delta}\frac{\sin^2 v}{v^2}\,dv < \frac{\epsilon}{\pi}\int_0^{+\infty}\frac{\sin^2 v}{v^2}\,dv = \frac{\epsilon}{2}.$$

(For the evaluation of the improper integral, cf. Ex. 14, § 1409.)

For the remaining part of the proof we use the following inequality (with $|f(x)| < K$):

(12) $$\left|\frac{1}{n}\int_\delta^\pi \frac{\sin^2 \frac{1}{2}nu}{u^2} \phi(x, u)\, du\right| \leq \frac{1}{n}\int_\delta^\pi \frac{1}{\delta^2} \cdot 4K\, du.$$

This quantity is less than $\frac{1}{2}\epsilon$ if $n > 8\pi K/\delta^2\epsilon$.

Corollary. *If a trigonometric Fourier series converges at a point x where $f(x+)$ and $f(x-)$ exist, it converges to $\frac{1}{2}[f(x+) + f(x-)]$.*

Proof. Any convergent series is summable to the sum of the series (cf. Ex. 15, § 717).

★★1522. UNIFORM SUMMABILITY

Inasmuch as each $\sigma_n(x)$ defined in (2), § 1521, is everywhere continuous, it follows that whenever the Fourier series of a function $f(x)$ possessing one-sided limits is uniformly summable over some interval (in the sense that the limit (3), § 1521, is a uniform limit over that interval) the function $f(x)$ must be continuous over that interval. A converse of this fact is the following:

Theorem. *If $f(x)$ is continuous on an open interval J, and if I is any closed interval contained in J, then the convergence of the Cesàro means to the function $f(x)$ is uniform throughout I:*

(1) $$\sigma_n(x) \rightrightarrows f(x), \quad x \in I.$$

If $f(x)$ is everywhere continuous, the limit (1) is uniform for all x.

Proof. We may assume without loss of generality that the length of the interval J does not exceed π. Let I_1 be a closed interval within J and containing I in its interior, and let $\eta > 0$ be so small that if $x \in I$ and $0 \leq u < \eta$, then $x \pm u \in I_1$.

Turning our attention to the details of the proof in § 1521, we see that our task is to show that the limit (9), § 1521, is uniform for $x \in I$, since the estimate relating the quantities in (8) and (9), § 1521, is independent of x. Since the inequality (12), § 1521, is also independent of x, the only remaining detail is to show that the number δ, as used in (11), § 1521, can be chosen independently of x. But this follows from the uniform continuity of $f(x)$ on the closed interval I_1: corresponding to $\epsilon > 0$ there exists $\delta > 0$ such that $x_1, x_2 \in I_1, |x_1 - x_2| < \delta$ imply $|f(x_1) - f(x_2)| < \epsilon/\pi$; we require that $\delta < \eta$ and let $x \in I$; then, since $0 \leq u \leq \delta$, $x \pm u \in I_1$.

★★1523. WEIERSTRASS'S THEOREM

Weierstrass's theorem on the uniform approximation of a continuous function by means of polynomials (Ex. 22, § 911) is a simple consequence

DENSITY OF POLYNOMIALS

of the uniform summability theorem of § 1522. We give two forms of Weierstrass's theorem, the first for *trigonometric polynomials:*

Theorem I. *If $f(x)$ is continuous on a closed interval I of length less than $2p$, and if ϵ is an arbitrary positive number, there exists a* **trigonometric polynomial**

(1) $$T(x) = \frac{1}{2}a_0 + a_1 \cos \frac{\pi x}{p} + b_1 \sin \frac{\pi x}{p} + \cdots + a_n \cos \frac{n\pi x}{p} + b_n \sin \frac{n\pi x}{p}$$

such that for all $x \in I$,

(2) $$|f(x) - T(x)| < \epsilon.$$

Proof. If the interval I is $[a, b]$, extend the domain of $f(x)$ to the closed interval $[a, a + 2p]$ by linear interpolation in $(b, a + 2p)$, the value of f at $a + 2p$ being the same as that at a: $f(a + 2p) = f(a)$. Then extend the domain to $(-\infty, +\infty)$ by periodicity with period $2p$. Then $f(x)$ is everywhere continuous, so that the Cesàro means $\sigma_n(x)$ of the Theorem, § 1522 (with the appropriate adjustment from 2π to $2p$ as indicated in Note 2, § 1514), converge uniformly to $f(x)$ everywhere, and in particular on I. Since every $\sigma_n(x)$ is a trigonometric polynomial of the type required, the proof is complete.

Theorem II. *If $f(x)$ is continuous on a closed interval I and if ϵ is an arbitrary positive number, there exists a polynomial $P(x)$ such that for all $x \in I$,*

(3) $$|f(x) - P(x)| < \epsilon.$$

Proof. We first find a trigonometric polynomial $T(x)$ according to Theorem I, such that for all $x \in I$,

$$|f(x) - T(x)| < \tfrac{1}{2}\epsilon.$$

Then, since $T(x)$ is a finite sum of sines and cosines it is represented by its Maclaurin series, which converges uniformly on any bounded set and in particular on I. We can therefore find a partial sum $P(x)$ of this Maclaurin series such that on I,

$$|T(x) - P(x)| < \tfrac{1}{2}\epsilon.$$

In combination, these two inequalities give (3).

★★1524. DENSITY OF POLYNOMIALS

In any distance space S a subset A is said to be **dense** if and only if corresponding to an arbitrary member s of S and an arbitrary positive number ϵ there exists a member a of A such that $d(a, s) < \epsilon$. For the space of real numbers this definition is equivalent to that given in (v), § 112, as the reader may easily verify.

We shall use Weierstrass's uniform approximation theorems (§ 1523) to establish two density theorems. The first is phrased for simplicity for the space $R_{2\pi}{}^2$, but can easily be reformulated for the space $R_{2p}{}^2$. The details of the proofs will be given only for the second theorem. (For Theorem I, cf. Ex. 7, § 1528.)

Theorem I. *The trigonometric polynomials*

(1) $$T(x) = \tfrac{1}{2}a_0 + a_1 \cos x + b_1 \sin x + \cdots + a_n \cos nx + b_n \sin nx$$

are dense in the space $R_{2\pi}{}^2$. That is, if $f(x)$ is any periodic function with period 2π integrable on $[-\pi, \pi]$ and if $\epsilon > 0$, there exists a trigonometric polynomial $T(x)$ of the form (1) such that

(2) $$d(f, T) = \|f - T\| = \left\{ \int_{-\pi}^{\pi} [f(x) - T(x)]^2 \, dx \right\}^{\frac{1}{2}} < \epsilon.$$

Theorem II. *The polynomials*

(3) $$P(x) = a_0 + a_1 x + \cdots + a_n x^n$$

are dense in the space R^2, for any fixed interval $[a, b]$. That is, if $f(x)$ is integrable on $[a, b]$ and if $\epsilon > 0$, there exists a polynomial $P(x)$ such that

(4) $$d(f, P) = \|f - P\| = \left\{ \int_a^b [f(x) - P(x)]^2 \, dx \right\}^{\frac{1}{2}} < \epsilon.$$

Proof of Theorem II. We start the proof by means of two lemmas, the first of which is a repetition of Exercise 19, § 1511.

Lemma 1. *The space C is dense in the space R^2.*

Lemma 2. *The polynomials are dense in the space C.*

Proof. For a given function $g(x)$ continuous on $[a, b]$ and $\epsilon > 0$, by Weierstrass's theorem (Theorem II, § 1523) there exists a polynomial $P(x)$ such that $|P(x) - g(x)| < \min(\epsilon^2/(b-a), 1)$ on $[a, b]$, and hence

$$d(P, g) = \left\{ \int_a^b |g(x) - P(x)| \cdot |g(x) - P(x)| \, dx \right\}^{\frac{1}{2}}$$
$$\leq \left\{ \int_a^b |g(x) - P(x)| \, dx \right\}^{\frac{1}{2}} < \left\{ \int_a^b [\epsilon^2/(b-a)] \, dx \right\}^{\frac{1}{2}} = \epsilon.$$

The details of the proof of Theorem II are completed with the aid of the triangle inequality (§§ 1503, 1504). For a given $f \in R^2$ and $\epsilon > 0$, we first find $g \in C$ such that $d(f, g) < \tfrac{1}{2}\epsilon$, and then P such that $d(g, P) < \tfrac{1}{2}\epsilon$. Then $d(f, P) \leq d(f, g) + d(g, P) < \tfrac{1}{2}\epsilon + \tfrac{1}{2}\epsilon = \epsilon$.

★★1525. UNIFORM CONVERGENCE

The question of uniform convergence of a (trigonometric) Fourier series does not play the important role that one might expect since, as will be shown in the following section, *any Fourier series can be integrated term by term*, whether it converges uniformly or not or, indeed, whether it con-

§ 1526] INTEGRATION OF FOURIER SERIES 573

verges or not. The theorem that follows is not easy to prove. Hints on details of the proof are given in Exercises 9-11, § 1528.

Theorem. Jordan Uniform Convergence Theorem. *Assume that $f(x)$ is periodic with period 2π, and of bounded variation on $[-\pi, \pi]$, and that $f(x)$ is continuous on an open interval J. If I is any closed interval contained in J, then the convergence of the Fourier series of $f(x)$ is uniform throughout I:*

$$(1) \qquad S_n(x) \rightrightarrows f(x), \quad x \in I.$$

If $f(x)$ is everywhere continuous (as well as being periodic with period 2π and of bounded variation on $[-\pi, \pi]$), then the limit (1) is uniform for all x.

★★1526. INTEGRATION OF FOURIER SERIES

The question of term by term integration of a Fourier series has a simple answer: *It is always legitimate.* The precise formulation is given in Theorem V, below. We start with three preliminary theorems dealing with integration, leaving the proofs to the student in Exercises 12-14, § 1528.

Theorem I. *If $g(x)$ is integrable on a closed interval $[a, b]$ and if the function $G(x)$ is defined on $[a, b]$ by the formula:*

$$(1) \qquad G(x) = \int_a^x g(t)\, dt,$$

then $G(x)$ is continuous and of bounded variation on $[a, b]$.

Theorem II. *If $g(x)$ and $G(x)$ are subject to the conditions of Theorem I, and if $f(x)$ is integrable on $[a, b]$, then f is Riemann-Stieltjes integrable with respect to $G(x)$, and*

$$(2) \qquad \int_a^b f(x)\, dG(x) = \int_a^b f(x)\, g(x)\, dx.$$

Theorem III. *If $f(x)$ and $g(x)$ are integrable on $[a, b]$, and if $F(x)$ and $G(x)$ have the form*

$$(3) \qquad F(x) = \int_a^x f(t)\, dt + C_1, \quad G(x) = \int_a^x g(t)\, dt + C_2,$$

then the following integration-by-parts formula holds:

$$(4) \quad \int_a^b F(x)\, g(x)\, dx + \int_a^b f(x)\, G(x)\, dx$$
$$= \int_a^b F(x)\, dG(x) + \int_a^b G(x)\, dF(x) = F(b)\, G(b) - F(a)\, G(a).$$

Theorem IV. *If $f \in RI_{2\pi}$, with Fourier series*

$$(5) \qquad f \sim \tfrac{1}{2}a_0 + a_1 \cos x + b_1 \sin x + a_2 \cos 2x + b_2 \sin 2x + \cdots,$$

and if

$$F(x) \equiv \int_0^x [f(t) - \tfrac{1}{2}a_0]\, dt, \tag{6}$$

for all real x, then $F(x)$ is continuous and periodic with period 2π, and of bounded variation on $[-\pi, \pi]$, and

$$\begin{cases} \int_{-\pi}^{\pi} f(x) \cos nx\, dx = n \int_{-\pi}^{\pi} F(x) \sin nx\, dx, \\ \int_{-\pi}^{\pi} f(x) \sin nx\, dx = -n \int_{-\pi}^{\pi} F(x) \cos nx\, dx, \end{cases} \tag{7}$$

$n = 1, 2, \cdots$.

Proof. The periodicity property is true by:

$$F(a + 2\pi) - F(a) = \int_a^{a+2\pi} [f(t) - \tfrac{1}{2}a_0]\, dt = \int_{-\pi}^{\pi} f(t)\, dt - a_0\pi = 0.$$

The remaining conclusions follow, by integration by parts, from Theorems I, II, III; for example:

$$\int_{-\pi}^{\pi} f(x) \cos nx\, dx = \int_0^{2\pi} \cos nx\, [f(x) - \tfrac{1}{2}a_0]\, dx = \int_0^{2\pi} \cos nx\, dF(x)$$
$$= \Big[\cos nx\, F(x)\Big]_0^{2\pi} + n \int_0^{2\pi} F(x) \sin nx\, dx.$$

Theorem V. *Any Fourier series can be integrated term by term. That is, if $f \in RI_{2\pi}$, with Fourier series (5), and if a and b are any two real numbers,*

$$\int_a^b f(x)\, dx = \int_a^b \tfrac{1}{2}a_0\, dx + \int_a^b a_1 \cos x\, dx + \int_a^b b_1 \sin x\, dx + \cdots. \tag{8}$$

Proof. We assume without loss of generality that $a = 0$, replace b by x, rearrange the first term on the right of (8), and (with the notation of (6)) rewrite the equation to be established in the form

$$F(x) = \sum_{n=1}^{+\infty} \int_0^x [a_n \cos nt + b_n \sin nt]\, dt \tag{9}$$
$$= \sum_{n=1}^{+\infty} \frac{a_n \sin nx + b_n(1 - \cos nx)}{n}.$$

By Theorem IV, and Jordan's convergence theorem, $F(x)$ can be expanded in a Fourier series which converges everywhere to $F(x)$:

$$F(x) = \tfrac{1}{2}A_0 + A_1 \cos x + B_1 \sin x + A_2 \cos 2x + B_2 \sin 2x + \cdots, \tag{10}$$

where, by (7), $A_n = -b_n/n$, $B_n = a_n/n$, $n = 1, 2, \cdots$. Therefore (10) can be rewritten in the form

$$F(x) = \tfrac{1}{2}A_0 + \sum_{n=1}^{+\infty} \frac{a_n \sin nx - b_n \cos nx}{n}. \tag{11}$$

Substitution of $x = 0$ in (11), gives the value $\tfrac{1}{2}A_0 = \sum_{n=1}^{+\infty} \frac{b_n}{n}$, and the desired result (9). This completes the proof.

★★1527. THE GIBBS PHENOMENON

In 1899 J. W. Gibbs (1839-1903, U.S.A.) reported in a letter to *Nature* a remarkable property of the convergence of a trigonometric Fourier series in the neighborhood of a jump discontinuity of the given function. He observed that near such a point the peaks of the approximating graphs on the high side of the discontinuity are higher than one would normally expect from a superficial examination of the graph of the function, and that the troughs on the low side are unexpectedly deep. Furthermore, he showed that these maximum and minimum points approach points respectively above and below the one-sided limit points of the graph of the function, and separated from those points by a fixed fraction (approximately 9 per cent) of the total jump of the function. (See Fig. 1505.)

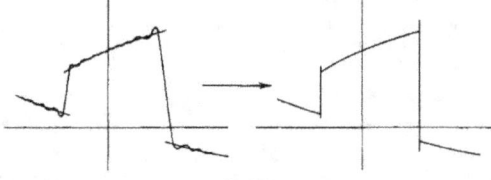

FIG. 1505

This peculiarity—in sharp distinction to the *uniform* convergence that holds throughout an interval strictly inside an interval of continuity—is known as the **Gibbs phenomenon**. To demonstrate its validity for any function of $BV_{2\pi}$ that is continuous throughout a deleted neighborhood of a point of jump discontinuity, it is sufficient to prove it for any *particular* example, since a suitable translation and change of scale with such a function can be used to "erase" the discontinuity and produce uniform convergence (cf. Ex. 15, § 1528).

A convenient special function for establishing the Gibbs phenomenon is the odd function of period 2π equal to 1 for $0 < x < \pi$ (cf. Ex. 1, § 1511). The student is asked in Exercises 16 and 17, § 1528, to show the following two facts regarding the approximating sums (cf. Fig. 1506)

$$S_n(x) = \frac{4}{\pi}\left[\frac{\sin x}{1} + \frac{\sin 3x}{3} + \cdots + \frac{\sin (2n-1)x}{2n-1}\right];$$

(i) The absolute maximum y_n of $S_n(x)$ for $0 \leq x \leq \frac{1}{2}\pi$ is given by $x = \pi/2n$;

(ii) $$\lim_{n \to +\infty} y_n = \frac{2}{\pi} \int_0^\pi \frac{\sin x}{x}\, dx.$$

576 FOURIER SERIES; ORTHOGONAL FUNCTIONS [§ 1528

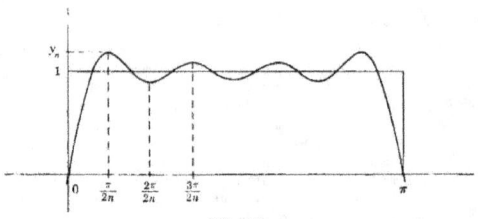

FIG. 1506

The value of the limit in (ii) is approximately equal to $\frac{2}{\pi}(1.852)$† $= 1.179$, so that its excess above 1 is 0.179, or about 9 per cent of the total jump of 2.

For further discussion, and graphs, see the Carslaw reference at the end of the chapter.

★★1528. EXERCISES

★★**1.** Prove the identity (3), § 1516, with the aid of the identity $2 \cos A \sin B = \sin (A + B) - \sin (A - B)$.

★★**2.** Prove (2), § 1517, directly by means of (3), § 1516.

★★**3.** Prove that the spaces BV (for an interval $[a, b]$) and $BV_{2\pi}$ are linear spaces.

★★**4.** Prove that the space described in (iii), § 1519, is a subspace of BV, but is not a linear space.

★★**5.** Prove that the assumptions (2), § 1520, can be made without loss of generality, as claimed.

★★**6.** Prove the identity given between (6) and (7), § 1521.

★★**7.** Prove Theorem I, § 1524.

★★**8.** Prove that the space $BV_{2\pi}$, subject to the condition $f(x) = \frac{1}{2}[f(x+) + f(x-)]$ at every point x, as a subspace of $R_{2\pi}{}^{2}$, is a metric space.

★★**9.** Prove the **Generalized Riemann-Lebesgue Theorem**: If $f \in RI_{2\pi}$, if $G(x)$ is continuously differentiable for $0 \leq x \leq \pi$, if $0 \leq a \leq b \leq \pi$, and if $\phi(x, u) = f(x + u) + f(x - u) - 2f(x)$, then

$$\int_a^b \sin \lambda u G(u) \phi(x, u) \, du \rightrightarrows 0 \quad \text{as} \quad \lambda \to +\infty,$$

uniformly for all x. Hints: Use integration by parts to prove that $\int_a^b \sin \lambda u G(u) f(x + u) \, du \rightrightarrows 0$ in case $f(x)$ is a broken line function, continuous and of period 2π. Then prove this same result if $f \in RI_{2\pi}$. Finally, establish similar results with factors $f(x - u)$ and $-2f(x)$.

† For the value of the integral see the book by Jahnke and Emde, p. 8, in the references at the end of the chapter.

§ 1529]   APPLICATIONS OF FOURIER SERIES   577

10. If I and J are open intervals such that $\bar{I} \subset J$ and $\bar{J} \subset (a, b)$ where $b < a + 2\pi$, and if f is continuous and monotonically increasing in (a, b), prove that corresponding to $\epsilon > 0$ there exists $\delta > 0$ such that for all $x \in \bar{I}$ and for all λ:

$$\left| \int_0^\delta \frac{\sin \lambda u}{u} [f(x+u) - f(x)] \, du \right| < \epsilon, \quad \left| \int_0^\delta \frac{\sin \lambda u}{u} [f(x-u) - f(x)] \, du \right| < \epsilon.$$

Hint: Use the Bonnet form for the Second Mean Value Theorem for Riemann integrals (Ex. 29, § 518).

11. Prove the Jordan Uniform Convergence Theorem (§ 1525). *Hints:* First use the Generalized Riemann-Lebesgue Theorem (Ex. 9) to reduce the proof to showing

(1) $$\int_0^\pi \frac{\sin \lambda u}{u} \phi(x, u) \, du \rightrightarrows 0,$$

for $x \in I$. Next show that if $f \in BV_{2\pi}$, and if a is any number, then $f(x)$ can be written as $g(x) - h(x)$, where g and h are monotonically increasing in $[a, a + 2\pi)$, and continuous there wherever f is, and g and $h \in BV_{2\pi}$ (cf. Ex. 41, § 518). Finally, use Exs. 9 and 10.

12. Prove Theorem I, § 1526.

13. Prove Theorem II, § 1526. *Hints:* First show that it can be assumed without loss of generality that $f(x) \geqq 0$. Then squeeze $g(x)$ between stepfunctions.

14. Prove Theorem III, § 1526.

15. Prove that in establishing the Gibbs phenomenon (§ 1527) it can be assumed without loss of generality that the function under consideration is that of Exercise 1, § 1511. Illustrate with an example.

16. Establish (i), § 1527. *Hints:* Show that $S_n'(x) = 2 \sin 2nx/\pi \sin x$, sketch the graph of $S_n'(x)$, and observe that $S_n(x) = \int_0^x S_n'(t) \, dt$.

17. Establish (ii), § 1527. *Hint:* Write

$$y_n = \frac{2}{\pi} \left[\frac{\sin \frac{\pi}{2n}}{\frac{\pi}{2n}} + \frac{\sin \frac{3\pi}{2n}}{\frac{3\pi}{2n}} + \cdots + \frac{\sin \frac{(2n-1)\pi}{2n}}{\frac{(2n-1)\pi}{2n}} \right] \frac{\pi}{n}.$$

(Cf. Exs. 19, 24, § 503.)

18. By integrating the Fourier sine series for the function $\pi/4$ (cf. Ex. 1, § 1511) an odd number of times, evaluate $\zeta(2n)$, for a few values of n. (Cf. Exs. 2-4, 20, 22, § 1515.)

*1529. APPLICATIONS OF FOURIER SERIES. THE VIBRATING STRING

We discuss in this section one of the more basic applied problems for which the techniques of Fourier analysis are especially appropriate. This is the problem of the *vibrating string*, of considerable significance in the history of Fourier series. (For a more complete discussion, see the

578 FOURIER SERIES; ORTHOGONAL FUNCTIONS [§ 1529

Assume that a stretched string of uniform composition is vibrating with fixed endpoints and in a fixed plane. As a matter of convenience, we shall take units and a coordinate system such that the string is vibrating in the xy-plane, with its endpoints fixed at the two points $(0, 0)$ and $(\pi, 0)$ on the x-axis. The motion of the string is determined when the displacement y of the point of the string whose abscissa is x is known for any instant of time t:

(1) $$y = u(x, t).$$

The general problem of the vibrating string is to determine a function (1) that satisfies certain prescribed conditions or restrictions.

Our first assumption is that the function (1) satisfies a certain partial differential equation, whose derivation is based on physical considerations (and will not be given here),† known as the **wave equation**:

(2) $$\frac{\partial^2 u}{\partial t^2} = a^2 \frac{\partial^2 u}{\partial x^2},$$

where a is a positive constant depending on the string.

We shall now give detailed consideration to a particularization of the vibrating string problem, known as the *plucked string* problem. For this we assume that the string is initially deformed to fit a prescribed graph:

(3) $$y = f(x),$$

and then instantaneously released. Taking $t = 0$ to be the instant that the string is released from rest, and expressing the fact that the endpoints remain fixed, we have the following four conditions:

(4) $\quad u(0, t) = 0,$ ⎫
(5) $\quad u(\pi, t) = 0,$ ⎬ Boundary conditions

(6) $\quad u(x, 0) = f(x),$ ⎫
(7) $\quad u_2(x, 0) = 0,$ ⎬ Initial conditions

for $0 \leq x \leq \pi$ and $t \geq 0$, and where $u_2(x, t)$ means $\partial u/\partial t$. We now seek a solution of the differential equation (2), subject to the boundary conditions (4) and (5) and the initial conditions (6) and (7).

A useful technique, often employed in the solving of partial differential equations, is to find solutions of the differential equation that have a particularly simple form, namely that of the product of factors each of which is a function of precisely one of the independent variables. In the present circumstance we are looking for a function of the form

(8) $$u = X(x) \, T(t),$$

where $X(x)$ is a function of x alone, and $T(t)$ is a function of t alone, that satisfies (2). Substitution gives the equation

---

† See the book by Widder, p. 344, listed in the references at the end of the chapter, for a derivation.

APPLICATIONS OF FOURIER SERIES

(9) $$T''X = a^2 TX''.$$

Proceeding formally, and not concerning ourselves with minor obstacles like dividing by 0 (cf. Ex. 3, § 1531), we have the identity

(10) $$\frac{X''}{X} = \frac{T''}{a^2 T}$$

Since the left-hand member of (10) is independent of t, and the right-hand member is independent of x, they must both be equal to the same constant, which might conceivably be positive, zero, or negative. The student is asked in Exercise 4, § 1531, to show that *positive* values for this constant have no significance for this particular problem.

Let us see now why the value *zero* must be rejected. The main idea ahead of us, after we have found solutions of (2) of the factored form XT, is to combine such solutions in a fashion that gives a new solution that satisfies the boundary and initial conditions as well. The first step in this direction is to *impose the boundary conditions* (4) *and* (5) *and the initial condition* (7) *on the solutions in factored form*. Now, if the left-hand member of (10) were identically equal to zero, the function X would have the form $X = ax + b$. But then, the boundary conditions (4) and (5) would require a and b to be zero, and only the trivial function identically zero would result.

We therefore set the two members of (10) equal to a *negative* constant, and obtain the two equations:

(11) $$X''(x) + \lambda^2 X(x) = 0, \quad T''(t) + \lambda^2 a^2 T(t) = 0,$$

where $\lambda > 0$.

Solving (11) (by elementary differential equations theory), and forming the product, we have the following *solutions of* (2) *in factored form*:

(12) $$u = (A \cos \lambda x + B \sin \lambda x)(C \cos \lambda at + D \sin \lambda at).$$

If these solutions are to be nontrivial, the boundary conditions (4) and (5) require that $A = 0$ and $\lambda = 1, 2, 3, \cdots$. The initial condition (7) requires that $D = 0$. Our conclusion is that the nontrivial solutions of (2) in factored form that satisfy (4), (5), and (7) are multiples of

(13) $$\sin nx \cos nat,$$

where n is a positive integer.

We are now ready to put things together. Because of the form of the differential equation (2) (it is "linear") and that of the conditions (4), (5), and (7) it follows readily (check the details) that any finite linear combination of functions of the form (13) is again a solution of (2), (4), (5), and (7). It is not too much to hope that the same may be said for a suitable *limit* of such finite linear combinations, in the form of an *infinite series*:

(14) $$u(x, t) = \sum_{n}^{+\infty} b_n \sin nx \cos nat.$$

580 FOURIER SERIES; ORTHOGONAL FUNCTIONS [§ 1530

Assuming that this is possible, we are led to the following form for the remaining initial condition (6):

(15) $$u(x, 0) = f(x) = \sum_{n=1}^{+\infty} b_n \sin nx.$$

In other words, a solution (14) is obtained, at least formally, by expanding the prescribed function $f(x)$ in a sine series. If the function $f(x)$ is sufficiently well-behaved, the legitimacy of such operations as term by term differentiation is not difficult to supply. For uniqueness of the solution, see Exercises 5, 6, § 1531.

Examples for detailed study are given in Exercises 1, 2, § 1531. The student is asked in Exercise 7, § 1531, to work out similar details for the problem of the *struck string*.

NOTE. The values of λ, $\lambda = \lambda_n = n = 1, 2, 3, \cdots$, found in the solution of the differential equation $X'' + \lambda^2 X = 0$, subject to the boundary conditions of the problem, are called **characteristic values** or **eigenvalues**. The corresponding functions $\sin nx$, $n = 1, 2, 3, \cdots$, are called **characteristic functions** or **eigenfunctions**. The set of all characteristic values is called the **spectrum**. The characteristic values of λ give the *resonant frequencies:* if the string produces musical tones, the value $\lambda = 1$ corresponds to the *fundamental*, $\lambda = 2$ to the *first octave*, $\lambda = 3$ the *fifth* above the octave, and higher values of λ to higher *harmonics* or *overtones*.

★1530. A HEAT CONDUCTION PROBLEM

Consider a thin rod located on the x-axis for $0 \leq x \leq \pi$, with an initial temperature distribution

(1) $$\text{initial temperature} = f(x)$$

along the rod. Furthermore, assume that the rod is insulated except at the endpoints, which are maintained at a fixed temperature which can be assumed to be 0 (for a suitable temperature scale). If the temperature at the point x at the time t is denoted

(2) $$\text{temperature} = u(x, t),$$

the physical theory of heat conduction demands that this function satisfy the **heat equation**:

(3) $$\frac{\partial u}{\partial t} = a^2 \frac{\partial^2 u}{\partial x^2},$$

where $a > 0$.

The problem as stated calls for a solution of the partial differential equation (3) subject to the boundary and initial conditions:

(4) $u(0, t) = 0,$
(5) $u(\pi, t) = 0,$ } **Boundary conditions**

(6) $\quad u(x, 0) = f(x)$, **Initial condition**

for $0 \leq x \leq \pi$ and $t \geq 0$.

Adopting the method of the preceding section, with $u = X(x)\, T(t)$, we find

(7) $$\frac{X''}{X} = \frac{T'}{a^2 T} = -\lambda^2,$$

where $\lambda > 0$, with the result that any nontrivial solutions of (3) in factored form that satisfy (4) and (5) are multiples of

(8) $$\sin nx \, e^{-n^2 a^2 t},$$

where n is a positive integer.

Finally, if $f(x)$ is a sufficiently well-behaved function, a solution of the system (3), (4), (5), (6) exists having the form

(9) $$u(x, t) = \sum_{n=1}^{+\infty} b_n \sin nx \, e^{-n^2 a^2 t},$$

where the coefficients b_n are those of the sine series expansion of $f(x)$:

(10) $$u(x, 0) = f(x) = \sum_{n=1}^{+\infty} b_n \sin nx.$$

For uniqueness of the solution, and examples for detailed study, see Exercises 9–11, § 1531.

NOTE. The technique of solutions in factored form used in this and the preceding section is not applicable to all applied vibrational and heat-flow problems, even fairly simple ones. Other methods, such as integral equations, will be found in the references listed at the end of the chapter. For further discussion of applications, see in particular the books by Churchill, Courant and Hilbert, Hildebrand, and Jackson.

★1531. EXERCISES

★1. Solve the problem of the plucked string if the initial condition (6), § 1529, is given by the function

$$f(x) = \begin{cases} 2px/\pi, & 0 \leq x \leq \tfrac{1}{2}\pi, \\ 2p(\pi - x)/\pi, & \tfrac{1}{2}\pi \leq x \leq \pi. \end{cases}$$

★2. Discuss the vibrations of a plucked string if the initial shape is that of a single arch of a sine curve: $f(x) = p \sin x$. Show that the string retains the same general shape. Complete the discussion by considering functions of the form $f(x) = p \sin kx$, and their linear combinations.

★3. Prove that any solution of (2), § 1529, that has the form (8), § 1529, and is not identically zero, must be such that $X(x)$ and $T(t)$ satisfy equations of the form $X'' = kX$ and $T'' = a^2 kT$ simultaneously. *Hint:* If $X(x)$ is not identically zero, let x_1 and x_2 be such that $X''(x_1) = k_1 X(x_1)$ and $X''(x_2) = k_2 X(x_2)$, where $k_1 \neq k_2$ and $X(x_1) X(x_2) \neq 0$. Then $T(t)$ must vanish identically.

★4. Prove that the constant of (10), §1529, cannot be positive if the boundary conditions of the problem are satisfied.

★5. Prove uniqueness of the solution of the plucked string problem of §1529, assuming continuous existence of the second order partial derivatives involved, as follows: First assume without loss of generality that $f(x) = 0$, identically. For each t expand the solution in a sine series: $u(x, t) = \sum \phi_n(t) \sin nx$, where

(1) $$\phi_n(t) = \frac{2}{\pi} \int_0^\pi u(x, t) \sin nx \, dx.$$

Differentiate (1) twice using Leibnitz's rule (§1226), then with the aid of (2), (4), (5), §1529, obtain the relation $\phi_n''(t) = -n^2 a^2 \phi_n(t)$, whence $\phi_n(t) = A_n \cos nat + B_n \sin nat$. Finally, from (1), $\phi_n(0) = \phi_n'(0) = 0$, and hence $A_n = B_n = 0$.

★6. Show that the change of variables $r = x + at$, $s = x - at$ transforms the wave equation (2), §1529, to $\partial^2 u / \partial r \partial s = 0$. Conclude that any solution must have the form (cf. Ex. 22, §1216):

(2) $$u(x, t) = \phi(x + at) + \psi(x - at).$$

Show that the initial conditions (6) and (7), §1529, with $f(x) = 0$, become $\phi(x) + \psi(x) = \phi'(x) - \psi'(x) = 0$. Use this to construct a new proof of the uniqueness theorem of Exercise 5.

★7. The problem of the **struck string** is that of solving the differential equation (2), §1529, subject to the boundary conditions (4) and (5), §1529, and the two initial conditions

(3) $$u(x, 0) = 0,$$
(4) $$u_t(x, 0) = f(x),$$

for $0 \leq x \leq \pi$ and $t \geq 0$. Assuming such legitimacy as term by term differentiation, obtain the solution

$$u = \sum_{n=1}^{+\infty} c_n \sin nx \sin nat, \quad \text{where} \quad c_n = \frac{2}{na\pi} \int_0^\pi f(x) \sin nx \, dx.$$

Show that this solution is unique (cf. Exs. 5, 6).

★8. Solve the problem of the struck string (Ex. 7) for the function $f(x)$ that is equal to the positive constant p when x is in the interval $[\frac{1}{2}\pi - h, \frac{1}{2}\pi + h]$ and identically zero otherwise.

★9. Solve the heat conduction problem subject to (3), (4), (5), (6), §1530, for the function $f(x)$ of Exercise 1.

★10. Solve the heat conduction problem subject to (3), (4), (5), (6), §1530, for the function

$$f(x) = A \sin 2x + B \sin 7x.$$

★11. State and prove a uniqueness theorem for the solution of the heat conduction problem as described in §1530. (Cf. Ex. 5.)

★12. The problem of heat conduction in a *rod with insulated ends* is that of solving the differential equation (3), §1530, subject to the initial condition (6), §1530, and the two boundary conditions

(5) $$u_1(0, t) = 0,$$
(6) $$u_1(\pi, t) = 0.$$

for $t \geq 0$. Assuming legitimacy of the operations involved, find the general solution, and show that it is unique (cf. Ex. 11).

★**13.** Solve the heat conduction problem of Exercise 12 for the case of the function $f(x)$ given in Exercise 1.

★**14.** Discuss the solution of the heat conduction problem of a rod whose endpoints are maintained at constant, but distinct, temperatures as follows: Show that a solution of the differential equation (3), § 1530, subject to the conditions

(7) $\qquad u(0, t) = \alpha, \quad u(\pi, t) = \beta, \quad u(x, 0) = f(x)$

can be obtained by adding a solution $\sigma(x, t)$ of the **steady-state problem** (3), § 1530, with conditions

(8) $\qquad u(0, t) = \alpha, \quad u(\pi, t) = \beta, \quad u(x, 0) = L(x) \equiv \alpha + (\beta - \alpha)x/\pi$

and a solution $\tau(x, t)$ of the **transient problem**, (3), § 1530, with conditions

(9) $\qquad u(0, t) = 0, \quad u(\pi, t) = 0, \quad u(x, 0) = f(x) - L(x)$.

Show that these solutions are unique, and that $\sigma(x, t) = L(x)$. (Cf. Ex. 11.)

★**15.** Use Exercise 14 to solve the problem of a rod whose temperature distribution is initially that of the steady state with end temperatures of 10° and 20°, if these end temperatures are suddenly changed to 50° and 80°, respectively, and then maintained at these temperatures.

★★**1532. BASES. CLOSEDNESS. COMPLETENESS**

If

(1) $\qquad \phi_1, \phi_2, \cdots$

is an orthonormal sequence of functions in R^2, for an interval $[a, b]$, and if f is a given function integrable on $[a, b]$, the question of approximation of f (in the sense of least squares) by a linear combination of certain terms of the sequence $\{\phi_n\}$ was introduced in § 1505. It was found that for any finite subfamily the best approximation was given by the Fourier coefficients $\alpha_1, \alpha_2, \cdots$ as defined by (4), § 1505. Since the linear combination $\alpha_1\phi_1 + \cdots + \alpha_n\phi_n$ is one of the competing linear combinations $\beta_1\phi_1 + \cdots + \beta_m\phi_m$ for any $m > n$ (let $\beta_i = 0$ for $i > n$), it follows that if the number of terms of $\{\phi_n\}$ appearing in a linear combination approximating f is increased, the approximation is either unchanged or improved. In other words,

(2) $\qquad \|f - (\alpha_1\phi_1 + \cdots + \alpha_n\phi_n)\| \downarrow .$

This fact is also a direct consequence of the expansion (2) of the quantity (1), § 1505, when the β_i are the Fourier coefficients α_i, $i = 1, 2, \cdots$:

(3) $\qquad \|f - \sum_{i=1}^{n} \alpha_i\phi_i\|^2 = \|f\|^2 - \sum_{i=1}^{n} \alpha_i^2.$

Therefore the limit of the expression in (2) always exists and is nonnegative. The important question now is whether this limit is equal to zero. If it is equal to zero we say that the generalized Fourier series

584 FOURIER SERIES; ORTHOGONAL FUNCTIONS [§ 1532

$$\text{(4)} \qquad \sum_{n=1}^{+\infty} \alpha_n \phi_n = \alpha_1 \phi_1 + \alpha_2 \phi_2 + \cdots$$

of f with respect to the orthonormal sequence $\{\phi_n\}$, **converges in the mean** to the function f, and write

$$\text{(5)} \qquad f = \underset{n \to +\infty}{\text{l.i.m.}} \sum_{i=1}^{n} \alpha_i \phi_i$$

to mean

$$\text{(6)} \qquad \lim_{n \to +\infty} \left\| f - \sum_{i=1}^{n} \alpha_i \phi_i \right\| = 0$$

(the letters l.i.m. representing the phrase **limit in the mean**).

We shall see that convergence in the mean of the generalized Fourier series of any integrable function depends only on the orthonormal sequence (and not on the function itself). To put it crudely, convergence in the mean obtains if and only if there are "sufficiently many" terms in the orthonormal sequence, in a sense to be set forth presently. Convergence in the mean, for a function f, signifies that linear combinations of the sequence $\phi_1, \phi_2, \cdots$ can be found arbitrarily close to f according to *distance* in the space R^2. This type of convergence should not be confused with **point-wise convergence**, which means that for each point x of the interval $[a, b]$

$$\text{(7)} \qquad f(x) = \lim_{n \to +\infty} \sum_{i=1}^{n} \alpha_i \phi_i(x).$$

Point-wise convergence depends not only on the sequence $\{\phi_n\}$, but on the function f (cf. §§ 1510, 1520).

To facilitate discussion we introduce two important concepts:

Definition. *An orthonormal sequence $\{\phi_n\}$ in the space R^2, for an interval $[a, b]$, is **closed** in R^2 or, equivalently, is **a basis** for R^2 if and only if every f of R^2 is the limit in the mean of its generalized Fourier series with respect to $\{\phi_n\}$, in the sense of (5) and (6). An orthonormal sequence $\{\phi_n\}$ is **complete** in R^2 if and only if it is maximal in the sense that the only orthonormal family Ψ containing $\{\phi_n\}$ is $\{\phi_n\}$ itself.*[†]

For closed and complete orthonormal sequences we have the following three theorems:

Theorem I. *Each of the following three conditions on an orthonormal sequence $\{\phi_n\}$ of R^2, for an interval $[a, b]$, is equivalent to each of the other two:*

[†] Variation in the use of terms for these two concepts is great. Some German writers (cf. the Courant and Hilbert reference at the end of the chapter) interchange the meanings of the two words *closed* and *complete*. Some authors retain the words *basis* and *complete*, but avoid the word *closed*. The two words *fundamental* and *total* are sometimes used to mean *closed* and *complete*, respectively, as defined here.

BASES. CLOSEDNESS. COMPLETENESS

(i) $\{\phi_n\}$ is closed (equivalently, a basis);
(ii) the finite linear combinations of the members of $\{\phi_n\}$ are dense in R^2;
(iii) Parseval's equation holds for every $f \in R^2$:

(8) $$\sum_{i=1}^{+\infty} \alpha_n^2 = \|f\|^2.$$

Proof. We give the proof by establishing a cyclic set of implications.

(i) implies (ii): Assume equation (6) holds. Then for any $\epsilon > 0$ there exists a number n such that $\|f - \sum_{i=1}^{n} \alpha_i \phi_i\| < \epsilon$, and $\sum_{i=1}^{n} \alpha_i \phi_i$ is a linear combination of $\phi_1, \phi_2, \cdots, \phi_n$.

(ii) implies (iii): For a given $\epsilon > 0$ assume that there are numbers $\beta_1, \beta_2, \cdots, \beta_N$ such that $\|f - \sum_{i=1}^{N} \beta_i \phi_i\| < \sqrt{\epsilon}$. Then, by the minimizing property of the Fourier coefficients (§ 1505), $\|f - \sum_{i=1}^{N} \alpha_i \phi_i\| < \sqrt{\epsilon}$, and by (3), $0 \leq \|f\|^2 - \sum_{i=1}^{N} \alpha_i^2 < \epsilon$. Therefore $n > N$ implies $\left|\|f\|^2 - \sum_{i=1}^{n} \alpha_i^2\right| < \epsilon$.

(iii) implies (i): If the right-hand member of (3) $\to 0$, so does the left-hand member.

Theorem II. *Each of the following three conditions on an orthogonal sequence $\{\phi_n\}$ of R^2, for an interval $[a, b]$, is equivalent to each of the other two:*

(i) $\{\phi_n\}$ is complete;
(ii) a member f of R^2 is orthogonal to every ϕ_n if and only if it has zero norm, $\|f\| = 0$;
(iii) two members, f and g, of R^2 have identical Fourier coefficients with respect to $\{\phi_n\}$ if and only if they are a zero distance apart, $\|f - g\| = 0$.

Proof. (i) implies (ii): If $\|f\| = 0$, by the Schwarz inequality $|(f, \phi_n)| \leq \|f\| \cdot 1 = 0$, and every Fourier coefficient of f vanishes. Conversely, if every $\alpha_n = (f, \phi_n) = 0$, f is orthogonal to every ϕ_n. If $\|f\| > 0$, let $\psi = f/\|f\|$. Then the family obtained by adjoining ψ to $\{\phi_n\}$ is an orthonormal family containing $\{\phi_n\}$ but not equal to it. (Contradiction.)

(ii) implies (iii): If $\|f - g\| = 0$, then by (ii) $f - g$ is orthogonal to every ϕ_n, so that $(f - g, \phi_n) = (f, \phi_n) - (g, \phi_n) = 0$, so that f and g have identical Fourier coefficients. Conversely, if $(f, \phi_n) = (g, \phi_n)$ for every n, $(f - g, \phi_n) = 0$ and $f - g$ is orthogonal to every ϕ_n, so that by (ii) $\|f - g\| = 0$.

(iii) implies (i): If $\{\phi_n\}$ is not complete there exists a larger orthonormal system containing $\{\phi_n\}$, and therefore a ψ such that $\|\psi\| = 1$ and ψ is orthogonal to every ϕ_n. Then both ψ and the function 0 (identically equal to 0) have identical Fourier coefficients, but $\|\psi - 0\| \neq 0$. (Contradiction.)

Theorem III. *Every basis is complete.*

Proof. If a basis $\{\phi_n\}$ were not complete, there would exist, by Theorem II, an integrable function f with positive norm orthogonal to every ϕ_n. Therefore, since every Fourier coefficient is zero, while $\|f\| > 0$, Parseval's equation could not hold for f, in contradiction to Theorem I.

NOTE. In the space R^2 not every complete orthonormal sequence is a basis. This fact is not an elementary one to demonstrate. For those familiar with Lebesgue theory a proof is given in the author's paper, "Completeness and Parseval's Theorem," in the *American Mathematical Monthly*, Vol. 65, No. 5 (May, 1958), pp. 343-345. When the Riemann integral is replaced by the Lebesgue integral an orthonormal sequence is complete *if and only if* it is a basis, and all six properties of Theorems I and II are equivalent. For further discussion see the second Halmos reference and that of Courant and Hilbert at the end of the chapter.

As a consequence of Theorem I, § 1524, on the density of the trigonometric polynomials in the space $R_{2\pi}{}^2$, and of the Theorems of this section, we have:

Theorem IV. *In the space $R_{2\pi}{}^2$, the orthonormal sequence*

(9) $$\frac{1}{\sqrt{2\pi}}, \frac{\cos x}{\sqrt{\pi}}, \frac{\sin x}{\sqrt{\pi}}, \frac{\cos 2x}{\sqrt{\pi}}, \frac{\sin 2x}{\sqrt{\pi}}, \ldots$$

(Example 2, § 1504) *is a basis and therefore satisfies all six properties of Theorems I and II.*

For algebraic polynomials we have the corresponding result, for the space R^2, for any interval $[a, b]$ (give the proof in Ex. 1, § 1538):

Theorem V. *In the space R^2 any orthonormal sequence of polynomials containing, for every $n = 0, 1, 2, \cdots$, a polynomial of degree n,† is a basis and therefore satisfies all six properties of Theorems I and II.*

★★1533. LINEAR DEPENDENCE AND INDEPENDENCE

In a linear function space a concept of basic importance is given in the following definition:

Definition. *The functions of a set $f_1, f_2, \cdots, f_n$ are **linearly dependent** on $[a, b]$ if and only if there exist constants $\alpha_1, \alpha_2, \cdots, \alpha_n$, not all zero, such that the equation*

(1) $$\alpha_1 f_1(x) + \alpha_2 f_2(x) + \cdots + \alpha_n f_n(x) = 0$$

holds identically on $[a, b]$. If no such constants $\alpha_1, \alpha_2, \cdots, \alpha_n$ exist, that is,

† The convention for degrees of constant polynomials is that the degree is 0 if the constant is nonzero, and $-\infty$ if the constant is 0.

§ 1534]   THE GRAM-SCHMIDT PROCESS   587

if (1) holds (*identically in x*) *only for constants all of which are zero, then the functions* $f_1, f_2, \cdots, f_n$ *are* **linearly independent** *on* $[a, b]$.

NOTE. A necessary and sufficient condition for a set of functions to be linearly dependent is that at least one of them is a linear combination of the others. A necessary and sufficient condition for a set of functions to be linearly independent is that any particular linear combination of them has a unique representation as a linear combination. A single function is linearly dependent if and only if it is identically zero. Two functions are linearly dependent if and only if they are proportional (the constant of proportionality possibly being zero). Any subset of a set of linearly independent functions are linearly independent. (Cf. Ex. 4, § 1538.)

A relationship between linear independence and orthogonality is:

Theorem. *Any finite orthogonal set of functions each of which has nonzero norm are linearly independent. Any finite orthonormal set of functions are linearly independent.*

Proof. Assume that $f_1, f_2, \cdots, f_n$ are orthogonal, that $\|f_i\| \neq 0$, $i = 1, 2, \cdots, n$, and that an equation of the form (1) holds, where at least one of the constants, say α_k, is nonzero. We can obtain a contradiction by forming the inner product

$$(\alpha_1 f_1 + \alpha_2 f_2 + \cdots + \alpha_k f_k + \cdots + \alpha_n f_n, f_k).$$

On the one hand, by equation (1), this is equal to 0, but on the other hand, by the orthogonality assumed, and the linearity property of the inner product (Theorem I, § 1504), this is equal to

$$\alpha_1(f_1, f_k) + \cdots + \alpha_k(f_k, f_k) + \cdots + \alpha_n(f_n, f_k) = \alpha_k \|f_k\|^2,$$

which is not equal to 0. (Contradiction.)

★★1534. THE GRAM-SCHMIDT PROCESS

We shall describe in this section a process of converting a sequence $\{f_n\}$ of functions every finite subset of which are linearly independent, on an interval $[a, b]$, into an orthonormal sequence $\{\phi_n\}$. It will be assumed that the functions f_n are contained in a linear space S of functions integrable on $[a, b]$, where S is a *metric* subspace $((v), \S 1503)$ of R^2. The procedure, known as the **Gram-Schmidt orthonormalization process**, is inductive in nature. (Let us note in passing that the assumptions on the space S, together with the Note, § 1533, imply that $\|f_n\| \neq 0$ for every $n = 1, 2, 3, \cdots$.)

In the first place, we define

(1) $$\phi_1 \equiv \frac{\pm 1}{\|f_1\|} f_1,$$

with either choice of sign.

Our objective, next, is to construct a linear combination of ϕ_1 and f_2,

(2) $$\psi_2 = a_{21}\phi_1 + f_2,$$

orthogonal to ϕ_1:

(3) $$(\psi_2, \phi_1) = a_{21} + (f_2, \phi_1) = 0.$$

Thus, with $a_{21} = -(f_2, \phi_1)$, the function ψ_2, defined by (2), is orthogonal to ϕ_1. Furthermore, since f_1 and f_2 are linearly independent, $\psi_2 \neq 0$† and, since S is assumed to be a metric space, $\|\psi_2\| \neq 0$. Therefore ψ_2 can be normalized:

(4) $$\phi_2 = \frac{\pm 1}{\|\psi_2\|}\psi_2$$

(with either choice of sign). At this point it should be observed that the set of all linear combinations of ϕ_1 and ϕ_2 is identical with the set of all linear combinations of f_1 and f_2 (show this).

For the next step we define

(5) $$\psi_3 = a_{31}\phi_1 + a_{32}\phi_2 + f_3,$$

subject to the orthogonality conditions

(6) $$\begin{cases}(\psi_3, \phi_1) = a_{31} + (f_3, \phi_1) = 0, \\ (\psi_3, \phi_2) = a_{32} + (f_3, \phi_2) = 0.\end{cases}$$

With $a_{31} = -(f_3, \phi_1)$ and $a_{32} = -(f_3, \phi_2)$, the function ψ_3, defined by (5), is orthogonal to ϕ_1 and ϕ_2. Furthermore, since f_3 is not a linear combination of ϕ_1 and ϕ_2, $\psi_3 \neq 0$, and can be normalized:

(7) $$\phi_3 = \frac{\pm 1}{\|\psi_3\|}\psi_3.$$

As before, the linear combinations of ϕ_1, ϕ_2, and ϕ_3 are the same as those of f_1, f_2, and f_3.

The general inductive procedure should now be clear. With $\phi_1, \phi_2, \cdots, \phi_{n-1}$ determined, we define

(8) $$\psi_n = a_{n1}\phi_1 + a_{n2}\phi_2 + \cdots + a_{n\,n-1}\phi_{n-1} + f_n,$$

subject to the orthogonality conditions

(9) $$(\psi_n, \phi_i) = a_{ni} + (f_n, \phi_i) = 0, \ i = 1, \cdots, n-1.$$

Thus prescribed, $\psi_n \neq 0$, and we define

(10) $$\phi_n = \frac{\pm 1}{\|\psi_n\|}\psi_n,$$

the linear combinations of $\phi_1, \cdots, \phi_n$ being the same as those of $f_1, \cdots, f_n$. The sequence $\{\phi_n\}$ is orthonormal.

† The statement $\psi_2 \neq 0$ means that ψ_2 is not the zero function, and is therefore equivalent to the statement that the function ψ_2 is not *identically* zero.

§ 1535]   LEGENDRE POLYNOMIALS   589

Note. At each stage the Gram-Schmidt orthonormalization process is unique except for sign.

Specific examples illustrating the Gram-Schmidt process for the sequence $1, x, x^2, \cdots$ are considered in the remaining three sections and in the Exercises, § 1538.

As a corollary to Theorem V, § 1532, we have:

Theorem. *The orthonormal sequence produced by applying the Gram-Schmidt process to the sequence $1, x, x^2, x^3, \cdots$ on any interval $[a, b]$ is a basis.* (Cf. Ex. 5, § 1538.)

★★1535. LEGENDRE POLYNOMIALS

If the sequence of functions

(1) $$1, x, x^2, x^3, \cdots, x^n, \cdots$$

on the interval $[-1, 1]$ is subjected to the Gram-Schmidt process (§ 1534; cf. Ex. 5, § 1538), and if the signs are chosen at each stage to produce a polynomial with *positive* leading coefficient, the important closed orthonormal sequence (cf. § 1534) of **normalized Legendre polynomials**, $\{p_n(x)\}$, results. The first few of these are (derive these formulas in Ex. 7, § 1538):

(2) $$\begin{cases} p_0(x) = \tfrac{1}{2}\sqrt{2} & p_3(x) = \tfrac{1}{4}\sqrt{14}(5x^3 - 3x), \\ p_1(x) = \tfrac{1}{2}\sqrt{6}\,x & p_4(x) = \tfrac{1}{16}\sqrt{18}(35x^4 - 30x^2 + 3), \\ p_2(x) = \tfrac{1}{4}\sqrt{10}(3x^2 - 1), & p_5(x) = \tfrac{1}{16}\sqrt{22}(63x^5 - 70x^3 + 15x). \end{cases}$$

Legendre polynomials (without the adjective "normalized") differ from (2) by constant factors. These polynomials, denoted $P_0(x), P_1(x), \cdots, P_n(x), \cdots$ are introduced and studied principally by means of the following four formulas:

(*i*) the **formula of Rodrigues**:

(3) $$P_n(x) = \frac{1}{2^n n!} \frac{d^n}{dx^n} (x^2 - 1)^n;$$

(*ii*) the **recurrence relation**:

(4) $$(n + 1)P_{n+1}(x) - (2n + 1)xP_n(x) + nP_{n-1}(x) = 0;$$

(*iii*) the **differential equation**:

(5) $$(1 - x^2)P_n''(x) - 2xP_n'(x) + n(n + 1)P_n(x) = 0;$$

(*iv*) the **generating function**:

(6) $$(1 - 2xr + r^2)^{-\frac{1}{2}} = P_0(x) + P_1(x)r + P_2(x)r^2 + \cdots.$$

The first few Legendre polynomials are

(7) $\begin{cases} P_0(x) = 1, & P_3(x) = \tfrac{1}{2}(5x^3 - 3x), \\ P_1(x) = x, & P_4(x) = \tfrac{1}{8}(35x^4 - 30x^2 + 3), \\ P_2(x) = \tfrac{1}{2}(3x^2 - 1), & P_5(x) = \tfrac{1}{8}(63x^5 - 70x^3 + 15x). \end{cases}$

As a consequence of the closedness of the orthonormal sequence of normalized Legendre polynomials, the **Legendre series** of an arbitrary function f integrable on $[-1, 1]$,

$$\text{(8)} \qquad \sum_{n=0}^{+\infty} a_n p_n(x), \quad a_n = \int_{-1}^{1} f(x) p_n(x) \, dx,$$

converges in the mean to the function f (§ 1532).

Legendre series have many interesting convergence properties, some of which bear a strikingly close resemblance to similar properties of Fourier series. For a more detailed discussion, and further references, see the book by Jackson in the references at the end of the chapter.

★★1536. ORTHOGONALITY WITH RESPECT TO A WEIGHT FUNCTION

For many problems in both pure and applied mathematics it has been found fruitful to generalize the concept of orthogonality according to the following definition:

Definition. *Let $\rho(x)$ be a nonnegative function integrable on an interval $[a, b]$. Then two functions $f(x)$ and $g(x)$ integrable on $[a, b]$ are orthogonal there with respect to the **weight function** $\rho(x)$ if and only if*

$$\text{(1)} \qquad (f, g)_\rho \equiv \int_a^b f(x) g(x) \rho(x) \, dx = 0.$$

The definition of orthogonality given originally in § 1504 can be considered as a special case of the present definition, the weight function being identically equal to 1. A substantial part of the discussion in this chapter having to do with generalized Fourier series needs little more than a change of wording in order to apply to orthogonality with respect to a more general weight function. In particular, the concepts of *norm* and *orthonormal sequence*, and the Gram-Schmidt process (§ 1534), are readily adapted to accommodate the new definition (cf. Exs. 9-10, § 1538). Furthermore, the proper Riemann integral can often be replaced by an improper integral. Specific examples of important orthonormal sequences of polynomials produced by the Gram-Schmidt process, for various intervals and weight functions, are given in the following section. For further reading, see the references at the end of the chapter.

★★1537. OTHER ORTHOGONAL SYSTEMS

If the Gram-Schmidt process, as mentioned in the preceding section, is applied to the sequence of polynomials $1, x, x^2, \cdots$, for the following combinations of intervals and weight functions, important families of polynomials result:

Interval	Weight function	Polynomials
$[-1, 1]$	1	*Legendre*
$[-1, 1]$	$(1 - x^2)^{-\frac{1}{2}}$	*Tchebycheff*
$[0, 1]$	$x^{\alpha-1}(1 - x)^{\beta-1}\,(\alpha, \beta > 0)$	*Jacobi*
$(-\infty, +\infty)$	e^{-x^2}	*Hermite*
$[0, +\infty)$	e^{-x}	*Laguerre*

In conclusion, let us call the attention of the reader to one of the most significant families of functions for which the concept of orthogonality with respect to a weight function is important, namely, the *Bessel functions* (which are neither algebraic nor trigonometric polynomials). For a discussion, see in particular the books by Courant and Hilbert, Jackson, Watson, and Whittaker and Watson in the references at the end of the chapter.

★★1538. EXERCISES

★★1. Prove Theorem V, § 1532.

★★2. Prove the following form of Parseval's equation, where $\{\phi_n\}$ is a basis and f and g are arbitrary functions in R^2:
$$f \sim \alpha_1\phi_1 + \alpha_2\phi_2 + \cdots, \quad g \sim \beta_1\phi_1 + \beta_2\phi_2 + \cdots$$
imply
$$(f, g) = \alpha_1\beta_1 + \alpha_2\beta_2 + \cdots.$$
Hint: Expand $(f - [\alpha_1\phi_1 + \cdots + \alpha_n\phi_n], g - [\beta_1\phi_1 + \cdots + \beta_n\phi_n])$.

★★3. Prove that each of the following is a basis of the space R^2 for the interval $[0, \pi]$:

(i) $1/\sqrt{\pi}, \sqrt{2/\pi}\cos x, \sqrt{2/\pi}\cos 2x, \sqrt{2/\pi}\cos 3x, \cdots$,

(ii) $\sqrt{2/\pi}\sin x, \sqrt{2/\pi}\sin 2x, \sqrt{2/\pi}\sin 3x, \cdots$.

★★4. Prove the statements of the Note, § 1533.

★★5. Prove that any finite set of polynomials no two of which have the same degree are linearly independent on every interval. In particular, the polynomials $1, x, x^2, \cdots, x^n$ are linearly independent.

★★6. Let $f_1, f_2, \cdots, f_n$ be functions in a metric linear subspace S of R^2, for an interval $[a, b]$. Prove that these functions are linearly dependent if and only if their **Gramian** determinant

(1) $$\begin{vmatrix} (f_1, f_1) & (f_1, f_2) & \cdots & (f_1, f_n) \\ (f_2, f_1) & (f_2, f_2) & \cdots & (f_2, f_n) \\ \cdots & & & \end{vmatrix}$$

592 FOURIER SERIES; ORTHOGONAL FUNCTIONS [§ 1538

vanishes. *Hints:* "Only if": If one function is a linear combination of the other functions, one row of (1) is a linear combination of the other rows. "If": If (1) vanishes there exist numbers $a_1, \cdots, a_n$, not all 0, such that $\sum a_i(f_i, f_j) = 0$. Hence $\sum (f_i, a_i f_j) = (f_i, \sum a_i f_j) = (a_i f_i, \sum a_i f_j) = (\sum a_i f_i, \sum a_i f_j) = \|\sum a_i f_i\|^2 = 0$.

★★7. Derive formulas (2), § 1535.

★★8. Find the first four polynomials with positive leading coefficients resulting from the Gram-Schmidt process applied to the polynomials $1, x, x^2, x^3, \cdots$ on the interval $[0, 1]$.

★★9. Define the norm of a function f with respect to a weight function ρ, for an interval $[a, b]$:

$$(2) \qquad \|f\|_\rho = \|f\sqrt{\rho}\| = \{(f, f)_\rho\}^{\frac{1}{2}} = \left\{ \int_a^b [f(x)]^2 \rho(x)\, dx \right\}^{\frac{1}{2}}.$$

Prove that this norm has all four properties of norm of Theorem II, § 1504. Prove that the space RI with the distance function $d(f, g)_\rho = \|f - g\|_\rho$ is a distance space (§ 1503).

★★10. Discuss the Gram-Schmidt orthonormalization process with respect to a weight function, obtaining formulas similar to those of § 1534.

In Exercises 11-13, find the first four polynomials of the specified orthonormal sequence, with a positive leading coefficient. In each case a suitable reduction formula (cf. § 509) will facilitate the details.

★★11. Tchebycheff.   ★★12. Hermite.   ★★13. Laguerre.

★★14. Find the first three Jacobi polynomials except for the normalization factor. Instead of normalizing, find the orthogonal polynomials having a constant term equal to 1. Find the norm of the first one of these only. (Cf. Ex. 22, § 510.)

REFERENCES

BOHR, H., *Almost Periodic Functions* (New York, Chelsea Publishing Co., 1947).

CARSLAW, H. S., *Fourier's Series and Integrals*, 3rd ed. (New York, Dover Publications, 1930).

CHURCHILL, R. V., *Fourier Series and Boundary Value Problems* (New York, McGraw-Hill Book Co., 1941).

———, *Modern Operational Mathematics in Engineering* (New York, McGraw-Hill Book Co., 1944).

COURANT, R., and HILBERT, D., *Methods of Mathematical Physics* (New York, Interscience Publishers, 1953.)

FRANKLIN, P., *A Treatise on Advanced Calculus* (New York, John Wiley and Sons, 1940).

HALMOS, P. R., *Finite-Dimensional Vector Spaces* (Princeton, D. Van Nostrand Co., 1958).

———, *Introduction to Hilbert Space* (New York, Chelsea Publishing Co., 1951).

HARDY, G. H., and ROGOSINSKI, W. W., *Fourier Series* (Cambridge University Press, 1944).

HILDEBRAND, F. B., *Advanced Calculus for Engineers* (Englewood Cliffs, N. J., Prentice-Hall, 1949).

———, *Methods of Applied Mathematics* (Englewood Cliffs, N. J., Prentice-Hall, 1952).

HOBSON, E. W., *The Theory of Functions* (Washington, Harren Press, 1950).
JACKSON, D., *Fourier Series and Orthogonal Polynomials* (Carus Mathematical Monographs, 1941).
JAHNKE, E., and EMDE, F., *Tables of Functions* (New York, Dover Publications, 1943).
ROGOSINSKI, W. W., *Fourier Series* (New York, Chelsea Publishing Co., 1950).
SZEGÖ, G., *Orthogonal Polynomials* (New York, American Mathematical Society Colloquium Publications, No. 23, 1939).
TITCHMARSH, E. C., *The Theory of Functions* (Oxford Press, 1932).
WATSON, G. N., *Theory of Bessel Functions*, 2nd ed. (New York, Macmillan Co., 1944).
WHITTAKER, E. T., and WATSON, G. N., *Modern Analysis*, 4th ed. (Cambridge University Press, 1940).
WIDDER, D. V., *Advanced Calculus* (Englewood Cliffs, N. J., Prentice-Hall, 1947).
ZYGMUND, A., *Trigonometrical Series* (New York, Dover Publications, 1955).

Answers

§ 111, page 18

12. $x > a + |b|$.
13. $-1 < x < 5$.
14. $x \leq -5$ or $x \geq -1$.
15. $x > 2$.
16. $x < 3$.
17. No values.
18. All values.
19. $-\sqrt{3} \leq x \leq -1$ or $1 \leq x \leq \sqrt{3}$.
20. $-3 < x < 5$.
21. No values.
22. $x < -\frac{1}{2}$ or $x > 5$.
23. $4 < x < 6$.
24. $2 < x < 4$ or $6 < x < 12$.
25. $|x| > |a|$.
26. $|x| < |a|$ or $x = a(a \neq 0)$.
27. $|x - 1| > 2$.
28. If $a = b$, no values; if $a < b$: $-|b| < x < 0$ or $x > |b|$; if $a > b$: $0 < x < |b|$ or $x < -|b|$.

§ 113, page 20

17. $ad = bc$.

§ 203, page 33

11. $\dfrac{2n}{2n-1}$.
12. $\dfrac{(-1)^{n+1}}{n^2+2}$.
13. $\dfrac{(-1)^{n-1}}{(n-1)!}$.
14. $(2n-2)!$.
15. $1 \cdot 3 \cdot 5 \cdots (2n+1) = \dfrac{(2n+1)!}{2^n n!}$.
16. $a_{2n} = 2,\ a_{4n-1} = 3,\ a_{4n-3} = 1$.
17. $a_{3n} = a_{3n-1} = n,\ a_{3n-2} = -n$.
18. n even, $2^n n!$; n odd, $\dfrac{(2n)!}{2^n n!}$.
19. 2.
20. 1.
21. $\frac{1}{2}$.
22. $+\infty$.
23. ∞.
24. $-\infty$.
25. $N(\epsilon) = 1$.
26. $N(\epsilon) = 1/\epsilon$.
27. $N(\epsilon) = 1/\epsilon$.
28. $N(B) = B$.
29. $N(B) = \frac{1}{2}(7B+2)$.
30. $N(B) = |B| + 10$.

§ 208, page 45

21. 4.
22. $\frac{7}{3}$.
23. $\frac{1}{3}$.
24. $-\frac{1}{4}$.
25. $3a^2$.
26. ma^{m-1}.

ANSWERS

33. (a) $-\infty$; (b) $+\infty$; (c) ∞; (d) $+\infty$; (e) $-\infty$; (f) ∞.
34. (c) 0; (e) 1; (f) 0; (a), (b), (d): there is no limit.
37. $\tfrac{3}{5}$. **38.** $+\infty$. **39.** $-\infty$.
40. $+\infty$. **41.** 0. **42.** $-\infty$.

59. 6; $\delta(\epsilon) = \tfrac{\epsilon}{3}$. **60.** 9; $\delta(\epsilon) = \min\left(1, \tfrac{\epsilon}{7}\right)$.

61. 28; $\delta(\epsilon) = \min\left(1, \tfrac{\epsilon}{22}\right)$. **62.** $-\tfrac{1}{5}$; $\delta(\epsilon) = \min(1, 20\epsilon)$.

63. $\tfrac{3}{7}$; $\delta(\epsilon) = \min\left(1, \tfrac{\epsilon}{5}\right)$. **64.** 6; $\delta(\epsilon) = \min\left(\tfrac{1}{8}, \tfrac{\epsilon}{42}\right)$.

65. 0; $N(\epsilon) = \tfrac{5}{\epsilon}$. **66.** 0; $N(\epsilon) = -\max\left(1, \tfrac{1}{\epsilon}\right)$.

67. 3; $N(\epsilon) = \tfrac{17}{\epsilon}$. **68.** $\tfrac{5}{3}$; $N(\epsilon) = \max\left(1, \tfrac{1}{3\epsilon}\right)$.

69. $+\infty$; if $B > 0$, $\delta(B) = \min\left(1, \tfrac{1}{B}\right)$; otherwise $\delta(B) = 392$.

70. $-\infty$; if $B < 0$, $\delta(B) = \min\left(1, \tfrac{1}{-2B}\right)$; otherwise $\delta(B) = 1$.

§ 216, page 57

32. $\delta(\epsilon) = \min(3, \epsilon)$. **33.** $\delta(\epsilon) = \min\left(1, \tfrac{\epsilon^2}{5}\right)$.

34. $\delta(\epsilon) = \min\left(1, \tfrac{\epsilon}{5}\right)$. **35.** $\delta(\epsilon) = \min(5, \epsilon)$.

36. $\delta(\epsilon) = \min(5, 6\epsilon)$. **37.** $\delta(\epsilon) = \min\left(\tfrac{7}{8}, \tfrac{\epsilon}{2}\right)$.

§ 305, page 66

19. 1; 0. **20.** $+\infty$; $-\infty$.
21. 1; 0. **22.** 1; 0.
29. $+\infty$; $-\infty$. **30.** 1; -1.
31. 1; -1. **32.** 1; -1.
33. 1; -1. **34.** 1; -1.

§ 306, page 73

3. $\delta(\epsilon) = \tfrac{\epsilon}{2}$. **4.** $\delta(\epsilon) = \tfrac{\epsilon}{4}$.
5. $\delta(\epsilon) = 2\epsilon$. **6.** $\delta(\epsilon) = \epsilon^2$.
7. $\delta(\epsilon) = \epsilon$. **8.** $\delta(\epsilon) = \tfrac{\epsilon^2}{4}$.
13. $\delta(\epsilon) = \min\left(\tfrac{x_0}{2}, \tfrac{x_0^2 \epsilon}{2}\right)$. **14.** $\delta(\epsilon) = \min\left(1, \tfrac{\epsilon}{1 + 2x_0}\right)$.

§ 404, page 91

1. $2x - 4$. **2.** $3x^2$.

3. $-\tfrac{2}{x^3}$. **4.** $-\tfrac{22}{(5x-4)^2}$.

5. $\tfrac{1}{2\sqrt{x}}$. **6.** $\tfrac{1}{3\sqrt[3]{x^2}}$.

ANSWERS

15. Yes. **16.** Yes. **17.** No.
18. No. **19.** No. **20.** Yes.
21. $n > 1;\ n > 1$. **22.** $n > k;\ n > k$.
23. $f'(x) = nx^{n-1}\sin\frac{1}{x} - x^{n-2}\cos\frac{1}{x},\ x > 0;\ f'(0) = 0,\ n > 1.\ n > 1.\ n > 2.$
24. $f''(x) = [n(n-1)x^{n-2} - x^{n-4}]\sin\frac{1}{x} - 2(n-1)x^{n-3}\cos\frac{1}{x},\ x > 0;$
$f''(0) = 0,\ n > 3.\ n > 3.\ n > 4.$

§ 408, page 100

1. $\pi, 2\pi$, or 3π. **2.** $2 - \sqrt{2}$.
3. $e - 1$. **4.** $\frac{1}{2}(a + b)$.
5. $\dfrac{1}{e-1}$. **6.** $\frac{1}{2}$.
7. $\dfrac{a+b}{2}$. **8.** $\frac{1}{3}^{\frac{1}{3}}$.
9. $1 - \sqrt[n+1]{1-b}$. **10.** $2/\sqrt[3]{\ln 27}$.

§ 412, page 111

1. Relative maximum $= 9$; relative minimum $= 5$; increasing on $(-\infty, 1]$ and $[3, +\infty)$; decreasing on $[1, 3]$.
2. Absolute maximum $= 1$; absolute minimum $= -1$; increasing on $[-1, 1]$; decreasing on $(-\infty, -1]$ and $[1, +\infty)$.
3. Relative maximum $= 3\sqrt[5]{20}/25$; relative minimum $= 0$; increasing on $[0, \frac{2}{3}]$; decreasing on $(-\infty, 0]$ and $[\frac{2}{3}, +\infty)$.
4. Relative maximum $= 2\sqrt{3}/9$; absolute minimum $= 0$; increasing on $[0, \frac{1}{3}]$ and $[1, +\infty)$; decreasing on $[\frac{1}{3}, 1]$.
5. Maximum $= \frac{1}{2}$; minimum $= -1$.
6. Maximum $= -4$; minimum $= -109/16$.
7. Maximum $= 2$; minimum $= -9/8$.
8. No maximum; minimum $= -1/e$.
15. (a) $x = 20$; (b) $x = 25$; (c) $x = 21$; (d) no profit possible.
16. (a) $x = 20$; (b) $x = 37.4$; (c) full speed, $x = 60$.
19. If $t \leq s,\ x = b$; if $t > s,\ x = \min(b, as/\sqrt{t^2 - s^2})$.
20. $r \leq -1$: no minimum; $-1 < r \leq -\frac{1}{2}$: minimum $= -a^2/4(r+1)$; $-\frac{1}{2} \leq r \leq 0$: minimum $= ra^2$; $r \geq 0$: minimum $= 0$.
33. $\epsilon(\Delta x) = (6x^2 - 5)\Delta x + 4x\Delta x^2 + \Delta x^3$.
34. $\epsilon(\Delta x) = \Delta x/x^2(x + \Delta x)$.
35. $\epsilon(\Delta x) = -\sqrt{x}\,\Delta x/2(x + \sqrt{x^2 + x\Delta x})^2$.
37. 10.48810. **38.** h. **39.** -0.06.
40. x. **41.** x. **42.** $\dfrac{1}{2} - \dfrac{\sqrt{3}}{2}\left(x - \dfrac{\pi}{3}\right)$.
43. x. **44.** $1 + \dfrac{h}{n}$. **45.** 0.
46. h. **47.** $1 + x$. **48.** 0.

§ 417, page 121

1. $\frac{1}{3}$. **2.** $\frac{1}{2}$. **3.** $\frac{1}{4}$.

ANSWERS

4. $-\frac{\pi}{2}$.   **5.** $\frac{7}{24}$.   **6.** 1.
7. $\frac{1}{17}$.   **8.** 1.   **9.** $\frac{\ln a}{\ln b}$.
10. 2.   **11.** $\frac{3}{4}$.   **12.** 1.
13. 1.   **14.** Meaningless.   **15.** $\frac{1}{2}$.
16. 0.   **17.** 0.   **18.** $+\infty$.
19. 0.   **20.** $-\frac{2a}{\pi}$.   **21.** -1.
22. $\frac{1}{2}$.   **23.** 1.   **24.** e^3.
25. e.   **26.** e^2.   **27.** 0.
28. 1.   **29.** e^{-2}.   **30.** 0.

§ 503, page 143

14. $0; 3; \frac{1}{2}(n^2 - n)$.
25. $\frac{\pi}{4}$.   **26.** $\frac{\pi}{6}$.   **27.** $\frac{\ln 2}{6}$.

§ 506, page 156

3. 0.   **4.** $\sin x^3$.
5. $-\sin x^2$.   **6.** $3x^2 \sin x^6$.
7. $4x^3 \sin x^8 - 3x^2 \sin x^6$.
8. $2x \cos x^2 \sin (\sin^2 x^2) - \sin 2x \sin (\sin^2 x)$.

§ 508, page 160

3. No. No.

§ 510, page 162

13. $-\frac{1}{4}\sin^3 x \cos x - \frac{3}{24}\sin^3 x \cos x - \frac{3}{16}\sin x \cos x + \frac{3}{16}x + C$.
14. $\frac{1}{5}\sin x \cos^4 x + \frac{4}{15}\sin x \cos^2 x + \frac{8}{15}\sin x + C$.
15. $-\frac{5}{4}\cot^4 \frac{x}{5} + \frac{5}{2}\cot^2 \frac{x}{5} + 5 \ln \sin \frac{x}{5} + C$.
16. $\frac{1}{4}\sec^3 x \tan x + \frac{3}{24}\sec^2 x \tan x + \frac{3}{16}\sec x \tan x + \frac{3}{16}\ln |\sec x + \tan x| + C$.
17. $\frac{1}{4}[(4x^3 - 6x)\sin 2x + (-2x^4 + 6x^2 - 3)\cos 2x] + C$.
18. $\frac{1}{3}(x^2 + 4x)^{\frac{3}{2}} - (x + 2)\sqrt{x^2 + 4x} + 4 \ln |x + 2 + \sqrt{x^2 + 4x}| + C$.
19. $\frac{1}{3}x^3 - \frac{1}{4}(x + \frac{1}{2})\sqrt{x^2 + x + 1} + \frac{3}{8}\ln |x + \frac{1}{2} + \sqrt{x^2 + x + 1}| + C$.
20. $-\frac{1}{80}(4x^2 + 21x + 105)(6x - x^2)^{\frac{3}{2}} + \frac{3}{2}x^2 (x - 3)\sqrt{6x - x^2}$
$\qquad + \frac{1701}{8} \operatorname{Arc\,sin} \frac{x - 3}{3} + C$.

§ 515, page 171

1. $\frac{\pi}{2}$.   **2.** 0.   **3.** Divergent.
4. 2.   **5.** 2.   **6.** Divergent.
7. $\frac{\pi}{2}$.   **8.** $\frac{1}{(k-1)(\ln 2)^{k-1}}, k > 1$; divergent, $k \leq 1$.   **9.** 2.   **10.** $\frac{\pi}{2}$.
15. Convergent.   **16.** Divergent.

ANSWERS 599

17. Convergent.
18. Convergent.
19. Divergent.
20. Convergent.
21. Convergent.
22. Convergent.
23. Divergent.
24. Convergent.

§ 518, page 184

1. $\frac{2}{3}$.   **2.** 3.   **3.** $\frac{\pi}{4}$.
4. $-(1 + e + e^2)$.   **5.** $e + e^{-1} - 2$.   **6.** $\pi - 2$.
35. $v(x) = x + [x];\ p(x) = x;\ n(x) = [x]$.
36. $v(x) = 3 + 2x - x^2,\ -1 \leqq x \leqq 1;\ 5 - 2x + x^2,\ 1 \leqq x \leqq 2$;
$p(x) = 3 + 2x - x^2,\ -1 \leqq x \leqq 1;\ 4,\ 1 \leqq x \leqq 2$;
$n(x) = 0,\ -1 \leqq x \leqq 1;\ 1 - 2x + x^2,\ 1 \leqq x \leqq 2$.

§ 606, page 196

1. $\frac{1}{5}\ln|\sec 5x| + C$.
2. $\frac{1}{4}\ln|\sec 4x + \tan 4x| + C$.
3. $\operatorname{Arc\,sin}(x/\sqrt{2}) + C,\ x^2 < 2$.
4. $\frac{1}{4}(x + 2)\sqrt{x^2 + 4x} - 2\ln|x + 2 + \sqrt{x^2 + 4x}| + C,\ x \geqq 0\ \text{or}\ x \leqq -4$.
5. $\frac{1}{2}\ln|2x - 1 + \sqrt{4x^2 - 4x + 5}| + C$.
6. $\frac{6x - 1}{12}\sqrt{x - 3x^2} + \frac{\sqrt{3}}{72}\operatorname{Arc\,sin}(6x - 1) + C,\ 0 < x < \frac{1}{3}$.
7. $\frac{1}{\sqrt{5}}\ln|(\sqrt{5 - 2x^2} - \sqrt{5})/x| + C,\ |x| < \sqrt{5/2}$.
8. $\sec\sqrt{x}\tan\sqrt{x} + \ln|\sec\sqrt{x} + \tan\sqrt{x}| + C,\ x > 0$.
9. $\frac{1}{\sqrt{109}}\ln\left|\frac{6x + 5 - \sqrt{109}}{6x + 5 + \sqrt{109}}\right| + C$.
10. $\frac{2}{\sqrt{59}}\operatorname{Arc\,tan}\frac{6x + 5}{\sqrt{59}} + C$.

§ 609, page 200

1. $3\sinh 3x$.   **2.** $\sinh 2x$.
3. $-\operatorname{sech}^2(2 - x)$.   **4.** $\coth x^2 - 2x^2\operatorname{csch}^2 x^2$.
5. $2\coth 2x$.   **6.** $e^{ax}(a\cosh bx + b\sinh bx)$.
7. $\frac{4}{\sqrt{1 + 16x^2}}$.   **8.** $\frac{e^x}{\sqrt{e^{2x} - 1}},\ x > 0$.
9. $\frac{2x}{1 - x^2},\ |x| < 1$.   **10.** $-\csc x,\ x \neq n\pi$.
11. $\frac{1}{6}\ln\cosh 6x + C$.   **12.** $\ln(\sinh e^x) + C$.
13. $\sinh x + \frac{1}{3}\sinh^3 x + C$.   **14.** $x - \frac{1}{10}\tanh 10x + C$.
15. $\frac{1}{2}\sinh 2x - \frac{1}{2}x + C$.   **16.** $\frac{1}{4}[(x^2 - 1)\tanh^{-1} x + \frac{1}{2}x^2 + x] + C$.
17. $\operatorname{Cosh}^{-1}\frac{x}{\sqrt{2}} + C$.   **18.** $\frac{1}{2}\sinh^{-1}\frac{2x - 1}{2} + C$.
19. $-\frac{2}{\sqrt{109}}\coth^{-1}\frac{6x + 5}{\sqrt{109}} + C$.   **20.** $\frac{2}{\sqrt{109}}\tanh^{-1}\frac{6x + 5}{\sqrt{109}} + C$.

ANSWERS

§ 707, page 211

1. $\frac{n}{n+1}$; 1.
2. $\frac{n}{2n+1}$; $\frac{1}{2}$.
3. $1 - (\frac{1}{2})^n$; 1.
4. $\frac{n(n+1)}{2}$; $+\infty$.
5. $2 + \frac{1}{2} + \frac{1}{6} + \frac{1}{12}$; $\sigma_1 = 2$, $a_n = \frac{1}{n(n-1)}$, $n > 1$.
6. $1 + 2 - 2 + 2$; $a_1 = 1$, $a_n = 2(-1)^n$, $n > 1$.
7. $1.3 - 0.39 + 0.117 - 0.0351$; $1.3 (-0.3)^{n-1}$.
8. $2 - 0.7 - 0.21 - 0.063$; $a_1 = 2$, $a_n = -0.7(0.3)^{n-2}$, $n > 1$.
9. $\frac{31}{15}$.
10. Divergent.
11. 0.3.
12. $\frac{10{,}201}{201}$.
13. $\frac{1}{2}$.
14. $\frac{3333}{1000}$.
15. $\frac{343}{9}$.
16. $\frac{1}{3}$.
17. Divergent.
18. Divergent.
19. Convergent.
20. Convergent.
21. Convergent.
22. Divergent.
23. Divergent.
24. Convergent.
25. Convergent.
26. Convergent.

§ 711, page 219

1. Divergent.
2. Divergent.
3. Convergent.
4. Convergent.
5. Convergent.
6. Divergent.
7. Divergent.
8. Convergent.
9. Convergent.
10. Convergent.
11. Convergent.
12. Divergent.
13. Convergent.
14. Convergent.
15. Convergent for $\alpha > \frac{1}{2}$; divergent for $\alpha \leq \frac{1}{2}$.
16. Convergent.
17. Divergent.
18. Convergent for $\alpha > 1$; divergent for $\alpha \leq 1$.
19. Convergent.
20. Convergent for $0 < r < 1$; divergent for $r \geq 1$ unless $\alpha = k\pi$.

§ 713, page 223

1. Divergent.
2. Convergent.
3. Divergent.
4. Convergent for $p > 2$; divergent for $p \leq 2$.

§ 717, page 229

1. Conditionally convergent.
2. Divergent.
3. Conditionally convergent.
4. Absolutely convergent.
5. Absolutely convergent.
6. Absolutely convergent.
7. Absolutely convergent for $x > 0$; divergent for $x \leq 0$.
8. Absolutely convergent for $|r| < 1$; divergent for $|r| \geq 1$.
9. Absolutely convergent for $p > 1$; conditionally convergent for $p \leq 1$.
10. Absolutely convergent for $p > 2$; conditionally convergent for $0 < p \leq 2$; divergent for $p \leq 0$.

§ 721, page 237

13. 2.718.
14. 0.6931.
15. 1.202.
16. 0.1775.
17. 0.7854.
18. 0.4055.
22. $0.5770 < C < 0.5774$.

ANSWERS

§ 802, page 243

1. Absolutely convergent for $-\infty < x < +\infty$.
2. Absolutely convergent for $|x| < 1$; conditionally convergent for $x = \pm 1$.
3. Absolutely convergent for $-\infty < x < +\infty$.
4. Absolutely convergent for $x = 0$.
5. Absolutely convergent for $-2 \leq x \leq 0$.
6. Absolutely convergent for $-\infty < x < +\infty$.
7. Absolutely convergent for $4 < x < 6$; conditionally convergent for $x = 4$.
8. Absolutely convergent for $0 < x < 2$.
9. Absolutely convergent for $|x| < 1$; conditionally convergent for $x = 1$.
10. Absolutely convergent for $|x| \leq 1$.
11. Absolutely convergent for $x < 2$ or $x > 4$; conditionally convergent for $x = 2$.
12. Absolutely convergent if $|x| \neq \frac{2n+1}{2}\pi$; conditionally convergent if $|x| = \frac{2n+1}{2}\pi$.

§ 811, page 261

1. $1 + \frac{x^2}{2!} + \frac{x^4}{4!} + \frac{x^6}{6!} + \cdots$.
2. $1 - x^2 + x^4 - x^6 + \cdots$.
3. $x - \frac{x^3}{3} + \frac{x^5}{5} - \frac{x^7}{7} + \cdots$.
4. $1 + \frac{x}{1!\cdot 2} + \frac{x^2}{2!\cdot 4} + \frac{x^3}{3!\cdot 8} + \cdots$.
5. $1 - \frac{x^4}{2!} + \frac{x^8}{4!} - \frac{x^{12}}{6!} + \cdots$.
6. $\ln 2 + \frac{3x}{1\cdot 2} - \frac{9x^2}{2\cdot 4} + \frac{27x^3}{3\cdot 8} - \cdots$.
7. $1 + \frac{1}{2}x^2 + \frac{1\cdot 3}{2\cdot 4}x^4 + \frac{1\cdot 3\cdot 5}{2\cdot 4\cdot 6}x^6 + \cdots$.
8. $2 + \frac{2}{8}x - \frac{2\cdot 1}{8\cdot 16}x^2 + \frac{2\cdot 1\cdot 3}{8\cdot 16\cdot 24}x^3 - \frac{2\cdot 1\cdot 3\cdot 5}{8\cdot 16\cdot 24\cdot 32}x^4 + \cdots$.
9. $2\left[x + \frac{x^3}{3} + \frac{x^5}{5} + \frac{x^7}{7} + \cdots\right]$.
10. $x - \frac{x^3}{3\cdot 3!} + \frac{x^5}{5\cdot 5!} - \frac{x^7}{7\cdot 7!} + \cdots$.
11. $x - \frac{x^3}{3} + \frac{x^5}{2!\cdot 5} - \frac{x^7}{3!\cdot 7} + \cdots$.
12. $\frac{x^3}{3} - \frac{x^7}{3!\cdot 7} + \frac{x^{11}}{5!\cdot 11} - \frac{x^{15}}{7!\cdot 15} + \cdots$.
13. $\frac{1}{2} - \frac{x}{4} + \frac{x^3}{48} - \frac{x^5}{480}$.
14. $1 + x - \frac{x^2}{3} - \frac{x^4}{6} - \frac{x^5}{30}$.
15. $x - \frac{1}{3}x^3 + \frac{2}{15}x^5 - \frac{17}{315}x^7$.
16. $1 + x + \frac{x^2}{2} + \frac{x^3}{2} + \frac{3x^4}{8} + \frac{37x^5}{120}$.
17. $-\frac{x^2}{6} - \frac{x^4}{180} - \frac{x^6}{2835}$.
18. $-\frac{x^2}{2} - \frac{x^4}{12} - \frac{x^6}{45} - \frac{17x^8}{2520}$.
19. $\cos a \left[1 - \frac{(x-a)^2}{2!} + \frac{(x-a)^4}{4!} - \cdots\right]$
    $- \sin a \left[(x-a) - \frac{(x-a)^3}{3!} + \frac{(x-a)^5}{5!} - \cdots\right]$.
20. $\frac{\sqrt{2}}{2}\left[1 - \left(x - \frac{\pi}{4}\right) - \frac{\left(x - \frac{\pi}{4}\right)^2}{2!} + \frac{\left(x - \frac{\pi}{4}\right)^3}{3!} + - - \cdots\right]$.
21. $1 + \frac{x-e}{e} - \frac{(x-e)^2}{2e^2} + \frac{(x-e)^3}{3e^3} - \cdots$.
22. $1 + m(x-1) + \frac{m(m-1)}{2!}(x-1)^2 + \cdots$.
23. $(1-x)^{-1}$.

25. $(1 + x^2)^{-1}$.
26. $3 \ln \frac{3}{2} - 1$.
35. $B_0 = 1$, $B_1 = -\frac{1}{2}$, $B_2 = \frac{1}{6}$, $B_4 = -\frac{1}{30}$, $B_6 = \frac{1}{42}$, $B_8 = -\frac{1}{30}$, $B_{10} = \frac{5}{66}$.

§ 814, page 266

1. $\frac{1}{2}$.　　**2.** $-\frac{1}{2}$.　　**3.** $\frac{3}{4}$.　　**4.** $\frac{1}{4}$.
5. $-\frac{1}{7!}$.　　**6.** 1.　　**7.** $\frac{1}{2}$.　　**8.** -2.
9. $2e^3$.　　**10.** 0.
11. 7.38906.　　　　　　**12.** 1.64872.
13. 1.0986.　　　　　　**14.** 0.8776.
15. 2.15443.　　　　　**16.** 0.10033.
17. 0.4931.　　　　　　**18.** 0.0976.
19. 0.4920.　　　　　　**20.** 0.7468.

§ 816, page 268

3. $b_1 = 1$, $b_2 + a_1 = 0$, $b_3 + 2a_1b_2 + a_2 = 0$, $b_4 + 3a_1b_3 + (a_1^2 + 2a_2)b_2 + a_3 = 0$.

§ 904, page 275

37. $\dfrac{\ln \epsilon}{\ln \sin (\frac{1}{2}\pi - \eta)}$.　　**38.** $\dfrac{b}{\eta}$.

39. $\dfrac{1}{e\eta}$.　　**40.** $\dfrac{2b}{\epsilon^2}$.

41. $\max \left(\dfrac{2}{\eta}, \dfrac{6}{e\eta}\right)$.　　**42.** $\dfrac{\ln \epsilon}{\ln (1 - \eta)}$.

§ 911, page 290

7. $\ln 4 - 1$.　　**8.** $\frac{1}{4}\pi - \frac{1}{2}\ln 2$.

§ 1005, page 299

1. (a) and (b) no, (c) yes, (d)-(g) no, (h) and (i) $a \leq x \leq b$, $c \leq y \leq d$, (j) $a < x < b$, $c < y < d$, (k) the four edges of the rectangle, including the corners.
2. (a) yes, (b)-(g) no, (h) and (i) the entire plane, (j) the set itself, (k) the two coordinate axes.
3. (a) no, (b)-(d) yes, (e)-(g) no, (h) and (i) the set itself, (j) $0 < \rho^2 < \cos 2\theta$, (k) $\rho^2 = \cos 2\theta$.
4. (a)-(g) no, (h), (i), and (k) the set together with all (x, y) such that $x = 0$ and $-1 \leq y \leq 1$, (j) empty.
5. (a)-(g) no, (h) and (i) all (x, y) such that $y \leq [x]$, (j) all points of the set except for x integral and $y \geq x - 1$, (k) all (x, y) such that $y = [x]$ together with all (x, y) such that x is integral and $x - 1 \leq y < x$.
6. (a)-(g) no, (h) all (x, y) of the form (m, n) or $\left(m + \dfrac{1}{p}, n\right)$ or $\left(m, n + \dfrac{1}{q}\right)$, (i) and (k) the set together with (h), (j) empty.

§ 1011, page 309

1. No.　　**2.** Yes. 0.　　**3.** Yes. 0.　　**4.** No.
5. No.　　**6.** No.　　**7.** Yes. 0.　　**8.** Yes. 0.
9. Yes. 0.　　**10.** Yes. 0.

ANSWERS

§ 1015, page 314

11. $\delta(\epsilon) = \epsilon\eta$.
12. $\delta(\epsilon) = \epsilon/b$.
13. $\delta(\epsilon) = \epsilon/\max(|a|, |b|)$.
14. $N(\epsilon) = K/\epsilon$.

§ 1021, page 322

13. $\delta(\epsilon) = \epsilon/2$.
14. $\delta(\epsilon) = \epsilon/2e^a$.

§ 1023, page 327

15. False.
16. False.
17. True.
18. True.
19. False.
20. True.

§ 1110, page 352

1. $\int_0^2 \sqrt{1 + 9x^4}\, dx$.
2. $\sqrt{17} + \frac{1}{4}\ln(4 + \sqrt{17})$.
3. πa.
4. $\frac{k}{2}\left[\alpha\sqrt{1 + \alpha^2} + \ln(\alpha + \sqrt{1 + \alpha^2})\right]$.
5. $10n\pi$.
6. $6a$.
7. $6a$.
8. $a \sinh \frac{b}{a}$.
9. $\sqrt{2} + \ln(\sqrt{2} + 1)$.
10. $\sqrt{2}(e^{\pi/2} - 1)$.
11. $x = r\cos(s/r),\ y = r\sin(s/r)$.
12. $x = 2\,\text{Arc}\cos\frac{4-s}{4} - \frac{(4-s)\sqrt{8s - s^2}}{8},\ y = \frac{8s - s^2}{8},\ 0 \le s \le 8$.
13. $x = [(3 - 2s)/3]^{3/2},\ y = [2s/3]^{3/2},\ 0 \le s \le \frac{3}{2}$.
14. $\theta = \ln[(s + \sqrt{2})/\sqrt{2}],\ \rho = (s + \sqrt{2})/\sqrt{2}$.

§ 1113, page 357

1. $(3, 0, 2),\ (3, 3\sqrt{3}, -1)$.
2. $(4, \pi, 1),\ (4\sqrt{3}, \pi/6, 0)$.
3. $(-7, 0, 0),\ (-3, 3\sqrt{3}, -6\sqrt{3})$.
4. $(5, \pi/2, 3\pi/2),\ (\sqrt{3}, \text{Arc}\cos 3^{-1/2}, \pi/4)$.
5. Cylinder, half-plane, cylinder.
6. Sphere, half-cone, surface of revolution.
7. $x^2 + y^2 - z^2 = 0$.
8. $\rho = \cos\phi$.
9. $x^2 + y^2 + z^2 = 9$.
10. $\phi = \pi/3,\ 2\pi/3$.
13. $\sqrt{a^2\lambda^2 + \mu^2}\cdot|t_2 - t_1|$.
14. $\sqrt{a^2\lambda^2 + \mu^2}\cdot|t_2 - t_1|$.
15. $\dfrac{t\sqrt{a^2 + \mu^2 + a^2\lambda^2 t^2}}{2} + \dfrac{a^2 + \mu^2}{2a\lambda}\ln\dfrac{a\lambda t + \sqrt{a^2 + \mu^2 + a^2\lambda^2 t^2}}{\sqrt{a^2 + \mu^2}}$.
16. $\dfrac{\nu t\sqrt{1 + \lambda^2(\sin^2\alpha)t^2}}{2} + \dfrac{\nu}{2\lambda\sin\alpha}\ln[\lambda(\sin\alpha)t + \sqrt{1 + \lambda^2(\sin^2\alpha)t^2}]$.

§ 1117, page 363

1. $R = \frac{1}{2};\ (0, \frac{1}{2})$.
2. $R = \frac{1}{2}\sqrt{5};\ (-4, \frac{1}{2})$.
3. $R = \frac{1}{2}\sqrt{13};\ (\frac{1}{2}, -\frac{1}{2})$.
4. $R = \frac{1}{2}\sqrt{5};\ (\frac{1}{\sqrt{2}}, \frac{1}{2})$.
5. $R = 6;\ (3, -4)$.
6. $R = 4;\ (\pi, -2)$.
7. $R = \frac{1}{2};\ (\frac{1}{2}, \pi)$.
8. $R = \frac{1}{2};\ (\pm\frac{1}{2}, 0)$.
14. $3\sqrt{2}/2$.
19. $\alpha = (3y^2 + 2p^2)/2p,\ \beta = -y^3/p^2$.

604 ANSWERS

20. $\alpha = 2x^2, \beta = -2y^2$. **21.** $\alpha = -t^3 - \tfrac{3}{2}t^2, \beta = \tfrac{3}{2}t + 4t^3$.
22. $\alpha = \tfrac{3}{2}(1 + 2t^2 - t^4), \beta = -4t^3$.
23. $\alpha = x + 2\sqrt{y}(x + y), \beta = y + 2\sqrt{x}(x + y)$.
24. $2\alpha = 3\cos\theta - \cos 3\theta, 2\beta = 3\sin\theta + \sin 3\theta$.
29. $x = \cos s + s \sin s, y = \sin s - s \cos s$.

§ 1204, page 369

1. $u_x = \sin(x+y) + (x-y)\cos(x+y)$,
$u_y = -\sin(x+y) + (x-y)\cos(x+y)$.
2. $u_x = e^y + yx^2, u_y = xe^y + e^x$.
3. $f_x = x/\sqrt{x^2+y^2+z^2}, f_y = y/\sqrt{x^2+y^2+z^2}, f_z = z/\sqrt{x^2+y^2+z^2}$.
4. $x_r = \sin\phi\cos\theta, x_\phi = r\cos\phi\cos\theta, x_\theta = -r\sin\phi\sin\theta$.
5. $f_{xx} = 2A, f_{xy} = 2B, f_{yy} = 2C$.
6. $z_{xx} = -\dfrac{2y}{(x+y)^3}, z_{xy} = \dfrac{x-y}{(x+y)^3}, z_{yy} = \dfrac{2x}{(x+y)^3}$.
7. $u_{xx} = 0, u_{yy} = xe^y, u_{zz} = -y\sin z, u_{xy} = e^y, u_{xz} = 0, u_{yz} = \cos z$.
8. $u_{xx} = \dfrac{2xyz}{(x^2+y^2)^2}, u_{yy} = \dfrac{-2xyz}{(x^2+y^2)^2}, u_{zz} = 0$,
$u_{xy} = \dfrac{(y^2-x^2)z}{(x^2+y^2)^2}, u_{xz} = \dfrac{-y}{x^2+y^2}, u_{yz} = \dfrac{x}{x^2+y^2}$.

§ 1209, page 378

1. $0; 6t$. **2.** $3r^2e^{3z}, 3r^2e^{3z}; 4r, 0$.
9. $f'(g(x) + h(x))(g'(x) + h'(x))$. **10.** $f'(g(x)h(x))(g(x)h'(x) + g'(x)h(x))$.
13. $x^x(\ln x + 1); x^{x^x-1}[x \ln x(\ln x + 1) + 1]$.

§ 1211, page 384

1. $x = -ar, y = b, z = \tfrac{1}{2}\pi c + cr$ (r a parameter); $2ax - 2cz + \pi c^2 = 0$.
2. $x = -\tfrac{1}{2}\pi ar, y = \tfrac{1}{2}\pi b + br, z = \tfrac{1}{2}\pi c + cr$ (r a parameter);
$\pi ax - 2by - 2cz + \pi b^2 + \pi c^2 = 0$.
3. $x = 0, y = t, z = -t; y = z$.
4. $\dfrac{x-x_1}{Ax_1 + Fy_1 + Ez_1} = \dfrac{y-y_1}{Fx_1 + By_1 + Dz_1} = \dfrac{z-z_1}{Ex_1 + Dy_1 + Cz_1}$;
$Ax_1x + By_1y + Cz_1z + D(y_1z + z_1y) + E(x_1z + z_1x) + F(x_1y + y_1x) = G$.
7. $2\sqrt{61}; \tfrac{3}{5}$. **8.** $\sqrt{41}; -7/\sqrt{3}$. **10.** $2\sqrt{5}, -2 - \sqrt{3}$.

§ 1216, page 392

1. $\theta = \tfrac{1}{2}$.
2. $\dfrac{e}{2} = e^\theta\left[\sin\dfrac{\theta\pi}{6} + \dfrac{\pi}{6}\cos\dfrac{\theta\pi}{6}\right]$.
3. $\theta = \tfrac{1}{2}$.
4. $1 = 0 + \ln(1 + \theta e - \theta) + \dfrac{2(\theta e - \theta)}{1 + \theta e - \theta} - \dfrac{1}{2}\left(\dfrac{\theta e - \theta}{1 + \theta e - \theta}\right)^2$.
5. $2\tfrac{1}{12}$ sq. ft.; $\tfrac{1}{2}$ sq. in.
6. 1.387 in.; 0.003 in.
7. $a > 0$, minimum value $-a^2$; $a = 0$, no extremum; $a < 0$, maximum value $-a^2$.
8. Maximum value b/e; minimum value $\min(a, 0)/e$.
9. $1 + x + y + \dfrac{x^2}{2} + xy + \dfrac{y^2}{2} + \dfrac{x^3}{6} + \dfrac{x^2y}{2} + \dfrac{xy^2}{2} + \dfrac{y^3}{6}$.

ANSWERS

10. $(x-1) + (y-1) - \frac{(x-1)^2}{2} - \frac{(y-1)^2}{2}$

12. $1 + \frac{1}{2}(x-y); y.$ 15. $6 = 0 + 6\theta^0 + 6\theta^0 + 6\theta^0.$ 21. $f(x,y) = ax + by + c.$

22. (a) $f(x,y) = g(x) + h(y);$ (b) $f(x,y) = axy + bx + cy + d;$ (c) It does not exist.

§ 1219, page 397

1. $-\frac{2x+y}{x+6y}.$
2. $-\left(\frac{y}{x}\right)^{\frac{1}{3}}.$

3. $-\frac{\sin y + y\cos x}{\sin x + x\cos y}.$
4. $(a - cx + acy)^{-1}.$

5. $-\frac{Cy+Bz}{Bx+Ay}, -\frac{Cx+Az}{Bx+Ay}.$
6. $\frac{x^2-ayz}{axy-z^2}, \frac{y^2-axz}{axy-z^2}.$

7. $\frac{z}{x(1+yz)}, \frac{-z^2}{1+yz}.$
8. $\frac{1}{x(\cot z - 1)}, \frac{1}{y(1-\cot z)}.$

11. $\frac{4x-6x}{15u^4}, \frac{8yz-9y^2}{15u^4}.$
12. $\frac{3x^2-6x^2z+6xz^2}{(1-2z)(1-3u^2)}, \frac{3y^2-6y^2z+6yz^2}{(1-2z)(1-3u^2)}.$

13. $\frac{2-\cos z}{(1-\cos x)(1-e^w)}, \frac{2-\cos z}{(1-\cos x)(1-e^w)}.$

14. $\frac{f_2g_1-f_1g_2}{f_2g_2}, \frac{f_2g_1-f_1g_2}{f_2g_2}.$
16. $-\frac{f_2}{f_1}, \frac{1-f_2}{f_1}, \frac{f_1}{1-f_2}.$

17. $\frac{g_1-f_3}{f_1}, \frac{g_2}{f_1-g_1}, \frac{f_1}{g_2}.$
18. $\frac{f_1-2y}{2x-f_1}, \frac{f_2}{2y-f_2}, \frac{2x-f_1}{f_2}.$

19. $\frac{f_1+xzf_2}{f_1+yzf_2}, \frac{1-f_1-xyf_2}{f_1+xzf_2}, \frac{f_1+yzf_2}{1-f_1-xyf_2}.$
20. $\frac{g'-f_1}{f_2+f_3g'}.$

§ 1221, page 400

1. The circle $x^2 + y^2 = 9.$
2. The parabola $x^2 + 4y = 0.$
3. There is no envelope.
4. The parabola $y^2 = x + \frac{1}{4}$ if $t \geq \frac{1}{2}$; no envelope if $t < \frac{1}{2}.$
5. The lines $y = \pm 2x.$
6. The semicubical parabola $27x^2 = 4y^3.$
8. $27y^2 = 32.$
9. The parabola $2y = x^2.$
10. The line $y = 0$ and the circle $2x^2 + 2y^2 = y.$
11. The hypocycloid $x^{\frac{2}{3}} + y^{\frac{2}{3}} = a^{\frac{2}{3}}.$
12. The parabola $(x-y)^2 - 2c(x+y) + c^2 = 0.$
13. Two conjugate rectangular hyperbolas, $xy = \pm k.$
14. The hypocycloid $x^{\frac{2}{3}} + y^{\frac{2}{3}} = c^{\frac{2}{3}}.$
15. The square $|x| + |y| = c.$
16. The lemniscate $(x^2 + y^2)^2 = 4(x^2 - y^2).$

§ 1225, page 411

1. $\frac{du}{dx} = \frac{x^2u-xv^2}{u^3+v^3}, \frac{dv}{dx} = \frac{-xu^2-x^2v}{u^3+v^3}.$

2. $\frac{\partial u}{\partial x} = \frac{e^v-1}{e^{u+v}-1}, \frac{\partial u}{\partial y} = \frac{-e^v-1}{e^{u+v}-1}, \frac{\partial v}{\partial x} = \frac{e^u-1}{e^{u+v}-1}, \frac{\partial v}{\partial y} = \frac{e^u+1}{e^{u+v}-1}.$

ANSWERS

3. $\dfrac{du}{dx} = \dfrac{u(u-2x)}{(u-v)(u-w)}$, $\dfrac{dv}{dx} = \dfrac{v(v-2x)}{(v-u)(v-w)}$, $\dfrac{dw}{dx} = \dfrac{w(w-2x)}{(w-u)(w-v)}$.

4. If $D = 1 + \cos u \cos v \cos w$, $\dfrac{\partial u}{\partial x} = 1/D$, $\dfrac{\partial u}{\partial y} = -\cos v\, D$, $\dfrac{\partial v}{\partial x} = -\cos u \cos v/D$,
$\dfrac{\partial v}{\partial y} = 1/D$, $\dfrac{\partial w}{\partial x} = -\cos u/D$, $\dfrac{\partial w}{\partial y} = -\cos u \cos v\, D$.

5. $\left(\dfrac{\partial u}{\partial x}\right)_y = \dfrac{z-x}{u-z}$, $\left(\dfrac{\partial u}{\partial x}\right)_z = \dfrac{y-x}{u-y}$, $\left(\dfrac{\partial x}{\partial u}\right)_y = \dfrac{u-z}{z-x}$, $\left(\dfrac{\partial x}{\partial u}\right)_z = \dfrac{u-y}{y-x}$.

6. $\left(\dfrac{\partial u}{\partial x}\right)_{yz} = \dfrac{v-x}{u-v}$, $\left(\dfrac{\partial u}{\partial x}\right)_{yv} = \dfrac{z-x}{u-z}$, $\left(\dfrac{\partial u}{\partial x}\right)_{vz} = \dfrac{y-x}{u-y}$.

10. $-(x^2+y^2)^{-2}$.

34. x': $2x+y=0$, y': $x-2y=0$, $4x'^2+9y'^2=36$.

36. $\sqrt{6}$, 2, $\sqrt{2}$.

§ 1227, page 416

1. $\dfrac{2 \sin x^2}{x}$. $\quad\quad$ 2. $\dfrac{4}{x}\sinh x^4$.

§ 1231, page 428

8. $u^2+v^2=1$. $\quad\quad$ 9. $u^2-v^2=4$.
10. $u^2+v^2-w^2=0$. $\quad\quad$ 11. $u+2v-w^2=0$.

§ 1310, page 448

1. $\displaystyle\int_0^1\int_{x^2}^{x^3} f(x,y)dy\,dx$, $\displaystyle\int_0^1\int_{y^{1/3}}^{y^{1/2}} f(x,y)dx\,dy$, $\tfrac{1}{20}$.

2. $\displaystyle\int_0^a\int_{x^2/a}^{\sqrt{2ax-x^2}} f(x,y)dy\,dx$, $\displaystyle\int_0^a\int_{a-\sqrt{a^2-y^2}}^{\sqrt{ay}} f(x,y)dx\,dy$, $\dfrac{a^3}{12}(3\pi-4)$.

3. $\displaystyle\int_{-5}^{-4}\int_{-\sqrt{25-x^2}}^{\sqrt{25-x^2}} f(x,y)dy\,dx + \int_{-4}^{4}\int_{3x/4}^{\sqrt{25-x^2}} f(x,y)dy\,dx$,

$\displaystyle\int_{-3}^{3}\int_{-\sqrt{25-y^2}}^{4y/3} f(x,y)dx\,dy + \int_{3}^{5}\int_{-\sqrt{25-y^2}}^{\sqrt{25-y^2}} f(x,y)dx\,dy$, $25\pi/2$.

4. $\displaystyle\int_{-3}^{-2}\int_{x^2}^{\frac{1}{2}(15+\sqrt{129+40x})} f(x,y)dy\,dx + \int_{-2}^{1}\int_{\frac{1}{2}(15-\sqrt{129+40x})}^{\frac{1}{2}(15+\sqrt{129+40x})} f(x,y)dy\,dx$

$+ \displaystyle\int_{1}^{4}\int_{x^2}^{\frac{1}{2}(15+\sqrt{129+40x})} f(x,y)dy\,dx$, $\displaystyle\int_{1}^{4}\int_{\frac{1}{10}(y^2-15y+24)}^{\sqrt{y}} f(x,y)dx\,dy$

$+ \displaystyle\int_{4}^{9}\int_{-\sqrt{y}}^{\sqrt{y}} f(x,y)dx\,dy + \int_{9}^{10}\int_{\frac{1}{10}(y^2-15y+24)}^{\sqrt{y}} f(x,y)dx\,dy$, $58\tfrac{1}{4}$.

9. $\displaystyle\int_{-2}^{1}\int_{-y}^{\sqrt{2-y}} f(x,y)dx\,dy + \int_{1}^{2}\int_{-\sqrt{2-y}}^{\sqrt{2-y}} f(x,y)dx\,dy$.

ANSWERS

10. $\int_0^8 \int_{x^2/32}^{\sqrt{x}} f(x,y)\,dy\,dx$.

11. $\int_{\frac{1}{2}}^1 \int_1^{2x} f(x,y)\,dy\,dx + \int_1^2 \int_x^{2x} f(x,y)\,dy\,dx + \int_2^4 \int_x^4 f(x,y)\,dy\,dx$.

12. $\int_0^1 \int_0^1 f(x,y)\,dx\,dy + \int_1^{1+e} \int_{\phi(y)}^1 f(x,y)\,dx\,dy$, where $x = \phi(y)$ is the inverse of $y = x + e^x$.

13. $\frac{1}{12}$. 14. 45. 15. $\dfrac{a^4}{8}$.

16. $\frac{1}{5}(19 - 18a - 18a^2 - 10a^3 - 3a^4)$, where a is the (positive) real root of $e^{-x} = x$.

§ 1312, page 453

1. $\int_{-4}^4 \int_{-\sqrt{16-z^2}}^{\sqrt{16-z^2}} \int_7^{23-z^2-y^2} f(x,y,z)\,dz\,dy\,dx$, 128π.

2. $\int_{-1}^1 \int_{-1+|x|}^{1-|x|} \int_{-1+|x|+|y|}^{1-|x|-|y|} f(x,y,z)\,dz\,dy\,dx$, $\tfrac{2}{3}$.

5. $\int_0^2 \int_0^{\sqrt{2x-x^2}} \int_0^{\sqrt{2x-x^2-y^2}} f(x,y,z)\,dz\,dy\,dx$,

$\int_{-1}^1 \int_{1-\sqrt{1-y^2}}^{1+\sqrt{1-y^2}} \int_0^{\sqrt{2x-x^2-y^2}} f(x,y,z)\,dz\,dx\,dy$,

$\int_0^2 \int_0^{\sqrt{2x-x^2}} \int_{-\sqrt{2x-x^2-z^2}}^{\sqrt{2x-x^2-z^2}} f(x,y,z)\,dy\,dz\,dx$,

$\int_0^1 \int_{1-\sqrt{1-z^2}}^{1+\sqrt{1-z^2}} \int_{-\sqrt{2x-x^2-z^2}}^{\sqrt{2x-x^2-z^2}} f(x,y,z)\,dy\,dx\,dz$,

$\int_{-1}^1 \int_0^{\sqrt{1-y^2}} \int_{1-\sqrt{1-y^2-z^2}}^{1+\sqrt{1-y^2-z^2}} f(x,y,z)\,dx\,dz\,dy$,

$\int_0^1 \int_{-\sqrt{1-z^2}}^{\sqrt{1-z^2}} \int_{1-\sqrt{1-y^2-z^2}}^{1+\sqrt{1-y^2-z^2}} f(x,y,z)\,dx\,dy\,dz$.

6. $\tfrac{1}{10}$.

§ 1315, page 459

1. $\pi a^4/16$. 2. $5\pi a^4/128$.

3. $\dfrac{1}{5}\int_0^{\pi/4} \sec^5\theta\,d\theta = [7\sqrt{2} + 3\ln(\sqrt{2}+1)]/40$.

4. $16\pi/3$. 5. $3\pi/2$. 6. $\pi a^3/8$.

7. $(\tfrac{1}{3}\pi + \tfrac{1}{2}\sqrt{3})a^2$. 8. $\pi - \tfrac{3}{2}\sqrt{3}$. 9. $\tfrac{4}{3}a^3$.

ANSWERS

10. $\frac{2}{3}(3\pi - 4)a^2$. 11. $8\pi(10\sqrt{5} - 19)/3$. 12. $\pi/2$.
13. $\int_0^a \int_0^{\frac{\pi}{4}} F(\rho, \theta)\rho\, d\theta\, d\rho + \int_a^{\sqrt{2}a} \int_{\text{Arccos }(a/\rho)}^{\frac{\pi}{4}} F(\rho, \theta)\rho\, d\theta\, d\rho$.

§ 1318, page 461

1. $\bar{x} = \frac{1}{2}b$, $\bar{y} = \frac{1}{3}\sqrt{2bp}$.
2. $\bar{x} = \bar{y} = \frac{256a}{315\pi}$.
3. $\bar{x} = 0$, $\bar{y} = \frac{8\pi + 3\sqrt{3}}{4\pi + 6\sqrt{3}}a$.
4. $\bar{x} = \bar{y} = \frac{128a}{105\pi}$.
5. $\bar{x} = \frac{3}{5}$, $\bar{y} = \frac{1}{4}$.
6. $\bar{x} = \bar{y} = \frac{8a}{5\pi}$.
7. $\bar{x} = \frac{6a}{5}$, $\bar{y} = 0$.
8. $\bar{x} = \bar{y} = \frac{16a}{35}$.
9. $\frac{1}{4}\pi ab^3 = \frac{1}{4}b^2 A$.
10. $\frac{(3\pi + 8)a^4}{48} = \frac{3\pi + 8}{48}a^2 A$.
11. $\frac{bh^3k}{5} = \frac{3}{5}h^2 M$.
12. $\frac{63\pi k}{20}a^5 = 1.89\, a^2 M$.

§ 1322, page 466

1. $\frac{1}{3}\pi(16\sqrt{2} - 17)$.
2. $V = \frac{1}{3}\pi a^3(\sqrt{2} - 1)$, $\bar{z} = \frac{3}{8}a(\sqrt{2} + 1)$.
3. $\frac{1}{6}\pi(13\sqrt{13} - 41)$.
4. $\frac{1}{2}\pi a^2 hk = \frac{1}{2}a^2 M$.
5. $\bar{x} = \frac{3}{8}a$, $\bar{y} = \frac{3\pi}{16}a$, $\bar{z} = \frac{3\pi}{32}ab$.
6. $\frac{1}{15}\pi a^4 hk = \frac{2}{15}a^2 M$.
7. Centroid on axis of symmetry, distance $\frac{3}{8}a$ from center (a = radius), $I = \frac{4}{15}\pi a^5 k = \frac{2}{5}a^2 M$.
8. $M = \frac{2}{3}\pi a^3(1 - \cos\alpha)k$, $\bar{z} = \frac{3}{8}a(1 + \cos\alpha)$.
9. $M = \frac{\pi^2 a^4 k}{8}$, $\bar{z} = \frac{16a}{15\pi}$, $I_z = \frac{\pi^2 a^6 k}{16} = \frac{a^2}{2}M$.
10. $M = \pi a^4 k$; $I = \frac{4a^2}{9}M$.
11. $M = abck$, $I_{yz} = \frac{a^2 M}{3}$, $I_x = \frac{(b^2 + c^2)M}{3}$, $I_0 = \frac{(a^2 + b^2 + c^2)M}{3}$.
12. $(\bar{x}, \bar{y}, \bar{z}) = \left(\frac{a}{4}, \frac{b}{4}, \frac{c}{4}\right)$, $I_z = \frac{a^2 + b^2}{10}M$.
13. $\bar{z} = \frac{3(b^4 - a^4)}{8(b^3 - a^3)}$.
14. $M = \frac{1}{3}[\sqrt{2} + \ln(1 + \sqrt{2})]a^3 k$, $I = \frac{1}{30}[7\sqrt{2} + 3\ln(1 + \sqrt{2})]a^5 k$.
16. Centroid: distance $4a/3\pi$ from center; volume: $2\pi^2 a^2 b$.

§ 1325, page 471

1. $\frac{GMh}{(a^2 + h^2)^{\frac{3}{2}}}$.
2. $\frac{2GM}{a^2}\left(1 - \frac{h}{\sqrt{a^2 + h^2}}\right)$.
3. $\frac{GM}{2ab}\ln\left[\frac{c + a}{c - a} \cdot \frac{b + \sqrt{(c - a)^2 + b^2}}{b + \sqrt{(c + a)^2 + b^2}}\right]$.

ANSWERS

4. $\dfrac{2GM}{a^2 h}[h + \sqrt{a^2 + (c-h)^2} - \sqrt{a^2 + c^2}]$.

5. $\dfrac{GM}{a^2}(\sqrt{2} - 1)$.
6. $\dfrac{6GM}{a^2}\left(1 - \dfrac{h}{\sqrt{a^2 + h^2}}\right)$.

§ 1330, page 488

1. $\tfrac{1}{15}\pi$.
2. $\dfrac{\sqrt{3}\,\pi}{18}$.
3. $\tfrac{1}{6}(b-a)(d-c)$.
4. $\tfrac{1}{4}(e^2 - e)$.
5. $\tfrac{1}{3}[\sqrt{2} + \ln(1 + \sqrt{2})]$.
6. n odd, 0; n even, $\dfrac{1\cdot 3\cdot 5 \cdots (n-1)}{2\cdot 4\cdot 6 \cdots (n+2)} \cdot 2\pi(a^2 + b^2)^{n/2}$.
7. $-\tfrac{1}{12}$.
8. $\bar{x} = \bar{y} = \bar{z} = \tfrac{1}{2}$.
19. π. 20. -4π. 21. Divergent. 22. 2π.
23. $\tfrac{1}{3}\pi$. 24. -8π. 25. 8π. 26. Divergent.

§ 1403, page 499

1. Absolutely convergent.
2. Absolutely convergent.
3. Divergent.
4. Conditionally convergent.
5. Conditionally convergent, $0 < p < 1$.
6. Absolutely convergent, $0 < p < 2$, conditionally convergent, $-1 < p \leq 0$.
7. Absolutely convergent, $p < 1$, conditionally convergent, $1 \leq p < 2$.
8. Absolutely convergent, $1 < p < 3$.

§ 1415, page 523

9. $\tfrac{1}{2}\sqrt{\pi}$. 10. $\sqrt{\pi/5}$.
11. $\sqrt{2\pi}$. 12. $\pi/8$.

§ 1511, page 556

1. $\dfrac{4}{\pi}\left[\dfrac{\sin x}{1} + \dfrac{\sin 3x}{3} + \dfrac{\sin 5x}{5} + \cdots\right]$; $1 - \tfrac{1}{3} + \tfrac{1}{5} - \tfrac{1}{7} + \cdots = \dfrac{\pi}{4}$.

2. $\dfrac{1}{2} + \dfrac{2}{\pi}\left[\dfrac{\sin x}{1} + \dfrac{\sin 3x}{3} + \dfrac{\sin 5x}{5} + \cdots\right]$; $1 - \tfrac{1}{3} + \tfrac{1}{5} - \tfrac{1}{7} + \cdots = \dfrac{\pi}{4}$.

3. $\dfrac{1}{4} + \dfrac{1}{\pi}\sum_{n=1}^{+\infty}\left[\dfrac{1}{n}\sin\tfrac{1}{2}n\pi\cos nx + \dfrac{1}{n}(1 - \cos\tfrac{1}{2}n\pi)\sin nx\right]$;

$\sum_{n=1}^{+\infty}(-1)^{n+1}/(2n-1) = \pi/4$.

4. $\dfrac{1}{4} + \dfrac{1}{\pi}\sum_{n=1}^{+\infty}\left[-\dfrac{1}{n}\sin\tfrac{1}{2}n\pi\cos nx + \dfrac{1}{n}(\cos\tfrac{1}{2}n\pi - \cos n\pi)\sin nx\right]$;

$\sum_{n=1}^{+\infty}(-1)^{n+1}/(2n-1) = \pi/4$.

5. $\dfrac{\pi}{4} + \sum_{n=1}^{+\infty}\left[-\dfrac{1 + (-1)^{n+1}}{n^2\pi}\cos nx + \dfrac{(-1)^{n+1}}{n}\sin nx\right]$; $\sum_{n=1}^{+\infty}(2n-1)^{-2} = \pi^2/8$.

6. $\dfrac{\pi}{16} + \sum_{n=1}^{+\infty}\left\{\left[\dfrac{1}{2n}\sin\tfrac{1}{2}n\pi - \dfrac{1}{n^2\pi}(1 - \cos\tfrac{1}{2}n\pi)\right]\cos nx \right.$

$\left. + \left[-\dfrac{1}{2n}\cos\tfrac{1}{2}n\pi + \dfrac{1}{n^2\pi}\sin\tfrac{1}{2}n\pi\right]\sin nx\right\}$; $\sum_{n=1}^{+\infty}(2n-1)^{-2} = \pi^2/8$.

ANSWERS

§ 1515, page 562

2. $b_n = (-1)^{n+1}\dfrac{2\pi^2}{n} + (-1)^n\dfrac{12}{n^3}$.

3. $a_0 = \dfrac{2\pi^4}{5}$, $a_n = (-1)^n\dfrac{8\pi^2}{n^2} + (-1)^{n+1}\dfrac{48}{n^4}$, $n > 0$.

4. $b_n = (-1)^{n+1}\dfrac{2\pi^4}{n} + (-1)^n\dfrac{40\pi^2}{n^3} + (-1)^{n+1}\dfrac{240}{n^5}$.

5. $\dfrac{1}{2} + \dfrac{2}{\pi}\sum_{n=1}^{+\infty}(-1)^n\cos(2n-1)x/(2n-1)$; $\dfrac{2}{\pi}\sum_{n=1}^{+\infty}\dfrac{1}{n}(\cos\tfrac{1}{2}n\pi - \cos n\pi)\sin nx$.

6. $\dfrac{\pi}{4} - \dfrac{8}{\pi}\left[\dfrac{\cos 2x}{2^2} + \dfrac{\cos 6x}{6^2} + \dfrac{\cos 10x}{10^2} + \cdots\right]$; $\dfrac{4}{\pi}\left[\dfrac{\sin x}{1^2} - \dfrac{\sin 3x}{3^2} + \dfrac{\sin 5x}{5^2} - \cdots\right]$.

7. $\cos x$; $\dfrac{4}{\pi}\left[\dfrac{2\sin 2x}{2^2-1} + \dfrac{4\sin 4x}{4^2-1} + \dfrac{6\sin 6x}{6^2-1} + \cdots\right]$.

8. $\dfrac{2}{\pi} - \dfrac{4}{\pi}\left[\dfrac{\cos 2x}{2^2-1} + \dfrac{\cos 4x}{4^2-1} + \dfrac{\cos 6x}{6^2-1} + \cdots\right]$; $\sin x$.

9. $\dfrac{4}{\pi}\left[\dfrac{\sin \pi x}{1} + \dfrac{\sin 3\pi x}{3} + \dfrac{\sin 5\pi x}{5} + \cdots\right]$.

10. $\dfrac{4}{\pi}\left[\dfrac{\sin \tfrac{1}{2}\pi x}{1} - \dfrac{\sin \tfrac{3}{2}\pi x}{2} + \dfrac{\sin \tfrac{5}{2}\pi x}{3} - \cdots\right]$.

11. $1 + \dfrac{4}{\pi}\sum_{n=1}^{+\infty}\left\{-\dfrac{1+(-1)^{n+1}}{n^2\pi}\cos\dfrac{n\pi x}{4} + \dfrac{(-1)^{n+1}}{n}\sin\dfrac{n\pi x}{4}\right\}$.

12. $\dfrac{3}{\pi} + \dfrac{6}{\pi}\sum_{n=1}^{+\infty}\left\{\left[\dfrac{1}{2n}\sin\tfrac{1}{2}n\pi - \dfrac{1}{n^2\pi}(1-\cos\tfrac{1}{2}n\pi)\right]\cos\dfrac{n\pi x}{6}\right.$
$\left. + \left[-\dfrac{1}{2n}\cos\tfrac{1}{2}n\pi + \dfrac{1}{n^2\pi}\sin\tfrac{1}{2}n\pi\right]\sin\dfrac{n\pi x}{6}\right\}$.

15. $a_n = \dfrac{2}{p}\int_0^p f(x)\cos\dfrac{n\pi x}{p}dx$, $b_n = \dfrac{2}{p}\int_0^p f(x)\sin\dfrac{n\pi x}{p}dx$.

21. $B_0(x) = 1$, $B_1(x) = x - \tfrac{1}{2}$, $B_2(x) = x^2 - x + \tfrac{1}{6}$, $B_3(x) = x^3 - \tfrac{3}{2}x^2 + \tfrac{1}{2}x$, $B_4(x) = x^4 - 2x^3 + x^2 - \tfrac{1}{30}$.

§ 1531, page 581

1. $u = \dfrac{8p}{\pi^2}\sum_{n=1}^{+\infty}(-1)^{n+1}\dfrac{\sin(2n-1)x\cos(2n-1)at}{(2n-1)^2}$.

8. $u(x,t) = \dfrac{4p}{\pi a}\sum_{n=1}^{+\infty}\left[\dfrac{(-1)^{n+1}\sin(2n-1)h}{(2n-1)^3}\sin(2n-1)x\sin(2n-1)at\right]$.

9. $u(x,t) = \dfrac{8p}{\pi^2}\sum_{n=1}^{+\infty}(-1)^{n+1}\dfrac{\sin(2n-1)x\,e^{-(2n-1)^2a^2t}}{(2n-1)^2}$.

10. $u(x,t) = A\sin 2x\,e^{-4a^2t} + B\sin 7x\,e^{-49a^2t}$.

12. $u(x,t) = \tfrac{1}{2}a_0 + \sum_{n=1}^{+\infty}a_n\cos nx\,e^{-n^2a^2t}$, $a_n = \dfrac{2}{\pi}\int_0^\pi f(x)\cos nx\,dx$, $n = 0, 1, 2, \cdots$.

13. $u(x,t) = \dfrac{p}{2} - \dfrac{16p}{\pi^2}\sum_{n=1}^{+\infty}\dfrac{\cos(4n-2)x\,e^{-(4n-2)^2a^2t}}{(4n-2)^2}$.

15. $50 + \dfrac{30x}{\pi} - \dfrac{200}{\pi}\sum_{n=1}^{+\infty}\dfrac{\sin(2n-1)x\,e^{-(2n-1)^2a^2t}}{(2n-1)} + \dfrac{40}{\pi}\sum_{n=1}^{+\infty}\dfrac{\sin 2nx\,e^{-4n^2a^2t}}{2n}$.

ANSWERS

§ 1538, page 591

6. $1, \sqrt{3}(2x-1), \sqrt{5}(6x^2-6x+1), \sqrt{7}(20x^3-30x^2+12x-1)$.

11. $\sqrt{\dfrac{1}{\pi}}, \sqrt{\dfrac{2}{\pi}}x, \sqrt{\dfrac{8}{\pi}}(x^2-\tfrac{1}{2}), \sqrt{\dfrac{32}{\pi}}(x^3-\tfrac{3}{4}x)$.

12. $\sqrt[4]{\dfrac{1}{\pi}}, \sqrt[4]{\dfrac{4}{\pi}}x, \sqrt[4]{\dfrac{9}{\pi}}(x^2-\tfrac{1}{3}), \sqrt[4]{\dfrac{16}{9\pi}}(x^3-\tfrac{3}{5}x)$.

13. $1, x-1, \tfrac{1}{2}x^2-2x+1, \tfrac{1}{6}x^3-\tfrac{3}{2}x^2+3x-1$.

14. $1, 1-\dfrac{\alpha+\beta}{\alpha}x, 1-2\dfrac{\alpha+\beta+1}{\alpha}x+\dfrac{(\alpha+\beta+1)(\alpha+\beta+2)}{\alpha(\alpha+1)}x^2;\ \sqrt{B(\alpha,\beta)}$.

642

References For Further Reading In Undergraduate Real Analysis

This recommended reading section is going to be quite a bit longer then I usually write for a Blue Collar Scholar text because analysis is a make or break subject for mathematics students. Therefore, I want to recommend as many potential study sources as possible. We'll discuss several standard texts as well as some nonstandard ones at this level. Brand new editions of most of these standard texts are prohibitively expensive. But not discussing them would be committing educational malpractice because these are the top standard analysis books at both the basic and intermediate levels. You really need to at least look at them if you're serious about learning undergraduate analysis beyond calculus. Since Olmsted is a very broad text for all students of undergraduate analysis, we'll be suggesting texts at both the beginning and intermediate levels.

Let's begin with "baby real variable" texts which can well supplement the early chapters of Olmsted. A great-and now cheap-text on single variable advanced calculus/honors calculus/elementary real analysis by a master teacher at MIT is Arthur Mattuck's book (1).I was absolutely floored to learn this incredible book was reissued in a fantastically cheap paperback last year after becoming exceedingly scarce-and correspondingly expensive-in its' original hardcover edition. For bare beginners, analysis should look like calculus, only much more careful. The author takes great pains to give proofs in great detail, all done by specific calculation methods at first, which slowly give way to more general arguments as theorems are established throughout the book. The Completeness property is given in terms of Cauchy convergence. The author writes beautifully and clearly, with many deep insights that are usually omitted as obvious in not only analysis courses, but calculus courses. For example, he goes into some detail on why subtracting inequalities is illegal. He also can be wryly amusing at times. There are tons of excellent problems-best of all, they come with complete solutions, which will make the book incredibly useful for self-study. Mattuck has written an outstanding textbook that all students of mathematics regardless of level can learn from. It will make a terrific companion to Olmsted, which covers a number of topics beyond Mattuck.[1]

Some other excellent books at about the same level are Ross (2) and Liebl (3). Ross is exactly what the title says it is and it's by a master analyst. There's no better book for a student beginning real analysis with a weak calculus background who needs to get up to speed, which sadly is all too common these days. I used the first edition in my student days. There's currently a considerably enlarged second edition with many added topics like metric spaces, but the first edition is quite good and much cheaper.[2] Another book you could try which is very good for this, but considerably

---

[1] Several other books described here will make excellent companions depending on the level student and the ultimate goal of learning. For example, Olmsted combined with one of the functions of several variables texts listed later will provide the basis for a complete undergraduate course on analysis of one and several variables.

[2] But if you're lucky enough to afford the second edition, by all means, go for it.

more difficult, is Michael Spivak's classic (4). Don't let the title fool you-this is a rigorous presentation of one variable calculus with thorough discussions of sequences, convergence, functions and limits. It's beautifully written with many examples-but it's considerably harder than either Mattuck or Ross, especially the exercises. It's also sadly quite expensive. Lebl covers essentially the same topics as Mattuck and is quite inexpensive. In fact, it's available online for free at the author's website. The paperback is cheap, so you should be able to get a second hand copy. It's quite a bit more concise in exposition and lacks the depth of either Mattuck or Spivak. But the author writes quite well and the low price makes it a good option. Another good reason to buy the book besides the cost is that the author has continued to revise and refine the text as he's continued to teach out of it. The book also has a sequel volume on functions of several variables, but it's nowhere near as good in my opinion.

The famous text by Rudin (5) has been, of course, the All Hollowed Holy Ordained Text of Undergraduate Analysis for over half a century. It's more abstract, concise and covers less ground then Olmsted. It was the first modern analysis text on abstract metric spaces and it's still the prototype for such courses. It's notorious for its legendarily difficult problem sets, which have been a rite of passage for mathematics majors for generations. To be honest, I'm not a big fan of it and never really have been. I know, the purists in the audience want to lynch me right now. As I said earlier, calculus classes that usually preceded Rudin in the curricula just ain't what they used to be. But even if that wasn't the case, I wouldn't love this book. When I was a student learning analysis, I had an old beat up copy that I scraped together enough to buy. When I was done with it, I couldn't wait to ditch it. It's dry as dust and reads almost like a legal document because it completely lacks motivation. There's the barest minimum exposition with almost no explanation and few examples. I know, a lot of people like it for that reason because it forces students to wrestle with it and build their own examples and exposition. I certainly understand that approach. But honestly, unless you're dealing with superior students, is that really a reasonable expectation in a first brush with a subject as difficult as real analysis? Even with well-prepared honors students, they really won't get as much out of the course using it simply because it's so pristine and terse. Sure, they'll be able to do a lot of the problems and understand definitions and theorems. They'll get what they need out of it. But unless they have a professional analyst who's also a first rate teacher, I'll bet most of them walk out of the course not having a clue why it's necessary to build all this abstract machinery to support the structures of analysis/calculus. It'll definitely teach them how to do analysis, but will they really understand it after that? I have my doubts.

 The long-standing competitor to Rudin as the standard intermediate level analysis course-and the book I much preferred when first learning the subject-is the book by Tom

Apostol (6).³ Apostol's book provides an interesting contrast to Rudin. While it covers more or less the same ground as Rudin and at the same level-that is to say, it develops intermediate level analysis on metric spaces-it is less terse and far more readable. In fact, it is about the same difficulty level as Olmsted, although it is somewhat more advanced. It takes a fully set-theoretic approach to analysis and the topology of metric spaces, which allows Apostol to define everything much more precisely and in a detailed manner, with many examples. Rudin often skims or skips over many of these fine points, which is one of the things that makes the book confusing to a beginner. Unfortunately, the exercises are nowhere near as deep or interesting as Rudin's and that hurts the book. ⁴However, Apostol's exposition is second to none and demonstrates why the late Apostol was an award winning teacher in calculus, analysis and differential equations, particularly for the gifted incoming freshmen at the California Institute of Technology, for so many years.

A book I much prefer to Rudin-and is superior to Apostol in some ways -is Charles Chapman Pugh's (7). It's based on the honors analysis course Pugh taught at The University of California at Berkeley for over 30 years. I affectionately

---

³ I actually used the 1957 first edition, which is considerably different in organization from the more well-known second edition. The first edition contains considerable material on classical vector analysis and Fourier analysis limited only to Riemann integrable functions. both of which I found immensely useful as a beginner. The second edition removes the vector analysis material and instead develops the Lebesgue integral in terms of step functions and then gives a fully general treatment of Fourier series. The second edition also has cleaner and more modern notation, developing basic point set topology more fully. However, I'll always love the first edition and still heartily recommend it to anyone who can't afford the far more expensive current edition for learning analysis.

⁴ In fact, the best advice I gave analysis students for many years was to use both Apostol **and** Rudin: Study the subject out of Apostol and do the exercises in Rudin. Although this approach lead to some unavoidable repetition, in general, the 2 books complemented each other quite well.

refer to it as Rudin Done Right. The book is pitched at the same level as Rudin, with exercises just as difficult if not more so. So what's different about it? Firstly, Pugh writes in a beautiful, conversational yet very concise style. It's immensely personable, contains many references to the mathematical literature and yet is amazingly curt and direct. There's very little chatter. So how is Pugh able to be concise like Rudin and still be very informative and illuminating? Pugh has a remarkable gift as a textbook author: He seems to know instinctively exactly how many words it takes to explain something completely-not one word more, not one word less. For example, when defining the Cantor set, he refers to it as "Cantor dust". This phrase is extremely insightful in picturing the Cantor set for the first time. This intentional parsimony, where he so carefully chooses every word and constructs every sentence is one way he does it. The other is the book is full of pictures-there's literally a picture on every other page. But what's amazing here is that none of the pictures are purely decorative, throwaways or space fillers-every single picture is very specifically designed to make a point. For example, when proving that in a general metric space, every open ball is itself an open set, he presents it alongside a picture of an open disk in the plane and how a smaller open ball can be constructed anywhere inside an open ball. It also covers certain topics vastly better then Rudin or other texts-his chapters on function spaces and their applications, differential equations and multivariable analysis are considerably better than Rudin's. Indeed, the chapter on functions of several variables is an excellent short course that's far superior to the very confusing treatment in *Principles*. In particular, it has a brief treatment of differential forms that is far superior and can set the stage for a follow up course in functions of several variables. It will act as a perfect complement to Olmsted by supplying the few modern topics missing from it.

While we're on the subject, sadly, my favorites among the standard texts on functions of several variables at this level are quite expensive. Therefore, I hesitate to recommend any here. But this is a suggested bibliography of undergraduate analysis texts and we've already decided not to put a price limit on these recommended texts. So I feel if I don't make several recommendations, I'll be performing academic malpractice. The standard books for learning this material are Spivak (8) and Munkres (9). Both contain roughly the same topics: the topology of Euclidean space, differential calculus of several variables, multiple integrals, line and path integrals, multilinear algebra, manifolds and differential forms in Euclidean space and the general Stokes' theorem. Spivak's book is basically a problem course with quite a few pictures. It's rough going, but it's worth the effort if you've got the patience. Munkres is more of a standard textbook and covers the same material with much more detail. If you're looking for a rapid introduction that'll cover the material in minimum space and you've already had a standard multivariable calculus course, Spivak really will fill your needs best if you're willing to work through it. Indeed, the entire point of the book is to build to careful proofs of the variants of Stokes' theorem. Munkres sacrifices brevity for clarity, so if you need more detail, that's a better choice. Be warned, though- Munkres has a bit more in the way of prerequisites then Spivak. Spivak basically assumes only good courses in linear algebra and rigorous $\epsilon-\delta$ single variable calculus. By contrast, Munkres assumes in addition, a good working knowledge of basic topology in metric spaces. Both books work mostly in R^n and discuss abstract manifolds at the end in passing. Both are still really good choices-again, it depends on how much time you have to invest.

Lastly, one can recommend the classic (10). To the joy of all who love great mathematics, it was finally republished in an inexpensive softcover in 2014. This was after over 35 years of inaccessibility unless one wanted to pay a king's ransom for it. (I know, I did and then had to break my own heart selling the book to help pay for groceries when my father had cancer.) The book is famous as the first textbook to develop out of Harvard University's notorious genius freshman-destroying Math 55 course and was designed to give a completely modern development of vector analysis. It's the most comprehensive and advanced text on multivariable calculus ever written. But make no mistake-it is not an easy book to learn from by any means. It's more difficult than either Spivak or Munkres, pitched about midway in difficulty between Rudin and a full blown graduate analysis text. It presents limits of functions of several variables, partial derivatives, the differential,

differential equations, integration of several variables using Jordan content (an old fashioned idea of measure that works well for relatively non-pathological functions with connected domains) differential forms and manifolds, all modelled on Banach spaces. The book contains both abstract theory and many calculation topics, such as the Jacobian matrix and iterated integrals. It's far broader and more thorough in scope then Cartan's texts, with many, many topics not even brushed on in most of the books listed here. It also contains examples in Euclidean spaces-not as many as I'd prefer, but a substantial number. It also covers modern and abstract presentations of differential equations, calculus of variations, classical potential theory and modern classical mechanics! Lastly, it has many strong and very difficult exercises developing entire tangential topics such as Fourier analysis. It's a remarkable book for any serious analysis student who wants a modern presentation of multivariable calculus. It also will find a lot of combat action as a reference for graduate students of analysis.

Most second courses in analysis focus on metric spaces. The unusual and beautiful text by Hoffman (11) is an important exception to this rule. Its' an affordable alternative at about the same level as Rudin or Apostol, but it focuses mostly on Euclidean normed spaces. Since most of the important topological and algebraic properties of calculus are identical in Banach spaces as in Euclidean spaces, Hoffman makes for excellent complementary text to books on metric spaces.

More importantly, Hoffman gives a beautifully deep and insightful presentation of standard classical analysis topics that emphasizes how calculus on real and complex Euclidean and abstract normed spaces compares to that on the real line-and more important for the purposes of modern analysis, how Euclidean and abstract normed spaces compare in properties. Hoffman's unique approach is to use classical analysis on normed spaces to prepare students for how functional analysis is done on abstract normed spaces. Topics covered are sequences and series in both real and complex normed spaces, convergence and compactness theorems, Fourier series, completion of normed spaces, Lebesgue integration via completion, quotient spaces and much, much more. There are also many deep and wonderful exercises. After a dedicated study of Hoffman and the appropriate algebra, a mathematics student will be more than ready to tackle a serious modern analysis course.

While we're on the topic, the first textbook that fully presented analysis entirely on normed spaces was Dieudonne's classic (12), which is now available in an inexpensive edition. This book had a seminal influence on many analysts who were coming up in the 1960's, particularly those trained in Europe. It's comprehensive, uncompromisingly modern and abstract. It contains a great deal of graduate level material, including the Lebesgue measure and integral, Hilbert spaces and an introduction to the spectral theory of linear operators. Indeed, it was originally written in the late 1950's as a first year graduate text, although I doubt it could be used as the main text for such a course today unless it is extensively supplemented. I'd agree the book is historically important as representative of the Bourbakian approach to classical analysis and it does have a wealth of interesting material. That being said, I'd consider looking at it after a Rudin-level analysis course or as a sequel to Olmsted's book. I believe it would be a perfect sequel to Hoffman. Approach with caution, for serious advanced users only!

Another famous text that presents analysis on normed spaces at the intermediate level that's at a somewhat higher level than Olmsted is Henri Cartan's 2 volume set (13) and (14). This is an experimental advanced course in calculus on normed spaces that's set entirely in Banach spaces, which has the interesting effect that functions of one and several values become unified. The first volume covers the differential calculus in normed spaces via the Frechet operator as well as applications such as local extrema. It also contains an in-depth study of the theory of differential equations on Banach spaces requiring only basic analysis and linear algebra. The second volume covers multilinear algebra and differential forms on Banach spaces as well path integrals and the general Stokes' theorem, basic differential geometry as well as a very influential treatment of the calculus of variations with applications to mechanics. The first volume was out of print for nearly 4 decades and was incredibly expensive to obtain a used copy of.[5] Strangely, the second volume was republished by the venerable Dover Books 12 years ago in a very inexpensive edition. Now both are once again available easily and cheaply. The pair makes a terrific and geometric follow up to an intermediate Olmsted/Rudin-level analysis course, particularly for students with an interest in topology and geometry in higher spaces.

Another very good choice I like a great deal is the second edition of G. Krantz's excellent book (15).[6] Krantz is overall a good textbook author and I like all his books, but I personally think this is his best. It's an interesting book, pitched at about the same level as Olmsted and not quite at the same level as Rudin. The book covers all the same topics as Rudin's and the influence of the book on Krantz is clear. But there is *far* more explanation, examples and details and those are chosen with the greatest of care so the student is not spoon-fed. One of my favorite parts of the book are brief "glimpses" of important advanced topics, such as applications to differential equations and physics, functions

---

[5] Which is why once I got permission, it was the first book I republished. However, I've been informed by some users they found the small print of my preface disconcerting. I'll have to modify the book at some point to rectify this. Otherwise- I'm very proud this was the first book BCS republished.

[6] **Warning: This recommendation is for the 2nd edition.** At this writing (early 2019), there have been 4 editions of Krantz' book. Why am I recommending the 2nd edition specifically? Well, while the first edition is ok, it's full of errors and doesn't cover as much. While the later 2 editions are very solid, they remove most of the additional topics discussed above as too difficult for a course at this level. I think this was a mistake by the author. Even though he's right that full, detailed treatments of these topics are more suited to graduate courses, I feel he does a very good job giving short primers to math majors that will both pique their interest and prepare them for graduate analysis courses. In addition, the 2nd edition can be obtained a lot cheaper than the newer ones. This is why I prefer the 2nd edition and I think it's better for students.

of several variables, differential forms, harmonic analysis and wavelets at the undergraduate level. There is also a set of wonderful, tough exercises that nevertheless, unlike Rudin, are all doable by the average mathematics major. Best of all, the book's level of difficulty gradually increases towards the advanced topics at the end. This is a modern Rudin well suited for today's students: It educates them at a level appropriate for their preparation, but it still succeeds by the end in giving the student a deep, challenging course in the wondrous world of real analysis.

My favorite intermediate level analysis text is probably Korner's (16). This is a bit of an oddball text pitched at a level between Olmstead and Rudin. But it's one of the best all-around textbooks on analysis out there. Firstly, it's written by Tom Korner, who besides being one of the top research analysts in the world at the University of Cambridge, is one of the best currently active math textbook writers. Second, the major themes of the book are 1) why a rigorous approach to analysis is needed and 2) how can it be constructed. Korner asks and answers many deep related questions about classical analysis at several different levels of generality-the real line, abstract metric spaces, normed spaces, functions of one and several variables, etc. Some are: *Is the intermediate value theorem obvious? Are the real numbers unique? How does one define powers on infinite sequences and series? Why isn't the Riemann integral sufficient for most analysis? How do norms create a connection between the algebraic structure on vector spaces and abstract metric spaces?* Korner has a wonderfully literate and humorous style. He intersperses many fascinating historical footnotes on the history of the subject and unorthodox references that together with his mathematical expertise make the book an absolute pleasure to read. He also has a lot of original presentations of classical results that are very insightful. For example, Korner explains the bisection algorithm that plays such an important role in the classical proofs of the intermediate value theorem by calling it "lion hunting" i.e. creating a smaller and smaller pen around the lion's jungle until the lion is captive in a fixed location. There are also lots of additional topics that aren't presented in most calculus or analysis books, such as interchanging limits in multiple sums and integrals, general limits via directed sets, constructive analysis, the contraction mapping theorem and differential equations and a careful presentation of the logarithm function. Best of all, there are quite literally hundreds of exercises of all kinds-half interspersed throughout the text and the other half in a lengthy appendix-ranging in difficulty from simple one word answers to substantial brain teasers culled from the Tripos exams that'll challenge the very best students. Plus, Korner is one of the best texts to complement with Cartan's text(s) since the books overlap and complement each other in surprising ways and on surprising topics, such as differential equations and the calculus of variations. Seriously, if you can only afford one of these standard texts, Korner's is the one you need to get. Try and get a second hand copy or even better, an international softcover editions as it's much cheaper.

Rosenlicht's (17) for many years was one of the standard collateral readings with Rudin. It gives a very clear, careful and inexpensive analysis course on metric spaces. It has many nice exercises, none too hard. It reads like a very good set of lecture notes. However, this is both its strength and its weakness. Like a set of lecture notes, it's very sparse in detail and limited to the most essential facts and topics. It's also quite dry. Still, it's very clear and it'll certainly be a helpful supplement to any of the texts listed above.

The book by the late Robert Ash (18) is considerably more readable and broader in range then Rosenlicht. Although the choice of topics is fairly standard for such a course, Ash develops both the topological and analytical aspects of both Euclidean and metric spaces with considerable detail, intuition and many pictures. Ash's teaching skill and background in electrical engineering both shine through on every page to illuminate the subject. Although it's not as original or well written as either Pugh or Korner, its low price and excellent clarity make it a very good choice as an analysis text. Best of all, it's the only textbook on analysis I know that comes with complete solutions, making it ideal as a self-study text. It'll make a wonderful low cost supplement to Olmsted. Ash wrote a number of equally solid advanced mathematics textbooks on algebra, probability and analysis for courses as well as self-study students-all are available in inexpensive paperbacks from Dover. Sadly, there will be no more. Ash was tragically killed while taking his dog for a walk by being run over in 2015.

(Believe me, I wish that was a joke in poor taste.)

A recent text I've become very enamored with is the wonderfully idiosyncratic 2 volume text by Terence Tao (19),(20). This beautiful text is based on Tao's lecture notes for the honors analysis course at UCLA and demonstrates that Tao is that rarest and most precious of individuals in academia: A preeminent researcher who is also an extremely talented teacher. Such individuals bring a richness and insight to their instruction from their work at the forefront of the subject that many lecturers utterly lack and which the best of students can immensely profit from. The unique quality of Tao's presentation of single and multivariable analysis in these books is his emphasis on foundational concepts. The structure of the first volume really exemplifies Tao's approach. The first half of volume 1 centers around a full construction of the number systems beginning with naïve set theory and the Peano axioms of the natural numbers through the real numbers as convergent sequences of rational numbers. Along the way, Tao not only defines the fundamental tools and objects of mathematics such as sets, relations, functions, cardinality, equivalence relations, inequalities, etc., he also shows many beautiful and careful examples. Many mathematicians hesitate to give an explicit construction of the reals in an analysis course because it's quite time consuming.⁷ It's a daring approach and Tao argues in the preface that it was well worth the time investment because students then have no problem applying the properties of the real numbers in explicit limit calculations.⁸ Tao's course brims with this kind of educational as well as mathematical insight. I heartily recommend both these books for strong students and teachers of analysis as well as a follow-up to Olmsted.

I have one last suggested book, one that's brand new and I've been very impressed with. It's the textbook by Cummings (21). At the website for the text, Cummings claims he wrote his book from the lecture notes he's produced for his students in real analysis in order to have a book which was deliberately wordy and inexpensive. It was written for complete beginners in rigorous analysis who may have never seen careful mathematics before. In many ways, the book is the anti-Rudin. It's pitched at about the same level as Olmsted, although it doesn't cover quite as many topics. But its' written in the most detailed and gentle manner possible so that students can learn analysis by reading the book and working the many exercises once with maximum clarity and minimal confusion. It covers more or less the standard topics of analysis on the real line: the real

---

[7] Remember, I originally considering including such a construction in this text and ultimately relented because I didn't think I could do such an important construction justice in a single chapter. Which shows why Tao is a far better mathematician then I am.

[8] My own experiences as a student support this. I struggled mightily with epsilons, deltas and using inequalities to compute limits until I constructed the reals from scratch. After that, I rarely had any problems working with inequalities because the construction internalizes the properties of all the number systems.

numbers, basic set theory and cardinality, limits of sequences and series, functions, the basic topology of the real line, continuity and differentiation of functions, Riemann integration and sequences and series of functions. It also contains 2 terrific appendices: The first giving a brief outline of the construction of the real numbers via Dedekind cuts beginning with the natural numbers. The second gives an overview of many of the strange counterexamples of analysis such as the Cantor set and the Devil's staircase. But there are lots of pictures and examples and accompanying each section are lengthy explanations, comments and even jokes. The book is not only immensely informative and gentle, it's genuinely fun to read-something one would think a real variables course would be incapable of being. While Cummings' book certainly could be used as a classroom text, I think it would serve students and teachers even better as a study guide supplementing a standard text-a kind of Schaum's Outline for basic analysis. I have no doubt any student would be thrilled to have Cummings' book as a supplement to their first course in analysis-it will greatly assist any student by accompanying any of the standard texts here. It would also be one of the very best choices for students in other fields of science for self-study in real analysis. This is an absolute must-have for any student or teacher of analysis to put on their shelf. Get a copy along with this copy of Olmsted. Or any real analysis text, for that matter. You won't regret it, believe me.

These are my suggestions as study texts for undergraduate analysis courses. There are so many others I wish I could have listed, but many are as expensive as Rudin. So it's simply beyond the means of students to purchase more than one of them-and most of my target audience will barely be able to buy the most inexpensive here. This is most tragic as many are truly wonderful and most mathematics is effectively learned by studying multiple sources.

But then that's indicative of the central problem of our modern world that's driving my publishing company,isn't it?

Bibilography

(1) Mattuck, Arthur, *An Introduction To Analysis*, CreateSpace Independent Publishing Platform,2013

(2) Ross, Kenneth, *Elementary Analysis: The Theory of Calculus*, Springer-Verlag, 1st ed, 1980, 2nd ed, 2013

(3) Lebl, Jeri, *Basic Analysis I: Introduction to Real Analysis*, Createspace   Independent Publishing Platform, 5th ed. 2018

(4) Spivak, Michael, *Calculus,* Publish or Perish Inc., 4th ed, 2008

(5) Rudin, Walter, *Principles of Mathematical Analysis*, McGraw-Hill Education, 3rd ed,1976

(6) Apostol, Tom, *Mathematical Analysis*, Pearson, 2nd ed, 1974

(7) Pugh, Charles Chapman, *Real Mathematical Analysis*, Springer-Verlag, 2nd edition, 2016

(8) Spivak, Michael, *Calculus On Manifolds*, New Ed., CRC Press, 2018

(9) Munkres, James R., *Analysis On Manifolds*, Westview Press, 1997

(10) Loomis, Lynn, Sternberg, Schlomo, *Advanced Calculus Revised Edition*, World Scientific, 2014

(11) Hoffmann, Kenneth, *Analysis In Euclidean Spaces*, Dover Publishing, 2007

(12) Dieudonne, Jean, *Foundations Of Modern Analysis*, Hesperides Press, 2006

(13) Cartan, Henri, Maestro, Karo. *Differential Calculus On Normed Spaces: A Course In Analysis*, Createspace Independent Publishing Platform, 2nd ed, 2017

(14) Cartan, Henri, *Differential Forms*, Dover Publishing, 2006

(15) Krantz, Steven G., *Real Analysis And Foundations*, CRC Press, 2nd ed, 2004

(16) Korner, Tom, *A Companion To Analysis: A First Second And A Second First Course In Analysis*, American Mathematical Society, 2003

(17) Rosenlicht, Maxwell, *Introduction To Analysis*, Dover Publishing, 1985

(18) Ash, Robert B., *Real Variables With Basic Metric Space Topology*, Dover Publishing, 2009

(19) Tao, Terence, *Analysis I*, 3rd ed, Springer-Verlag, 2015

(20) Tao, Terence, *Analysis II*, 3rd ed, Springer-Verlag, 2016

(21) Cummings, Ray, *Real Analysis: A Long Form Mathematics Textbook*, Createspace Independent Publishing Platform, 2018

INDEX

(The numbers refer to pages)

Abel's theorem, 288
Abel test for improper integrals, 496, 499 (Ex. 11)
   uniform convergence, 504, 513 (Ex. 30)
Abel test for series, 230 (Exs. 22, 23)
   uniform convergence, 277 (Exs. 49, 50)
Absolute convergence of improper integrals, 174 (Ex. 44), 496
Absolute convergence of infinite products, 531 (Ex. 1)
Absolute convergence of series, 228
Absolute maximum (minimum), 105
Absolute value, 17
Addition of numbers, 3
Addition of series, 233
Algebraic function, 202
Algebraic number, 201
Almost everywhere, 153 (Ex. 54), 437
Almost periodic functions, 550
Alternating harmonic series, 226
Alternating integrals, 496
Alternating p-series, 228
Alternating series, 225
Analytic functions, 267-269
Antiderivative, 155
Approximations by differentials, 109-111, 387
Arc, 340
   simple, 341
   space-filling, 342
Archimedean property, 11 (Ex. 24), 24
Arc length, 343, 356
   existence of, 345
   independence of parametrization, 346
   integral form, 347, 356
Area, 431
   independence of coordinates, 431
   inner, 434
   outer, 434
Area of a surface, 492 (Ex. 9)
Associative law, 2, 3, 8, 25, 327 (Exs. 1, 2)
Asymptotes, 123
Attraction, 468

Axiom of completeness, 22
Axiom of induction, 7, 27 (Ex. 16)
Axioms, closure, 335
   open set, 335
   Peano, 1
Axioms of order, 5
Axioms of the basic operations, 2, 3
Axioms of the real numbers, 27

Basis, 504
Bernoulli numbers, 263 (Ex. 35), 269 (Ex. 9)
Bernoulli polynomials, 563 (Exs. 21-22)
Bessel functions, 591
Bessel's inequality, 548
Beta function, 171, 521-524
Between, 7
Bicontinuous mapping, 317
Big O notation, 169, 213
Binary number, 20 (Ex. 4)
Binomial series, 252
Binomial theorem, 12 (Ex. 35), 252
Bliss's theorem, 149 (Ex. 42), 339, 450 (Ex. 37)
Bolzano-Weierstrass Theorem, 82 (Ex. 13), 323 (Ex. 22)
Bonnet form of the second mean value theorem, 186 (Ex. 29)
Bound, lower, 23
   upper, 22
Boundary, 297
   of a surface, 493 (Ex. 15)
Boundary conditions, 578, 580
Bounded convergence, 152 (Ex. 49)
Bounded function, 54
Bounded sequence, 35, 320
Bounded set, 23, 297
Bounded variation, 175-179, 566, 573
Bracket function, 29, 51
Bray (see Helly-Bray), 533-534

Cancellation law, 4 (Exs. 2, 9)
Category, 332 (Ex. 43)

INDEX

Cauchy criterion, for functions, 65, 315 (Ex. 19)
  for improper integrals, 174 (Ex. 43), 495, 503
  for infinite products, 532 (Ex. 4)
  for sequences, 62-63, 277 (Ex. 45), 320
  for series, 230 (Ex. 17), 277 (Ex. 46)
  for uniform convergence, 277 (Exs. 45, 46), 315 (Ex. 19), 503, 532 (Ex. 4)
Cauchy form of the remainder, 248
Cauchy inequality, 13 (Ex. 43), 147 (Ex. 29), 220 (Ex. 26), 229 (Ex. 14), 543
Cauchy principal value, 173
Cauchy-Riemann differential equations, 370 (Ex. 23), 413 (Ex. 25)
Cauchy sequence, 63, 320
Center of curvature, 359
Centroid, 460, 465, 468
Cesàro summability, 229 (Ex. 15), 230 (Ex. 16), 277 (Ex. 44), 568, 570
Chain rule, 87, 374, 407
Change of variables, 87, 374, 407
  in multiple integrals, 472-492
Characteristic equation, 414 (footnote)
Characteristic function, 580
Characteristic value, 580
Circle of curvature, 359
Circle, osculating, 364 (Ex. 27)
Classification of numbers and functions, 201-205
Closed curve, 341
Closed disk, 295
Closed interval, 16
Closed orthonormal sequence, 584
Closed region, 298
Closed set, 75, 296
Closure axioms, 335
Closure of a set, 82 (Ex. 18), 298
Commutative law, 3, 9, 25, 325
Compact region, 298
Compact set, 76, 297, 320, 334, 336
Comparison tests for improper integrals, 168-171, 495, 502
  for series, 212-215, 274
Complement of a set, 75, 325
Complete metric space, 334, 338 (Ex. 33)
Completeness, axiom of, 22
Complete ordered field, 27
Complete orthonormal sequence, 584
Component of a set, 332 (Ex. 37)
Composite number, 11 (Ex. 22)
Composition of functions, 87, 374, 407
Composition of ordinates, 123
Computation of series, 235-239
Computations with series, 264-267
Conditional convergence, improper integrals, 174 (Ex. 45), 496
  series, 228

Connected set, 77, 320, 321, 322
Constant function, 28
Constant of integration, 156
Content, 431
Continued fraction, 21 (Ex. 15)
Continuity, 49, 306, 316
  and integrability, 133, 137, 153 (Ex. 54), 432
  and uniform convergence, 278, 312, 505
  from the right or left, 50
  modulus of, 287
  negation of, 53 (Ex. 14)
  of inverse function, 57, 81, 321
  sectional, 159
  uniform, 71, 318
Continuity theorems, 52-55, 57, 80, 81, 137, 309, 320
Continuous image, 317
Continuous nondifferentiable function, 86, 286 (Ex. 41)
Convergence, bounded, 152 (Ex. 49)
  dominated, 175 (Ex. 47)
  interval of, 240-241
  in the mean, 584
  of improper integrals, 164-175, 486
  of sequences, 31, 319, 328 (Ex. 23)
  of series, 206
  uniform, of improper integrals, 500
  uniform, of sequences, 270
  uniform, of series, 273
Convergence theorem for Fourier series, 555, 567
Convex closure, 332 (Ex. 39)
Convex hull, 332 (Ex. 39)
Convex set, 332 (Ex. 38)
Coordinate transformation, 406
Cosine series, 559
Covering of a set, 83 (Ex. 28)
Cracked china definition, double integral, 432
Criterion, Cauchy (see Cauchy criterion)
Criterion for continuity, sequential, 64
Critical point, 105, 390
Critical value, 105, 390
Curvature, 357
  circle of, 359
  radius of, 359
Curve, 341
  closed, 341
  simple, 341
  smooth, 349
Curve, Jordan, 476, 477 (footnote)
Curve tracing, 123-128
Cylindrical coordinates, 354, 462

Decimal expansions, 20 (Ex. 5), 220 (Exs. 28-30)

INDEX

Dedekind's theorem, 26 (Ex. 11)
Definite integral, 131-154
Deleted neighborhood, 41, 294
De Morgan's laws, 328
Dense, 19, 541
Density of polynomials, 571
Denumerable set, 21 (Ex. 12)
Dependence, functional, 410
Dependence, linear, 586
Dependent variable, 28
Derivative, 85
    directional, 380, 384 (Ex. 9)
    one-sided, 88-91
    partial, 366, 367
Derived set, 328
Determinant, derivative of, 379 (Ex. 15
Determinant, functional, 403
Diameter of a set, 331 (Ex. 33), 433
Differentiable function, 85, 372
Differential, 107, 373
    approximations by, 109-111, 387
Differentiation, 85-130, 366-428
    and uniform convergence, 281-283, 313, 506
    implicit, 394, 401, 419
    of power series, 259
    under the integral sign, 416, 506
Directional derivative, 380, 384 (Ex. 9)
Dirichlet's integral, 564
Dirichlet's theorem, 232
Discontinuity, jump, 52
    removable, 51
Discontinuous derivative, 90
Discrete topology, 336
Disjoint sets, 76, 325
Disk, closed, 295
    open, 295
Distance between a point and a set, 82 (Ex. 19), 324 (Ex. 26)
Distance between two sets, 82 (Ex. 21), 324 (Ex. 26)
Distance function, 333
Distance in a function space, 539
Distance space, 540
Distributive law, 3, 9, 326, 327 (Exs. 3, 9)
Divergence theorem, 493 (Ex. 12)
Division of numbers, 3
Domain of definition, 28, 60
Dominance, 168, 212, 274, 502
Dominant terms, 123
Dominated convergence, 175 (Ex. 47)
Double integral, 430, 433
    in polar coordinates, 455
    transformation theorem, 473-486
Duhamel's principle, 339

e, 60 (Ex. 40), 190 (Ex. 5)
Eigenfunctions, 580

Eigenvalues, 580
Elementary functions, 203
Empty set, 75
End-points, 16
Enumerable set, 21 (Ex. 12)
Envelope of a family, 398
Equality of mixed partial derivatives, 368
Equivalence of neighborhoods, 294
Euclid, fundamental theorem of, 11 (Ex. 25), 14 (Ex. 15)
Euclidean spaces, 16, 293-301
Euler's constant, 239 (Exs. 19, 22), 527
Euler's theorem, 378, 380 (Ex. 31)
Evaluation of improper integrals, 508-512
Evaluation of series (see Zeta-function)
Even function, 144 (Exs. 7-13)
Evolute, 361
Exact differential, 491 (Ex. 3)
Existence theorem for implicit functions, 419
Existence theorem for inverse transformations, 424
Exponential function, 188-191, 292

F_σ set, 332 (Ex. 40)
Factor, 11 (Ex. 22)
    highest common, 14 (Ex. 13)
Factorial, 12 (Ex. 34)
Fejér summability theorem, 568
Field, 20 (Ex. 8)
    complete ordered, 27
    ordered, 21 (Ex. 11)
Finitely decomposable compact region, 491 (Ex. 4), 493 (Ex. 11)
First category, 332 (Ex. 43)
First derivative test, 106
First mean value theorem for integrals, 137, 143 (Exs. 5, 6), 186 (Ex. 25)
Fourier coefficients, 546
Fourier series, 537-593
    convergence theorem, 555, 567
    cosine series, 559
    generalized, 547
    integration of, 573
    Jordan convergence theorem, 567
    sine series, 559
    trigonometric, 551
    uniform convergence of, 572
Fresnel integrals, 513
Frontier, 297
Fubini's theorem, 446, 452
Function, 28, 60, 301, 315
    algebraic, 202
    analytic, 267-269
    beta, 171, 521-524
    binomial, 252

INDEX

bounded, 54
bracket, 29, 51
constant, 28
continuous, 49, 306
differentiable, 85, 372
direct, 57, 60
elementary, 203
even, 144 (Exs. 7-13)
gamma, 171, 518-521
greatest integer, 29, 51
harmonic, 370 (Ex. 18)
homogeneous, 377
inverse, 57, 60
locally homogeneous, 380 (Ex. 31)
monotonic, 55
nondifferentiable, 86, 286 (Ex. 41)
odd, 144 (Exs. 8-13)
rational, 202
real-valued, 28
signum, 42, 52
single-valued, 28, 60
transcendental, 202
Functional dependence, 410, 426
Functional determinant, 403
Functions, hyperbolic, 197-201
 trigonometric, 191-194, 292, 351
Functions of several variables, 293-338
Function space, 538
Fundamental increment formula, 371
Fundamental theorem for double integrals, 444
 for triple integrals, 452
Fundamental theorem of integral calculus, 154-155, 158 (Exs. 19, 20)
Fundamental theorem on bounded sequences, 61, 320
Fundamental theorems on mappings, 320

G_δ set, 332 (Ex. 40)
Gamma function, 171, 518-521
Gauss's test, 224 (Ex. 7)
Geometric series, 208
Gibbs phenomenon, 575
Gradient, 382, 384 (Ex. 9)
Gramian determinant, 591 (Ex. 6)
Gram-Schmidt process, 587, 592 (Ex. 8)
Graph of a function, 60, 324 (Ex. 30)
Greatest integer function, 29, 51
Greatest lower bound, 23
Green's theorem in the plane, 491 (Ex. 5)
Group, 20 (Ex. 7)
Grouping of terms, 231

Half-open interval, 16
Harmonic function, 370 (Ex. 18)
Harmonic series, 210
Hausdorff space, 337 (Ex. 15), 338 (Exs. 29, 30)

Heat conduction, 580-583
Heat equation, 580
Heine-Borel theorem, 83 (Ex. 28), 324 (Exs. 24, 25), 337 (Ex. 11)
Helly-Bray convergence theorem, 533-534
Hermite polynomials, 591
Highest common factor, 14 (Ex. 13)
Hölder inequality, 415 (Exs. 38, 40)
Homeomorphism, 317
Homogeneous function, 377
 locally, 380 (Ex. 31)
l'Hospital's rule, 115-121
Hyperbolic functions, 197-201
Hypergeometric series, 224 (Ex. 6)

Identity transformation, 408
Image, 317
 continuous, 317
 inverse, 317
Implicit differentiation, 394, 401, 419
Implicit function theorem, 419
Improper integrals, 164-175, 495-536
 alternating, 496
 Cauchy principal value, 173
 evaluation of, 508-512
Improper integrals of infinite series, 515
Improper multiple integrals, 486
Improper Riemann-Stieltjes integrals, 532
Indefinite integral, 155
Independence, linear, 586
Independence of parametrization, 346
Independence of path, 491 (Ex. 3), 492 (Ex. 8)
Independent variable, 28
Indeterminate forms, 115-121, 264
Index set, 325
Induction, mathematical, 7, 24, 27 (Ex. 16)
Inductive set, 27 (Ex. 16)
Inequality, Bessel's, 548
 Hölder, 415 (Exs. 38, 40)
 Minkowski, 13 (Ex. 44), 147 (Ex. 30), 229 (Ex. 14), 415 (Exs. 39, 40), 543
 Schwarz (Cauchy), 13 (Ex. 43), 147 (Ex. 29), 220 (Ex. 26), 229 (Ex. 14), 543
 triangle, 17, 301, 333, 540, 543
Infinite derivative, 86
Infinite interval, 16
Infinite limit, of a function, 42, 302
 of a sequence, 31
Infinite products, 524-527, 530, 563 (Ex. 19)
Infinite sequence, 30
Infinite series (see Series, infinite), 206-292

INDEX

Initial conditions, 578, 581
Inner area, 434
Inner product, 541
Integer, negative, 13
    positive, 7, 13, 27 (Ex. 16)
Integral, definite, 131-154
    double, 430, 433, 455
    Fresnel, 513
    improper, 164-175, 495-536
    indefinite, 155
    iterated, 444, 452, 457, 463, 465, 506, 513
    line, 490-494
    lower, 150 (Ex. 44), 450 (Ex. 38)
    multiple, 429-494
    proper, 132, 165
    Riemann, 131-154
    Riemann-Stieltjes, 179-187, 573
    surface, 490-494, 493 (Ex. 10)
    triple, 451, 462, 463
    upper, 150 (Ex. 44), 450 (Ex. 38)
Integral form of the remainder, 247
Integral points, 296
Integral test, 209
Integration, 131-187, 429-494
    and uniform convergence, 279-281, 312, 506
    by parts, 157 (Exs. 10-13), 160 (Ex. 7), 181, 573
    by substitution, 156, 185 (Ex. 18)
    of Fourier series, 573
    of power series, 259
Interior, 297
Interior point, 17, 297
Intermediate value property, 55, 103 (Ex. 47), 309, 322
Intersection of sets, 325
Interval, 16
Interval of convergence, 240-241
Intrinsic property, 324 (Ex. 32)
Invariance of area, 431
Inverse function, 57, 60
    continuity of, 57, 81, 321
    differentiability of, 88, 98
Inverse image, 317, 337 (Ex. 16)
Inverse mapping, 317, 406, 424
Inverse transformation, 317, 406, 424
Inversion of power series, 269 (Exs. 7, 8)
Involute, 361
Irrational number, 13
Irrationality of e, 202, 263 (Ex. 34)
Irrationality of π, 202
Iterated integral, 444, 452, 457, 463, 465, 506, 513
Iterated limits, 304, 310 (Ex. 32)
Iterated suprema, 515

Jacobian, 403
    and multiple integral, 472, 490

Jacobi polynomials, 591
Jordan content, 431
Jordan convergence theorem, 567
Jordan curve, 476, 477 (footnote)
Jump discontinuity, 52

Kummer's test, 221

Lagrange form of the remainder, 247
Lagrange multipliers, 413 (Ex. 28)-414 (Ex. 37)
Laguerre polynomials, 591
Lamina, 459
Landau, E., 1
Laplace's equation, 370 (Ex. 18), 409, 413 (Exs. 26, 27)
Law, associative, 2, 3, 8, 25, 327 (Exs. 1, 2)
    cancellation, 4 (Exs. 2, 9)
    commutative, 3, 9, 25, 325
    distributive, 3, 9, 326, 327 (Exs. 3, 9)
    transitive, 6 (Exs. 2, 19)
    trichotomy, 6 (Ex. 3)
Law of cosines, 558 (Ex. 24)
Law of the mean, 94, 385
    extended, 99, 386
    generalized, 95
Leading coefficient, 12 (Ex. 42)
Least squares, 545
Least upper bound, 23
Lebesgue measure zero, 153 (Exs. 52-54)
Lebesgue theorem on bounded convergence, 152 (Ex. 49)
Lebesgue theorem on dominated convergence, 175 (Ex. 47)
Legendre polynomials, 589
Legendre series, 590
Leibnitz's rule, 93 (Ex. 25), 416, 417
l'Hospital's rule, 115-121
Limit, along a path, 303
    in the mean, 584
    iterated, 304, 310 (Ex. 32)
    of a function, 40, 301
    of a sequence, 31, 319, 328 (Ex. 23)
    of a variable, 44
    one-sided, 41
    uniform, 270, 311
Limit inferior, 67 (Ex. 16), 68 (Ex. 24), 323 (Ex. 17), 328 (Ex. 21)
Limit point, of a sequence, 40 (Exs. 20-22)
    of a set, 75, 296
Limit superior, 67 (Ex. 16), 68 (Ex. 24), 323 (Ex. 17), 328 (Ex. 21)
Limit theorems, for functions, 44
    for sequences, 34
Line integral, 490-494
Linear combination, 538

INDEX

Linear dependence and independence, 536
Linear space, 538
Little o notation, 169, 213
Locally exact differential, 492 (Ex. 7)
Locally homogeneous function, 380 (Ex. 31)
Logarithm, 188-191, 292
Lower bound, 23
Lower integral, 150 (Ex. 44), 450 (Ex. 38)
Lower semicontinuity, 69 (Ex. 38)
Lowest terms, 11 (Ex. 23)

Maclaurin series, 245-253
   for $\cos x$, 251
   for e^x, 250
   for $\ln(1+x)$, 251
   for $\sin x$, 250
   for $(1+x)^m$, 252
Mapping, 315
   bicontinuous, 317
   inverse, 317, 406, 424
   topological, 317
Mass, 459, 465, 468
Mathematical induction, 7, 24, 27 (Ex. 16)
Maxima and minima, 105-107, 388-391, 413 (Ex. 28)-414 (Ex. 37)
Maximal orthonormal sequence, 584
Maximum of a function, 54, 80
Mean value theorem for derivatives, 94, 385
   extended, 99, 386
   generalized, 95
Mean value theorem for integrals, first, 137, 143 (Exs. 5, 6), 186 (Ex. 25)
   second, 157 (Ex. 14), 186 (Exs. 27-29)
Mean value theorem for multiple integrals, 440
Measure zero, 153 (Ex. 52)
Membership notation, 327
Mertens, theorem of, 239 (Ex. 20)
Metric space, 333, 540
Minimum of a function, 54, 80
Minkowski inequality, 13 (Ex. 44), 147 (Ex. 30), 229 (Ex. 14), 415 (Exs. 39, 40), 543
Mixed partial derivatives, 367
   equality of, 368
Modulus of continuity, 287
Moments, 460, 465, 468
Monotonic function, 55
Monotonic sequence, 37
Monotonically decreasing sequence of sets, 83 (Ex. 26)
Moore-Osgood theorem, 285 (Ex. 37), 313

Multiple integrals, 429-494
   and Jacobians, 472-490
   improper, 486
Multiplication of numbers, 3
Multiplication of series, 233

Natural numbers, 1, 27 (Ex. 16)
Negation of continuity, 53 (Ex. 14)
Negation of uniform continuity, 73
Negation of uniform convergence, 272, 311, 501
Negative integer, 13
Negative number, 5
Negative of a number, 3
Neighborhood, 17, 294, 336
   circular, 293
   cubical, 299
   deleted, 41, 294
   of infinity, 32
   spherical, 299
   square, 294
Nested intervals theorem, 83 (Ex. 26)
Net, 131, 429, 451
Nondenumerability of the reals, 220 (Ex. 30)
Nondifferentiable function, 86, 286 (Ex. 41)
Norm in a linear space, 542
Norm of a net, 131, 430
Norm of a partition, 433
Normal derivative, 383
Normal direction, 383, 412 (Ex. 18)
Normalization, 544
Normalized Legendre polynomials, 589
Normal line, 383
Normal plane, 383
Normed linear space, 543
Notational pitfalls, 396, 404
Nowhere dense, 332 (Ex. 42)
Number, algebraic, 201
   binary, 20 (Ex. 4)
   composite, 11 (Ex. 22)
   integral, 13
   irrational, 13
   natural, 1, 27 (Ex. 16)
   negative, 5
   positive, 5
   prime, 11 (Ex. 22)
   rational, 13
   real, 1-27
   transcendental, 202

Odd function, 144 (Exs. 8-13)
One, 3
One-sided derivative, 88-91
One-to-one correspondence, 60
Open covering, 83 (Ex. 28)
Open disk, 295

INDEX

Open interval, 16
Open mapping, 425
Open set, 75, 295
Open set axioms, 335
Order of magnitude, 169, 213
Order relation, 5
Ordered field, 21 (Ex. 11)
Ordered pair, 60
Orientable surface, 494 (Ex. 16)
Orthogonal functions, 537-593
Orthogonality, 541
Orthogonality with respect to a weight function, 590
Orthogonal projection, 318, 547
Orthonormal family, 544
Orthonormal sequence, closed, 584
   complete, 584
   maximal, 584
Oscillation of a function, 84 (Ex. 30)
Osculating circle, 364 (Ex. 27)
Outer area, 434

Parametric equations, 125
Parseval's equation, 553, 559, 591 (Ex 2)
Partial derivatives, 366, 367
Partial differentiation, 366-428
Partial fractions, 16 (Ex. 22), 562 (Ex 16)
Partition, 432, 451
Parts, integration by, 157 (Exs. 10-13), 160 (Ex. 7), 181, 573
Path, limit along, 303
Peano axioms, 1
Peano curve, 342
Periodic functions, 549
   almost, 550
Point set algebra, 324
Point sets, 74-84, 294-301, 324-338
Point transformation, 405
Polynomial, 46 (Ex. 19)
Polynomial equation, 201
Polynomials, Hermite, 591
   Jacobi, 591
   Laguerre, 591
   Legendre, 591
   Tchebycheff, 591
Positive integer, 7, 13, 27 (Ex. 16)
Positive number, 5
Power series, 240-269
   differentiation of, 259
   integration of, 259
   inversion of, 269 (Exs. 7, 8)
Preimage, 317
Prime numbers, 11 (Ex. 22)
   relatively, 11 (Ex. 23)
Primitive, 155
Principal value, Cauchy, 173 (Ex. 30)

Products, infinite, 524-527, 530
Product series, 233
Product space, 60
Projection, 318, 547
Proper integral, 132, 165
p-series, 210
   alternating, 228
Pseudo-metric space, 540
Pythagorean theorem, 544

Quotient of numbers, 3
Quotient of polynomials, 14 (Ex. 15)

R^1, 541
R^2, 544
R^p, 557 (Ex. 21)
Raabe's test, 222
Radius of convergence, 241, 242
Radius of curvature, 359
Range of values, 28, 60
Ratio test, 215, 220 (Ex. 31), 228
Rational function, 47 (Ex. 20), 48 (Ex. 35), 202
Rational number, 13
Rational points, 296
Real number, 1-27
Real-valued function, 28
Rearrangement of terms, 231
Reciprocal of a number, 3
Rectifiable arc, 344
Reducible covering, 83 (Ex. 28)
Reduction formulas, 161-163
Region, 298, 331 (Ex. 35)
   closed, 298
   compact, 298
   simply-connected, 492 (Ex. 8)
Regular point, 348, 382
Relative maximum (minimum), 105, 388
Remainder, 14 (Ex. 15) (see Taylor)
Removable discontinuity, 51
Representation by series, 248
Riemann integral, 131-154
Riemann-Lebesgue theorem, 565, 576 (Ex. 9)
Riemann's lemma, 548
Riemann-Stieltjes integral, 179-187, 532
Riemann Zeta-function, 214, 236, 556, 559, 560, 562 (Exs. 2-4), 563 (Exs. 20, 22), 577 (Ex. 18)
Riesz-Fischer theorem, 554
Rodrigues formula, 589
Rolle's theorem, 93
Root test, 218, 220 (Ex. 32)
Roots of numbers, 55

Scalar, 539
Schwarz inequality, 13 (Ex. 43), 147

INDEX

(Ex. 29), 220 (Ex. 26), 229 (Ex. 14), 543
Second category, 332 (Ex. 43), 338 (Ex. 33)
Second derivative test, 106
Second mean value theorem for integrals, 157 (Ex. 14), 186 (Exs. 27-29)
Sectional continuity, 159
Sectional smoothness, 159
Semicontinuity, 69 (Ex. 38)
Separated sets, 77
Sequence, 30, 319, 328 (Ex. 23)
Sequence of distinct points, 78
Sequential criterion for continuity, 64
Sequential criterion for limits, 64
Series, infinite, 206-292
   addition of, 233
   alternating, 225
   alternating harmonic, 226
   basic definitions, 206
   computation of, 235-239
   computations with, 264-267
   Fourier, 537-593
   geometric, 208
   harmonic, 214
   Maclaurin, 245-253
   multiplication of, 233
   p-series, 210, 214
   power, 240-269
   product, 233
   representation by, 248
   subtraction of, 233
   Taylor, 244-253, 391
   uniform convergence of, 273
Set, bounded, 23, 297
   closed, 75, 296
   compact, 76, 297, 320, 334, 336
   connected, 77, 320-322
   denumerable, 21 (Ex. 12)
   open, 75, 295
Sets of points, 74-84, 294-301, 324-338
Signum function, 42, 52
Simple arc, 341
Simple compact region, 491 (Ex. 4), 493 (Ex. 11)
Simple continued fraction, 21 (Ex. 15)
Simple curve, 341
Simply-connected region, 492 (Ex. 8), 494 (Ex. 19)
Simpson's rule, 146 (Ex. 23)
Sine series, 559
Single-valued function, 28, 60
Singular point, 348, 382
Smooth curve, 349
Smoothness, sectional, 159
Smooth surface, 382
Space, Euclidean, 16, 293-301
   metric, 333
   product, 60
   topological, 335
Space-filling arc, 342
Spectrum, 580
Spherical coordinates, 355, 463
Square root, 7, 55
Steady-state problem, 583 (Ex. 14)
Step-function, 138, 439
Stirling's formula, 528
Stokes's theorem, 494 (Ex. 17)
Subsequence, 32
Substitution of power series, 255-259
Subtraction of numbers, 3
Subtraction of series, 233
Subtraction of sets, 327
Summability of series, 229 (Ex. 15), 230 (Ex. 16), 568
   uniform, 277 (Ex. 44), 570
Summation notation, 12 (Ex. 36)
Surface, 382, 492 (Ex. 9)
   orientable, 494 (Ex. 16)
   smooth, 382
Surface area, 492 (Ex. 9)
Surface integral, 490-494, 493 (Ex. 10)

Tangent line, 383, 412 (Ex. 17)
Tangent plane, 383
Taylor series, 244-253, 391
Taylor's formula with a remainder, 246-248, 386
   Cauchy form, 248
   integral form, 247
   Lagrange form, 247
Tchebycheff polynomials, 591
Terminating decimal, 20 (Ex. 5)
Topological equivalence, 317
Topological mapping, 317
Topological spaces, 335
Topology, 335
   discrete, 336
   Hausdorff, 337 (Ex. 15), 338 (Exs. 29, 30)
   trivial, 336
Total differential, 373
Total variation, 175-177
Transcendental function, 202
Transcendental number, 202
Transformation, 315
   coordinate, 406
   identity, 408
   inverse, 317, 406, 424
   point, 405
Transformation theorem for double integrals, 473-486
Transient problem, 583 (Ex. 14)
Transitive law, 6 (Exs. 2, 19)
Trapezoidal rule, 145 (Ex. 22)

INDEX

Triangle inequality, 17, 301, 333, 540, 543
Trichotomy, 6 (Ex. 3)
Trigonometric Fourier series, 551
Trigonometric functions, 191-194, 292, 351
Trigonometric series, 551
Triple integral, 451, 462, 463
  in cylindrical coordinates, 462
  in spherical coordinates, 463
Trivial topology, 336

Uniform approximation theorem of Weierstrass, 291 (Ex. 22), 570
Uniform continuity, 71, 318
Uniform convergence, 270-292, 311-314
  and continuity, 278, 312, 505
  and differentiation, 281-283, 313, 506
  and integration, 279-281, 312, 506
Uniform convergence of Fourier series, 572
Uniform convergence of improper integrals, 500
Uniform convergence of infinite products, 532 (Ex. 4)
Uniform limits, 311
Uniform summability, 277 (Ex. 44), 570
Union of sets, 325
Unique factorization theorem, 11 (Ex. 29)
Unity, 3
Unwinding string, 364 (Ex. 28)
Upper bound, 22
Upper integral, 150 (Ex. 44), 450 (Ex. 38)
Upper semicontinuity, 69 (Ex. 38)

Variable, dependent, 28
  independent, 28
  real, 28
Variation, bounded, 175-179, 566, 573
  negative, 177
  positive, 177
  total, 175-177
Vector, 381, 539
Vector space, 539
Venn diagram, 325-327
Very nearly everywhere, 437
Vibrating string, 577-582
  plucked, 578
  struck, 582 (Ex. 7)
Volume, 431, 451

Wallis's formulas, 174 (Ex. 36)
Wallis's infinite product for π, 526
Wave equation, 578
Weierstrass continuous nondifferentiable function, 86
Weierstrass infinite product for $1/\Gamma(\alpha)$, 530
Weierstrass M-test, 274, 502, 532 (Ex. 5)
Weierstrass uniform approximation theorem, 291 (Ex. 22), 570
Weight function, 590
Well-ordering principle, 8
Without loss of generality, 129
Wronskian determinant, 104 (Ex. 51)

Zero, 3
Zeta-function of Riemann, 214, 236, 556, 559, 560, 562 (Exs. 2-4), 563 (Exs. 20, 22), 577 (Ex. 18)

About the Author

John M. H. Olmsted (1911-1997) was part of the pre-WW II generation of American mathematics students who were the first native teachers of modern mathematics trained by the European-trained missionaries at top programs. He was a graduate student at Princeton University in the 1930's and got his PhD in 1940 under Salomon Bochner. His research was mostly focused on integration theory and function spaces, but he was known from the beginning for his passion for teaching. It was really this skill as a teacher for which he was known. Olmsted taught most of his career at the University of Southern Illinois At Carbondale with periods at the University of Minnesota and a fellowship at the Princeton Institute for Advanced Study. He was Department Chair at USIC for many years, He was known for personally designing and teaching virtually all analysis courses there ranging from freshman calculus to graduate analysis. In his memory, USIC has named a department award for the best Masters' student in the department-The John M. H. Olmsted Memorial Award. In addition to his teaching and research, Olmstead authored a number of textbooks sharing the same range as the courses they're based on i.e. from basic freshman to first year graduate level which became standard texts in the 1970's: *Solid Analytic Geometry, Calculus With Analytic Geometry, Prelude To Calculus and Linear Algebra, A Second Course In Calculus, The Real

Number System, *Advanced Calculus*, *Intermediate Analysis* and most famously the standard reference *Counterexamples In Analysis* co-authored with Bernard Gelbaum as well as the more general follow up volume with the same co-author, *Theorems And Counter-examples In Mathematics*-and of course, the volume you hold in your hands. This new edition of *Real Variables* is a fine legacy for his reputation as a teacher and an excellent addition to the line of textbooks published by Blue Collar Scholar.

About the Publisher/ Editor:

Karo Maestro aka The Mathemagician is the *nome-de-plume* of a former graduate student in mathematics. His true identity will remain a secret for now, but one day soon will be revealed to all. Some things this online enigma can reveal: He was a distinguished undergraduate student as a double major in mathematics and biochemistry whose poor health prevented completing graduate studies. Unbowed and undaunted, he plans to return to ultimately obtain a PhD before dying. Partially to that end, he is building the publishing company Blue Collar Scholar, committed to making high quality sources of mathematics-both original works and reprints-available widely and inexpensively to students of all backgrounds. He also hopes to fall in love with and marry one of the most beautiful women on Earth before he dies, but frankly the doctorate is far more likely. He has 3 bachelors' degrees, in philosophy, physical chemistry and mathematics as well as minors in biochemistry and psychology. He is also a past reviewer of textbooks for the Mathematical Association Of America. Among his more recent achievements are the website, TULOOMATH (www.tuloomath.com) , which is designed to be a one stop hub for free downloadable lecture notes and online textbooks in university mathematics from high school algebra to PhD level topics. He's also the author of *Tables,Chairs And Beermugs (* https://tableschairsandbeermugsmathemagician.blogspot.com/) , the associated blog for the website where he reviews textbooks and vents on matters mathematical and academic. He'll soon be beginning 2 more blogs on progressive politics and pop culture. He's painfully blunt, opinionated and has a comment on just about everything in creation-from math textbooks to progressive politics to how to make a perfect cup of organic green tea. He loves JRR Tolkien, science fiction, comic books and movies of all kinds, the comedy of the late great George Carlin, playing with his wonderful nieces, real barbeque, creative burgers and fresh cut French fries, tall curvaceous women and popular music, particularly Bon Jovi and U2. Among his current musical favorites are Alysia Cara, Adele, Imagine Dragons and Ed Sheeran. He is currently working on 2 original books on how to earn a PhD in pure mathematics by self-study alone.

(No, he's not insane-at least he doesn't think so.)

677

677

**And he still has no clue what people see in Ariana Grande.
Not a clue.**

www.ingramcontent.com/pod-product-compliance
Lightning Source LLC
Chambersburg PA
CBHW081552220526
45468CB00010B/2641